HYDRODYNAMICS OF HIGH-SPEED MARINE VEHICLES

Hydrodynamics of High-Speed Vehicles discusses the three main categories of high-speed marine vehicles, vessels supported by submerged hulls, air cushions, or foils. The wave environment, resistance, propulsion, seakeeping, sea loads, and maneuvering are extensively covered based on rational and simplified methods. Links to automatic control and structural mechanics are emphasized. A detailed description of waterjet propulsion is given, and the effect of water depth on wash, resistance, sinkage, and trim is discussed. Chapter topics include resistance and wash; slamming; air cushion–supported vessels, including a detailed discussion of wave-excited resonant oscillations in air cushion; and hydrofoil vessels. The book contains numerous illustrations, examples, and exercises.

Odd M. Faltinsen received his Ph.D. in naval architecture and marine engineering from the University of Michigan in 1971 and has been a Professor of Marine Hydrodynamics at the Norwegian University of Science and Technology since 1974. Dr. Faltinsen has experience with a broad spectrum of hydrodynamically related problems for ships and sea structures, including hydroelastic problems and slamming. He has published more than 200 scientific papers, and his textbook *Sea Loads on Ships and Offshore Structures*, published by Cambridge University Press in 1990, is used at universities worldwide.

Hydrodynamics of High-Speed Marine Vehicles

ODD M. FALTINSEN
Norwegian University of Science and Technology

CAMBRIDGE UNIVERSITY PRESS
Cambridge, New York, Melbourne, Madrid, Cape Town, Singapore,
São Paulo, Delhi, Dubai, Tokyo, Mexico City

Cambridge University Press
32 Avenue of the Americas, New York, NY 10013-2473, USA

www.cambridge.org
Information on this title: www.cambridge.org/9780521178730

First published 2005
Reprinted 2007, 2010
First paperback edition 2010

A catalog record for this publication is available from the British Library.

Library of Congress Cataloging in Publication Data

Faltinsen, O. M. (Odd Magnus), 1944–
Hydrodynamics of high-speed marine vehicles / Odd M. Faltinsen.
p. cm.
Includes bibliographical references and index.
ISBN 0-521-84568-8 (hardback)
1. Motorboats. 2. Ships – Hydrodynamics. 3. Hydrodynamics. 4. Hydrofoil boats.
I. Title.
VM341.F35 2005
623.8'1231-dc22 2005006328

ISBN 978-0-521-84568-7 Hardback
ISBN 978-0-521-17873-0 Paperback

Contents

Preface

Writing a book on the hydrodynamics of high-speed marine vehicles was challenging because I have had to cover all areas of traditional marine hydrodynamics, resistance, propulsion, seakeeping, and maneuvering. However, there is a need to combine all aspects of hydrodynamics in the design of which high-speed vessels are very different from conventional ships, depending on whether they are hull supported, air cushion supported, foil supported, or hybrids.

High-speed vessels are a fascinating topic, and I have been deeply involved in research on high-speed vessels since a national research program under the leadership of Kjell Holden started in Norway in 1989. We also started the International Conference on Fast Sea Transportation (FAST), which has a much broader scope than marine hydrodynamics. I have also benefited from being the chairman of the Committee of High-Speed Marine Vehicles of the International Towing Tank Conference (ITTC) from 1990 to 1993. Further, this book would not have been possible without the work done by the many doctoral students who I have supervised. Their theses are referenced in the book. Parts of the book have been taught to the fourth year, master of science students and doctoral students at the Department of Marine Technology, Norwegian University of Science and Technology (NTNU).

My philosophy in writing the book has been to start from basic fluid dynamics and to link this to practical issues for high-speed vessels. Mathematics is a necessity, but I have tried to avoid this when physical explanations can be given. Knowledge of calculus, including vector analysis and differential equations, is necessary to read the book in detail. The reader should also be familiar with dynamics and basic hydrodynamics of potential and viscous flow of an incompressible fluid.

Computational fluid dynamics (CFD) are commonly used nowadays, but my emphasis is on giving simplified and rational explanations of fluid behavior and its interaction with the vessel. This is beneficial in planning and interpreting experiments and computations. I also believe that examples and exercises are important parts of the learning process.

Automatic control and structural mechanics of high-speed marine vehicles are two disciplines that rely on hydrodynamics. These links are emphasized in the book and are also important aspects of the Centre for Ships and Ocean Structures, NTNU, where I participate.

My presentation of the material is inspired by the book *Marine Hydrodynamics* by Professor J. N. Newman.

I am thankful to Professor Newman for reading through the manuscript and offering suggestions for improvement. Dr. Svein Skjørdal spent a lot of time giving detailed comments on different versions of the manuscript. He was also helpful in seeing the topics from a practical point of view. Sun Hui also did a great job in confirming all my calculations and providing solutions to all exercises. I have benefited from Professor K. J. Minsaas' expertise in propulsion and hydrodynamic design of hydrofoil vessels. Many other people should be thanked for their critical reviews and contributions, including Dr. Tony Armstrong, Professor Tor Einar Berg, J. Bloch Helmers, Professor Lawrence Doctors, Dr. Svein Ersdal, Lars Flæten, Professor Thor I. Fossen, Dr. Chunhua Ge, Dr. Marilena Greco, Dr. Martin Greenhow, Dr. Ole Hermundstad, Egil Jullumstrø, Dr. Toru Katayama, Professor Katsuro Kijima, Professor Spyros A. Kinnas, Dr. Kourosh Koushan, David Kristiansen, Professor Claus Kruppa, Dr. Jan Kvaalsvold, Dr. Burkhard Müller-Graf, Professor Dag Myrhaug, Professor Makoto Ohkusu, Professor Bjørnar Pettersen, Dr. Olav Rognebakke, Renato Skejic, Dr. Nere Skomedal, Professor Sverre Steen, Gaute Storhaug, Professor Asgeir Sørensen, Professor Ernest O. Tuck, and Dr. Frans van Walree.

The artwork was done by Bjarne Stenberg. Anne-Irene Johannessen and Keivan Koushan were helpful in drawing figures. Jorunn Fransvåg organized and typed the many versions of the manuscript in an accurate and efficient way, which required a tremendous amount of work.

The support from the Centre of Ships and Ocean Structures and the Department of Marine Technology at NTNU is appreciated.

List of symbols

A	area; planform area of foil
A_D	developed area, propeller blades
A_E	expanded area, propeller blades
A_{jk}	3D added mass coefficient in the jth mode due to kth motion
a_{jk}	2D added mass coefficient
A_O	area of propeller disc
AP	after perpendicular
A_R	rudder area
A_W	waterplane area
AHR	average hull roughness
B	beam
b	beam of section
BAR	blade area ratio
B_{cr}, b_{cr}	critical damping
B_{jk}	3D damping coefficient in jth mode due to kth motion
b_{jk}	2D damping coefficient
c	chord length; half wetted length in 2D impact; speed of sound
C_B	block coefficient, ship
C_D	drag coefficient
C_f	friction coefficient
C_F	frictional force coefficient
CFD	computational fluid dynamics
C_H	head coefficient
C_{jk}	restoring force coefficient in jth mode due to kth motion
C_L	lift coefficient
$C_{L\beta}$	lift coefficient for planing vessel
C_{L0}	$C_{L\beta}$ at zero deadrise angle
$C(k_f)$	Theodorsen function
C_M	midship section coefficient; mass coefficient in Morison's equation
COG	center of gravity
C_p	pressure coefficient
$C_{p\min}$	minimum pressure coefficient
C_P	longitudinal prismatic coefficient
C_R	residual resistance coefficient
C_T	propeller thrust-loading coefficient; total resistance coefficient
C_W	wave-making resistance coefficient
C_{WP}	wave pattern resistance coefficient

C_Q	capacity coefficient
D	draft; drag force; propeller diameter
DNV	Det Norske Veritas
D_T	transom draft
E	Young's modulus of elasticity
EI	flexural rigidity of a beam
E_k	kinetic fluid energy
E(t)	energy
f	frequency (Hz); maximum camber
F	densimetric Froude number; fetch length
Fn	Froude number $U/\sqrt{gL}$
Fn_B	beam Froude number
Fn_D	draft Froude number
Fn_h	depth or submergence Froude number
Fn_T	transom draft Froude number
FP	forward perpendicular
F_v	volumetric Froude number
g	acceleration of gravity
$G(x,y,z;\xi,\eta,\zeta)$	Green function
$\overline{GM}$	transverse metacentric height
$\overline{GM_L}$	longitudinal metacentric height
$\overline{GZ}$	moment arm in heel (roll) about COG
h	water depth; submergence
h_j	height of the center of the jet at station S_7 (see Figure 2.54) above calm free surface
H	wave height; head
$H_{1/3}$	significant wave height
i	imaginary unit
I_{jk}	moment or product of inertia
i, j, k	unit vectors along x, y and z-axis, respectively
IVR	inlet velocity ratio
J	advance ratio of propeller
k	wave number; roughness height; form factor
KC	Keulegan-Carpenter number
k_f	reduced frequency
$\overline{KG}$	height of COG above keel
K_T	thrust coefficient
K_Q	torque coefficient
L	length of ship; lift of a foil; hydrodynamic roll moment in maneuvering
L_C	chine wetted length
LCB	longitudinal center of buoyancy
lcg	longitudinal center of gravity measured from the transom stern
LCG	longitudinal center of gravity
L_K	keel wetted length
L_{OA}	length, overall
L_{OS}	length, overall submerged

l_p	longitudinal position of the center of pressure measured along the keel from the transom stern
L_{PP}	length between perpendiculars
L_{WL}	length of the designer's load waterline
M	mass; moment; hydrodynamic pitch moment in maneuvering
M	fluid momentum vector
m	mass per unit length
M_{jk}	components of mass matrix
n	propeller revolutions per second
n	surface normal vector positive into the fluid
N	normal force; hydrodynamic yaw moment in maneuvering
O	origin of coordinate system
$O(\varepsilon)$	order of magnitude of ε
P	power; pitch of propeller; probability
p	pressure; roll component of angular velocity; half of the distance between the center lines of the demihulls of a catamaran; stagger between foils
p_a	atmospheric pressure
P_D	delivered power
p_o	ambient pressure; static excess pressure
p_v	vapor pressure of water
Q	propeller torque; volume flux; source strength
q	pitch component of angular velocity
r	yaw component of angular velocity
R	radius; resistance
R_{AA}	added resistance in air and wind
R_{AW}	added resistance in waves
r_{jj}	radius of gyration in rigid body mode j
RMS	root mean square
Rn	Reynolds number
R_R	residual resistance
R_S	spray resistance
R_T	total resistance
R_V	viscous resistance
R_W	wave-making resistance
s	span length of foil
S	area of wetted surface; cross-sectional area
S_B	body surface
$S(\omega)$	wave spectrum
t	time; thrust-deduction coefficient; maximum foil thickness
T	period; propeller thrust
T_0	modal or peak period
T_1	mean wave period
T_2	mean wave period
T_e	encounter period
T_n	natural period
T_S	surface tension

U	forward velocity of vessel
U_I	mean velocity at the most narrow cross-section of the waterjet inlet
U_S	propeller slip stream velocity
u	x-component of vessel velocity
v	y-component of vessel velocity
vcg	vertical distance between COG and the keel
V_g	group velocity
V_p	phase velocity
v^*	wall friction velocity
V	water entry velocity
W	weight
w	wake fraction; z-component of vessel velocity; vertical deflection
Wn	Weber number
x, y, z	Cartesian coordinate system. Moving with the forward speed in seakeeping analysis. Body-fixed in maneuvering analysis.
X	x-component of hydrodynamic force in maneuvering
X_E, Y_E, Z_E	Earth-fixed coordinate system
x_T	x-coordinate of transom
x_s	$L_K - L_C$
Y	y-component of hydrodynamic force in maneuvering
Z	z-component of hydrodynamic force in maneuvering

Greek symbols

α	angle of attack
α_c	Kelvin angle
α_f	flap angle
α_i	ideal angle of attack
α_0	angle of zero lift
β	wave propagation angle; deadrise angle; drift angle
Γ	circulation; gamma function; dihedral angle
γ	vortex density; sweep angle; ratio of specific heat for air
δ	boundary layer thickness; rudder angle; flap angle
δ^*	displacement thickness
Δ	vessel weight
ε	angle
ζ	surface elevation
ζ_a	wave amplitude
η	overall propulsive efficiency
η_H	hull efficiency
η_J	jet efficiency
η_k	wave-induced vessel motion response, where $k = 1, 2, 3, \ldots, 6$ refers to surge, sway, heave, roll, pitch, and yaw, respectively
η_P	propeller efficiency; pump efficiency
η_R	relative rotative efficiency
η_S	sinkage
η_T	thrust power efficiency
θ	pitch angle; momentum thickness

Λ	aspect ratio of foil
Λ_L	ratio between full scale and model length
λ	wavelength
λ_w	mean wetted length-to-beam ratio
μ	dynamic viscosity coefficient
ν	kinematic viscosity coefficient
ξ	ratio between damping and critical damping
ρ	mass density of fluid (water)
ρ_a	mass density of air
σ	cavitation number; source density; standard deviation
σ_i	cavitation inception index
σ_o	propeller cavitation number
$\sigma_{0.7}$	propeller cavitation number defined at 0.7 R
τ	trim angle in radians; $\omega_e U/g$
τ_{deg}	trim angle in degrees
τ_{ij}	Newtonian stress relations
τ_w	frictional stress on hull surface
ϕ	heel (roll) angle
φ	velocity potential
ψ	yaw angle
ω	circular frequency in radians per second
$\boldsymbol{\omega}$	vorticity vector; vector of rotational vessel motion
ω_n	natural frequency
ω_e	frequency of encounter
ω_o	frequency of waves in an Earth-fixed coordinate system
$\boldsymbol{\Omega}$	vector of rotational vessel velocity
Ω	volume

Special symbols

∇	displaced volume of water; vector differential operator
∇^2	$\frac{\partial^2}{\partial x^2} + \frac{\partial^2}{\partial y^2} + \frac{\partial^2}{\partial z^2}$

1 Introduction

Baird (1998) defines a high-speed vessel as a craft with maximum operating speed higher than 30 knots, whereas hydrodynamicists tend to use a Froude number $Fn = U/\sqrt{Lg}$ larger than about 0.4 to characterize a fast vessel supported by the submerged hull, such as monohulls and catamarans. Here, U is the ship speed, L is the overall submerged length L_{OS} of the ship, and g is acceleration of gravity. The pressure carrying the vessel can be divided into hydrostatic and hydrodynamic pressure. The hydrostatic pressure gives the buoyancy force, which is proportional to the submerged volume (displacement) of the ship. The hydrodynamic pressure depends on the flow around the hull and is approximately proportional to the square of the ship speed. Roughly speaking, the buoyancy force dominates relative to the hydrodynamic force effect when Fn is less than approximately 0.4. Submerged hull–supported vessels with maximum operating speed in this Froude number range are called displacement vessels. When $Fn > 1.0$–1.2, the hydrodynamic force mainly carries the weight, and we call this a planing vessel. Vessels operating with maximum speed in the range 0.4–$0.5 < Fn < 1.0$–1.2 are called semi-displacement vessels. This means that high-speed submerged hull–supported vessels denote vessels in which the buoyancy force is not dominant at the maximum operating speed.

Ship speeds of about 50 knots represent an important barrier for a high-speed vessel. At this speed, cavitation typically starts to be a problem, for instance, on the foils and the propulsion system. Cavitation means that the pressure somewhere on the upper side (suction side) of the foil becomes equal to the vapor pressure. This is only 0.012 times the atmospheric pressure at 10°C. If a large part of the suction side of the foil is cavitating, the lift is clearly reduced relative to a noncavitating foil at the same speed. For instance, the lift of a supercavitating 2D flat foil in infinite fluid is only 25% of the lift of a noncavitating 2D flat foil at the same speed and the same orientation of the foil relative to the forward speed (Newman 1977). Supercavitation means that the suction side of the foil is not wetted. Partial cavitation may also cause damage to the foil structure in terms of implosion of bubbles. In addition, ventilation may occur, for instance, as a consequence of cavitation. Ventilation means that there is a connection or an air tunnel between the air and the foil surface. Occurrence of ventilation also leads to significant drop in lifting capacity of a foil. Supercavitating foils and propellers are used to increase the speed barrier substantially beyond 50 knots. Such foil shapes have a sharp leading edge to initiate cavitation.

Minimization of the hull weight with consideration of the structural strength is important for all high-speed vessels. One early foil catamaran design resulted in too-heavy foils and struts. The consequence was reduced payload and unsatisfactory transport economy.

The 35th edition (2002–2003) of *Jane's High-Speed Marine Transportation* refers to four major limitations for future market developments of fast "ro-pax" vessels carrying passengers and allowing roll-on roll-off payloads (most often in terms of cars):

- Limited seakeeping ability
- Reliability of the main propulsion machinery
- Cost of the higher-grade fuel used
- Limited freight-carrying ability

Wave generation, that is, wash, is also an issue for further market expansion. The decay of the generated waves perpendicular to the ship's course is important from a coastal engineering point of view. When the waves enter shallow water, the wavelength decreases and the wave amplitude increases, resulting in breaking waves on a beach. This may happen when the ship is out of sight, surprising swimmers. The reflection of the generated waves from vertical walls, such as a quay, may also be a problem and a safety issue. The total wave amplitude will be twice the incident amplitude, and water may flow over the quay. The wash also affects the environment, for instance, in terms of erosion. There is no simple universal criterion in terms of maximum wave amplitude that quantifies the wash effect. The criterion must be different if the waves are affecting the seashore or affecting other ships. If, for instance, the effect on other ships is analyzed, the ship response due to wash of a passing ship must be studied. Ferry operators in the United Kingdom must prepare a route assessment with regard to wash that must be approved by the Maritime and Coastguard Agency (Whittaker and Elsässer 2002).

There is a broad variety of high-speed vessels in use, with very different physical features. The vessels differ in the way the weight is supported. The vessel weight can be supported by:

- Submerged hulls
- Hydrofoils
- Air cushions
- A combination of the above

Figure 1.1, used in the announcement of the FAST'91 Conference in Trondheim, Norway, illustrates a fictitious high-speed vessel using air cushion, foils, and submerged hulls to support the vessel weight. The air cushion is enclosed between the side hulls and by seals in the forward and aft end of the vessel. The main types of high-speed vessels are discussed below.

Figure 1.1. Fictitious high-speed vessel with air cushion, foils, and SWATH effects. (Artist: Bjarne Stenberg)

Submerged hull–supported vessels

Examples of semi-displacement and planing vessels are presented. Figure 1.2 shows a SWATH (small waterplane area twin hull) vessel. As the name says, this vessel is characterized by a small waterplane area and two demihulls. A SWATH has higher natural periods in heave and pitch and generally lower vertical wave excitation loads than a similarly sized catamaran. The explanation is similar to that of a semi-submersible platform (Faltinsen 1990). The consequence is better seakeeping behavior of a SWATH compared with the catamaran in head sea conditions. However, if the sea state, speed, and heading cause resonant vertical motions of the SWATH, it may not have good seakeeping behavior. Wetdeck slamming is then a danger. Further, if motion control surfaces are not used, a SWATH is dynamically unstable in the vertical plane beyond a certain speed. A SWATH is often not classified as a high-speed vessel.

The most common type of high-speed vessel is the catamaran. The catamaran is often equipped with an automatic motion control system, such as foils, which minimize wave-induced motions. Catamaran designs include the wave-piercing (Figure 1.3) and semi-SWATH types of hulls. Trimarans and pentamarans (Figure 1.4) with one large center hull combined with smaller outrigger hulls are other types of multihull vessels.

The beam-to-draft ratio of semi-displacement monohulls with lengths longer than approximately 50 m may vary from around 5 to more than 7 which is very different from displacement ships. Large monohulls are often equipped with automatic motion control devices similar to the ones used for catamarans. Stern flaps and roll fins are commonly used. A pronounced increase in the length of a submerged hull is generally favorable for wave-induced vertical motion and acceleration. It means that a relatively long monohull with the same displacement as a catamaran has an advantage relative to the catamaran. However,

Figure 1.2. SWATH (small waterplane area twin hull). (Artist: Bjarne Stenberg)

Figure 1.3. "Wave-piercing" catamaran. (Artist: Bjarne Stenberg)

Figure 1.4. Pentamaran. (Artist: Bjarne Stenberg)

Figure 1.5. Planing vessel. (Artist: Bjarne Stenberg)

attention has to be paid to roll motion and dynamic stability of monohull vessels.

Planing vessels (Figure 1.5) are typically smaller vessels used as patrol boats, sportfishing vessels, and service craft, and for sport competitions. Dynamic stability, cavitation, and ventilation are of concern for planing vessels.

Foil-supported vessels

Hydrofoil-supported monohulls with either fully submerged or free surface–piercing foils are shown in Figures 1.6 and 1.7. The first commercial high-speed vessels were the monohull hydrofoil boats with free surface–piercing foils. If the flap angle of the foils and the trim of the vessel are held constant, the foil lifting capacity increases approximately with the square of the vessel's speed until cavitation occurs. Because the foil lift is approximately proportional to the projection of the foil area onto the mean free surface, the inclined free surface–piercing foils need a larger foil area than that required by fully submerged foils for a given weight and design speed. The free surface–piercing foil is self-stabilizing with respect to vertical position, heel, and trim.

In the beginning of the 1990s, foil catamarans were a promising concept, having small resistance and good seakeeping behavior. Fully submerged

Figure 1.6. Hydrofoil vessel with fully submerged foil system. (Artist: Bjarne Stenberg)

Figure 1.7. Hydrofoil vessel with free surface–piercing foils. (Artist: Bjarne Stenberg)

horizontal foil systems were used. A control system that activates foil flaps is needed to stabilize the heave, roll, and pitch of a hydrofoil boat with fully submerged foils in the foilborne condition. Another important design consideration is sufficient power and efficiency of the propulsor system to lift the vessel to the foilborne condition. This is of special concern when waterjet propulsion is used because of its decreased efficiency at lower speeds. Another concern is the ventilation along one of the two forward struts during maneuvering, which may ventilate the forward foil system and cause loss of the lift force.

Foil cavitation limits the vessel's speed to about 50 knots. Proper design to delay cavitation on the aft foil system requires evaluation of the wake from the forward foil system. An important effect is caused by roll-up of tip vortices originating from the forward foil system. The wake from the forward foil causes an angle of attack that varies along the span of the aft foil, which can be counteracted by using a twisted aft foil that is adapted to the inflow. One foil catamaran experienced problems with foil cavitation during operation, which were resolved by drilling holes in the aft part of the foils to provide communication between the flow on the pressure and suction sides of the foils.

Very precise and smooth foil surfaces are needed from a resistance, lift, and cavitation point of view. These surfaces require special fabrication procedures and frequent cleaning during operation. The high production and maintenance costs are important reasons why few foil catamarans have been built. There also exist hydrofoil-assisted catamarans in which the foils only partially lift the vessel.

Air cushion–supported vessels

Surface effect ships (SES) or air-cushion catamarans of lengths less than 40 m were frequently built for commercial use until the mid-1990s. An air cushion is enclosed between the two side hulls and by flexible rubber seals in the bow and aft end (Figure 1.8). The skirt in the front end is easily worn out.

The excess pressure in the air cushion is produced by a fan system that lifts the vessel, thereby carrying about 80% of the weight. The excess pressure reduces the metacentric height, but the static stability is still good. It also causes a mean depres-

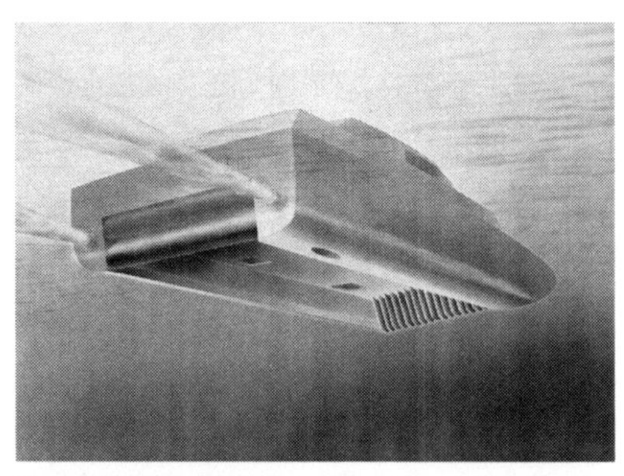

Figure 1.8. Artist's fish-eye view of an SES (surface effect ship) illustrating the air cushion with flexible skirts in the bow and a flexible bag in the aft end used to enclose the air cushion between two catamaran hulls. Fans are used to create an excess pressure in the air cushion that lifts the vessels. (Artist: Bjarne Stenberg)

sion of the free surface inside the cushion that results in waves and wave resistance. However, because the hull wetted surface is diminished, the total calm water resistance is small relative to a catamaran of similar dimensions. The lifting up of the SES also causes an increase in air resistance. Because resistance is proportional to the mass density of the fluid and the air density is only about 1/1000 of the water density, the air resistance is smaller than water resistance. The ship speed can be up to 50 knots in low sea states.

Resonance oscillations in the air cushion cause "cobblestone" oscillations with a dominant frequency around 2 Hz for a 30 to 40 m–long vessel. The word *cobblestone* is associated with the feeling of driving a car on a road with badly layed cobblestones. The highest natural period is the result of a mass-spring system in which the compressibility of the air in the cushion acts like a spring. The mass is related to the total weight of the SES. The damping is small and caused by air leakage and the lifting fans. The excitation is induced by volume changes in the air cushion due to incident waves. The resonant oscillations require incident wave energy at a frequency of encounter close to the natural frequencies of the cobblestone oscillations, which occurs in very small sea states. The resulting vertical accelerations are of concern from a comfort point of view. Damping of the cobblestone oscillations can be increased by an active control system introducing air leakage through louvers. If special attention is not

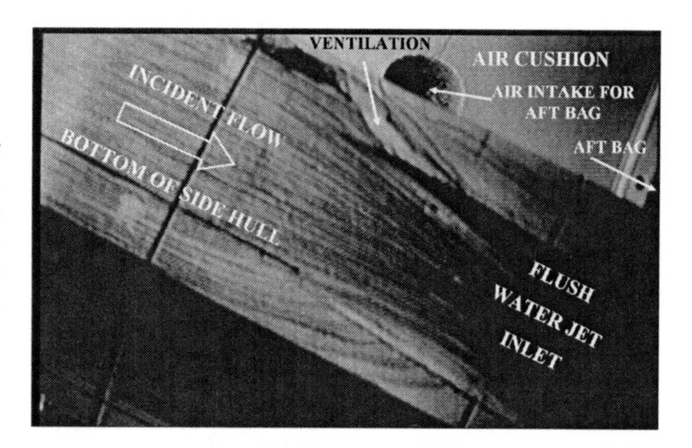

Figure 1.9. Fish-eye view of the bottom of a side hull of an SES with the waterjet inlet. A tube with air (white part) coming into the waterjet inlet can be seen; this is ventilation. The air intake for the aft bag shown in the figure is in the roof of the air cushion.

paid to scaling laws, the cobblestone phenomenon will not be detected in model tests that are based on Froude scaling. If the SES is on cushion and no cobblestone oscillations occur, the vessel has vertical accelerations that are generally lower than those of a similarly sized catamaran in head seas.

When the SES is on cushion, there is a small distance from a waterjet inlet at the hull bottom to the air cushion, which can easily cause ventilation of the waterjet inlet in a seaway. Because the waterjet inlet flow acts similarly to a flow sink, cross-flow occurs in the vicinity of the inlet. If the hull cross section has a small radius of curvature in the inlet area, very high local velocities and low pressures occur, increasing the danger of ventilation even in calm water. Figure 1.9 illustrates model tests of the occurrence of ventilation to the waterjet inlet in calm water conditions. Fences on the cushion side of the side hulls have been proposed to deal with this problem.

Figure 1.10. Air-cushion vehicle (ACV). (Artist: Bjarne Stenberg)

An SES experiences a more significant involuntary speed loss than that of a similarly sized catamaran in a seaway. The relative vertical motions between the vessel and the waves cause air leakage, which decreases the air cushion pressure when the lifting power is kept constant. The resulting sinkage implies higher resistance. If the fan system does not have sufficient power to maintain air cushion pressure, significant speed loss may occur, even in moderate sea states.

The air-cushion vehicle (ACV) shown in Figure 1.10 is the oldest type of air cushion–supported vessel. Because a flexible seal system is used for the air cushion, the ACV is amphibious. It also implies that air propellers are used, which may represent a noise problem. Because there is no submerged hull to provide hydrostatic restoring moments in roll and pitch, static stability in these modes of motion needs attention during the design stage.

Air lubrication technology (ALT) uses air caverns that run for approximately half the length of a hull in the aft part of the vessel. An air cushion can facilitate the lifting to the airborne condition of Ekranoplanes or wing-in-ground (WIG) vehicles. The air cushion is part of the Hoverwing design (see Figure 1.11 and Fischer and Matjasic 1999). A small portion of the propeller slip stream is used to create an air cushion with an excess pressure between the two floats (catamaran hulls) and the flexible textile skirts at the front and aft ends. The WIG flies close to the water surface. This gives extra lift (see Figure 6.46 and accompanying text). The Hoverwing cruises at a speed of 180 km/hour (90 knots) and is claimed to have high maneuvrability and short stopping distance. Low noise emission at all speeds is also an important issue.

Figure 1.11. Artist's impression of a WIG vehicle based on the Hoverwing techology by Fischer-Flugmechanick. An air-cushion effect is generated between the floats during takeoff. (Artist: Bjarne Stenberg)

Papanikolaou (2002) has systematically presented the many types of high-speed marine vehicles that exist today. He explains the many different acronyms used, together with his view on the advantages and disadvantages of the different types of vessels.

There are also sailboats that can be categorized as high-speed marine vessels. The current world speed sailing record is 46.52 knots, set by *Yellow Pages Endeavour* in 1993. Our detailed discussion of the flow around lifting surfaces and hulls is relevant in this context. The keels, the rudder, and the sails are all lifting surfaces from a fluid dynamics point of view. The fluid dynamics of sailboats are handled in the books by Larsson and Eliasson (2000), Marchaj (2000), Garrett (1987), and Bethwaite (1996).

1.1 Operational limits

Operational limits are set by

- Safety, comfort, and workability criteria
- Structural loading and response
- Machinery and propulsion loading and response

Seakeeping criteria typically used for conventional ships are presented in Tables 1.1 and 1.2.

Table 1.1. *General operability limiting criteria for ships (NORDFORSK 1987).*

	Merchant ships	Naval vessels	Fast small craft
Vertical acceleration at forward perpendicular (RMS value)	0.275 g ($L \leq 100$ m) 0.05 g ($L \geq 330$ m)[a]	0.275 g	0.65 g
Vertical acceleration at bridge (RMS value)	0.15 g	0.2 g	0.275 g
Lateral acceleration at bridge (RMS value)	0.12 g	0.1 g	0.1 g
Roll (RMS-value)	6.0°	4.0°	4.0°
Slamming criteria (probability)	0.03 ($L \leq 100$ m) 0.01 ($L \geq 300$ m)[b]	0.03	0.03
Deck wetness criteria (probability)	0.05	0.05	0.05

[a] The limiting criterion for lengths between 100 and 330 m varies almost linearly between the values $L = 100$ m and $L = 330$ m, where L is the length of the ship.
[b] The limiting criterion for lengths between 100 and 300 m varies linearly between the values $L = 100$ m and 300 m.

Table 1.2. *Criteria (root mean square) with regard to accelerations and roll (NORDFORSK 1987).*

Vertical acceleration	Lateral acceleration	Roll	Description
0.20 g	0.10 g	6.0°	Light manual work
0.15 g	0.07 g	4.0°	Heavy manual work
0.10 g	0.05 g	3.0°	Intellectual work
0.05 g	0.04 g	2.5°	Transit passengers
0.02 g	0.03 g	2.0°	Cruise liner

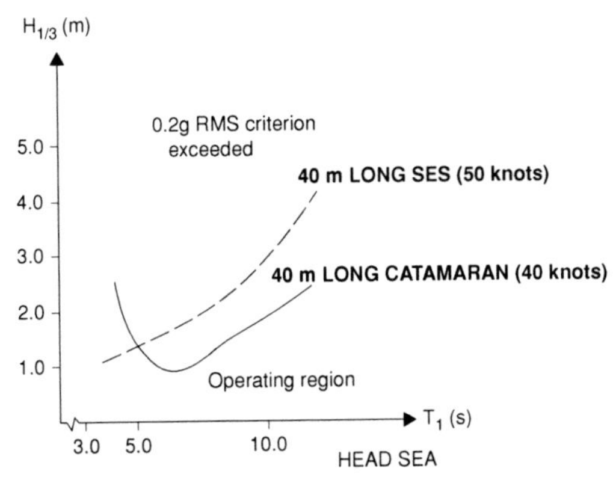

Figure 1.12. Calculated operational limits of similarly sized catamaran and SES in head sea long-crested waves with different significant wave heights ($H_{1/3}$) and mean wave periods (T_1). The 0.2 g RMS value of vertical acceleration at the center of gravity (COG) is used as a criterion. Involuntary speed loss due to wind resistance and added resistance in waves are considered.

Those criteria are related to slamming, deck wetness, RMS values of roll, and lateral and vertical accelerations. RMS values mean root mean square values or standard deviation. The rightmost column of Table 1.2 includes a brief description of what the criteria relate to. *Light manual work* means work carried out by people adapted to ship motions. This work is not tolerable for longer periods, and causes fatigue quickly. *Heavy manual work* means work, for instance, on fishing vessels and supply ships. *Intellectual work* relates to work carried out by people not so well adapted to ship motions, such as scientific personnel on an ocean research vessel. *Transit passenger* means passengers on a ferry exposed to the acceleration level for about two hours. *Cruise liner* refers to older passengers on a cruise liner.

The criteria can be used to determine voluntary speed loss and operability of vessels in different sea areas. For example, Figure 1.12 illustrates the calculated operational limits of a 40 m–long catamaran and a 40 m–long SES for head sea conditions. No active motion control systems are used in the calculations. The criterion used was RMS value of vertical acceleration at COG equal to 0.2 g. However, other criteria as well as other headings must be considered. Generally speaking, the catamaran has the lowest operational limits in Figure 1.12, but these can be improved by an active control system. The reason the SES has the lowest operational limit for small sea states (small mean wave periods, T_1) is the outset of cobblestone oscillations.

Faltinsen and Svensen (1990) have pointed out the relatively large variation in published criteria, which may lead to quite different predictions of voluntary speed reduction and operational limits. For high-speed vessels, other criteria are also needed, such as operational limits in a seaway due to the propulsion and engine system. Meek-Hansen (1990, 1991) presented service experience with a 37 m–long SES equipped with diesel engines and waterjet propulsion. An example with significant wave height, $H_{1/3}$, around 2 m, head sea, and 35 knots speed shows significant engine load fluctuations at intervals of 6 to 12 seconds (Figure 1.13). These fluctuations result in increased thermal loads in a certain time period,

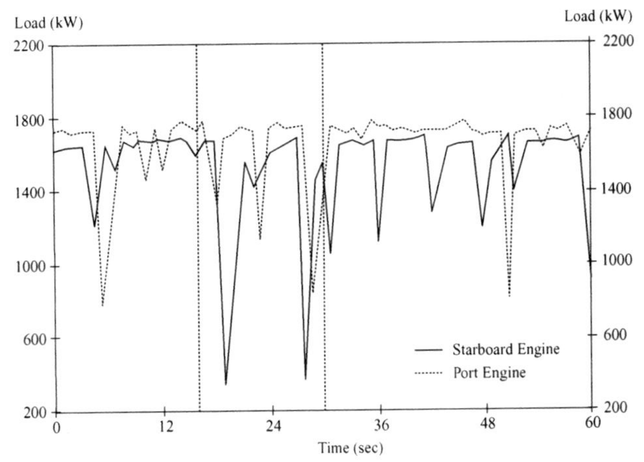

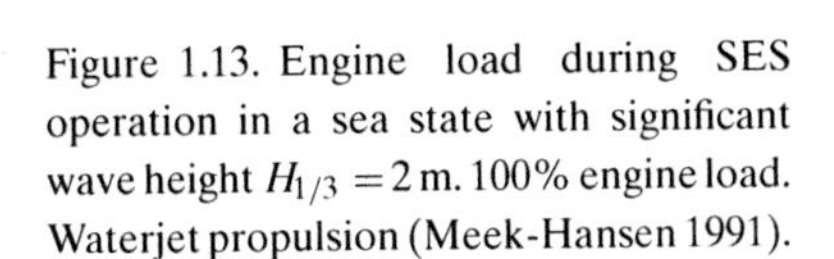
Figure 1.13. Engine load during SES operation in a sea state with significant wave height $H_{1/3} = 2$ m. 100% engine load. Waterjet propulsion (Meek-Hansen 1991).

caused by a very high fuel-to-air ratio. These high thermal loads may lead to engine breakdowns.

Possible reasons for the engine load fluctuations are believed to be:

- Exposure of the waterjet inlet to free air
- Flow separation in front of and inside the inlet
- Ventilation and penetration of air from the free water surface or from entrained air in the boundary layer

The phenomenon mentioned above often interacts in a complicated way; for example, separation may be one of the causes for onset of ventilation and cavitation. Under certain conditions, a cavity may be penetrated and filled with air. Separation and cavitation are primarily dependent on the pressure distribution in and near the waterjet inlet. For a given inlet geometry, this distribution depends mainly on the speed and thrust (resistance) of the ship.

Exposure of the waterjet inlet to free air is a result of the relative vertical motions between the vessel and the seawater. An operational limit may be related to the probability of exceeding a certain limit of the relative vertical motion amplitude between the vessel and the waves at the waterjet inlet. In particular, with an SES equipped with flush inlets, the exposure to free air represents a problem even for small sea states. The reason is the small distance between the inlet and the calm water surface inside the air cushion.

The seasickness criterion according to NS-ISO 2631/3 is commonly used for the assessment of passenger comfort in high-speed vessels (see Figure 1.14). It gives limits for RMS (root mean square) values of the accelerations as a function of frequency. This criterion needs some explanation. It refers to the a_z or a human's head–to-foot component of the acceleration. For a broadband spectrum, frequency f_c in Figure 1.14 means the average frequency of a one-third–octave band, defined as the frequency interval between f_1 and f_2, where $f_2 = 2^{1/3} f_1$. Further, the center frequency f_c of the one-third–octave band is $(f_1 f_2)^{1/2}$. This means $f_1 = f_c/2^{1/6}$ and $f_2 = f_c 2^{1/6}$. A broadband spectrum should be divided into one-third–octave bands, and the RMS value should be evaluated separately for each of the one-third–octave bands. Each RMS value should be compared with the limits given in Figure 1.14 for different exposure periods. Because the motion

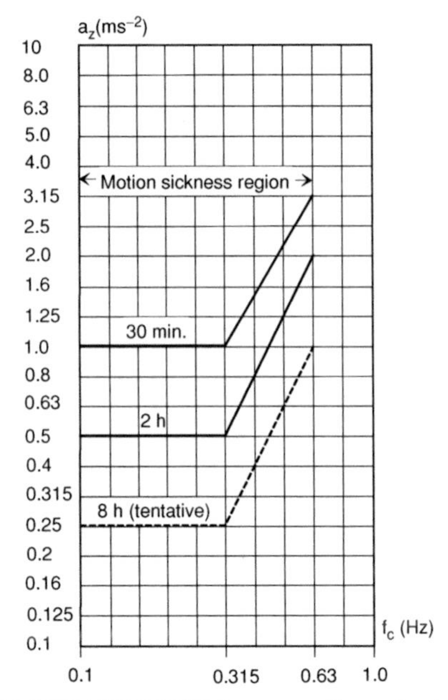

Figure 1.14. NS-ISO 2631/3 – severe discomfort boundaries (1. ed. Nov. 1985). a_z is the RMS value of human's head–to-foot component of acceleration in a one-third–octave band of a spectrum with center frequency f_c.

sickness region in Figure 1.14 is from 0.1 to 0.63 Hz, it implies that the cobblestone effect of an SES does not cause motion sickness. According to ISO 2631/1, there are other criteria for accelerations in the frequency range from 1 to 80 Hz, which are related to workability or human fatigue. An example is shown in Figure 1.15 that expresses the limits of the RMS value of the a_z-component of the acceleration as a function of frequency. This figure should be interpreted in the same way as Figure 1.14. In addition, by multiplying the acceleration values in Figure 1.15 by 2, one gets boundaries related to health and safety, and by dividing the acceleration values by 3.15, one gets boundaries for reduced comfort.

Operational studies should ideally take into account that the shipmaster may change speed and heading. It may sound wrong, but a semi-displacement vessel equipped with foils may improve the seakeeping behavior by increasing the speed. The reason is that the heave and pitch damping of a foil increases with forward speed. In particular, the roll motion magnitude is important for monohull vessels. However, if the ship is equipped with roll stabilization means, high-speed conditions should be of minor concern.

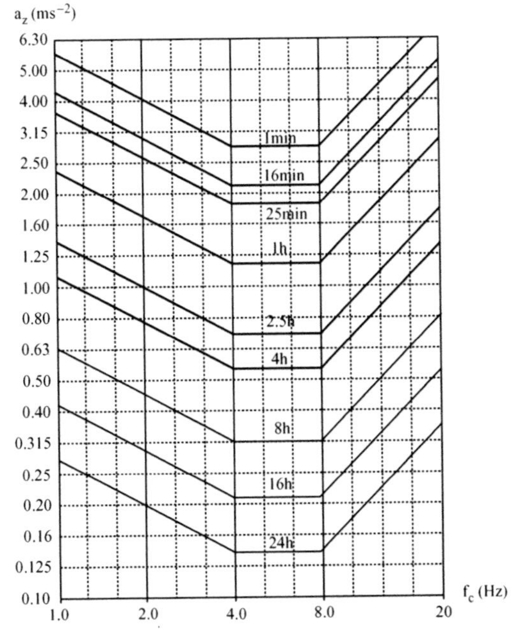

Figure 1.15. ISO 2631/1 – fatigue-decreased proficiency boundaries. a_z is the RMS value of a human's head–to-foot component of acceleration in a one-third–octave band of a spectrum with center frequency f_c.

There is a need to establish better seakeeping criteria for wetdeck slamming and the behavior of the propulsion and machinery systems in a seaway. The wetdeck is the underside of the deck structure between the side hulls of multihull vessels, that is, the deck part facing the water.

Because cavitation and ventilation of foils mean that the foils become less efficient as damping devices and cause an increase in the vessel motions and accelerations, these effects should be accounted for in operational studies. However, knowledge about these issues is still in its infancy.

It is important to investigate different vessel headings relative to the wave propagation direction. For instance, a catamaran in following regular waves may have a speed close to the phase speed of the waves, that is, the speed of the propagating geometry of the waves. Further, if the wavelength is of the order of the vessel's length, the catamaran can assume a position relative to the waves so that the fore part of the vessel dives into a wave crest. The slender fore part may not have sufficient buoyancy, and the more-voluminous aft part will be lifted up by the waves. The result is a significant amount of water over the fore deck.

The loss of steady heel moment with forward speed of semi-displacement round-bilge monohulls is an important safety issue. When the Froude number is larger than 0.6 to 0.7 in calm water, the vessel may suddenly lean over to one side. At higher speeds, this may cause dangerous "calm water broaching" and is the main reason round-bilge hulls are unsuitable for Froude numbers above 1.2 (Lavis 1980).

Directional instability in following seas with the subsequent risk of the vessel becoming broadside to the waves and eventually capsizing, is a well-known phenomenon of monohulls. This is referred to as "broaching" and may occur under conditions similar to those in a "dive-in." Because a multihull semi-displacement vessel has good static stability in roll and is very difficult to capsize in waves, broaching is less important for catamarans. However, large sway and yaw motions as well as steering problems may also occur for catamarans in following and quartering sea.

Quasi-steady stability in the roll of monohulls in following seas with a small frequency of encounter should also be considered. This is of particular concern if the local waterplane area, that is, local width of the hull at the hull/water line intersection, clearly changes as a function of local draft (i.e., large flare). The hydrostatic transverse stability should then be calculated as a function of different frozen incident wave shapes along the ship. These frozen conditions in following seas should also be considered as structural load cases for the hull girder. When calculating hydrostatic stability, the increased importance of steady hydrodynamic pressure on the hull with increasing speed relative to hydrostatic pressure should be recognized. This is an implicit consequence of being a "semi-displacement" vessel.

The propulsion unit, rudders, stabilization fins in faulty position, cavitation, and ventilation may also influence stability. A scenario might be two supercavitating propellers, one of which suddenly ventilates, causing an asymmetry in thrust with resulting directional instability.

If the ship is in a planing condition, that is, the Froude number is larger than one, special dynamic instability problems may occur. Examples are "chine-walking" (dynamic roll oscillations), "porpoising" (dynamic coupled pitch-heave oscillations), and "cork-screwing" (pitch-yaw-roll oscillations). However, the major part of the commercial high-speed vessel fleet does not operate in planing conditions.

Müller-Graf (1997) has given a comprehensive presentation of the many different dynamic stability problems of high-speed vessels. This work includes design features and factors influencing dynamic instabilities. Recommendations are given on how to minimize dynamic instabilities of monohulls.

1.2 Hydrodynamic optimization

A ship is often hydrodynamically optimized in calm water conditions. Because good seakeeping behavior is an important feature of a high-speed vessel, optimization in calm water conditions may lead to unwanted behavior in a seaway. Both wave resistance and wave radiation damping are caused by the ship's ability to generate waves. Because low wave resistance may imply low wave radiation damping in heave and pitch, the result may be unwanted large resonant vertical motions of a semi-displacement vessel. This relationship was illustrated by a project with first-year students knowing little about hydrodynamics. A catamaran design was proposed in which each of the two side hulls had a very small beam-to-draft ratio. This hull form was fine for resistance, but the vessel jumped out of the water during seakeeping tests when the wave periods were in resonant heave-and-pitch conditions. This extreme behavior could have been counteracted at high speed if the vessel were equipped with damping foils.

Another example is the recent designs of passenger cruise vessels with very shallow local draft and nearly horizontal surfaces in the aft part of the ship. These designs were the result of hydrodynamic optimization studies in calm water. One does not need to be a hydrodynamicist to understand that this caused slamming (water impact) problems. Aft bodies with shallow draft should also be of concern for directional stability and for ventilation of waterjet inlets in waves. Hydrodynamic optimization studies must therefore consider resistance, propulsion, maneuvering, and seakeeping. There obviously are also constraints of a nonhydrodynamic character. For instance, minimalization of ship motions may lead to higher global structural loads.

1.3 Summary of main chapters

This textbook focuses on high-speed vessels. However, some of the text on semi-displacement vessels is also relevant for conventional ships. Further, the discussion of slamming (water impact) is important in many other marine applications, including offshore structures.

Chapter 2 considers resistance and propulsion in calm water conditions. The two most important resistance components of semi-displacement vessels and SES are viscous resistance and wave resistance. Viscous resistance is important for hydrofoil-supported vessels, but induced drag due to trailing vortices should also be considered.

The waterjet is the most common propulsion system for high-speed vessels. We use conservation of fluid momentum and kinetic fluid energy to derive the thrust and efficiency of the waterjet system. The possibility of cavitation at the waterjet inlet is also discussed.

Chapter 3 presents linear wave theory and a stochastic description of the waves. This is necessary background for later chapters that describe wave-induced motions and loads on high-speed vessels. Linear wave theory is also used to describe wave resistance and wash in detail. This is done in Chapter 4.

Chapter 4 considers wave resistance of semi-displacement vessels and air cushion–supported vessels. Ship waves are traditionally classified as divergent and transverse waves. The transverse waves have crests nearly perpendicular to the ship's track. The dominant wave picture far away from the ship is normally the result of divergent bow waves. The divergent waves are a major source for the wave resistance of a semi-displacement vessel at the maximum operating speed. The effects of finite water depth on monohull vessels, including the effect on trim and sinkage, is also discussed.

Chapter 5 concentrates on SES. However, the issues presented also have relevance for other air cushion–supported vessels. The chapter explains how the air cushion causes a depression of the free surface and affects the roll metacentric height. The air cushion typically carries 80% of the weight of an SES. Details are given about the seal system of the air cushion. Resistance and propulsion in calm water are covered in Chapters 2 and 4. This chapter discusses cobblestone oscillations and the added resistance and speed loss in waves.

Chapter 6 discusses foil-supported vessels. Relevant hydrodynamic foil theory is presented. The chapter starts out describing a boundary element method (BEM) based on source and dipole distributions that may account for nonlinearities, 3D flow, interaction between foils and struts, and free surface effects. Thereafter is a presentation of linear theory. The advantage of a linear theory is that we can more easily show how the angle of attack, foil camber, foil flaps, and three-dimensionality of the flow influence lift and drag of the foil. It is also shown how the free surface and interaction between tandem foils affect the steady lift and drag of a foil. Unsteady flow conditions due to incident waves and vessel motions are also handled. This discussion is used in Chapter 7 to estimate damping of vertical motions of a semi-displacement vessel due to an attached foil.

Chapter 7 describes the wave-induced motions and global wave loads on semi-displacement vessels. The effects of foil damping and hydrodynamic hull-hull interaction on multihull vessels are also considered. Added resistance in waves and dynamic stability are other issues.

We discuss local and global slamming effects in *Chapter 8*. This is an important structural loading mechanism for all high-speed vessels. The local slamming analysis may need a local hydroelastic analysis. This is shown to be important when the local angle between the impacting free surface and the body surface is less than about five degrees. Global hydroelastic response (springing and whipping) due to wave effects is also discussed in Chapter 8. Springing is a steady-state response, whereas whipping is associated with transient response, such as that caused by water impact (slamming) on the wetdeck, bow flare slamming. or stern slamming.

Chapter 9 discusses both steady and unsteady flow effects around planing vessels. The steady lift and trim moment can, to a large extent, be explained by potential flow theory.

The hydrodynamic performance of prismatic planing hulls in calm water is discussed by examples. Instabilities may play an important role for a planing boat. One example is porpoising, which is unstable heave-and-pitch motions. This is discussed in detail.

Wave-induced vertical motions of planing vessels are also discussed. It is demonstrated that nonlinear effects are more important for planing vessels than for semi-displacement vessels.

Chapter 10 considers maneuvering of a ship in water of infinite horizontal extent. A slender-body theory for a monohull at Froude numbers smaller than approximately 0.2 is presented. This theory can also be applied to a catamaran and an SES. Directional stability, automatic motion control, and viscous effects are other items considered. It is shown that the directional stability changes with forward speed. Further, a maneuvering analysis of a high-speed vessel must, in general, consider motions in six degrees of freedom. A derivation of Euler's equation of motion is given.

2 Resistance and Propulsion

2.1 Introduction

The power of the installed propulsion machinery is an indirect measure of the maximum resistance of a vessel. However, the actual amount of this power that can be transformed into thrust to counteract the resistance depends on the efficiency of the propulsion device. For an ACV and an SES, power is also needed to lift the vessel. For an SES, this is about 10% to 20% of the power needed for propulsion. Casanova and Latorre (1992) have collected data on installed horsepower in different types of high-speed marine vehicles (HSMV).

Our focus in this chapter is on resistance and propulsion in calm water. When we consider a ship with constant speed on a straight course in calm water conditions, the balance of forces is simple: the ship resistance must be equal to the thrust delivered by the propulsion unit.

It is most common in model tests and in numerical calculations to consider the ship without an integrated propulsion system. The resistance is therefore evaluated without the presence of the propulsion unit. We will follow this approach. This means the ship resistance R_T is defined as the force that is needed to tow the ship in calm water with a constant velocity U on a straight track (of course, the towing unit must not affect the flow around the ship). The power needed to tow the vessel is:

$$P_E = R_T U \qquad (2.1)$$

This is not true when wind and waves are present. In that case, added resistance in wind and waves has to be accounted for when required engine power is estimated. Ship maneuvering will also increase the resistance. Even sticking to our assumption of a straight course in calm water, important issues to consider are the efficiency of the propulsion system and how the resistance is affected by the propulsion system. For instance, the flow in the vicinity of a waterjet inlet on a ship hull affects the trim and sinkage of a high-speed vessel, which will then influence the resistance. On the other hand, the flow along the ship hull will affect the inflow conditions to the propulsion unit and hence the thrust. So ideally, we should not have considered resistance and propulsion as separate issues.

We can divide the calm water resistance into

- Viscous water resistance
- Air resistance
- Spray and spray rail resistance
- Wave resistance

Actually, part of the spray resistance is viscous water resistance, whereas the pressure part of spray resistance is difficult to distinguish clearly from the total wave resistance obtained by pressure integration. Each component is discussed in the following text, with the main focus on semi-displacement monohulls and catamarans. However, SES and hydrofoil vessels are also addressed. Additional details on resistance of hydrofoil vessels are given in Chapter 6. Planing hulls are discussed in detail in Chapter 9. More in-depth studies of wave resistance of semi-displacement vessels and SES are considered in Chapter 4.

The resistance is influenced by the trim angle, and trim devices are used on semi-displacement and planing vessels to optimize the trim angles. Examples are interceptors (see Figure 2.2), trim tabs (stern flaps) (see Figure 7.4), and transom wedges, which start forward of the transom and end at the transom. The entire wedge is under the hull and is a local abrupt modification in the buttock lines aft of station $19\frac{1}{2}$ (Cusanelli and Karafiath 1997).

There is ongoing research on how to reduce the ship resistance. One example is by injecting microbubbles into the turbulent boundary layer. Latorre et al. (2003) report that microbubble drag reduction (MBDR) has the potential of reducing the local skin friction by 15%. However, MBDR will not be considered in this text.

As said above, to properly analyze the ship resistance, the latter must be considered in conjunction with the vessel propulsion system. Waterjet propulsion is the most common type of propulsion for high-speed vessels of nonplaning type. Different types of propulsion systems for planing hulls are illustrated in Figure 2.1 and discussed by Savitsky (1992). The most common

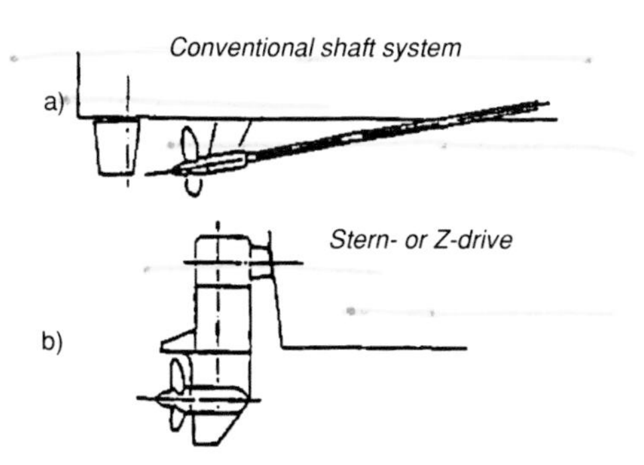

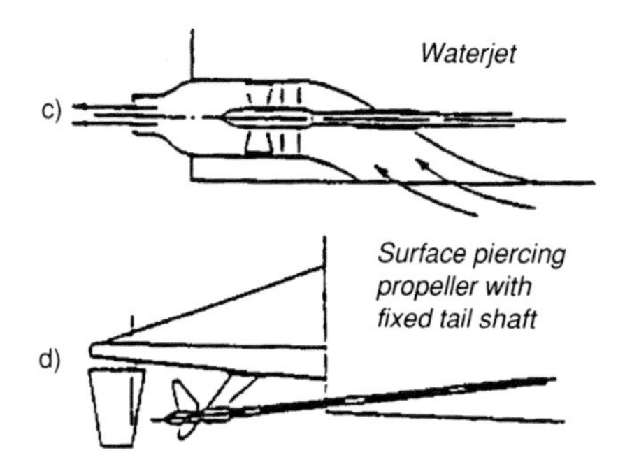

Figure 2.1. Various propulsors for high-speed vessels (Savitsky 1992).

propulsion is a subcavitating or partial cavitating propeller in combination with an inclined shaft (Figure 2.2). The appendage drag due to struts, shaft, and rudder becomes important at higher speeds. The unsteady forces on the inclined propeller may lead to undesirable vibrations. When the maximum speed is higher than 40 knots, free surface–piercing propellers are sometimes used.

The two other types of propulsion systems shown in Figure 2.1 are waterjet propulsion and stern drive propulsion. Stern drive propulsion and outboard engines are used mainly for pleasure and recreation craft. Outboard engines up to 300 hp are made today. In some cases, you might find up to four of these on one boat. In practice, outboard engines are not often used on boats longer than 40 feet. The outboard engines also include a version of the waterjet referred to as a jet drive. These may be used on recreation craft running on very shallow waters, where there is a risk for a propeller to be damaged. Another possibility, then, is to use propeller tunnels. Design of propeller tunnels for high-speed craft is discussed by Blount (1997).

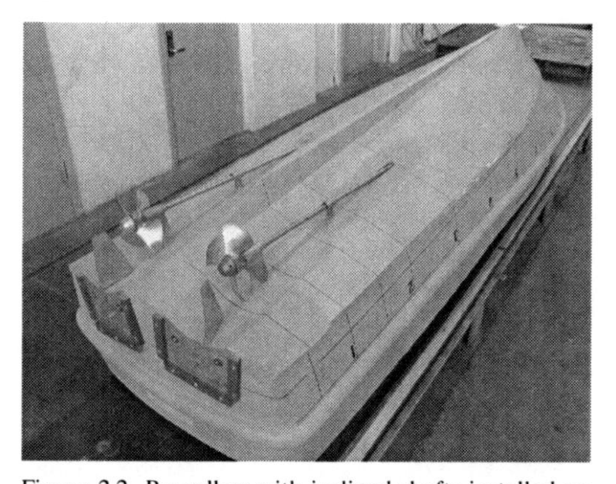

Figure 2.2. Propellers with inclined shafts installed on a model of a planing vessel with hard chines. Propeller tunnels are used to minimize the shaft angle. The rudders are twisted and adopted to the propeller slip stream. Two interceptors are placed at the transom to control the trim angle (see section 7.1.3 and Figure 7.5 for more details about interceptors). (Photo by K.A. Hegstad)

Oblique-flow conditions, locally concentrated wake peaks, and high loading density at high speeds make it difficult to avoid cavitation on a propeller. Oblique flow occurs, for instance, when the propeller shaft has an angle relative to the vessel velocity (see Figure 2.1a). Propeller tunnels are beneficial in this context. Cavitation has been discussed by van Beek (1992), who considers thrust breakdown and cavitation at a propeller blade root as limiting criteria for the application of conventional high-speed propellers.

If oblique flow can be avoided, conventional propellers may run with little cavitation, even at 45 knots. This has been demonstrated by tractor propellers in conjunction with right-angle drives installed in catamarans and foil catamarans (Halstensen and Leivdal 1990).

Our way of treating resistance and propulsion of high-speed vessels follows the traditional route in ship hydrodynamics. However, an interesting question is: What can we learn about resistance and propulsion from aquatic animals? Despite potential payoffs, relatively little work has been done to answer this question. An introduction to this field is given by Triantafyllou and Triantafyllou (1995) and Sfakiotakis et al. (1999).

2.2 Viscous water resistance

A main resistance component is caused by the friction force on the wetted hull. Pressure loads acting perpendicularly to the hull surface matter, but have less importance. Boundary layer theory

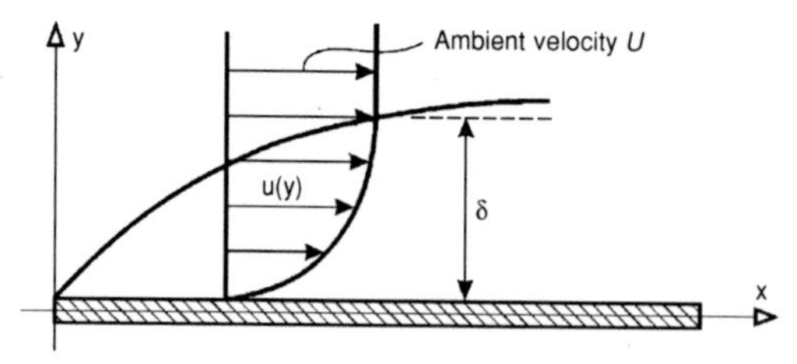

Figure 2.3. Boundary layer along a flat plate with incident (ambient) flow velocity U along the x-axis. $\delta =$ boundary-layer thickness.

may be used to describe the effect of fluid viscosity. It means that the viscosity only matters in a thin layer close to the hull surface. The two-dimensional boundary layer along a flat plate can be used to describe important characteristics of the viscous flow. We can approximate the wetted hull surface as a flat plate. If we look at the flow from a reference frame following the ship, the forward speed of the ship appears as an incident flow with velocity U on a stationary hull, as shown in Figure 2.3.

One important characteristic is that the water must adhere to the plate, that is, there is no slip. That means the flow velocity is zero on the plate. At a short perpendicular distance $\delta(x)$ from the plate (function of the longitudinal distance x from the leading edge of the plate), the flow velocity is equal to U.

The viscous flow is laminar for Reynolds number $Rn_x = Ux/\nu$ less than $\approx 10^5$. Here ν is the kinematic viscosity coefficient with $1.35 \cdot 10^{-6}\,\mathrm{m^2s^{-1}}$ for salt water at 10°C (see Table A.2 in the Appendix). The transition to turbulent flow occurs for Rn_x between $2 \cdot 10^5$ and $3 \cdot 10^6$. Turbulent flow is characterized by a velocity and a pressure that vary irregularly with a high frequency. Laminar flow means that the flow is well organized in layers. It is steady when the incident velocity is steady. One can make the analogy between laminar flow and a school class marching orderly in a parade. Every pupil keeps his or her position relative to the others so that a clear structure with rows and columns appears. Then things get out of order and the pupils run everywhere without an apparent system except that they have a mean forward motion. This is like a turbulent flow. This analogy between hydrodynamics and human beings is used in simulating evacuation of passengers from passenger vessels during catastrophic events. Hinze (1987) gives the following definition of turbulence:

"Turbulent fluid motion is an irregular condition of flow in which the various quantities show a random variation with time and space coordinates, so that statistically distinct values can be discerned." Turbulence frequencies may vary between 1 and 10,000 $\mathrm{s^{-1}}$, and turbulent fluctuations are roughly 10% of average velocity (Hinze 1987). The upper and lower bounds of the turbulence frequencies depend on the field of application. Consider, for instance, cross-flow past a circular cylinder at a high Reynolds number. This is associated with a vortex shedding frequency that is described by the Strouhal number as a function of the Reynolds number (Faltinsen 1990). Depending on the cylinder diameter and the ambient flow, a vortex shedding frequency can be 1 Hz in marine applications. This frequency cannot be considered a turbulence frequency. If the frequency range around a vortex shedding frequency was filtered out by an averaging process, one would lose important information on vortex-induced vibrations of structures.

Figure 2.4 illustrates how the flow changes from being laminar to turbulent along a smooth flat plate. The laminar 2D flow becomes unstable at a critical Reynolds number Rn_{crit}. If there is negligible turbulence intensity in the incident flow, this corresponds to $Ux/\nu = 2 \cdot 10^5$. Rn_{crit} can be found by a linear stability analysis (Schlichting 1979). The unstable 2D waves shown in Figure 2.4 are called Tollmien-Schlichting (T/S) waves. As the amplitudes of the T/S waves grow, three-dimensional instabilities occur. Fully turbulent flow occurs at the transition Reynolds number Rn_{tr}. If there is negligible turbulence intensity in the incident flow, Rn_{tr} is $3 \cdot 10^6$.

The horizontal velocity distribution in Figure 2.3 is representative of a laminar boundary layer in the case of a 2D flow along a flat plate. This can be described by the Blasius theory. The frictional stress (longitudinal force per unit area) on the plate is

$$\tau_w = \mu \left. \frac{\partial u}{\partial y} \right|_{y=0} \tag{2.2}$$

Here μ is the dynamic viscosity coefficient, which is related to the kinematic viscosity coefficient ν by $\nu = \mu/\rho$, where ρ is mass density of the fluid. Eq. (2.2) is also applicable to turbulent boundary layer flow, but then u means in practice a velocity that has been time averaged on the time scale of turbulence.

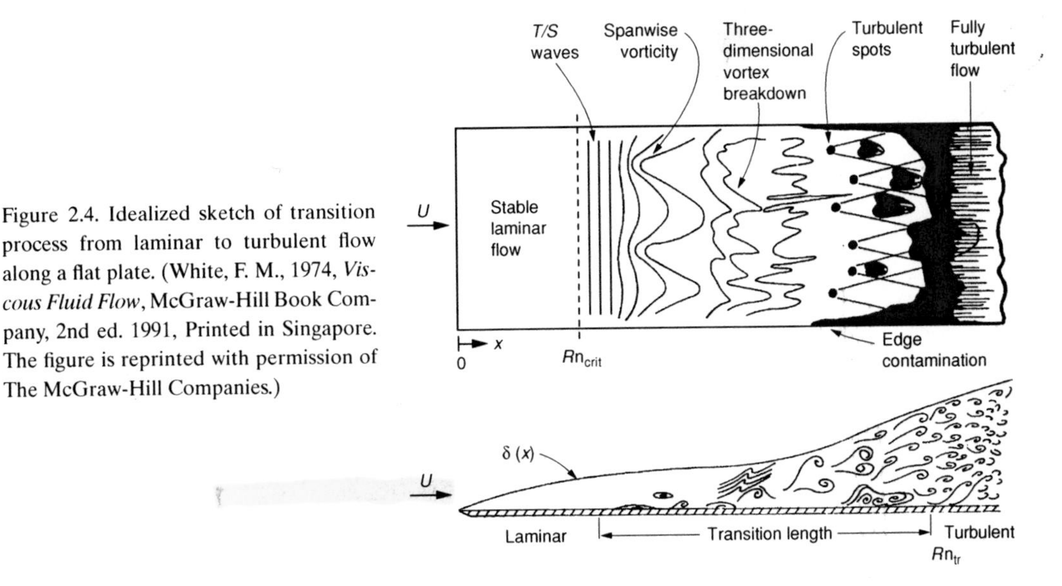

Figure 2.4. Idealized sketch of transition process from laminar to turbulent flow along a flat plate. (White, F. M., 1974, *Viscous Fluid Flow*, McGraw-Hill Book Company, 2nd ed. 1991, Printed in Singapore. The figure is reprinted with permission of The McGraw-Hill Companies.)

The velocity gradient $\partial u/\partial y$ at the plate is very different for laminar and turbulent flows (Figure 2.5). Because turbulent flow implies much a larger exchange of fluid momentum in the y-direction than laminar flow does, both the boundary layer thickness δ and $\partial u/\partial y$ at the plate are much larger for turbulent flow than for laminar flow.

We cannot avoid turbulent flow along the hull surface of a full-scale ship, and we must ensure turbulent flow along the hull surface during model tests to be able to scale the results to full scale. Because we do not have sufficient knowledge yet on how to model turbulence theoretically, empiricism has to be partly used. We discuss this in more detail in section 2.2.4. The empirical formulas for frictional resistance are amazingly simple. They express the viscous resistance as

$$R_V = 0.5\rho C_F S U^2, \tag{2.3}$$

where S is the wetted surface area. It is common to estimate S at zero speed. However, S changes in reality as a result of the free surface elevation along the hull. Further, the transom stern of a semi-displacement vessel becomes dry for Froude numbers $Fn = U/\sqrt{Lg}$ higher than approximately 0.4, and an SES on cushion causes a lower free surface elevation inside the cushion than outside the cushion. Here L is the overall submerged ship length L_{OS} and g is the acceleration of gravity. The International Towing Tank Conference (ITTC) 1957 model–ship correlation line expresses the friction coefficient C_F for a smooth hull surface as

$$C_F = \frac{0.075}{(\log_{10} Rn - 2)^2}, \tag{2.4}$$

where $Rn = UL/\nu$ is the Reynolds number. Eq. (2.4) agrees well with experimental results for turbulent flow along a smooth flat plate.

Figure 2.6 illustrates how C_F changes going from a laminar boundary layer to a turbulent boundary layer along a flat plate. The Blasius solution is used for laminar flow, and the Prandtl–von Karman

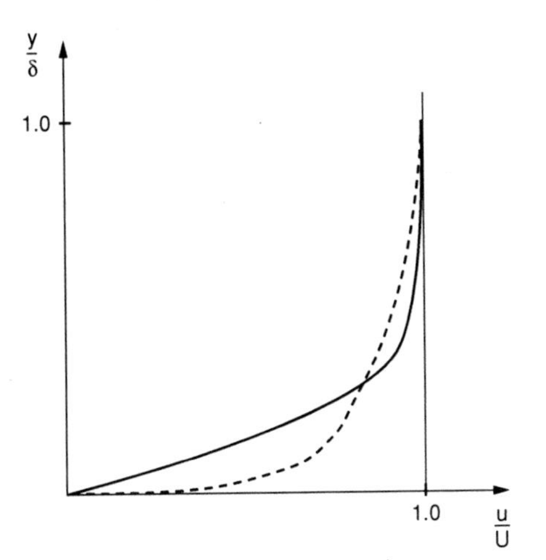

Figure 2.5. Laminar (——) and mean turbulent (- - - - -) velocity profiles for the boundary-layer flow along a flat plate.

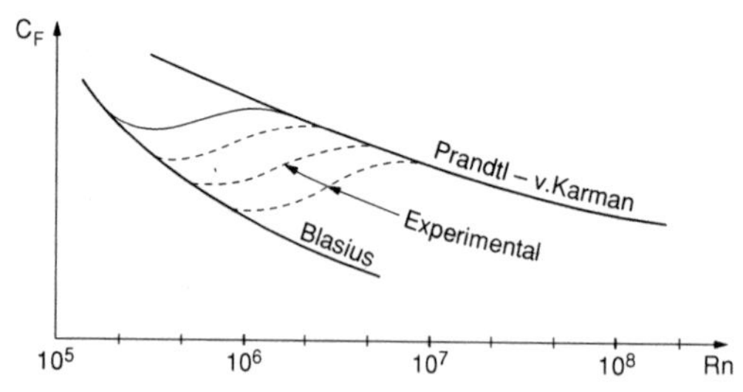

Figure 2.6. Friction coefficients C_F for flow along a flat plate as a function of Reynolds number Rn (Walderhaug 1972).

expression is applied for turbulent flow. Later on, we will see that different empirical formulas exist for turbulent flows. The Blasius solution for laminar flow may be expressed as

$$C_F = \frac{1.328}{Rn^{1/2}}. \tag{2.5}$$

The Prandtl–von Karman expression is

$$C_F = \frac{0.072}{Rn^{1/5}}. \tag{2.6}$$

In Figure 2.6, several curves (experimental data) for C_F are indicated in the transition between laminar and turbulent flows. These depend on the turbulence intensity T in the inflow velocity, which may be expressed as

$$T = \sqrt{\overline{u'^2}}/U, \tag{2.7}$$

u' being the turbulent part of the longitudinal component of the inflow velocity u, that is, $u = U + u'$. Further, $\overline{u'^2}$ means the time average over the turbulence time scale of u'^2. When $T < 0.001$, there is no influence of T and the transition Reynolds number is $2.8 \cdot 10^6$. However, if $T = 0.03$, the transition Reynolds number becomes 10^5.

Because a hydrofoil has a Reynolds number much smaller than that of a submerged hull-supported vessel, one may be tempted to design laminar foil shapes, which are used in connection with gliders. Schlichting (1979) gives examples of foil shapes and their C_F values. The "laminar effect" reduces the drag of normal airfoils by 30% to 50% in the Reynolds number range of $2 \cdot 10^5$ to $3 \cdot 10^7$. When $Rn > 5 \cdot 10^7$, the laminar effect is lost and the flow is fully turbulent. If we consider as an example a hydrofoil with velocity $U = 20\ \text{ms}^{-1}$ and use $\nu = 10^{-6}\text{m}^2\text{s}^{-1}$, we see that foils with chord lengths less than 1.5 m may benefit from the laminar effect. However, we should note that the results are for zero incidence, that is, the foils do not cause lift. The presence of lift implies a change in the pressure gradient along the foil relative to no lift. This influences the drag force and the transition. Further, the presence of surface roughness may change the results. This is discussed in section 2.2.6. Attention must also be given to cavitation inception and the effect of nonuniform inflow in the hydrofoil design.

We will present a theoretical basis for the resistance formula for turbulent flow along a smooth flat plate. To better understand this, we need first to present Navier-Stokes equations.

2.2.1 Navier-Stokes equations

The flow around a ship is governed by the Navier-Stokes equations, so to study the vessel resistance, such equations should be solved for the problem of interest. Navier-Stokes equations are presented in many textbooks of fluid mechanics, such as Newman (1977), Schlichting (1979), and White (1974). We will limit ourselves to two-dimensional flow of an incompressible fluid and refer to the above-mentioned textbooks for a detailed and general derivation of Navier-Stokes equations. For our applications, water can be considered incompressible, that is, sound waves do not matter. The flow around a ship is, of course, three-dimensional, but empirical formulas for viscous resistance are to a large extent based on two-dimensional flow for a flat plate. Because the boundary layer in which viscosity matters generally has a small thickness δ relative to the local radii of curvature of the hull surface, we can justify that the hull surface appears locally flat and that the 2D flat plate flow represents a first approximation. We later will see how we correct empirically for three-dimensional flow around a ship hull by introducing form factors.

The two-dimensional Navier-Stokes equations for an incompressible fluid without gravity can be written as

$$\frac{\partial u}{\partial t} + u\frac{\partial u}{\partial x} + v\frac{\partial u}{\partial y} = -\frac{1}{\rho}\frac{\partial p}{\partial x} + \nu\left(\frac{\partial^2 u}{\partial x^2} + \frac{\partial^2 u}{\partial y^2}\right) \tag{2.8}$$

$$\frac{\partial v}{\partial t} + u\frac{\partial v}{\partial x} + v\frac{\partial v}{\partial y} = -\frac{1}{\rho}\frac{\partial p}{\partial y} + \nu\left(\frac{\partial^2 v}{\partial x^2} + \frac{\partial^2 v}{\partial y^2}\right). \tag{2.9}$$

The continuity equation is

$$\frac{\partial u}{\partial x} + \frac{\partial v}{\partial y} = 0. \tag{2.10}$$

Here we have used a Cartesian coordinate system (x, y) as in Figure 2.3. u and v are the x- and y-components of the fluid velocity, t is the time variable, and p is the pressure. We have three equations and three unknowns: u, v, and p. In order to solve eqs. (2.8) to (2.10), we need a set of initial and boundary conditions. The body boundary condition requires that the fluid adheres to the body surface (no-slip).

Eqs. (2.8) and (2.9) follow by analyzing the motion inside an arbitrary fluid volume and enforcing that the time rate of change of momentum inside the fluid volume is equal to the sum of forces acting on the fluid volume, that is, Newton's second law. These forces are the result of hydrodynamic pressure and viscous stresses. Concerning the hydrodynamic pressure contribution, the force per unit area due to the pressure p acts perpendicularly to a surface element as $-p\mathbf{n}$. Here $\mathbf{n}$ is the surface unit normal vector with positive direction outward from the fluid volume. To introduce the viscous stresses, we consider a two-dimensional rectangular fluid volume with sides parallel to the x- and y-axes (Figure 2.7). On the top side AB of the volume, we have the viscous stress components τ_{xy} and τ_{yy} along the x- and y-axes, respectively. They can be expressed as

$$\tau_{xy} = \mu \left(\frac{\partial u}{\partial y} + \frac{\partial v}{\partial x} \right) \tag{2.11}$$

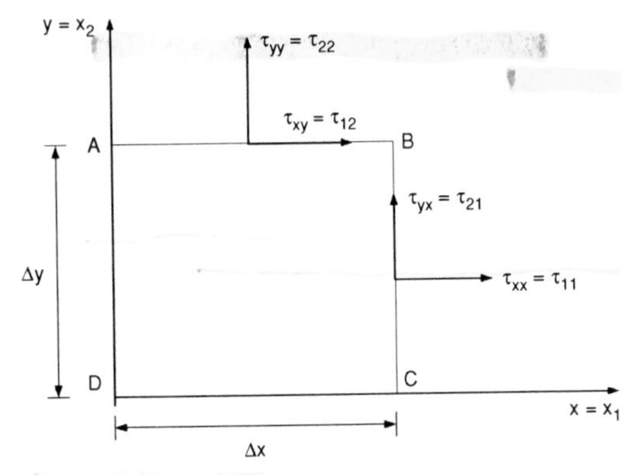

Figure 2.7. Rectangular fluid volume ABCD and viscous stresses τ_{ij} acting on AB and BC sides.

and

$$\tau_{yy} = 2\mu \frac{\partial v}{\partial y} \tag{2.12}$$

The pressure force per unit area on AB is $-p$ and acts along the y-axis. This means the total hydrodynamic force per unit area on AB consists of the components τ_{xy} and $(-p + \tau_{yy})$ along the x and y-axes, respectively.

The viscous stress components on the vertical side BC are

$$\tau_{yx} = \mu \left(\frac{\partial v}{\partial x} + \frac{\partial u}{\partial y} \right) \tag{2.13}$$

and

$$\tau_{xx} = 2\mu \frac{\partial u}{\partial x}. \tag{2.14}$$

Here τ_{yx} and τ_{xx} are viscous stress components directed along the y- and x-axes, respectively. The total hydrodynamic force per unit area on BC consists of the components $-p + \tau_{xx}$ and τ_{yx} along the x and y-axes, respectively. In order to express the viscous stresses in a more abbreviated and general way in three dimensions, we change notation so that $x = x_1$, $y = x_2$, $z = x_3$, $u = u_1$, $v = u_2$, $w = u_3$ and introduce τ_{ij}, where i or j is equal to 1, 2, or 3 when referring to x, y, and z, respectively. This notation is depicted in Figure 2.7 for the two-dimensional case. Here we have also included a third dimension by introducing the z-coordinate of the Cartesian coordinate system $(x, y\ z)$ and the velocity component $w = u_3$ along the z-axis. We define a surface element with an outward unit normal vector $\mathbf{n} = (n_1, n_2, n_3)$. If this surface element belongs to a side of a fluid volume as in Figure 2.7, then $\mathbf{n}$ is pointing outward from the fluid volume. If we consider a surface element on a body surface, then the normal direction is into the fluid domain. The viscous stress (force per unit area) in the ith direction is then

$$\tau_{i1}n_1 + \tau_{i2}n_2 + \tau_{i3}n_3, \tag{2.15}$$

where

$$\tau_{ij} = \mu \left(\frac{\partial u_i}{\partial x_j} + \frac{\partial u_j}{\partial x_i} \right). \tag{2.16}$$

We note the symmetry $\tau_{ij} = \tau_{ji}$. The justification of this linear relationship between viscous stresses and derivatives of velocity components is, for instance, discussed in Newman (1977). The Newtonian stress relations given by eq. (2.16) assume an incompressible fluid.

We will demonstrate that eqs. (2.15) and (2.16) are consistent with eq. (2.11). For example, let us consider the top side AB in Figure 2.7, where $n_1 = 0$, $n_2 = 1$, and $n_3 = 0$. This means that we have the viscous stress components $\mu(\partial u_1/\partial x_2 + \partial u_2/\partial x_1)$ and $2\mu\partial u_2/\partial x_2$, that is, the same as eqs. (2.11) and (2.12). With the same procedure at the bottom side CD, where $n_1 = 0$, $n_2 = -1$, and $n_3 = 0$, we find the viscous stress components $-\mu(\partial u_1/\partial x_2 + \partial u_2/\partial x_1)$ and $-2\mu\partial u_2/\partial x_2$ directed along the x- and y-axes, respectively. These expressions are similar to eqs. (2.11) and (2.12), but with opposite signs. This has to be kept in mind for our next derivation of eqs. (2.8) and (2.9). As said, they follow from Newton's second law. We shall focus on eq. (2.8) and have in mind the fluid volume in Figure 2.7. The sides Δx and Δy are assumed small so that all quantities can be approximated by the lowest-order terms in a Taylor expansion about the center of the volume. We first evaluate the forces acting on the volume. The resultant viscous force component in the x-direction acting on AD and BC can then be approximated as

$$\Delta x \frac{\partial}{\partial x}\left(2\mu \frac{\partial u}{\partial x}\right) \Delta y. \tag{2.17}$$

Further, the viscous force component in the x-direction along AB and DC becomes

$$\Delta y \frac{\partial}{\partial y}\left[\mu\left(\frac{\partial u}{\partial y} + \frac{\partial v}{\partial x}\right)\right] \Delta x. \tag{2.18}$$

The sum of eqs. (2.17) and (2.18) can finally be rewritten as

$$\mu\left(\frac{\partial^2 u}{\partial x^2} + \frac{\partial^2 u}{\partial y^2}\right) \Delta x \Delta y \tag{2.19}$$

by means of the continuity equation (2.10). By a similar Taylor expansion, the pressure force on the surface of the fluid can be approximated as

$$-\frac{\partial p}{\partial x} \Delta x \Delta y. \tag{2.20}$$

Then we consider the time rate of change of fluid momentum in the x-direction of the volume. Part of this is the result of momentum flux through AB, BC, CD, and DA. The momentum flux through a surface element that is not moving is

$$\rho \mathbf{u}\,(\mathbf{u} \cdot \mathbf{n})\, dS, \tag{2.21}$$

where $\mathbf{u} = (u, v, w)$ and dS is the area of the surface element. Once more making a Taylor expansion, we find that the momentum flux in x-direction through AD and BC can be approximated as

$$\rho \Delta x \frac{\partial}{\partial x}\left(u^2\right) \Delta y. \tag{2.22}$$

The momentum flux in x-direction through AB and CD reduces to

$$\rho \Delta y \frac{\partial}{\partial y}(uv)\, \Delta x. \tag{2.23}$$

The sum of eqs. (2.22) and (2.23) can by means of the continuity eq. (2.10) be rewritten as

$$\rho\left(u\frac{\partial u}{\partial x} + v\frac{\partial u}{\partial y}\right) \Delta x \Delta y. \tag{2.24}$$

Then we have to add the term

$$\rho \frac{\partial u}{\partial t} \Delta x \Delta y \tag{2.25}$$

to get all the contributions to the time rate of change of the fluid momentum in the x-direction inside the volume. These must be balanced by the forces acting on the volume. By doing this, we find the following equation:

$$\begin{aligned} &\rho\left(\frac{\partial u}{\partial t} + u\frac{\partial u}{\partial x} + v\frac{\partial u}{\partial y}\right) \Delta x \Delta y \\ &= -\frac{\partial p}{\partial x} \Delta x \Delta y + \mu\left(\frac{\partial^2 u}{\partial x^2} + \frac{\partial^2 u}{\partial y^2}\right) \Delta x \Delta y, \end{aligned} \tag{2.26}$$

as a first order equation valid for small Δx and Δy. By dividing by $\rho \Delta x \Delta y$ on both sides and then letting Δx and Δy go to zero, we see that this leads to eq. (2.8).

2.2.2 Reynolds-averaged Navier-Stokes (RANS) equations

In principle, we can directly use Navier-Stokes equations and solve them numerically to study the turbulent flow. However, a very small time and spatial discretization are needed for the numerical solution. Available computer technology limits the possibilities. Reynolds-averaged Navier-Stokes (RANS) formulations are commonly used instead.

The RANS formulation means that we decompose the variables of interest as $u = \bar{u} + u'$, $v = \bar{v} + v'$, $p = \bar{p} + p'$, where u', v', and p' are varying on the time scale of turbulence and $\bar{u}$, $\bar{v}$, and

$\bar{p}$ are time averaged over the time scale of turbulence. Then we insert this into eqs. (2.8) to (2.10) and time average the equations over the time scale of turbulence. Let us show this procedure for the convective acceleration term $u\partial u/\partial x + v\partial u/\partial y$ in eq. (2.8). By using the continuity equation, this contribution can be rewritten as

$$\frac{\partial u^2}{\partial x} + \frac{\partial uv}{\partial y}. \tag{2.27}$$

This result was already shown when we rewrote eqs. (2.22) and (2.23) into eq. (2.24). We can now write eq. (2.27) as

$$\frac{\partial \bar{u}^2}{\partial x} + \frac{\partial \bar{u}\bar{v}}{\partial y} + 2\frac{\partial \bar{u}u'}{\partial x} + \frac{\partial}{\partial y}(\bar{u}v' + u'\bar{v}) + \frac{\partial u'^2}{\partial x} + \frac{\partial u'v'}{\partial y}. \tag{2.28}$$

Then we time average eq. (2.28). The two first terms remain the same. Because the time averages $\overline{u'}$ and $\overline{v'}$ are zero, the third and fourth terms give zero contribution. However, the time averages of the two last terms are not zero. Because turbulence is 3D for 2D inflow conditions (see Figure 2.4), we should actually have done the time averaging by starting with the 3D Navier-Stokes equations (see Schlichting 1979). Because the $\partial u/\partial t$-term in eq. (2.8) also refers to time variations on a time scale larger than that for turbulence, we must include the effect of this term. Eq. (2.8) can eventually be expressed as

$$\frac{\partial \bar{u}}{\partial t} + \bar{u}\frac{\partial \bar{u}}{\partial x} + \bar{v}\frac{\partial \bar{u}}{\partial y} = -\frac{1}{\rho}\frac{\partial \bar{p}}{\partial x} + \nu\left(\frac{\partial^2 \bar{u}}{\partial x^2} + \frac{\partial^2 \bar{u}}{\partial y^2}\right) - \frac{\partial \overline{u'^2}}{\partial x} - \frac{\partial \overline{u'v'}}{\partial y}. \tag{2.29}$$

The term proportional to ν in eq. (2.29) is the result of viscous stresses. The two last terms on the right-hand side of eq. (2.29) are the result of what is called turbulent stresses or Reynolds stresses. The challenge in solving eq. (2.29) is that we have introduced several new unknowns, such as $\overline{u'^2}$ and $\overline{u'v'}$; therefore, we need new equations. In practice, these are empirical, that is, we need guidance from experiments.

Turbulence modeling and numerical computations based on RANS, particularly for 2D flow around bodies, are extensively covered by Cebeci (2004). Efforts are also made to use large-eddy simulations (LES). Empirical relationships are then only needed for the small-scale turbulent flow. CFD (computational fluid dynamics) methods relevant to ship resistance and flow are discussed by Larsson and Baba (1996).

2.2.3 Boundary-layer equations for 2D turbulent flow

Our problem deals with boundary-layer flows, that is, we are interested in the turbulent flow in a narrow region near the hull surface. Therefore we can further approximate eq. (2.29). The boundary-layer thickness δ in Figure 2.3 is small relative to the distance x from the leading edge. The mean velocity varies rapidly across the boundary layer from zero on the body to the free stream velocity U_e at $y = \delta$. This implies that $\partial\bar{u}/\partial y$ is much larger than $\partial\bar{u}/\partial x$. The consequence is that the $\partial^2\bar{u}/\partial x^2$ term in eq. (2.29) can be neglected relative to the $\partial^2\bar{u}/\partial y^2$ term. It is implicit from Figure 2.3 that the flow must vary both with x and y. If we neglect one of the terms in the continuity equation $\partial\bar{u}/\partial x + \partial\bar{v}/\partial y = 0$, this will not be true. Because $\partial\bar{v}/\partial y$ is the order of $\bar{v}$ divided by δ and $\partial\bar{u}/\partial x$ is the order of $\bar{u}$, $\bar{v}$ is the order of $\bar{u}\cdot\delta$, that is, $\bar{v}$ is smaller than $\bar{u}$. This implies that both terms $\bar{u}\partial\bar{u}/\partial x$ and $\bar{v}\partial\bar{u}/\partial y$ in the convective acceleration of eq. (2.29) are of the same order. From eq. (2.9), it follows that $\partial\bar{p}/\partial y$ is of the order of $\bar{u}\cdot\delta$. This means that, as a first approximation, in eq. (2.29) we can set $\partial\bar{p}/\partial y = 0$. This gives that $\bar{p}$ in eq. (2.29) is the same as $\bar{p}$ at $y = \delta$. Thus as long as the boundary layer has a small thickness δ, $\bar{p}$ can be calculated from the flow outside the boundary layer. There, the fluid is accurately described by the potential flow theory, that is, the fluid can be modeled as inviscid and in irrotational motion. This estimate of $\bar{p}$ can be done by neglecting the boundary layer and finding the tangential velocity U_e at the body surface. The steady version of eq. (2.29) based on potential flow gives $\rho U_e dU_e/dx = -d\bar{p}/dx$. We can also neglect the term $\partial\overline{u'^2}/\partial x$ in eq. (2.29). In this way, we end up with the following steady 2D boundary layer equations for turbulent flow:

$$\bar{u}\frac{\partial \bar{u}}{\partial x} + \bar{v}\frac{\partial \bar{u}}{\partial y} = U_e\frac{dU_e}{dx} + \frac{\partial}{\partial y}\left(\nu\frac{\partial \bar{u}}{\partial y} - \overline{u'v'}\right) \tag{2.30}$$

$$\frac{\partial \bar{u}}{\partial x} + \frac{\partial \bar{v}}{\partial y} = 0. \tag{2.31}$$

In the case of steady flow along a flat plate, we will have $U_e = U$ and $dU_e/dx = 0$. Further, the last term in eq. (2.30) can be expressed as

$$\frac{1}{\rho}\frac{\partial}{\partial y}\left(\mu\frac{\partial \bar{u}}{\partial y} - \rho\overline{u'v'}\right) = \frac{1}{\rho}\frac{\partial}{\partial y}\left(\tau_l + \tau_t\right),$$

where

$$\tau_l = \mu\frac{\partial \bar{u}}{\partial y} \quad (2.32)$$

is the viscous (also called laminar) shearing stress and

$$\tau_t = -\rho\overline{u'v'} \quad (2.33)$$

is the turbulent stress.

We have pointed out earlier (see eq. (2.16)) that many stress components exist. Eq. (2.32) is a boundary-layer approximation of τ_{xy} given by eq. (2.11). This means that τ_t is also a longitudinal force per unit area on a horizontal surface like AB in Figure 2.7. Measurements show that there is a domain very close to the body surface where $\tau_l \gg \tau_t$. We can understand this by noting that τ_t is zero on the body surface, which is a consequence of the body boundary condition, that is, $u' = v' = 0$ on the body surface. However, τ_l is not zero on the body surface, as we see from the velocity distribution in Figure 2.5. The domain in which τ_l dominates is called the viscous sublayer.

2.2.4 Turbulent flow along a smooth flat plate. Frictional resistance component

Instead of proceeding with finding a numerical solution to the boundary-layer equations, we will follow a very different way to find the shear stress on a flat plate, the velocity distribution in the boundary layer, and the boundary-layer thickness.

The first step is to define three layers of fluid next to the surface of the flat plate:

Inner layer or viscous sublayer:	Viscous shear τ_l dominates
Outer layer:	Turbulent shear τ_t dominates
Overlap layer:	Both types of shear are important

Why we can state that the turbulent shear dominates in the outer layer is a consequence of experimental results.

The inner layer is very thin relative to the boundary-layer thickness δ. Then on the scale of the inner layer, the outer layer is very far away. It could just as well be at infinity. Therefore, the mean longitudinal velocity $\bar{u}$ in the inner layer will not be a function of δ. To see what parameters $\bar{u}$ depends on, we start with a Taylor expansion of $\bar{u}$ about $y = 0$, that is, the surface of the plate. We can write

$$\bar{u} = \bar{u}|_{y=0} + y\frac{\partial \bar{u}}{\partial y}\bigg|_{y=0} + \frac{1}{2}y^2\frac{\partial^2 \bar{u}}{\partial y^2}\bigg|_{y=0} + \frac{1}{6}y^3\frac{\partial^3 \bar{u}}{\partial y^3}\bigg|_{y=0} + O(y^4). \quad (2.34)$$

The body boundary condition gives $\bar{u}|_{y=0} = 0$, and eq. (2.2) gives $\partial\bar{u}/\partial y|_{y=0} = \tau_w/\mu$. If we apply eq. (2.30) at $y = 0$ and neglect turbulent stresses, we find that

$$-U_e\frac{dU_e}{dx} = \nu\frac{\partial^2 \bar{u}}{\partial y^2}\bigg|_{y=0}. \quad (2.35)$$

This means $\partial^2\bar{u}/\partial y^2|_{y=0} = 0$ for the steady flow along a flat plate. Further, if we differentiate eq. (2.30) with respect to y, we find that $\partial^3\bar{u}/\partial y^3|_{y=0} = 0$. In this way, we have shown that $\bar{u}$ in the inner layer can be expressed as

$$\bar{u} = y\frac{\tau_w}{\mu} + O(y^4), \quad (2.36)$$

where $O(\,)$ means order of magnitude. This means $\bar{u}$ is a function of y, τ_w, and μ, but it also will be a function of ρ as a consequence of the fact that the laminar stresses τ_l decelerate the fluid particles. This is expressed by eq. (2.30). Similar to Prandtl (1933), this gives

$$\text{Inner law: } \bar{u} = f(\tau_w, \rho, \mu, y). \quad (2.37)$$

We do not know the function f in the whole inner layer but only very close to the surface of the flat plate, as expressed by eq. (2.36).

von Karman (1930) deduced that in the outer layer, we can write

$$\text{Outer law: } U - \bar{u} = f(\tau_w, \rho, y, \delta). \quad (2.38)$$

We have used the same symbol f in eqs. (2.37) and (2.38) to indicate a function, but obviously it is not the same function in the two expressions. Because laminar stresses τ_l do not matter in the outer layer, we can understand why eq. (2.38) does not depend on μ. The presence of τ_w in eq. (2.38) expresses the fact that the wall retards the flow in the outer layer.

In the overlap layer, we expect that the outer law and the inner law match, or that both eqs. (2.37) and (2.38) are valid. Before proceeding with the matching, we will introduce nondimensional variables using the Pi-theorem. The Pi-theorem is due to Buckingham (1915) and was elaborated in detail by Rouse (1961).

The Pi-theorem states:
Let a physical law be expressed in terms of n physical quantities, and let k be the number of fundamental units needed to measure all quantities. Then the law can be re-expressed as a relation among (n-k) dimensionless quantities.

Both eqs. (2.37) and (2.38) contain five physical quantities and three fundamental units (mass, length, and time). This means that according to the Pi-theorem, eqs. (2.37) and (2.38) can be re-expressed in terms of $5-3=2$ dimensionless variables. The expressions are

$$\text{Inner law: } \frac{\bar{u}}{v^*} = f\left(\frac{yv^*}{\nu}\right) \tag{2.39}$$

$$\text{Outer law: } \frac{U-\bar{u}}{v^*} = g\left(\frac{y}{\delta}\right). \tag{2.40}$$

Here

$$v^* = \sqrt{\frac{\tau_w}{\rho}} \tag{2.41}$$

is called the wall friction velocity. In order to check that v^* has the units of ms^{-1}, it is noted that τ_w has the units of Nm^{-2} or $\mathrm{kgm}^{-1}\mathrm{s}^{-2}$ and that ρ has the units of kgm^{-3}. This means that τ_w/ρ has the units of $\mathrm{m}^2\mathrm{s}^{-2}$. We can find the function f in eq. (2.39) very near the wall by using eq. (2.36). We then divide both sides of eq. (2.36) by $\sqrt{\tau_w/\rho}$, that is, v^*. Using $\mu = \nu\rho$, this gives

$$\frac{\bar{u}}{v^*} \approx \frac{y\tau_w}{\nu\rho\sqrt{\tau_w/\rho}} = \frac{yv^*}{\nu}. \tag{2.42}$$

We now apply eqs. (2.39) and (2.40) in the overlap region. This means

$$\text{Overlap law: } \frac{\bar{u}}{v^*} = f\left(\frac{yv^*}{\nu}\right) = \frac{U}{v^*} - g\left(\frac{y}{\delta}\right). \tag{2.43}$$

Differentiating eq. (2.43) with respect to y gives

$$f'(y^+)\frac{v^*}{\nu} = -g'(\eta)\frac{1}{\delta}. \tag{2.44}$$

Here

$$y^+ = \frac{yv^*}{\nu} \tag{2.45}$$

and

$$\eta = \frac{y}{\delta}. \tag{2.46}$$

We then multiply eq. (2.44) by y and get the separated variables form:

$$f'(y^+)y^+ = -g'(\eta)\eta. \tag{2.47}$$

Let us now consider y^+ and η as independent variables. The only way to satisfy eq. (2.47) is for both the left- and right-hand sides to be equal to the same constant, which we denote $1/\kappa$. This means

$$\begin{aligned} f'(y^+)\,y^+ &= 1/\kappa \\ g'(\eta)\,\eta &= -1/\kappa \end{aligned}.$$

Integrating these two equations gives

$$\begin{aligned} f(y^+) &= \tfrac{1}{\kappa}\ln(y^+) + B \\ g(\eta) &= -\tfrac{1}{\kappa}\ln(\eta) + A \end{aligned},$$

where A and B are constants. This means that in the overlap layer, we can write either

$$\frac{\bar{u}}{v^*} = \frac{1}{\kappa}\ln\left(\frac{yv^*}{\nu}\right) + B \tag{2.48}$$

by using inner layer variables or

$$\frac{U-\bar{u}}{v^*} = -\frac{1}{\kappa}\ln\left(\frac{y}{\delta}\right) + A \tag{2.49}$$

by using outer-layer variables. The constants κ, B, and A have to be experimentally determined and are found to be $\kappa = 0.4$ and $B = 5.5$ according to Nikuradse (1930). Schultz-Grunow (1940) found that $A \approx 2.35$. The overlap region corresponds to $35 < yv^*/\nu < 350$. We should note that eq. (2.48) cannot be valid in the whole inner layer. It does not agree with eq. (2.42) and will actually give infinite value of $\bar{u}$ for $y = 0$. Further, eq. (2.49) cannot be valid in the whole outer layer. We see that it does not give $\bar{u} = U$ when $y = \delta$.

Based on experimental results, it is possible to construct a composite representation of $\bar{u}$ for both the outer layer and the overlap layer (White 1974). For turbulent flow along a flat plate, we can write

$$\frac{\bar{u}}{v^*} = 2.5\ln\left(\frac{yv^*}{\nu}\right) + 5.5 + 2.5\sin^2\left(\frac{\pi}{2}\frac{y}{\delta}\right). \tag{2.50}$$

When y/δ is small, that is, in the overlap layer, we get eq. (2.48). If we substitute $y = \delta$ in eq. (2.50), we get a relationship between U and δ.

In order to determine $\bar{u}$ as a function of y and x based on eq. (2.50), we need to know $v^* = \sqrt{\tau_w/\rho}$ and δ. Because eq. (2.50) does not apply in the

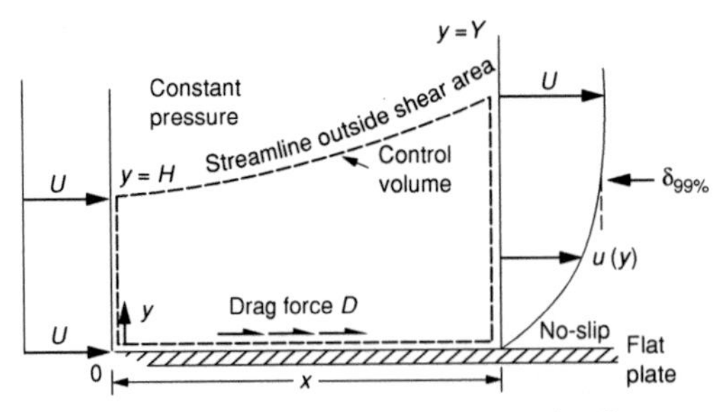

Figure 2.8. Definition of control volume for flow past a flat plate. (White, F. M., 1974, *Viscous Fluid Flow*, McGraw-Hill Book Company, 2nd ed. 1991, Printed in Singapore. The figure is reprinted with permission of The McGraw-Hill Companies.)

inner layer, we cannot determine τ_w based on eqs. (2.2) and (2.50). However, it is possible to find an expression for τ_w based on conservation of fluid momentum. We use a control volume, as shown in Figure 2.8. The control volume has a horizontal extent from the leading edge of the plate to x. The upper boundary of the control volume is outside the shear area or boundary layer. We need to consider forces due to pressure, viscous, and Reynolds (turbulent) stresses on the control volume. This will be based on boundary-layer theory. This means the pressure does not vary with y. Because $\partial p/\partial x$ is zero for flow along a flat plate, the pressure does not vary along the sides of the control volume. The force on the control volume due to pressure is therefore zero. We use eqs. (2.15) and (2.16) in combination with Figure 2.7 to consider the viscous stresses on the control volume. There, a viscous stress $\tau_{11} = \tau_{xx} = 2\mu\partial\bar{u}/\partial x$ acts on the vertical side parallel to the y-axis at x. This is negligible according to the boundary-layer theory. The only viscous stress acting on the control volume is at the side coinciding with the surface of the plate from the leading edge to x. This means that on the control volume, a longitudinal force D acts where D is the frictional (drag) force on the plate. D follows from integrating τ_w given by eq. (2.2) from the leading edge to x. τ_w also follows from eqs. (2.15) and (2.16) by noting the difference in normal vector $\mathbf{n} = (n_1, n_2, n_3)$ that applies.

Then we have to consider the turbulent stresses. The turbulent stress $-\rho\overline{u'v'}$ acting along the flat plate is zero. This is a simple consequence of the fact that u' and v' are zero on the plate. The longitudinal component of turbulent stresses on the vertical side of the control volume placed at the horizontal position x is $-\rho\overline{u'^2}$. This gives a negligible effect according to the boundary-layer theory. We can also see this from eq. (2.30), where no stress effect comes from a term like that.

The conservation of fluid momentum in the x-direction can then be expressed as

$$-D = \rho\int_0^Y \bar{u}^2\,dy - \rho\int_0^H U^2\,dy. \qquad (2.51)$$

The first and second terms on the right-hand side of eq. (2.51) correspond to the momentum flux through the vertical side at x and at the leading edge, respectively. The integration limits Y and H are defined in Figure 2.8. Because eq. (2.30) follows from conservation of fluid momentum, we could obviously have integrated this equation over the control volume and obtained eq. (2.51). We can rewrite eq. (2.51) by using conservation of mass for the control volume, that is,

$$UH = \int_0^Y \bar{u}\,dy. \qquad (2.52)$$

This gives

$$D = \rho\int_0^Y \bar{u}(U - \bar{u})\,dy. \qquad (2.53)$$

This means that D can be expressed in terms of the momentum thickness

$$\theta = \int_0^Y \frac{\bar{u}}{U}\left(1 - \frac{\bar{u}}{U}\right)dy, \qquad (2.54)$$

that is,

$$D = \rho U^2\theta. \qquad (2.55)$$

Eq. (2.54) is a general definition of momentum thickness, whereas eq. (2.55) applies only to our considered boundary-layer flow along a flat plate. We define the friction coefficient

$$C_f = \frac{\tau_w}{0.5\rho U^2} \qquad (2.56)$$

expressing the frictional stress on the plate. By using that $D = \int_0^x \tau_w dx$, i.e. $\tau_w = dD/dx$, we have

$$C_f = 2\frac{d\theta}{dx}. \qquad (2.57)$$

Because eq. (2.50) is valid everywhere in the boundary layer except in the viscous sublayer, which is a very small fraction of the boundary

Table 2.1. *Total drag computation for turbulent flow along a smooth flat plate*

Rn	C_F, "Exact" (White 1974, Table 6.6)	C_F, ITTC	Error, %	C_F (eq. 2.66)	Error, %	C_F (Hughes eq. 2.67)	Error, %
10^6	0.004344	0.004688	7.9	0.004210	−3.1	0.004188	−3.6
10^7	0.003015	0.003000	0.5	0.003030	0.5	0.002672	−11.4
10^8	0.002169	0.002083	−3.9	0.002181	0.5	0.001852	−14.6
10^9	0.001612	0.001531	−5.0	0.001569	−2.6	0.001359	−15.7
10^{10}	0.001236	0.001172	−5.2	0.001129	−8.7	0.001039	−15.9

layer, we can use eq. (2.50) as a good approximation in calculating θ given by eq. (2.54). We then set $Y = \delta$. This gives (see White 1974)

$$\frac{\theta}{\delta} = \frac{3.75}{\lambda} - \frac{24.778}{\lambda^2}, \tag{2.58}$$

where

$$\lambda = \frac{U}{\upsilon^*} \equiv \sqrt{\frac{2}{C_f}}. \tag{2.59}$$

We can also express λ by using eq. (2.50) at $y = \delta$. This gives

$$\lambda = 2.5 \ln\left[\frac{U\delta}{\nu\lambda}\right] + 8. \tag{2.60}$$

Eliminating δ between eqs. (2.58) and (2.60) gives

$$Rn_\theta \equiv \frac{U\theta}{\nu} = \lambda\left(\frac{3.75}{\lambda} - \frac{24.778}{\lambda^2}\right) e^{0.4(\lambda-8)}. \tag{2.61}$$

By using that C_f can be expressed in terms of λ by eq. (2.59) and by curve-fitting, White (1974) found that eq. (2.61) can be approximated as

$$C_f \approx 0.012 Rn_\theta^{-1/6}. \tag{2.62}$$

By substituting eq. (2.62) into (2.57), we find

$$Rn_x \equiv \frac{Ux}{\nu} = \frac{1}{0.006}\int_0^{Rn_\theta} Rn_\theta^{1/6}\, dRn_\theta$$

or

$$Rn_\theta = 0.0142 Rn_x^{6/7}. \tag{2.63}$$

Substituting eq. (2.63) into eq. (2.62) gives

$$C_f = 0.0244 Rn_x^{-1/7}. \tag{2.64}$$

According to White (1974), a more correct formula is

$$C_f = 0.027 Rn_x^{-1/7}. \tag{2.65}$$

From the expression above, C_f is infinite at $x = 0$. This means the local behavior near the leading edge is wrong, but this causes a negligible error in predicting the drag on the plate. By integrating eq. (2.65), we find the frictional force coefficient C_F or

$$C_F = \frac{\int_0^L \tau_w dx}{0.5\rho U^2 L} = \frac{1}{L}\int_0^L C_f(x)\, dx$$

$$= \frac{1}{Rn}\int_0^{Rn} C_f(Rn_x)\, dRn_x,$$

where $Rn = UL/\nu$. This means

$$C_F = 0.0303 Rn^{-1/7}. \tag{2.66}$$

In Table 2.1, this formula is compared with what White (1974) considered the "exact" solution of C_F in a broad Reynolds number range. The lower Reynolds numbers are typical for ship model testing, whereas the higher Reynolds numbers are representative for full-scale ships. C_F-values based on the ITTC formula and given by eq. (2.4) are also included in the table. Also, other formulas not considered here exist for C_F for turbulent flow along a smooth flat plate. We must also mention the Hughes (1954) formula that was commonly used in ship model testing:

$$C_F = \frac{0.066}{(\log_{10} Rn - 2.03)^2}. \tag{2.67}$$

Calculations by eq. (2.67) are also presented in Table 2.1 and show that the ITTC formula is in general a better approximation than the Hughes formula. However, eq. (2.66) generally gives the results closest to what White (1974) considers the correct expression.

In order to estimate the boundary-layer thickness δ, we first find a relationship between C_f and δ, for instance, by using eq. (2.60). By curve fitting, White (1974) found that

$$C_f \approx 0.018 Rn_\delta^{-1/6},$$

where $Rn_\delta = U\delta/\nu$. Using eq. (2.65) gives

$$\frac{U\delta}{\nu} = 0.11\left(\frac{Ux}{\nu}\right)^{6/7}.$$

However, using eq. (2.64) gives a relatively different result, that is,

$$\frac{U\delta}{\nu} = 0.16\left(\frac{Ux}{\nu}\right)^{6/7}$$

or

$$\delta = \frac{0.16x}{(Rn_x)^{1/7}}. \tag{2.68}$$

Let us consider the case $U = 20$ ms^{-1}, $x = 100$ m and use $\nu = 10^{-6}$ m^{2}s^{-1}. This gives the boundary-layer thickness as 0.75 m. This has relevance for the boundary-layer thickness at the aft end of a 100 m–long monohull at a speed of 20 ms^{-1}. Let us consider the corresponding thickness at model scale and assume a model length $L_M = 4$ m. The ratio between full-scale ship length L_S and L_M is 25. Model testing is done by Froude scaling. This means that the model speed is $(L_M/L_S)^{0.5}$ times the full-scale speed, or 4 ms^{-1} in this case. Assuming turbulent flow in model scale and using $\nu = 10^{-6}$ m^{2}s^{-1} and eq. (2.68) gives that the boundary layer thickness is equal to 0.06 m at the aft end of the ship model.

Eq. (2.68) is a geometrical measure of the boundary-layer thickness. There are also other measures of the boundary-layer thickness. One is the momentum thickness θ defined by eq. (2.54). Another is the displacement thickness δ^*. We will introduce this by means of Figure 2.8. From continuity of fluid mass of an incompressible fluid, it follows that

$$\begin{aligned} UH &= \int_0^Y u\,dy = \int_0^Y (U + \bar{u} - U)\,dy \\ &= UY + \int_0^Y (\bar{u} - U)\,dy. \end{aligned} \tag{2.69}$$

Here $Y = H + \delta^*$ so that δ^* expresses how much the streamline at $y = H$ at the leading edge has moved outward with respect to the plate at the location x (Figure 2.8). From eq. (2.69), it follows that

$$U(Y - H) = U\delta^* = \int_0^Y (U - \bar{u})\,dy$$

or

$$\delta^* = \int_0^{Y\to\infty} \left(1 - \frac{\bar{u}}{U}\right) dy. \tag{2.70}$$

We can use eq. (2.50) to calculate δ^*. We introduce then $\eta = y/\delta$ as an integration variable and integrate from $y = 0$ until $y = \delta$ instead of until infinity. Eq. (2.70) can be rewritten as

$$\frac{\delta^*}{\delta} = \frac{v^*}{U}\int_0^1 \left(\frac{U}{v^*} - \frac{\bar{u}}{v^*}\right) d\eta.$$

Further, eq. (2.50) can be expressed as

$$\frac{\bar{u}}{v^*} = 2.5\ln\eta + 2.5\ln\frac{\delta v^*}{\nu} + 5.5 + 2.5\sin^2\left(\frac{\pi}{2}\eta\right).$$

This means

$$\frac{U}{v^*} = 2.5\ln\frac{\delta v^*}{\nu} + 8.$$

Further, integrating and using eq. (2.59) gives

$$\frac{\delta^*}{\delta} = 3.75\sqrt{C_f/2}.$$

We have already found δ and C_f as a function of Rn_x (see eqs. (2.64) and (2.68)). This gives

$$\delta^* = 0.066\,x/Rn_x^{3/14}. \tag{2.71}$$

We note that δ^* is clearly smaller than δ. This thickness parameter can be used to measure how much the flow outside the boundary layer is affected by the boundary layer. As shown in Figure 2.8, the slope of the streamline is $d\delta^*/dx$. Because the flow is parallel to a streamline and U is the dominant velocity outside the boundary layer, we find that there is a vertical velocity

$$\bar{v} = U\frac{d\delta^*}{dx} \tag{2.72}$$

at the outer part of the boundary layer. This represents then a boundary condition for the potential flow outside the boundary layer. Eq. (2.72) expresses that there in the potential flow appears a flow coming out from the plate. A consequence of this is that there exists a pressure gradient in the x-direction. However, when calculating the effect of this pressure gradient on the viscous flow, we must also correct the boundary-layer equation.

This effect is not negligible in general, but the analysis will not be pursued here.

In order to find a measure of the thickness of the viscous sublayer we use the information about the velocity distribution given by eq. (2.50), which is valid in the outer and overlap layers but not in the viscous sublayer. The lowest y value for which eq. (2.50) is valid corresponds to $yv^*/\nu = 35$. We then need to determine the friction velocity $v^* = \sqrt{\tau_w/\rho}$, where $\tau_w = 0.5\rho U^2 C_f$. Using eq. (2.65) for C_f gives

$$v^* = 0.114\, U/Rn_x^{1/14}. \tag{2.73}$$

We define the thickness δ_{VS} of the viscous sublayer by $\delta_{VS} v^*/\nu = 35$. This means

$$\delta_{VS} = 307 \frac{x}{Rn_x^{13/14}}. \tag{2.74}$$

Using the previous example with $U = 20\,\mathrm{ms}^{-1}$, $x = 100\,\mathrm{m}$, and $\nu = 10^{-6}\,\mathrm{m}^2\mathrm{s}^{-1}$ gives $\delta_{VS} = 70 \cdot 10^{-6}$ m or only $0.9 \cdot 10^{-4}$ times the boundary-layer thickness we found.

The above-discussed formulas for the frictional force constitute only one part of the total viscous resistance effect for the ship. We consider other effects in the following section.

2.2.5 Form resistance components

Experimental results show that eq. (2.4) has to be modified to describe the viscous resistance of high-speed monohulls and catamarans. A form resistance component exists because of the interaction between the ship's three-dimensional shape and the viscosity. Wave resistance is also a function of the ship's three-dimensional and finite transverse-dimensional shape. However, viscosity does not have an important influence on wave resistance, or at least it is common to assume this. This means wave resistance is not considered as a part of the form resistance. The form resistance can be associated with the following three effects:

- Frictional resistance
- Viscous pressure resistance
- Flow separation

We discuss these different effects in the following text. When we derived the formulas for viscous resistance of a flat plate, we used the ship speed as the tangential velocity outside the boundary layer. However, the ship's three-dimensional form affects (and in general increases) the tangential velocity just outside the boundary layer. As a consequence, the frictional stress on the hull generally will increase along the ship. When calculating the contribution due to viscous resistance, the frictional stress must be resolved into a component parallel to the longitudinal coordinate of the ship. However, this effect is small, particularly for slender hull forms. We should also note that the ship's three-dimensional shape causes a pressure gradient along the hull. This influences the velocity distribution in the boundary layer and therefore the frictional stress at the hull surface. Another important aspect is that the flow in the boundary layer is 3D and not 2D as we assumed in the analysis of turbulent flow along the flat plate. Further, we have implicitly assumed a thin boundary layer, which may be less appropriate in the aft end of the ship.

The second main contribution to form resistance is the viscous pressure resistance. We explain this by referring to a situation in which viscous resistance is dominant and wave resistance does not matter; this means for Froude numbers $U/(Lg)^{0.5}$ less than ≈ 0.15. In this case, there is negligible free-surface motion and the normal velocity on the mean free surface can be set equal to zero. Let us look upon the flow from a reference frame following the ship. This means the forward speed of the ship appears as an incident flow velocity U along the longitudinal x-axis pointing toward the stern of the ship (Figure 2.9). The free-surface condition and the horizontal direction of the incident flow make the flow around the ship the same as the flow around a double body consisting of the submerged part of the ship and its image about the mean free surface. This is a consequence of the fact that the steady flow around the double body is symmetric about the xy-plane. This means zero normal velocity on $z = 0$ outside the body,

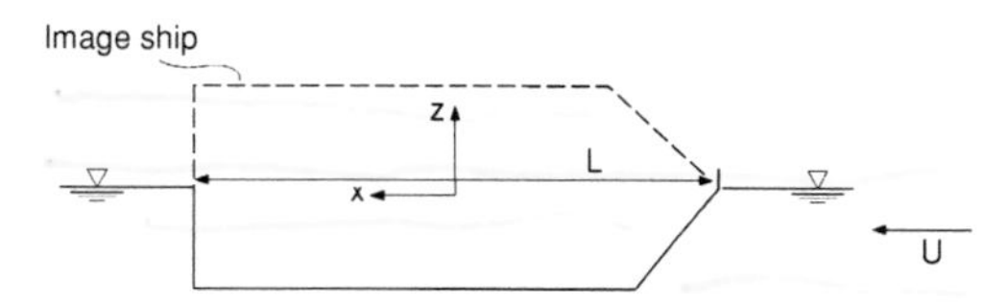

Figure 2.9. Double-body approximation. For Froude number $Fn = U/\sqrt{Lg} < \approx 0.15$ the flow around a ship with speed U can be represented by the flow around a double body consisting of the submerged ship and its image about the mean free surface.

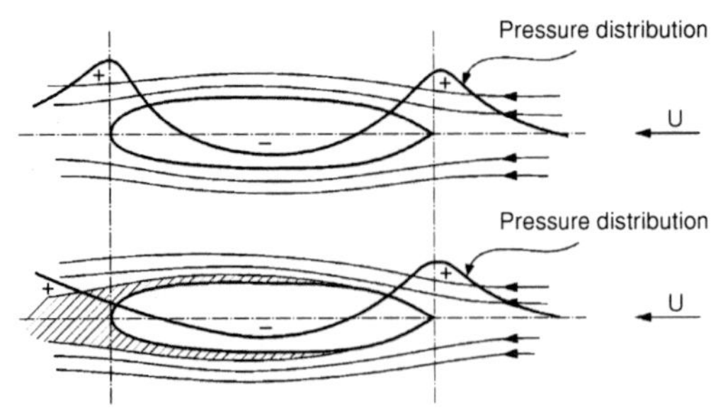

Figure 2.10. The flow and pressure distributions around a ship when $Fn < \approx 0.15$ (see Figure 2.9) Ambient pressure is excluded. The upper figure does not account for viscosity. The shaded area in the lower figure is the boundary layer and wake where viscosity matters (Walderhaug 1972).

that is, a similar condition that we have specified in the free-surface condition for the ship problem.

Having now created the equivalent to the double-body problem, we can use knowledge about the flow around a body in infinite fluid. If viscosity is neglected, the flow at the waterline and the pressure force distribution look like those in the upper drawing of Figure 2.10. Remember that the spacing between the streamlines is an implicit expression of the flow velocity, with high velocities in regions with narrow spacings. Because increasing velocity means decreasing pressure, we see that the pressure increases in regions with wider spacing. Because the ambient pressure is constant in space and gives zero force, its effect on the pressure force distribution is not included in Figure 2.10. The resulting force on the ship due to the pressure force distribution in the upper drawing in Figure 2.10 is zero. This is the well-known D'Alembert paradox, that is, there is no hydrodynamic force acting on a body in infinite fluid due to steady potential flow without circulation. However, the pressure influenced by the boundary layer changes this situation. The shaded area in the lower drawing of Figure 2.10 indicates the boundary layer. Because of this viscosity region next to the hull, the pressure force in the bow part does not cancel the pressure force in the aft part of the ship, so the boundary layer affects the pressure distribution. We discussed this previously in connection with eq. (2.72). Where the boundary layer ends at the stern, a wake forms behind the ship, where turbulent stresses are important. The pressure approaches ambient pressure (or zero in Figure 2.10) at some distance downstream in the wake not shown in Figure 2.10. Actually, the pressure has not yet reached its maximum value in the small part of the wake presented in Figure 2.10. From this figure, we see that the effect of the boundary layer on the pressure is negligible in the bow part, where the boundary layer is thin relative to that in the aft part. The lower drawing in Figure 2.10 illustrates clearly that there is a viscous pressure resistance.

The third main cause of form resistance is flow separation. If the flow separates from the hull, we get a larger domain aft of the separation line, where viscosity matters. This implies a larger influence on the pressure distribution and increased form resistance. Cross-flows past a circular cylinder and a sphere are classical examples of separated flow. When the Reynolds number is larger than $\approx 10^3$ for circular cylinders, the major part of the drag forces is the result of the pressure.

If a surface has a sharp edge, the flow will separate from the sharp edge when there is a cross-flow past the edge. However, the flow may also separate from a surface without sharp corners, as we have seen, for example, in bluff bodies such as spheres and circular cylinders. We illustrate how flow separation starts for a 2D flow situation by means of Figure 2.11. If there is a point S on the body surface where $\partial u/\partial y = 0$ and there is backflow aft of the point S, we get flow separation from point S. If $\partial u/\partial y = 0$ also aft of S, we do not get flow separation from S. This situation is beneficial because $\partial u/\partial y = 0$ means zero shear stress τ_w on the wall. This can be obtained by a proper design of the hull surface (Tregde 2004) and is referred to as Stratford (1959) flow. The position of point S depends on the pressure gradient $\partial p/\partial x$ along the hull surface and on the flow conditions (laminar or turbulent) in the boundary layer ahead of the separation point. An adverse pressure

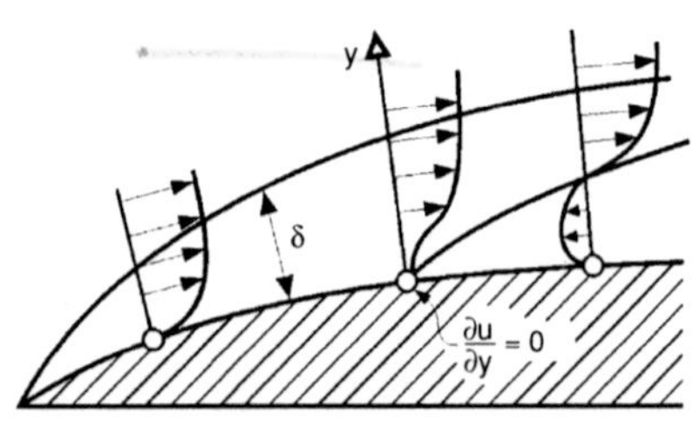

Figure 2.11. 2D flow with a boundary layer of thickness δ. Illustration of conditions for flow separation, that is, $\partial u/\partial y$ is zero at the surface at S and there is a backflow near the surface aft of S. The flow will then separate at point S (Walderhaug 1972).

gradient, that is, $\partial p/\partial x$ is positive, is necessary for flow separation according to boundary-layer theory. We can see this from eq. (2.30) by first noting that $\partial p/\partial x = -\rho U_e \, dU_e/dx$. Applying eq. (2.30) for $y = 0$, that is, on the body surface, gives

$$\frac{\partial p}{\partial x} = \mu \frac{\partial^2 u}{\partial y^2} \quad \text{for } y = 0. \qquad (2.75)$$

We recall from calculus that the condition $du/dy = 0$ for flow separation is also a condition for $u\,(y)$ to have either a local maximum or minimum value. Because in our case we obviously have a minimum value, it follows that $\partial^2 u/\partial y^2$ is positive at the separation point S. It then follows from eq. (2.75) that a necessary condition for separation to occur is that $\partial p/\partial x$ is positive on the hull surface at the separation point. Because $\partial p/\partial x = 0$ for flow along a flat plate, flow separation will not occur in this case.

Cross-flow past a circular cylinder is a classical case of flow separation. Figure 2.12 illustrates the different regimes of boundary-layer flow for 2D cross-flow past a circular cylinder at transcritical flow. The instability point of the laminar boundary-layer flow is Reynolds-number dependent. No instability will occur ahead of the separation point for subcritical flow. The subcritical flow regime is for Reynolds number $Rn = UD/\nu$ less than $\approx 2 \cdot 10^5$ for flow around a smooth circular cylinder with diameter D in steady incident flow with velocity U. The critical flow regime is for $\approx 2 \cdot 10^5 < Rn < \approx 5 \cdot 10^5$. The supercritical flow regime is for $\approx 5 \cdot 10^5 < Rn < 3 \cdot 10^6$, and the transcritical flow is for Reynolds numbers larger than $3 \cdot 10^6$.

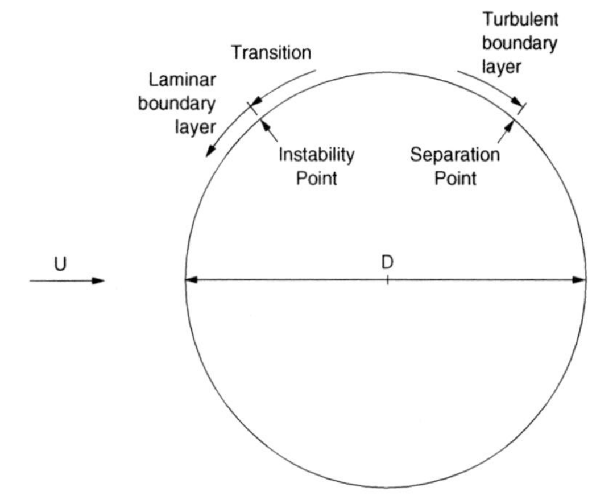

Figure 2.12. Schematic view of flow domains in the boundary layer of a circular cylinder with steady-state cross-flow at transcritical flow. The incident flow velocity U is constant. The instability and separation points depend on the Reynolds number UD/ν.

If the flow is turbulent ahead of the separation point (line in 3D case) S, separation occurs further downstream than if the flow is laminar ahead of the separation point (line in 3D case). The reason is that the large exchange of fluid momentum occurring in turbulent flows causes less deceleration of the boundary-layer flow relative to laminar flows. Even if the boundary-layer flow stays laminar up to separation, the wake flow will become turbulent for Reynolds numbers of practical interest. The mean drag on a circular cylinder will decrease as a consequence of turbulent separation. This is related to the fact that the wake behind the cylinder is narrower than that for laminar separation. The frictional drag on the cylinder at high Reynolds number will be lower for laminar boundary layer than for turbulent boundary layer, as we have seen for the flow along a flat plate. However, the frictional drag is a small part of the drag on a circular cylinder for Reynolds numbers of practical interest to us.

When the flow separation occurs from sharp corners and shear forces have secondary importance relative to pressure loads, the Reynolds number effect is not important in scaling drag forces from model to full scale. An example is the cross-flow past the rectangular cross section illustrated in Figure 2.13. The model is 2.63 m high and of constant cross-sectional form with side lengths 0.5 m and 0.42 m. The model was towed vertically in the middle of a towing tank of breadth 10 m and depth 5.5 m with the longest cross-sectional side in the towing direction. The submergence was 2.5 m, and the constant carriage velocity was 0.3 ms^{-1}. This is a sufficiently low velocity to avoid the influence of surface waves. Further, the cross-flow at the free surface can be approximated as two-dimensional. Even though the towing speed is constant, the flow is unsteady relative to the body. Pictures of the flow were taken by covering the water surface with confetti and using an exposure time of 1 s.

Figure 2.13 shows an instantaneous picture when the flow is symmetric about the center plane of the model. The separated vortices from the two leading edges are much more oval in form than the two vortices downstream of the body. Later on, an asymmetry is created in the wake, causing alternate vortex shedding from the edges. The consequence is an oscillatory lift force perpendicular to the towing direction. The drag force contains a

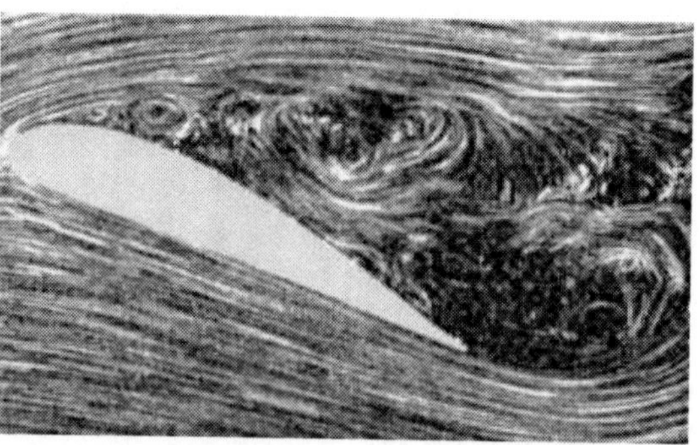

Figure 2.13. The photo on the left shows visualization of initial instantaneous flow around a rectangular cross section that is towed with constant velocity in the left direction. The flow is unsteady relative to the body and will later be asymmetric about the center plane, resulting in alternate vortex shedding from the edges. The photo on the right is from Prandtl (1956) and shows separated flow around an airfoil at a high angle of attack, resulting in stalling.

steady part and an oscillatory part. If the structure has natural frequencies in the vicinity of vortex shedding frequencies, strong dynamic oscillations of the structure may occur. Vortex-induced vibrations are a classical problem associated with cross-flow past cylindrical structures (Faltinsen 1990).

Flow separation should not occur at continuously curved surfaces of a properly designed ship on a straight course. However, we cannot avoid it at the transom stern. When the Froude number is higher than about 0.4, this flow separation causes a dry transom stern. The flow separation at the transom stern has an important effect on the hydrodynamic lift on a planing vessel (see Chapter 9). Generally speaking, the flow separation at the transom stern is not beneficial at low Froude numbers, but has clear advantages for the hydrodynamic performance at high speed. In Chapter 10, we discuss the effect of cross-flow separation during maneuvering of a ship.

When the angle of attack of the flow around a foil is not small, there is also the possibility of flow separation near the nose of the foil ("stalling") (see Figure 2.13). This decreases the mean lift force on the foil relative to no flow separation. Flow separation will then also affect cavitation inception.

2.2.6 Effect of hull surface roughness on viscous resistance

In the previous sections, we assumed a smooth hull surface. Actually, the frictional resistance is affected by the roughness of the hull surface. The influence of roughness depends in principle on many parameters, such as the height, form, and distribution of the roughness. Figure 2.14 illustrates both idealized types of roughness and roughness due to paint. If we look by microscope at the flow around the roughness structure on the scale of the roughness dimensions, the flow we see has similarities to flow around macroscale structures. For instance, we have the possibility of flow separation. If we look at the sinusoidal roughness in Figure 2.14, this will depend on the roughness height k and the wavelength λ as defined in the

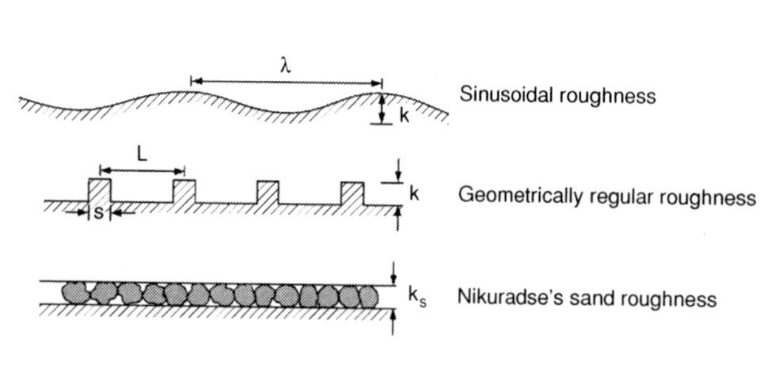

Figure 2.14. Visualization of different types of roughness.

figure. The larger k/λ, the larger the possibility of flow separation. We are sure that the flow will separate around a single rectangular type of roughness. If there is a geometrically regular system of rectangular roughness elements, the flow will depend on k, the horizontal length s of each rectangle, and the distance L between two subsequent rectangular elements, as defined in Figure 2.14. Even if the flow around the roughness structure does not separate, there will be a form resistance on the roughness structure similar to that discussed in the previous section. The details of the flow become very complicated to describe. We must also realize the irregular shape of realistic roughness, for instance, because of paint, as shown in Figure 2.14. If these were irregular sea waves, we would use a sea spectrum to characterize the "signal" (See Chapter 3).

A common practice is to use an equivalent sand-grain height to characterize the influence of roughness on viscous resistance. This is based on Nikuradse's (1933) experiments with closely spaced sand grains of equal height k_S as shown in Figure 2.14. We will come back to that later. For the time being, we assume that the roughness is uniformly distributed over the plate, like sand grains of the same size and constant density. The roughness height k is used as the only roughness parameter.

We can derive expressions for the mean longitudinal velocity $\bar{u}$ on the boundary layer of a rough plate the same way we did for the flow along a smooth plate. This means, by dimensional analysis, we can establish a logarithmic dependence in an overlap layer. The coefficients in the expression are then experimentally determined. Once the $\bar{u}$ behavior in the boundary layer is identified, we can proceed as we did for a smooth plate to obtain frictional drag on the plate.

Let us show the procedure for finding $\bar{u}$ in some detail. We first assume that the roughness height k is larger than the thickness of the viscous sublayer for the smooth plate. This means kv^*/ν is larger than about 35. It implies that there is no viscous sublayer, but we can postulate that there is an inner layer, where

$$\bar{u} = f(\tau_w, \rho, k, y). \tag{2.76}$$

This has similarities to eq. (2.37) for a smooth plate. The difference is that μ is substituted with k. Using dimensionless variables as we did for a smooth plate, we can re-express eq. (2.76) as

$$\frac{\bar{u}}{v^*} = f\left(\frac{y}{k}\right), \tag{2.77}$$

where $v^* = \sqrt{\tau_w/\rho}$. The outer law is the same for smooth and rough plates (see eqs. (2.38) and (2.40) in dimensional and dimensionless forms, respectively). By defining an overlap layer and following the same procedure that leads to eqs. (2.48) and (2.49), we find the law

$$\frac{\bar{u}}{v^*} = \frac{1}{\kappa}\ln\left(\frac{y}{k}\right) + \text{const} \tag{2.78}$$

for the overlap region in fully rough flow. Experiments show that $1/\kappa$ is about the same value as that for a smooth plate, that is, 2.5. White (1974) showed that it is possible to construct a formula for $\bar{u}$ valid for any roughness, including a smooth plate. The expression for closely spaced sand grains is

$$\frac{\bar{u}}{v^*} = 2.5\ln\left(\frac{yv^*}{\nu}\right) + 5.5 - 2.5\ln\left(1 + 0.3\frac{k_s v^*}{\nu}\right). \tag{2.79}$$

We have set $k = k_s$ to indicate that it is valid for closely spaced sand grains. If $k_s = 0$, eq. (2.79) agrees with the expression for a smooth plate. If $k_s v^*/\nu$ is larger than 60, which corresponds to fully rough flow, then $1 + 0.3k_s v^*/\nu \approx 0.3k_s v^*/\nu$ and eq. (2.79) becomes

$$\frac{\bar{u}}{v^*} = 2.5\ln\left[\left(\frac{yv^*}{\nu}\right)\left(\frac{\nu}{k_s v^*}\right)\right] + 5.5 - 2.5\ln(0.3),$$

that is,

$$\frac{\bar{u}}{v^*} = 2.5\ln\left(\frac{y}{k_s}\right) + 8.5. \tag{2.80}$$

This agrees with eq. (2.78). We are interested in finding the frictional force on the rough plate. This can be achieved following the same procedure we used for a smooth plate; however, this is not done here. Instead we present the final formulas for the frictional force coefficient C_F as given by Schlichting (1979). These corresponding results are shown in Figure 2.15, in which curves with either Uk_s/ν or L/k_s constant are indicated. Here L is the plate length. We note that C_F along curves with constant L/k_s becomes independent of Reynolds number beyond the broken line; this is called fully rough flow. The C_F-value can then be expressed as

$$C_F = \left(1.89 + 1.62\log_{10}\left(\frac{L}{k_s}\right)\right)^{-2.5} \tag{2.81}$$

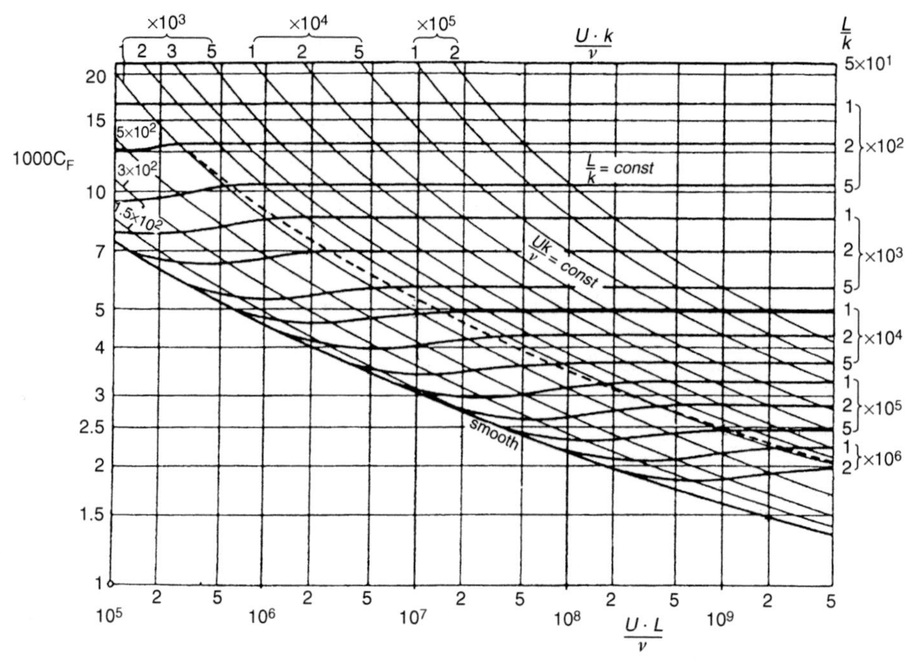

Figure 2.15. Resistance formula of sand-roughened plate; coefficient of total skin friction. L = length of flat plate, k = roughness height. (ISBN 0-07-055334-3, Schlichting, H., 1979, *Boundary-LayerTheory*, 7th ed., New York: McGraw-Hill Book Company. The figure is reprinted with permission of The McGraw-Hill Companies).

for $10^2 < L/k_s < 10^6$. Let us try to estimate when the flow can be considered fully rough by using $k_s v^*/\nu > 60$ as a criterion. We use eq. (2.73) for v^*. Because this applies to a smooth plate, the results would not be exactly the same as those in Figure 2.15. Further, because v^* changes with x, the criterion for fully rough flow will depend on x. We have just set $x = L$ in our estimate. This gives that a minimum value k_{rough} of k_s for fully rough flow corresponds to either

$$\frac{L}{k_{rough}} = 1.9 \cdot 10^{-3} Rn^{13/14} \qquad (2.82)$$

or

$$\frac{k_{rough} U}{\nu} = \frac{10^3}{1.9} Rn^{1/14}. \qquad (2.83)$$

Results from these formulas are presented in Table 2.2. They agree reasonably well with the dashed line in Figure 2.15.

We note that the results in Figure 2.15 become nearly independent of roughness when Uk_s/ν is smaller than the order of 100. We can use this to define an admissible roughness height k_{adm} given that there is no increase in drag compared with a smooth plate. Schlichting (1979) proposes that

$$k_{adm} = 100 \frac{\nu}{U}. \qquad (2.84)$$

Because Froude scaling is used in ship model testing and the model speed is lower than full-scale speed, we see that it is easier to obtain a hydraulically smooth surface in model condition than in full scale. As an example, if we consider a model speed of 5 ms^{-1} and use $\nu = 10^{-6}$ m^{2}s^{-1}, this leads to $k_{adm} = 20\ \mu$m (or $20 \cdot 10^{-6}$ m). Let us say that

Table 2.2. *Estimates of lowest sand roughness height k_{rough} for fully rough flow along a flat plate of length L as a function of Reynolds number*[a]

Rn	$\frac{L}{k_{rough}}$	$\frac{Uk_{rough}}{\nu}$
10^6	$7 \cdot 10^2$	$1.4 \cdot 10^3$
10^7	$6 \cdot 10^3$	$1.7 \cdot 10^3$
10^8	$5 \cdot 10^4$	$2.0 \cdot 10^3$
10^9	$4 \cdot 10^5$	$2.3 \cdot 10^3$

[a] Estimates based on eqs. (2.82) and (2.83). The latter uses the wall friction velocity v^* for a smooth plate as a basis.

Table 2.3. *Effect of roughness on viscous frictional resistance*

L(m)	Rn	$\frac{L}{k_{rough}}$	$\frac{L}{k}$	$C_{F,ITTC}$	$C_{F,ROUGH}$	ΔC_F	ΔC_F (eq. 2.85)
50	10^9	$4.3 \cdot 10^5$	$3.3 \cdot 10^5$	0.00153	0.00259	0.00106	0.00045
100	$2 \cdot 10^9$	$8.2 \cdot 10^5$	$6.7 \cdot 10^5$	0.00141	0.00232	0.00091	0.00038
150	$3 \cdot 10^9$	$1.2 \cdot 10^6$	10^6	0.00134	0.00218	0.00084	0.00035

L = ship length, Rn = Reynolds number, k_{rough} = minimum roughness height for fully rough flow, k = roughness height = 150 μm, $C_{F,ITTC}$ = ITTC friction coefficient for smooth surface, $C_{F,ROUGH}$ = Schlichting's friction coefficient, $\Delta C_F = C_{F,ROUGH} - C_{FITTC}$, $U = 20\ \text{ms}^{-1}$, $\nu = 10^{-6}\ \text{m}^2\text{s}^{-1}$.

the corresponding full-scale speed is 20 ms^{-1}. This gives $k_{adm} = 5$ μm. A ship model is always made hydraulically smooth, but this is not possible for a full-scale ship hull, even for a newly built ship without marine fouling accumulated during service. This is connected to standard fabrication procedures and the low value of k_{adm}. However, efforts are made to fabricate hydraulically smooth foils.

It is worth keeping in mind that the results in Figure 2.15 are for a sand-roughened plate. The practical use of the results for our purposes requires that we define an equivalent sand roughness height k_s. This is, for instance, 1 μm for polished metal such as stainless steel used for foils and 60 μm for antifouling paint. Because roughness may vary significantly in size and density, it may be difficult to find an equivalent sand roughness height.

Example: Effect of hull roughness on viscous resistance

We consider an average hull roughness height $k = 150 \cdot 10^{-6}$ m or 0.15 mm, which in practice is an upper limit for a newly built ship. We assume a ship speed $U = 20\ \text{ms}^{-1}$. If $\nu = 10^{-6}\ \text{m}^2\text{s}^{-1}$, this corresponds to $Uk/\nu = 3 \cdot 10^3$. Let us for simplicity set k equal to k_s and use eq. (2.83) to estimate if the flow is fully rough. This depends on the Reynolds number or the ship length. We consider only fully rough flow and use eq. (2.81) to calculate C_F and define ΔC_F so that $C_F = C_{F,ITTC} + \Delta C_F$, where $C_{F,ITTC}$ is the ITTC formula for a smooth hull (see eq. (2.4)). We have already pointed out that there are many formulas for C_F of a smooth plate and that they do not give the same results. For instance, C_F for a smooth plate presented in Figure 2.15 does not give the same results as the $C_{F,ITTC}$ formula. However, because $C_{F,ITTC}$ is commonly used for ship hulls, we prefer such a formula in this context. Results for different ship lengths are presented in Table 2.3. In the same table, ΔC_F values from

$$\Delta C_F = \left[111\,(AHR \cdot U)^{0.21} - 404\right] C_{F,ITTC}^2 \tag{2.85}$$

are also given. Here AHR means the average hull roughness in micrometers. This has been set equal to 150 μm. U should be given in units of meter per second (ms^{-1}). Eq. (2.85) is relevant only for full-scale ships and is used by MARINTEK (Minsaas, private communication). We will compare ΔC_F for ship lengths equal to 50, 100, and 150 m. Because L/k_{rough} in all cases is larger than L/k, fully rough flow can be assumed. We note that ΔC_F for fully rough flow is about twice the value given by the MARINTEK formula. Why is that so? The reason is simply that AHR is not the same as Nikuradse's sand roughness height k_s. In a way, it accounts for differences in the details of the roughness as shown in Figure 2.14. Values of AHR for newly built ships may vary between 75 and 150 μm, and ΔC_F between 0.0002 and 0.0008 are commonly used. From Table 2.3 we note that ΔC_F is not small relative to C_F for a smooth surface. Other empirical formulas exist for the roughness effect. The formula

$$10^3 \Delta C_F = 44\left[(AHR/L)^{1/3} - 10 Rn^{-1/3}\right] + 0.125 \tag{2.86}$$

by Bowden and Davison (1974) accounts for correlation between model tests and full scale and includes the effect of roughness. In this case, AHR has dimension meter.

2.2.7 Viscous foil resistance

Special formulas are used for the viscous resistance of a foil. Before we discuss details about

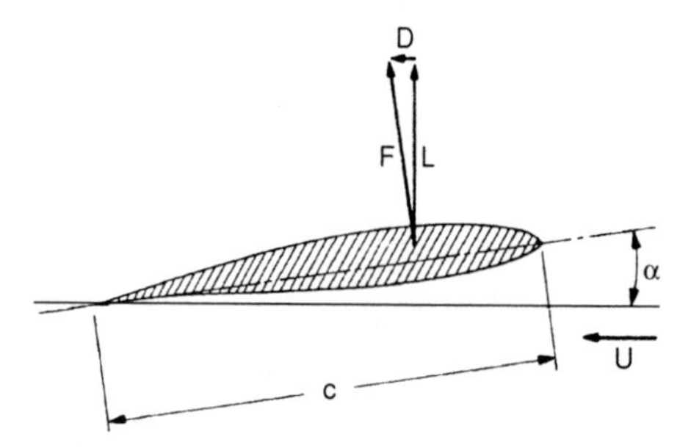

Figure 2.16. Flow past a cambered hydrofoil section. The angle of attack α is generally defined with respect to the "nose-tail line," between the center of the minimum radius of curvature of the leading edge and the sharp trailing edge. L and D denote the lift and drag components of the total force F, and are defined respectively to be perpendicular and parallel to the free-stream velocity vector. (Newman, J. N., 1977, *Marine Hydrodynamics*, Cambridge: The MIT Press. The figure is reprinted with the permission of The MIT Press)

it, we need some definitions. Let us refer to the two-dimensional case shown in Figure 2.16. The length c is called the chord length. The lateral extent s of the hydrofoil in three dimensions is called the span. The incident flow with velocity U has an angle of attack α as defined in the figure. The lift force L and the drag force D act perpendicular and parallel, respectively, to the free-stream velocity vector. The planform area A is defined as the projected area of the foil in the direction of the lift force for zero angle of attack α. This means that A in the two-dimensional case shown in Figure 2.16 is the chord length c. Further, A for a thin foil is close to half the wetted surface. The aspect ratio Λ is defined as

$$\Lambda = \frac{s^2}{A}. \tag{2.87}$$

A large-aspect ratio means that the flow is close to two-dimensional at each cross section of the foil, as shown in Figure 2.16. A planar foil means that the angle of attack is the same at each cross section. A foil is uncambered when the foil is symmetric about the "nose-tail line," shown by the dot–long-dashed line in Figure 2.16. The camber line is the mean line between the upper and lower surfaces.

Lift and drag coefficients C_L and C_D for a foil are defined as

$$C_L = \frac{L}{\frac{\rho}{2}U^2 A} \tag{2.88}$$

and

$$C_D = \frac{D}{\frac{\rho}{2}U^2 A}, \tag{2.89}$$

respectively.

Figures 2.17 and 2.18 show examples of lift and drag coefficients as a function of the angle of attack α for steady flow past a 2D foil in infinite fluid with turbulent boundary-layer flow conditions. The ambient flow velocity or the foil speed U is assumed small relative to the speed of sound, that is, the fluid may be considered incompressible. Abbot and von Doenhoff (1959) have presented comprehensive experimental results for C_L and C_D. If linear theory and an inviscid fluid are

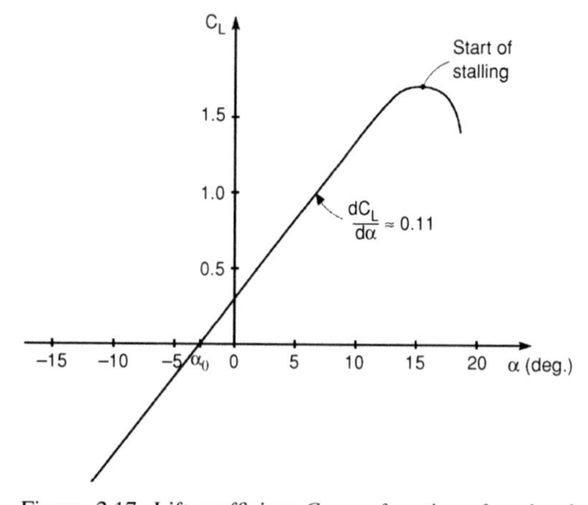

Figure 2.17. Lift coefficient C_L as a function of angle of attack α for a 2D foil with turbulent boundary layer in infinite fluid and steady inflow conditions. Linear theory predicts a lift slope $dC_L/d\alpha = 0.11$ when α is measured in degrees.

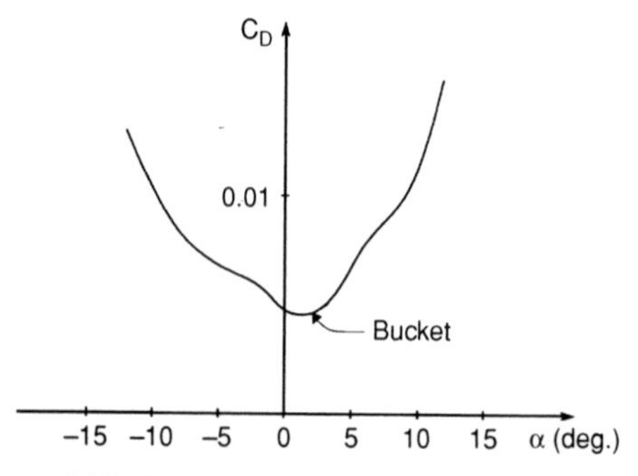

Figure 2.18. Example of drag coefficient C_D as a function of angle of attack α for a 2D foil with turbulent boundary layer in infinite fluid and steady inflow conditions.

assumed, the lift coefficient can be expressed as $2\pi(\alpha - \alpha_0)$, where α is in radians. If the foil section is uncambered and there is no flap or zero flap angle, α_0 is zero (see Chapter 6). If α is measured in degrees, linear theory predicts $dC_L/d\alpha = 2\pi(\pi/180) = 0.11$. The comparison with experimental results shows good agreement for a broad range of angles of attack and for many foil shapes. Good agreement means less than a 5% difference. Less good agreement may, for instance, occur for large foil thickness-to-chord ratio t/c, that is, $t/c > \approx 0.14$.

When the flow separation occurs from the leading edge (see Figure 2.13), the lift coefficient will start decreasing with increasing α. This is called stalling. Flow separation is affected by the boundary-layer flow and occurs much easier in laminar than in turbulent flow. This means a higher maximum value of C_L is obtained for turbulent boundary-layer flow. A necessary requirement for flow separation is an adverse pressure gradient (see Section 2.2.5). Because this occurs at the aft part of the foil forward of the trailing edge, flow separation may also happen there.

Inviscid theory for steady flow past a 2D foil in an infinite and incompressible fluid predicts zero drag. The drag forces illustrated in Figure 2.18 are therefore the result of viscous effects and are Reynolds-number dependent. The C_D coefficient presented as a function of α has a bucket form with a minimum drag coefficient. It is obviously desirable from resistance and operational points of view, with a wide bucket causing lowest possible C_D values for the broadest range of angles of attack.

Many aspects of viscous foil resistance have implicitly been covered in the previous sections. The first step is to assume a smooth flat plate and use the ITTC formula given by eq. (2.4) to calculate frictional resistance. However, the viscous resistance depends on the pressure distribution around the foil, which is a function of the angle of attack α. Because the difference between the pressures on the pressure and suction sides gives the dominant contribution to the lift force, we may say that the lift force affects the viscous resistance. Further, the viscous boundary layer flow will affect the pressure distribution, as illustrated in Figure 2.10.

We can estimate the viscous resistance of a 2D foil by using information from Hoerner (1965). If we disregard the influence from lift, we can write the drag coefficient defined in eq. (2.89) as

$$C_D = 2C_F[1 + 2(t/c) + 60(t/c)^4], \qquad (2.90)$$

where t/c is the thickness-to-chord ratio of the foil and C_F is given by eq. (2.4). This means that eq. (2.90) applies to a smooth surface in turbulent flow condition. The formula may also be applied to a strut.

If the aspect ratio is large, we can use strip theory to obtain the viscous drag force on the foil. This means the foil is divided into strips of small lengths and with cross sections as depicted in Figure 2.16. The flow is assumed two-dimensional at each strip. A contribution to the viscous resistance is then found by using eq. (2.90) with correction for the angle of attack on each strip (see Figure 2.18) and simply adding together the results obtained for each strip. In addition, we must add the induced pressure drag due to the trailing vortex sheet (see eq. 2.98). This is a 3D flow effect that increases with decreasing aspect ratio. The strongest effect is caused by tip vortices. Figure 2.19 illustrates this together with other strong trailing vortex systems present behind a submarine. The tip vortices behind the wings of a big airplane can cause strong rotating flow. This requires that a small airplane coming after a big airplane must wait some time before takeoff. The effect of tip vortices can partly be counteracted by using winglets (Figure 2.20) on hydrofoils. Winglets are used on many airplanes. They force the flow to be more two-dimensional at the foils, which means smaller induced drag due to the tip vortices.

The vorticity in the trailing vortex sheet downstream from the foil comes from the boundary layer along the foil and in the numerical flow simulations it is often assumed are concentrated in a thin shear layer downstream from the foil. Actually, both the boundary layer and the free shear layer are assumed to have zero thickness in the calculations. If we look upon the flow in the free shear layer on the scale of thickness of the free shear layer, we find a rapid change of the flow across the free shear layer. On the other hand, looking upon the flow from outside the shear layer, we will just see a jump in the tangential velocity across the shear layer. The pressure is continuous through the thin free shear layer in a similar way as we deduced for the boundary layer, that is, as

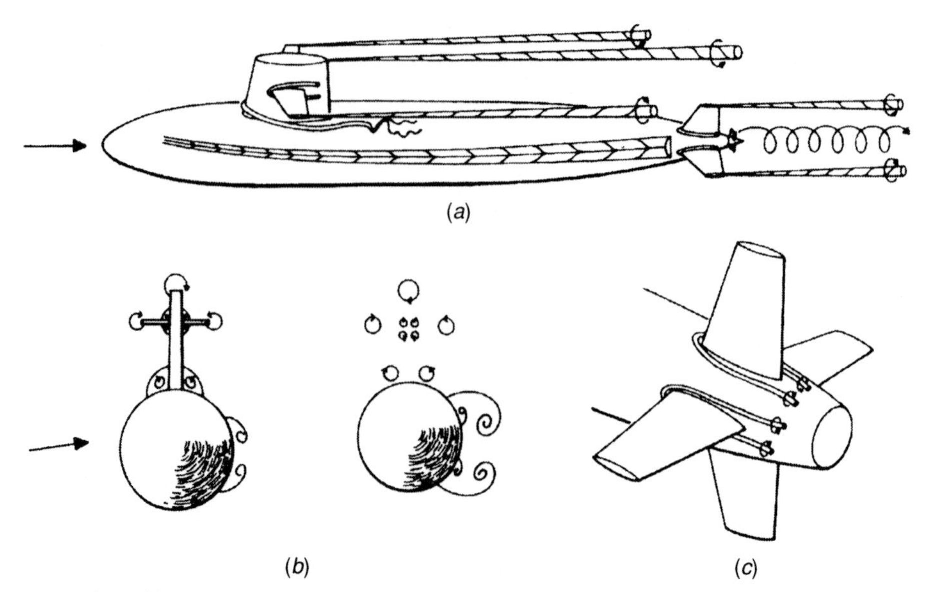

Figure 2.19. (a) The major vortex configurations around a submarine at a small angle of attack in a slight turn. (b) Two cross sections of the hull at and behind the sail. In the latter case, the local angle of attack (drift angle) is larger. (c) Necklace vortices around the control surfaces at the stern (Lugt 1981).

a first approximation, the pressure does not vary across the boundary layer.

There is zero vorticity outside the boundary layer and the free shear layer. Vorticity ω is a vector defined by

$$\omega = \nabla \times \mathbf{u} = \mathbf{i}\left(\frac{\partial w}{\partial y} - \frac{\partial v}{\partial z}\right) + \mathbf{j}\left(\frac{\partial u}{\partial z} - \frac{\partial w}{\partial x}\right) + \mathbf{k}\left(\frac{\partial v}{\partial x} - \frac{\partial u}{\partial y}\right), \tag{2.91}$$

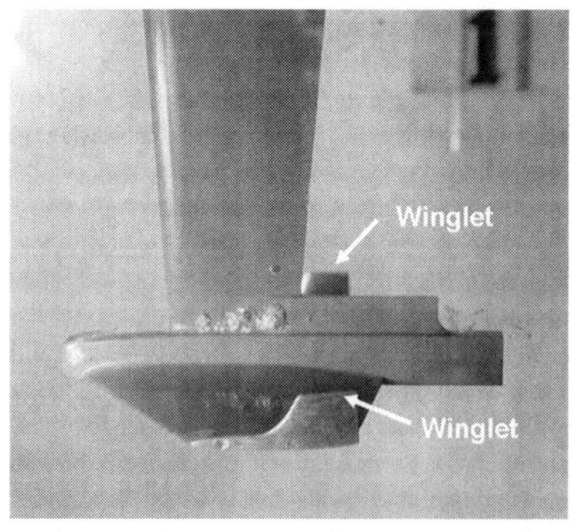

Figure 2.20. Winglets on one of the two low-aspect ratio foils in the front strut-foil system of a foil catamaran tested by MARINTEK at Technical University of Berlin (side view).

where $\mathbf{u} = (u, v, w)$ is the fluid velocity, (x, y, z) is a Cartesian coordinate system, and $\mathbf{i}$, $\mathbf{j}$, and $\mathbf{k}$ are unit vectors along the x-, y-, and z-axes, respectively.

Because the vorticity is zero outside the boundary and free shear layers, a velocity potential φ can be introduced so that

$$\mathbf{u} = \nabla\varphi = \mathbf{i}\frac{\partial\varphi}{\partial x} + \mathbf{j}\frac{\partial\varphi}{\partial y} + \mathbf{k}\frac{\partial\varphi}{\partial z}. \tag{2.92}$$

We can now combine eq. (2.92) with the assumption that the fluid is incompressible, that is, use the continuity equation

$$\frac{\partial u}{\partial x} + \frac{\partial v}{\partial y} + \frac{\partial w}{\partial z} = 0.$$

This gives

$$\frac{\partial^2\varphi}{\partial x^2} + \frac{\partial^2\varphi}{\partial y^2} + \frac{\partial^2\varphi}{\partial z^2} = 0. \tag{2.93}$$

Eq. (2.93) expresses the fact that the velocity potential satisfies the 3D Laplace equation. In order to solve this equation, we need boundary conditions on the foil and along the free shear layer. This is just one way to solve the potential flow problem. We also could use the Euler equations that correspond to the Navier-Stokes equations with a zero viscosity coefficient. If we substitute eq. (2.92) into Euler's equations and integrate these equations in space, we get Bernoulli's equation for the pressure p, that is,

$$p + \rho\frac{\partial\varphi}{\partial t} + \frac{\rho}{2}|\nabla\varphi|^2 + \rho g z = \text{Constant}. \quad (2.94)$$

Here we have included gravitational acceleration g in Euler equations and assumed that the z-axis is vertical and positive upward. The terms $-\rho g z$ and $-\rho\partial\varphi/\partial t - 0.5\rho|\nabla\varphi|^2$ will be referred to as hydrostatic and hydrodynamic pressure terms, respectively.

Another way to represent the potential flow solution is by means of Biot-Savart's law (see section 6.4.2). This expresses the influence of the vorticity on the fluid velocity in the potential flow domain. If the foil is thin, this represents the complete contribution to the fluid velocity. We will not pursue these different ways of finding the potential flow solution in this context, but use eq. (2.91) to express the vorticity in the boundary and free shear layers.

We start with the boundary layer and use (x, y, z) as a local coordinate system, with the y-axis normal to the foil surface. Consistent with our derivation for a 2D boundary layer, we assume that $\partial u/\partial y$ and $\partial w/\partial y$ are much larger than $\partial v/\partial z$, $\partial u/\partial z$, $\partial w/\partial x$, and $\partial v/\partial x$. Eq. (2.91) can then be approximated as

$$\omega = \mathbf{i}\frac{\partial w}{\partial y} - \mathbf{k}\frac{\partial u}{\partial y}. \quad (2.95)$$

By integrating eq. (2.95) across the boundary layer and noting that u and w are zero on the foil surface, we get

$$\int_0^\delta \omega\, dy = \mathbf{i}w - \mathbf{k}u, \quad (2.96)$$

where δ is the boundary layer thickness and u and w are tangential velocity components on the foil surface as calculated by potential flow theory.

As for the free shear layer, the integration of ω across the free shear layer gives $\mathbf{i}\Delta w - \mathbf{k}\Delta u$, where Δu and Δw are the jumps in the tangential velocity components u and w across the free shear layer. Δu and Δw can be obtained from the potential flow solution.

What we have tried to illustrate in the previous text is that the induced drag due to a thin trailing vortex sheet downstream from a foil can be calculated by potential flow theory. This is possible because vorticity generated in the boundary layer is convected downstream in the fluid in a thin free shear layer and does not diffuse much into the rest of the fluid.

As an example, consider a thin foil with zero camber and an elliptical planform. The aspect ratio is then given by $\Lambda = 4s/\pi\ell_0$, where s is the span and ℓ_0 is the chord at midspan. If the aspect ratio is high, the angle of attack α is small, and we consider steady incident flow in infinite fluid, we can use Prandtl's lifting theory as described in section 6.7.1. The lift and drag coefficients defined as in eqs. (2.88) and (2.89) become then

$$C_L = \frac{2\pi\alpha}{1 + 2/\Lambda} \quad (2.97)$$

and

$$C_D = \frac{4\pi\alpha^2\Lambda}{(\Lambda + 2)^2}, \quad (2.98)$$

respectively. Here eq. (2.98) represents the induced pressure drag due to the trailing vortex sheet. If Λ goes to infinity, we get the two-dimensional results, that is, $C_L = 2\pi\alpha$ and $C_D = 0$.

2.3 Air resistance component

The air resistance with no wind present may be expressed as

$$R_{AA} = 0.5\rho_a C_D A U^2, \quad (2.99)$$

where ρ_a is the mass density of the air and A is the area of the above-water hull form projected onto a transverse plane of the vessel. We note the different form between eqs. (2.3) and (2.99); we used a surface area S in eq. (2.3) and a projected area A in eq. (2.99). It is logical to use S when frictional forces dominate, whereas it is logical to use A when pressure forces dominate. Efforts are made to design streamlined superstructures for high-speed vessels in order to minimize C_D. Wind tunnel tests are commonly used to determine C_D. Typical values of C_D are between 0.5 and 0.7. Because ρ_a is only 1.25 kgm^{-3} for dry air at 10°C whereas ρ for salt water at 10°C is 1026.9 kgm^{-3}, the air resistance makes a small contribution; however, it should not be neglected a priori. For instance, an SES on cushion has a small water resistance relative to a similarly sized catamaran. The air resistance will then be more significant for the SES than for the catamaran. The airflow can also influence trim and sinkage, which again affect the resistance.

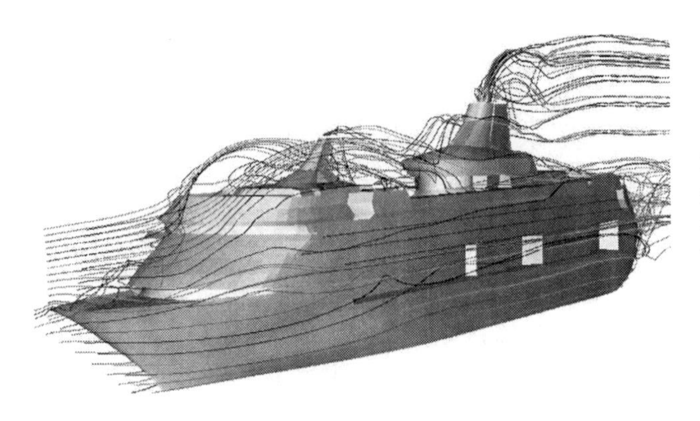

Figure 2.21. Exhaust gas tracing on a cruise ship by means of computational fluid dynamics (CFD). (The figure is reprinted with the permission of CFD Norway AS.)

If a model is made with complete superstructure and is mounted below the main part of the towing tank carriage in a model tank, the towing carriage will affect the airflow and cause false effects, which can be avoided by mounting the model in front of the main part of the towing carriage.

Wind-tunnel tests and computional fluid dynamics (CFD) are commonly used to estimate the air resistance. Figure 2.21 illustrates the use of CFD to trace exhaust gas on a cruise ship by drawing streamlines. Calculations of the pressure distribution and drag coefficient are an integrated part of the analysis.

2.4 Spray and spray rail resistance components

When the Froude number is larger than approximately 0.5, the occurrence of spray increases strongly with speed. Figure 2.22 shows an example of large spray formation due to a round-bilge hull. Müller-Graf (1991) divides the spray resistance R_s into two components, that is,

$$R_S = R_{SP}(Fn) + R_{SF}(Rn, Wn), \quad (2.100)$$

where the spray pressure resistance R_{SP} is a function of the Froude number and the spray frictional resistance R_{SF} is a function of the Reynolds number Rn and the Weber number

$$Wn = \frac{\rho V_{SR}^2 d_{SR}}{T_S}. \quad (2.101)$$

Here V_{SR} is the spray velocity, d_{SR} is the spray thickness, and T_S is the surface tension at the water-air interface. A representative value of T_S is 0.073 Nm^{-1}.

Müller-Graf (private communication, 2004) gives the following explanation of R_{SP}: "Due to the high stagnation pressure and large pressure gradients at the hull of the forebody near below the free surface, the spray root, a sheet of green water breaks violently out of the water surface, causing hereby a fully turbulent flow in the spray root. The rear part part of the spray root climbs the hull sides up and aft. Because of the development of Helmholtz-Taylor instabilities at the surface of the spray root (Birkhoff and Zarantonello 1957), which are initiated by the turbulence condition, the outward thrown part of the spray root bursts into a white mixture of water droplets and air. This phenomenon starts at the top and on the front side of the spray root. R_{SP} is caused by the generation of the spray root." When calculations or experiments are done, one cannot clearly separate this effect from the total pressure distribution on the

Figure 2.22. Development of spray at a semi-displacement round-bilge hull (Müller-Graf 1991).

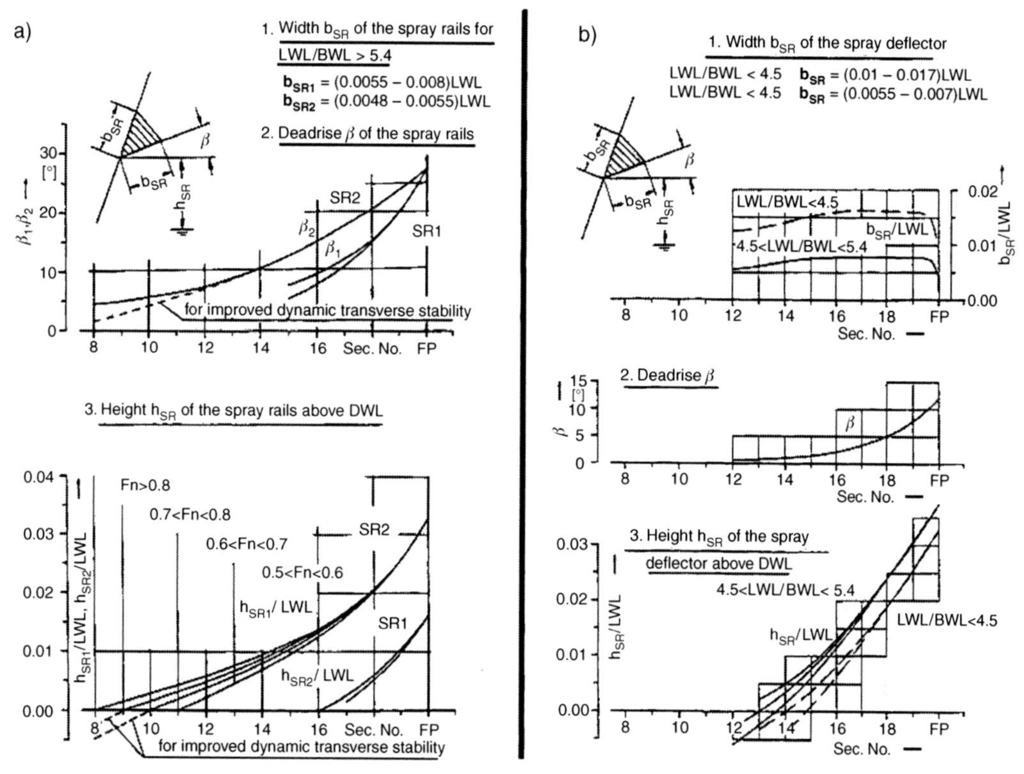

Figure 2.23. (a) Optimal dimensions and arrangement of the Advanced Spray Rail System for round-bilge monohulls and round-bilge and hard-chine catamaran hulls. (b) Optimal dimensions and arrangement of spray deflectors at round-bilge and hard-chine monohulls (Müller-Graf 1994).

hull. The spray frictional resistance is caused by the friction of the spray on the above-water hull form. Because the velocity of the coherent water sheet of the spray at the hull is difficult to estimate and by this the frictional resistance coefficients are unknown, the spray frictional resistance is hard to calculate. The main direction of flow in the spray sheet and the direction of the resulting component of the frictional spray resistance are also undefined. These quantities can be determined reliably only by full-scale tests.

In order to vary the spray pressure resistance, one must change the hull form in the bow part. The spray frictional resistance can be reduced by using spray rails separating the spray sheet from the hull and reducing the spray wetted area. However, introducing spray rails causes spray rail resistance R_{SR}. Müller-Graf (1991) divides R_{SR} into two components, that is,

$$R_{SR} = R_{SRP} + R_{SRF}, \qquad (2.102)$$

where the spray rail pressure resistance R_{SRP} is caused by generation of hydrodynamic lift due to deflection of the spray sheet in the longitudinal and transversal directions at the bottom of the rail. The spray rail frictional resistance R_{SRF} is caused by frictional force on the rail. This resistance component is negligible relative to the spray frictional resistance R_{SF}. We cannot calculate R_{SR} in a simple way. The hydrodynamic lift on the spray rails will affect the trim of the vessel and thereby the hull resistance.

Müller-Graf (1991, 1997) describes the results of systematic model tests with semi-displacement round-bilge hulls that are equipped with spray rails. The Froude number range is between 0.3 and 1.0. Number, position, length, and cross-sectional shape of the rails are varied, and optimum configurations are derived (Figure 2.23a and b). The optimal set of spray rails, which consists of the two staggered external spray rails with triangular cross sections was introduced under the name

Advanced Spray Rail System (ASRS). In order to combine reduced wetted area and best possible running trim, both transom wedge and spray rails have to be installed at the same time. The effect of the spray rails is Froude number dependent. They can reduce the full-scale hull resistance of the naked hull by up to about 12% for Froude numbers larger than 0.5. However, at $Fn = 0.3$, the resistance can increase by up to about 20% because of the spray rails and the transom wedge. In a seaway, rails will reduce deck wetness. If long enough and extended down to the waterline, or even below, spray rails will increase dynamic transverse stability and dampen roll motions. As Müller-Graf (1999a, 1999b) demonstrated, the Advanced Spray Rail System has been proven to function very well, not only for round-bilge monohulls but also for round-bilge catamarans and trimarans as well as for hard-chine catamaran hulls.

Müller-Graf (private communication, 2004) states, "At very small length-to-beam ratios, $L/B < 5.0$, which are common for the daughter boats of rescue-, police-, and custom craft, the blunt forebody lines generate a very thick spray which climbs straight upwards to the deck. To separate the spray from the hull and to throw it sidewards away from the hull, the optimum bottom width of the spray deflectors had to be approximately three times the spray rail width of hulls having length-to-beam ratios $L/B > 5.4$. It was found, that only one spray deflector at each hullside is necessary. The dimensions and arrangement of the spray deflectors as well as their heights above the waterline depending on their longitudinal position at the hull are given in Figure 2.23."

2.5 Wave resistance component

The wave resistance R_W is caused by the waves the vessel generates following a straight course with constant speed U in calm water conditions. This means there are no incident waves. By waves we mean both the local wave elevation along the hull and the far-field waves. In Chapter 4, we discuss this in more detail and show that the wave resistance is influenced by the wetted hull form and the Froude number. Further, the air cushion of an SES generates waves which causes wave resistance. If the ship is in shallow water, R_W can be strongly influenced by the water depth h. A depth Froude number $Fn_h = U/(hg)^{0.5}$ is used to characterize this effect. Large changes in wave resistance occur around the critical Froude number $Fn_h = 1$ when h/L is small. There is no simple formula for wave resistance, like the one we have for viscous resistance.

An important part of wave resistance is associated with the energy in the far-field waves caused by the ship. The wave elevation can be measured along longitudinal cuts parallel to the ship's track, and the associated wave resistance can be calculated by assuming small wave slopes. This is called wave pattern resistance R_{WP}, but it does not account for the fact that the wave slopes can be large or that the waves break near the ship. The wave breaking resistance R_{WB} due to breaking bow waves of the blunt forms of tankers has been extensively studied by Baba (1969). R_{WB} can make an important contribution. However, the bow waves of high-speed vessels, particularly with flare, may also overturn and impact on the underlying free surface. Further, the overturning bow waves are associated with spray and it is common to talk specifically about spray resistance for high-speed vessels (see the previous discussion). The behavior of the pressure at the spray root causing the spray is an integrated part of the pressure causing wave resistance. Strong flow interaction effects due to the demihulls of a catamaran may cause steep and breaking waves between the two hulls in the aft part of the catamaran. Further, the waves may break aft of the transom stern of monohull and multihull vessels.

2.6 Other resistance components

Drag forces on appendages have to be considered. Examples of appendages for a hydrofoil vessel are inclined propeller shafts, rudders not included in the foil system, propeller nacelles, and waterjet intakes (pods). Similar appendages may also be used for other high-speed vessels. Empirical formulas may be found in van Walree (1999) and Hoerner (1965).

The skirt of an SES will cause a resistance component. The skirt may be considered a high-aspect ratio planing surface. The forward jet flow that is generated will cause a drag force, as described in section 9.2.4 for a 2D rigid planing surface. However, a skirt is in reality flexible and vibrating, even in calm water conditions. The percentage of additional resistance related to the flexible vibrations is hard to estimate. The viscous resistance on the skirt can be approximated by considering it as a flat plate.

A planing vessel will have an important resistance component associated with the hydrodynamic pressure causing lift force and trim moment on the vessel. This is considered in more detail in section 9.2.

All the previous resistance components refer to a ship with constant speed on a straight course in calm water. Added resistance that is caused by waves, wind and ship maneuvering should also be considered. Added resistance in waves is often misunderstood and believed to be wave resistance. The added resistance R_{AW} in waves is a consequence of interaction between incident waves and the ship. It is particularly large when the relative vertical motion between the bow of the vessel and the waves is large. The physical reasons for added resistance are different for a semi-displacement vessel and an SES. For a semi-displacement vessel, it is caused by diffraction of the incident waves by the ship and by radiation of waves due to wave-induced ship motions. A dominant effect for an SES is associated with the leakage from the air cushion caused by the relative vertical motions between the SES and the waves. If the lifting power of the fans for the cushion is unchanged, the air cushion pressure drops and the SES sinks to a lower position with a larger wetted surface. The calm water resistance in this lower position explains the major part of added resistance for an SES in a seaway.

When the wind resistance is calculated, the wind velocity is time averaged over one hour and the average wind velocity u_{10} at 10 m over free surface is used as a reference value. The incident horizontal wind velocity $\bar{u}$ will vary with height z above the mean free surface, as illustrated in Figure 2.24. The logarithmic horizontal velocity profile in the overlap layer of a fully rough boundary layer flow is often used. One version is to set the constant in eq. (2.78) equal to zero. This means $\bar{u}/v^* = (1/\kappa) \cdot \ln(z/k)$ and $u_{10}/v^* = (1/\kappa) \cdot \ln(10/k)$. Dividing those two equations gives

$$\frac{\bar{u}}{u_{10}} = \frac{\ln(z/k)}{\ln(10/k)}. \tag{2.103}$$

In order to find a measure of the roughness height k, we can relate eq. (2.103) to another formula

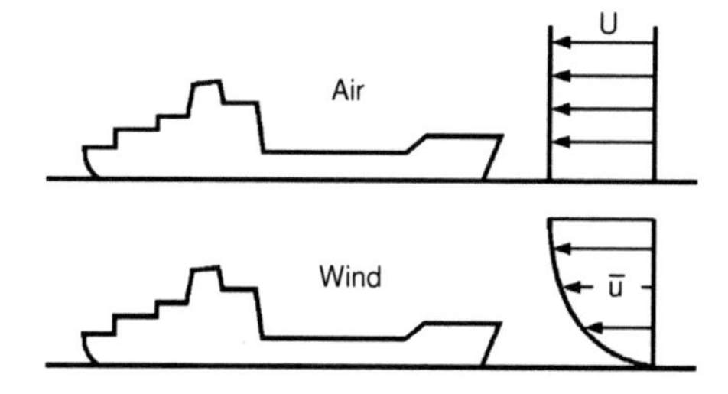

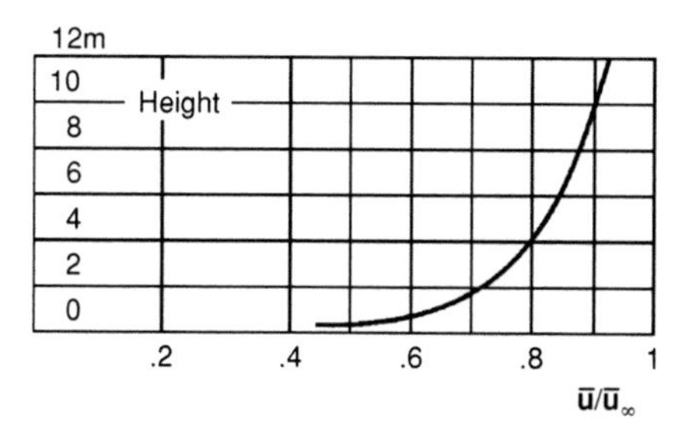

Figure 2.24. Incident flow velocity as a function of height above mean free surface when air and wind resistances are considered (Walderhaug 1972).

often used for $\bar{u}$, that is,

$$\frac{\bar{u}}{u_{10}} = \left(\frac{z}{10}\right)^{1/7}. \tag{2.104}$$

Using $k = 0.006$ m gives almost the same velocity distribution as in eq. (2.103), from around $z > k$ to about 100 or 200 m. In fully developed seas, u_{10} can be related to the sea state, as we shall see in Table 3.5.

Figure 2.24 illustrates that the incident flow velocity will not vary with height when air resistance is considered. We can in principle use a drag formula such as eq. (2.99) to obtain the influence of wind resistance. If head winds are considered, we should add U and an average wind velocity over the superstructure. In this context, a challenge in the C_D prediction is that this is influenced by the incident flow variation with height above mean free surface.

2.7 Model testing of ship resistance

Model testing is the standard procedure to predict the resistance of a ship. A model that is geometrically similar to the ship is manufactured. However, it is normally not equipped with appendages such as a rudder and propulsion system. The reason is scaling problems. The scaling of resistance from model scale to full scale is by means of nondimensional resistance coefficients and knowledge about what flow parameters (Reynolds and Froude numbers) matter. Let us illustrate what we mean by nondimensional resistance using eq. (2.3). We can write

$$\frac{R_V}{0.5\rho S U^2} = C_F,$$

which is a nondimensional quantity. Let us see the consequence of requiring that

$$\frac{R_{VM}}{0.5\rho_M S_M U_M^2} = \frac{R_{VS}}{0.5\rho_S S_S U_S^2}. \quad (2.105)$$

Here the subscripts M and S refer to model and full-scale ships, respectively. Let us assume that the hull surface is smooth and the flow at the hull surface is turbulent in both the model and full scales. We will use eq. (2.4), which states that if eq. (2.105) is true, then

$$\frac{U_M L_M}{\nu_M} = \frac{U_S L_S}{\nu_S},$$

that is, we must have the same Reynolds number in model and full scales. Let us say that $L_S =$ 100 m, $L_M = 5$ m, $U_S = 20$ ms^{-1}, and the kinematic viscosity coefficient $\nu_M = \nu_S$. This leads to $U_M = 400$ ms^{-1}. It is not difficult to understand that this is not the right procedure. It is impossible in practice to have a towing carriage with a speed like that. Because the nondimensional wave resistance is a function of Froude number and $U_M = 400$ ms^{-1} would lead to very different Froude numbers in model and full scales, we have in practice no way to scale wave resistance. Further, new physical phenomena associated with cavitation will occur with a model speed like this.

The procedure that is followed in practice is based on Froude's hypothesis. The total water resistance coefficient C_T is divided into two parts, that is,

$$C_T = \frac{R_T}{0.5\rho S U^2} = C_F + C_R, \quad (2.106)$$

where C_F is defined as in eq. (2.3) and assumed to be only a function of Reynolds number for geometrically similar models. Further,

$$C_R = \frac{R_R}{0.5\rho S U^2}, \quad (2.107)$$

where R_R is the residual resistance. A main component of R_R is the wave resistance. C_R is assumed to be only a function of Froude number for geometrically similar models. This procedure assumes that the air resistance is either negligible or is corrected for when the total resistance is measured. If the flow around the superstructure separates from sharp corners, the Reynolds number dependence of the drag coefficient associated with the airflow will be small. This means the air resistance will Froude scale in cases like that.

The model speed is obtained by Froude scaling. That means we require

$$\frac{U_M}{\sqrt{L_M g}} = \frac{U_S}{\sqrt{L_S g}}$$

or

$$U_M = U_S \sqrt{\frac{L_M}{L_S}}. \quad (2.108)$$

Repeating the previous example with $L_S =$ 100 m, $L_M = 5$ m, and $U_S = 20$ ms^{-1} gives $U_M =$ 4.47 ms^{-1}. This leads to a very different Reynolds numbers in model and full scales. We must then know how C_F depends on Rn. For instance, if the ITTC 1957 model-ship correlation line expressed by eq. (2.4) is used, then we know how to extrapolate C_F from model to full scale. Because this formula is based on turbulent flow, we must ensure that the flow along the ship model is turbulent. We will use the results for two-dimensional flow along a smooth flat plate as a basis. Transition to turbulence depends on the turbulence intensity of the inflow. If we neglect the effect of turbulence in the inflow, the transition to turbulence occurs when $Ux/\nu = 3 \cdot 10^6$. Here x is the longitudinal distance from the leading edge. Using $U = 4.47$ ms^{-1} and $\nu = 10^{-6}$ m^{2}s^{-1} gives that turbulence is ensured for longitudinal distances from the bow larger than 0.67 m. To avoid errors in our extrapolation, we want transition to turbulence to happen closer to the bow. This is achieved by using turbulence stimulators at the bow of the model. For instance, Molland et al. (1996) used trip studs of 3.2 mm diameter and 2.5 mm height at a spacing of 25 mm. The studs were situated 37.5 mm aft of the stem on 1.6 m–long models. White (1974) gives

$$Uk/\nu = 826 \quad (2.109)$$

as the criterion for a wire of diameter k to be fully effective in causing turbulence. We see from eq. (2.109) that it is the lowest tested speed that determines k. Let us say that this corresponds to $Fn = 0.15$. If a 5 m–long model is used, this means $U = 1.05$ ms^{-1} and $k = 0.0008$ m

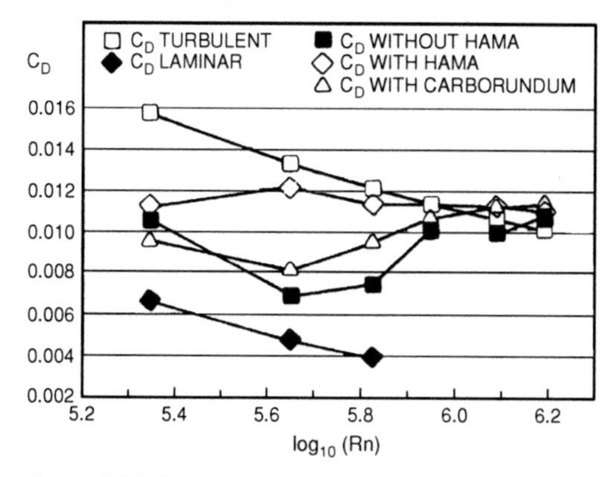

Figure 2.25. Drag coefficient C_D at zero angle of attack for NACA 16012 section as a function of Reynolds number Rn (van Walree and Yamaguchi 1993).

with $\nu = 10^{-6}\,\text{m}^2\text{s}^{-1}$. If the diameter is too large, it will affect the global flow. This is of concern, for instance, in testing the foils and struts of a hydrofoil vessel.

van Walree and Yamaguchi (1993) carried out a series of experiments with a NACA 16012 foil in order to investigate possible scale effects on the lift and drag. The tests were performed with a vertical surface-piercing strut at two different immersions (1.0 m and 0.5 m), with the intention of eliminating free-surface effects. With a chord length of 250 mm, the foil was tested in the Reynolds number range $0.2 \leq Rn \cdot 10^{-6} \leq 1.5$, with and without turbulence stimulation. The results for zero lift angle are presented in Figure 2.25. They clearly reveal the effect of turbulence stimulation, particularly at low Reynolds numbers. However, neither the Carborundum size (60 μm, coverage about 40% over 5% of the chord length from the leading edge) nor the Hama strip thickness (0.18 mm, 5% from the leading edge extending over 8% of the chord length) seems sufficient to trigger fully turbulent boundary-layer flow at low Reynolds numbers. Carborundum is a trade name. It consists of silicone carbide. The basis of a Hama strip is a tape (Hama et al. 1956, Hama 1957). A sawtooth shape is made on the upstream edge by means of a scissors.

An advantage of the Hama strip seems to be the almost constant drag coefficients that can be obtained for relatively low Reynolds numbers in the transition from laminar to turbulent boundary-layer flow. This makes extrapolation procedures more simple to apply. The results for untripped foils and foils with Carborundum are less well defined in this region and therefore result in undefined scale effects.

If C_F is known, as in eq. (2.4), Figure 2.26 illustrates how the calm water resistance of a ship in full scale can be obtained by scaling model test results to full scale. Because the model tests are based on a smooth model without appendages, we have to add a resistance coefficient C_A that accounts for hull roughness, air resistance, and appendage resistance.

Another method that was commonly used to establish viscous resistance for a smooth hull surface is the Hughes (1954) method. This expresses the viscous resistance as

$$C_F = C_{F0}\,(1 + k)\,, \qquad (2.110)$$

where

$$C_{F0} = \frac{0.066}{(\log_{10} Rn - 2.03)^2}. \qquad (2.111)$$

The form factor k must be experimentally determined and is assumed to be independent of Reynolds number. Estimation of k requires that model tests be done for small Froude numbers where wave resistance is negligible, let us say $Fn < \approx 0.15$.

A difficulty with the Hughes method is that it requires measurements at small speeds. The flow at the hull surface may then be laminar at the bow part of the ship model. Further, because the resistance decreases with decreasing speed, measurement accuracy becomes important. Typical values of $(1 + k)$ for displacement ships are between 1.2 and 1.4. If the flow separates, $(1 + k)$ can be as high as 1.8.

We can speculate what $(1 + k)$ represents. Because Hughes assumes that C_{F0} is the friction coefficient for a flat plate, it is logical to say that $(1 + k)$ accounts for the three-dimensionality in

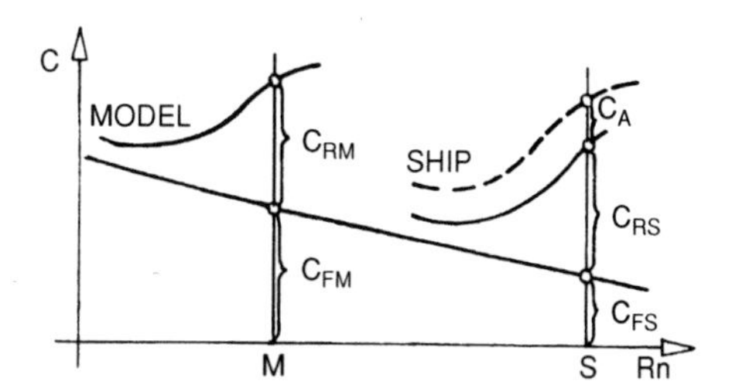

Figure 2.26. Scaling of ship resistance from model (M) to full scale (S) (Walderhaug 1972).

the flow along the hull surface and for the effect of pressure drag. The latter effect is particularly important when flow separation occurs. However, we must be sure that C_{F0} as given by Hughes is a correct value for 2D turbulent flow along a flat plate. The results in Table 2.1 do not indicate that. Actually, C_F given by eq. (2.4) is closer to the 2D flow formula. If we calculate C_F/C_{F0}, where C_F is given by eq. (2.4), and interprete this in terms of $(1+k)$, it gives $(1+k)$-values between 1.12 and 1.13 for Rn between 10^6 and 10^{10}. Molland et al. (1996) instead use C_F given by eq. (2.4) as a basis and multiply this by $(1+k)$ to predict viscous resistance. They obtain experimentally $(1+k)$-values for high-speed monohulls between 1.22 and 1.45. The $(1+k)$-values for high-speed catamarans are higher when compared with the demihull in isolation. The highest $(1+k)$-value found was 1.65.

One possible reason why the $(1+k)$-values for high-speed mono- and multihull vessels are high is flow separation at the transom stern. When the form factor is determined, the Froude number is small and the transom stern is wet. The sharp edge of the transom stern ensures flow separation, which causes a pressure drag force. This is called base drag, where the word *base* now refers to the transom stern. What happens is that the vortex shedding due to flow separation alters the pressure distribution on the transom stern. Hoerner (1965) has presented base drag coefficients for longitudinal ambient flow along axisymmetric bodies with a large length-to-beam ratio in infinite fluid. We can express the base drag as $0.5\rho C_D A U^2$, where A is the base area.

Because the same form factor is applied to viscous resistance for all Froude numbers, a contradiction occurs when the flow separation causes a dry transom stern for $Fn > 0.3$–0.4. The base drag coefficient should be influenced by this.

2.7.1 Other scaling parameters

In the previous discussion, we assumed that only Froude and Reynolds numbers are the parameters to consider when scaling from model to full scale. However, cavitation number may matter for hydrofoils. Cavitation may lead to ventilation. The cavitation number is defined as

$$\sigma = \frac{p_0 - p_V}{0.5\rho U^2}, \qquad (2.112)$$

where p_0 is the ambient pressure at the position of the foil, p_V is the vapor pressure, and U is the foil velocity. We can write p_0 as $p_a + \rho g h$, where p_a is atmospheric pressure and h is the submergence of the foil. How the vapor pressure depends on temperature is presented in Table A.3 in the Appendix. Because U in practice will be Froude scaled and the major contribution to $p_0 - p_V$ is p_a, we see that σ is not the same in model and full scales. Because cavitation may be a problem for hydrofoils and planing vessels, we should be concerned about this. The only possibility is then to use a depressurized model tank. However, this is not possible in most ship model basins.

If we want to get a correct spray picture of a model, we must scale surface tension correctly. The Weber number Wn expresses the influence of surface tension. It can be written as

$$Wn = \frac{\rho U^2 L}{T_S}, \qquad (2.113)$$

where T_S is the surface tension, which is nearly the same in model and full scales. A representative value is 0.073 Nm^{-1}. Because the ship speed is Froude scaled, that is,

$$U_M = \sqrt{\frac{L_M}{L_S}}\, U_S,$$

we find by assuming the same value of T_S/ρ in model and full scale that

$$(Wn)_M = \frac{\frac{L_M}{L_S} U_S^2 \frac{L_M}{L_S} L_S}{\frac{T_S}{\rho}}.$$

$$= \left(\frac{L_M}{L_S}\right)^2 (Wn)_S$$

This means we will not obtain the same Weber number in model and full scales. However, the difference in the spray picture will have a small effect on the pressure and frictional forces on the hull. It means that the Weber number is not important in scaling model test results of resistance to full scale.

2.8 Resistance components for semi-displacement monohulls and catamarans

Molland et al. (1996) presented systematic results for calm water resistance, trim, and sinkage of monohulls and catamarans. The monohulls correspond to demihulls of the catamarans. The length L of the still waterline is 1.6 m, block coefficient is $C_B = 0.397$, longitudinal prismatic coefficient is $C_P = 0.693$, and midship section coefficient is $C_M = 0.565$. Here

$$C_B = \frac{\nabla}{L_{PP} \cdot B \cdot D}$$

$$C_P = \frac{\nabla}{A_M L_{PP}}$$

$$C_M = \frac{A_M}{B \cdot D},$$

where L_{PP} is the length between perpendiculars. The forward perpendicular (FP) is a vertical line through the intersection of the designer's load waterline (DLWL) and the fore side of the stem. The after perpendicular (AP) is a vertical line passing through the DLWL and the rudder post or the transom profile. This implies that there is an ambiguity in the definition of AP. When the vessel has a transom stern and no rudder, AP is at the transom according to the previous description. This means L_{PP} is the same as the length of DLWL. Further, ∇ = displaced volume of water, A_M = area of midship section, and B and D are the beam and draft, respectively. $LCB = -6.4\%L$ was the same for all models. Here $LCB = -6.4\%L$ means that the position of the center of buoyancy is 6.4% of the length between perpendiculars aft of midships for all models. The length-to-beam ratio L/B of the demihulls varied between 7 and 15.1, and the beam-to-draft ratio B/D of the demihulls varied between 1.5 and 2.5. The ratio $2p/L$ varied between 0.2 and 0.5. Here $2p$ is the distance between two demihull centerlines.

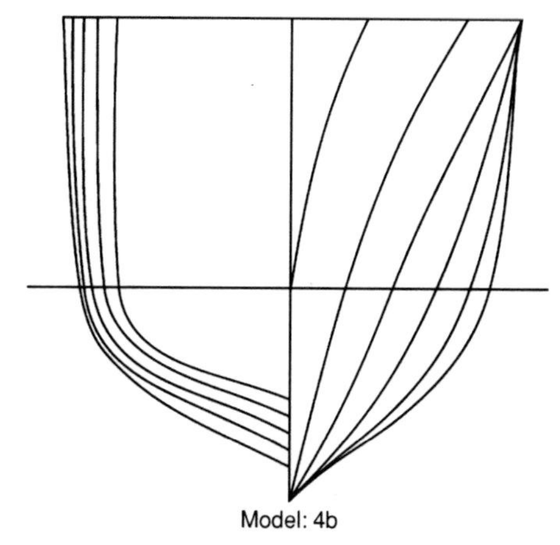

Figure 2.27. Monohull or demihull of a catamaran in experiments by Molland et al. (1996). Length L = 1.6 m. Length-to-beam ratio $L/B = 9$. Beam-to-draft ratio $B/D = 2$. $L/\nabla^{1/3} = 7.41$. $C_B = 0.397$. $C_P = 0.693$. $C_M = 0.565$. Wetted hull surface area at zero speed $S = 0.338$ m^2. $LCB = -6.4\%L$.

The Molland group found that the length to displacement ratio $L/\nabla^{1/3}$ was the most significant hull parameter causing decreasing resistance with increasing $L/\nabla^{1/3}$. The effect of B/D on resistance was not large.

We will present their results for models of a type denoted 4b. Main characteristics and the body plan are shown in Figure 2.27. Figure 2.28 gives the residual resistance coefficient C_R as a function of Froude number for the monohulls and catamarans with $2p/L = 0.2, 0.3, 0.4, 0.5$. C_R is obtained by subtracting C_F given by the ITTC 1957 model-ship correlation line (see eq. (2.4)) from the total

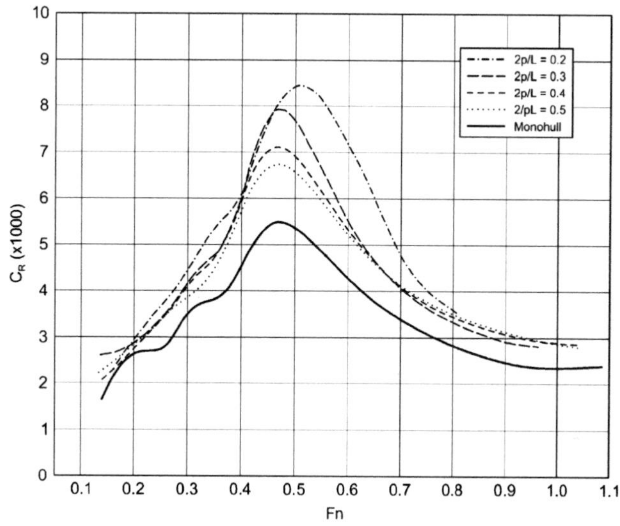

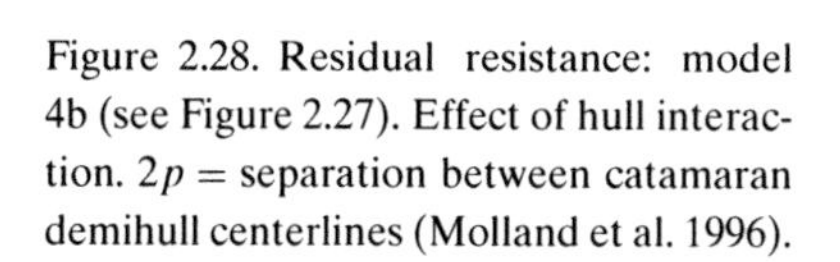
Figure 2.28. Residual resistance: model 4b (see Figure 2.27). Effect of hull interaction. $2p$ = separation between catamaran demihull centerlines (Molland et al. 1996).

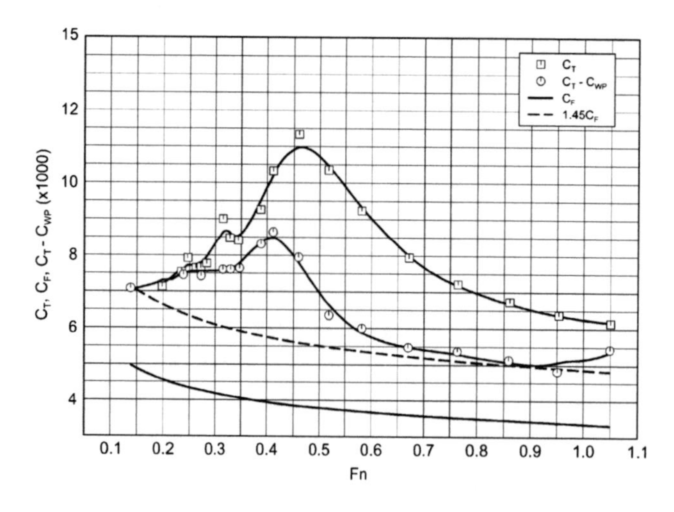

Figure 2.29. Resistance components: model 4b (see Figure 2.27). $2p/L = 0.4$. $2p$ = separation between catamaran demihull centerlines (Molland et al. 1996).

resistance coefficient C_T. The static wetted surface area is used in calculating C_T, C_R, and C_F. The results show that the catamarans have resistance higher than twice the resistance of a monohull. There is a tendency for the C_R for the catamarans to approach the C_R for the monohull when $2p/L$ increases. However, there is still a clear difference when $2p/L = 0.5$, showing that hull interaction still matters. We note that C_R is not zero for a small Froude number, which means that C_F obtained by the ITTC 1957 line is not sufficient to explain the viscous resistance. This is further clarified in Figure 2.29, in which C_T, $C_T - C_{WP}$, and C_F are presented for Fn between 0.15 and 1.05. Here C_{WP} means the wave pattern resistance coefficient, that is, the contribution to the wave resistance from the far-field waves, as described earlier in the text. C_{WP} is also nondimensionalized by the static wetted surface area. There is also a curve $1.45C_F$ in Figure 2.29 that fits the total resistance for the lowest Froude number when the wave resistance is believed negligible. This means the form factor $(1 + k)$, with ITTC 1957 line as a basis, is 1.45. If Hughes's C_{F0} had been used as a basis, $(1 + k)$ would be $1.45 \cdot 1.12 = 1.62$. The figure shows that the wave pattern resistance cannot explain the total wave resistance if we now interpret C_{WP} to be $C_T - 1.45\ C_F$. The results show that, generally speaking, viscous resistance and wave resistance are of equal importance. However, viscous resistance dominates for the smallest and highest Froude numbers.

Figures 2.30 and 2.31 present the trim angle and sinkage for the catamarans and the monohull, respectively. Positive trim angle means bow up and sinkage is positive downward. The trim angle and sinkage have a clear influence of Froude number when Fn is larger than 0.3 to 0.4. The catamarans

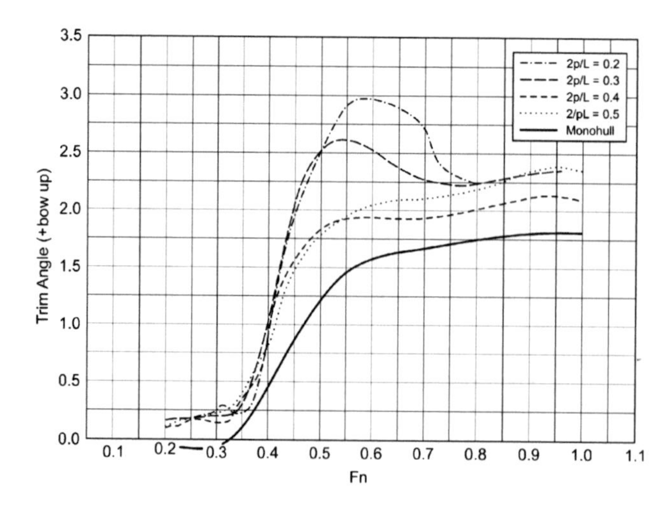

Figure 2.30. Running trim: model 4b (see Figure 2.27; Molland et al. 1996).

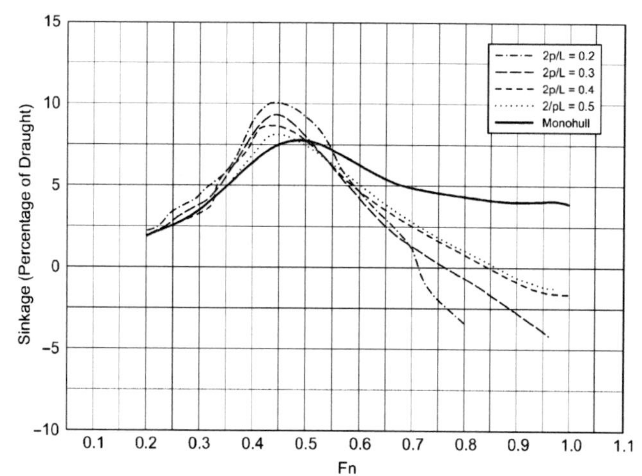

Figure 2.31. Running sinkage: model 4b (see Figure 2.27; Molland et al. 1996).

display significantly higher running trim angles than the monohull, but generally approach the monohull angle as $2p/L$ increases. The increased trim with speed will in practice be counteracted by trim tabs and/or interceptors (see section 7.1.3) in order to minimize the resistance.

2.9 Wake flow

A wake flow, for instance, is important in the analysis of the inflow to a propeller behind a ship. It is affected by both potential and viscous flow effects. In Chapter 6, we show that the wake generated by an upstream foil provides an important inflow condition to an aft foil of a hydrofoil vessel with a fully submerged foil system. Our analysis in this section is idealized and concentrates on viscous flow effects.

We consider a vertical strut with forward speed U and analyze 2D flow in a horizontal cross-sectional plane. The effect of the free surface is neglected, and the incident flow has zero angle of attack. The flow is described in a coordinate system following the strut (see Figure 2.32). There is then an incident flow with velocity U along the positive x-axis. A wake is generated behind the strut. Even if the flow is laminar in the boundary layer of the strut, the flow in the wake will be turbulent at a small distance behind the foil. The reason is that a laminar wake profile becomes unstable more easily than does a laminar boundary-layer profile (Schlichting 1979). Figure 2.32 illustrates the time-averaged longitudinal velocity profile $\bar{u}$ in the wake at some distance behind the foil. We express $\bar{u}$ as

$$\bar{u} = U - u_1, \tag{2.114}$$

where u_1 is positive in the wake and zero outside the wake. The wake will be analyzed far away from the strut. This means $u_1/U \ll 1$. The wake flow can be described by the boundary-layer equations given by eqs. (2.30) and (2.31). However, the boundary conditions differ. It is meaningless, of course, to require a nonslip condition as we would on the strut surface. Further, the turbulent stress τ_t expressed by eq. (2.33) dominates over laminar stress τ_l given by eq. (2.32). This is a consequence of measurements and is documented later in the text. In order to solve the boundary-layer equations, we need to relate τ_t to the mean velocity. This will be done by expressing τ_t as

$$\tau_t = \mu_t \frac{\partial \bar{u}}{\partial y}. \tag{2.115}$$

The form of this expression is the same as the one for laminar stress given by eq. (2.32). We assume μ_t is a constant. The actual value of μ_t has to be experimentally determined and is dealt with later.

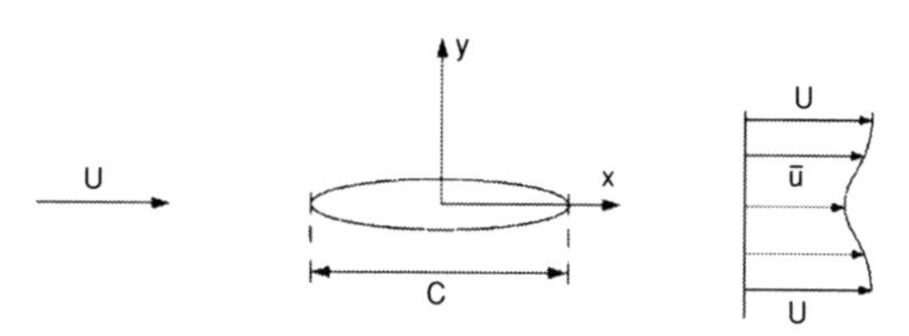

Figure 2.32. 2D mean wake flow with velocity $\bar{u}$ behind a strut with incident flow velocity U.

The pressure can be assumed to be constant and equal to the ambient pressure far away from the strut. This is the same as saying that the velocity outside the wake is equal to the incident flow velocity U far away from the strut. Because $u_1/U \ll 1$, eq. (2.30) can be linearized. This gives

$$U\frac{\partial u_1}{\partial x} = \nu_t \frac{\partial^2 u_1}{\partial y^2}, \qquad (2.116)$$

where $\nu_t = \mu_t/\rho$. The solution of eq. (2.116) for our wake problem can be found in Schlichting (1979), that is,

$$u_1 = A_1 U x^{-1/2} \exp(-0.25\eta^2), \qquad (2.117)$$

where

$$\eta = y\left(\frac{U}{\nu_t x}\right)^{1/2}. \qquad (2.118)$$

Eq. (2.117) shows that the wake goes exponentially to zero when $|y| \to \infty$. The constant A_1 is determined below. The x-dependence of u_1 at $y = 0$ according to eqs. (2.117) and (2.118) is $x^{-1/2}$. Other x-dependency may also satisfy eq. (2.116), but it is only the x-dependence in eqs. (2.117) and (2.118) that is consistent with the conservation of fluid momentum. We will show this and choose a control volume as in Figure 2.33. The strut surface and the surfaces B, C, D, and E in the fluid far away from the strut are enclosing the considered fluid volume. The flow velocity at B, C, and E is equal to the incident flow velocity, whereas the flow velocity at D is given by eqs. (2.117) and (2.118). We can then write the momentum flux into the control volume as

$$-\rho\int_B U^2\,dy + \rho\int_D \bar{u}^2\,dy. \qquad (2.119)$$

This must balance the longitudinal force acting on the control volume. The pressures at B, C, D, and E are equal to the ambient pressure. Because this is a constant, it gives zero total pressure force on the sum of the surfaces B, C, D, and E. There is a Reynolds (turbulent) stress $-\rho\overline{u'^2}$ that in principle gives a force on surface D. However, this is negligible. This means the force acting on the control volume is minus the drag force R acting on the strut, that is,

$$R = \rho\int_B U^2\,dy - \rho\int_D \bar{u}^2\,dy. \qquad (2.120)$$

We can rewrite the expression for R by using continuity of fluid mass, that is,

$$\rho\int_B U\,dy = \rho\int_D \bar{u}\,dy. \qquad (2.121)$$

This gives

$$R = \rho\int_D \bar{u}(U - \bar{u})\,dy. \qquad (2.122)$$

We now introduce u_1 defined by eq. (2.114), let C and E tend to infinity, and use the fact that $u_1/U \ll 1$ to linearize the expression for R. The result is

$$R = \rho U\int_{-\infty}^{\infty} u_1\,dy. \qquad (2.123)$$

Using eqs. (2.117) and (2.118) gives then

$$R = \rho U^2 A_1 x^{-1/2}\sqrt{\frac{\nu_t x}{U}}\int_{-\infty}^{\infty}\exp\left(-\frac{1}{4}\eta^2\right)d\eta.$$
$$= \rho U^2 A_1 2\sqrt{\pi}\sqrt{\nu_t/U} \qquad (2.124)$$

This means R is independent of x as it should be. We can also express R as

$$R = 0.5\rho C_D U^2 c, \qquad (2.125)$$

where C_D is known as in eq. (2.90) and c is the chord length, as shown in Figure 2.32. Using eqs. (2.124) and (2.125) gives

$$A_1 = \frac{C_D c}{4\sqrt{\pi}}\sqrt{\frac{U}{\nu_t}}. \qquad (2.126)$$

Further, according to experiments,

$$\nu_t = 0.0222 U C_D c. \qquad (2.127)$$

This means

$$A_1 = 0.95\sqrt{C_D c}. \qquad (2.128)$$

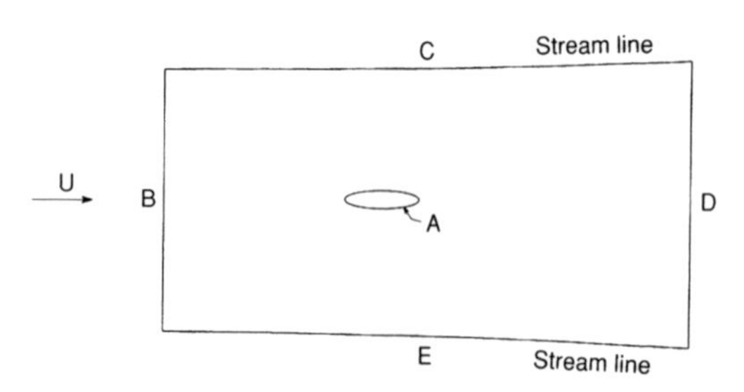

Figure 2.33. Surfaces A, B, C, D, and E enclosing the fluid volume used in application of the conservation of fluid momentum to express the drag force on the strut.

Using eqs. (2.127) and (2.128) in eqs. (2.117) and (2.118), we find that

$$\frac{\bar{u}}{U} = 1 - 0.95\left(\frac{x}{C_D c}\right)^{-1/2} e^{-0.25\eta^2}, \quad (2.129)$$

where

$$\eta = y\,(0.0222 C_D c x)^{-1/2}. \quad (2.130)$$

Let us evaluate the relative importance between turbulent and laminar stresses in the far-field wake. This can be expressed by ν_t/ν. Using eq. (2.127) gives

$$\frac{\nu_t}{\nu} = 0.0222 C_D Rn. \quad (2.131)$$

As an example, let us consider a strut with zero thickness, that is, a flat plate. We see from eq. (2.90) that $C_D = 2C_F$, where C_F is given by eq. (2.4). This means ν_t/ν is equal to 208 and 1332 for $Rn = 10^6$ and 10^7, respectively. So we confirm that Reynolds stresses dominate over laminar stresses in the wake.

We can use the factor $\exp(-0.25\eta^2)$ in eq. (2.129) to define a measure of the thickness of the wake. The y-value corresponding to $\exp(-0.25\eta^2) = 0.5$ is called $b_{1/2}$ (Schlichting 1979). From eq. (2.130), it follows that

$$b_{1/2} = 0.25\sqrt{C_D c x}. \quad (2.132)$$

This means $b_{1/2}$ is proportional to $\sqrt{x}$. Because the wake velocity decays exponentially with y, we cannot define geometrically the thickness of the wake. However, instead of using $2b_{1/2}$ as a measure of the wake thickness, we could introduce, for instance, the y-value corresponding to $\exp(-0.25\eta^2) = 0.01$ and call it $b_{1/100}$. The value $2b_{1/100}$ is obviously a better measure of the wake thickness than $2b_{1/2}$. Whatever value of $\exp(-0.25\eta^2)$ we use as a basis to define the wake thickness, we will find that the wake thickness is proportional to $\sqrt{x}$.

There is nothing in the previous derivation of the far-field wake that requires the body to be a strut. It could just as well be a circular cylinder or any other 2D body that is symmetric about the x-axis and in which the incident steady flow is along the x-axis. However, an important feature of the flow around a bluff body is flow separation and resulting vortex shedding. This causes additional time-dependent flow velocities and forces that are not accounted for in the previous application of conservation of fluid momentum.

As long as the body is streamlined and no flow separation occurs, we could in principle use the boundary-layer equations given by eqs. (2.30) and (2.31) from the leading edge to the far-field wake for a body that is symmetric about the x-axis and in which the incident steady flow is along the x-axis. However, this requires knowledge about the Reynolds stresses, which cannot be expressed as simply as we did in the far-field wake.

Our derivation is based on $u_1/U \ll 1$. How close to the body one can apply the wake solution requires comparisons with model tests. However, one can get a qualitative understanding by using eq. (2.129) for $y = 0$, and calculating u_1/U as a function of x. Blevins (1990) was able to use the far-field mean wake solution very close to a circular cylinder by redefining the origin of the coordinate system.

The wake solution may be used to define the inflow velocity to a 2D body in the wake of another 2D body. As long as the downstream body is not too close to the upstream body, there is little influence from the downstream body on the upstream body. However, the upstream body may have an important effect on the downstream body.

In the following section, we see, for instance, that the wake behind a ship hull provides an important inflow to the propeller. However, this highly 3D flow phenomenon cannot be described by a simplified procedure. It requires either model tests or CFD (computational fluid dynamics) simulations with proper turbulence modeling.

2.10 Propellers

Hydrodynamics of ship propellers is a speciality by itself and there are textbooks and lecture notes dealing comprehensively with the topic (e.g., Breslin and Andersen 1994, Carlton 1994, Kerwin 1991). Our presentation is of an introductory nature. We first discuss open-water propeller characteristics, meaning how the propeller performs when the ship does not influence the propeller flow. This is the converse of what we assumed to analyze resistance. In that case, we assumed the propeller was not present. Finally, we discuss how to correct for hull-propeller interaction.

Figure 2.34 shows a typical propeller drawing and definitions of commonly used parameters. Most propellers are fixed-pitch propellers. However, some have adjustable blades and are

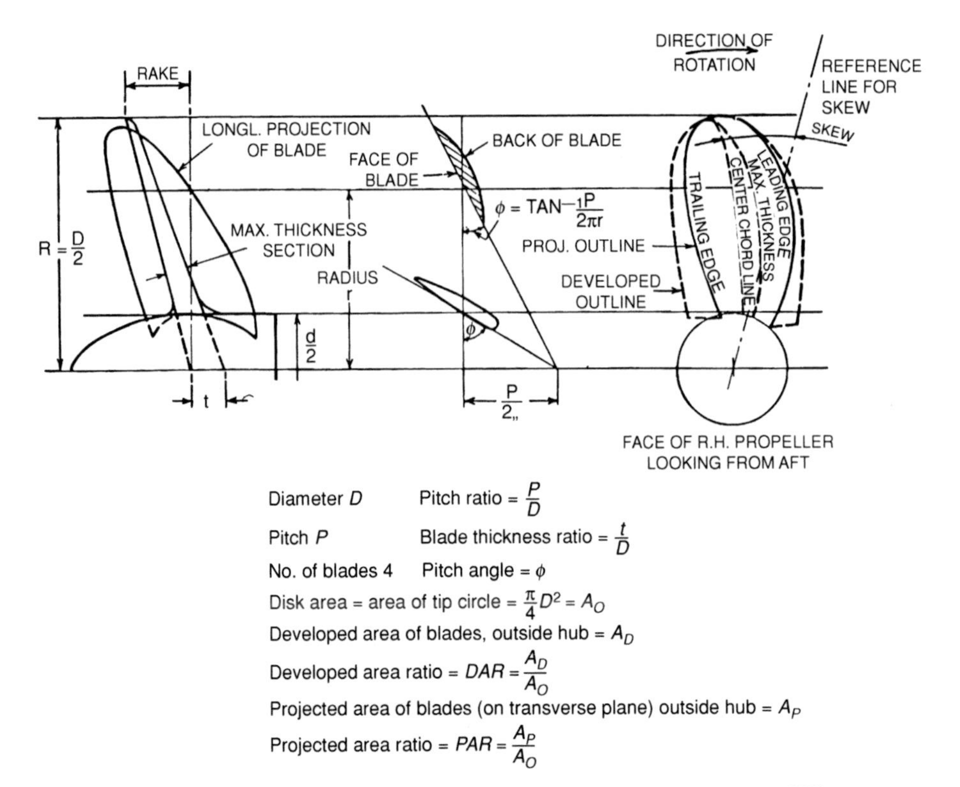

Figure 2.34. Typical propeller drawing (van Manen and van Oossanen 1988).

called controllable-pitch propellers. If the propeller viewed from aft of the propeller turns clockwise, it is called right-handed. A left-handed propeller rotates counterclockwise. If a ship is equipped with two propellers (see Figure 2.2), the starboard and port propellers are normally right-handed and left-handed, respectively. The face of the propeller is the propeller surface seen from aft of the propeller. The other side is called the back.

Different areas are used to characterize the propeller. The propeller disc area A_0 is equal to $\pi D^2/4$, where D is the propeller diameter. The expanded area A_E of the propeller is obtained by considering different circular cylinders with axis coinciding with the propeller shaft axis and with different radius r between r_h and the propeller radius R. Here r_h is the radius of the root section, that is, the hub. The intersection between the cylinder surfaces and a propeller blade defines propeller blade sections. These are indicated by lines in Figure 2.35. We then unfold the cylinder surface with the propeller blade section so that the section becomes planar. The chord length $c\,(r)$ of this section is the length of the "nose-tail line" as

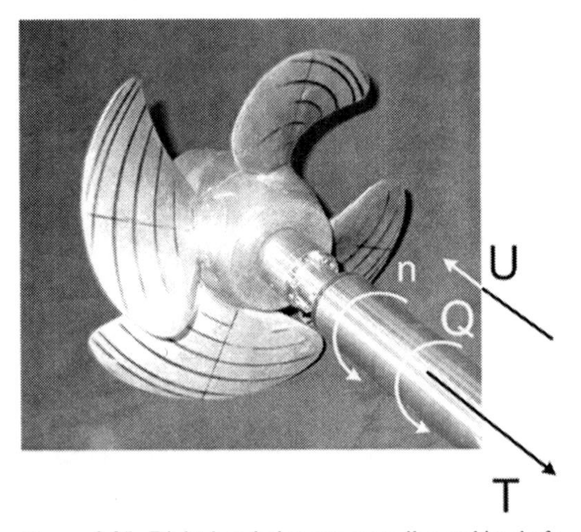

Figure 2.35. Right-handed screw propeller and its shaft in uniform inflow velocity U corresponding to the ship speed. n = shaft revolutions per second, Q = shaft torque, T = propeller thrust. There are lines drawn on the propeller with constant radial distance from the shaft axis. Each line defines a propeller blade section. (Photo by K. A. Hegstad)

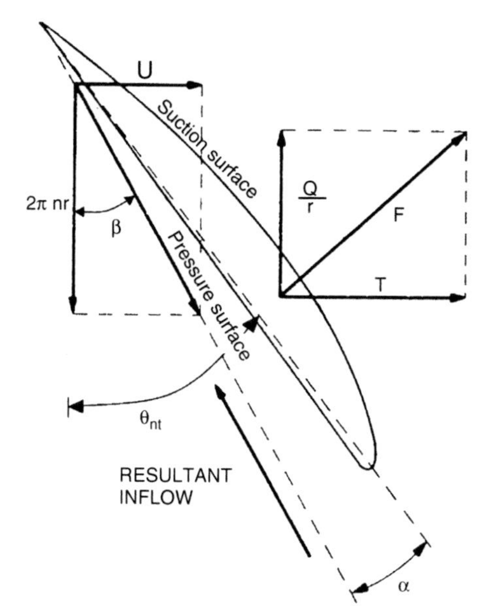

Figure 2.36. Two-dimensional view of a propeller blade section, moving to the right with velocity U and in the peripheral direction with velocity $2\pi nr$. Here r is the radial distance from the shaft axis. The local angle of attack α is shown here as the difference between the pitch angle of the blade and the inflow angle $\tan^{-1}(U/2\pi nr)$. The resultant force F contains an axial component (thrust), and a peripheral component (torque/radius). (Newman, J. N., 1977, *Marine Hydrodynamics*, Cambridge: The MIT Press. The figure is reprinted with the permission of The MIT Press.)

shown in Figure 2.16. The expanded area is defined as

$$A_E = Z \int_{r_h}^{R} c(r)\,dr,$$

where Z is the number of propeller blades.

A developed area A_D can also be defined (van Manen and van Oossanen 1988). This is more elaborate to evaluate. However, in practice, A_D and A_E are close (Carlton 1994). The blade area ratio (BAR) is either A_E/A_O or A_D/A_O.

We use Figures 2.35 and 2.36 to give a simplified picture of how a propeller works. Infinite fluid is assumed. There is a shaft torque Q delivered by the engine and a constant number n of shaft revolutions per second. Assuming a steady inflow to the propeller, Q must balance a steady hydrodynamic torque on the propeller blades and the hub. Actually, the hub torque is negligible. Seen from a propeller blade section as in Figure 2.36, there is an axial inflow velocity U equal to the ship's speed. This is a consequence of temporarily neglecting hull-propeller interaction. The propeller revolution causes an inflow velocity $2\pi nr$ in the peripheral direction. Here r is the radial distance from the propeller shaft axis to the propeller blade section. The resultant inflow velocity vector causes an angle of attack α relative to the blade section. This angle of attack is equal to the difference between the pitch angle θ_{nt} of the blade section and the inflow angle $\beta = \tan^{-1}(U/2\pi nr)$, that is,

$$\alpha = \theta_{nt} - \tan^{-1}(U/2\pi nr). \qquad (2.133)$$

The pitch angle θ_{nt} is the angle between the plane of rotation and the chord line, that is, the nose-tail line shown in Figure 2.16. The word *pitch* is related to a helical curve with angle θ_{nt}, as shown in Figure 2.37.

Let us consider a point A on a cylinder surface of radius r. The point rotates around the cylinder axis and moves axially with constant velocity. The starting and ending positions after one complete revolution are denoted A_0 and A_1 in the figure. The axial motion during this time period is the pitch P. If the cylindrical surface between A_0 and A_1 is cut open along A_0 and A_1 and unfolded into a plane, we get the right picture in the figure. We see from this picture, that the relationship between the pitch and the pitch angle is $P = 2\pi r \tan\theta_{nt}$. The propeller pitch is defined in the literature in different ways. Instead of using the nose-tail line to define the pitch angle, a face-pitch line is also used (Carlton 1994). Further, the propeller pitch may be defined as the local pitch at r equal to either $0.7R$ or 0.75R, where R is the propeller radius.

The flow situation in Figure 2.36 is similar to that in Figure 2.16. In both cases, there is a flow with an angle of attack past a 2D foil. When the foil is symmetric about the nose-tail line, that is, uncambered, and the angle of attack is zero, there is zero lift according to potential flow theory for 2D steady flow past a foil in infinite fluid (see Chapter 6). According to the same theoretical assumptions, there is also zero drag. However, drag occurs, for instance, as a result of 3D flow and viscous effects. There are in general both a lift force and a drag force acting on the propeller blade section. Figure 2.16 shows that the lift force dominates. This is normally the case. Because a sharp trailing edge is essential in developing a large lift force, the way the propeller rotates matters. It implies that the propeller is less efficient when going astern.

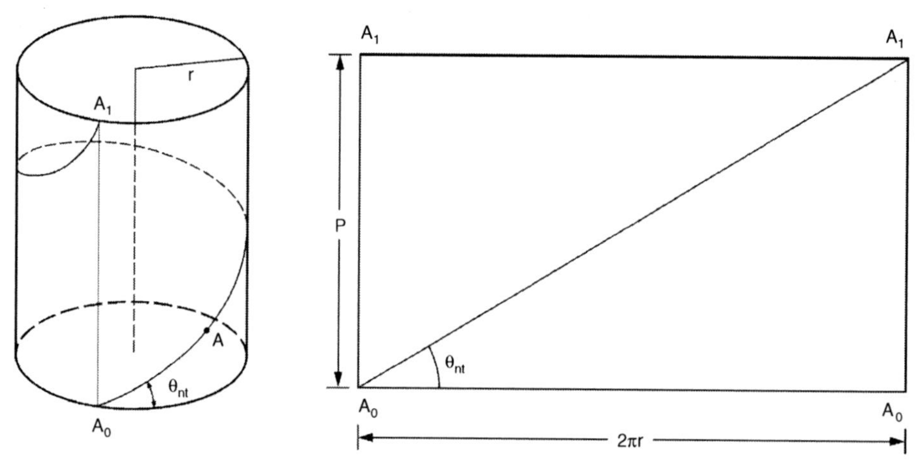

Figure 2.37. Left figure: helical curve described by point A on a cylinder surface with radius r. Point A rotates with constant velocity around the cylinder axis and moves axially with constant velocity. Right picture: the cylinder surface is unfolded into a plane. P = pitch, θ_{nt} = pitch angle.

Let us try to give a simplified description of the flow around a foil. We then disregard the details of the boundary layer. The lift is closely related to the circulation Γ around the foil, where $\Gamma = \oint_C \mathbf{u} \cdot d\mathbf{s}$, $\mathbf{u}$ is the fluid velocity, C is a closed curve enclosing the foil. The circulation implies higher velocities on the suction side than on the pressure side of the foil and is a consequence of the flow leaving tangentially from the trailing edge of the foil (Figure 2.38). Figure 2.38 also illustrates that the flow velocity is higher on the top side (suction side) of the foil than on the lower side. We can see this by using the conservation of fluid mass, noting that there is no flow through streamlines and that the distance between streamlines is shorter on the top side than on the lower side.

The fact that the flow leaves tangentially from the trailing edge is referred to as the Kutta condition in mathematical foil theory. The Kutta condition is a restriction imposed on the flow based on physical flow observations. If a Kutta condition was not imposed and there was no circulation around the foil, it would lead to zero force on a foil in steady potential flow in infinite fluid (D'Alembert's paradox). The corresponding unphysical theoretical flow would imply cross-flow with infinite velocity at the trailing edge of the foil (see Figure 2.39). This suggests that a vortex will be shed in reality from the trailing edge. It is the reason for the development of circulation around the foil. The flow adjusts itself so that it leaves tangentially from the trailing edge. The reader should refer to Chapter 6 for details about foil theory.

The resultant force vector $\mathbf{F}$ in Figure 2.36 is decomposed into an axial force component T that contributes to the propeller thrust and a peripheral component Q/r that contributes to the propeller torque.

We said that this explanation was simplified. So what are the simplifications used? Well we have basically used a strip theory approach, which implies that the flow is two-dimensional at the blade section. If the aspect ratio of the propeller blade is large, the strip theory represents a good approximation. However, by using a definition of aspect ratio like eq. (2.87) for a propeller blade,

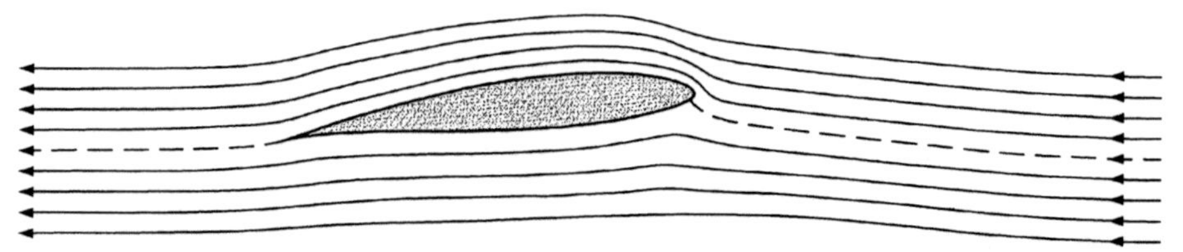

Figure 2.38. Assumed flow past a foil in which the flow leaves tangentially from the trailing edge (Kutta condition). The lines represent streamlines. (Newman, J. N., 1977, *Marine Hydrodynamics*, Cambridge: The MIT Press. The figure is reprinted with the permission of The MIT Press.)

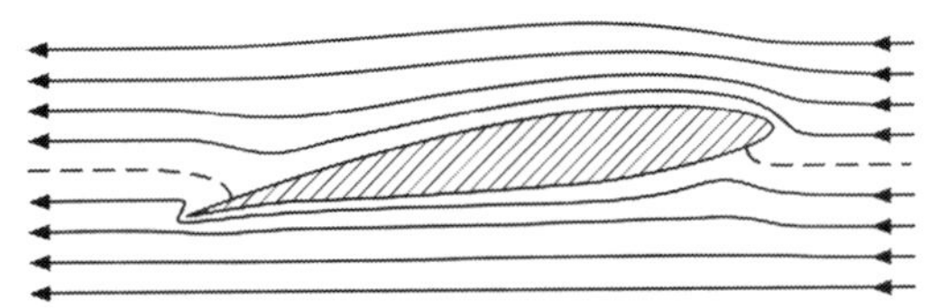

Figure 2.39. Flow past a foil without circulation. (Newman, J. N., 1977, *Marine Hydrodynamics*, Cambridge: The MIT Press. The figure is reprinted with the permission of The MIT Press.)

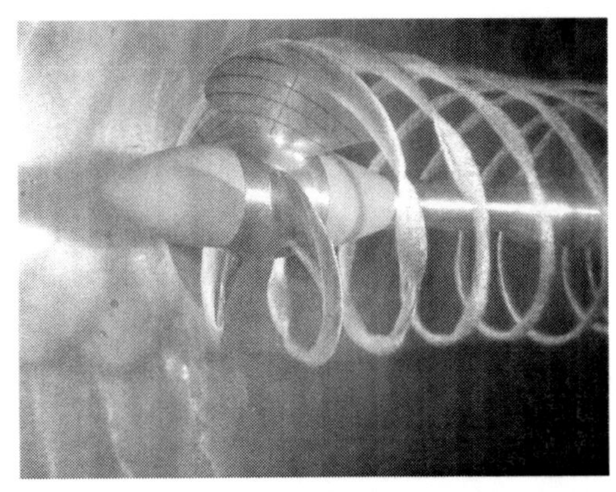

Figure 2.40. Tip vortex sheet cavitation of a four-bladed propeller used for a high-speed vessel. The tests have been done in the cavitation tunnel at the Marine Technology Centre, Trondheim. $P/D = 1.2$, $A_E/A_0 = 0.71$. No shaft inclination. (Photo by F. Bolstad.)

we find that some propellers can have an aspect ratio of an order of one. This means that the 3D flow effects cannot be ignored. One consequence of the 3D flow is the tip vortices from the propeller blades (Figure 2.40). As a first approximation, they form helical paths that can be determined by noting that vorticity is convected with the fluid velocity. We then use a propeller-fixed coordinate system. As first approximation, the longitudinal (axial) and peripheral velocities at the propeller tip are, respectively, U and $\pi n D$. We can construct a helical curve for a tip vortex as in Figure 2.37. The radius R is $0.5D$. The angle θ_{nt} is replaced with the inflow angle $\tan^{-1}(U/\pi n D)$.

The tip vortices will influence the inflow to the propeller blade section in Figure 2.36. However, there are also influences from other parts of the trailing vortex sheet and other sections of the blade, as well as other blades. If the aspect ratio of a propeller blade is high, the influence is only caused by the trailing vortex sheets of the different blades. The inflow velocity to a propeller blade can then be represented as in Figure 2.41, which shows axial (U_a) and tangential (U_t) corrections to the inflow velocity due to 3D flow effects. Similar 3D flow effects are discussed in connection with high-aspect ratio lifting surfaces in section 6.7.

Propeller slip stream

The propeller slip stream can as a first approximation be estimated by considering the propeller as an actuator disc in infinite fluid. This means that we let the number of propeller blades go to infinity and the cross-dimensions of each blade go to zero in such a way that there is flow through the propeller disc. We will limit our discussion to an axially symmetric flow (see e.g., Lewis 1996). The inflow velocity far upstream is assumed equal to

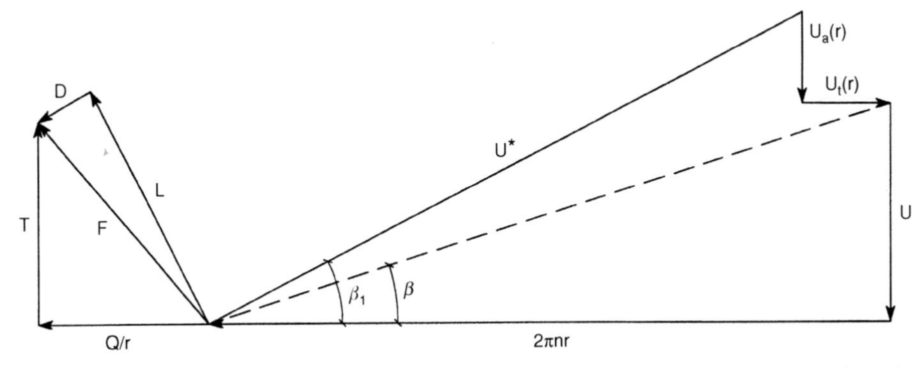

Figure 2.41. Incident flow velocity components $U_t(r)$ and $U_a(r)$ due to trailing vortex sheet from the propeller blades modify the incident flow velocity to a propeller blade section relative to the two-dimensional flow pictured in Figure 2.36. There is a lift force L and a drag force D perpendicular and parallel, respectively, to the incident flow velocity with magnitude U^*. The resultant force F contains an axial component (thrust T) and a peripheral component (torque/radius Q/r).

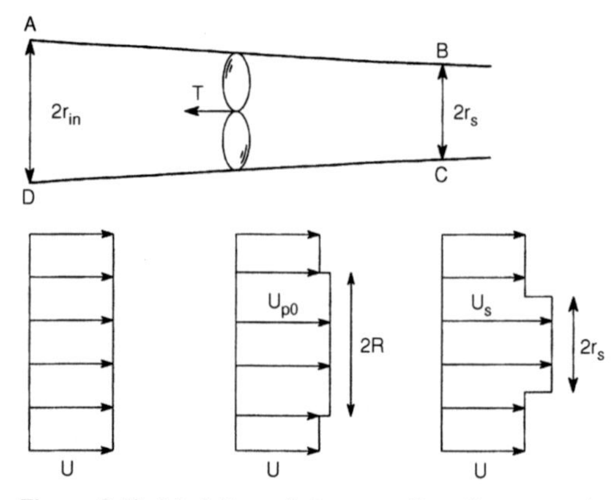

Figure 2.42. Modeling of the propeller slip stream in infinite fluid by means of an actuator disc model. U = the inflow flow velocity far upstream, U_{p0} = through-flow velocity at the actuator disc, U_S = the slip stream velocity far downstream. AB and DC are parts of streamlines.

the ship velocity U. The flow picture is illustrated in Figure 2.42 with a stream surface consisting of streamlines AB and DC that touch the propeller blade tips. The flow velocity is everywhere in the axial direction, that is, the swirling flow associated with the trailing vortex sheets behind the propeller is neglected. The velocity outside the stream surface shown in Figure 2.42 is equal to the ship velocity U. The flow velocity inside the stream surface varies only in the longitudinal direction and is equal to U_{p0} and U_S at the propeller plane and far downstream, respectively.

An incompressible and inviscid fluid is assumed. The propeller thrust force T may be expressed by conservation of fluid momentum and is equal to the difference in momentum flux through the cross-plane at BC and AD in Figure 2.42, that is,

$$T = \rho\pi r_S^2 U_S^2 - \rho\pi r_{in}^2 U^2. \qquad (2.134)$$

Here r_{in} and r_s are the radial distances from the propeller axis to the stream surface at AD and BC, respectively. Continuity of fluid mass gives

$$\rho\pi r_{in}^2 U = \rho\pi R^2 U_{p0} = \rho\pi r_S^2 U_S. \qquad (2.135)$$

Here R is the propeller radius. Using eq. (2.135) in eq. (2.134) gives

$$T = \rho\pi R^2 U_{p0}(U_S - U). \qquad (2.136)$$

The propeller thrust may also be expressed by direct pressure integration. This gives consistent with our simplified model that

$$T = (p_2 - p_1)\pi R^2, \qquad (2.137)$$

where $(p_2 - p_1)$ is the pressure rise across the actuator disc representing the propeller. The pressure can be related to the flow velocity by means of Bernoulli's equation (see eq. (2.94)). Neglecting gravity and applying Bernoulli's equation separately on the upstream and downstream side of the actuator disc gives

$$p_1 + \frac{\rho}{2}U_{p0}^2 = p_{amb} + \frac{\rho}{2}U^2 \qquad (2.138)$$

$$p_2 + \frac{\rho}{2}U_{p0}^2 = p_{amb} + \frac{\rho}{2}U_S^2. \qquad (2.139)$$

Here p_{amb} means the ambient pressure. This means that eq. (2.137) can be expressed as

$$T = \frac{\rho}{2}(U_S^2 - U^2)\pi R^2. \qquad (2.140)$$

Comparing eqs. (2.136) and (2.140) gives $U_{p0}(U_S - U) = 0.5(U_S^2 - U^2)$, that is,

$$U_{p0} = \frac{1}{2}(U_S + U). \qquad (2.141)$$

This expresses that the through-flow velocity U_{p0} at the actuator disc is the average of the inflow velocity U and the slip stream velocity U_S far downstream of the propeller.

Eq. (2.140) can be rewritten as

$$U_S = U\sqrt{1 + C_T}, \qquad (2.142)$$

where

$$C_T = \frac{T}{0.5\rho U^2 \pi R^2}. \qquad (2.143)$$

This determines the slip stream velocity U_S in terms of the propeller thrust-loading coefficient C_T. The radius r_S associated with the far downstream slip velocity can be determined by eqs. (2.135) and (2.141), giving

$$r_S = R\sqrt{0.5\left(1 + \frac{U}{U_S}\right)}. \qquad (2.144)$$

Eqs. (2.142) and (2.144) are commonly used in expressing the inflow velocity to a rudder behind a propeller (see Figure 10.30 and associated text). Even though U_S and r_S are asymptotic values far downstream of the actuator disc, they represent good approximations when the longitudinal distance between the rudder and the propeller is the order of the propeller radius (Söding 1982). If we want to evaluate the propeller slip stream at a much larger downstream distance, turbulent stresses must be considered. This can be analyzed by considering the slip stream as a turbulent axissymmetric jet flow. Details are given by Schlichting (1979). A more sophisticated approach must obviously account for the presence of viscous propeller forces, swirling flow, the free surface, and the vessel. However, we must still keep in mind that representing the propeller as an actuator disc is an approximation.

2.10.1 Open-water propeller characteristics

Even if three-dimensionality matters in the flow around the propeller, it does not change the fact that the inflow angle $\tan^{-1}(U/2\pi nr)$ is an important parameter. A global measure of the inflow angle is the advance ratio

$$J = \frac{U}{nD}, \tag{2.145}$$

where D is the propeller diameter. If Reynolds number and cavitation effects are disregarded, then the nondimensional thrust and torque, expressed by

$$K_T = \frac{T}{\rho n^2 D^4} \tag{2.146}$$

and

$$K_Q = \frac{Q}{\rho n^2 D^5}, \tag{2.147}$$

respectively, will depend only on the advance ratio J for geometrically similar propellers. However, we should recall that we have assumed infinite fluid, that is, no free surface effects. Figure 2.43 gives examples from Gawn (1953) on the thrust coefficient K_T and the torque coefficient K_Q as a function of J for three-bladed propellers with different pitch-to-diameter ratios. The blade area ratio in this case is 0.65, whereas Gawn (1953) also presented results for other blade area ratios from 0.2 to 1.1. The propeller efficiency η_p is also plotted. This is the ratio of the propulsive power UT done by the propeller in developing a thrust force, divided by the shaft power $2\pi nQ$ required to overcome the shaft torque. The latter is also the power delivered by the vessel machinery. This means

$$\eta_p = \frac{UT}{2\pi nQ} = \frac{J}{2\pi}\frac{K_T}{K_Q}. \tag{2.148}$$

It is obviously an advantage to have high efficiency, and Figure 2.43 shows that a maximum efficiency of up to 0.8 can be achieved with the studied propellers. Maximum values of K_T and K_Q occur for $J = 0$. However, η_p is then zero. This follows from eq. (2.148). We note that K_T and K_Q tend to zero for approximately the same J-value for a given pitch-to-diameter ratio. We will explain this by the simplified picture in Figure 2.36, that is, we assume first a 2D flow condition. We will neglect viscous forces. There is then only a lift force L perpendicular to the inflow velocity vector. This can be expressed as

$$L = \frac{\rho}{2}(U^2 + 4\pi^2 n^2 r^2)c\frac{dC_L}{d\alpha}(\alpha - \alpha_0), \tag{2.149}$$

where c is the chord length of the blade section. $dC_L/d\alpha$ is a constant and depends on the foil profile. It is 2π according to linear theory for a 2D foil in an infinite and incompressible fluid when the angle of attack α (see Figure 2.16) is given in radians. When $\alpha = 0$, there still will be a lift for a cambered section. This is expressed by α_0 in eq. (2.149). α_0 will be negative relative to the flow situation in Figure 2.16. Actually most of the lift on the propeller blade at design speed is the result of the foil camber.

The decomposition of eq. (2.149) in axial and peripheral directions gives contributions to the thrust and torque equal to $L\cos\beta$ and $rL\sin\beta$, respectively. Here $\cos\beta = 2\pi nr/(U^2 + 4\pi^2 n^2 r^2)^{1/2}$ and $\sin\beta = U/(U^2 + 4\pi^2 n^2 r^2)^{1/2}$. We then have to add contributions from all blade sections and blades to find propeller thrust and torque. If a representative blade section with r equal to 0.7 times the propeller radius R is chosen, this gives

$$\alpha = \theta_{nt} - \tan^{-1}\left(\frac{J}{0.7\pi}\right). \tag{2.150}$$

We must require that $\alpha - \alpha_0$ is positive. This gives the thrust and torque directions as in Figure 2.36. If $\alpha - \alpha_0$ were negative, then the hydrodynamic torque on the propeller would act in the same direction as the propeller shaft torque delivered by the engine. This means we cannot balance the torque and the inflow will drive the propeller. Another matter is that it would lead to negative

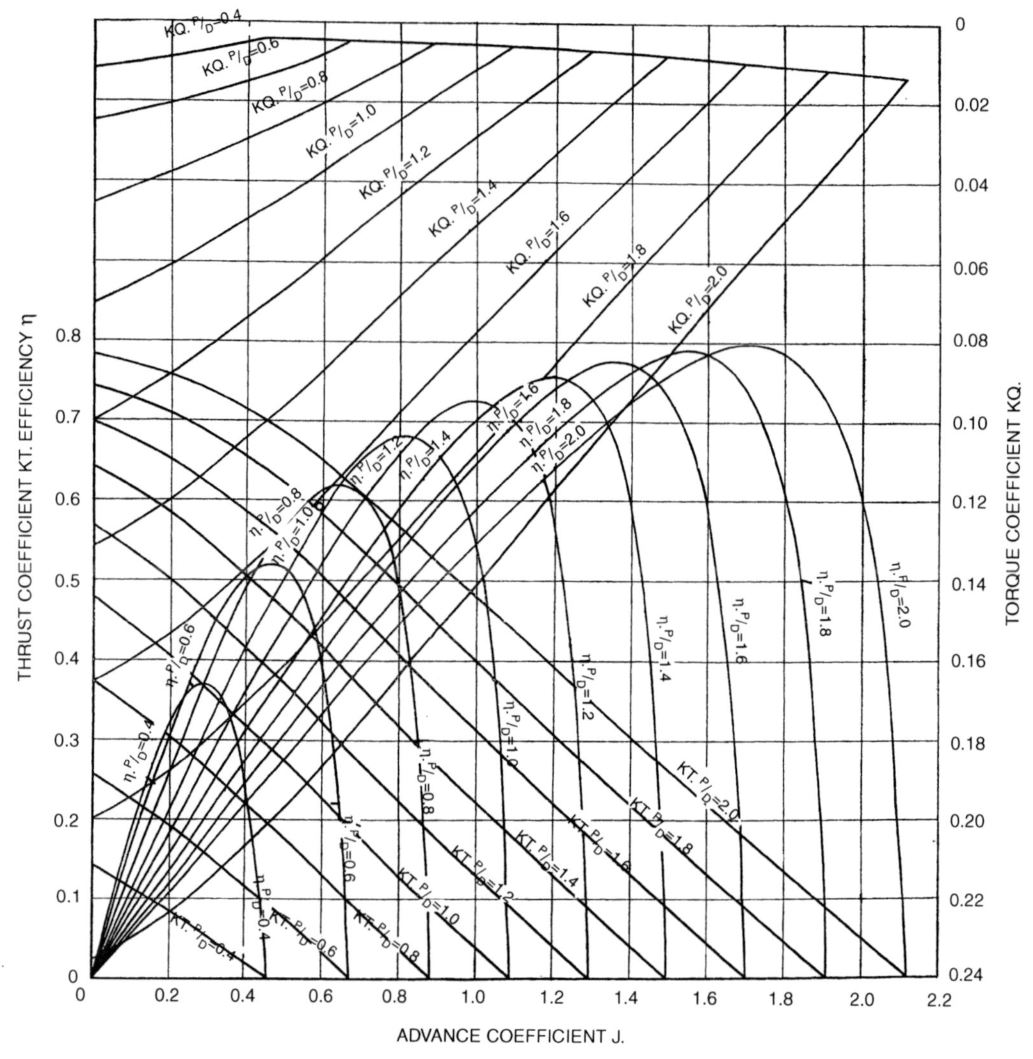

Figure 2.43. Thrust, torque, and efficiency coefficients for a series of three-bladed propellers with pitch-to-diameter ratios 0.4 to 2.0. Blade area ratio (BAR) = 0.65 (Gawn 1953).

thrust. Three-dimensional flow effects will reduce the angle of attack (see Figure 2.41), how much is a matter of detailed investigation. Let us now roughly say that $C_L = 2\pi\alpha$, that is, the effect of α_0 is offset by 3D flow effects. Because the thrust and torque are proportional to α in this simplified analysis, by means of eq. (2.150) we see that K_T and K_Q become zero for the same J-value for a given pitch-to-diameter ratio. Eq. (2.150) shows that the J-value that corresponds to zero K_T and K_Q values increases with increases in the pitch angle θ_{nt}, that is, pitch-to-diameter ratio. This agrees with Figure 2.43. We can estimate what this J-value is from eq. (2.150). We then use Figure 2.37 showing that $\tan\theta_{nt} = P/(2\pi r)$ and use the fact that the propeller pitch corresponds to $r = 0.7R$. This shows that zero α corresponds to a propeller pitch-to-diameter ratio equal to the advance ratio J. That is in reasonable agreement with Figure 2.43.

If the propeller is approximated as an actuator disc and an axially symmetric flow is assumed, the propeller thrust, torque, and efficiency can be expressed in a simple way. The propeller thrust can be written as either eq. (2.136) or eq. (2.140). The shaft power is equal to the difference in kinetic energy flux through the cross-planes at BC and AD in Figure 2.42 (see part B of exercise 2.12.4), that is,

$$2\pi n Q = \frac{1}{2}\rho\pi r_S^2 U_S^3 - \frac{1}{2}\rho\pi r_{in}^2 U^3. \quad (2.151)$$

Using eq. (2.138) gives

$$2\pi n Q = \frac{1}{2}\rho\pi R^2 U_{p0}\left(U_S^2 - U^2\right). \quad (2.152)$$

Eqs. (2.148), (2.136), and (2.152) imply that the propeller efficiency can be expressed as

$$\eta_P = \frac{2}{1 + U_S/U}. \quad (2.153)$$

This can be rewritten as

$$\eta_P = \frac{2}{1 + \sqrt{1 + C_T}} \quad (2.154)$$

by means of eq. (2.142). The propeller thrust-loading coefficient C_T in eq. (2.154) can be expressed as

$$C_T = \frac{8}{\pi}\frac{K_T}{J^2} \quad (2.155)$$

by using the definitions of C_T, J, and K_T given by eqs. (2.143), (2.145), and (2.146), respectively. Eq. (2.154) is called the *ideal efficiency*. The efficiency of various propulsion devices will for a given thrust-loading coefficient always be lower than the ideal efficiency. One reason is that our simplified actuator disc model for the propeller does not account for viscous drag forces on the propeller and the kinetic energy associated with the swirling flow behind the propeller. However, the ideal efficiency represents a measure in judging the qualities of different propulsion devices (see examples in Breslin and Andersen 1994).

2.10.2 Propellers for high-speed vessels

The previous discussion does not account for the possibility of propeller cavitation, which is of concern for high-speed vessels. A broad and in-depth coverage of cavitation is given by Brennen (1995). Kato (1996) reviewed the modeling, analysis, and computational methods of cavitation for hydrofoils and marine propellers. Generally speaking, cavitation occurs when the total pressure becomes equal to vapor pressure p_v somewhere on the propeller surface. p_v is only 0.012 times the atmospheric pressure at 10°C (see Table A.3 in the Appendix). This means practically zero pressure. The pressure can, by the way, never be negative in a fluid. The total pressure is the sum of atmospheric pressure p_a and hydrostatic pressure at a given instantaneous position on the propeller and the hydrodynamic pressure on the propeller. The latter is the cause of the lift expressed by eq. (2.149) and therefore of the propeller thrust and torque. If we continue using the flow picture in Figure 2.38, there will be a suction pressure on the curved "upper" side of the blade section. So except for some details around the nose of the blade section, it is the difference between the higher pressure on the nearly flat "lower" (face) side and the lower pressure on the "upper" (back) side that causes lift. It is the suction pressure that is of concern from a cavitation point of view. Because the lift increases with the square of the inflow velocity, the hydrodynamic pressure will do the same. It implies that by increasing the ship's speed, the load on the propeller must increase, thus the possibility of cavitation will increase. Further, the hydrostatic pressure also matters. This implies that the closer a propeller blade section comes to the free surface, the larger is the possibility of cavitation on the low-pressure areas of the section. The scenario may then be a varying cavitation volume on the outer part, that is, near the tip, of the propeller blade as the propeller blade passes near the free surface. When the cavitation disappears and bubbles implode, it may lead to propeller erosion. Face cavitation is of most concern in this context. Further, the varying cavity volume on a propeller blade as it passes near the free surface acts similarly to a fluid source generating pulsating pressures on the hull (Huse 1972). These pressure pulses may excite undesirable hull vibrations.

If partial cavitation is unavoidable by the propeller design, a supercavitating propeller design may be concidered. If the propeller is to be used on a high-speed vessel with a speed above the order of 50 knots, cavitation will occur. A supercavitating propeller means that the entire suction side of the foil is contained within a cavity. The blade section of a supercavitating propeller typically has a sharp leading edge. The flow around this sharp edge causes high velocities, which means low pressure. This facilitates cavitation inception. A supercavitating blade section has a much lower lift coefficient C_L and lift-to-drag ratio than does a noncavitating blade section. A pioneering analysis was made by Tulin (1953). Two-dimensional linear steady supercavitating flow around a flat plate

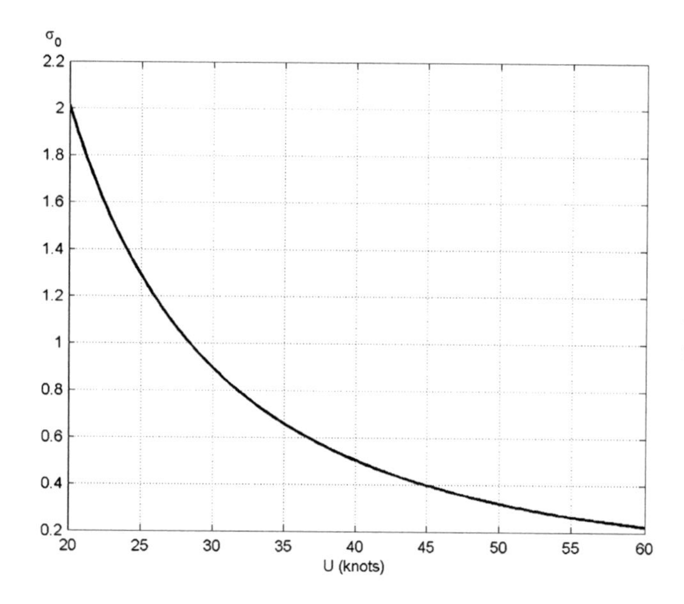

Figure 2.44. Propeller cavitation number σ_0 (see eq. (2.157)) as a function of vessel speed U. Ambient pressure is based on 1 m submergence of the propeller shaft.

in infinite fluid at small angles of attack α was studied as a function of the cavitation number. Geurst (1960) extended this theory to cambered foils. If we consider a flat plate at zero cavitation number, then C_L is $0.5\pi\alpha$ for supercavitating flow compared with $2\pi\alpha$ for a noncavitating flat plate. This has obvious consequences for propeller thrust and torque.

The cavitation number is an important parameter. For a blade section, this can be expressed as

$$\sigma = \frac{p_a + \rho g h - p_v}{0.5\rho[U^2 + (2\pi n r)^2]}, \tag{2.156}$$

where h is the minimum instantaneous submergence of the blade section during the propeller rotation. The higher the cavitation number, the smaller the probability of cavitation.

The propeller cavitation number σ_0 follows by inserting $r = 0$ in eq. (2.156), that is,

$$\sigma_0 = \frac{p_a + \rho g h - p_v}{0.5\rho U^2}. \tag{2.157}$$

Figure 2.44 illustrates how σ_0 decreases with increasing speed when the propeller shaft is 1 m submerged. Newton and Rader (1961) have presented extensive model test results showing how the thrust coefficient K_T, the torque coefficient K_Q, and the propeller efficiency η_P depend on σ_0 and the advance ratio J for uniform axial inflow. The Newton-Rader series consists of three bladed propellers and covers a wide range of pitch-diameter and blade area ratios ($1.0 \le P/D \le 2.0$, $0.5 \le A_E/A_O \le 1.0$). Here A_E means the expanded blade area and A_O is the propeller disc area. The parent propeller ($P/D = 1.25$, $A_E/A_O = 0.71$) is illustrated in Figure 2.45. Figure 2.46 presents K_T, K_Q, and η_P as a function of σ_0 and J for the parent propeller.

A radial distance r corresponding to 0.7 times propeller radius (R) is typically used in eq. (2.156)

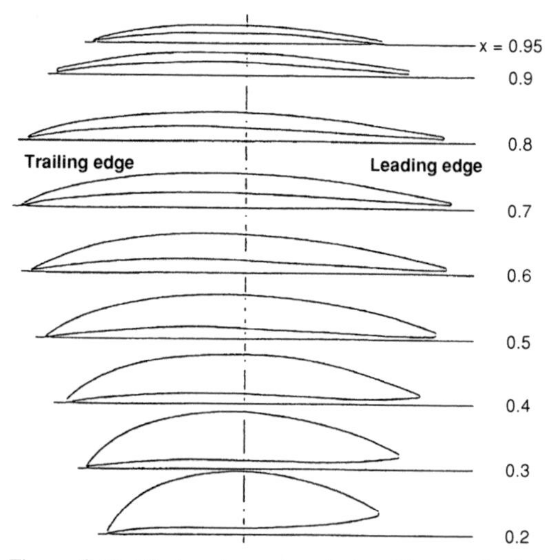

Figure 2.45. Blade elements of the Newton–Rader series parent propeller. x means radial distance from propeller shaft divided by propeller radius. Presented by Kruppa (1990) based on Newton and Rader (1961).

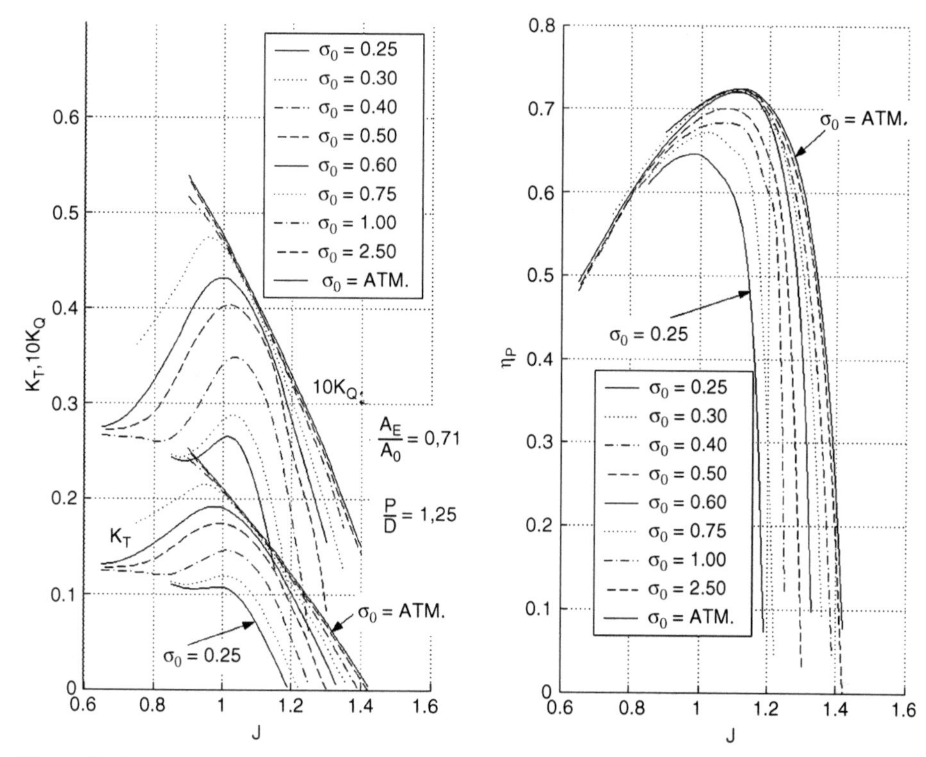

Figure 2.46. Hydrodynamic characteristics of the Newton-Rader series parent propeller. J = advance ratio. K_T = thrust coefficient, K_Q = torque coefficient, η_P = propeller efficiency, σ_0 = propeller cavitation number defined in eq. (2.157), A_E = expanded blade area, A_0 = propeller disc area, P = propeller pitch, D = propeller diameter.

to characterize the occurrence of cavitation for the propeller. This gives a cavitation number $\sigma_{0.7}$ that can be expressed as

$$\sigma_{0.7} = \frac{p_a + \rho g h - p_v}{0.5\rho U^2} \cdot \frac{1}{1 + \left(\frac{\pi \cdot 0.7}{J}\right)^2}. \quad (2.158)$$

It is possible to design a propeller in uniform axial inflow so that cavitation is avoided when $\sigma_{0.7} > 0.12$ (Kruppa 1990). This criterion is illustrated in Figure 2.47 by presenting J as a function of U when minimum submergence of the blade section at $r = 0.7R$ is 1 m. The results show that increasing speed requires increasing minimum advance ratio to avoid cavitation.

Uniform axial inflow can be achieved in calm water conditions when the combination of a tractor propeller and a right-angle drive (Z-drive) is used. A Z-drive is illustrated in Figure 2.1b. A tractor propeller is rotated 180° about a vertical axis relative to the propeller in Figure 2.1b and faces an incident flow very close to the forward speed of the

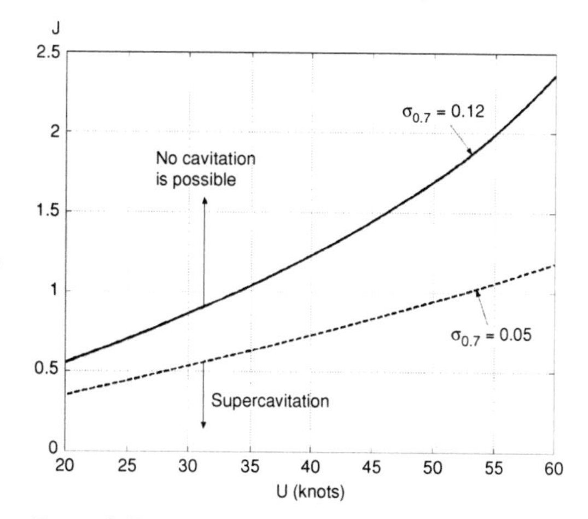

Figure 2.47. Cavitation domains as a function of vessel speed U and advance ratio J. $\sigma_{0.7}$ = local cavitation number at radial distance 0.7 times the propeller radius. Ambient pressure is based on 1 m submergence. Criteria are according to Kruppa (1990).

Figure 2.48. Model testing of a tractor propeller that is integrated in the aft foil-strut system of a foil catamaran (see Figure 6.4). Only the strut part is seen in the figure (Minsaas, unpublished).

vessel (see also Figures 2.48 and 2.49), similar to a propeller on an airplane. Propellers with negligible cavitation can then operate beyond 40 knots. A propeller such as this was designed by MARINTEK and reported by Halstensen and Leivdal (1990). This SpeedZ propulsion system uses controllable pitch propellers with four blades. Pitch-to-diameter ratios of $P/D = 1.80$ and advance coefficient $J = 1.50$ ensure that no thrust breakdown occurs at a vessel speed of 45 knots.

Propellers fitted to inclining shafts are sometimes used for planing vessels (see Figures 2.1a and 2.2) and hydrofoil vessels. The inflow to the propeller blade sections is then nonuniform. Local wake peaks occur because of struts and shafting. The criterion $\sigma_{0.7} > 0.12$ is no longer valid, and cavitation is of concern.

Fully cavitating conditions can be achieved when $\sigma_{0.7} < 0.05$. The cavity starts then at the leading edge and extends beyond the trailing edge. The pressure at the suction side of the propeller is then equal to the vapor pressure. Even if fully cavitating conditions can be achieved for smaller ship speeds (see Figure 2.47), the allowable J-values will give unacceptable efficiencies. Kruppa (1990) states, "It is felt that the proper field of application of fully cavitating propellers is beyond 40 knots." Supercavitating propellers are also discussed by Venning and Haberman (1962) and Todd (1967).

In model testing propellers, it is important that the cavitation number σ is the same as in full scale. If tests are done in a cavitation tunnel without a free surface, we do not have to worry about the Froude number. One tries to obtain as high an axial inflow velocity U as possible, say 15 ms^{-1}, to maximize the Reynolds number to obtain turbulent boundary-layer conditions. The correct cavitation number is obtained by lowering the ambient pressure. This can be lowered to about 0.1 bar (10 kNm^{-2}) in MARINTEK's facilities. If model tests are done with a ship model equipped with a propeller and with free-surface effects, we can use a depressurized towing tank, such as the one at MARIN in Wageningen, to properly model cavitation. There are, however, few facilities like that. Another possibility is to use a cavitation tunnel with a free surface.

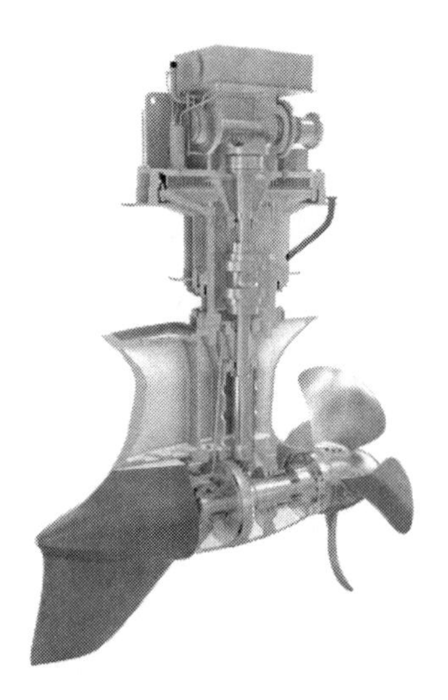

Figure 2.49. Azimuth thrusters by Rolls-Royce that can operate up to 24 knots. The figure on the left illustrates the Z-drive. The right picture shows the two thrusters working as tractor propellers.

If the propeller tip gets too close to the free surface, there is danger of ventilation. That means that air is drawn into the suction side of the propeller blade. The higher the propeller loading, that is, the lower the suction pressure on the propeller blade, the larger the possibility of ventilation. A rough guide is that ventilation should be considered when $h/R \leq 1.5$. Here R is the propeller radius and h is the vertical distance between the free surface and the propeller shaft. Relative vertical motions between the ship and the waves in a seaway increase the possibility of ventilation (Minsaas et al. 1986).

Free surface–piercing propellers are also used for high-speed vessels operating at maximum speeds higher than 40 knots. When the propeller blade exits and enters the free surface, unsteady lifting effects occur. Because the propeller extends from the transom, there is no appendage drag. Some systems may be used for steering and trim adjustments. Rose and Kruppa (1991) presented model test results from a methodical series of four-bladed surface-piercing propellers. Optimum propeller dimensions derived from these tests were published earlier by Kruppa (1990). The paper addresses aspects of testing techniques for surface-piercing propellers. It demonstrates that cavitation number affects propeller performance adversely unless fully ventilated propeller flow exists. Complete suction-side ventilation, however, is usually neither achieved nor desirable if one aims at maximum possible efficiency from a design point. Design charts for surface-piercing propellers should therefore contain the cavitation number as a parameter. The same is necessary for data on vertical and transverse propeller forces. Further requirements for design and testing of partially submerged propellers are formulated by Kruppa (1991, 1992). Rose et al. (1993) have presented secondary force coefficients for the Rolla Propeller series of four-bladed surface-piercing propellers studied by Rose and Kruppa (1991). The data can be used to estimate vertical and side forces and to understand the influence of propeller dimensions and position. The data also provide a basis for calculating shaft stresses.

Example: Determination of propeller characteristics

A planing vessel is considered. In sections 9.2 and 9.3, it is shown how to predict the resistance of a planing vessel. We assume that a resistance of 110 kN has been estimated. A right-angle drive with a tractor propeller is assumed. The inflow velocity to the propeller can then be approximated as the ship speed. Further, the effect of the propeller on the flow around the ship hull will be neglected. The propeller shaft is 1 m submerged. The maximum operating vessel speed U is set equal to 20.7 ms^{-1}. This gives a propeller cavitation number σ_0 equal to 0.5. A Newton-Rader propeller with $P/D = 1.25$ and $A_E/A_0 = 0.71$ is selected as a first try. We must assume a propeller diameter in order to find the advance ratio J. Let us say $D = 1.42$ m.

K_T can be rewritten as

$$K_T = \frac{T}{\rho n^2 D^4} = \frac{T}{\rho U^2 D^2} \cdot \frac{U^2}{n^2 D^2} = \frac{T}{\rho U^2 D^2} J^2. \tag{2.159}$$

In eq. (2.159), K_T is known as a function of J. This follows from the propeller diagram. Further, $T/(\rho U^2 D^2)$ is known from the values $T = 110$ kN, $\rho = 1025$ kgm^{-3}, $U = 20.7$ ms^{-1} and $D = 1.42$ m. The solution for J can, for instance, be found graphically by plotting K_T and $T/(\rho U^2 D^2) J^2$ as a function of J (Figure 2.50). The intersection point between the two curves corresponds to $J = U/(nD) = 1.1$. This gives $n = 13.2$ propeller revolutions per second. The K_Q-value at $J = 1.1$ is equal to 0.0376 according to the propeller diagram. The corresponding propeller

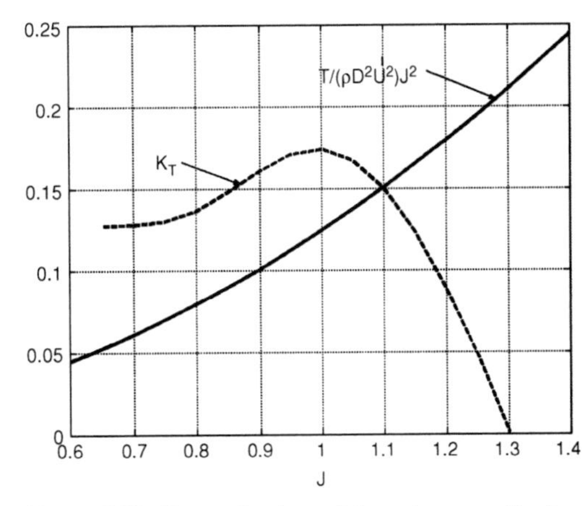

Figure 2.50. Determination of the advance ratio J as the intersection between K_T and $T/(\rho U^2 D^2) J^2$. $U = 20.7$ ms^{-1}, $D = 1.42$ m, $T = 110$ kN, $\rho = 1025$ kgm^{-3}. The data for K_T are from the Newton–Rader propeller series with $P/D = 1.25$, $A_E/A_0 = 0.71$, and $\sigma_0 = 0.5$.

efficiency η_P is 0.699. The delivered power by the propeller is

$$P_D = 2\pi n Q = 2\pi n \rho n^2 D^5 K_Q = 3235\,kW.$$

We note that this differs from the needed effect $TU = 2273\ kW$. The previous procedure must now be repeated for several propeller diameters and with consideration of other propeller diagrams in order to also find balance between P_D and TU.

The resistance of a high-speed vessel may not be the largest at maximum operating speed. A hump in the resistance may occur at a lower speed. This happens, for instance, in planing vessels and monohull hydrofoil vessels before takeoff to the foilborne condition (see Figure 6.15), and SES with a small length-beam ratio. One then has to ensure sufficient propulsion power and thrust for the hump speed.

Figure 2.47 shows that partial cavitation will occur for $J = 1.1$ at $U = 20.7\ \text{ms}^{-1}$ (40.2 knots). However, no face cavitation was present in the experiments by Newton and Rader (1961) for the selected propeller at $J = 1.1$ and $\sigma_0 = 0.5$. This suggests that cavitation will not be destructive for the propeller blades.

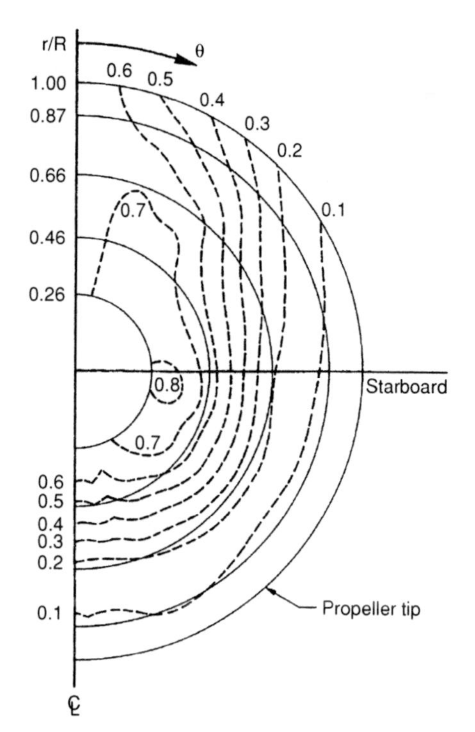

Figure 2.51. Curves of constant wake fraction w for a 125-m cargo ship model. The axial velocity that the propeller sees is $U(1 - w)$ (Breslin and Andersen 1994).

2.10.3 Hull-propeller interaction

If a propeller operates in the wake behind the ship, the inflow velocity to the propeller is affected. Figure 2.51 shows an example of the longitudinal wake fraction $w(r, \theta)$ at the position of a propeller located aft of the stern of a displacement vessel. Here (r, θ, x) are polar coordinates with x in the longitudinal direction of the ship. Zero radial distance r corresponds with the propeller shaft axis. The axial velocity, which the propeller sees, is $U(1 - w)$. If we consider a blade section as in Figure 2.36, the θ-dependence of w means that the blade section sees a time-dependent axial inflow velocity. This causes time-dependent hydrodynamic loads on the propeller that can, for instance, cause propeller shaft vibrations. However, this is not our concern in the present context of propulsion. This means we should average w over θ. In practice, one also averages w over r and uses an averaged wake fraction $\bar{w}$ that may vary between 0 and 0.4 for a displacement vessel, depending on the hull form. However, the averaged wake fraction may be negative for a semi-displacement vessel. This is Froude number dependent and caused by the waves generated by the ship.

We can determine w in model tests. Because there is also a Reynolds number effect, there are problems in how to scale w to full scale. We have already discussed this in the context of boundary-layer flow along the ship hull. For instance, the ratio between boundary-layer thickness and ship length will not be the same in model tests and full scale. The wake is just a continuation of the viscous boundary-layer flow.

We can account for the wake by redefining the advance ratio J in the open-water propeller characteristics as $J = U(1 - \bar{w})/(nD)$. However, because $\bar{w}$ is an average wake effect over the whole propeller area, we should not expect that this could be correct both for K_T and K_Q. So in addition, K_Q is corrected as follows

$$(K_Q)_{SP} = (K_Q)_{OW}/\eta_R. \tag{2.160}$$

Here the subscripts SP and OW refer to a self-propelled model and open-water propeller conditions, respectively. Further, η_R is called the *relative rotative efficiency*. Typical values of η_R are between 1.0 and 1.1. A *self-propelled test* uses a model fitted with and propelled by a scale-model propeller. Because the total resistance coefficient C_T is larger in the model test than in full scale (see Figure 2.26), additional thrust is provided by the towing carriage.

The propeller will affect the flow around the ship. This means the resistance on the hull is not the same with and without the propeller. The suction due to the propeller will reduce the pressure at the stern and therefore increase the resistance. This can be expressed as

$$R_T = (1 - t)\,T, \tag{2.161}$$

where R_T is the resistance without the propeller and t is called the *thrust-deduction coefficient*. Typical values of t are lower than 0.2.

The relative rotative efficiency, the thrust deduction coefficient, and the wake fraction can be determined by combining self-propelled tests, open-water propeller characteristics, and towing tests without the propeller (Newman 1977). If the wake fraction is obtained in this way, it is called *Taylor wake* w_T.

We can now define an overall propulsive efficiency

$$\eta = \frac{R_T U}{2\pi n Q_{SP}} = (1-t)\left(\frac{U}{2\pi n D}\right)\frac{T_{SP} D}{Q_{OW}}\eta_R. \tag{2.162}$$

We can express this in terms of $(K_T)_{OW}$ and $(K_Q)_{OW}$, which now must be considered a function of the modified advance ratio $U(1-\bar{w})/(nD)$. Then, we will have similarity between the open-water and self-propelled thrust coefficients, that is, $(K_T)_{OW} = (K_T)_{SP}$. The propeller efficiency η_P given by eq. (2.148) is then

$$\eta_P = \frac{U(1-\bar{w})}{2\pi n D}\frac{K_T}{(K_Q)_{OW}}. \tag{2.163}$$

This means eq. (2.162) becomes

$$\eta = \frac{(1-t)}{(1-\bar{w})}\eta_P\eta_R. \tag{2.164}$$

Here

$$\eta_H = \frac{1-t}{1-\bar{w}} \tag{2.165}$$

is called the *hull efficiency*. Because η_H is typically between 1 and 1.2, the propulsive efficiency of a propeller behind a hull will be larger than the open-water efficiency. We can determine the ship's speed when we know the hull interaction coefficients, the resistance, and the machinery and propeller characteristics. This follows by the fact that the delivered torque from the machinery should balance the hydrodynamic torque on the propeller and by equalizing the ship's resistance with the propeller thrust. The analysis will implicitly determine the number of shaft revolutions n. Eventually we can, for instance, determine the overall propulsive efficiency given by eq. (2.164).

2.11 Waterjet propulsion

Waterjet propulsion is the most common propulsion system for high-speed vessels of nonplaning type. A comprehensive and practically oriented article on waterjet propulsion is presented by Allison (1993). Cavitation is easier to avoid for a waterjet than for a propeller at high speed.

If the engine power and the resistance curve of the vessel are known, we can easily make a preliminary selection of a waterjet propulsion system with diagrams provided by the manufacturer. There are different series with different outlet nozzle diameters and blade pitch angles. Let us illustrate this with an example. Figure 2.52 shows, for one particular waterjet system, different lines of constant brake power in kW (BKW) as a function of vessel speed. The brake power is the power delivered by the propulsion machinery. The vertical axis is the net thrust accounting for mechanical losses. In Figure 2.52, we have also plotted the resistance of the vessel together with the waterjet propulsion data. The intersection between the resistance and BKW-curves determines the needed power for a given vessel speed. The vessel should operate less than 10% of the time in zone 2, according to the diagram. This is associated with pump cavitation. Figure 2.52 also shows a zone 3, where a vessel must be only less than 1% of the operating time. However, Figure 2.52 shows that the studied vessel will never be in zone 3 with this particular waterjet propulsion system. Different loading conditions, that is, resistance curves, of the vessel must be investigated.

The power and shaft speed must be consistent with the engine and the gear. Figure 2.53 shows the connection between BKW and shaft speed for the same waterjet propulsion system as the one studied in Figure 2.52. The manufacturer makes

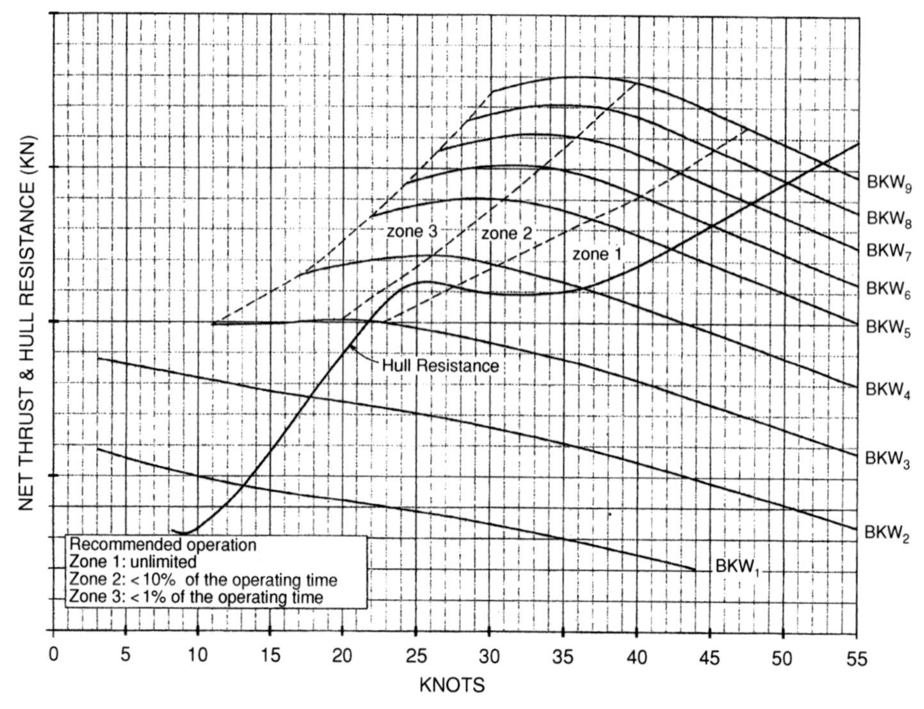

Figure 2.52. Combination of the resistance curve of a vessel with waterjet propulsion data. This determines the brake power in kilowatts (BKW) as a function of vessel speed.

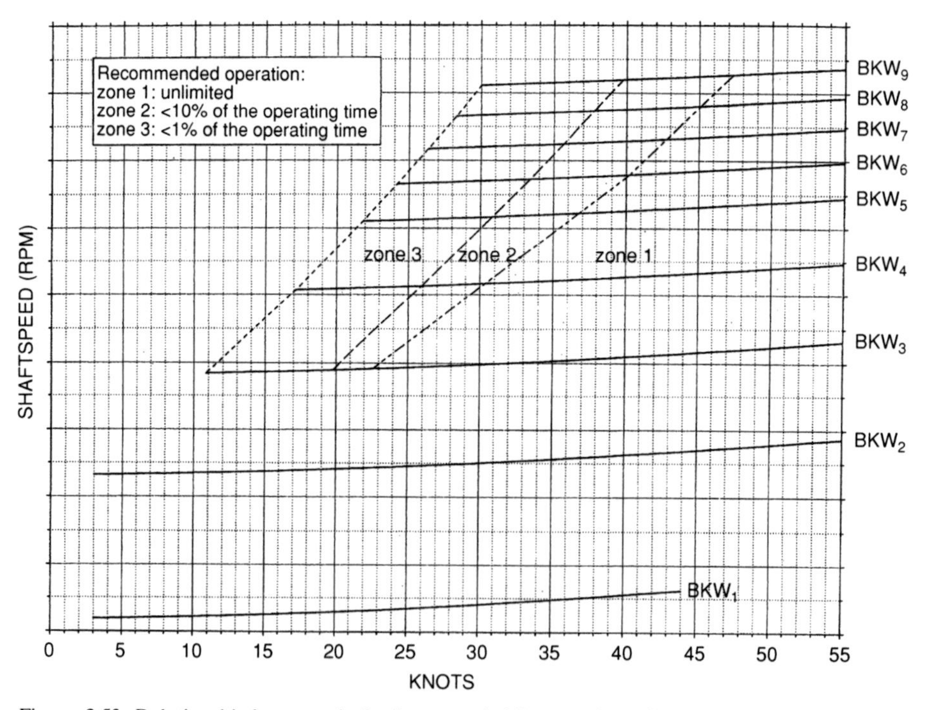

Figure 2.53. Relationship between the brake power in kilowatts (BKW) and the shaft speed in revolutions per minute (RPM) as a function of vessel speed for a waterjet propulsion system.

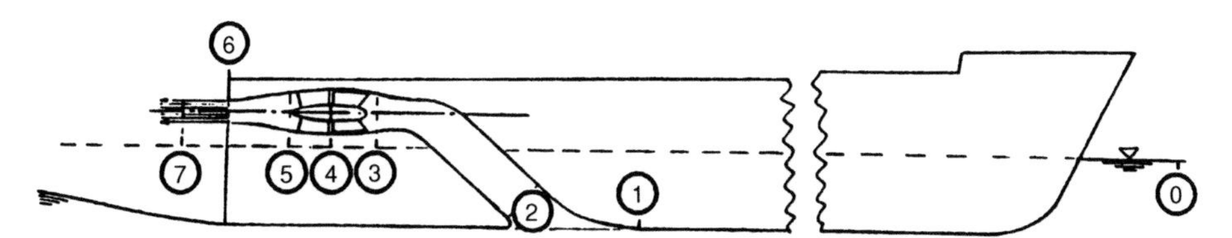

Figure 2.54. Waterjet propulsion system. The numbers in the figure correspond to:

Station no.	Location
0	In undisturbed flow far ahead of the vessel
1	Far enough in front of the intake ramp tangency point, before inlet losses occur
2	Normal to the internal flow at the aft lip of the intake
3	Just ahead of the pump
4	Between pump and stator
5	Behind stator
6	At the nozzle outlet plane
7	Behind the nozzle outlet plane where the pressure in the jet is equal to atmospheric pressure (vena contracta)

the final choice regarding the outlet nozzle diameter and the blade pitch angle in order to optimize performance and minimize fuel consumption.

Figure 2.54 presents an overview of all parts of a waterjet system with flush inlet. Details between stations 3 and 5 in Figure 2.54 are illustrated in Figure 2.55. The pump with the impeller (propeller) and the stator are on the right and left sides of the figure, respectively. Figure 2.56 demonstrates how the waterjet direction can be changed in order to stop, reverse, and steer the vessel.

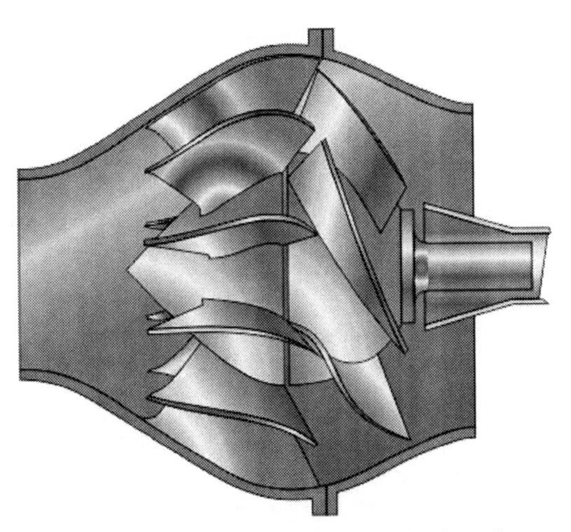

Figure 2.55. Details of the pump with the impeller on the right-hand side and the stator on the left-hand side of a Kamewa waterjet propulsion system by Rolls-Royce.

A procedure using Figures 2.52 and 2.53 does not, from a physical point of view, provide insight into how a waterjet works. The following presentation emphasizes this.

We start by describing how the thrust and efficiency of a waterjet propulsion system can be determined by model tests. The theoretical basis for doing this is the use of conservation of fluid mass, momentum, and energy. We will show how the thrust can be expressed by conservation of fluid momentum and how the effect delivered by the impeller can be expressed by conservation of kinetic energy for a viscous flow.

Finally, we discuss how the details of the inlet area and inflow velocity affect the possibility of cavitation.

2.11.1 Experimental determination of thrust and efficiency by model tests

The measurement of thrust and power for a waterjet system is more complex than for a ship equipped with a propeller. As outlined in the previous section, it is possible to separate the effect of a conventional propeller and the ship's hull. This is difficult for a waterjet system, as shown in Figure 2.54. By model testing, we cannot identify which part of the thrust is produced by the pump and what is the force due to the flow in the

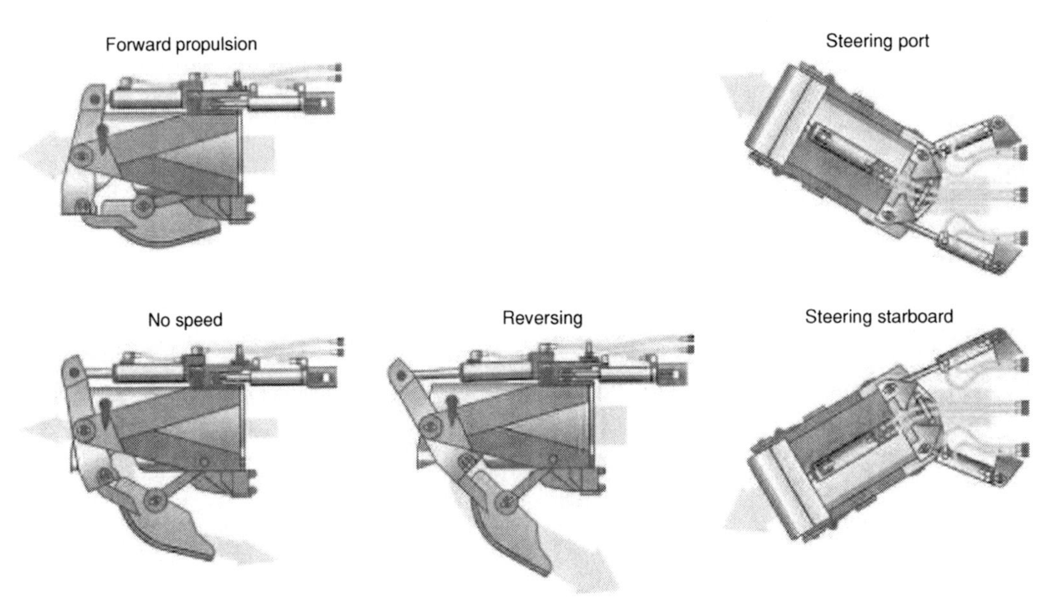

Figure 2.56. Illustration of how the waterjet direction of a Kamewa waterjet propulsion system by Rolls-Royce can be changed in order to stop, reverse, and steer the vessel.

internal ducting. Further, the flow into the waterjet affects the flow at the outer hull and causes trim and sinkage of the vessel. Specialist committees of the ITTC are still working on how to measure thrust and power from a waterjet system.

Let us consider a ship model equipped with a waterjet system at forward speed U. The first question is, What is the amount of water captured into the waterjet system?

We can answer this by considering the continuity of fluid mass and by measuring at station ⑦ in Figure 2.54 the amount of water going out of the waterjet system. The volume flux at station ⑦ can be expressed as

$$Q_7 = \int_{S_7} u_1 \, dA. \tag{2.166}$$

Here S_7 is the cross-sectional surface of the jet flow at station ⑦. Further, we define a Cartesian coordinate system (x, y, z) fixed relative to the ship (Figure 2.57). u_1 in eq. (2.166) is then the x-component of the fluid velocity at station ⑦, where the pressure is atmospheric. u_1 can be measured.

In the description of station ⑦ in Figure 2.54, the term *vena contracta* is used. This means that the jet flow has a smaller cross-sectional area at section ⑦ than at the nozzle outlet. The effect is small for a waterjet outlet. However, this is not the case for a jet flow escaping from a small opening in a container.

We then have to define a capture area of the fluid going into the waterjet system. We imagine a stream tube through the waterjet system as in Figure 2.57, in which stations ① and ⑦ are also shown. We select a new station (1a) that is not affected by the local flow at the inlet. In practice, the distance between stations ① and (1a) can be chosen as one inlet width, as recommended by the 23rd ITTC.

We now have to define the shape of the capture area at station (1a). The 21st ITTC proposed the use of a rectangular capture area with width b_1 that is 30% wider than the inlet width. The height of the capture area can then be determined by using the continuity of fluid mass. This requires that fluid velocities are measured at capture area

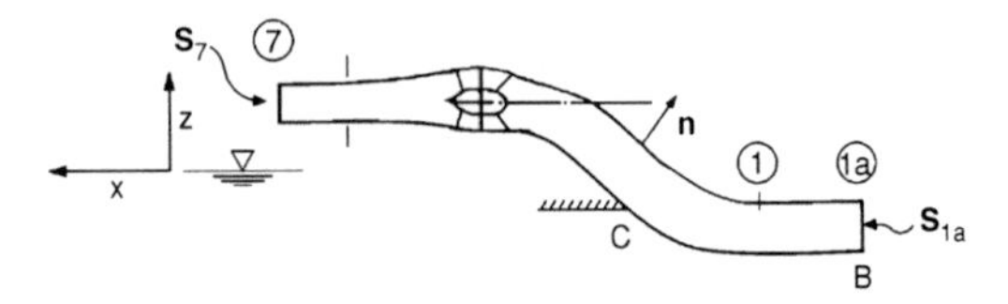

Figure 2.57. The control volume used in the calculation of the forward-acting force (thrust) on the ship due to the waterjet propulsion system and the effect that the impeller has on the flow. Conservation of fluid momentum and kinetic fluid energy are used. The bounding surface of the control volume is divided into a number of surfaces (see eq. (2.168)). The bounding surfaces at station (1a) and ⑦ are perpendicular to the x-axis.

S_{1a}. Let us, for simplicity, assume that this area is perpendicular to the x-axis and u_1 varies only with z. This means that the height h_{1a} of the capture area is determined by

$$b_1 \int_{-d-h_{1a}}^{-d} u_1(z)\, dz = Q_7. \quad (2.167)$$

Here d is the draft of the vessel at station ⓵a. There is no specific physical reason why the capture area should be rectangular. The important point is that the thrust and power given by the subsequent analysis are not sensitive to the detailed shape of the capture area.

Because the boundary layer at the inlet has a thickness of the order of the inlet duct diameter, u_1 in eq. (2.167) is affected by viscosity. We can estimate the boundary-layer thickness δ by eq. (2.68). This is valid for turbulent flow along a smooth flat plate. The use of eq. (2.68) illustrates a scaling problem. Let us call L the length from the ship's bow to the waterjet inlet and use subscripts M and S to indicate model and full scale. We then get

$$\delta_s = \frac{0.16 L_s}{\left(\frac{U_s L_s}{\nu}\right)^{1/7}}.$$

By now, using Froude scaling for the velocity, that is,

$$U_M = U_S \left(\frac{L_M}{L_S}\right)^{\frac{1}{2}},$$

and assuming the same value of the kinematic viscosity coefficient ν in model and full scales, we get

$$\frac{\delta_M}{\delta_S} = \left(\frac{L_M}{L_S}\right)^{11/14}.$$

According to Froude scaling, δ_M/δ_S should be L_M/L_S. This means we have scaling problems. Practical procedures for how to deal with this are presented by the ITTC. In addition, roughness will influence the full-scale boundary layer. We will not pursue this here, but rather continue with the estimation of the thrust by model tests. We assume that the vessel has only one waterjet. However, the procedure can easily be generalized to several waterjets on one vessel.

Thrust by conservation of fluid momentum

The control volume illustrated in Figure 2.57 is now considered. The bounding surface S is divided into several parts, that is,

$$S = S_{1a} + S_7 + S_H + S_{IM} + S_{ST} + S_{SH} + S_{WAT} + S_{AIR}. \quad (2.168)$$

Here S_{1a} and S_7 mean the surface at stations ⓵a and ⑦, respectively, having areas A_{1a} and A_7. S_{IM}, S_{ST}, and S_{SH} are, respectively, the surfaces of the impeller, the stator, and the shaft. S_H means the part of the remaining enclosing surface S that is connected to the vessel. It consists of the surfaces of the inlet, ducts, and outlet. S_{WAT} and S_{AIR} are the remaining part of the enclosing surface S. S_{WAT} is in the water, that is, like BC in Figure 2.57, and S_{AIR} is bounded by air, that is, aft of station ⑥ in Figure 2.54.

Conservation of fluid momentum expresses:

Sum of forces acting on the control volume

=

Rate of change of fluid momentum in the control volume

We will consider the longitudinal forces in the x-direction. The rate of change of fluid momentum in the x-direction can be expressed by the difference in the momentum flux at stations ⑦ and ⓵a, that is,

$$\Delta M = \rho \iint_{S_7} u_1^2\, dS - \rho \iint_{S_{1a}} u_1^2\, dS. \quad (2.169)$$

The longitudinal forces acting on S_{AIR} and S_7 are the result of the atmospheric pressure p_a. Because a constant pressure acting on the complete bounding surface S will not cause a net force on the control volume, we can just as well neglect the force due to p_a on S_{AIR} and S_7 and operate with forces due to the difference in total pressure and atmospheric pressure on the other parts of S.

The sum of the longitudinal forces acting on the control volume at S_{IM}, S_{ST}, S_{SH}, and S_H causes an opposite force on the vessel. This is the law of actio and reactio. Because positive x-direction is opposite the forward motion of the ship and positive thrust is in the direction of the forward motion of the ship, the thrust T is the sum of longitudinal forces acting on S_{IM}, S_{ST}, S_{SH}, and S_H. Because the flow inside the waterjet system is turbulent, these forces are caused not only by pressures.

There are also longitudinal forces acting on S_{1a} and S_{WAT}. However, Masilge (1991) states that this contribution to the thrust is small.

Neglecting the contribution from longitudinal forces on S_{1a} and S_{WAT} leads to the following commonly used expression for the thrust:

$$T = \rho Q_f (U_j - U_\theta). \tag{2.170}$$

Here Q_f is the same as the volume flux Q_7 defined by eq. (2.166), that is, the flow rate ingested by the waterjet. The subscript j denotes jet and the subscript θ is used because of the resemblance to how the momentum thickness of a boundary layer is defined (see eq. (2.54)). Further,

$$\rho Q_f U_\theta = \rho \iint_{S_{1a}} u_1^2 \, dS \tag{2.171}$$

$$\rho Q_f U_j = \rho \iint_{S_7} u_1^2 \, dS. \tag{2.172}$$

Eq. (2.170) can also be expressed as

$$T = \rho Q_f (U_j - U(1 - \bar{w}_f)), \tag{2.173}$$

where

$$\bar{w}_f = 1 - \frac{U_\theta}{U}. \tag{2.174}$$

Here $\bar{w}_f$ is a mean wake fraction that accounts for both approach flow losses in the boundary layer and potential flow effects outside the boundary layer. The longitudinal flow velocity U_1 outside the boundary layer at S_{1a} differs from the forward speed U. For instance, it will be influenced by the waves generated by the vessel.

Impeller effect by conservation of kinetic fluid energy

The effect that the impeller has on the flow is needed in expressing the efficiency of the waterjet system. This can be done by using conservation of kinetic fluid energy in the control volume illustrated in Figure 2.57. A derivation is part of exercise 2.12.4.

The effect P_D given by the impeller to the flow can be expressed as

$$P_D = \iint_{S_7} u_1 \left[\frac{\rho}{2}\mathbf{u}\cdot\mathbf{u} + p + \rho g z\right] dS - \iint_{S_{1a}} u_1 \left[\frac{\rho}{2}\mathbf{u}\cdot\mathbf{u} + p + \rho g z\right] dS + \textit{losses due to turbulent and viscous stresses} \tag{2.175}$$

Here $\mathbf{u} = (u_1, u_2, u_3)$ is the fluid velocity vector. It means that $0.5\rho \int_{S_7}\int u_1\mathbf{u}\cdot\mathbf{u}\, dS$ and $-0.5 \int_{S_{1a}}\int u_1\mathbf{u}\cdot\mathbf{u}\, dS$ in eq. (2.175) represent kinetic energy flux out of and into the control volume, respectively. Further, the z-coordinate in eq. (2.175) is defined in Figure 2.57, with $z = 0$ corresponding to the mean free surface. Because there are mechanical losses, the power delivered by the propulsion machinery (brake power) is not the same as the power P_D delivered by the impeller.

If we are outside the boundary layer at S_{1a}, the Bernoulli equation given by eq. (2.94) can be used to express the pressure p. Our problem is steady and the constant in eq. (2.94) can be expressed by considering a point on the free surface far upstream of the ship. Here the fluid velocity is equal to the ship's speed in our reference frame. This gives

$$p = -\rho g z + \frac{\rho}{2}(U^2 - u_1^2 - u_2^2 - u_3^2) + p_a \tag{2.176}$$

everywhere in the water where potential flow theory applies. Here p_a is the atmospheric pressure. We also use the approximation that the pressure does not vary across a thin boundary layer when gravity is neglected. By assuming that the hydrostatic pressure part $-\rho g z$ is valid through the boundary layer, the pressure on S_{1a} can be expressed as:

$$p = -\rho g z + \frac{\rho}{2}\left[U^2 - U_1^2 - U_2^2 - U_3^2\right] + p_a. \tag{2.177}$$

Here (U_1, U_2, U_3) is the fluid velocity just outside the boundary layer.

Assuming U_2 and U_3 are small relative to U_1 at S_{1a} gives

$$p = -\frac{\rho}{2}U_1^2 - \rho g z + p_a + \frac{\rho}{2}U^2. \tag{2.178}$$

The pressure at S_7 is p_a. Because

$$\iint_{S_7} u_1 \, dS = \int_{S_{1a}} u_1 \, dS, \tag{2.179}$$

the atmospheric pressure will not contribute to the sum of the integrals over S_{1a} and S_7 in eq. (2.175). Let us also approximate $\mathbf{u}$ as $u_1\mathbf{i}$ at S_{1a} and S_7 and assume u_1 is a constant jet velocity U_j at S_7. Approximating $\mathbf{u}$ as $u_1\mathbf{i}$ at S_7 means that we neglect the rotatory flow caused by the impeller, that is, as we see in terms of tip vortices in Figure 2.40. We could argue that the stator counteracts the rotatory flow caused by the impeller.

Further, $\int_{S_7}\int u_1 z\, dS$ is approximated as $Q_f h_j$, where h_j is the height of the center of the jet at S_7 above the mean free surface. This gives

$$P_D = \rho Q_f\left(0.5U_j^2 + g h_j\right) - \frac{\rho}{2}\iint_{S_{1a}} u_1\left[u_1^2 - U_1^2 + U^2\right] dS. \qquad (2.180)$$

$+$ *losses due to turbulent and viscous stresses*

Eq. (2.180) expresses that the effect delivered by the impeller to the flow is used to

- Accelerate the water from station (1a) to station (7)
- Lift the water to a vertical distance h_j above the mean free surface
- Overcome the losses due to Reynolds and viscous stresses in the waterjet system

It is common to study separately the effect of the pump or what is happening between stations (3) and (5) in Figure 2.54. The pump efficiency η_P, when the pump is not installed in the waterjet system and there is a homogenous inflow to the pump (see position 3 in Figure 2.54), is introduced. We can then write

$$P_D = \frac{\rho g Q_f H}{\eta_P \eta_R}, \qquad (2.181)$$

where η_R is a relative rotative efficiency that accounts for the irregular inflow to the pump. Eqs. (2.180) and (2.181) give an expression for the head H:

$$H = 0.5\frac{U_j^2}{g}\left(1+\zeta_{ex}\right) + h_j - 0.5\frac{U^2}{g}\left(1 - \tilde{\bar{w}}_f^2 - \zeta_{in}\right), \qquad (2.182)$$

where

$\zeta_{ex} =$ nozzle loss coefficient in terms of $0.5\rho U_j^2$. This accounts for losses after station (5) in Figure 2.54.

$\zeta_{in} =$ inlet and internal flow loss coefficient in terms of $0.5\rho U^2$. This accounts for losses between stations (1) and (3) in Figure 2.54.

These loss coefficients can be experimentally determined. Further, $\tilde{\bar{w}}_f^2$ in eq. (2.182) is defined as

$$\tilde{\bar{w}}_f^2 = \frac{1}{U^2 Q_f}\iint_{S_{1a}} u_1\left(U_1^2 - u_1^2\right) dS. \qquad (2.183)$$

Efficiency

The thrust power efficiency is

$$\eta_T = \frac{TU}{P_D} = \eta_P \eta_R \frac{\left[U_j - U\left(1-\bar{w}_f\right)\right]U}{gH}. \qquad (2.184)$$

Further, an overall propulsive efficiency is defined as

$$\eta = \frac{R_{BH}U}{P_D}. \qquad (2.185)$$

Here R_{BH} is the bare hull resistance of the vessel. It is related to a condition when the waterjet inlet is closed and the vessel has the same trim and sinkage that it would have had with a working waterjet system. We can write

$$T = \frac{R_{BH}}{1-t}, \qquad (2.186)$$

where t is a thrust deduction coefficient. Introducing eq. (2.186) gives

$$\eta = (1-t)\eta_T. \qquad (2.187)$$

A jet efficiency is introduced as

$$\eta_J = \frac{\left[U_j - U\left(1-\bar{w}_f\right)\right]U}{gH}. \qquad (2.188)$$

Eq. (2.184) can then be written as

$$\eta_T = \eta_P \eta_R \eta_J. \qquad (2.189)$$

Using eq. (2.182) gives

$$\eta_J = \frac{2\left[\left(\frac{U_j}{U}\right) - \left(1-\bar{w}_f\right)\right]}{\left(\frac{U_j}{U}\right)^2\left(1+\zeta_{ex}\right) + \frac{2gh_j}{U^2} - \left(1 - \tilde{\bar{w}}_f^2 - \zeta_{in}\right)}. \qquad (2.190)$$

Realistic values for the parameters in eq. (2.190) are $\bar{w}_f = 0.02$, $\tilde{\bar{w}}_f^2 = 0.04$, $\zeta_{ex} = 0.04$. Further, ζ_{in} may vary from 0.1 to 0.25. Higher values may, for instance, occur as the result of marine growth. We have exemplified η_J as a function of U_j/U in Figure 2.58 by using these values together with $U = 25\ \mathrm{ms}^{-1}$ and $h_j = 0.5$ m. Maximum jet efficiency occurs at about $U_j = 1.3U$ for $\zeta_{in} = 0.1$ and at about $U_j = 1.5U$ for $\zeta_{in} = 0.25$. Figure 2.59 illustrates how maximum jet efficiency $\eta_{J\max}$ decreases and the corresponding value of jet velocity–ship velocity ratio $(U_j/U)_{\max}$ increases with increasing $\zeta = \zeta_{in} + 2gh_j/U^2 + \tilde{\bar{w}}_f^2$ when $\zeta_{ex} = 0.04$ and $\bar{w}_f = 0$.

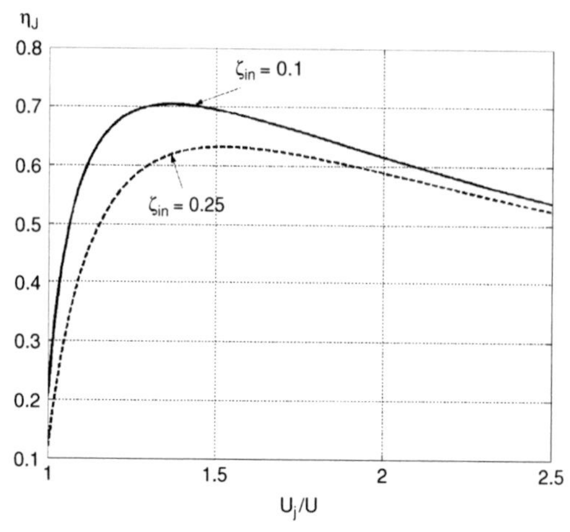

Figure 2.58. Jet efficiency η_J as a function of the ratio between jet velocity U_j and ship velocity U for a waterjet system with $\bar{w}_f = 0.02$, $\tilde{\bar{w}}_f^2 = 0.04$, $\zeta_{ex} = 0.04$, $U = 25\ \mathrm{ms}^{-1}$, $h_j = 0.5$ m.

Additional parameters for the pump have to be introduced in order to find the pump efficiency. Figure 2.60 presents an example of a pump diagram showing pump efficiency η_P and head coefficient $C_H = gH/(nD)^2$ as a function of the capacity coefficient $C_Q = Q_f/(nD^3)$. Here D means the impeller diameter and n is the number of shaft revolutions per second. Typical values of n are between 5 and 10 revolutions per second. As long as the cavitation and Reynolds number effects do not matter, C_H and η_P as a function of C_Q are sufficient to characterize a given pump.

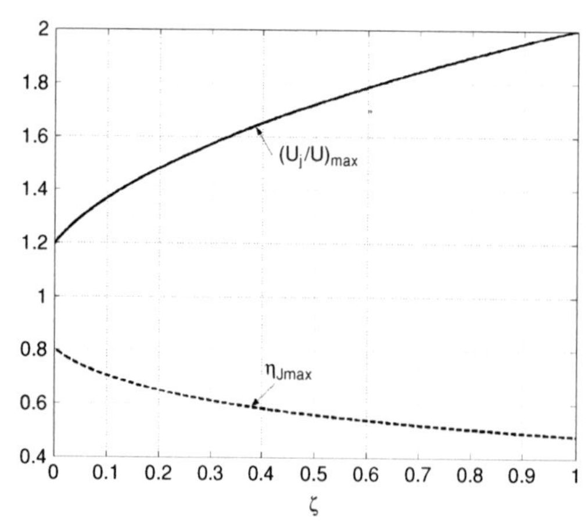

Figure 2.59. Maximum jet efficiency $\eta_{J\,\max}$ and corresponding value of jet velocity–ship velocity ratio $(U_j/U)_{\max}$ as a function of $\zeta = \zeta_{in} + 2gh_j/U^2 + \tilde{\bar{w}}_f^2$, $\zeta_{ex} = 0.04$, $\bar{w}_f = 0$.

The analogue parameters for a propeller will be the propeller efficiency η_P and either the thrust coefficient K_T or the torque coefficient K_Q as a function of the advance ratio J.

Various pump types exist, such as centrifugal pumps, mixed-flow pumps, axial-flow pumps, and inducers. Mixed-flow pumps are typically used for waterjet systems. The nondimensional specific speed

$$n_S = \frac{nQ_f^{1/2}}{(gH)^{3/4}} \tag{2.191}$$

characterizes what pump type should be used (Carlton 1994).

An important consideration for a pump is cavitation. The nondimensional specific suction velocity

$$S = n\sqrt{Q_f}/\left(gH_{SV}\right)^{0.75} \tag{2.192}$$

plays a role for a pump similar to the one the cavitation number plays for a propeller. Here H_{SV} is called the net positive suction head. It can be expressed as

$$H_{SV} = \frac{p_{SV}}{\rho g}, \tag{2.193}$$

where p_{SV} is the difference between the pressure (including atmospheric pressure) at station ③ in Figure 2.54 and the vapor pressure. Cavitation may occur when S is larger than 0.6 to 0.9.

Let us illustrate how both a high pump efficiency and jet efficiency can be obtained. We do not consider the possibility of cavitation of the pump in this context. The procedure is divided into different steps. Only one waterjet system is assumed.

Step 1

We consider maximum operating speed of the vessel and assume the bare hull resistance R_{BH} of the vessel is known. The needed thrust can then be estimated by eq. (2.186). A typical thrust deduction coefficient is 0.02.

Step 2

We determine the ratio between the jet velocity U_j and the vessel speed U by considering the jet efficiency η_J as a function of U_j/U (see Figure 2.58 as an example). This is done by using eq. (2.190). It requires that loss coefficients ζ_{ex} and ζ_{in} as well as wake factors $\bar{w}_f$ and $\tilde{\bar{w}}_f^2$ are known from experiments. There exists a value

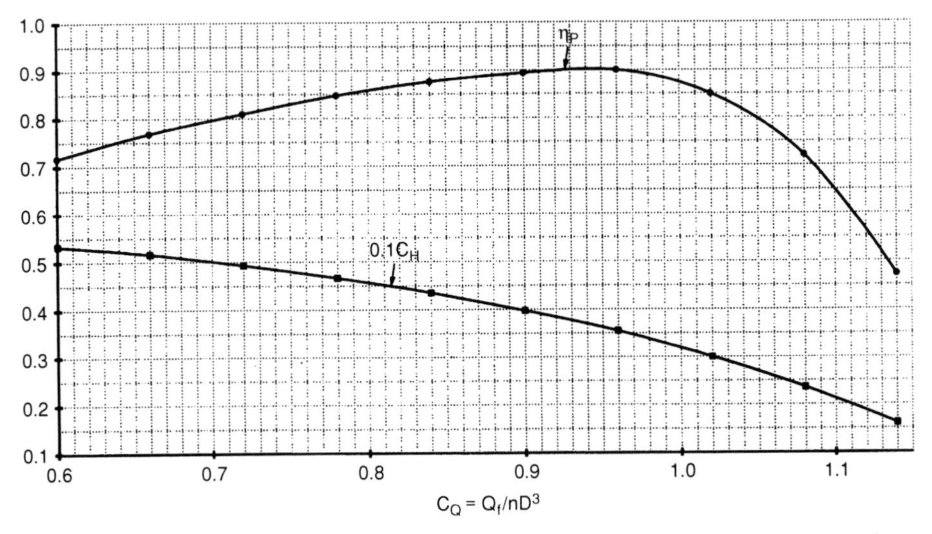

Figure 2.60. Example of pump diagram showing pump efficiency η_P and head coefficient $C_H = gH/(nD)^2$ as a function of the capacity coefficient $C_Q = Q_f/(nD^3)$. n = shaft revolutions per second, D = impeller diameter (Minsaas, unpublished).

$(U_j/U) = (U_j/U)_{\max}$ where maximum η_J occurs (see Figure 2.58). Because η_J decays rapidly for U_j/U less than $(U_j/U)_{\max}$, one may choose $U_j/U = 1.1\,(U_j/U)_{\max}$ to ensure large values of η_J and small sensitivity between η_J and (U_j/U).

Step 3

Because all variables except Q_f are now known in eq. (2.173), we can use this equation to determine Q_f. Because $Q_f = 0.25\pi D_j^2 U_j$, where D_j is close to the nozzle exit diameter, D_j is also determined. D_j must of course represent a realistic value. A typical value of D_j is of the order of 0.5 m. A large value of D_j implies a large added weight of the vessel due to the water in the waterjet system. This is not desirable. One can change Q_f and D_j by changing U_j. However, it should be kept in mind that we want a large η_J as described in Step 2.

Step 4

We must now consider a pump diagram as illustrated in Figure 2.60. We then vary Q_f/nD^3 by considering realistic combinations of n and impeller diameter D. The value of $C_H = gH/(nD)^2$ must be consistent with $C_Q = Q_f/nD^3$. H can be calculated by eq. (2.182). Different pump diagrams in which, for instance, the pitch of the impeller is varied, may have to be considered. Obviously we want Q_f/nD^3 to correspond to a large pump efficiency η_P. Figure 2.60 illustrates that η_P can be as high as 0.9. However, values up to 0.93 have been achieved.

Step 5

We can now evaluate the thrust power efficiency η_T by eq. (2.189). A typical value of the relative rotative efficiency is 0.98. The overall propulsive efficiency is obtained by eq. (2.187).

The procedure outlined above is meant to illustrate how a waterjet system can be selected. It is not necessarily the way chosen by a waterjet manufacturer. Further, pump diagrams are not always presented as in Figure 2.60.

The previously described procedure to determine thrust and efficiency relies strongly on model tests. An important issue is how to scale the model test results. There are Reynolds-number effects associated with the inflow to the waterjet inlet as we have already mentioned. Further, part of the thrust is the result of viscous effects, but our procedure of using conservation of fluid momentum does not tell us to what extent forces due to viscous and Reynolds stresses in the waterjet system affect the thrust. Also, the losses in the waterjet system are Reynolds-number dependent. Scaling procedures are an important issue to solve by the ITTC. van Terwisga (1991, 1992) has described in detail the model test procedures used at MARIN.

An alternative or supplement to model tests is, of course, the use of CFD. Taylor et al. (1998)

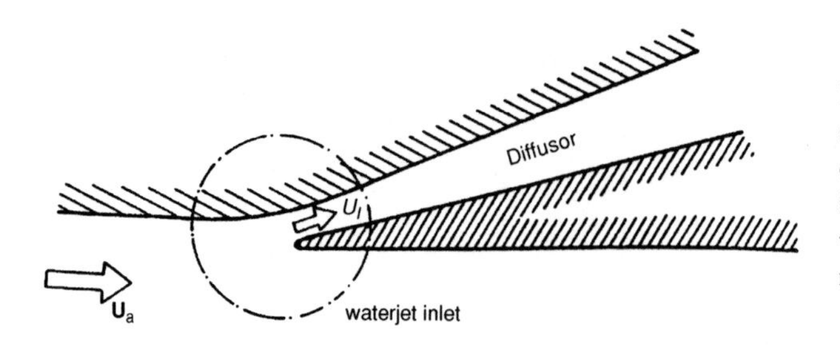

Figure 2.61. Carstensen's (1983) local 2D steady flow analysis of inlet area. The details of the flow within the circle is examined. U_a = ambient flow velocity, U_I = inlet velocity defined as the mean velocity at the most narrow cross section of the inlet.

have presented a waterjet pump design procedure. This includes analysis with a technique that couples a lifting surface program for blade-row calculations with an axi-symmetric Reynolds-averaged Navier–Stokes (RANS) viscous flow solver for computation of the pump through-flow. An example of application of CFD to the analysis of waterjet inlets is given by Førde et al. (1991).

2.11.2 Cavitation in the inlet area

The inlet area requires special attention in design to avoid engine load fluctuations, which may occur because of:

1. Exposure of the waterjet inlet to the free air
2. Flow separation in front of and inside the inlet
3. Cavitation inside the inlet
4. Ventilation and penetration of air from the free surface (see Figure 1.9) or from entrained air in the boundary layer

The phenomena mentioned above are often coupled in a complicated way. As an example, separation may be one of the requirements for onset of ventilation. Cavitation occurs in connection with separation. Under given conditions, a cavity will be penetrated and filled with air. Separation and cavitation are, first of all, dependent on the pressure distribution in and near the entrance. For a given shape, this distribution depends mainly on the speed and thrust (resistance) of the ship.

In a seaway, the power is often of the same order of magnitude as in calm water, but the speed will be reduced, which generally decreases the pressure near the inlet. With an inlet shape optimized for the maximum speed, this may lead to cavitation and separation. Further, the relative vertical motions between the vessel and the water in a seaway may lead to exposure of the waterjet inlet to the free air. The consequence in terms of engine load fluctuations is illustrated in Figure 1.13. In the design, one should therefore pay attention to off-design conditions.

Carstensen (1983) has studied theoretically the possibility of cavitation in the inlet area by using a two-dimensional steady-potential flow model. The possibility of flow separation was not examined. Only the details of the inlet area are included in the numerical model. The focus was on the local flow within the circle shown in Figure 2.61. The impeller, stator, and outlet appear far away on the scale of the local flow, and the effect on the local flow is in terms of a water mass-flux ρQ_f. This is used to define an inlet velocity

$$U_I = \frac{Q_f}{A_I}, \qquad (2.194)$$

where A_I is the most narrow cross section of the inlet. The ambient longitudinal flow velocity is denoted U_a. This flow velocity is unaffected by the local hull and the waterjet system. It represents an upstream inflow velocity to the local flow at the inlet. U_a can be set equal to U_1, that is, the longitudinal flow just outside the boundary layer at station ⓵a.

Faltinsen et al. (1991a) also presented a numerical method to calculate the pressure distribution in a flush-type waterjet inlet. This method is also based on a local steady two-dimensional potential flow model in a longitudinal cross section. In principle, it is straightforward to generalize the method to three dimensions. However, viscous effects ought to be considered. We know the water mass-flux through the waterjet system from the previously outlined power-prediction method. This condition and zero normal velocity through the hull are imposed. The far-field of the local flow is composed of two parts, that is, the x-component of the velocity at the waterjet inlet as if the waterjet were not there and a sink flow with a strength determined from the water mass-flux through the waterjet. The first far-field component

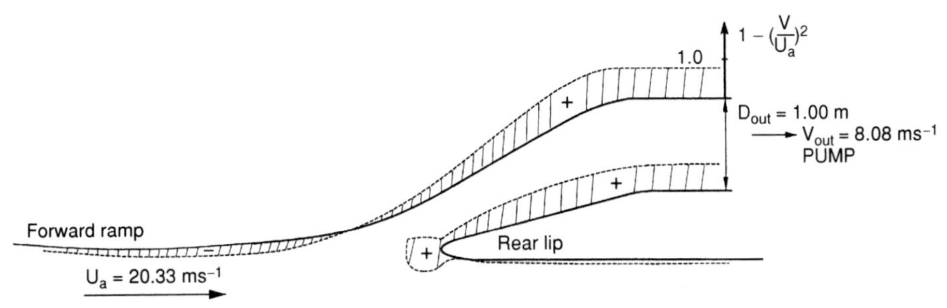

Figure 2.62. Longitudinal cross section of a waterjet inlet. The figure shows predicted nondimensional hydrodynamic pressure distribution: $1-(V/U_a)^2$. V = local velocity relative to the vessel, U_a = ship speed.

can be determined from a global steady-flow analysis. A boundary element method based on Green's second identity (see section 6.4) is used to find the velocity potential in the local flow at the waterjet inlet. Bernoulli's equation determines the pressure distribution. From this, we can investigate the possibility of cavitation and separation (see items 2 and 3 above).

We will illustrate this by the example shown in Figure 2.62. A longitudinal cross section of an inlet ahead of the pump with diameter $D_{out} = 1$ m is shown together with predicted velocity distribution V on the hull surface. The total steady pressure is the sum of atmospheric pressure, hydrostatic pressure, and hydrodynamic pressure $0.5\rho U_a^2(1-(V/U_a)^2)$, where $U_a = 20.33$ ms^{-1} in this case. If the total pressure is equal to vapor pressure, cavitation occurs. The lower the hydrodynamic pressure, the larger the possibility of cavitation. This does not occur in this example. However, the results show the lowest hydrodynamic pressure on the forward ramp and the lower surface of the lip. There is a stagnation pressure on the front of the lip and high pressures ahead of the pump.

In this context, an important parameter is the inlet velocity ratio (IVR), defined as

$$IVR = \frac{U_I}{U}. \tag{2.195}$$

Typical values of IVR are between 0.7 and 1.2. The inlet velocity defined by eq. (2.194) is in the case presented in Figure 2.62:

$$U_I = \frac{8.08 \cdot 1}{0.5} = 16.16 \text{ ms}^{-1}.$$

Here the area A_I per unit length at the most narrow cross section of the inlet is 0.5 m. This gives IVR = 0.79 if we set $U_a = U$. A restriction on IVR follows from eq. (2.173) for the thrust. In order to have positive thrust it is necessary that

$$U_j - U(1-\bar{w}_f) > 0.$$

The continuity equation gives $U_j A_7 = U_I A_I$, that is,

$$IVR\, A_I > A_7(1-\bar{w}_f). \tag{2.196}$$

Using the thrust and continuity equations gives the following relationship between thrust and IVR:

$$\frac{T}{\rho U^2 A_I} = IVR\left(IVR\frac{A_I}{A_7} - (1-\bar{w}_f)\right). \tag{2.197}$$

Carstensen (1983) presented systematic results of the pressure coefficient C_p along the inlet surface as a function of IVR based on $U_a = U$. The pressure coefficient is defined as

$$C_p = \frac{p-p_a}{0.5\rho U^2}, \tag{2.198}$$

where p is the total pressure and p_a is the ambient pressure. Because $p = p_a + 0.5\rho U^2 - 0.5\rho V^2$, C_p can be expressed as $1-(V/U)^2$, similar to the formula in Figure 2.62. The pressure coefficient at the forward ramp is always negative upstream. The IVR value determines when C_p becomes positive and continues to increase within the diffuser on the forward ramp. A suction peak is generally present near the tip of the rear lip. The general tendency is for the suction peak to occur on the outside of the rear lip when IVR is small. When IVR is sufficiently high, the suction peak on the rear lip occurs inside the inlet.

The minimum pressure coefficient $C_{p\min}$ can be used to assess the occurrence of cavitation. $-C_{p\min}$ as a function of IVR is presented in Figure 2.63 based on Carstensen's calculations for an inlet with a 10° diffuser angle and a 25° diffuser axis.

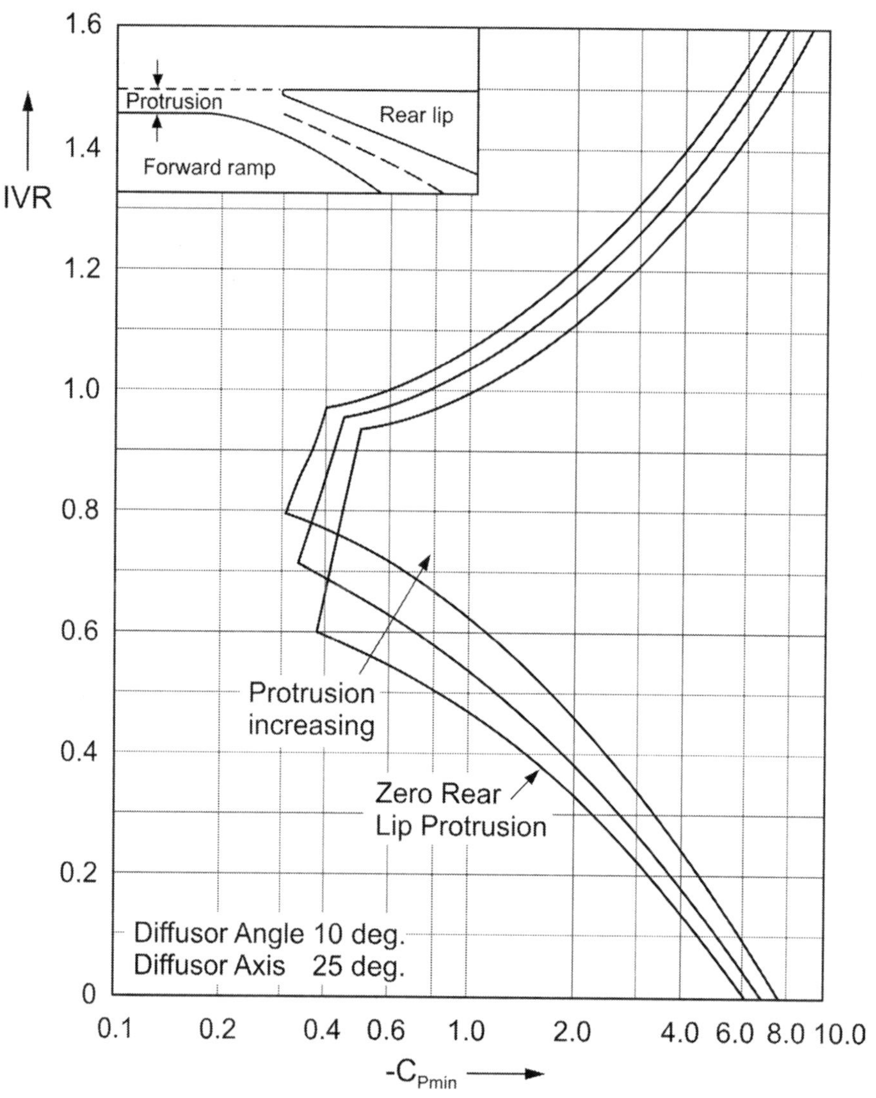

Figure 2.63. Bucket curves for flush and protruding inlets, $C_{p\min}$ = minimum pressure coefficient on the forward ramp and rear lip. $IVR = U_I/U$ with $U = U_a$ (see Figure 2.61). Presented by Kruppa (1990) based on Carstensen (1983).

The effect of rear lip protrusion is also shown. The presented curves are called bucket curves. The upper and lower branches of the bucket correspond to minimum pressure conditions of the inner and outer surfaces of the rear lip, respectively. These curves may be used as follows to assess the occurrence of cavitation. We set as a criterion for onset of cavitation that $p = p_v$, where p_v is the vapor pressure. We may then express $C_{p\min}$ as $(p_v - p_a)/(0.5\rho U^2)$. This is the same as minus the cavitation number σ. This means the curves in Figure 2.63 tell us at what cavitation number onset of cavitation will occur as a function of IVR.

As an example, let us assume that the outer surface of the rear lip has a draft $d = 1$ m. We set the ambient pressure equal to the sum of the atmospheric pressure and the hydrostatic pressure $\rho g d$. If $U \approx 40$ knots, this gives $\sigma = 0.5$. Let us then consider the bucket curve for zero rear lip protrusion in Figure 2.63. Onset of cavitation will occur outside the bucket, that is, onset of cavitation will not occur for IVR between approximately 0.55 and

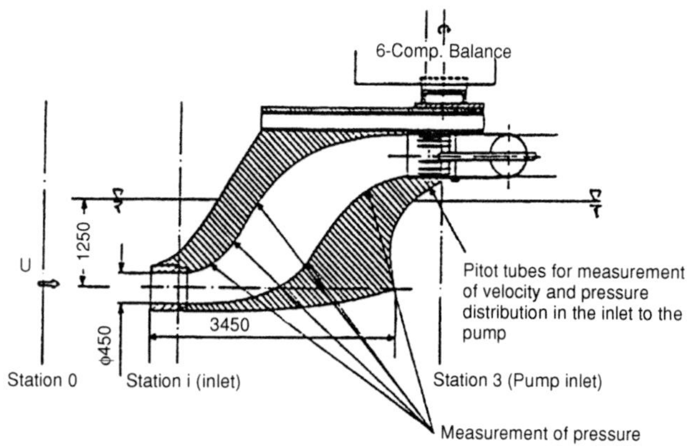

Figure 2.64. Model test setup of a ram (scoop) inlet tested in a cavitation tunnel with free surface (Minsaas 1996).

0.93 for the examined condition. If $\sigma < 0.3$, cavitation will occur for all IVR-values for the cases presented in Figure 2.63. These results are based on 2D calculations, but nowadays similar calculations can easily be done in three dimensions. Further, model tests with pressure measurements in the inlet are commonly done.

There are other types of waterjet inlets than the one described above. Waterjets with ram (scoop, pilot) inlets are used, for instance, on hydrofoil vessels with a fully submerged foil system. Figure 2.64 illustrates the model test setup of a ram inlet tested in a cavitation tunnel with free surface. Part of the study was to investigate the cavitation pattern at the inlet corresponding to a full-scale speed of 50 knots at different IVR-values. Bubble cavitation was detected outside the inlet, as illustrated in Figure 2.65. This can cause both noise problems and erosion of the material. Sheet cavitation was detected at the inside of the inlet at certain IVR-values shown in Figure 2.65. This will cause blockage of the flow to the waterjet system and reduce the efficiency. The sheet cavitation detected at the entrance may also lead to cavitation at other places in the waterjet system.

2.12 Exercises

2.12.1 Scaling

a) Use the Pi-theorem to show that nondimensional water resistance

$$\frac{R}{0.5\rho U^2 L^2}$$

is a function of Reynolds number and Froude number for geometrically similar submerged hull forms. Start by assuming that R is a function of ρ, g, μ, U, L, and other submerged hull dimensions.

How would you generalize this to account also for Weber number?

b) Use the Pi-theorem to express the fact that the nondimensional thrust and torque coefficients K_T and K_Q in open-water conditions without free-surface effects are functions of the advance ratio and Reynolds number for geometrically similar propellers.

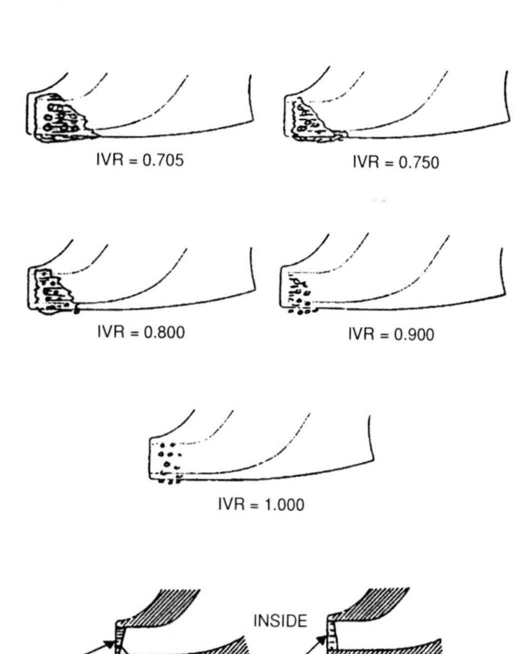

Figure 2.65. Model tests of a ram inlet corresponding to 50 knots in full scale. Influence of the inlet-velocity ratio (IVR) on cavitation pattern. The ram inlet is shown in Figure 2.64 (Minsaas 1996).

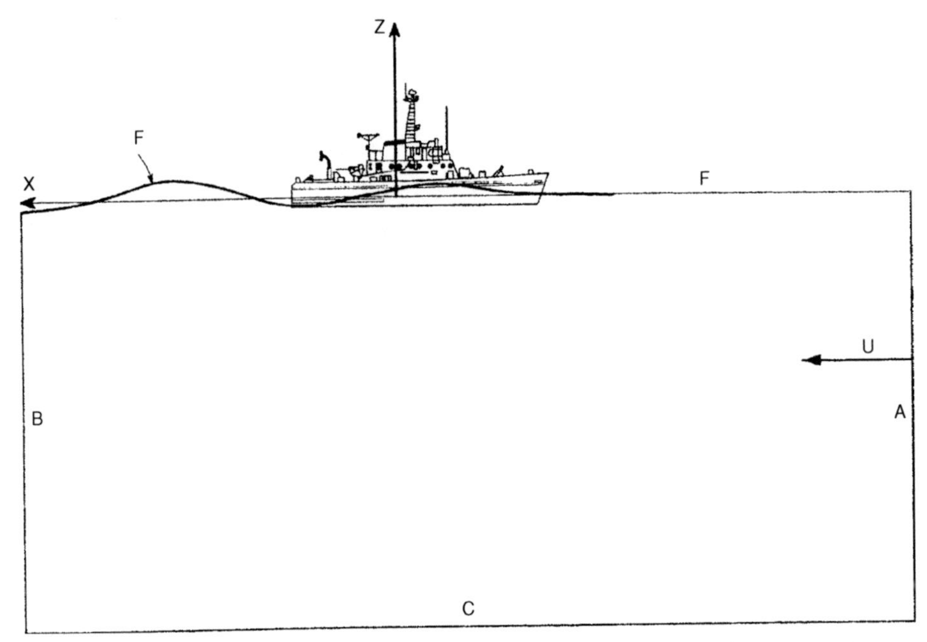

Figure 2.66. Control surfaces in applying conservation of fluid momentum to estimate ship resistance.

c) Start with the Navier-Stokes equations expressed by eqs. (2.8), (2.9), and (2.10), and assume steady mean flow along a smooth flat plate of length L. Make the equations and boundary conditions nondimensional by introducing

$$u^* = \frac{u}{U}, v^* = \frac{\text{v}}{U}, x^* = \frac{x}{L}, y^* = \frac{y}{L}.$$

Explain the fact that the nondimensional resistance of the plate is a function only of Reynolds number.

2.12.2 Resistance by conservation of fluid momentum

The following exercise is partly based on Ogilvie (1969a).

a) We are going to express water resistance R of a ship by means of steady RANS equations and conservation of fluid momentum. We choose a body of fluid bounded by the following surfaces (Figure 2.66).

A:	$x = x_A$,	far ahead of the ship
B:	$x = x_B$,	behind the ship
C:	$z = z_C$,	far below the ship
D:	$y = y_D$,	far to starboard of the ship ($y_D > 0$)
E:	$y = y_E$,	far to port of the ship ($y_E < 0$)
F:	$z = \zeta(x, y)$,	the free surface
H:		the hull surface

The normal $\mathbf{n}$ to the surface enclosing the fluid volume is positive outward of the fluid volume. The fluid velocity at A is $U\mathbf{i}$, where U is the ship's speed and $\mathbf{i}$ is the unit vector along the x-axis.

Neglect viscous and Reynolds stresses on F and show by means of continuity of fluid momentum and fluid mass that

$$R = \iint_B dS\left[2\mu\frac{\partial \bar{u}}{\partial x} - \rho\overline{u'^2} - (\bar{p} - p_a)\right] + \iint_A dS(\bar{p} - p_a) - \iint_B dS\rho\bar{u}\,(U + \bar{u}) \quad (2.199)$$

Here

$U + \bar{u} =$ time-averaged longitudinal fluid velocity

$-\rho\overline{u'^2} =$ Reynolds stress component, with u' being the turbulent part of longitudinal velocity

$\bar{p} =$ time-averaged pressure

$p_a =$ atmospheric pressure

b) Express $\bar{p}$ at A and discuss why the contribution from integration of $\bar{p}$ over A is cancelled by a similar contribution from the pressure integral over B.

c) We consider the wave resistance problem and neglect viscosity. We can now use Bernoulli's equation. By using eq. (2.199) and Bernoulli's equation, show that the wave resistance R_W can be expressed as

$$R_w = -0.5\rho \iint_B (u^2 - v^2 - w^2)\, dS + 0.5\rho g \int_B \zeta^2 (x, y)\, dy. \qquad (2.200)$$

Here the fluid velocity has the components $U + u$, v, and w along, respectively, the x-, y-, and z-axes, and $z = \zeta$ is the free-surface elevation.

2.12.3 Viscous flow around a strut

We consider a vertical strut with forward speed $U = 15\ \text{ms}^{-1}$ and analyze 2D flow in a horizontal cross-sectional plane. The effect of the free surface is neglected. The chord length is 5 m, and the thickness-chord length ratio is 0.10. Assume $\nu = 1.35 \cdot 10^{-6}\ \text{m}^2\text{s}^{-1}$ and that the flow is turbulent from the leading edge.

a) What is the drag force in Nm^{-1} if the strut surface is smooth?

b) What is the drag force in Nm^{-1} if the strut surface has a roughness height of 150 μm?

c) Estimate the boundary-layer thickness at the trailing edge by using a formula for a smooth flat plate.

d) Assume that the strut surface is smooth, and plot the velocity distribution in the wake at a longitudinal distance 50 m from the center of the strut.

e) Consider two smooth struts in a tandem arrangement. The strut dimensions and the speed are as above. The longitudinal distance between the centers of the two struts is 50 m. Assume the upstream strut is not influenced by the flow around the downstream strut. However, the downstream strut is influenced by the wake of the upstream strut.

Estimate the drag force on the downstream strut.

(Hint: Use wake solution at the center of the downstream strut to define a modified inflow velocity to the downstream strut.)

f) Generalize the arrangement in question e) to a staggered arrangement, and discuss the behavior of the drag force on the downstream strut.

2.12.4 Thrust and efficiency of a waterjet system

A. Thrust by conservation of fluid momentum

In this part of the exercise, we study an expression for the thrust of a waterjet that is more detailed mathematically and physically than the one in section 2.11.1. We start by expressing the equation for conservation of momentum $\mathbf{M}(t) = (M_1, M_2, M_3)$ in the fluid in a general way. Let S be a closed surface that encloses a fluid volume Ω. The momentum inside S can be written as

$$\mathbf{M}(t) = \iiint_\Omega \rho \mathbf{u}\, d\Omega, \qquad (2.201)$$

where $\mathbf{u} = (u_1, u_2, u_3)$ is the fluid velocity. The enclosing surface S does not need to follow the fluid motion.

By using the definition of a derivative and noting that both the volume and the velocity may change with time (Figure 2.67), show that

$$\frac{d\mathbf{M}}{dt} = \rho \iiint_\Omega \frac{\partial \mathbf{u}}{\partial t} d\Omega + \rho \iint_S \mathbf{u} U_{Sn}\, ds. \qquad (2.202)$$

Here U_{Sn} is the normal component of the velocity of the surface S. Note that here we have defined the positive normal direction out of the fluid. The last integral is the effect of integrating over the shaded area in Figure 2.67 and letting Δt be small (i.e., go to zero). The volume integral in eq. (2.202) may

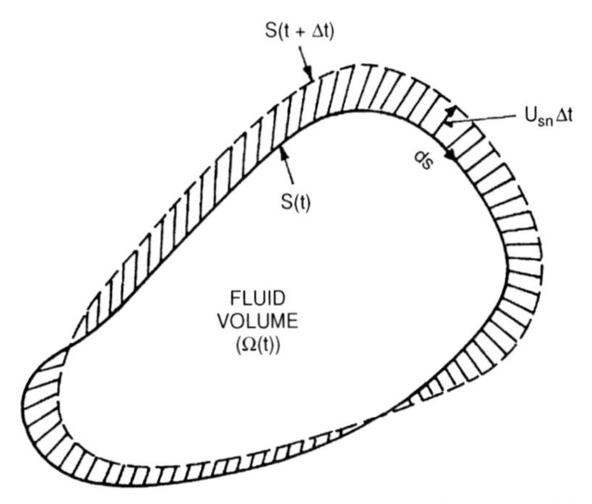

Figure 2.67. Illustration of how the control volume $\Omega(t)$ changes in a time increment Δt. U_{Sn} = normal component of the velocity of the surface S.

be rewritten by expressing $\partial \mathbf{u}/\partial t$ by the Navier–Stokes equation. This, for an incompressible fluid, can be written as

$$\frac{\partial \mathbf{u}}{\partial t} + \mathbf{u} \cdot \nabla \mathbf{u} = -\nabla \left(\frac{p}{\rho} + gz \right) + \nu \nabla^2 \mathbf{u}, \quad (2.203)$$

where z is a vertical coordinate and the z-axis is positive upward, with $z = 0$ in the mean free surface.

Using vector algebra, show that

$$\nabla \cdot (\mathbf{uu}) = \mathbf{u} \cdot \nabla \mathbf{u} \quad (2.204)$$

for an incompressible fluid.

The volume integral can be reduced to a surface integral by using a generalized Gauss theorem that states that

$$\iiint_{\Omega} \nabla \circ X \, d\Omega = \iint_{S} \mathbf{n} \circ X \, ds. \quad (2.205)$$

Here X may be a scalar, vector, or tensor, and $\circ$ denotes a dot, a cross, an ordinary multiplication, or nothing. It is assumed that X has continuous derivatives in Ω.

Show that

$$\frac{d\mathbf{M}}{dt} = -\iint_{S} p\mathbf{n} \, dS - \rho g \iint_{S} z\mathbf{n} \, dS$$
$$(2.206)$$
$$-\rho \iint_{S} \mathbf{u}(u_n - U_{Sn}) \, dS + \mu \iint_{S} \mathbf{n} \cdot \nabla \mathbf{u} \, dS$$

Here u_n is the normal component of the fluid velocity at the surface S.

The second term on the right-hand side of eq. (2.206) can be rewritten by eq. (2.205). Show that this gives

$$-\rho g \iint_{S} z\mathbf{n} \, dS = -\rho g \Omega \mathbf{k}, \quad (2.207)$$

where $\boldsymbol{k}$ is the unit vector along the z-axis.

The control volume illustrated in Figure 2.57 is now considered. The bounding surface S is divided into several parts, as described by eq. (2.168).

On which surfaces are U_{Sn} and u_n zero?

What is the relationship between U_{Sn} and u_n on the impeller surface S_{IM}?

Eq. (2.206) is also valid for turbulent flow, which is the real case for our application. Average eq. (2.206) over the time scale of turbulence in a way similar to the one in which the Reynolds-averaged Navier-Stokes equations were derived. Illustrate that this will lead to terms similar to those of the Reynolds stress terms.

If there is steady homogenous inflow to the impeller, the time average of $d\mathbf{M}/dt$ over the scale of turbulence is zero. However, a steady nonhomogenous inflow in the rotational direction of the impeller will cause unsteady flow effects. We will disregard these effects and set the time average of $d\mathbf{M}/dt$ equal to zero.

When using the time average of eq. (2.206) to find expression for the thrust provided by the impeller, the longitudinal force in the x-direction due to pressure, Reynolds, and viscous stresses acting on S_{IM}, S_{ST}, S_{SH}, and S_H must be considered, as we discussed in section 2.11.1.

Show that we can formally express the thrust T by the waterjet system as

$$T = \rho \iint_{S_7} u_1^2 \, dS - \rho \iint_{S_{1a}} u_1^2 \, dS$$
$$+ \int\limits_{S_7 + S_{1a} + S_{WAT} + S_{AIR}} p n_1 \, dS \quad (2.208)$$

– *Longitudinal force due to Reynolds and viscous stresses.*

Here the fluid velocity and pressure are time averaged over the scale of turbulence.

The pressure forces on the right-hand side of eq. (2.208) acting on S_{1a} and S_{WAT} will now be considered. Express the pressure by eq. (2.177).

Show that the contribution from the pressure part $-\rho gz$ to the longitudinal force on $S_{1a} + S_{WAT}$ can be neglected.

(Hint: Express the longitudinal force part as

$$\rho g \int\limits_{S_{1a} + S_{WAT}} z n_1 dS. \quad (2.209)$$

We introduce the closed surface $S_{1a} + S_{WAT} + S_{HI}$, where S_{HI} is close to horizontal and consists of the waterjet inlet area and part of the hull surface between S_{1a} and the inlet area. Because n_1 is approximately zero on S_{HI}, eq. (2.209) can be approximated as

$$\rho g \int\limits_{S_{1a} + S_{WAT} + S_{HI}} z n_1 \, dS. \quad (2.210)$$

Show by using the generalized Gauss theorem given by eq. (2.205) that the expression in eq. (2.210) is zero.)

The contribution to the thrust from the pressure term $-0.5\rho(U^2 - U_1^2 - U_2^2 - U_3^2)$ on $S_{1a} + S_{WAT}$ is more difficult to ignore. Discuss this by noting that the pressure varies strongly at the inlet.

B. Impeller effect by conservation of kinetic fluid energy

We will now consider in more detail the expression for the effect that the impeller gives to the flow. This will be done by using conservation of kinetic fluid energy. We start by expressing this in a general way as we did for conservation of fluid momentum.

The kinetic fluid energy inside the closed surface in Figure 2.67 can be expressed as

$$E_k(t) = \iiint_\Omega \frac{\rho}{2}\left(u_1^2 + u_2^2 + u_3^2\right) d\Omega = \iiint_\Omega \frac{\rho}{2} \mathbf{u} \cdot \mathbf{u}\, d\Omega. \tag{2.211}$$

Show in a way similar to that used in deriving eq. (2.206) that

$$\frac{dE_k(t)}{dt} = \rho \iiint_\Omega \mathbf{u} \cdot \frac{\partial \mathbf{u}}{\partial t} d\Omega + \frac{\rho}{2} \iint_S \mathbf{u} \cdot \mathbf{u}\, U_{Sn}\, dS. \tag{2.212}$$

We then rewrite $\partial \mathbf{u}/\partial t$ by means of Navier–Stokes equations for an incompressible fluid. This can be expressed in component form as

$$\frac{\partial u_i}{\partial t} = -u_j \frac{\partial u_i}{\partial x_j} - \frac{1}{\rho}\frac{\partial}{\partial x_i}(p + \rho g x_3) + \frac{1}{\rho}\frac{\partial \tau_{ij}}{\partial x_j}, \tag{2.213}$$

where the viscous stress components τ_{ij} are given by eq. (2.16) and $x_1 = x, x_2 = y$ and $x_3 = z$. Further, a conventional summation convention is used. This means

$$\frac{\partial \tau_{ij}}{\partial x_j} = \frac{\partial \tau_{i1}}{\partial x_1} + \frac{\partial \tau_{i2}}{\partial x_2} + \frac{\partial \tau_{i3}}{\partial x_3}. \tag{2.214}$$

Based on Landau and Lifschitz (1959) we can write

$$\begin{aligned} \rho \mathbf{u} \cdot \frac{\partial \mathbf{u}}{\partial t} &= -\rho \mathbf{u} \cdot (\mathbf{u} \cdot \nabla)\mathbf{u} - \mathbf{u} \cdot \nabla (p + \rho g z) \\ &\quad + u_i \frac{\partial \tau_{ij}}{\partial x_j}. = -\rho(\mathbf{u} \cdot \nabla)\left(\frac{1}{2}\mathbf{u} \cdot \mathbf{u} + \frac{p}{\rho} + gz\right) \\ &\quad + \nabla \cdot (\mathbf{u} \cdot \boldsymbol{\tau}) - \tau_{ij}\frac{\partial u_i}{\partial x_j} \end{aligned} \tag{2.215}$$

Here $\mathbf{u} \cdot \boldsymbol{\tau}$ is a vector with components $u_1\tau_{1j}$, $u_2\tau_{2j}$, $u_3\tau_{3j}$. Show eq. (2.215) and express the term $\tau_{ij}\partial u_i/\partial x_j$ without using the summation convention. Explain why we can express eq. (2.215) as

$$\begin{aligned} \rho \mathbf{u} \cdot \frac{\partial \mathbf{u}}{\partial t} &= -\nabla \cdot \left[\rho \mathbf{u}\left(\frac{1}{2}\mathbf{u} \cdot \mathbf{u} + \frac{p}{\rho} + gz\right) - \mathbf{u} \cdot \tau\right] \\ &\quad - \tau_{ij}\frac{\partial u_i}{\partial x_j}. \end{aligned} \tag{2.216}$$

Show by using eq. (2.212), eq. (2.216) and the generalized Gauss theorem (see eq. (2.205)) that

$$\begin{aligned} \frac{dE_k}{dt} &= -\iint_S \left[(u_n - U_{Sn})\frac{1}{2}\rho \mathbf{u} \cdot \mathbf{u}\right] dS \\ &\quad - \iint_S \rho u_n \left(\frac{p}{\rho} + gz\right) dS \\ &\quad + \iint_S \mathbf{n} \cdot (\mathbf{u} \cdot \tau) dS - \iiint_\Omega \tau_{ij}\frac{\partial u_i}{\partial x_j} d\Omega \end{aligned} \tag{2.217}$$

Eq. (2.217) will now be averaged over the time scale of turbulence. The time average of dE_k/dt is assumed to be zero. The argument is similar to the one for the fluid momentum.

The fluid volume in Figure 2.67 is considered, and S is divided into surfaces as in eq. (2.168). Which terms in eq. (2.217) express the effect P_D given by the impeller to the flow?

Show that eq. (2.217) leads to eq. (2.175).

2.12.5 Steering by means of waterjet

Figure 2.56 illustrates how the waterjet can be used to steer the vessel. Express longitudinal and transverse thrust on the vessel by means of conservation of fluid momentum and fluid mass in a way similar to that used in the main text for the thrust.

3 Waves

3.1 Introduction

Before we can describe wave resistance in detail, we need to introduce wave theory. This theory is also needed in the description of wave-induced motions and loads on a high-speed vessel. We first present linear wave theory in regular harmonic waves in deep and finite water depths. This includes analysis of wave refraction. An irregular sea state can be represented as a sum of regular waves of different frequencies and wave propagation directions. Recommended wave spectra that describe the frequency content in irregular waves are then given.

Linear wave theory assumes that the wave slope is asymptotically small. Not all waves occurring in reality can be described by linear wave theory; an extreme example is breaking waves. Figure 3.1 illustrates plunging, breaking waves generated in a wave flume. We see breaking waves on a beach, but they can also occur in the open sea in deep water. A strong current in the opposite direction of the wave propagation steepens the waves. This is a phenomenon known in connection with the Agulhas current off the east coast of Africa. A typical feature of a nonlinear wave is that the vertical distance between the wave crest and the mean water level is larger than the distance between the mean water level and the wave trough.

Scatter diagrams of significant wave heights and mean wave periods are needed in operational and design studies. These diagrams describe the probability of occurrence of different sea states for a given operational area. An example on scatter diagram recommended for high-speed vessels is presented.

3.2 Harmonic waves in finite and infinite depth

Our assumptions are irrotational motions so that $\nabla \times \mathbf{u} = 0$, where $\mathbf{u}$ is the fluid velocity. It follows then that the fluid velocity can be expressed by the velocity potential Φ as

$$\mathbf{u} = \nabla\Phi = \mathbf{i}\frac{\partial\Phi}{\partial x} + \mathbf{j}\frac{\partial\Phi}{\partial y} + \mathbf{k}\frac{\partial\Phi}{\partial z}, \qquad (3.1)$$

where if the fluid is incompressible, that is, $\nabla \cdot \mathbf{u} = 0$, then Φ satisfies the Laplace equation

$$\frac{\partial^2\Phi}{\partial x^2} + \frac{\partial^2\Phi}{\partial y^2} + \frac{\partial^2\Phi}{\partial z^2} = 0. \qquad (3.2)$$

We will choose the Cartesian coordinate system (x, y, z) so that z is a vertical coordinate with positive coordinate upward. Further, $z = 0$ represents the mean free surface (Figure 3.2).

3.2.1 Free-surface conditions

For the time being, we neglect surface tension and derive free-surface conditions based on potential flow. Surface tension matters only for linear propagating waves when the wavelength is less than 5 cm. There are two free-surface conditions. One requires that the fluid pressure be equal to the pressure in the air on the free surface. This is the *dynamic free-surface condition*. The other one is the *kinematic free-surface condition* and requires that fluid particles remain on the free surface.

We shall now derive a linear expression for the dynamic free-surface condition. We start with an expression for the fluid pressure p which follows from Bernoulli's equation

$$p + \rho\frac{\partial\Phi}{\partial t} + \frac{\rho}{2}|\nabla\Phi|^2 + \rho g z = C. \qquad (3.3)$$

Here Φ is the velocity potential for the flow, t is the time variable, and ρ is the mass density of the fluid. The constant C is derived by expressing eq. (3.3) on the free surface where the air pressure is atmospheric and there is no ambient wave motion or fluid velocity due to the ship. Before doing that we reformulate eq. (3.3) and express

$$\Phi = Ux + \varphi. \qquad (3.4)$$

This is, for instance, a description of the flow when we consider a ship with constant speed U on a straight course and observe the flow from a reference frame following the ship velocity U. The ship's velocity then appears as an incident flow to the ship with velocity U along an x-axis in the mean free surface with increasing x-values from the bow

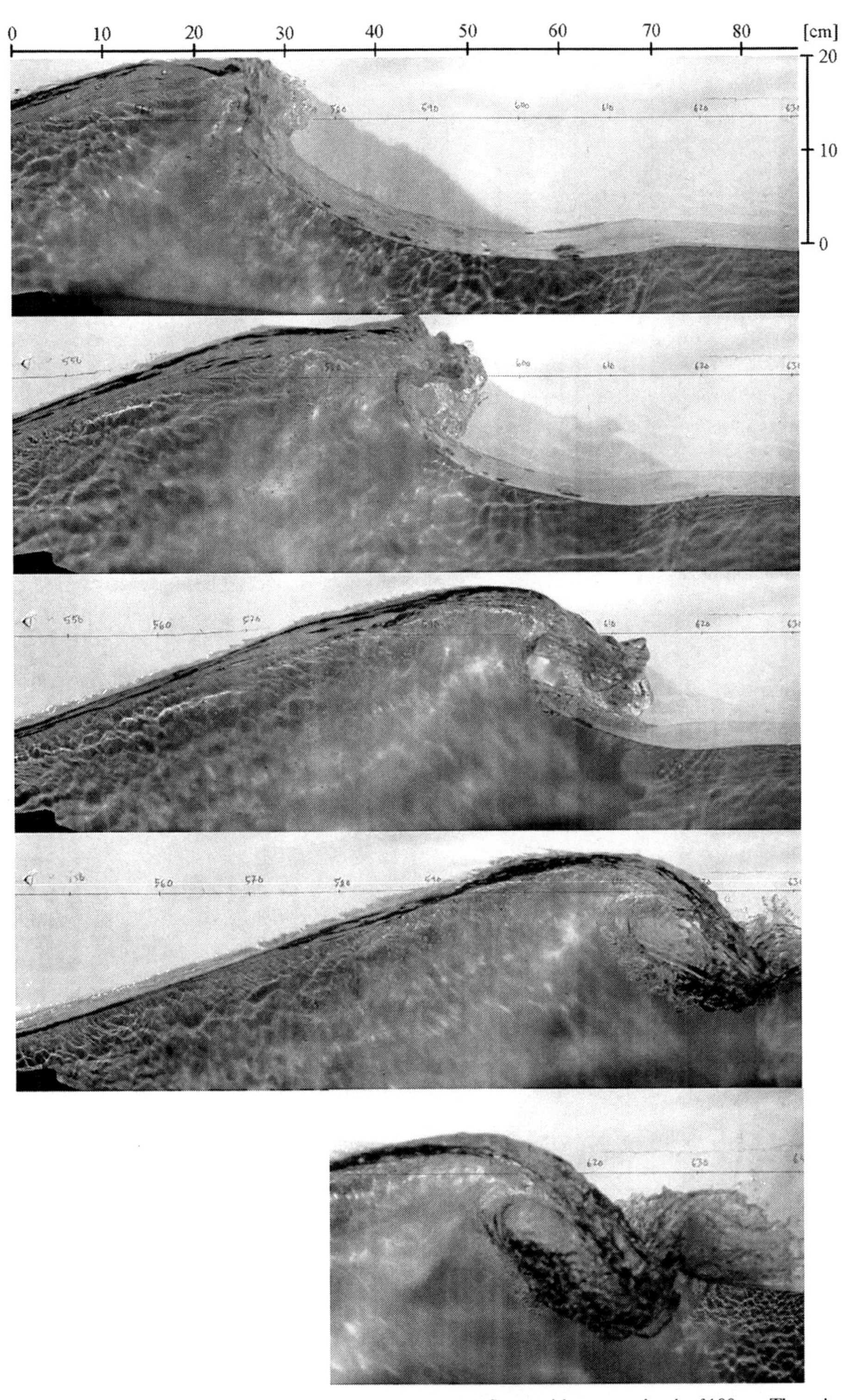

Figure 3.1. Plunging breaker generated in a narrow wave flume with a water depth of 100 cm. The ruler at the top and top right of the figure shows the physical dimensions in centimeters. The bottom picture gives a detailed view of the plunging breaker after impact on the underlying water. The impact causes reflection of the water, resulting in spray formation. (Photos by Pål Lader and Olav Rognebakke.)

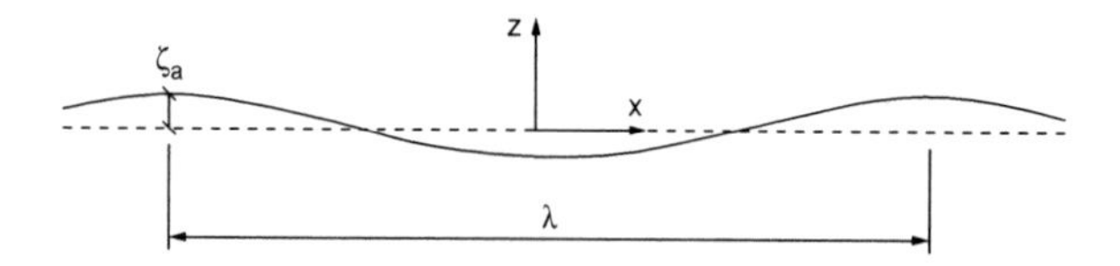

Figure 3.2. Coordinate system for linear harmonic long-crested waves.

toward the stern (Figure 3.3), or it could represent the flow of a river with a small slope past a fixed object.

We now determine C for the ship problem. We then go far away from the ship, where the ship causes zero disturbance. However, there is an inflow velocity U. Then we express eq. (3.3) at $z = 0$. This gives $p_a + 0.5\rho U^2 = C$, where p_a is the atmospheric pressure. Substituting eq. (3.4) into eq. (3.3) leads to

$$p = -\rho\frac{\partial\varphi}{\partial t} - \frac{\rho}{2}\left[\left(U + \frac{\partial\varphi}{\partial x}\right)^2 + \left(\frac{\partial\varphi}{\partial y}\right)^2 + \left(\frac{\partial\varphi}{\partial z}\right)^2\right] - \rho g z + p_a + \frac{\rho}{2}U^2. \quad (3.5)$$

We now linearize eq. (3.5) by assuming $|\nabla\varphi| \ll U$ and keep linear terms in φ. This gives

$$p = -\rho\frac{\partial\varphi}{\partial t} - \rho U\frac{\partial\varphi}{\partial x} - \rho g z + p_a. \quad (3.6)$$

Then we impose the dynamic free-surface condition by once more doing a linearization. This means that $\partial\varphi/\partial t$ and $\partial\varphi/\partial x$ in eq. (3.6) are Taylor series expanded about $z = 0$ and only linear terms are kept. Another way of saying this is that φ is assumed constant from $z = 0$ to the instantaneous free-surface elevation. The dynamic free-surface condition $p = p_a + p_0$ on $z = \zeta(x, y, t)$ can now be expressed as

$$p_0 = -\rho\frac{\partial\varphi}{\partial t} - \rho U\frac{\partial\varphi}{\partial x} - \rho g\zeta \quad \text{on } z = 0. \quad (3.7)$$

Here it is possible to deal with an SES that has an excess pressure p_0 in the air cushion.

Next we consider the kinematic free-surface condition, which can be expressed as

$$\frac{D}{Dt}(z - \zeta(x, y, t)) = 0 \quad \text{on } z = \zeta(x, y, t). \quad (3.8)$$

Here $D/Dt \equiv \partial/\partial t + \nabla\Phi \cdot \nabla$ is the substantive or material derivative, which expresses the time rate of change when we follow a fluid particle as it moves in the space. Eq. (3.8) states that a fluid particle on the free surface always has the property $z - \zeta(x, y, t) = 0$, that is, it stays on the free surface. We now linearize eq. (3.8) as we did in deriving eq. (3.7) and note that $Dz/Dt = \nabla\Phi \cdot \nabla z = \partial\varphi/\partial z$. After a Taylor expansion about $z = 0$ and linearization we find

$$\frac{\partial\zeta}{\partial t} + U\frac{\partial\zeta}{\partial x} = \frac{\partial\varphi}{\partial z} \quad \text{on } z = 0. \quad (3.9)$$

We can now combine eqs. (3.7) and (3.9) by first performing the operation $\partial/\partial t + U\partial/\partial x$ on

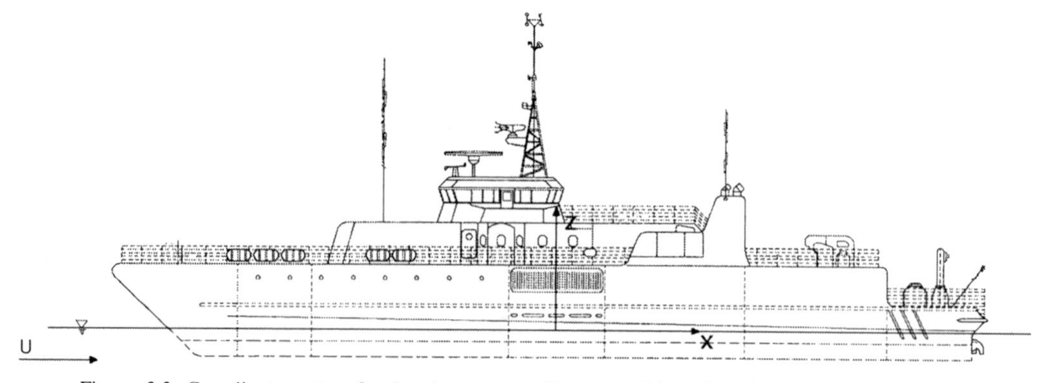

Figure 3.3. Coordinate system fixed to the mean oscillatory position of a ship with forward speed U. The forward speed appears from this coordinate system as a flow with velocity U in the x-direction. (Based on Figure 5.1. Copyright holder: Royal Norwegian Navy).

Table 3.1. *Velocity potential, dispersion relationship, wave profile, velocity, and acceleration for regular sinusoidal propagating waves in finite and infinite water depth according to linear theory (Faltinsen 1990)*

	Finite water depth	Infinite water depth
Velocity potential	$\varphi = \frac{g\zeta_a}{\omega}\frac{\cosh k(z+h)}{\cosh kh}\cos(\omega t - kx)$	$\varphi = \frac{g\zeta_a}{\omega}e^{kz}\cos(\omega t - kx)$
Connection between wave number k and circular frequency ω (dispersion relationship)	$\frac{\omega^2}{g} = k\tanh kh$	$\frac{\omega^2}{g} = k$
Connection between wavelength λ and wave period T	$\lambda = \frac{g}{2\pi}T^2\tanh\frac{2\pi}{\lambda}h$	$\lambda = \frac{g}{2\pi}T^2$
Wave profile	$\zeta = \zeta_a\sin(\omega t - kx)$	$\zeta = \zeta_a\sin(\omega t - kx)$
Hydrodynamic pressure	$p_D = \rho g\zeta_a\frac{\cosh k(z+h)}{\cosh kh}\sin(\omega t - kx)$	$p_D = \rho g\zeta_a e^{kz}\sin(\omega t - kx)$
x-component of velocity	$u = \omega\zeta_a\frac{\cosh k(z+h)}{\sinh kh}\sin(\omega t - kx)$	$u = \omega\zeta_a e^{kz}\sin(\omega t - kx)$
z-component of velocity	$w = \omega\zeta_a\frac{\sinh k(z+h)}{\sinh kh}\cos(\omega t - kx)$	$w = \omega\zeta_a e^{kz}\cos(\omega t - kx)$
x-component of acceleration	$a_1 = \omega^2\zeta_a\frac{\cosh k(z+h)}{\sinh kh}\cos(\omega t - kx)$	$a_1 = \omega^2\zeta_a e^{kz}\cos(\omega t - kx)$
z-component of acceleration	$a_3 = -\omega^2\zeta_a\frac{\sinh k(z+h)}{\sinh kh}\sin(\omega t - kx)$	$a_3 = -\omega^2\zeta_a e^{kz}\sin(\omega t - kx)$

$\omega = 2\pi/T$, $k = 2\pi/\lambda$, T = wave period, λ = wavelength, ζ_a = wave amplitude, g = acceleration of gravity, t = time variable, x = direction of wave propagation, z positive upward, $z = 0$ mean water level, h = average water depth, ρ = mass density of the fluid. Total pressure in the fluid: $p_D - \rho g z + p_a$ (p_a= atmospheric pressure).

eq. (3.7) and then using eq. (3.9). This gives

$$\frac{\partial^2\varphi}{\partial t^2} + 2U\frac{\partial^2\varphi}{\partial x\partial t} + U^2\frac{\partial^2\varphi}{\partial x^2} + g\frac{\partial\varphi}{\partial z} = -\frac{1}{\rho}\left(\frac{\partial p_0}{\partial t} + U\frac{\partial p_0}{\partial x}\right) \quad \text{on } z = 0. \tag{3.10}$$

3.2.2 Linear long-crested propagating waves

We consider linear long-crested waves propagating in the x-direction in a fluid with infinite horizontal extent and no obstacles present. The mass density and temperature are assumed constant in the fluid domain, that is, there is no stratification. Assuming harmonic oscillations with circular frequency ω (rad/s), no mean flow ($U = 0$), atmospheric pressure on the free surface ($p_0 = 0$), and linearity in terms of a small wave slope leads by using eq. (3.10) to the following combined dynamic and kinematic free-surface condition

$$-\omega^2\varphi + g\frac{\partial\varphi}{\partial z} = 0 \quad \text{on } z = 0. \tag{3.11}$$

The boundary condition on the sea floor $z = -h$ expresses no flow through the sea bottom, that is,

$$\frac{\partial\varphi}{\partial z} = 0 \quad \text{on } z = -h. \tag{3.12}$$

Further, φ satisfies the two-dimensional Laplace equation in x and z. The solution to this boundary-value problem can be found in many textbooks dealing with water waves and will not be derived here. We use the results by Faltinsen (1990), which are presented in Table 3.1. We note that the wave profile, the dynamic pressure $p_D = -\rho\partial\varphi/\partial t$, the fluid velocity, and acceleration are linearly dependent on the wave amplitude ζ_a.

A practical fact that will be used several times later is that the fluid motion for deep-water waves is negligible from half a wavelength λ down in the fluid. This results from the exponential factor $\exp(kz) = \exp(2\pi z/\lambda)$ in the deep-water results in Table 3.1. For instance, if $z = -0.5\lambda$, $\exp(kz) = 0.043$ or if $z = -\lambda$, $\exp(kz) = 0.002$.

According to linear theory, a fluid particle moves in a circle for deep water and in an ellipse for finite water depth (see exercise 3.5.1). The

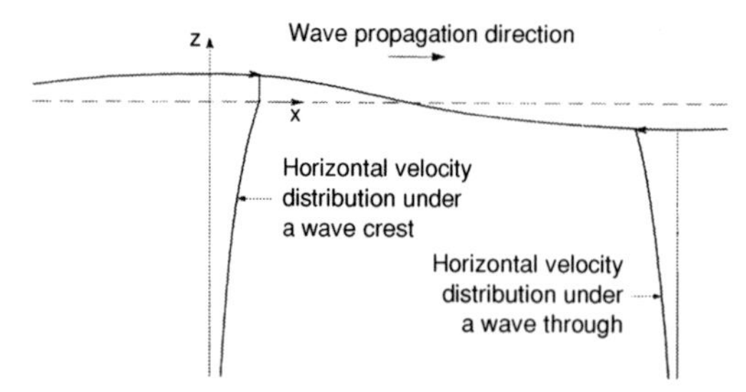

Figure 3.4. Horizontal velocity distribution under a wave crest and a wave trough according to linear wave theory. (The x- and z-axes have different scales.)

circle radius is equal to the wave amplitude for a fluid particle on the free surface. The two semi-axes of the elliptical motion for finite depth are horizontal and vertical with the horizontal semi-axis being the larger. The vertical semi-axis is equal to the wave amplitude for a fluid particle on the free surface. When the fluid depth is very shallow, the horizontal fluid velocity is much larger than the vertical fluid velocity. This can be seen by Taylor expanding the finite-depth expressions of fluid velocity at $z = -h$ and assuming kh to be small. Further, we can show that the total pressure is hydrostatic relative to the instantaneous free-surface elevation when $kh \to 0$.

It should be noted that the linear theory assumes the velocity potential and fluid velocity to be constant from the mean free surface to the actual free-surface level. This was assumed when the free-surface conditions were formulated. The horizontal velocity distribution shown in Figure 3.4 for the flow under a wave crest is consistent with linear theory. Figure 3.4 also shows the velocity under a wave trough, where we have used the analytical velocity distribution up to the free-surface level. It is then implicitly assumed that the difference between the horizontal velocity at the wave trough and the velocity at $z = 0$ is small compared with the velocity itself.

Figure 3.5 shows how the pressure varies with depth both under a wave crest and a wave trough. The "hydrostatic" pressure "$-\rho g z$" cancels the dynamic pressure $-\rho \partial\varphi/\partial t|_{z=0}$ at the free surface. This is the linear dynamic free-surface condition, which is exactly satisfied at the wave crest in Figure 3.5, whereas there is a higher-order error under the wave trough. By "higher-order error" we mean that the error is approximately proportional to $(\zeta_a/\lambda)^n$, where $n \geq 2$. This means that linear theory is correct to $O(\zeta_a/\lambda)$, where $O(\ )$ means order of magnitude. We should note that the dynamic pressure $-\rho\partial\varphi/\partial t$ half a wavelength down in the fluid is only 4% of its value at $z = 0$.

Linear theory represents a first-order approximation in satisfying the free-surface conditions. It can be improved by introducing higher-order terms in a consistent manner – a Stokes expansion. The next approximation would solve the problem to second order in the parameter ζ_a/λ characterizing the wave amplitude/wavelength ratio of the linear (first-order) solution. Second-order theory means that we keep in a consistent way all terms proportional to $O((\zeta_a/\lambda)^2)$ and $O(\zeta_a/\lambda)$. For sinusoidal unidirectional progressive *deep-water* waves for which the solution in Table 3.1

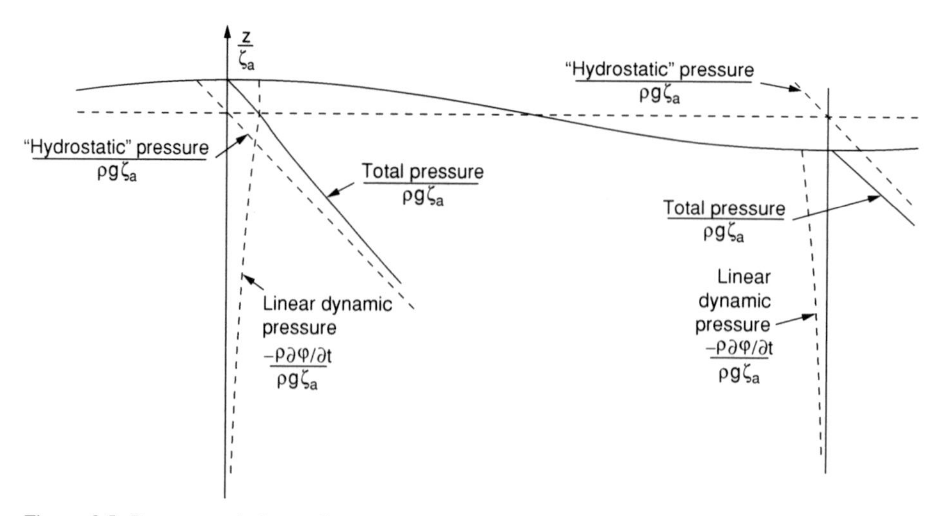

Figure 3.5. Pressure variation under a wave crest and a wave trough according to linear wave theory.

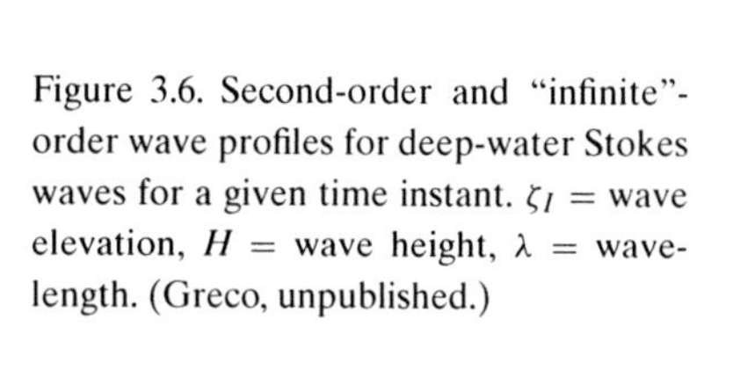
Figure 3.6. Second-order and "infinite"-order wave profiles for deep-water Stokes waves for a given time instant. ζ_I = wave elevation, H = wave height, λ = wavelength. (Greco, unpublished.)

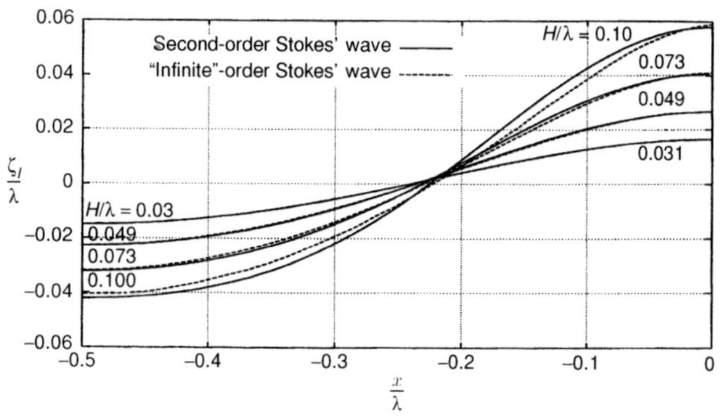

represents the first-order (linear) solution, it is possible to show that the second-order velocity potential is zero, and that the second-order wave elevation ζ_2 is

$$\zeta_2 = -\frac{1}{2}\zeta_a^2 k \cos[2(\omega t - kx)]. \quad (3.13)$$

By combining this with the first-order solution $\zeta_a \sin(\omega t - kx)$, we see that the second-order solution sharpens the wave crests and makes the trough more shallow. We leave it as a part of exercise 3.5.3 to show eq. (3.13).

In Figure 3.6, second-order wave profiles are compared to "infinite"-order wave profiles for four different wave steepnesses H/λ, where H is the wave height. The infinite-order wave profile for $H/\lambda = 0.10$ is given by Schwartz (1974), whereas the theory presented by Bryant (1983) is used to determine the wave profile for the other H/λ-values. The wave elevation is symmetric about $x = 0$ in Figure 3.6 and cannot describe plunging breakers as illustrated in Figure 3.1

The wave profiles computed by second-order theory and infinite-order theory compare very well for $H/\lambda = 0.031$ and $H/\lambda = 0.049$. For $H/\lambda = 0.073$, the deviation is larger, but still the relative error for the maximum wave elevation is less than 0.8%. When the wave steepness is increased to 0.1, the relative difference between the two different wave profiles becomes more significant. The exact wave profile is more peaked at the crest and flatter at the troughs than the second-order profile. This indicates that linear- and second-order wave theory is not sufficient to describe the wave properly for steep waves. However, in the following chapters, linear theory is to a large extent used to describe the incident wave elevation and kinematics.

When the ratio between the water depth and the wavelength becomes small, Boussinesq equations (Mei 1983) are commonly used to describe the effects of nonlinearity and dispersion. One example of application is ship waves in shallow water, which is discussed in section 4.5.

The expressions in Table 3.1 can be generalized to any wave propagation direction. We can show that by using the two coordinate systems (X, Y, Z) and (x, y, z) shown in Figure 3.7 The X-axis is the wave propagation direction and has an angle β relative to the x-axis. This means that by following the notation in Table 3.1, the wave elevation is $\zeta = \zeta_a \sin(\omega t - kX)$. Then we make a coordinate transformation to the (x, y, z) system, that is,

$$x = X\cos\beta - Y\sin\beta \quad X = x\cos\beta + y\sin\beta$$
$$y = X\sin\beta + Y\cos\beta \quad Y = -x\sin\beta + y\cos\beta.$$

This gives

$$\zeta = \zeta_a \sin(\omega t - kx\cos\beta - ky\sin\beta). \quad (3.14)$$

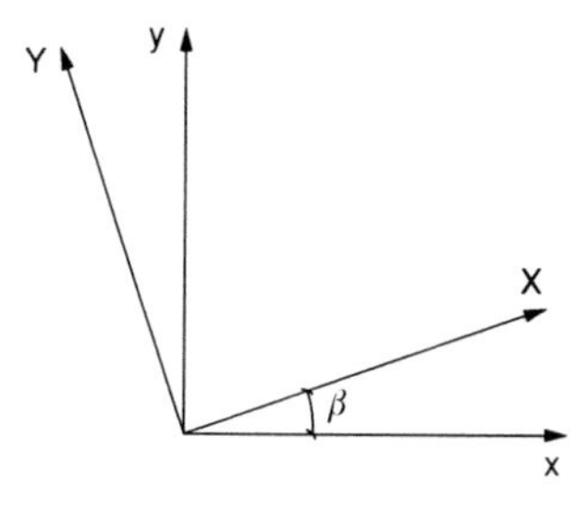

Figure 3.7. Coordinate systems used to derive expressions for waves propagating along the X-axis with an angle β relative to the x-axis.

3.2.3 Wave energy propagation velocity

The wave energy propagation velocity is important, for instance, in explaining ship waves and how waves are modified near a beach. We start out with general formulas presented, for instance, by Newman (1977, pp. 260–6). The total energy E in a fluid volume Ω consists of kinetic and potential energy. It can be written as

$$E(t) = \rho \iiint_{\Omega} \left(\frac{1}{2}V^2 + gz\right) d\Omega, \quad (3.15)$$

where $d\Omega$ is used as a symbol for volume integration. V is the fluid velocity, and ρ is the mass density of the fluid. Using the divergence theorem, the time derivative of the energy can be written as

$$\frac{dE(t)}{dt} = -\rho \iint_{S} \left(\frac{\partial\varphi}{\partial t}\frac{\partial\varphi}{\partial n} - \left(\frac{p - p_a}{\rho} + \frac{\partial\varphi}{\partial t}\right) U_n\right) ds, \quad (3.16)$$

where S is the boundary surface to Ω and $\partial/\partial n$ is the derivative along the normal vector $\boldsymbol{n}$ to S. The positive direction is into the fluid domain. (Note that Newman uses the opposite positive direction of $\boldsymbol{n}$.) U_n means the normal velocity of S, and p_a is the atmospheric pressure.

Let us assume two-dimensional flow. The volume of fluid Ω has two fixed vertical sides S_A and S_B at $x = A$ and $x = B$, bounded above by the free surface S_F and below by the bottom surface S_0 at $z = -h$. The boundary surface S in eq. (3.16) consists then of S_A, S_B, S_F, and S_0. We can write

$$U_n = \partial\varphi/\partial n \quad \text{on } S_F$$

$$U_n = \partial\varphi/\partial n = 0 \quad \text{on } S_0$$

$$U_n = 0 \quad \text{on } S_A \quad \text{and } S_B$$

$$p = p_a \quad \text{on } S_F.$$

Using eq. (3.16) gives that the time rate of change of E per unit length in y-direction of the energy in Ω is

$$\frac{dE(t)}{dt} = -\rho \int_{-h}^{\zeta_A} \left.\frac{\partial\varphi}{\partial t}\frac{\partial\varphi}{\partial x}\right|_{x=A} dz + \rho \int_{-h}^{\zeta_B} \left.\frac{\partial\varphi}{\partial t}\frac{\partial\varphi}{\partial x}\right|_{x=B} dz, \quad (3.17)$$

where ζ_A and ζ_B are the wave elevations at, respectively, $x = A$ and $x = B$. We can interpret the two terms in eq. (3.17) as energy flux through A and B, respectively. This means if we describe φ by linear theory as in Table 3.1, then the time-averaged energy flux per unit transverse length through a vertical plane perpendicular to the wave propagation direction is

$$\begin{aligned}\overline{\frac{dE}{dt}} &= -\rho \overline{\int_{-h}^{0} \frac{\partial\varphi}{\partial t}\frac{\partial\varphi}{\partial x} dz} \\ &= \frac{\rho g \zeta_a^2 \omega}{\sinh kh \cosh kh} \overline{\sin^2(\omega t - kx)} \\ &\quad \times \int_{-h}^{0} \cosh^2 k(z+h)\, dz. \\ &= \frac{1}{2}\rho g \zeta_a^2 \frac{\omega}{k}\left[\frac{1}{2} + \frac{kh}{\sinh 2kh}\right]\end{aligned} \quad (3.18)$$

The overbar in eq. (3.18) expresses time average over the period T. We note:

$$\overline{\sin^2(\omega t - kx)} = \frac{1}{T}\int_{0}^{T} \sin^2(\omega t - kx) dt = \frac{1}{2}.$$

We also note that the integration in eq. (3.18) is only to $z = 0$ and not to the instantaneous free-surface elevation as in eq. (3.17). This is consistent with using linear theory to describe φ. We can then only express dE/dt correctly to $O((\zeta_a/\lambda)^2)$. The integration from $z = 0$ to the instantaneous free surface gives terms of higher order than $O((\zeta_a/\lambda)^2)$. We need now to find the time-averaged wave energy $\overline{E(t)}$ per unit of horizontal area. So we consider a vertical column with unit cross-sectional area extending from the sea bottom to the free surface. The wave energy density E follows by using eq. (3.15) and excluding the potential energy $-(1/2)\rho g h^2$ of the fluid without any waves. This means

$$E = \frac{\rho}{2}\int_{-h}^{0} V^2\, dz + \frac{1}{2}\rho g \zeta^2. \quad (3.19)$$

Here ζ is the free-surface elevation given in Table 3.1. The kinetic energy is integrated only to $z = 0$. The argument is similar as for dE/dt, that is, we use linear theory to describe the flow and can describe E correctly only to $O((\zeta_a/\lambda)^2)$. We note that $V^2 = u^2 + w^2$, where the horizontal and vertical fluid velocities u and w follow from Table 3.1. The first step is to time average the kinetic energy E_k. We then use

$$\overline{\sin^2(\omega t - kx)} = \overline{\cos^2(\omega t - kx)} = \frac{1}{2}. \quad (3.20)$$

The next step is to integrate the expression with respect to z. This involves the integral

$$\int_{-h}^{0} (\cosh^2 k(z+h) + \sinh^2 k(z+h))\, dz = \frac{\cosh kh \sinh kh}{k},$$

which gives

$$\bar{E}_k = \frac{\rho}{4}\omega^2 \zeta_a^2/(k \tanh kh) = \frac{\rho g}{4}\zeta_a^2$$

by using the dispersion relationship between ω and k in Table 3.1. This is the same as the time average of the potential energy part of eq. (3.19). It then follows that

$$\bar{E} = \frac{\rho g}{2}\zeta_a^2, \quad (3.21)$$

which is independent of water depth. Using eq. (3.21) in combination with eq. (3.18) means we can interpret

$$V_g = \left(\frac{1}{2} + \frac{kh}{\sinh 2kh}\right)\frac{\omega}{k} \quad (3.22)$$

as the wave energy propagation velocity. Actually, V_g is also the group velocity (Newman 1977). This is defined as

$$V_g = \frac{d\omega}{dk} \quad (3.23)$$

and can be evaluated by using the dispersion relationship between ω and k. So actually there are several different velocities describing the waves. These give

- Fluid velocity (see Table 3.1)
- Wave energy propagation velocity, also denoted as group velocity (see eq. (3.22))
- Phase velocity (wave-shape velocity or celerity)

$$V_p = \frac{\omega}{k}. \quad (3.24)$$

The wave-shape velocity follows simply by looking at the expression for the wave elevation ζ in Table 3.1 and finding combinations of x and t that make $\sin(\omega t - kx)$ equal to any constant. This gives the phase $(\omega t - kx)$ equal to, say, a constant A, that is,

$$x = \frac{\omega}{k}t - \frac{A}{k},$$

which means that the wave elevation remains unchanged when we move with velocity ω/k. This results in eq. (3.24). The consequence of the fluid velocity and the phase velocity being different is that a fluid particle that at some instant is at a wave crest will not continue being at the wave crest. The fluid particle will stay at the free surface, however, and according to linear theory, move in a circular or elliptic path for infinite or finite water depth, respectively.

The fact that waves of different frequencies travel with different phase velocities is called dispersion. The relationship between frequency and wave number is called dispersion relationship. It is only in the limit of $kh \to 0$, that is, shallow water, that gravity waves are nondispersive. Let us show this by starting out with the dispersion relationship in Table 3.1 for finite water depth. It follows by expansion of $\tanh kh$ for small kh that

$$\frac{\omega^2}{g} = k^2 h \quad \text{when } kh \to 0.$$

This means

$$V_p = \frac{\omega}{k} = \sqrt{gh} \quad \text{when } kh \to 0,$$

which is independent of frequency.

Actually, we should note from eq. (3.22) that the wave energy propagation velocity is also equal to $\sqrt{gh}$ when $kh \to 0$. This means if we observe shallow-water waves, such as those outside the wave-breaking zone on a beach, we do not see that wave crests disappear, as they do at the front of a transient group of waves generated in a deep-water model basin. In the latter case, $V_g = 0.5V_p$ (see eq. (3.22)), so individual wave crests, traveling forward with V_p, outstrip the energy of the group that travels with half that velocity.

The fluid velocities will in general differ from both V_g and V_p. They depend on the wave amplitude, whereas V_p and V_g do not for linear waves. Because $V_p = g/\omega$ and $V_g = 0.5g/\omega$ for deep-water waves and small ω means long periods and wavelengths, this implies that longer waves travel faster than shorter waves in deep water. This can be used to concentrate wave energy or very steep waves at a given position in a ship model tank by gradually decreasing the forcing frequency of the wave maker. The first generated waves are then overtaken by the later generated waves. Some adjustments due to nonlinear effects are necessary.

3.2.4 Wave propagation from deep to shallow water

Our assumptions are

- Linear harmonic propagating waves based on potential flow for an incompressible fluid
- No surface tension
- No mean fluid velocity
- Fluid depth changes slowly on the length scale of the wavelength
- No wave reflection due to changing water depth
- Frequency of oscillation is constant
- No refraction, which means we consider a 2D problem in which the wave direction is normal to the bottom contours

We now use the previous theoretical results, exchange ζ_a with A, and use the subscript 0 for deep water. It follows from conservation of energy, see eq. (3.18), that

$$\left[\frac{1}{2} + \frac{kh}{\sinh 2kh}\right]\frac{\omega}{k}A^2 = \frac{1}{2}\frac{\omega}{k_0}A_0^2. \quad (3.25)$$

This means that

$$\frac{A}{A_0} = \left[\frac{k}{k_0}\left(\frac{1}{1+\frac{2kh}{\sinh 2kh}}\right)\right]^{1/2}, \quad (3.26)$$

where $k_0 = \frac{\omega^2}{g} = k \tanh kh$ or

$$k_0 h = kh \tanh kh \quad (3.27)$$

Further, by definition $k = 2\pi/\lambda$, $k_0 = 2\pi/\lambda_0$. We assume λ_0/h is given and solve eq. (3.27) for kh. We can calculate

$$\frac{\lambda}{\lambda_0} = \frac{k_0 h}{kh} \text{ as a function of } \frac{h}{\lambda_0}. \quad (3.28)$$

This is shown in Figure 3.8. We note that λ/λ_0 decreases with decreasing depth.

In Figure 3.9, we have plotted A/A_0 as a function of h/λ_0 We note that for $h/\lambda_0 > 0.05$, A/A_0 is less than one. However, A/A_0 will have a minimum for larger values of h/λ_0 than those shown in Figure 3.9. A/A_0 will then approach one when h/λ_0 goes to infinity. The value of A/A_0 increases very strongly for very small h/λ_0 and it is for instance two for $h/\lambda_0 = 0.0025$.

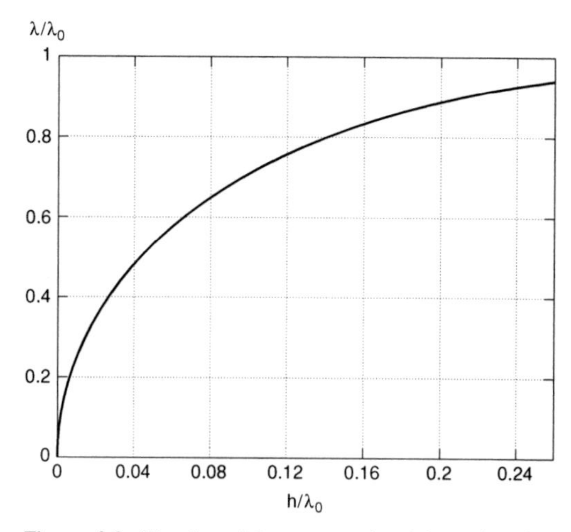

Figure 3.8. Wavelength λ at a water depth h as a function of wavelength λ_0 in deep water.

The steepness A/λ matters in evaluating qualitatively the importance of nonlinearities and the occurrence of breaking waves. Breaking waves imply both horizontal and vertical asymmetry of the wave profile. In deep water, it is common to set the wave-breaking limit as $H_0/\lambda_0 = 1/7$, where H_0 is the wave height. This is not appropriate for shallow water. Sorensen (1993) states that the wave-breaking limit depends on the relative depth h/λ and the beach slope. The ratio of the wave height to the water depth at breaking for common beach slopes and wave periods is between 0.8 and 1.2. It follows from Figure 3.8 and Figure 3.9 that $(A/\lambda)/(A_0/\lambda_0)$ increases with decreasing h/λ_0.

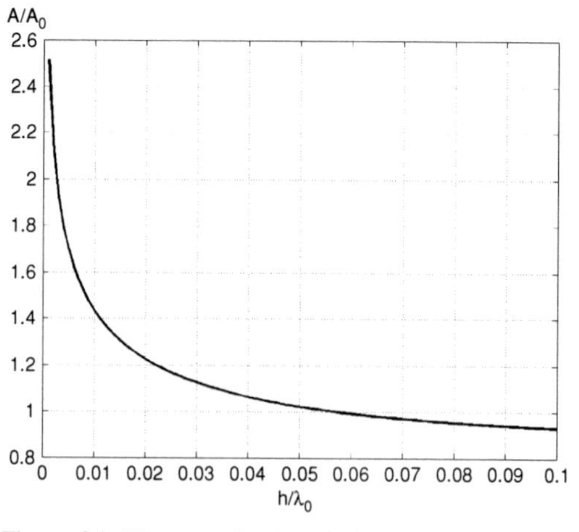

Figure 3.9. Wave amplitude ratio A/A_0 at a water depth h as a function of h/λ_0. A_0 and λ_0 are wave amplitude and wavelength in deep water.

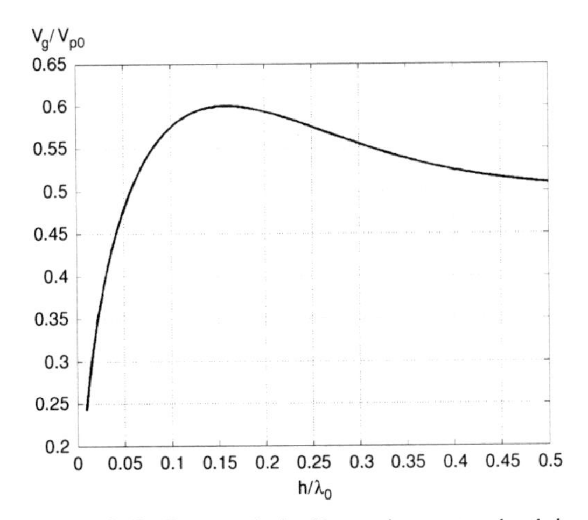

Figure 3.10. Group velocity V_g at a given water depth h divided by the deep-water phase velocity V_{p0} as a function of h/λ_0.

The group velocity V_g expresses the velocity of the wave front. This is presented nondimensionalized with respect to the phase velocity in infinite fluid depth in Figure 3.10. For small h/λ_0, we note that V_g decreases with depth.

Shallow water approximation

The previous expressions can be considerably simplified for shallow water, that is, $kh \to 0$. This follows by using a series expansion of the hyperbolic functions for small kh. Eq. (3.27) gives

$$k_0 h = (kh)^2 \quad \text{when } kh \to 0. \tag{3.29}$$

Eq. (3.26) gives then

$$\frac{A}{A_0} = \left(\frac{0.5}{(k_0 h)^{1/2}}\right)^{1/2} \quad \text{when } kh \to 0. \tag{3.30}$$

Eq. (3.28) gives

$$\frac{\lambda}{\lambda_0} = (k_0 h)^{1/2} \quad \text{when } kh \to 0. \tag{3.31}$$

Then what error do we make with these approximations? This is exemplified in Table 3.2. Roughly speaking, we may apply the shallow-water approximation for $h/\lambda_0 < 0.05$, that is, $h < \lambda_0/20$.

3.2.5 Wave refraction

Linear sinusoidal waves with a frequency ω in deep water are considered. There is no current. However, current will also have an influence (Mei 1983). When the waves approach a beach with a straight shoreline, the crests are nearly parallel to the shoreline even though the wave crests on deep water moving toward the beach may have a large angle relative to the coastline. This is called *wave refraction*. If we approximate the approaching waves as locally long crested and assume no wave reflection, we can explain the refraction by the phase speed V_p of two-dimensional (long-crested) linear harmonic waves. Because the number of crests reaching the beach per unit of time tends to be equal to the number approaching the coastline, we can assume the frequency ω to be a constant in the analysis. Using eq. (3.24) and that $k = 2\pi/\lambda$, gives then

$$\frac{V_p}{V_{p0}} = \frac{\lambda}{\lambda_0}. \tag{3.32}$$

Here the subscript 0 indicates deep water. Assuming as in section 3.2.4 that the fluid depth changes slowly on the length scale of the wavelength and there is no wave reflection, we can use eq. (3.32) along a ray orthogonal to the wave crest (Figure 3.11). Figure 3.8 shows by using eq. (3.32)

Table 3.2. *The validity of shallow-water approximation at different ratios between water depth h and wavelength λ_0 in deep water*

	$\frac{\lambda}{\lambda_0}$		$\frac{A}{A_0}$	
$\frac{h}{\lambda_0}$	Approximation	Exact	Approximation	Exact
0.1	0.79	0.71	0.79	0.93
0.05	0.56	0.53	0.94	1.02
0.025	0.40	0.39	1.12	1.17
0.01	0.25	0.25	1.41	1.43
0.005	0.18	0.18	1.68	1.69

A = wave amplitude. A_0 = wave amplitude in deep water. λ = wavelength.

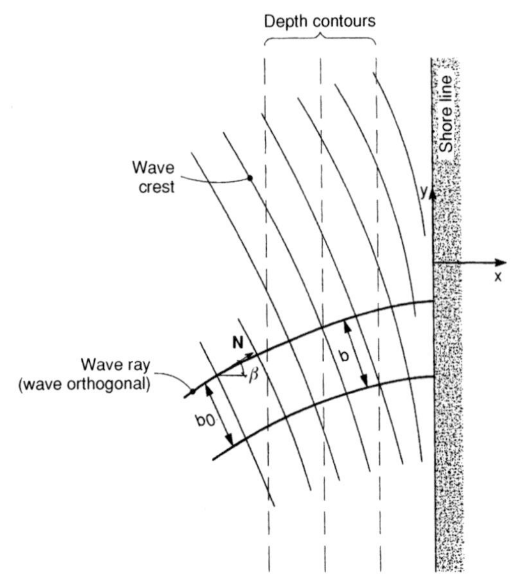

Figure 3.11. Wave refraction analysis. The depth contour is assumed to vary only with x in the figure, but the general analysis is not based on that. The distance b is measured along a wave crest between two nearby wave orthogonals (rays).

that V_p decreases as the water depth decreases. This means that those parts of a wave crest entering shallow water first are slowed down relative to those parts in deeper water. If we imagine the wave crest as a string, the deeper-water part of the string rotates relative to the shallow-water part and toward the beach.

Let us consider a wave refraction analysis in more detail. We express the wave elevation as in eq. (3.14), that is,

$$\zeta = A(x, y)\sin\left[\omega t - k(x, y)\left(x\cos\beta(x, y) + y\sin\beta(x, y)\right)\right], \tag{3.33}$$

where A, k, and β are slowly varying with x and y. A wave crest is defined by

$$\psi = \omega t - kx\cos\beta - ky\sin\beta \tag{3.34}$$

being equal to $(0.5 + 2n)\pi$, where n is an integer. If a line is described by $\psi(x, y) = c$ where c is a constant, we can show by vector analysis that $\nabla\psi$ is normal to the line. The proof is as follows. Let $\mathbf{r} = x\mathbf{i} + y\mathbf{j}$ be a position vector to any point on the line. We can write for x and y on the line that ψ does not vary, that is,

$$d\psi = \frac{\partial\psi}{\partial x}dx + \frac{\partial\psi}{\partial y}dy = 0. \tag{3.35}$$

This can also be expressed as $\nabla\psi \cdot d\mathbf{r} = 0$. Because $d\mathbf{r}$ in the limit $d\mathbf{r} \to 0$ is a vector tangentially to the line, this proves that $\nabla\psi$ is normal to the line. We can therefore define a normal vector $\mathbf{N}$ to a wave crest by

$$\mathbf{N} = \nabla\psi. \tag{3.36}$$

Assuming that $\partial k/\partial x$, $\partial k/\partial y$, $\partial\beta/\partial x$, and $\partial\beta/\partial y$ are small, we can write approximately

$$\mathbf{N} = -k(x, y)(\cos\beta(x, y)\mathbf{i} + \sin\beta(x, y)\mathbf{j}). \tag{3.37}$$

Then we use from vector analysis that $\nabla \times \nabla\psi = 0$ for any function ψ and approximate $\nabla\psi$ by eq. (3.37), that is,

$$\nabla \times (-k\cos\beta\mathbf{i} - k\sin\beta\mathbf{j}) = 0$$

or

$$\frac{\partial}{\partial x}\left(k(x, y)\sin\beta(x, y)\right) = \frac{\partial}{\partial y}\left(k(x, y)\cos\beta(x, y)\right). \tag{3.38}$$

In addition, it follows from the dispersion relationship (see Table 3.1) that

$$\frac{\omega^2}{g} = k(x, y)\tanh(k(x, y)h(x, y)). \tag{3.39}$$

Eq. (3.39) determines $k(x, y)$ for a given ω and $h(x, y)$. If we use the deep-water conditions as boundary (start) conditions, eq. (3.38) numerically determines $\beta(x, y)$.

The wave amplitude $A(x, y)$ follows from conservation of wave energy. Orthogonals (rays) are then constructed normal to the wave crests (Figure 3.11). The distance b measured along the wave crest between two nearby rays is assumed small. Because there is no energy flux through vertical planes coinciding with the rays, we can generalize the results in section 3.2.3 and write

$$\bar{E}V_g b = \bar{E}_0 V_{g0} b_0, \tag{3.40}$$

where, once more, the subscript 0 indicates deep water and $\bar{E}$ and V_g are given by eqs. (3.21) and (3.22), respectively.

Let us as a special case assume that the water depth is only a function of x. The wave propagation

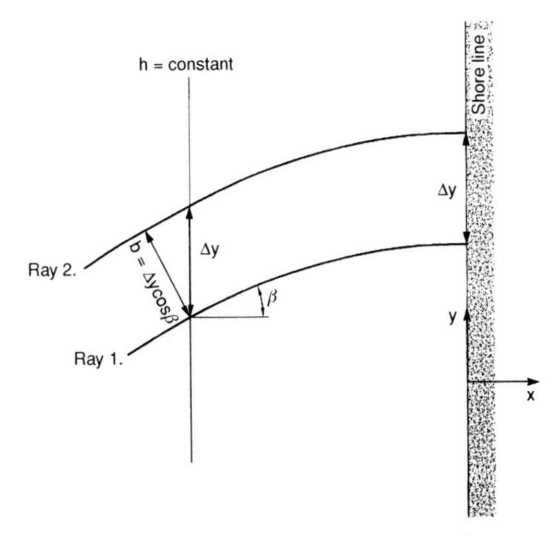

Figure 3.12. Sketch of two neighbor rays when the water depth varies only with x.

direction β will then also depend only on x. It follows from eq. (3.38) that

$$k(x)\sin\beta(x) = \text{constant}. \qquad (3.41)$$

Because V_p is equal to ω/k and ω is a constant, this gives

$$\frac{\sin\beta}{V_p} = \text{constant}. \qquad (3.42)$$

This is similar to Snell's law in optics. In order to find the wave amplitude, we can use eq. (3.40). We consider two neighbor rays as in Figure 3.12 and introduce the small distance Δy between the rays at the shore. Because there is no influence of y along a depth contour, Δy will also be the distance between the rays along a depth contour. Figure 3.12 illustrates that $b = \Delta y \cos\beta$. Using eq. (3.40) and eq. (3.21) with ζ_a equal to A gives

$$A^2 V_g \cos\beta = A_0^2 V_{g0} \cos\beta_0. \qquad (3.43)$$

Here V_g is given by eq. (3.22). Eq. (3.43) then determines A. If $\beta_0 = 0$, it has already been illustrated in Figure 3.9 how A depends on the water depth.

Figure 3.13 gives examples of wave refraction for more general depth contours where the shoreline is not a straight line. Cases A and C result in local concentration of wave energy density, whereas case B results in a decrease in the wave amplitude near the coastline.

A procedure such as this may be used to evaluate the effect of wash (waves) generated by a high-speed vessel (see Chapter 4). This implies that the dominant ship waves can be approximated as long-crested regular waves. Before starting the refraction analysis, we must transform the predicted wash from the body-fixed coordinate system to an Earth-fixed coordinate system. This transformation is discussed in Chapter 4. There is nothing in the previous procedure for harmonic waves that prohibits us from starting from a known condition in finite constant water depth instead of infinite water depth.

Because the previous analysis assumes linear waves, we can also consider an approaching sea composed of many frequency components. A first step then is to identify each frequency component and its wave propagation direction and then make a separate refraction analysis for each wave component. If we consider a stochastic sea described by a wave spectrum (see section 3.3), the wave

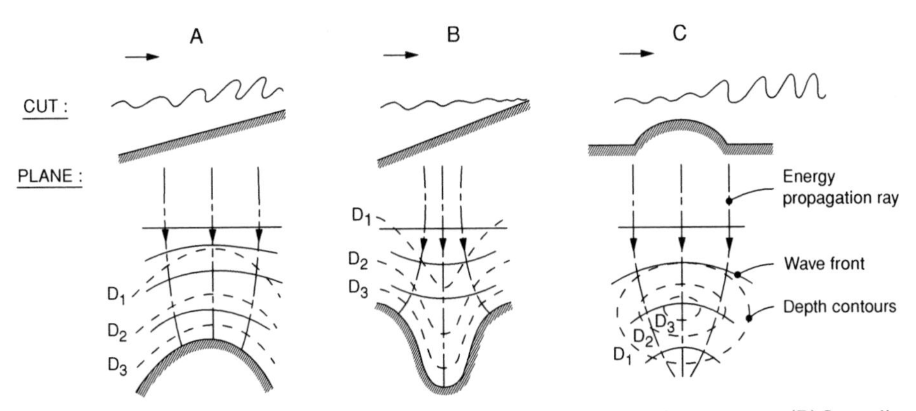

Figure 3.13. Three examples of depth refraction. (A) Focusing of rays and wave energy. (B) Spreading of rays and wave energy. (C) Focusing of rays and wave energy (Myrhaug 2004).

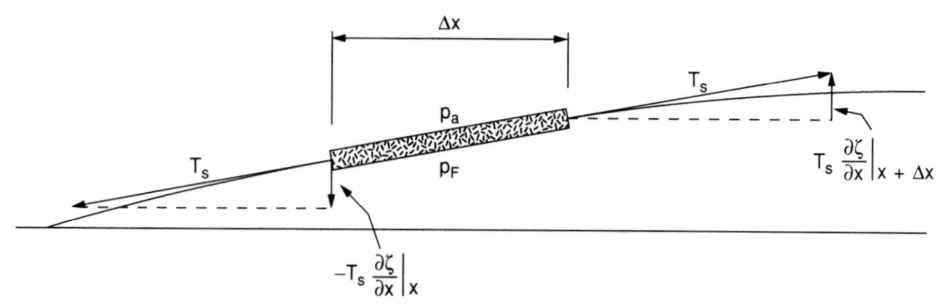

Figure 3.14. Vertical force component of the surface tension T_S acting on a free-surface element of length Δx and infinitesimal small thickness. A small wave slope $\partial\zeta/\partial x$ and 2D flow are assumed. p_F = fluid pressure. p_a = atmospheric pressure.

amplitude of each frequency component is determined by the wave spectrum. The phase of each regular wave component is random. The wave amplitude A for a given position (x, y) is found by

$$A^2 = \sum_{n=1}^{N} A_n^2. \tag{3.44}$$

Here N is the number of wave components and A_n is the wave amplitude predicted at position (x, y) for each wave component.

The linearity assumption obviously fails for breaking waves on a beach or when the wave rays focus and cause high waves outside the shoreline (see Figure 3.13c). A more accurate result for weakly nonlinear waves at small and moderate water depth–to-wavelength ratios h/λ can be obtained by directly solving the Boussinesq equations (Mei 1983 and Nwogu 1993). How large the h/λ we can apply to the Boussinesq equations depends on which version of these equations we use. For instance, the procedure by Nwogu (1993) may be applicable for h/λ up to approximately 0.3.

3.2.6 Surface tension

We shall assume a 2D flow situation in the x-z plane and study the effect of surface tension T_s per unit length when $U = 0$ and $p_0 = 0$. The wave slope $\partial\zeta/\partial x$ is assumed small. Figure 3.14 illustrates the vertical force due to the surface tension acting on a free-surface element of length Δx and infinitesimal thickness. The resultant vertical surface tension force is

$$T_s \left.\frac{\partial \zeta}{\partial x}\right|_{x+\Delta x} - T_s \left.\frac{\partial \zeta}{\partial x}\right|_{x} = T_s \frac{\partial^2 \zeta}{\partial x^2}\Delta x + O((\Delta x)^2).$$

The result must balance the sum of the pressure force $-p_a\Delta x$ due to atmospheric pressure p_a on the top of the free-surface element and the pressure force $p_F\Delta x$ due to hydrodynamic pressure p_F on the lower side of the free-surface element. This means

$$T_s \frac{\partial^2 \zeta}{\partial x^2} + p_F = p_a.$$

We express p_F by eq. (3.6) with $U = 0$ and linearize the free-surface condition as we did earlier. This leads to the dynamic free-surface condition

$$\frac{\partial \varphi}{\partial t} + g\zeta - \frac{T_s}{\rho}\frac{\partial^2 \zeta}{\partial x^2} = 0 \quad \text{on } z = 0. \tag{3.45}$$

The kinematic free-surface condition is similar to the one before and follows from eq. (3.9) with $U = 0$. We now combine the dynamic and kinematic free-surface condition by differentiating eq. (3.45) with respect to time and substitute $\partial\zeta/\partial t = \partial\varphi/\partial z$. This gives

$$\frac{\partial^2 \varphi}{\partial t^2} + g\frac{\partial \varphi}{\partial z} - \frac{T_s}{\rho}\frac{\partial^2}{\partial x^2}\frac{\partial \varphi}{\partial z} = 0 \quad \text{on } z = 0. \tag{3.46}$$

We now apply this to 2D propagating waves in deep water. The solution form for the velocity potential can be expressed in a way similar to the one in Table 3.1. This satisfies Laplace equation, decays with depth, and has the form of propagating waves. This means

$$\varphi = Be^{kz}\cos(\omega t - kx). \tag{3.47}$$

A difference now is that ω is not related to k as in Table 3.1. B can be expressed in terms of the wave amplitude from the use of eq. (3.45). We will instead focus on how ω is related to the wave

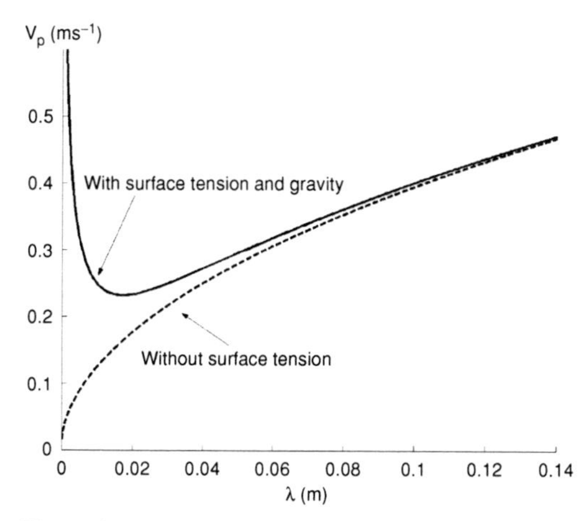

Figure 3.15. Effect of surface tension on 2D deep-water gravity waves. V_p = phase speed, λ = wavelength.

number $k = 2\pi/\lambda$. Substituting eq. (3.47) into eq. (3.46) gives

$$\omega^2 = gk + k^3 T_s/\rho. \tag{3.48}$$

When the wavelength is long, such that $k \ll \sqrt{g\rho/T_s}$, the effect of surface tension will be negligible. When the wavelength is very short, gravity may be neglected, that is,

$$\omega = \sqrt{T_s/\rho} \cdot k^{3/2}. \tag{3.49}$$

This corresponds to capillary waves or ripples. To exemplify when the surface tension matters, we have used eq. (3.48) and in Figure 3.15 plotted the phase speed $V_p = \omega/k$ as a function of λ for water with $T_s = 0.074$ Nm^{-1} and $\rho = 1000$ kgm^{-3}. The phase speed without surface tension is also plotted. Figure 3.15 demonstrates that surface tension does not really matter until λ is less than about 0.05 m. The group velocity for ripples follows from eq. (3.49) as

$$V_g = \frac{d\omega}{dk} = \frac{3}{2}\sqrt{\frac{T_s}{\rho}} k^{1/2}. \tag{3.50}$$

The phase velocity of ripples is

$$V_p = \frac{\omega}{k} = \sqrt{\frac{T_s}{\rho}} k^{1/2}. \tag{3.51}$$

This means that

$$V_g = \frac{3}{2} V_p \tag{3.52}$$

for 2D deep-water ripples. This is clearly different from deep-water gravity waves in which $V_g = 0.5V_p$. The consequence of this will be further discussed in Chapter 4 on steady ship waves.

3.3 Statistical description of waves in a sea state

In practice, linear theory is used to simulate irregular seas and to obtain statistical estimates. The wave elevation of a long-crested irregular sea propagating along the positive x-axis can be written as the sum of a large number of wave components, that is,

$$\zeta = \sum_{j=1}^{N} A_j \sin(\omega_j t - k_j x + \varepsilon_j). \tag{3.53}$$

Here, A_j, ω_j, k_j, and ε_j mean, respectively, the wave amplitude, angular frequency, wave number, and random phase angle of wave component number j. The random phase angles ε_j are uniformly distributed between 0 and 2π and constant with time. ω_j and k_j are related by the dispersion relationship (see Table 3.1). The wave amplitude A_j can be expressed by a wave spectrum $S(\omega)$ as

$$\frac{1}{2} A_j^2 = S(\omega_j)\Delta\omega, \tag{3.54}$$

where $\Delta\omega$ is a constant difference between successive frequencies. The instantaneous wave elevation is Gaussian distributed with zero mean and variance σ^2 equal to $\int_0^\infty S(\omega)d\omega$, which can be shown by using the definition of mean value and variance applied to the "signal" represented by eq. (3.53). We find, for instance, that $\sigma^2 = \sum_{j=1}^{N} A_j^2/2$. By using eq. (3.54) and letting $N \to \infty$ and $\Delta\omega \to 0$, we get $\sigma^2 = \int_0^\infty S(\omega)\, d\omega$. The relationship between a time-domain solution of the waves (that is, eq. (3.53)) and the frequency-domain representation of the waves by a wave spectrum $S(\omega)$ is illustrated in Figure 3.16.

The wave spectrum can be estimated from wave measurements (Kinsman 1965). It assumes that we can describe the sea as a stationary random process. This means in practice that we are talking about a limited time period in the range of one half to maybe ten hours. In the literature, this is often referred to as a short-term description of the sea.

Recommended sea spectra from the ISSC (International Ship and Offshore Structures Congress) and the ITTC (International Towing Tank Conference) are often used to calculate $S(\omega)$. For instance, for open-sea conditions, the 15th

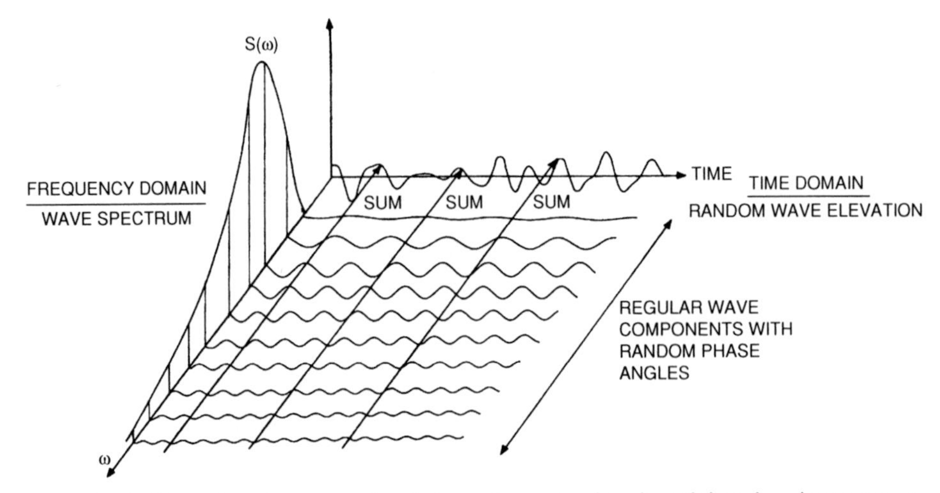

Figure 3.16. Illustration of the connection between frequency-domain and time-domain representations of waves in a long-crested short-term sea state. There are three "sum" lines drawn, illustrating how regular wave components with random phase angles at a given time add up to give the irregular wave elevation at that time instant.

ITTC recommended the use of ISSC spectral formulation for fully developed sea:

$$\frac{S(\omega)}{H_{1/3}^2 T_1} = \frac{0.11}{2\pi}\left(\frac{\omega T_1}{2\pi}\right)^{-5} \exp\left[-0.44\left(\frac{\omega T_1}{2\pi}\right)^{-4}\right], \tag{3.55}$$

where $H_{1/3}$ is the significant wave height defined as the mean of the one-third–highest waves and T_1 is a mean wave period defined as

$$T_1 = 2\pi m_0 / m_1.$$

where the spectrum moments, m_k, are given by

$$m_k = \int_0^\infty \omega^k S(\omega)\, d\omega; \mathrm{k} = 0, 1, 2 \ldots. \tag{3.56}$$

$H_{1/3}$ is often redefined as

$$H_{1/3} = 4\sqrt{m_0}, \tag{3.57}$$

giving a value that is usually close to the $H_{1/3}$ defined above.

Eq. (3.55) satisfies eq. (3.57). Strictly speaking, this relation is only true for a narrow-banded spectrum and when the instantaneous value of the wave elevation is Gaussian distributed.

The spectrum given by eq. (3.55) is the same as the modified Pierson-Moskowitz spectrum, in which it is more usual to use the mean wave period T_2 defined as

$$T_2 = 2\pi\,(m_0/m_2)^{1/2}. \tag{3.58}$$

The following relation exists between T_1 and T_2 for the spectrum given by eq. (3.55):

$$T_1 = 1.086 T_2. \tag{3.59}$$

The period T_0 corresponding to the peak frequency of the spectrum can be written as

$$T_0 = 1.408 T_2. \tag{3.60}$$

The peak period T_0 is also referred to as the modal period.

The spectrum formulation given by eq. (3.55) has little energy density when $\omega T_2/(2\pi)$ is less than 0.5 and larger than ≈1.5. For large frequencies, the wave spectrum decays like ω^{-5}.

The 17th ITTC recommended the following JONSWAP (Joint North Sea Wave Project) type spectrum for limited fetch:

$$S(\omega) = 155\frac{H_{1/3}^2}{T_1^4\omega^5}\exp\left(\frac{-944}{T_1^4\omega^4}\right)(3.3)^Y\ (\mathrm{m^2 s}), \tag{3.61}$$

where

$$Y = \exp\left(-\left(\frac{0.191\omega T_1 - 1}{2^{1/2}\sigma}\right)^2\right)$$

and

$$\sigma = 0.07 \quad \text{for } \omega \leq 5.24/T_1$$
$$= 0.09 \quad \text{for } \omega > 5.24/T_1.$$

The JONSWAP spectrum formulation may be used with the other characteristic periods by the substitution of

$$T_1 = 0.834T_0 = 1.073T_2. \quad (3.62)$$

We note that eq. (3.62) differs from the relationship between T_1, T_0, and T_2 given by eqs. (3.59) and (3.60) for the Pierson-Moskowitz spectrum.

The peak value of the modified Pierson-Moskowitz (ISSC) spectrum occurs at a different $(\omega T_2/2\pi)$-value than that for the JONSWAP spectrum. This can be seen from eqs. (3.60) and (3.62).

Tucker and Pitt (2001) have discussed how the parameters in the JONSWAP spectrum depend on limited fetch. We can, for instance, write

$$H_{1/3} = 0.0163F^{0.5}U_{10} \quad (3.63)$$

$$T_0 = 0.566F^{0.3}U_{10}^{0.4}. \quad (3.64)$$

Here U_{10} is the mean wind velocity in meter per second (ms^{-1}) 10 m above sea level. F is the fetch length in kilometers. The other dimensions are in SI units. We consider as an extreme case $U_{10} = 29.2\ \mathrm{ms}^{-1}$. If $F = 15$ km, this gives $H_{1/3} = 1.8$ m and $T_0 = 4.9$ s.

A good approximation to the probability density function for the wave amplitude maxima (peak values) A of the wave elevation can be obtained from the Rayleigh distribution given by

$$p(A) = \frac{A}{m_0} e^{-A^2/(2m_0)}, \quad (3.65)$$

where m_0 is related to $H_{1/3}$ by eq. (3.57). Strictly speaking, the Rayleigh distribution depends on the wave spectrum being narrow banded, which is an approximation for the spectra we have discussed. In deriving the Rayleigh distribution, it is also assumed that the instantaneous value of the wave elevation is Gaussian distributed.

We can simulate a seaway by using eq. (3.53), but this expression repeats itself after a time $2\pi/\Delta\omega$. Therefore, a large number N of wave components is needed to avoid this problem. A practical way to avoid this is to choose a random frequency in each frequency interval $(\omega_j - \Delta\omega/2, \omega_j + \Delta\omega/2)$ and calculate the wave spectrum with those frequencies. The number of wave components ought to be about 1000. This depends partly on the selection of the minimum and maximum frequency component. The minimum frequency component $\omega_{\min}$ is easier to select than the maximum frequency component $\omega_{\max}$. For instance, if a Pierson-Moskowitz spectrum is used, $\omega_{\min} \approx \pi/T_2$. The wave energy drops off more slowly for larger frequencies than for small frequencies. We should therefore investigate the results for different values of $\omega_{\max}$, to ensure that the results do not depend on the selection of $\omega_{\max}$.

We can use this procedure to simulate the wave elevation in a sea state with a given duration. The largest amplitude in each simulation (realization) is different because of the random selection of frequencies and phase angles. By selecting a large number of realizations, we will find that the extreme values have their own probability distribution. This was, for instance, discussed by Ochi (1982). In practice, the most probable largest value $A_{\max}$ is often used. This can be approximated as

$$A_{\max} = \left(2m_0 \ln \frac{t}{T_2}\right)^{1/2}, \quad (3.66)$$

where t is the time duration. We should note that $A_{\max}$ is the most probable largest value. With that we imply that there is a probability for $A_{\max}$ to be exceeded during the time t (Ochi 1982). The most probable maximum crest-to-trough wave height $H_{\max}$ during the same time is simply $2A_{\max}$.

The effect of short-crestedness may be important. A short-crested sea is often characterized by a two-dimensional wave spectrum, which in practice is often written as

$$S(\omega, \theta) = S(\omega)\, f(\theta), \quad (3.67)$$

where θ is an angle measuring the wave-propagating direction of elementary wave components in the sea. An example of $f(\theta)$ might be

$$f(\theta) = \begin{cases} \frac{2}{\pi}\cos^2\theta, & -\pi/2 \le \theta \le \pi/2 \\ 0; & \text{elsewhere} \end{cases}, \quad (3.68)$$

where $\theta = 0$ corresponds to the main wave-propagation direction. Other ways of representing a short-crested sea spectrum may be found in the report of the 10th ISSC. For a short-crested sea, eq. (3.53) can be generalized to

$$\zeta = \sum_{j=1}^{N}\sum_{k=1}^{K} \left(2S(\omega_j, \theta_k)\,\Delta\omega_j\Delta\theta_k\right)^{1/2} \times \sin\left(\omega_j t - k_j x\cos\theta_k - k_j y\sin\theta_k + \varepsilon_{jk}\right). \quad (3.69)$$

Table 3.3. *Classification of operation of high-speed craft according to DNV Rules for High-Speed and Light Craft, January 1999*

R0 Ocean	Long ocean–service restriction applies to craft on long international voyages when the craft is to be self-sustained without rescue assistance.
R1 Ocean	Short ocean–service restriction applies to craft on short international voyages on which the craft is assumed to be outside the range of rescue assistance from shore, other than from helicopter. The craft is assumed to be outside the possibility of seeking shelter if weather forces are above sea state 6.
R2 Offshore	Offshore service restriction applies to operation in water when weather conditions may change during the voyage, but in cases in which the possibility of seeking shelter at the coast exists if the weather forecast estimates more than sea state 6. Rescue assistance from shore is assumed.
R3 Coastal	Coastal service restriction applies to operation in waters in which the craft may be exposed to sea state 5 for part of the voyage, but in which the master can decide to change the voyage depending on weather conditions. Distance away from port of refuge is limited, and weather is not assumed to change between alternative places of refuge.
R4 Inshore	Inshore service restriction applies to operation in waters that are protected by islands and/or peninsulas. Rough sea is not expected, and close distance to place of refuge is assumed. Short voyages and good possibility of alternative places of refuge are assumed.
R5 Inland	Inland service restriction applies to operation on lakes, rivers, channels, and harbor areas where the sea is calm and the distance to refuge is short. Each voyage may stop in a few minutes, and alternative places of refuge are assumed to be numerous.

3.4 Long-term predictions of sea states

So far, we have discussed a "short-term" description of the sea, which means the significant wave height and the mean wave period are assumed constant during the time period considered. The significant wave height and mean wave period will vary in a "long-term" description of the sea. In order to construct a long-term prediction of the sea, we need to know the joint frequency of the significant wave height and the mean wave period. This discrete joint frequency distribution is referred to as a scatter diagram.

High-speed craft normally operate with a limitation on how far from shore they are allowed to go. This is expressed as an addition to the class notation issued by a classification society. DNV's service restrictions range from **R0** to **R5**. Typical operation for the different service restrictions is shown in Table 3.3. There is no direct link between the service restriction and the sea state a vessel may experience. A vessel operating at an open coastline may experience both high and low sea states, even if it is operating close to shore. It is, however, expected that a vessel designed for service restriction **R0** (ocean service) will during its lifetime experience higher sea states than will a vessel designed for service restriction **R3** (coastal service). This is reflected in the scatter diagrams. The scatter diagrams are "constructed" from theoretical parameters and are presented in DNV Report No. 97-0152.

We present in Table 3.4 the recommended scatter diagram for service restriction **R0**. The numbers in the table should first be summed up. This is 995, as we see in the table. Then, for instance, there is a number 44 for $H_{1/3} = 3$ m and $T_2 = 7$ s. Here $H_{1/3} = 3$ m means values between $H_{1/3} = 2.5$ m and 3.5 m. Further, $T_2 = 7$ s means values between $T_2 = 6.5$ s and 7.5 s. The probability of having a significant wave height between 2.5 m and 3.5 m and a mean wave period T_2 between 6.5 s and 7.5 s is then $44/995 = 0.044$. The footnote for Table 3.4 states that there is a 10% probability for exceedance of $H_{1/3} = 6$ m. We obtain this by using the values in the right column as follows. We use half the value for $H_{1/3} = 6$ m, that is, 26. Then we add the values for $H_{1/3} > 6$ m. The sum is then divided by the total number, 995. This means $(26 + 31 + 20 + 12 + 6 + 4)/995 = 0.1$ or 10% probability for exceedance of $H_{1/3} = 6$ m.

For the directional distribution of waves, the DNV rules say, "It may be assumed that the vessel's heading relative to the waves is equally

Table 3.4. *Scatter diagram for service restriction **R0** (ocean service)* [a]

$H_{1/3}$ (m)	T_2 (seconds)																
	1	2	3	4	5	6	7	8	9	10	11	12	13	14	15	16	Sum
1	0	0	1	22	67	84	63	35	16	6	2	1	0	0	0	0	297
2	0	0	0	2	17	48	60	44	23	10	4	1	0	0	0	0	209
3	0	0	0	0	4	22	44	45	29	13	5	2	1	0	0	0	165
4	0	0	0	0	1	8	25	34	27	15	6	2	1	0	0	0	119
5	0	0	0	0	0	3	12	22	21	13	6	2	1	0	0	0	80
6	0	0	0	0	0	1	6	12	14	10	6	2	1	0	0	0	52
7	0	0	0	0	0	0	2	6	9	7	4	2	1	0	0	0	31
8	0	0	0	0	0	0	1	3	5	5	3	2	1	0	0	0	20
9	0	0	0	0	0	0	0	2	3	3	2	1	1	0	0	0	12
10	0	0	0	0	0	0	0	1	1	2	1	1	0	0	0	0	6
11	0	0	0	0	0	0	0	0	1	1	1	1	0	0	0	0	4
12	0	0	0	0	0	0	0	0	0	0	0	0	0	0	0	0	0
13	0	0	0	0	0	0	0	0	0	0	0	0	0	0	0	0	0
Sum	0	0	1	24	89	166	213	204	149	85	40	17	7	0	0	0	995

[a] 10% probability for exceedence of $H_{1/3} = 6$ m.

distributed along the circle, i.e., an equal amount of time on each heading." It is also necessary to define running hours per year (see DNV Report No. 97-0152).

The sea state number is often used to classify the sea. This gives a more rough description of the relationship between $H_{1/3}$ and mean wave period. This is shown in Table 3.5 and also gives information about the wind speed.

3.5 Exercises

3.5.1 Fluid particle motion in regular waves

a) Use linear theory to show that the fluid particle motion $(x(t), z(t))$ in regular sinusoidal propagating waves in deep water can be expressed as

$$x(t) - x_0 = -\zeta_a e^{kz_0} \cos(\omega t - kx_0) \qquad (3.70)$$

$$z(t) - z_0 = \zeta_a e^{kz_0} \sin(\omega t - kx_0) \qquad (3.71)$$

when the fluid velocity is given as in Table 3.1. Here x_0 and z_0 are the coordinates of a fluid particle in the absence of any fluid motion.

b) Show that the fluid particle motion given by eqs. (3.70) and (3.71) represents a circular path with radius $\zeta_a \exp(kz_0)$. Discuss in what direction a fluid particle moves along this circular path and relate this motion to the free-surface elevation.

c) Make a numerical simulation of the fluid particle motion by using the linear expression for fluid velocity in deep-water regular waves at the instantaneous position of the fluid particle. Choose $\zeta_a = 2$ m, $\lambda = 100$ m and pick your own fluid particle.

d) Show both analytically and based on the result of question c) that a fluid-particle has a mean drift velocity

$$\bar{u}_D = \zeta_a^2 \omega k e^{2kz_0} \qquad (3.72)$$

in the wave propagation direction. Compare the magnitude of the fluid particle drift velocity, the fluid velocity amplitude at (x_0, z_0), the group and phase velocity for the case studied in question c). (Hint: You can show eq. (3.72) by Taylor expanding the linear fluid velocity at (x, z) about (x_0, z_0) and keeping terms of order $(\zeta_a/\lambda)^2$.)

e) Consider finite water depth and generalize eqs. (3.70) and (3.71). Show that a fluid particle follows the elliptical path given by

$$\left(\frac{x(t) - x_0}{a}\right)^2 + \left(\frac{z(t) - z_0}{b}\right)^2 = 1, \qquad (3.73)$$

where

$$a = \zeta_a \frac{\cosh k(z_0 + h)}{\sinh kh} \qquad (3.74)$$

$$b = \zeta_a \frac{\sinh k(z_0 + h)}{\sinh kh}. \qquad (3.75)$$

Discuss the fluid particle motion based on eq. (3.73) when $kh \to 0$.

Table 3.5. *Annual sea state occurrence in the open ocean of the North Atlantic and North Pacific (Lee and Bales 1984)*

	Significant wave height (m)		Sustained wind speed (knots)[a]		North Atlantic			North Pacific		
					Percentage probability of sea state	Modal wave period (s)		Probability of sea state (%)	Modal wave period (s)	
Sea state no.	Range	Mean	Range	Mean		Range[b]	Most probable[c]		Range[b]	Most probable[c]
0–1	0–0.1	0.05	0–6	3	0.70	–	–	1.30	–	–
2	0.1–0.5	0.3	7–10	8.5	6.80	3.3–12.8	7.5	6.40	5.1–14.9	6.3
3	0.5–1.25	0.88	11–16	13.5	23.70	5.0–14.8	7.5	15.50	5.3–16.1	7.5
4	1.25–2.5	1.88	17–21	19	27.80	6.1–15.2	8.8	31.60	6.1–17.2	8.8
5	2.5–4	3.25	22–27	24.5	20.64	8.3–15.5	9.7	20.94	7.7–17.8	9.7
6	4–6	5	28–47	37.5	13.15	9.8–16.2	12.4	15.03	10.0–18.7	12.4
7	6–9	7.5	48–55	51.5	6.05	11.8–18.5	15.0	7.00	11.7–19.8	15.0
8	9–14	11.5	56–63	59.5	1.11	14.2–18.6	16.4	1.56	14.5–21.5	16.4
>8	>14	>14	>63	>63	0.05	18.0–23.7	20.0	0.07	16.4–22.5	20.0

[a] Ambient wind sustained at 19.5 m above the surface to generate fully developed seas. To convert to another altitude, H_2, apply $V_2 = V_1(H_2/19.5)^{1/7}$.
[b] Minimum is 5 percentile and maximum is 95 percentile for periods given wave height range.
[c] Based on periods associated with central frequencies included in Hindcast Climatology.

3.5.2 Sloshing modes

Consider a rectangular tank with length a and breadth b. The mean water depth is h. Define a Cartesian coordinate system with origin in the center of the mean free surface in the tank. The z-axis is positive upward. The linear natural modes are found by solving the Laplace equation, using the linear free-surface conditions, and requiring the normal fluid velocity to be zero on the tank bottom and the sidewalls.

a) Show that the natural modes of the velocity potential can be expressed as

$$\varphi_{mn} = \cos\left[\frac{m\pi}{a}\left(x+\frac{a}{2}\right)\right]\cos\left[\frac{n\pi}{b}\left(y+\frac{b}{2}\right)\right] \times \cosh k_{mn}(z+h)\cos(\omega_{mn}t+\varepsilon_{mn}), \tag{3.76}$$

where

$$k_{mn}^2 = \pi^2\left(\frac{m^2}{a^2}+\frac{n^2}{b^2}\right) \tag{3.77}$$

and m and n are integers ranging from 0 to ∞. Further, show by neglecting surface tension that the natural frequencies ω_{mn} associated with the natural modes are

$$\omega_{mn}^2 = gk_{mn}\tanh k_{mn}h. \tag{3.78}$$

Express the free-surface elevation ζ_{mn} associated with a natural mode.

b) Consider now a 2D tank with length 20 m and water depth $h = 10$ m. Discuss when surface tension influences the natural modes.

c) Consider a square-base basin, that is, $a = b$. We define

$$f_m^{(1)}(x) = \cos\left[\frac{m\pi}{a}\left(x+\frac{a}{2}\right)\right],$$
$$f_n^{(2)}(y) = \cos\left[\frac{n\pi}{a}\left(y+\frac{a}{2}\right)\right]. \tag{3.79}$$

Why can

$$\left[f_1^{(1)}(x) \pm f_1^{(2)}(y)\right]\cos\omega t \tag{3.80}$$

be called diagonal standing waves?

Why can

$$f_1^{(1)}(x)\cos\omega t \pm f_1^{(1)}(y)\sin\omega t \tag{3.81}$$

be called swirling waves, that is, rotating waves that you may see if you excite your coffee cup?

What does ω mean in eqs. (3.80) and (3.81)?

Draw sketches of the 3D wave patterns in answering the questions. Discuss the meaning of $\pm$ in eqs. (3.80) and (3.81).

3.5.3 Second-order wave theory

We will consider regular sinusoidal propagating waves in infinite depth. The linear (first-order) results are presented in Table 3.1. We represent the velocity potential φ and the free-surface elevation ζ as

$$\varphi = \varphi_1 + \varphi_2 + \cdots \tag{3.82}$$
$$\zeta = \zeta_1 + \zeta_2 + \cdots, \tag{3.83}$$

where φ_1 and ζ_1 are the first-order (linear) results given in Table 3.1. φ_2 and ζ_2 are second-order approximations that are assumed proportional to the square of the linear incident wave amplitude divided by the incident wavelength.

a) Use the dynamic free-surface condition and show by a Taylor expansion about the mean free surface $z = 0$ that

$$g\zeta_2 = -\frac{\partial\varphi_2}{\partial t} - \frac{1}{2}\left[\left(\frac{\partial\varphi_1}{\partial x}\right)^2 + \left(\frac{\partial\varphi_1}{\partial z}\right)^2\right] - \zeta_1\frac{\partial^2\varphi_1}{\partial t\partial z} \quad \text{on} \quad z=0. \tag{3.84}$$

b) Express the kinematic free-surface condition correctly to second order. Combine this with the dynamic free-surface condition and show that

$$\frac{\partial^2\varphi_2}{\partial t^2} + g\frac{\partial\varphi_2}{\partial z} = -\frac{\partial}{\partial t}\left[\left(\frac{\partial\varphi_1}{\partial x}\right)^2 + \left(\frac{\partial\varphi_1}{\partial z}\right)^2\right] + \frac{1}{g}\frac{\partial\varphi_1}{\partial t}\frac{\partial}{\partial z}\left(\frac{\partial^2\varphi_1}{\partial t^2} + g\frac{\partial\varphi_1}{\partial z}\right) \quad \text{on} \quad z=0. \tag{3.85}$$

c) Show that the second-order potential is zero and that ζ_2 can be expressed as in eq. (3.13).

d) Draw pictures as in Figures 3.4 and 3.5 to show horizontal velocity distribution and pressure variation under a wave crest and wave trough that are consistent with the derivation of the second-order wave theory.

e) We consider two linear long-crested wave fields propagating in the same direction in deep water. The velocity potential is expressed as

$$\varphi_1 = \frac{ga_1}{\omega_1}e^{k_1 z}\cos(\omega_1 t - k_1 x + \delta_1) + \frac{ga_2}{\omega_2}e^{k_2 z} \times \cos(\omega_2 t - k_2 x + \delta_2). \tag{3.86}$$

Show that

$$\varphi_2 = \frac{2a_1a_2\omega_1\omega_2(\omega_1-\omega_2)}{-(\omega_1-\omega_2)^2 + g|k_1-k_2|}e^{|k_1-k_2|z} \times \sin[(\omega_1-\omega_2)t - (k_1-k_2)x + \delta_1 - \delta_2] \tag{3.87}$$

is a solution of the second-order boundary-value problem.

f) Consider then N linear wave components propagating in the same direction in deep water. Write up the expression for the second-order potential.

3.5.4 Boussinesq equations

Nwogu (1993) has presented extended Boussinesq-type equations. If we assume constant water depth h and linearize these equations, the following dispersion relationship can be derived for harmonic waves with frequency ω and wave number k:

$$\frac{\omega^2}{k^2} = gh\frac{1-\left(\alpha+\frac{1}{3}\right)(kh)^2}{1-\alpha\,(kh)^2}. \tag{3.88}$$

Here

$$\alpha = \frac{z_a}{h} + \frac{1}{2}\left(\frac{z_a}{h}\right)^2 \tag{3.89}$$

and z_a is a z-coordinate between 0 and $-h$ that is a variable that must be determined.

Use the dispersion relationship presented in Table 3.1 and the group velocity given in eq. (3.22) as the true values for finite depth. Make graphs of the relative error in using eq. (3.88) to predict the phase velocity and the group velocity as a function of the water depth–to-wavelength ratio h/λ for different choices of z_a/h. Which choice of z_a/h seems to be best if one wants to use these Boussinesq equations for $0 < h/\lambda < 0.3$?

3.5.5 Gravity waves in a viscous fluid

We consider linear long-crested propagating gravity waves in a viscous fluid with infinite depth. Laminar flow is assumed. A similar problem is considered by Landau and Lifschitz (1959).

a) Show based on the Navier-Stokes equations for an incompressible fluid that the following linearized equations apply:

$$\frac{\partial u}{\partial t} = -\frac{1}{\rho}\frac{\partial p}{\partial x} + \nu\left(\frac{\partial^2 u}{\partial x^2} + \frac{\partial^2 u}{\partial z^2}\right) \tag{3.90}$$

$$\frac{\partial w}{\partial t} = -\frac{1}{\rho}\frac{\partial p}{\partial z} + \nu\left(\frac{\partial^2 w}{\partial x^2} + \frac{\partial^2 w}{\partial z^2}\right) - g, \tag{3.91}$$

where (u, w) is the fluid velocity and p is the pressure. Further, the continuity equation is

$$\frac{\partial u}{\partial x} + \frac{\partial w}{\partial z} = 0. \tag{3.92}$$

The coordinate system is defined in Figure 3.2 What condition must be satisfied for the fluid behavior when $z \to -\infty$?

b) Show that the following solutions satisfy eqs. (3.90), (3.91) and (3.92):

$$u = e^{-i\omega t + ikx}\left(Ae^{kz} + Be^{mz}\right) \tag{3.93}$$

$$w = e^{-i\omega t + ikx}\left(-iAe^{kz} - \frac{ik}{m}Be^{mz}\right) \tag{3.94}$$

$$p/\rho = e^{-i\omega t + ikx}\cdot\frac{\omega A e^{kz}}{k} - gz + \frac{p_a}{\rho}, \tag{3.95}$$

where $m = (k^2 - i\omega/\nu)^{1/2}$ and p_a is the atmospheric pressure. Further, k and ω will be assumed real and complex, respectively.

c) Express the linear kinematic free-surface condition. Neglect surface tension and explain why the two dynamic free-surface conditions

$$-p + 2\mu\frac{\partial w}{\partial z} = p_a \quad \text{on } z = \zeta \tag{3.96}$$

$$\mu\left(\frac{\partial u}{\partial z} + \frac{\partial w}{\partial x}\right) = 0 \quad \text{on } z = 0 \tag{3.97}$$

are consistent with linear theory. Here ζ means the free-surface elevation. Differentiate eq. (3.96) with respect to time and apply the linear kinematic free-surface condition. Explain why we now, but not in eq. (3.96), can set $z = 0$ in this free-surface condition.

Show that the free-surface conditions lead to the following dispersion relationship:

$$\left(2 - \frac{i\omega}{\nu k^2}\right)^2 + \frac{g}{\nu^2 k^3} = 4\left(1 - \frac{i\omega}{\nu k^2}\right)^{1/2}. \tag{3.98}$$

Explain why the imaginary part of ω cannot be positive.

d) Assume that $\nu k^2 \ll (gk)^{1/2}$ and show that ω can be approximated as

$$\omega = \pm\sqrt{gk} - i2\nu k^2. \tag{3.99}$$

e) Consider water as the fluid and use eq. (3.99) to discuss how much the wave amplitude has decayed because of laminar viscous effects after 100 oscillation periods for wavelengths ranging from 5 cm to 100 m.

4 Wave Resistance and Wash

4.1 Introduction

In this chapter, we concentrate on wave resistance and ship-generated waves (wash) of high-speed vessels in calm water conditions. The focus is on semi-displacement vessels and air cushion–supported vehicles. Wave resistance of hydrofoils is discussed in section 6.8.

4.1.1 Wave resistance

Figure 4.1 shows the numerically calculated relative importance of resistance components of a 70 m–long catamaran in deep water. The main particulars are presented in Table 4.1. The ship speed U is 40 knots in calm water. The effect of head sea waves with different significant wave heights $H_{1/3}$ is also shown. The added resistance due to the incident waves and wind are accounted for by selecting a representative wind velocity and mean wave period for each $H_{1/3}$. Corresponding data for a 70 m–long surface effect ship (SES) are shown in Figure 4.2, with the main particulars given in Table 4.2. The ship power is kept constant in the calculations, which means increasing speed loss with increasing $H_{1/3}$. Voluntary speed reduction, for instance, as a result of excessive vertical accelerations, is not accounted for. The speed loss is most pronounced for the SES, as shown in Figure 4.2. The SES speed has dropped from 50 knots in calm water to about 35 knots in $H_{1/3} = 5$ m. The speed loss for the catamaran is not shown. However, because viscous frictional resistance is mainly proportional to U^2, the curve for frictional resistance as a function of $H_{1/3}$ shows that the speed loss is not large. Faltinsen et al. (1991b) report a study of an operational area with weather conditions similar to those of the northern North Sea. If the influence from different wave headings is neglected and the operational limits are disregarded, the mean vessel speed over one year for the 70 m–long catamaran is 37 knots, compared with 40 knots in calm water. Figure 4.1 for the catamaran shows that viscous frictional resistance is most important. The frictional resistance is divided into two components showing the influence of hull roughness. Even if the average hull roughness height is only 150 μm, the contribution matters. The air resistance is the smallest component. An obvious reason is that the resistance is proportional to the mass density of the fluid and the air density is the order of one thousandth of the water density. The added resistance in waves increases with $H_{1/3}^2$ for a given mean wave period and becomes significant for $H_{1/3} = 5$ m. The wave resistance, which is our main concern in this chapter, is more than 30% of the total resistance. The relative importance of the resistance components for an SES is different from that for a catamaran. The wave-making resistance (see Figure 4.2) is divided into two components. One is the result of the air cushion and the other of the side hulls. The excess pressure in the air cushion causes a mean depression of the free surface inside the hull relative to outside the hull. This results in generation of waves. Because the wave energy is directly related to the work done by the ship, this gives a wave-resistance component. The main effect of the side hulls is that the hulls push out the water in the bow and suck the water in at the stern. This causes changes in the free-surface elevation that result in far-field waves. Figure 4.2 shows that the wave-making resistance due to the air cushion is the most important resistance component in calm water. It is about 40% of total resistance. The significant drop in the air cushion–induced wave resistance as a function of $H_{1/3}$ is the result of air leakage caused by relative vertical wave-induced motions. The power of the fan-lifting system for the air cushion is assumed constant. The air leakage then implies lower excess pressure p_0 in the air cushion. We see this indirectly from the curve for cushion support in Figure 4.2. Cushion support expresses how much of the weight of the SES the air cushion carries. The other part is assumed to be caused by buoyancy of the side hulls. Because the importance of steady hydrodynamic force relative to hydrostatic pressure on the side hull increases with speed, this is an approximation. The cushion support drops from about 80% in calm water to less than 40% for $H_{1/3} = 5$ m. The wave resistance due to the air cushion is proportional to p_0^2. The

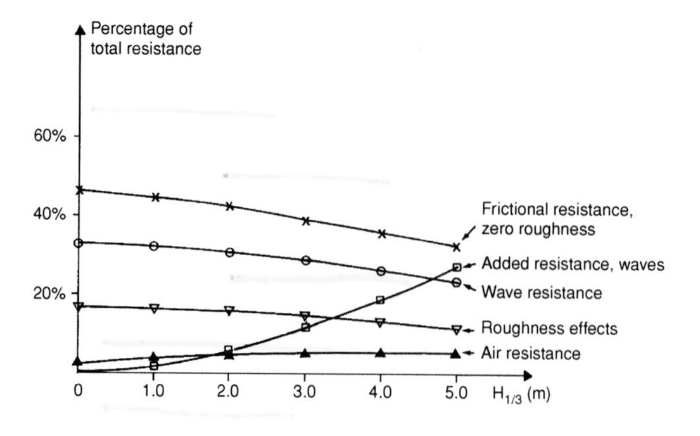

Figure 4.1. Relative importance of resistance components for a 70 m–long catamaran (see Table 4.1) in head sea waves. $H_{1/3}$ = significant wave height. Ship speed is 39.6 knots in calm water (Faltinsen et al. 1991b).

fact that lower p_0 causes lower air cushion wave resistance can also be understood from the fact that the mean depression of the free surface inside the air cushion is proportional to p_0.

A decrease in p_0 also means increased sinkage and wetted surface S of the vessel. The frictional resistance is proportional to S. The behavior of the frictional resistance as a function of $H_{1/3}$ can be explained by the increase in S and the decrease in U. Figure 4.2 shows that the air resistance is relatively more important for an SES than for a catamaran. The reason is related to the fact that the SES is lifted up. This means a larger projected transverse area in air than for the catamaran. It, of course, also has to do with the relative importance of the other resistance components. We note that the wave making due to the side hull of the SES is about 10% in calm water and clearly less than the wave resistance due to the air cushion. Roughly speaking, this has to do with the fact that the side hulls of the SES displace less water than does the air cushion. In this context, one should also remember that the buoyancy of the side hulls carries only about 20% of the weight in calm water.

The previous examples demonstrate that wave resistance is an important component. It is therefore of interest to understand what influences the wave resistance. This is dealt with later in the text. Another important effect of wave generation is that it influences the trim and sinkage of a high-speed vessel. Molland et al. (1996) did systematic experimental model tests with monohulls and catamarans showing a pronounced change in

Table 4.1. *Main particulars of a high-speed catamaran*

Length overall	70.0 m
Length of waterline	61.8 m
Beam of each hull	6.2 m
Draft of each hull	3.1 m
Distance between centerplanes of the hulls	14.0 m
Block coefficient of each hull	0.54
Pitch radius of gyration	15.0 m
Longitudinal position of center of gravity from FP	36.28 m
Projected area of above-water hull used in wind force calculations for head wind	125.0 m^2

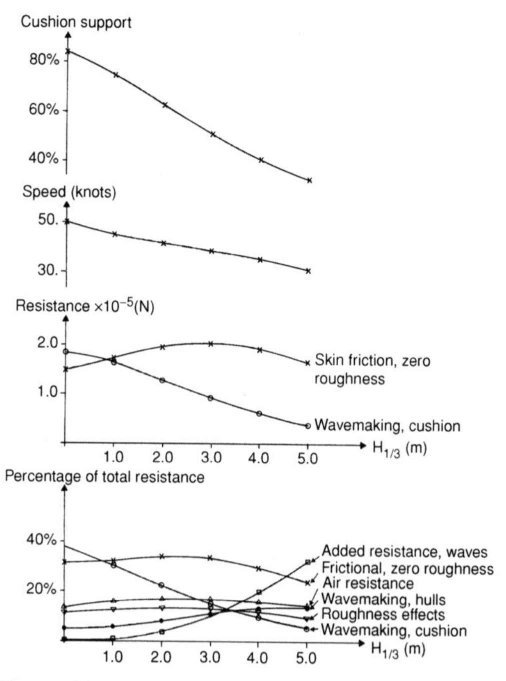

Figure 4.2. Cushion support and relative importance of resistance components for a 70 m–long SES (see Table 4.2) in head sea waves (Faltinsen et al. 1991b).

Table 4.2. *Main particulars of an SES*

Length overall	70.0 m
Length of waterline	60.9 m
Trim angle	0.82°
Longitudinal center of gravity from stern	26.5 m
Weight	1013.0 t
Cushion support, still water	83.3%
Cushion excess pressure, still water	0.0889 atm
Position of bow seal from stern	57.75 m
Pitch radius of gyration	18.2 m
Projected area of above-water hull used in wind force calculations for head wind	293.7 m^2

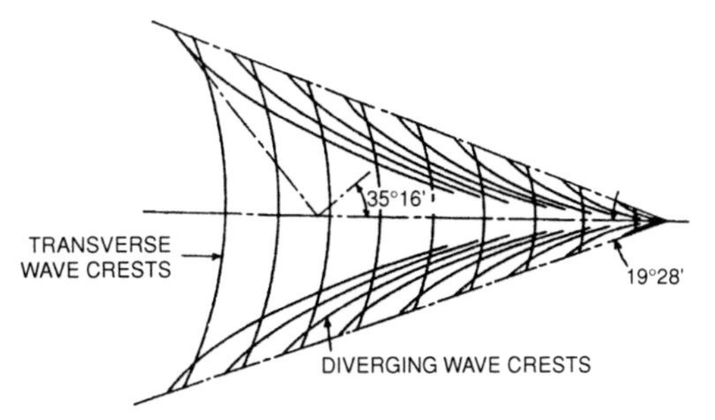

Figure 4.3. The Kelvin ship-wave pattern in deep water. The included half-angle 19°28′ of the waves is called the Kelvin angle and is affected by the water depth. (Newman, J. N., 1977, *Marine Hydrodynamics*, Cambridge: The MIT Press. The figure is reprinted with the permission of The MIT Press.)

trim and sinkage at Froude number $Fn = U/\sqrt{Lg}$ around 0.35. The trim and sinkage have a direct influence on the resistance. Trim and sinkage are also important factors to consider for wetdeck slamming on multihull vessels.

Theoretical predictions of wave resistance have a long history going back to Froude (1877) and Lord Kelvin (1887). The work by Michell (1898) is still very useful. Havelock (1908) studied how finite water depth affects ship waves. The many contributions by Havelock are collected in Havelock (1963). Reviews on the subject are given by Wehausen and Laitone (1962) and Wehausen (1973). More numerically oriented methods are needed in order to deal with important nonlinear aspects. This is discussed later in the text.

4.1.2 Wash

The waves (wash) caused by fast semi-displacement ships and SES are of concern in coastal and inland waters. A hydrofoil boat has an advantage in this respect. The wash may cause nearby small boats to capsize or ground or cause large moored ships to move and mooring lines to break. The waves may cause erosion or even collapse of a bank. When the waves approach a beach, the amplitudes increase and the waves break. This may happen when the ship is out of sight, surprise swimmers, and represent a risk factor.

The far-field waves in deep water can be classified as transverse and divergent waves (Figure 4.3). The angle between the boundary of the wave system and the ship course is 19°28′. This included half-angle of the waves is called the *Kelvin angle*. We will later see that it is affected by the water depth. The wavelength can be expressed as $\lambda = 2\pi U^2 \cos^2\theta/g$ where θ is the local wave propagation direction relative to the ship's course. It means that wavelength of the transverse waves along the ship's track can be expressed as

$$\frac{\lambda_T}{L} = 2\pi Fn^2. \tag{4.1}$$

A vessel is often categorized hydrodynamically to be a high-speed vessel if $Fn \geq 0.5$. This means that λ_T is larger than the ship length. The minimum θ-value for divergent waves occurs at the wave angle and is 35°16′ (see Figure 4.3). The corresponding wavelength λ_D is

$$\frac{\lambda_D}{L} = \frac{4\pi}{3} Fn^2. \tag{4.2}$$

The amplitude of the divergent waves is largest at the Kelvin angle. The transverse waves generated in the bow and at the stern of the ship may either tend to enforce each other or cancel each other behind the ship. This depends on λ_T/L and the particular ship. For instance, values around $\lambda_T/L = 0.5$ may mean large amplification of the wave height. This corresponds to $Fn = 0.28$. Large cancellation may happen around $\lambda_T/L = 1.0$ or $Fn = 0.4$. These strong amplification and cancellation effects cause humps and hollows in the wave resistance curve presented as a function of Froude number. However, the humps and hollows do not occur for $Fn > \approx 0.6 - 0.7$. Actually

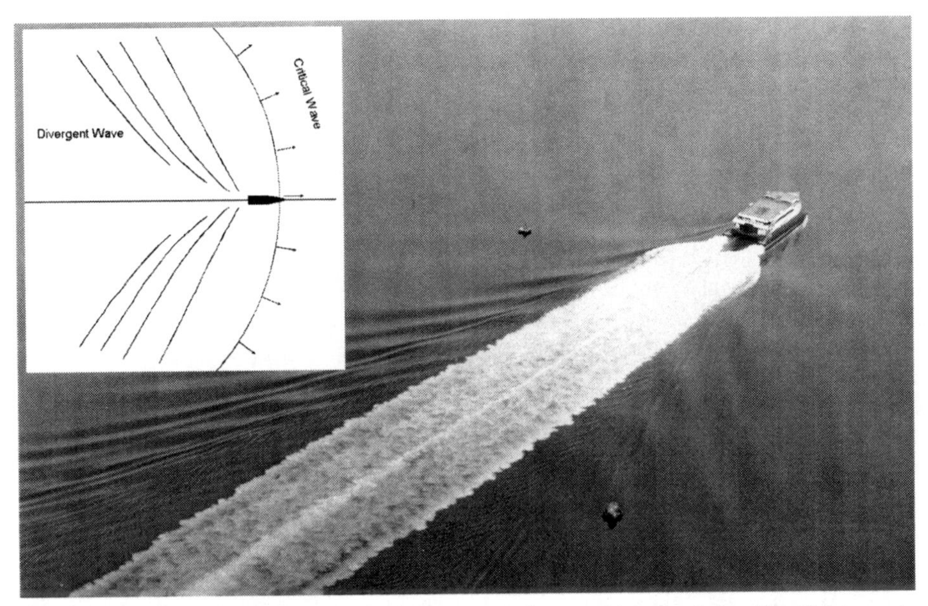

Figure 4.4. Waves near critical speed in shallow water due to a high-speed ship. The Kelvin angle (included half-angle of the waves) is close to 90°. Large waves are created at critical speed (Whittaker and Elsässer 2002).

the wave picture is dominated by divergent waves in the far field when Fn is larger than 0.6 to 0.7.

The wave decay along the distance $|y|$ perpendicular to the ship's track is important from a coastal engineering point of view. The waves at the Kelvin angle decay slowest with $|y|$. The decay rate is $|y|^{-1/3}$ in deep water. The leading waves in the wash of a high-speed vessel operating at maximum speed have long wavelengths. However, their amplitudes may not be large. When the waves come into shallow water, the wavelength decreases, and the wave amplitude increases. These waves may arrive unexpectedly at the shore and break after the fast vessel is out of sight.

When the ship is in finite water depth and the water depth–to-ship length ratio h/L is small, large changes occur near the critical depth Froude number $Fn_h = U/\sqrt{gh} = 1$. The wave resistance and the wave generation become large. Fast ships have sufficient power to overcome the critical speed. A first step in identifying which h/L and Fn_h cause large waves for a particular ship may be to study the ratio between wave resistance in finite and infinite depth. This was done by Yang (2002) by using linear theory. If either $Fn_h < \approx 0.6$ or $Fn_h > \approx 1.8$, the wave resistance ratio is close to 1. If $Fn_h \approx 1$, the ratios are larger than 3 when $h/L < \approx 0.2$. If $h/L < 0.13$ and $Fn_h \approx 1$, the ratios are larger than 15. The largest ratio predicted by Yang (2002) was 50. However, nonlinear and unsteady effects may then matter. Actually, large waves may be created upstream of the ship in confined water, such as in a channel.

If $h/L > \approx 0.4$, there is a very small effect of the water depth. This should be noted when doing model tests. For instance, at MARINTEK's facilities, the water depth is 5.5 m and the maximum towing speed is about $8\,\text{ms}^{-1}$. Then, the depth Froude number Fn_h may be larger than 1. Nevertheless if, for instance, a model length $L = 5.5$ m is used, then $h/L = 1$ and the model test results are not influenced by the tank bottom, even at critical- and supercritical-depth Froude numbers. This may seem surprising at first. It has to do with the fact that the lengths of the ship-generated waves are sufficiently small relative to the water depth. The ship then will not "feel" the tank bottom.

The ship generates both transverse and divergent waves at subcritical speed. However, the Kelvin angle, which is 19°28′ for deep water, is influenced by Fn_h. When $Fn_h < 0.5 - 0.6$, the value is practically the same as that for deep water. A rapid increase in the angle occurs for $Fn_h > 0.9$, and the angle is 90° for $Fn_h = 1$. We can imagine this from Figure 4.4, which shows a wave system near the critical speed. Only divergent waves exist for $Fn_h > 1$ when the waves "feel" the bottom.

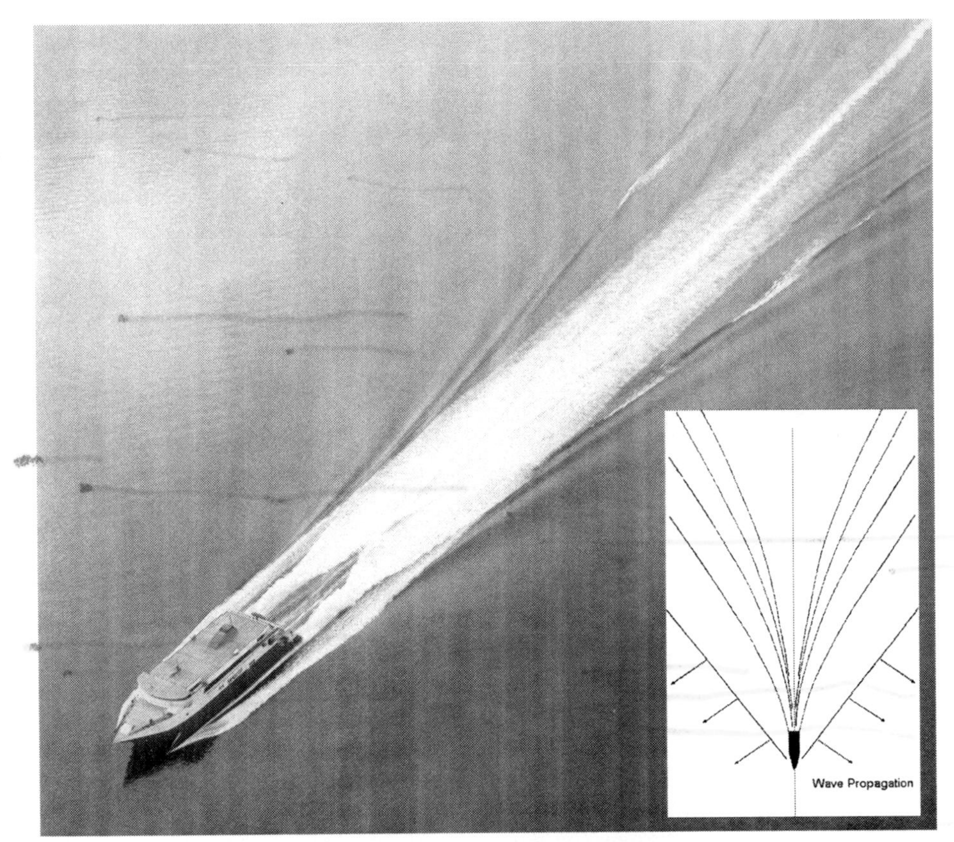

Figure 4.5. Waves due to a high-speed craft at supercritical speed in shallow water. There are no transverse waves at supercritical speed and the waves decay slowly perpendicular to the ship's track (Whittaker and Elsässer 2002).

Figure 4.5 illustrates a ship at supercritical speed. The decay rate for small h/L and supercritical speed is lower than for deep water waves at the wave angle. Decay rates as low as $|y|^{-0.2}$ were measured for shallow water by Doyle et al. (2001). Actually, the shallow-water theory by Tuck (1966) shows that the supercritical waves do not decay at all with $|y|$. Sections 4.4 and 4.5 give a detailed description of the effect of finite and shallow water depth.

There is no simple universal criterion in terms of maximum wave amplitude that quantifies the effect of wash. The criterion must be different for the effects on the seashore or if the effects on other ships are considered. If, for instance, the effect on other ships is analyzed, the ship response due to the wash of passing ships must be studied. This means that incident wavelengths, amplitudes, and relative directions all have to be considered. For instance, if the waves are very long relative to the ship length and they are not locally steep, the ship moves like a cork and the situation should not be of concern. The analysis for moored ships must include second-order mean and slowly varying drift forces (Faltinsen 1990). Proper knowledge about physical parameters influencing the wash will obviously enable the shipmaster to minimize the effect. Ferry operators in the United Kingdom must prepare a route assessment with regard to wash that must be approved by the Maritime and Coastguard Agency (Whittaker and Elsässer 2002).

In the following main text, we start out studying steady ship waves and wave resistance in deep water. We end up studying steady ship waves and wave resistance in finite and shallow water depth.

4.2 Ship waves in deep water

Figure 4.3 shows the wave pattern created by a ship moving by a constant speed on a straight course in a calm sea with infinite water depth and horizontal extent. There are two wave systems, categorized as divergent and transverse waves. The

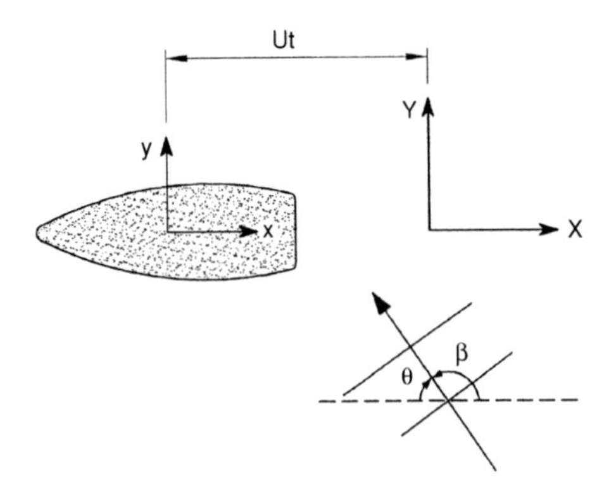

Figure 4.6. Ship-fixed coordinate system (x, y, z). Earth-fixed coordinate system (X, Y, Z). The z- and Z-axes are vertical upward. The origins of the coordinate systems are in the mean free surface. U = ship speed, β = wave propagation angle, θ = angle used in classification of ship waves, $-0.5\pi \leq \theta \leq 0.5\pi$. θ has a negative sign in the figure.

transverse waves have crests nearly perpendicular to the ship's course and have longer wavelengths relative to the divergent waves. The waves are confined in a domain aft of the ship. The angle between the outer boundary of the waves and the ship's course is 19°28′ (see Figure 4.3). The characteristics of the waves can be explained by representing the waves as an infinite sum of regular long-crested linear waves with different propagation directions. It is important to note that the wave system as observed from a ship-fixed coordinate system will be steady, that is, the wave system does not change with time. However, if the waves are observed in Earth-fixed coordinates, they are unsteady. Therefore, we define one ship-fixed coordinate system (x, y, z) and one Earth-fixed coordinate system (X, Y, Z) as shown in Figure 4.6. The x- and y-axes are in the mean free surface with the x-axis in the longitudinal aft direction of the ship. The z-axis is positive upward. We relate the coordinates as

$$x = X + Ut,\ y = Y,\ z = Z. \tag{4.3}$$

Here U is the ship's forward speed and t is the time variable. A regular wave profile with amplitude ζ_a propagating with an angle β relative to the X-axis can be expressed as

$$\zeta = \zeta_a \cos(kX\cos\beta + kY\sin\beta - \omega t - \varepsilon). \tag{4.4}$$

This follows from eq. (3.14), but one should not be confused that (X, Y, Z) has another meaning in deriving eq. (3.14). Further, we have added a constant phase angle ε in eq. (4.4) and changed sine with cosine. The wave number k in eq. (4.4) is related to the wavelength λ and the circular frequency ω by

$$k = \frac{2\pi}{\lambda} = \frac{\omega^2}{g}. \tag{4.5}$$

The phase velocity V_p, which expresses the velocity of the wave shape, is according to eq. (3.24):

$$V_p = \frac{\omega}{k} = \frac{g}{\omega}. \tag{4.6}$$

Further, eq. (3.23) gives that the wave energy propagation velocity or group velocity for deep water is

$$V_g = \frac{1}{2}\frac{\omega}{k} = \frac{1}{2}\frac{g}{\omega}. \tag{4.7}$$

V_g is the velocity of the front of the wave system. We will later use V_g to explain how far the waves generated by the ship can move relative to the ship.

Let us now transform the waves to the body-fixed coordinate system and also introduce an angle θ so that

$$\theta = \beta - \pi. \tag{4.8}$$

This gives

$$\begin{aligned}\zeta = \zeta_a \cos(&kx\cos\theta + ky\sin\theta \\ &+ (-kU\cos\theta + \omega)t + \varepsilon).\end{aligned} \tag{4.9}$$

The requirement that this wave form must be independent of time gives $kU\cos\theta = \omega$. We now use that $k = \omega^2/g$. This gives $\omega = 0$ as a possibility, but this is an uninteresting solution. We then get

$$\omega = \frac{g}{U\cos\theta}. \tag{4.10}$$

Because $\omega > 0$, this requires $\cos\theta$ to be positive or $-\pi/2 \leq \theta \leq \pi/2$. The angle θ is shown in Figure 4.6 together with β, which varies between $\pi/2$ and $3\pi/2$. The wave number k and the wavelength λ can now be expressed as

$$k = \frac{g}{U^2\cos^2\theta}, \quad \lambda = \frac{2\pi}{g}U^2\cos^2\theta. \tag{4.11}$$

Eq. (4.11) shows that $\lambda \to 0$ when $|\theta| \to \pi/2$. However, when λ becomes very small (less than 5 cm), surface tension matters (see Figure 3.15). These waves are so small they do not matter from a wave resistance point of view. If we approximate the divergent and transverse waves locally as

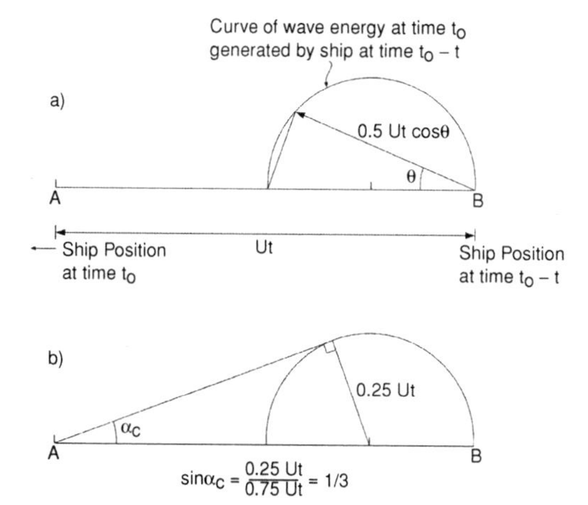

Figure 4.7. Simplified evaluation of the Kelvin angle α_C of a ship in deep water. $a_c = 19°28'$ as shown in Figure 4.3.

long-crested regular waves, the wavelength of transverse waves on the ship's track can be expressed as $\lambda_T = 2\pi U^2/g$. Nondimensionalizing λ_T by the ship length L gives

$$\lambda_T/L = 2\pi Fn^2. \tag{4.12}$$

This shows that the wavelength of the transverse waves is larger than the ship's length if $Fn > 0.4$. Eq. (4.11) also demonstrates that other wavelengths are shorter than λ_T. Eqs. (4.6), (4.7), and (4.10) make it possible for us to express the phase velocity V_p and group velocity V_g as

$$V_p = U\cos\theta, \quad V_g = \frac{1}{2}U\cos\theta. \tag{4.13}$$

4.2.1 Simplified evaluation of Kelvin's angle

Eq. (4.13) for the wave energy propagation velocity V_g now makes it possible to determine the Kelvin angle in a simple way. We first study Figure 4.7a. The ship is on a straight course with constant velocity U and is at position A at time t_0. We now look at the maximum distance the waves propagate from an arbitrary previous time $t_0 - t$ to time t_0. The distance can be determined by a circle, as shown in Figure 4.7a. We construct this circle by an arrow of length $V_g t = 0.5Ut\cos\theta$ out from B and note that the arrow length varies with θ. Then we draw a line from A that is tangent to the circle (Figure 4.7b). Because the time t is arbitrary, no waves due to the ship can be outside this line. It follows then by trigonometry and Figure 4.7b that the Kelvin angle α_c for deep water is

$$\sin\alpha_c = \frac{0.25Ut}{0.75Ut} = \frac{1}{3}. \tag{4.14}$$

So we note that all the waves are left behind the ship. A different situation would appear if we consider a very small model or actually small U or small wavelengths so that capillary waves dominate. We concentrate on the waves along the course of the model. Eq. (3.52) for the wave energy propagation velocity V_g for capillary waves shows this to be equal to $1.5V_p$, which is equal to $1.5U$. This means we see waves ahead of the model. You can do a simple experiment showing the effect. For instance, if you have a pencil at hand, put it vertically into water and draw it through the water.

4.2.2 Far-field wave patterns

Let us now focus on the complete gravity wave system generated by a ship. This is found by summing up all regular waves satisfying eqs. (4.9) to (4.11). This means

$$\zeta(x, y) = \int_{-\pi/2}^{\pi/2} |A(\theta)| \cos\left(\frac{g}{U^2\cos^2\theta} \times (x\cos\theta + y\sin\theta) + \varepsilon(\theta)\right) d\theta, \tag{4.15}$$

where $|A(\theta)|$ and $\varepsilon(\theta)$ are functions of the submerged ship shape. We will return to that later. We now write eq. (4.15) in complex form and introduce the polar coordinates (r, α) so that

$$x = r\cos\alpha, \quad y = r\sin\alpha. \tag{4.16}$$

This means

$$\zeta(r, \alpha) = \mathrm{Re}\int_{-\pi/2}^{\pi/2} A(\theta)\exp\left[i\frac{g}{U^2\cos^2\theta} r\cos(\theta - \alpha)\right] d\theta, \tag{4.17}$$

where now $A(\theta) = |A(\theta)|\exp(i\varepsilon)$ is complex and we can associate $A(\theta)\,d\theta$ with a complex wave amplitude for waves with propagation direction θ. When r is large relative to characteristic wavelength, eq. (4.17) results in the transverse and divergent wave system shown in Figure 4.3. How can we see that? This is not obvious. First of all we should note that the integrand of eq. (4.17) is highly oscillatory when r is large. This means

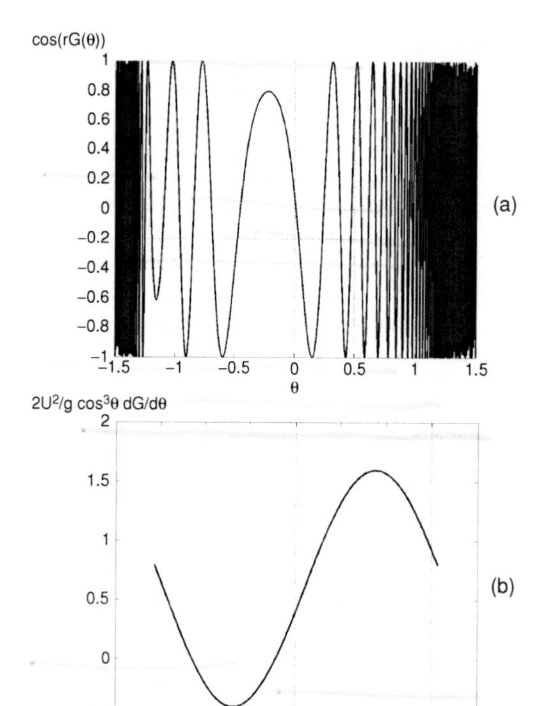

Figure 4.8. Determination of points of stationary phase for an integral with a highly oscillatory integrand.

there is a strong cancellation effect in the integration. We have illustrated this in Figure 4.8a by showing the integrand when $r/L = 10$, $\alpha = 0.2$ rad, $Fn = 0.5$, and $A(\theta) = 1$. Integrals like this can be evaluated by using the method of stationary phase, which is described in mathematical textbooks and in Stoker (1958) and Newman (1977).

If we express the exponential function in eq. (4.17) as $\exp[irG(\theta)]$, where

$$G(\theta) = \frac{g\cos(\theta - \alpha)}{U^2\cos^2\theta} \tag{4.18}$$

varies slowly, then the points of stationary phase correspond to solutions of $G'(\theta) = 0$. We can write

$$\begin{aligned}\frac{U^2}{g}\frac{dG}{d\theta} &= \frac{-\sin(\theta - \alpha)\cos^2\theta + 2\cos\theta\sin\theta\cos(\theta - \alpha)}{\cos^4\theta} \\ &= \frac{\sin(2\theta - \alpha) + 3\sin\alpha}{2\cos^3\theta}.\end{aligned} \tag{4.19}$$

This means that the points of stationary phase are the solution of

$$\sin(2\theta - \alpha) + 3\sin\alpha = 0. \tag{4.20}$$

Because $|\sin(2\theta - \alpha)| \leq 1$, the only possible solution to eq. (4.20) will be found for $|3\sin\alpha| \leq 1$, that is, that α is smaller than the Kelvin angle α_c. For $0 \leq \alpha \leq \alpha_c$, there exist two solutions to eq. (4.20), that is,

$$\begin{aligned}\theta_1 &= \frac{\alpha}{2} - \frac{1}{2}\sin^{-1}(3\sin\alpha) \\ \theta_2 &= -\frac{\pi}{2} + \frac{\alpha}{2} + \frac{1}{2}\sin^{-1}(3\sin\alpha).\end{aligned} \tag{4.21}$$

The function $dG(\theta)/d\theta$ is for the case in Figure 4.8 presented in Figure 4.8b. The zeros of $dG/d\theta$ are, according to eq. (4.21), $\theta_1 = -0.22$ rad and $\theta_2 = -1.15$ rad. The main contributions to the integral in eq. (4.17) come according to the method of stationary phase from the vicinity of θ_1 and θ_2. Further, by this method, eq. (4.17) can be approximated as

$$\zeta = \sum_{i=1}^{2} Re\left[A(\theta_i)\left(\frac{2\pi}{r|G''(\theta_i)|}\right)^{1/2} \times \exp\left\{i\left[rG(\theta_i) \pm \frac{\pi}{4}\right]\right\}\right],$$

where the $\pm$ sign corresponds to the sign of $G''(\theta_i)$. It follows by differentiating eq. (4.19) that

$$\frac{U^2}{g}\frac{d^2G}{d\theta^2}\bigg|_{\theta=\theta_i} = \frac{\cos(2\theta_i - \alpha)}{\cos^3\theta_i}.$$

This means

$$\frac{d^2G}{d\theta^2}\bigg|_{\theta=\theta_i} = \pm\frac{\sqrt{1 - 9\sin^2\alpha}}{|\cos^3\theta_i|}\frac{g}{U^2},$$

where + corresponds to θ_1 and – corresponds to θ_2. This means that for $0 \leq \alpha \leq \alpha_c$, the wave elevation is

$$\begin{aligned}\zeta = \sqrt{\frac{2\pi}{\nu r}}\left|1 - 9\sin^2\alpha\right|^{-1/4} & Re\left\{A(\theta_1)\,|\cos\theta_1|^{3/2} \times \exp\left[i\left[\nu r\frac{\cos(\theta_1 - \alpha)}{\cos^2\theta_1} + \frac{\pi}{4}\right]\right]\right. \\ & \left. + A(\theta_2)\,|\cos\theta_2|^{3/2} \times \exp\left[i\left[\nu r\frac{\cos(\theta_2 - \alpha)}{\cos^2\theta_2} - \frac{\pi}{4}\right]\right]\right\}\end{aligned} \tag{4.22}$$

where $\nu = g/U^2$. The solutions θ_1 and θ_2 given by eq. (4.21) are presented in Figure 4.9 as a function of the polar coordinate angle α for α between 0

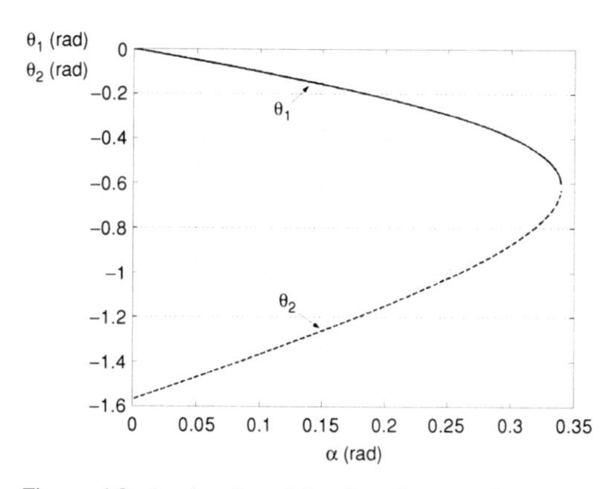

Figure 4.9. Angles θ_1 and θ_2 of stationary phase as a function of the polar coordinate angle α of a position inside the Kelvin angle $\alpha_c = 19°28' \approx 0.34$ rad. θ is defined in Figure 4.6. $-\theta_1$ and $-\theta_2$ are propagation angles for $y > 0$ of, respectively, transverse and divergent waves.

and the Kelvin angle α_c, that is, for positive y in Figure 4.6.

We note that θ_i is always negative and that $|\theta_1| < |\theta_2|$. $-\theta_1$ and $-\theta_2$ are propagation angles (see Figure 4.6) of respectively transverse and divergent waves. When $\alpha = \alpha_c$, $\theta_1 = \theta_2 = -\sin^{-1}(1/\sqrt{3}) = -35°16'$. This is the angle illustrated in Figure 4.3 for the divergent waves at the Kelvin angle. When $-\alpha_c < \alpha < 0$, we get similar solutions of θ_i, but with opposite signs. Eq. (4.22) shows that the wave amplitude decays like $r^{-1/2}$ when $0 \leq |\alpha| < \alpha_c$. When $|\alpha| = \alpha_c$, eq. (4.22) gives an infinite value for the amplitude. Actually, a separate analysis is needed in the close vicinity of $|\alpha| = \alpha_c$. This will show that the wave amplitude decays like $r^{-1/3}$. This decay factor or the fact that the wave amplitude decays as a function of horizontal distance $|y|$ from the ship's track is important from a coastal engineering point of view, that is, to assess the wash at the shore due to a passing ship. Because $y = r\sin\alpha$, the decay rate in terms of $|y|$ is the same as that in terms of r. One aspect is the decay rate, as discussed. Another aspect is how large the amplitudes of the waves are. We will later see that when $Fn > \approx 0.6$, the divergent waves will be dominant. The amplitude of the transverse waves is strongly dependent on the Froude number for $Fn < \approx 0.5$. This has to do with phasing between transverse waves generated in the bow and at the stern. Because the phasing has to do with the wavelength of the transverse waves, we can understand from eq. (4.12) that this is related to the Froude number. We explain this in more detail later. The conclusion of this discussion is that the divergent waves at the Kelvin angle are of primary concern from a wash point of view. It matters also from the seashore point of view that the wavelengths due to high-speed vessels are long. This can be illustrated by noting that the wavelength λ_c of the divergent waves at $\alpha = \alpha_c$ is according to eq. (4.11), given by

$$\frac{\lambda_c}{L} = \frac{4\pi}{3} Fn^2. \tag{4.23}$$

We discussed in sections 3.2.4 and 3.2.5 how waves are modified as they come close to the seashore.

We can find the wave crest positions inside the Kelvin angle as shown in Figure 4.3 by using eq. (4.22). If $A(\theta_i)$ are real, for the transverse waves they are given by

$$\nu r \frac{\cos(\theta_1 - \alpha)}{\cos^2\theta_1} + \frac{\pi}{4} = 2n\pi, \quad n = 0, 1, \ldots \tag{4.24}$$

Here θ_1 is given by eq. (4.21). Eq. (4.24) can be written as

$$r = \left(-\frac{\pi}{4} + 2n\pi\right) \frac{\cos^2\theta_1}{\nu\cos(\theta_1 - \alpha)}, \quad 0 \leq \alpha \leq \alpha_c \tag{4.25}$$

in polar coordinates (r, α). Similarly for the divergent waves, that is,

$$r = \left(\frac{\pi}{4} + 2n\pi\right) \frac{\cos^2\theta_2}{\nu\cos(\theta_2 - \alpha)}, \quad 0 \leq \alpha \leq \alpha_c \tag{4.26}$$

where θ_2, is given by eq. (4.21).

4.2.3 Transverse waves along the ship's track

We will now discuss how ship parameters influence the transverse waves along the ship's track. However, our theoretical model of how the ship generates waves is very simplified at this stage. The problem is analyzed from the ship reference system. This means that the ship speed appears as an incident flow with velocity U. We assume that the ship has a long parallel midpart with submerged cross-sectional area S. It is only at a small part in the bow and the stern where the ship cross section changes. Because a high-speed semi-displacement vessel has a transom stern and the flow separates from the transom stern with a resulting hollow in the water behind the ship, we may in this case consider the body to consist of the ship and the hollow.

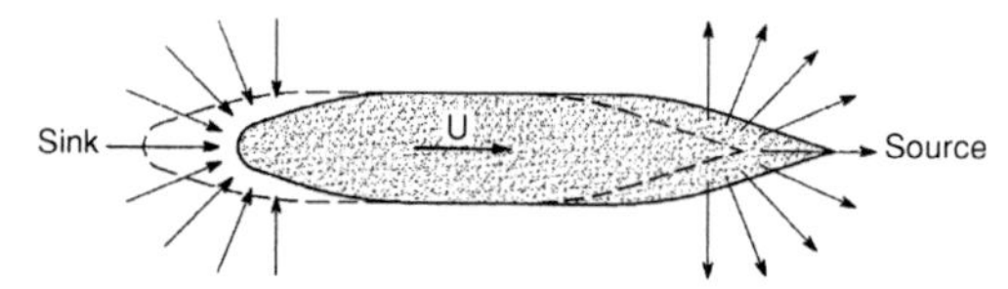

Figure 4.10. The ship moves the water. The ship has a long parallel midpart with submerged cross-sectional area S.

The effect of the ship will be that it pushes the water out in the bow, that is, it acts like a source, whereas the ship attracts the water in the stern, that is, it acts locally like a sink. This is illustrated from the Earth-fixed reference system in Figure 4.10. A source or a sink that properly satisfies the boundary conditions for this problem is a complicated function. The linearized free-surface condition follows from eq. (3.10) by setting p_0 and $\partial/\partial t$ equal to zero. This means

$$U^2 \frac{\partial^2 G}{\partial x^2} + g \frac{\partial G}{\partial z} = 0 \quad \text{on } z = 0, \tag{4.27}$$

where G is the velocity potential due to the source (sink). This is also called a Green function. Further, we must require that there is no flow at large water depths. A radiation condition ensuring that the ship waves are downstream of the ship is also needed. Newman (1987) has derived the following velocity potential for the source (sink):

$$G(x, y, z; \xi, \eta, \zeta)$$
$$= -\frac{Q}{4\pi} Re \left[\frac{1}{R} - \frac{1}{R_0} + \frac{2}{\pi} i\nu \int_{-\pi/2}^{\pi/2} \cos\theta \, e^{v} E_1(v) d\theta \right.$$
$$\left. + 4i\nu H(x - \xi) \int_{-\pi/2}^{\pi/2} \sec^2\theta \, e^{u} d\theta \right], \tag{4.28}$$

where

$$v = \nu[(z + \zeta)\cos^2\theta + |y - \eta| \cos\theta \sin\theta + i|x - \xi| \cos\theta]$$
$$u = \nu[(z + \zeta)\sec^2\theta + i|y - \eta| \sec^2\theta \sin\theta - i|x - \xi| \sec\theta]$$
$$\nu = g/U^2 \tag{4.29}$$
$$R = \sqrt{(x - \xi)^2 + (y - \eta)^2 + (z - \zeta)^2}$$
$$R_0 = \sqrt{(x - \xi)^2 + (y - \eta)^2 + (z + \zeta)^2}$$

This is based on the coordinate system (x, y, z) defined in Figure 4.6. (ξ, η, ζ) is the source point and (x, y, z) is a field point at which we want to evaluate the influence of the source (sink). E_1 is the complex exponential integral as defined in Abramowitz and Stegun (1964), and H is the Heaviside step function

$$H(x) = \left[\begin{array}{ll} 1 & \text{for } x \geq 0 \\ 0 & \text{for } x < 0 \end{array} \right.. \tag{4.30}$$

Further, Q is the source (sink) strength. To be precise, $Q > 0$ means a source and $Q < 0$ means a sink. The expression $-Q/(4\pi R)$ is a source (sink) in infinite fluid. The last integral term in eq. (4.28) represents the downstream waves, that is,

$$G \approx -i\frac{\nu}{\pi} Q \int_{-\pi/2}^{\pi/2} \sec^2\theta \, e^{u} \, d\theta \tag{4.31}$$

for large positive values of $(x - \xi)$. It can be shown that the source strength in the bow can be expressed as US, whereas the sink strength in the stern is $-US$. Here S is the submerged midship cross-sectional area. Later we show that this is correct. The source and sink coordinates are, respectively, $\xi = -0.5L$, $\eta = 0$, $\zeta = 0$ and $\xi = 0.5L$, $\eta = 0$, $\zeta = 0$. Here L is the ship length and the origin of the coordinate system is midships, and the x-axis points aft.

Let us call the velocity potential due to the source-sink pair φ. The resulting free-surface elevation follows from the dynamic free-surface condition, that is,

$$\zeta(x, y) = -\frac{U}{g} \frac{\partial \varphi(x, y, 0)}{\partial x} \tag{4.32}$$

(see eq. (3.7) with p_0 and $\partial/\partial t$ equal to zero). Let us start with the contribution ζ_{BS} from the bow source to the downstream wave elevation. We then use eq. (4.31) in approximating the velocity potential. This gives

$$\zeta_{BS} = \frac{U^2 S}{g\pi} \nu^2 \int_{-\pi/2}^{\pi/2} \sec^3\theta \, e^{i \frac{\nu}{\cos^2\theta} \cdot \left[|y| \sin\theta - \left| x + \frac{L}{2} \right| \cos\theta \right]} d\theta. \tag{4.33}$$

By noting that it is the real part of eq. (4.33) that has physical meaning and introducing $\theta' = -\theta$, we see that eq. (4.33) can be expressed in the form of eq. (4.15), that is,

$$\zeta_{BS} = \int_{-\pi/2}^{\pi/2} A(\theta') \cos\left[\frac{g}{U^2 \cos^2\theta'} \times \left(\left|x + \frac{L}{2}\right| \cos\theta' + |y| \sin\theta'\right)\right] d\theta', \tag{4.34}$$

where

$$A(\theta') = \frac{U^2 S}{g\pi} \nu^2 \sec^3\theta'. \tag{4.35}$$

The asymptotic analysis that leads to eq. (4.22) can then be used. A similar analysis can be followed for the contribution from the stern sink. We focus on the transverse waves along the ship's track. This gives

$$\zeta(x, 0) = S\sqrt{\frac{2\nu}{\pi}} \left[\sqrt{\frac{1}{x + L/2}} \cos\left(\nu(x + L/2) + \frac{\pi}{4}\right) - \sqrt{\frac{1}{x - L/2}} \cos\left(\nu(x - L/2) + \frac{\pi}{4}\right)\right]. \tag{4.36}$$

We should note that eq. (4.36) is similar to the θ_1-part of eq. (4.22) with $\theta_1 = 0$. Let us then make eq. (4.36) nondimensional, that is,

$$\frac{\zeta(x,0)L}{S} = \frac{1}{Fn}\sqrt{\frac{2}{\pi}} \left[\sqrt{\frac{1}{x/L + 0.5}} \times \cos\left(\frac{1}{Fn^2}\left(\frac{x}{L} + 0.5\right) + \frac{\pi}{4}\right) - \sqrt{\frac{1}{x/L - 0.5}} \cos\left(\frac{1}{Fn^2}\left(\frac{x}{L} - 0.5\right) + \frac{\pi}{4}\right)\right] \tag{4.37}$$

for large positive x. What large x means is difficult to quantify (see below).

Eq. (4.37) shows that the wave elevation is proportional to the submerged midships cross-sectional area S. Results are presented in Figure 4.11 as a function of x/L larger than 1 for Froude numbers 0.4, 0.5, and 0.8. The total wave elevation and the contributions from the bow source and the stern sink are shown. Looking only at the term outside the brackets in eq. (4.37), one may be tempted to believe that maximum values should decay like $1/Fn$ for increasing Fn. That this is not so has to do with the Froude number dependence of the term inside the brackets. For instance, looking at the maximum value for $Fn = 0.4$ shows it is relatively small compared with the results for $Fn = 0.5$. This has to do with cancellation effects of the wave systems from the bow source and the stern sink. This is also evident from Figure 4.11, which shows the contributions from the bow source and the stern sink

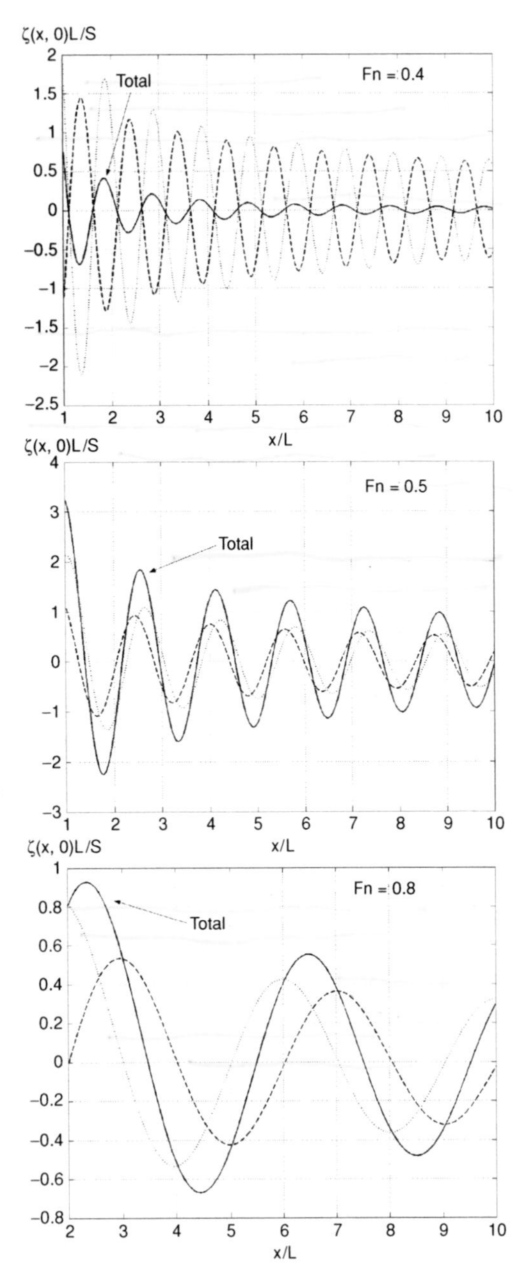

Figure 4.11. Wave elevation $\zeta(x, 0)$ along the track of a ship. The ship is represented by a source in the bow and a sink in the stern. The total wave elevation, together with the contributions from the bow source and the stern sink, is shown. L = ship length. S = midships submerged cross-sectional area. Effect of Froude number Fn.

separately. One can also explain it by the wavelength, which is $\lambda/L = 2\pi Fn^2 \approx 1$ for $Fn = 0.4$. Because the bow source and the stern sink are one wavelength apart and the effect of the source is 180° out of phase with the effect of the sink, we get a cancellation effect. However, the cancellation effect is not complete. This is because of the terms $(x + L/2)^{-1/2}$ and $(x - L/2)^{-1/2}$ associated with the source and the sink, respectively. It also means that the larger the x/L value, the stronger the cancellation effect. We note from Figure 4.11 that the contributions from the bow source and the stern sink are amplifying each other and causing a relatively large transverse wave amplitude when $Fn = 0.5$. A stronger amplification from a phasing point of view should occur when there is half a wavelength between the position of the source and the sink. This means $\lambda/L = 2\pi Fn^2 = 2$ or $Fn = 0.56$. There are also other Froude numbers for which either strong amplification or strong cancellation occurs because of the transverse wave systems from the bow source and stern sink. However, these amplifications or cancellations are for Froude numbers lower than those we have discussed.

Because the wave resistance arises directly from the far-field waves that the vessel generates, the amplification and cancellation tendency of the transverse waves causes humps (maxima) and hollows (minima) in the wave resistance as a function of Froude number. The magnitude of the humps and hollows depends on the relative contributions from transverse waves to the wave resistance. For instance, later we show that for specific cases, divergent waves cause nearly all the wave resistance when $Fn > \approx 0.8$.

4.2.4 Example

We consider a catamaran with length $L = 70$ m and $Fn = 0.5$, which means $U = 13.1\ \text{ms}^{-1}$. We use the results in Figure 4.11 as a basis by assuming no hydrodynamic interaction between the two catamaran hulls. This means S is two times the maximum cross-sectional area of one hull. We choose S equal to 25 m^2 and look at the results in Figure 4.11. However, there is a difficulty because we do not know for how small an x/L the results are applicable. However, we should not forget the simplifications made in using a source-sink pair and that our objective is to make an estimate only. We use the largest value in Figure 4.11 to estimate the amplitude A_0 of the waves. This means $A_0 L/S = 3.3$ or $A_0 = 1.2$ m.

4.3 Wave resistance in deep water

Newman (1977) shows by energy arguments how to relate the complex wave amplitude function $A(\theta)$ in eq. (4.17) to the wave resistance R_W, that is,

$$R_W = \frac{1}{2}\pi\rho U^2 \int_{-\pi/2}^{\pi/2} |A(\theta)|^2 \cos^3\theta \, d\theta. \quad (4.38)$$

The contribution to the integral from θ close to $\pm\pi/2$ corresponds to contributions for divergent waves with very small wavelength. This has negligible influence on the wave resistance.

There exist different computational methods to find the wave field around a ship and the wave resistance. A simple and still useful method for monohulls is the linear thin ship theory by Michell (1898). It represents the flow due to the ship as a source (sink) distribution along the center plane of the vessel, which requires a thin hull form. The local source strength is proportional to the longitudinal slope $\partial\eta(x, z)/\partial x$ of the mean submerged hull surface. Here

$$y = \pm\eta(x, z) \quad (4.39)$$

defines the hull surface.

Let us try to give some more detail about this. We start out with writing the total velocity potential as in eq. (3.4), where φ is the velocity potential caused by the ship. We must require that there be no flow through the hull surface. This means that

$$\frac{\partial}{\partial n}(Ux + \varphi) = 0$$
$$\text{on the hull surface } y = \pm\eta(x, z).$$

Here $\partial/\partial n$ is the derivative along the normal vector $\mathbf{n} = (n_1, n_2, n_3)$ to the hull surface. We choose the normal vector to be positive into the fluid domain. We can also write

$$\frac{\partial}{\partial n} = n_1\frac{\partial}{\partial x} + n_2\frac{\partial}{\partial y} + n_3\frac{\partial}{\partial z}.$$

This means the body boundary condition becomes

$$n_1\frac{\partial\varphi}{\partial x} + n_2\frac{\partial\varphi}{\partial y} + n_3\frac{\partial\varphi}{\partial z} = -Un_1$$
$$\text{on the hull surface.}$$

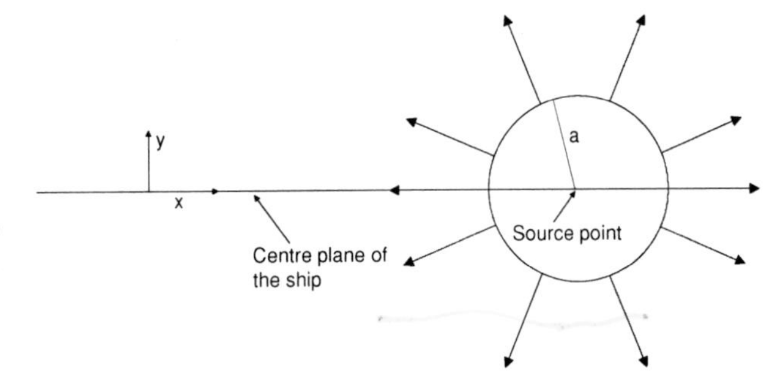

Figure 4.12. Single source in infinite fluid with source point on the centerplane of the ship. The figure shows the velocity vector due to the source on the surface of a sphere with radius a and center in the source point.

We now make assumptions about a slender hull. This means $n_1 \ll n_2$ and $n_1 \ll n_3$. Then we make the additional assumption of a thin ship. Then $n_3 \ll n_2$ and we can approximate $n_2 = \pm 1$ for respectively positive and negative y-values. We then Taylor series expand $\partial\varphi/\partial y$ about the center plane cp of the hull. The dominant term is then

$$\frac{\partial\varphi}{\partial y} = \pm U \frac{\partial\eta\,(x,z)}{\partial x} \quad \text{for, respectively, } y = 0^{+} \text{ and } y = 0^{-} \text{ on the } cp, \tag{4.40}$$

where we have expressed n_1 as $-\partial\eta/\partial x$ by geometric considerations.

Eq. (4.40) says the flow caused by the ship appears like a flow coming out from the ship with horizontal velocity $\pm U \cdot \partial\eta\,(x,z)\,/\partial x$ on the two sides of the hull. Water appears to either be pushed out or attracted, depending on the sign of $\partial\eta/\partial x$. Because $\partial\eta/\partial x$ is positive in the bow region, water is pushed out there. The opposite happens in the stern region. Figure 4.10 gives a simplified view of the flow caused by the ship. However, in our representation of the flow, this happens continually along the ship. So we can represent the flow caused by a ship as a source (sink) distribution along the ship's center plane. We will now show that the source strength follows by eq. (4.40).

A single source is expressed as in eq. (4.28). We concentrate first on the 1/R part, that is, the part corresponding to a source in infinite fluid. We have in Figure 4.12 sketched the velocity field due to the source when the source point is at the center plane of the hull. We find the velocity vector at the surface of a sphere with radius a and center in the source point. The velocity is radial to this surface. The magnitude of the velocity is constant on the spherical surface, and the flow appears as if it is coming from the center of the source. We note that the velocity along the center plane is tangential to the hull surface. So the source does not induce any velocity normal to the center plane outside the source point. However, at the source point, we must be careful. Taking the single source expression gives infinite velocity at the source point. We try to circumvent this problem by first calculating the mass flux through the spherical surface of radius a. The normal velocity of the infinite fluid source part of eq. (4.28) on the spherical surface is

$$\frac{Q}{4\pi} \cdot \frac{1}{a^2},$$

so the mass flux is Q. Because this is independent of a, we can make the radius very small. Before using this feature, we introduce a continuous source distribution with source density $\sigma(x,z)$, which means $\sigma(x,z)\,dx\,dz$ replaces Q. We concentrate on a given point (x,z) on the center plane. It is only the source with a center in this point that can introduce a normal velocity at that point.

Then we look at eq. (4.40), which says that a flow appears through an elemental area $dxdz$ with mass flux $2U\partial\eta/\partial x \cdot dx\,dz$ at a point (x,z) on the center plane. From the arguments above, this must be equal to $\sigma\,(x,z)\,dx\,dz$. This means

$$\sigma\,(x,z) = 2U\frac{\partial\eta}{\partial x}. \tag{4.41}$$

Have we forgotten something now? Let us express eq. (4.28) as $G = -(Q/4\pi)Re(1/R + f)$. We have argued as if the additional part f in eq. (4.28) for the source expression did not exist. However, this function has no singularities in the fluid domain such as those of $1/R$. Further due to the nature of a source, the flow induced by f must be symmetric about the x-z plane. Because f has no singularities in the fluid domain, the symmetry properties imply that $\partial f/\partial y$ is zero in the center plane. This means that eq. (4.41) is true and we have shown how to represent the flow due to the ship as a source distribution along the center plane of the hull. Before we proceed with this analysis, we should use eq. (4.41) to check that the source and sink strengths that we used previously in the single bow source and stern sink model are correct (see text after eq. (4.31) and Figure 4.10). Let us take the single bow source first. According to eq. (4.41), the source strength should be

$$Q = 2U \int_{-L/2}^{x_{max}} dx \int_{-D}^{0} \frac{\partial \eta}{\partial x} dz.$$

Here x_{max} is the x-coordinate of maximum cross-sectional area S. So we see that the integration gives $Q = US$ as previously claimed. A similar result holds for the stern sink.

Newman (1977) shows that the wave amplitude function $A(\theta)$ in eq. (4.38) can, according to Michell's thin ship theory, be expressed as

$$A(\theta) = \frac{2}{\pi} \nu \sec^3 \theta \times \iint_{cp} \frac{\partial \eta}{\partial x} \exp[\nu \sec^2 \theta (z + ix \cos \theta)]\, dz\, dx. \tag{4.42}$$

Here $\nu = g/U^2$, $\sec\theta = 1/\cos\theta$, and cp means the center plane of the hull. Because a high-speed vessel has a transom stern, Michell's expression needs some modifications. In this case, we can consider an elongated body consisting of the hull and the hollow in the water behind the transom stern. How to do this was studied by Doctors and Day (1997). We will leave the hollow out in the following examples on applications of Michell's theory.

4.3.1 Example: Wigley's wedge-shaped body

Wigley's wedge-shaped body has a constant draft D along the ship. Further, η which expresses the hull surface, is only a function of x. This means the ship has a flat bottom. By integrating eq. (4.42) in the z-direction, we get

$$A(\theta) = \frac{2}{\pi} \sec\theta \int_{-L/2}^{L/2} \eta_x (x, 0)\, e^{i\nu \sec\theta x} \left[1 - e^{-\nu \sec^2 \theta D}\right] dx. \tag{4.43}$$

The waterplane of the wedge-shaped body is shown in Figure 4.13. It is seen that the hull is symmetric about the y-z plane. It has a parallel midbody of length L_P and

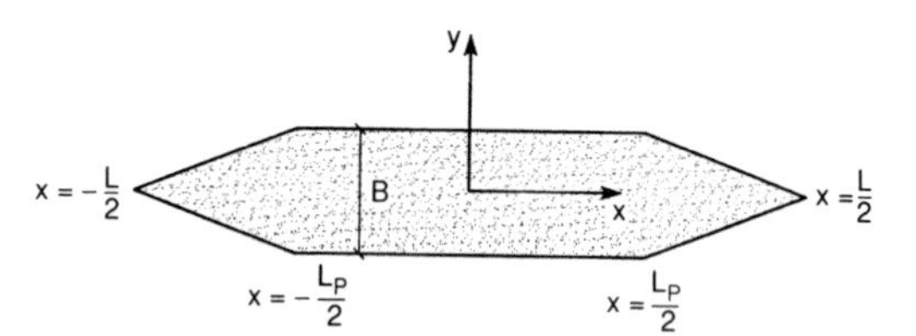

Figure 4.13. Wedge-shaped body with draft D.

$$\eta_x = \frac{B}{L - L_P} \equiv C \quad \text{from } x = -\frac{L}{2} \text{ to } -\frac{L_P}{2}$$
$$\eta_x = -C \quad \text{from } x = \frac{L_P}{2} \text{ to } \frac{L}{2}. \tag{4.44}$$

Then eq. (4.43) becomes

$$A(\theta) = \frac{4C}{\pi i \nu} \left(1 - e^{-\nu \sec^2 \theta D}\right) [\cos (0.5\nu \sec\theta \cdot L_P) - \cos (0.5\nu \sec\theta \cdot L)]. \tag{4.45}$$

We study $\theta = 0$ and write $L_P = L - 2L_B$, where L_B is the length of either the bow or stern part. This gives

$$A(0) = \frac{4C}{\pi i \nu}(1 - e^{-\nu D})[\cos(0.5\nu L)\cos(\nu L_B) + \sin(0.5\nu L)\sin(\nu L_B) - \cos(0.5\nu L)].$$

If L_B is small, that is, has a long parallel midbody, then $A(0)$ is approximately proportional to $\sin(0.5\nu L)$, which is zero for $0.5\nu L = n\pi, n = 1, 2, \ldots$ This means for $Fn = (0.5/n\pi)^{1/2}$ or for $Fn = 0.4, 0.28, 0.23, \ldots$ This is consistent with the results in Figure 4.11 showing small transverse wave amplitude along the ship's track. The maximum absolute values of $\sin(0.5\nu L)$ correspond to $0.5\nu L = \pi/2 + n\pi$. Here $n = 0$ gives $Fn = (1/\pi)^{1/2} = 0.56$, where Figure 4.11 also indicated large wave amplitude. Further, $n = 1$ gives $Fn = 0.33$. Because there is a direct connection between $A(\theta)$ and R_W through eq. (4.38) and the transverse waves are important contributors to $Fn < \approx 0.7$, local minimum value of $A(0)$ gives hollows in the wave resistance as a function of Froude number. Further, local maximum values of $A(0)$ give humps in the wave resistance curve.

4.3.2 Example: Wigley ship model

The equation for the hull surface for $y \geq 0$ for the Wigley's (1942) ship-shaped model is

$$\frac{y}{L} = \frac{B}{2L}\left(1 - \left(\frac{z}{D}\right)^2\right)\left(1 - \left(\frac{x}{0.5L}\right)^2\right). \tag{4.46}$$

Here B is the beam, D is the draft and L is the ship length. We use the same main particulars as Lunde (1951). This means $B/L = 0.75/8$ and $D/L = 1/16$. It is possible to analytically integrate the expression for $A(\theta)$. However, we do not show the details here. When $A(\theta)$ is obtained, we have to numerically integrate it into eq. (4.38) to find the wave resistance R_W. This can, for instance, easily be done by using the mathematical software Maple. The integrand is singular at $\theta = \pm\pi/2$ and rapidly oscillating in the vicinity of $\theta = \pm\pi/2$. However, this region has a negligible contribution to the wave resistance. Because this contribution is associated with divergent waves of very small wavelengths, we should also be able to understand this negligible influence from physical arguments. If it had some influence, we certainly would have to deal with the effect of surface tension, which matters for ripples. Mathematically, we can avoid the singular behavior by performing the integration numerically from $-\pi/2 + \delta$ to $\pi/2 - \delta$, where δ is a small positive number. The integration must be repeated to see that the results do not depend on δ.

Lunde (1951) presented interesting numerical studies with this hull form. He divided the contribution to the wave resistance into contributions from the transverse waves and divergent waves. The transverse waves correspond to $|\theta|$ between 0 and $\sin^{-1}(1/\sqrt{3})$ and the divergent waves correspond to $|\theta|$ between $\sin^{-1}(1/\sqrt{3})$ and $\pi/2$, as we discussed in the text after eq. (4.22). Lunde (1951) used the following definition of nondimensional wave resistance:

$$©_W = \frac{125}{\pi}\frac{R_W}{0.5\rho U^2 \nabla^{2/3}}, \tag{4.47}$$

where ∇ is the displacement, which is equal to $0.002604L^3$ in the present case. Further, the notations $©_{WD}$ and $©_{WT}$ were introduced to identify the contribution to $©_W$ from respectively divergent and transverse waves. $©_W$, $©_{WD}$, and $©_{WT}$ were presented as a function of the Froude number $Fn = U/\sqrt{Lg}$ between 0.15 and 0.6. We present similar results for Fn between 0.3 and 3.0 (Figure 4.14). It should be remembered that $Fn > \approx 1.0$ implies a planing craft. Because trim, sinkage, and nonlinear effects then are important and the hull shape would certainly not look like common semi-displacement and planing vessels, the practical relevance of doing calculations

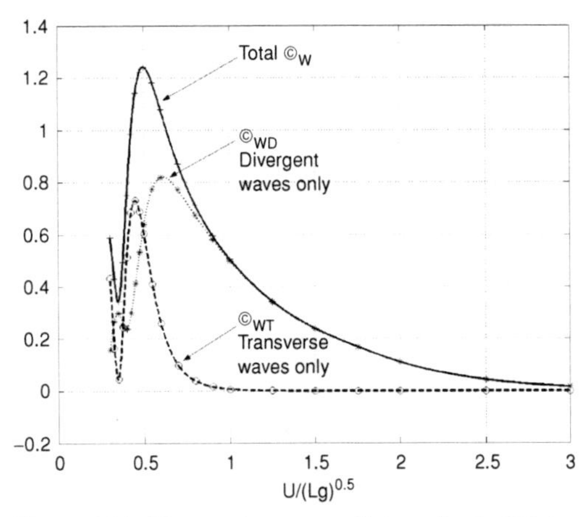

Figure 4.14. Wave resistance coefficients for the Wigley model studied by Lunde (1951). $B/L = 0.75/8$, $D/L = 1/16$, $\nabla/L^3 = 0.002604$. $©_W = \frac{125}{\pi}\frac{R_W}{0.5\rho U^2 \nabla^{2/3}}$. Similar for $©_{WD}$ and $©_{WT}$.

for such high Froude numbers is questionable. However, the results show some general features that do have practical relevance. The nondimensional wave resistance does not become negligible before $U/\sqrt{Lg} = 3.0$, contrary to common theoretical analysis of planing vessels that neglects gravity wave effects. One may also be misled by the nondimensionalization in terms of U^2. For instance, using the weight to nondimensionalize the resistance would illustrate better the relative contribution of wave resistance at high Froude numbers. Further, we note that the divergent waves give the main contribution to wave resistance for $Fn > \approx 0.8$. This has relevance when we later discuss what is referred to as 2.5D (2D+t) theory.

In the previous example, in which a ship was represented by a source in the bow and a sink in the stern, we emphasized that the superposition of the transverse waves originating from bow and stern caused amplification and cancellation of the resulting transverse waves behind the ship. This causes, respectively, maxima and minima in the contribution from transverse waves to the wave resistance. These maxima are for the Froude number range in Figure 4.14 and, for this case, occurring at $Fn \approx 0.3$ and $Fn \approx 0.5$. The minimum value occurs at $Fn \approx 0.35$. Because in this case we represent the ship by a continuous source distribution and the effect is dependent on the hull form, the maxima and minima will not be exactly

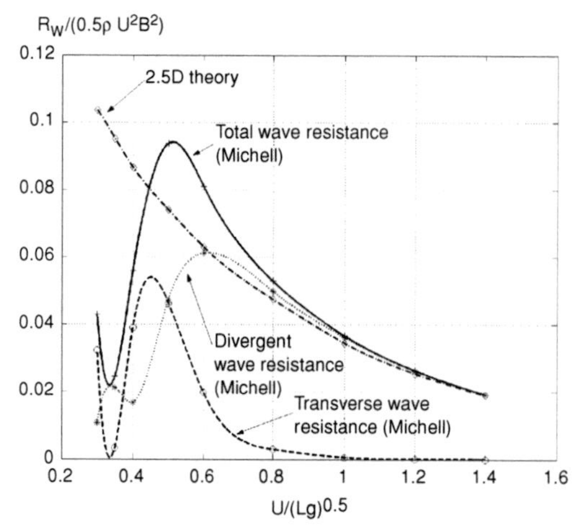

Figure 4.15. Wave resistance R_W of Tuck's parabolic strut. $D/L = 0.05$. B = beam, D = draft, L = length, U = forward speed.

the same for each case. This is also evident for the next example with a parabolic strut.

4.3.3 Example: Tuck's parabolic strut

The equation for the vertical hull surface for $y \geq 0$ is

$$\frac{y}{L} = \frac{B}{2L}\left(1 - \left(\frac{x}{0.5L}\right)^2\right). \tag{4.48}$$

The draft dependence of $A(\theta)$ is given by the term $(1 - \exp(-\nu \sec^2 \theta D))$ in eq. (4.43). This means the draft Froude number $Fn_D = U/\sqrt{gD}$ matters. Because transverse waves correspond to smaller θ than divergent waves, the dependence of the term $\exp(-\nu \sec^2 \theta D) = \exp(-\sec^2 \theta / Fn_D^2)$ on Fn_D matters for larger Fn_D for the transverse waves than for the divergent waves. There is a direct connection between $A(\theta)$ and R_W (see eq. (4.38)). This discussion is therefore relevant for the influence of Fn_D on the contribution from transverse and divergent waves to wave resistance. Because $A(\theta)$ is proportional to B for a parabolic strut, the wave resistance will be proportional to B^2. This means it is natural to use

$$\frac{R_W}{0.5\rho U^2 B^2} \tag{4.49}$$

as nondimensional wave resistance. Thus, the nondimensional wave resistance will not depend on B but is a function of $U/\sqrt{Lg}$ and $U/\sqrt{Dg}$ in this case. When studying the Wigley ship model, we used ©$_W$ given by eq. (4.47) to present nondimensional wave resistance. Using eq. (4.49) for that case also is believed to be better. This means we would have less significant B/L dependence. Another way to nondimensionalize R_W is to use

$$C_W = \frac{R_W}{0.5\rho U^2 S}, \tag{4.50}$$

where S is the wetted surface. This is natural to do for the viscous resistance R_V. The reason is that R_V for a slender ship is dominated by frictional forces that are proportional to S.

Tuck (1988) presented nondimensional wave resistance for a parabolic strut with $D/L = 0.05$. Both Michell's thin ship theory and a 2.5D (see section 4.3.4) thin ship theory were used. We recalculated the results by Michell's theory and separated the contributions from the transverse and divergent waves. The results are presented in Figure 4.15. Similar to the Wigley ship model presented in Figure 4.14, the transverse waves have a negligible influence for Froude numbers higher than approximately 0.8. We note that the 2.5D theory agrees well with the divergent part of Michell's theory for $Fn > \approx 0.6$. Note also that the 2.5D theory will clearly overpredict the wave resistance for very small Fn. A 2.5D theory is in practice used for $Fn > 0.4$. More details about the 2.5D theory are given in the next section.

Figure 4.16 shows nondimensional wave resistance for parabolic struts with D/L between 0.1

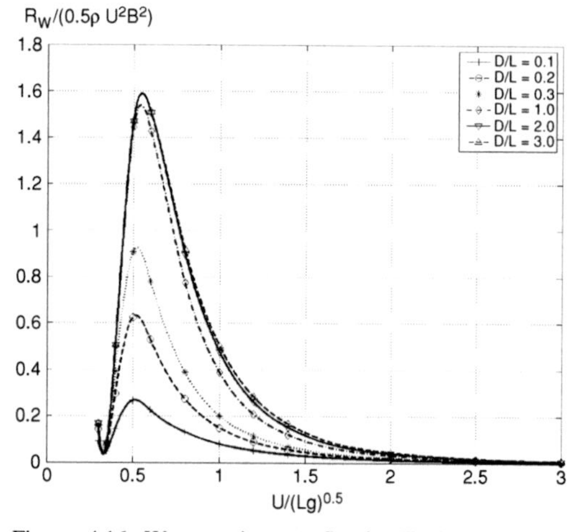

Figure 4.16. Wave resistance R_W for Tuck's parabolic strut calculated by Michell's thin ship theory. B = beam, D = draft, L = length, U = forward speed.

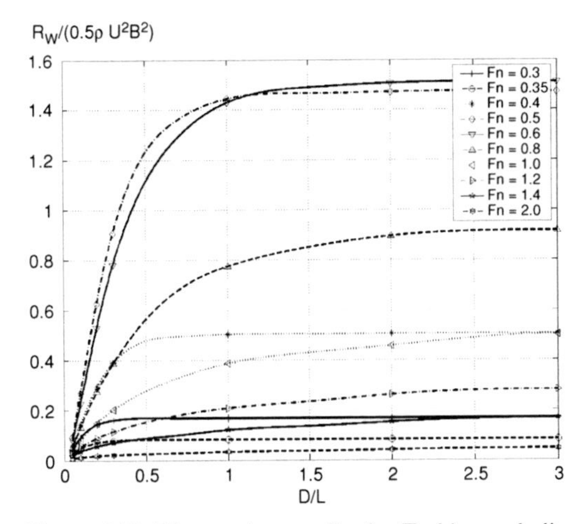

Figure 4.17. Wave resistance R_W for Tuck's parabolic strut calculated by Michell's thin ship theory. B = beam, D = draft, L = length, Fn = Froude number. The symbols represent calculated points.

and 3.0. The large values of D/L may, for instance, have relevance for the struts of a hydrofoil vessel. The Froude number range is from 0.3 to 3.0. The maximum nondimensional wave resistance occurs for $Fn \approx 0.5$. The nondimensional wave resistance for a given Fn increases with increasing D/L up to a value $(D/L)_{\min}$, beyond which there is a small dependence on D/L. This is also illustrated in Figure 4.17, in which nondimensional wave resistance is presented as a function of D/L for different Fn values. The general tendency is that the smaller Fn, the smaller $(D/L)_{\min}$. This can be explained by the fact that the wavelength for a given θ decreases with decreasing Fn and that the wave effect is negligible for a depth larger than half a wavelength. Figure 4.16 shows a small difference in the nondimensional wave resistance for D/L in the range from 2 to 3. It means that the results for $D/L = 3$ are close to the results for a parabolic strut with infinite draft.

4.3.4 2.5D (2D+t) theory

A 2.5D theory implies that the flow at a cross section of the ship is influenced only by the flow upstream of this section. Further, there is assumed to be no influence of the ship upstream of the bow. This means a numerical solution for the flow around the hull starts at the bow. The free-surface conditions are used to step the solutions of the free-surface elevation ζ and the associated velocity potential on the free surface in the longitudinal downstream direction of the hull. The velocity potential for each cross section is found by a two-dimensional analysis, that is, that the velocity potential due to the ship satisfies the 2D Laplace equation

$$\frac{\partial^2 \varphi}{\partial y^2} + \frac{\partial^2 \varphi}{\partial z^2} = 0. \tag{4.51}$$

This implies that changes of the flow in the x-direction are assumed small relative to changes in the cross-sectional plane. If the Froude number is small, the transverse wavelengths are small relative to the ship's length. This means longitudinal flow variations cannot be neglected relative to cross-sectional plane variations. The consequence is that we must assume high Froude numbers in the 2.5D theory. Ohkusu and Faltinsen (1990) showed by matched asymptotic expansions that a 2.5D theory only accounts for divergent waves. This is an implicit consequence of the results in Figure 4.15. The reason we call it a 2.5D theory is that the 2D Laplace equation is combined with 3D free-surface conditions.

Let us exemplify the procedure when the free-surface conditions can be linearized as in eqs. (3.7) and (3.9). We rewrite the dynamic and kinematic free-surface conditions as

$$\frac{\partial \varphi}{\partial x} = -g\frac{\zeta}{U} \quad \text{on } z = 0 \tag{4.52a}$$

$$\frac{\partial \zeta}{\partial x} = \frac{1}{U}\frac{\partial \varphi}{\partial z} \quad \text{on } z = 0. \tag{4.52b}$$

We start the procedure at the bow by setting $\varphi = 0$ and $\zeta = 0$ on the free surface. This means the upstream influence on the ship is neglected. Eqs. (4.52) express how φ and ζ change in the longitudinal direction. Let us consider an arbitrary transverse cross section. φ and ζ on $z = 0$ is then known from the upstream influence by using eq. (4.52). The problem in the cross-sectional plane is then solved by a numerical method such as the boundary element method (BEM). The body boundary condition is satisfied on the exact body boundary for $z \leq 0$. This is different from the thin-ship theory by Michell, in which body

Figure 4.18. Example of hollow in the water aft of the transom stern. The length Froude number is 0.47. The transom draft Froude number is 2.85. The length of the hollow increases with Froude number.

boundary conditions are transferred to the center plane. When the problem has been solved in the cross-sectional plane, we can evaluate $\partial\varphi/\partial z$ on $z = 0$. Then we know $\partial\zeta/\partial x$ from eq. (4.52b) and can, together with eq. (4.52a), continue to the next downstream cross section.

When the calculations come to the transom stern, the solution does not know that the pressure on the hull should be atmospheric. The procedure predicts a pressure different from the atmospheric pressure at the stern. This causes, in reality, an error in a small distance upstream of the transom stern. The flow for a high-speed vessel will, in reality, separate from the transom stern. This implies that the transom stern is dry and there is a hollow in the water behind the transom stern (Figure 4.18). This flow separation occurs for Froude numbers higher than approximately 0.4. However, Doctors (2003) showed by a systematic test series that the Froude number with the draft D_T at the transom as a length parameter, that is, $Fn_T = U/(D_T g)^{0.5}$, was a more important parameter. A representative lower value of Fn_T for a dry transom was approximately 2.5. Vanden-Broeck (1980) analyzed an idealized 2D problem that is relevant for the transom stern flow. The solution representing a dry transom gives a minimum $Fn_T = 2.26$ below which the downstream waves would exceed the theoretical breaking wave steepness limit ($H/\lambda = 0.141$) for Stokes waves (Schwartz 1974). Here H and λ are the wave height and wavelength, respectively.

Ogilvie (1972) derived a very simple 2.5D solution for the wave elevation ζ in the bow region along the surface of a symmetrical wedge with draft D and wedge half angle α. The body-boundary condition was transferred to the center-plane, and the linearized free-surface conditions given by eqs. (4.52) were used. The result was

$$Z(x) = \int_0^\infty d\mu \left(\frac{1 - e^{-\mu}}{\mu}\right)\left(\frac{\sin(\sqrt{\mu}X)}{\sqrt{\mu}}\right), \quad (4.53)$$

where

$$Z(x) = \frac{\pi}{2\alpha Fn_D}\frac{\zeta}{D} \quad (4.54)$$

and

$$X = \frac{x}{D \cdot Fn_D}. \quad (4.55)$$

Here Fn_D means the draft Froude number $U/\sqrt{gD}$ and x is a longitudinal coordinate with $x = 0$ corresponding to the edge of the bow. Eq. (4.53) is presented in Figure 4.19. The maximum value

$$\zeta_{\max} = 1.59D\frac{2\alpha}{\pi}Fn_D \quad (4.56)$$

of the wave elevation occurs when $x = 0.91D \cdot Fn_D$. Bow-wave elevation predictions by 2.5D theory for other hull shapes have been studied by Faltinsen (1983) and Fontaine et al. (2000).

We can illustrate the calculation procedure in a different way by using an Earth-fixed cross-sectional plane and letting the ship pass through this plane (Figure 4.20). The flow in this Earth-fixed plane is then time dependent. If the problem is linearized, the dynamic and kinematic free-surface conditions in the Earth-fixed cross-sectional plane are

$$\frac{\partial\varphi}{\partial t} + g\zeta = 0 \quad \text{on } z = 0. \quad (4.57a)$$

$$\frac{\partial\zeta}{\partial t} = \frac{\partial\varphi}{\partial z} \quad \text{on } z = 0. \quad (4.57b)$$

We see that eqs. (4.52) and (4.57) are the same if we make the coordinate transformation $x = Ut + x_0$ between the Earth-fixed and ship-fixed coordinate

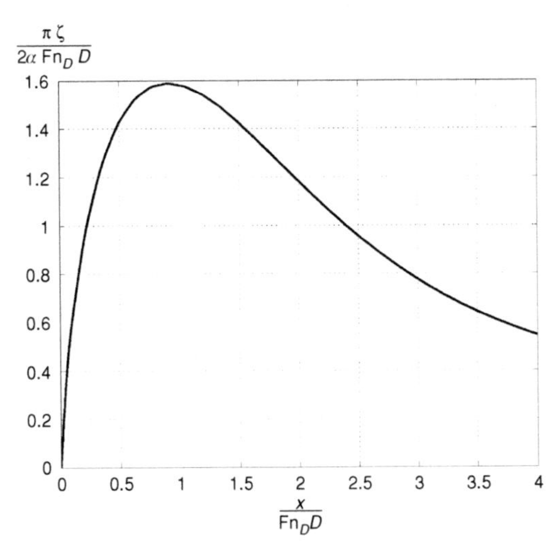

Figure 4.19. Bow-wave elevation ζ along the surface of a symmetrical wedge with draft D and wedge half-angle α as a function of the longitudinal coordinate x and draft Froude number $Fn_D = U/\sqrt{gD}$. $x = 0$ corresponds to the edge of the bow. The calculations are based on the linear 2.5D theory by Ogilvie (1972).

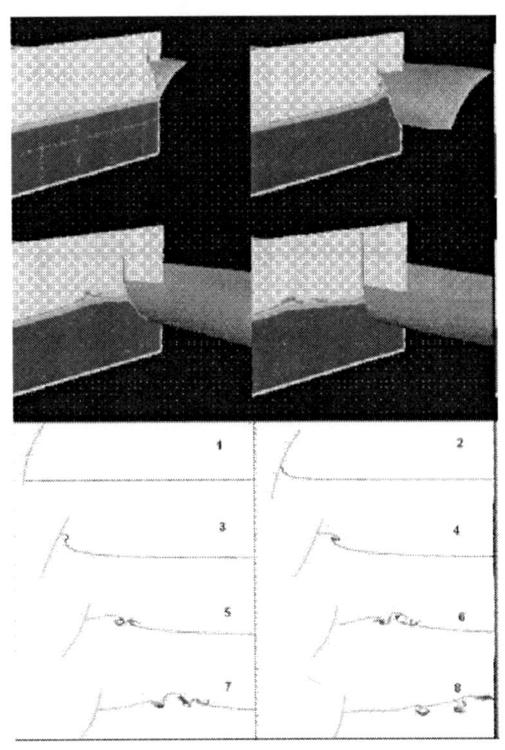

Figure 4.20. Numerical solution of the steady wave field by a 2D+t method. A 2D time-dependent problem is solved in an Earth-fixed cross-sectional plane. The ship passes through this cross-sectional plane. The simulations are illustrated in two different ways. In the last set of figures, we see clearly how the ship's cross section changes with time.

systems. Because we can formulate the problem as a 2D time-dependent problem in an Earth-fixed coordinate system, the 2.5D method is also referred to as a 2D+t method.

It is not necessary to assume linear flow. By using the nonlinear free-surface conditions, we can simulate breaking waves. However, a BEM breaks down when a plunging breaker impacts on the underlying fluid (see Figure 3.1). The SPH (smoothed particle hydrodynamics) method (Tulin and Landrini 2000) is then a more robust method. This is the method used in the calculations presented in Figure 4.20. The SPH method follows fluid particles. When a fluid particle in a plunging wave, as presented here, hits the underlying free surface, it is reflected. This causes a new splash flow, as illustrated both in Figures 4.20 and 3.1. This picture is, for instance, different from a waterfall, in which falling fluid particles penetrate the underlying free surface and create white water with entrained air. This can also be predicted by the SPH. The plunging wave impact simulations by the SPH create large gradients in the velocity tangential to the underlying free surface at the impact position. This is the same as saying that vorticity is generated.

Lugni et al. (2004) presented a comprehensive experimental and numerical study of the steady wave elevation around a semi-displacement monohull and catamaran. The ship models are described in Figure 4.21. Because one objective was to study the interaction between the demihulls of a catamaran, the submerged parts of the monohull and a demihull were identical. The chosen monohull has therefore a much smaller beam-to-draft ratio than typical semi-displacement monohulls. Both a nonlinear 2D+t method and a linear 3D Rankine panel method (RPM) were used. A panel method is the same as a boundary element method (BEM). Sometimes it is also referred to as a boundary integral method (BIM). In marine hydrodynamics, flow singularities such as sources, dipoles, and vortices are distributed over a boundary of the fluid to represent a potential flow satisfying Laplace equation in the fluid. The surface on which the singularities are distributed may consist of the total boundary enclosing the fluid or only part of it. The surface always includes the

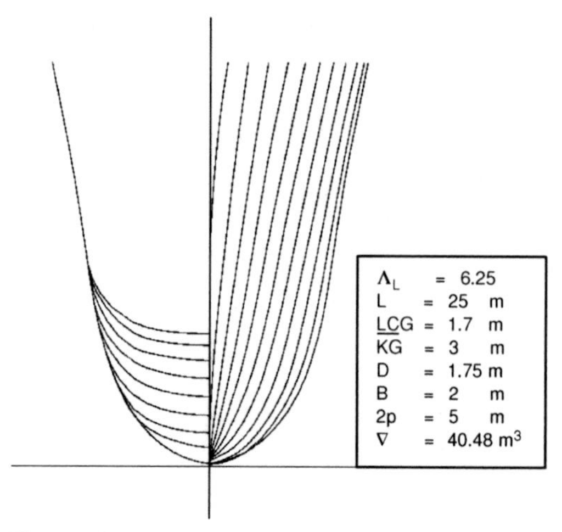

Figure 4.21. Body plan of the demihull and main particulars. Dimensions given in full-scale values. Λ_L is the ratio between the full-scale and model-scale lengths. LCG is relative to station 10. Station 20 is at the transom stern. D = draft, B = beam. 2p indicates the distance between the centerlines of the catamaran demihulls (Lugni et al. 2004).

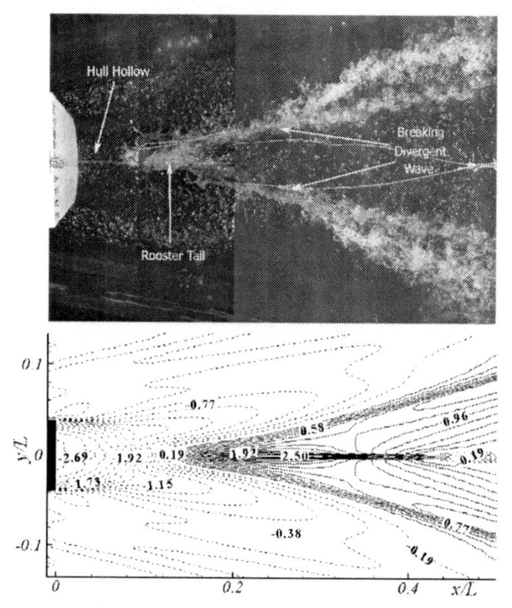

Figure 4.22. Transom stern wave field at $Fn = 0.5$ for the monohull described in Figure 4.21. Experimental picture (top) and contour lines of the steady wave pattern predicted by 2D+t theory (bottom). The numbers are $\zeta/L \cdot 10^2$, where ζ is the wave elevation and L is the ship length (Lugni et al. 2004).

wetted, or mean wetted surface, of the hull. How large a surface is used in the singularity distribution depends on what boundary conditions the flow singularities satisfy.

A BEM used to solve the 2D and 3D lifting problem of hydrofoils is described in detail in section 6.4. Lugni et al. (2004) apply a direct BEM formulation to solve the 2D fully nonlinear free-surface flow for each time step in their 2D+t method. The plunging breakers are numerically cut off to avoid impact with the underlying free surface. The 3D RPM used by Lugni et al. (2004) is similar to the numerical method described by Iwashita et al. (2000) but is based on the mathematical model proposed by Nakos (1990). The resulting solver does not account for flow separation from the transom stern. A Rankine panel method means that Rankine sources, that is, sources in infinite fluid, and Rankine dipoles in the surface normal direction are distributed over the mean wetted hull surface and a truncated part of the mean free surface around the ship model. Satisfaction of the body boundary condition, free-surface conditions, and the radiation condition requiring only far-field waves downstream of the vessel determine the source and dipole densities. When the 2D+t method was compared with the model tests, the experimentally determined sinkage and trim were used in the numerical calculations. This has an important effect when the Froude number (Fn) is larger than about 0.35. Froude numbers 0.3, 0.4, 0.5, 0.6, 0.7, and 0.8 were studied. The transom stern was dry for $Fn = 0.5$ and higher, which is consistent with Doctors (2003).

The 3D RPM simulations were able to capture the wave pattern along the hull for Fn smaller than 0.6. However, the stern and wake flows were not satisfactory for $Fn \geq 0.5$. This is because of the assumed wet transom and the linearization of the free-surface conditions. When $Fn \geq 0.5$, the nonlinearities are particularly important.

The nonlinear 2D+t results were not satisfactory for the smaller Froude numbers, 0.3 and 0.4. This is because of both the neglection of the transverse waves and the fact that a 2D+t method assumes a dry transom. When $Fn \geq 0.6$, the 2D+t method agrees quite well with the experiments. Figure 4.22 shows a picture of the experimental wave field behind the monohull as well as numerical results by the 2D+t method. The Froude number is 0.5. As we can see, the 2D+t formulation is able to reproduce the flow scenario behind the

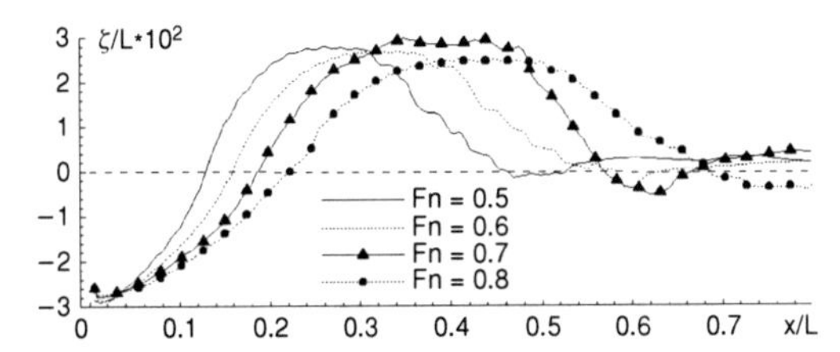

Figure 4.23. Cut of the steady wave pattern along the centerline of the transom stern for the monohull described in Figure 4.21. x is the longitudinal distance from the transom (Lugni et al. 2004).

transom: hull hollow, rooster tail, and incipient breaking divergent system. Nevertheless, because the plunging jet is cut to avoid the occurrence of impact on the underlying water, the energy of the wave system is focused close to the crest of the divergent wave. Differently, in the physical phenomenon, the breaking causes a spatial spread of the wave energy. Figure 4.23 shows the longitudinal wave cut along the centerline of the monohull transom as obtained by the 2D+t theory. The results show that the rooster tail height is not affected by the Froude number whereas both its horizontal width and the extension of the hollow increase with the speed. The hollow extension can be measured as the longitudinal distance between the transom position and the location where the free surface becomes zero. The wave elevation downstream from a catamaran demihull transom showed a quite different behavior. Except for the smaller speeds, showing an increase of the hollow extension with the Froude number, the hollow width is not particularly affected by the speed. The rooster tail height is lower than the corresponding value for the monohull. It shows a nonmonotonic but rather limited variation with the Froude number. This demihull wave behavior behind the transom is the result of the presence of the other demihull. The arrangement of the two demihulls causes three rooster tails downstream from the catamaran, respectively, in correspondence to the demihull transom sterns and the catamaran centerline.

Let us now return to our linearized steady problem in a ship-fixed coordinate system. The wave resistance is found by using a linearized Bernoulli equation for the pressure p, that is,

$$p = -\rho U \frac{\partial \varphi}{\partial x} - \rho g z. \tag{4.58}$$

Obviously we do not integrate the pressure over the dry transom stern. We should realize that the hydrostatic pressure $-\rho g z$ gives a longitudinal force. This is because of the flow separation at the transom stern. We use Figure 4.18 to demonstrate this. If the transom stern is wet, we know that the hydrostatic pressure integrated over the hull surface below $z = 0$ will not give a longitudinal force. We can either show this mathematically or simply appeal to Archimedes' principle:

$$\int_{S_B} \rho g z n_1 \, dS + \int_{S_T} \rho g z n_1 \, dS = 0. \tag{4.59}$$

Here n_1 is the x-component of the normal vector $\mathbf{n} = (n_1, n_2, n_3)$ to the hull surface. The positive normal direction is into the fluid domain. Further, S_B is the wetted body surface below $z = 0$ and S_T is the dry transom stern area below $z = 0$. By using eq. (4.59), we can write the longitudinal force F_1^{HS} due to hydrostatic pressure as

$$F_1^{HS} = \int_{S_B} \rho g z n_1 \, dS = -\int_{S_T} \rho g z dS. \tag{4.60}$$

If the transom stern has a rectangular cross section, which is not so in reality, we get that $F_1^{HS} = 0.5 \rho g D^2 B$, where D and B are the draft and beam at the transom stern.

Having obtained the pressure by eq. (4.58), we can also calculate the vertical force and the trim moment. This is needed in calculations of the sinkage and trim, which are important quantities affecting wave resistance (and viscous resistance) of high-speed vessels. Molland et al. (1996) showed by a systematic experimental series of different high-speed monohulls and catamarans that sinkage and trim started to be significantly different from their zero forward speed values when $Fn > \approx 0.35$. This is an implicit consequence of the increasing importance of the hydrodynamic pressure term $-\rho U \partial\varphi/\partial x$ relative to the hydrostatic pressure term $-\rho g z$ with increasing Froude number. Actually, the notation *semi-displacement vessel* means that the two terms are, roughly speaking, of equal importance for vertical steady loads. Trim tabs and/or interceptors are used to change the trim angle. This cannot be included in a rational way in the 2.5D theoretical procedure. We must add the effect of trim tabs and interceptors by a separate analysis in the calculation of trim. A description of trim tabs and interceptors is given in section 7.1.3.

In practice, a 2.5D theory is often used for $Fn > 0.4$. The results by thin ship theory (see e.g., Figures 4.14 and 4.15) show that the 2.5D theory is not really accurate before Fn becomes larger than about 0.8.

This section has shown that nonlinearities and breaking waves matter. The latter is a reason why the wave pattern resistance obtained by measuring the far-field wave elevation is different from the measured residual resistance. However, we should not conclude that the linear Michell thin ship theory is not useful in predicting the wave-pattern resistance. This type of analysis is very efficient relative to a nonlinear method and important at the predesign stage. Michell's thin ship theory can also be generalized and is useful for multihull vessels. This will be discussed in the next section.

4.3.5 Multihull vessels

A common way to calculate the wave resistance of a multihull vessel is to superimpose the waves generated by each hull as if they were individual hulls without the presence of other hulls (see e.g., Tuck and Lazauskas 1998 and Day et al. 2003). Let us describe this procedure. A local coordinate system (x_j, y_j, z_j) for hull number j is introduced (Figure 4.24). Similar to eq. (4.17), we can express the fact that hull number j creates a far-field wave system:

$$\zeta_j(x_j, y_j) = \mathrm{Re} \int_{-\pi/2}^{\pi/2} A_j(\theta) \exp\Big[i \frac{g}{U^2 \cos^2\theta} \times (x_j \cos\theta + y_j \sin\theta)\Big] d\theta. \tag{4.61}$$

Then we introduce the global coordinate system (x, y, z). The origin of the local coordinate system has coordinates (x_{j0}, y_{j0}, z_{j0}) in the global coordinate system. This means eq. (4.61) can be expressed as

$$\zeta_j(x, y) = \mathrm{Re} \int_{-\pi/2}^{\pi/2} A_j(\theta) \exp\Big[-i \frac{g}{U^2 \cos^2\theta} \times (x_{j0} \cos\theta + y_{j0} \sin\theta)\Big] \times \exp\Big[i \frac{g}{U^2 \cos^2\theta} (x \cos\theta + y \sin\theta)\Big] d\theta. \tag{4.62}$$

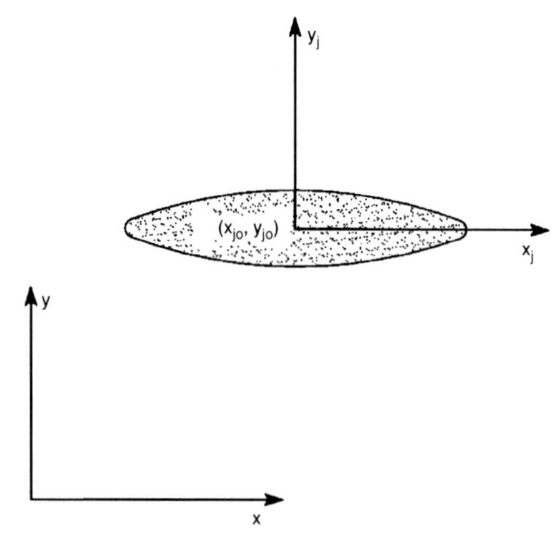

Figure 4.24. Global coordinate system (x, y, z) and local coordinate system (x_j, y_j, z_j) for hull number j.

This means the wave amplitude function $A(\theta)$ due to N hulls can be expressed as

$$A(\theta) = \sum_{j=1}^{N} A_j(\theta) \times \exp\Big[-i \frac{g}{U^2 \cos^2\theta} (x_{j0} \cos\theta + y_{j0} \sin\theta)\Big]. \tag{4.63}$$

Here $A_j(\theta)$ can be expressed by eq. (4.42) and use of the local coordinate system. Having obtained $A(\theta)$, we can use eq. (4.38) to calculate the wave resistance of the multihull vessel.

This means the wave resistance of a catamaran with nonstaggered identical side hulls can be expressed as

$$R_W = \frac{\pi}{2} \rho U^2 \int_{-\pi/2}^{\pi/2} |A(\theta)|_{SH}^2 \times 4 \cos^2\left(\frac{0.5(2p/L)}{Fn^2 \cos^2\theta} \sin\theta\right) \cos^3\theta \, d\theta, \tag{4.64}$$

where $|A(\theta)|_{SH}$ refers to the amplitude function for the side hull and 2p is the distance between the center lines of the demihulls. $A(\theta)$ can be evaluated by eq. (4.42). The hull interference function

$$F(\theta) = 4 \cos^2\left(\frac{0.5(2p/L)}{Fn^2 \cos^2\theta} \sin\theta\right) \tag{4.65}$$

determines then the amount of interference between the two hulls. This depends on the Froude number Fn, the ratio between the catamaran demihull centerlines 2p and the ship length L, as well as θ. The values θ that contribute to

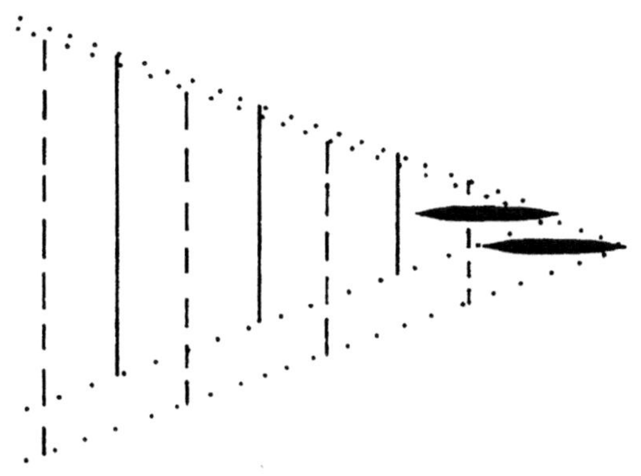

Figure 4.25. Schematic view of transverse wave crests generated by two staggered hulls. Superposition of the transverse wave system generated by each hull causes zero amplitude of the transverse waves in a broad area of the wake for the studied Froude number (Søding 1997).

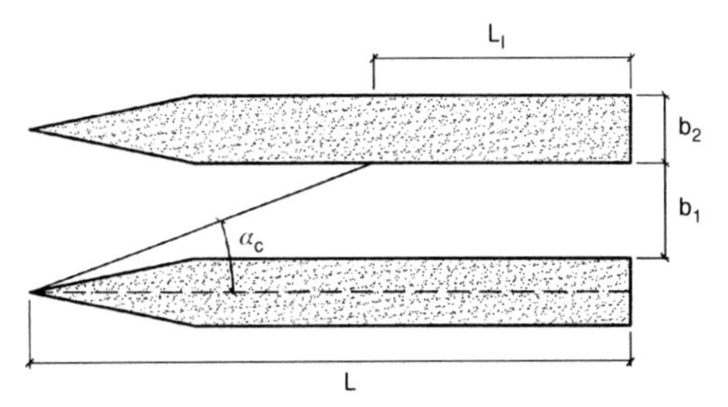

Figure 4.26. Hull interaction due to wave effects.

wave resistance can indirectly and qualitatively be seen from, for instance, Figure 4.14. When the Froude number is high, the divergent waves, that is, $|\theta| > \sin^{-1}(1/\sqrt{3})$, are most important. Generally speaking, the interference matters for divergent waves but is less important for transverse waves.

We will present another example showing the importance of the phasing of the waves generated by each hull. This is illustrated in Figure 4.25 with a staggered arrangement of the side hulls of a catamaran. The transverse waves generated by the two hulls are 180° out of phase, that is, the transverse wave amplitude of the catamaran is zero in a broad area of the wake. This cancellation effect depends on the Froude number and the longitudinal shift of the catamaran hull with respect to each other. For a Froude number of 0.4, this shift amounts to half the hull length. The cancellation effect of the transverse waves will have a beneficial effect on the wave resistance not only for the Froude number in which cancellation occurs but also in the vicinity of this Froude number (Søding, 1997). Tuck and Lazauskas (1998) have also theoretically optimized the placement of individual hulls in a multihull configuration in order to minimize the wave resistance.

However, the previous procedure does not account for the fact that waves generated by one hull will be incident to another hull and hence be diffracted (scattered) by the other hull. The diffracted waves include reflected and transmitted waves. This means the total wave system by a multihull vessel is not a superposition of waves generated by each hull as if the other hulls were not present. Let us discuss the hull interaction by considering a catamaran, which is the most common type of multihull vessel. To assess the importance of hydrodynamic hull interaction, we first assume there is no hydrodynamic hull interaction and then consider the Kelvin angle α_c for one hull and see if the waves inside the Kelvin angle are incident on the other hull and hence diffracted by the other hull. The length L_I of the aft part of this side hull that is affected by the other hull can, by using the notation in Figure 4.26, be expressed as

$$L_I = L - (b_1 + 0.5b_2)\cot\alpha_c.$$

By using $\sin\alpha_c = 1/3$, we find that

$$\frac{L_I}{L} = 1 - \left(\frac{b_1 + 0.5b_2}{L}\right)\sqrt{8}. \quad (4.66)$$

A representative example is $b_1/b_2 = 1.5$, $b_2/L = 1/12$. This means $L_I/L = 0.53$. We should note that it is the divergent waves at the Kelvin angle that cause the major part of the interaction. These incident waves oblique to the other hull will be diffracted by that hull. The transverse waves for $\theta = 0$ are not causing interaction effects. In the form presented here, the thin ship theory cannot predict this diffraction effect. However, a 2.5D or 3D theory can simultaneously account for the side hulls. Figure 4.27 shows calculations by a linear 3D Rankine panel method for the catamaran and monohull presented in Figure 4.21. Because the displacement of the catamaran is twice as large as for the monohull, we certainly expect a difference in the magnitude of the wave elevation for the two vessels. However, Figure 4.27 shows that the wave pattern around a catamaran is also quite different from that of a monohull. There are, for instance, large negative wave elevations occurring between the catamaran hulls in the aft part

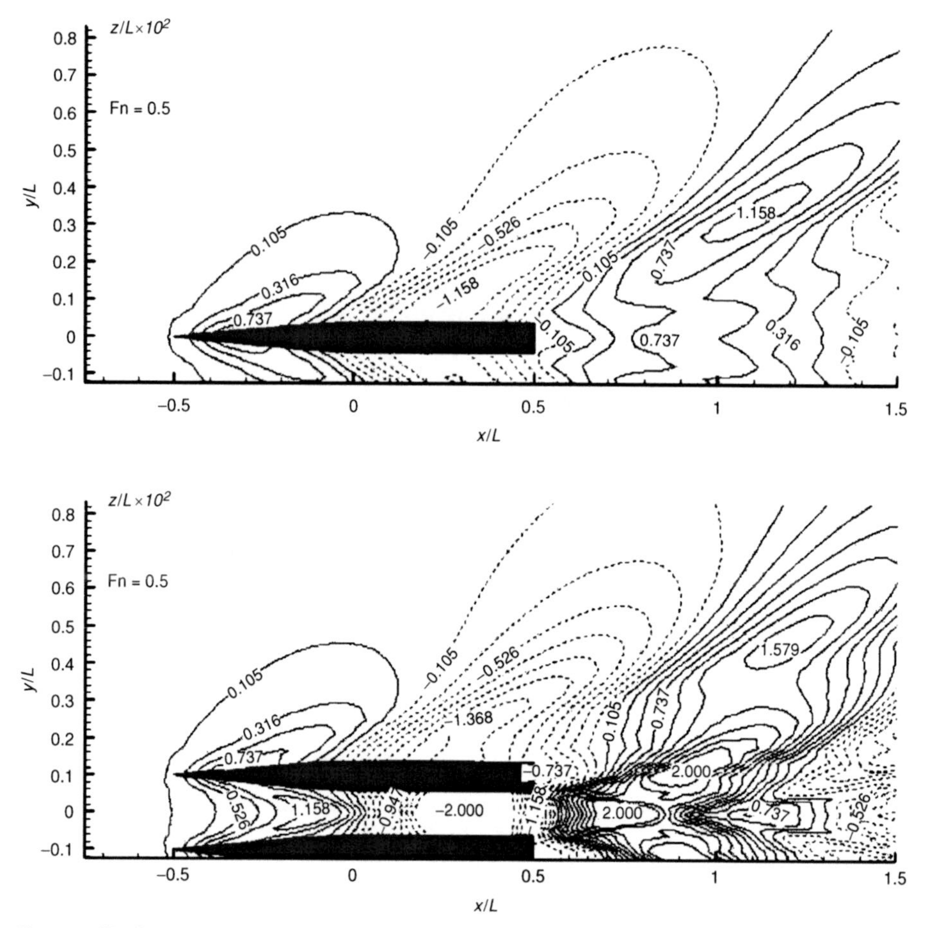

Figure 4.27. Comparison of numerically predicted linear wave elevation ζ around a monohull and a catamaran. Each demihull of the catamaran is identical to the monohull. The ship models are described in Figure 4.21. Calculations done with a 3D Rankine panel method (Lugni et al. 2004).

of the vessel. The effect of flow separation from the transom stern was not accounted for in the calculations.

We will illustrate the importance of hull interaction by another example from the experimental and numerical study by Lugni et al. (2004) with the catamaran model presented in Figure 4.21. Figure 4.28 shows the wave elevation along a longitudinal wave cut along the centerline of the catamaran for different Froude numbers. The 2D+t calculations are presented together with the experiments (for the tested speeds). In the plots, the curves with small squares give the 2D+t results obtained as the superposition of two monohull solutions, that is, the interaction between the demihulls is not accounted for, just their interference. From the results, the interference is not the only mechanism. The interaction between the two demihulls plays a fundamental role. This interaction is mainly nonlinear, as evidenced both by difficulties of the linear solution in capturing the first peak and by the phase shifting existing between the linear and nonlinear results accounting for the demihull interaction.

4.3.6 Wave resistance of SES and ACV

The excess pressure in the cushion of an SES or ACV deforms the free surface, causing surface waves when the vessel is moving forward. Tatinclaux (1975) has presented approximate analytical formulas, and Doctors (1992) has given a comprehensive survey article. Tuck and Lazauskas (2001) have studied choices of spatial pressure variations that minimize wave resistance of an ACV. The wave resistance due to the air cushion

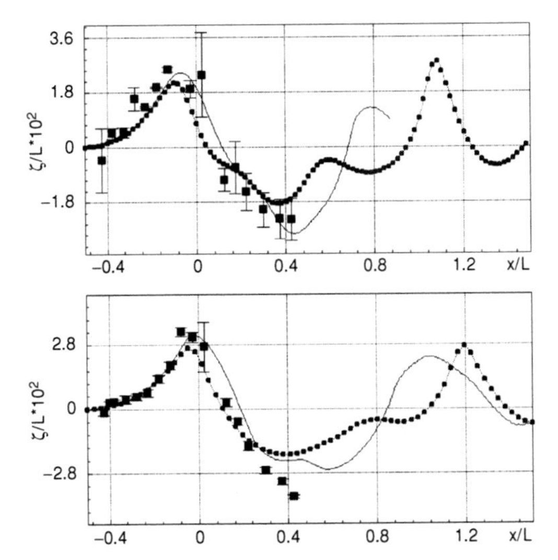

Figure 4.28. Catamaran: cut of the steady wave pattern along the catamaran centerline. Experimental data (large square symbols: mean value and error bar) and 2D+t theory (solid lines). The curves with small squares give the 2D+t results obtained as the superposition of the two monohull solutions, that is, the interaction between the demihulls is not accounted for. From top to bottom: $Fn = 0.5, 0.6$. The ship model is described in Figure 4.21 (Lugni et al. 2004).

of an SES and ACV can be calculated by eq. (4.38) with

$$|A(\theta)|^2 = \frac{4g^4}{\pi^2 U^8} \frac{(P^2 + Q^2)}{\cos^8 \theta} \tag{4.67}$$

and

$$P + iQ = \frac{1}{2\rho g} \iint_{A_b} p(x, y)\, e^{i \frac{g}{U^2 \cos^2 \theta}(x \cos\theta + y \sin\theta)}\, dx\, dy. \tag{4.68}$$

Here A_b is the horizontal cross-sectional area of the air cushion at the mean free surface and $p(x, y)$ is the excess pressure in the air cushion. If we assume a constant excess pressure p_0 and the cushion area A_b is rectangular with length L and breadth b, this gives a nondimensional wave resistance:

$$\frac{R_W}{\rho U^2 (p_0/\rho g)^2} = \frac{16}{\pi} \int_0^{\pi/2} \frac{\cos\theta}{\sin^2\theta} \sin^2\left(\frac{0.5}{Fn^2 \cos\theta}\right) \times \sin^2\left(\frac{0.5\,(b/L)\tan\theta}{Fn^2 \cos\theta}\right) d\theta. \tag{4.69}$$

This nondimensional wave resistance is presented in Figure 4.29 as a function of Fn for different

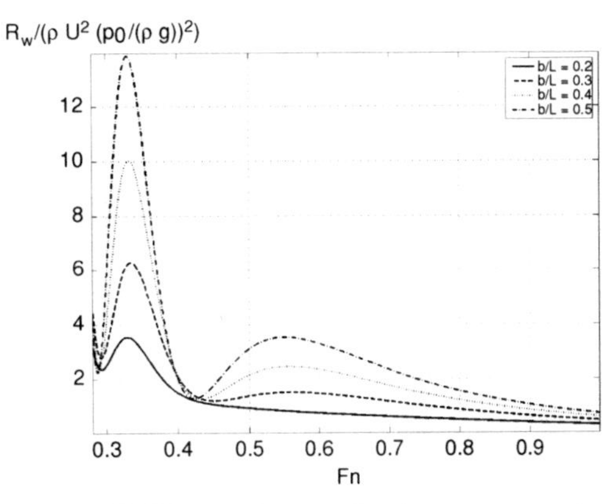

Figure 4.29. Wave resistance R_W due to rectangular cushion area with length L and breadth b presented as a function of Froude number $Fn = U/(Lg)^{0.5}$ for different b/L ratios. Constant excess pressure p_0.

b/L ratios. The general tendency is that nondimensional wave resistance increases with increasing b/L. The curves clearly show "humps and hollows" (maxima and minima), which are associated with the transverse wave system. The wave resistance is roughly proportional to b at the humps, whereas it is practically independent of b at the hollows. Doctors and Sharma (1972) discussed the influence of the falloff of the excess pressure at the edges of the cushion area. This has a pronounced effect for Froude numbers lower than those presented in Figure 4.29.

In addition, there is wave resistance due to the side hulls of an SES. In principle, there is an interaction between wave resistance due to the hulls and the air cushion.

4.4 Ship in finite water depth

We consider a ship in finite constant water depth h and assume infinite horizontal extent. The ship waves may be considerably modified because of the depth, particularly for shallow water.

The depth Froude number

$$Fn_h = \frac{U}{\sqrt{gh}} \tag{4.70}$$

plays an important role for ship waves and wave resistance in shallow water. Large changes occur at the critical depth Froude number equal to one. The ship waves are very different for subcritical, critical, and supercritical flows in shallow water. We return to this point later in the text but exemplify

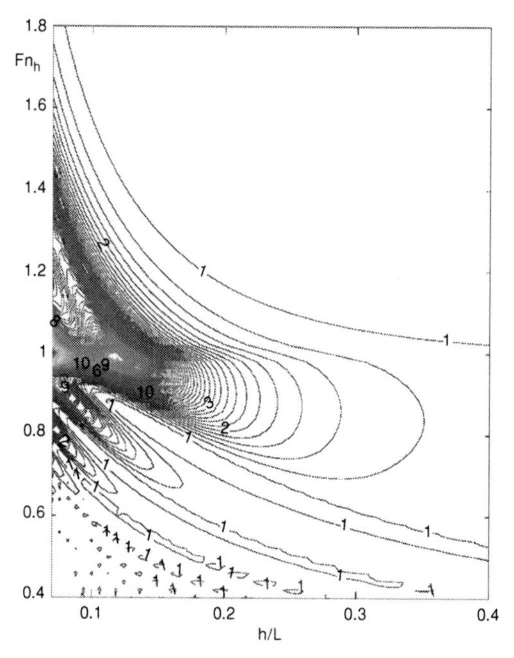

Figure 4.30. Shallow-water wave resistance ratio for a Wigley hull as a function of h/L and Fn_h. $B/L = 0.1$. $D/L = 0.0625$. See more details in Figure 4.31 (Yang 2002 and Yang et al. 2005).

the importance by presenting the ratio between wave resistance of a Wigley hull at depth h and infinite depth as a function of Fn_h and h/L, where L is the ship length. The equation of the Wigley hull model is given by eq. (4.46). Further, $B/L = 0.1$ and $D/L = 0.0625$ are used. The ratio between shallow-water and deep-water wave resistance is presented in Figures 4.30 and 4.31 and has been calculated using the linear thin ship theory, as previously outlined for infinite water depth. Even if it is not explicitly shown, this ratio can be less than 1. However, the ratio is generally larger than 1. When $h/L > \approx 0.4$, there is a very small effect of the depth. When Fn_h is around 1 and h/L is small, the shallow-water wave resistance ratio is very large. For instance if $h/L < 0.13$, the ratio is larger than 15 for $Fn_h \approx 1$. The largest ratio shown in Figure 4.31 is 50. Obviously, we should then question the linearity assumption of the theory. When $Fn_h < \approx 0.6$, there is negligible effect of the depth on the wave resistance for any h/L. When $Fn_h > \approx 2$, there is a small but not negligible effect.

Because there is a clear connection between the wave resistance and the waves generated by the ship, Figures 4.30 and 4.31 are also indicators of how the wash is influenced by Fn_h and h/L. However, because the wave decay is important from a coastal point of view, more detailed knowledge about the wave systems generated at subcritical, critical, and supercritical speeds is required. We deal with this in more detail later.

Another way of presenting the influence of water depth on wave resistance is shown in Figure 4.32. Experimental values for residual resistance R_R for a ship are shown as a function of Froude number for different h/L. The residual resistance is an approximation of the wave resistance. The ship model has a prismatic coefficient $C_P = 0.64$ and a beam-to-draft ratio of 3. The curves of R_R have very clear peaks for the presented h/L-values between 0.05 and 0.333. These peaks occur very close to a depth Froude number of 1. This can be seen from Table 4.3. Figure 4.32 illustrates that a ship needs extra power in shallow water to go through critical speed. Because Figure 4.32 also presents the residual resistance for infinite depth, we can compare the results with the trend in the ratio between the shallow-water and deep-water wave resistance for the Wigley hull presented in Figures 4.30 and 4.31. We see that this ratio at critical speed clearly increases with

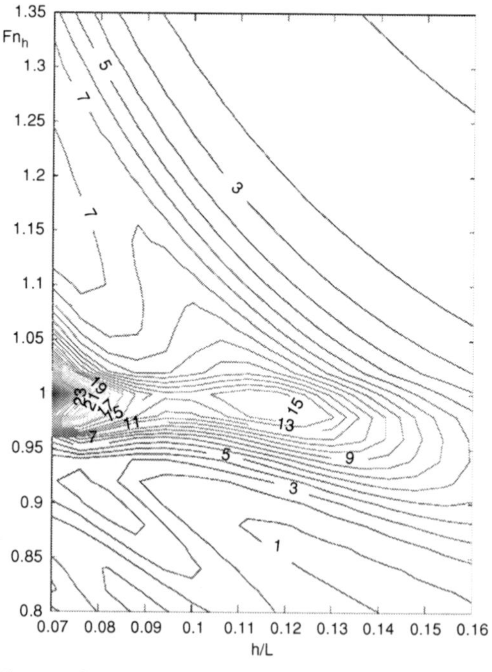

Figure 4.31. Shallow-water wave resistance ratio for a Wigley hull as a function of h/L and Fn_h. $B/L = 0.1$. $D/L = 0.0625$. Detailed view near critical speed and small h/L (Yang 2002 and Yang et al. 2005).

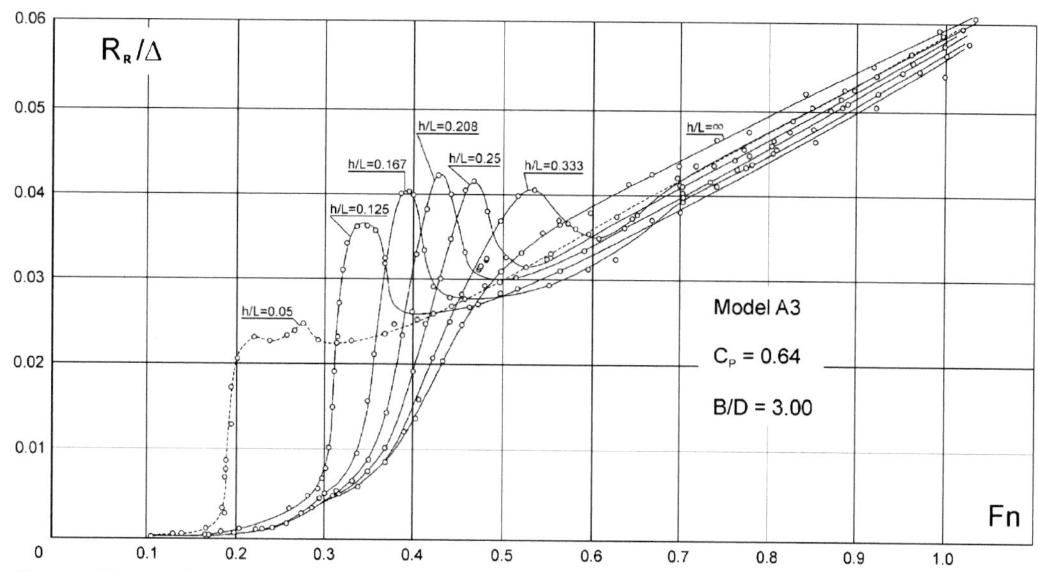

Figure 4.32. Experimentally obtained residual resistance R_R for a ship in deep and shallow water. The peak values occur at depth Froude number 1. See Table 4.3 (*Graff et al. 1964*).

decreasing h/L. The results also confirm that a relatively large range of supercritical speeds have a large influence on the wave resistance for small h/L. However, for all presented cases, the shallow-water wave resistance ratio approaches a value less than 1 for the highest presented supercritical speed. The trend for subcritical speed is that the wave resistance approaches the value for infinite depth when the speed decreases.

In order to translate the information in Figures 4.30 and 4.31 to ship speeds, water depth, and ship length, we first present in Figure 4.33 the relationship between ship speed and water depth at critical Froude number. For instance, Figure 4.30 shows that the shallow-water wave resistance ratio is less than 3 for $h/L \approx 0.2$. Using, for instance, $h/L = 0.2$ as an upper limit for significant influence of critical Froude number means that for a given h in Figure 4.33, ship lengths larger than 5 h would have a significant influence on critical speed effects. However, one should note that the discussion is based on the shallow-water wave resistance ratios, not on the values of shallow-water wave resistance. This means we must also have the wave resistance in infinite depth in mind. That depends on the length Froude number $Fn = U/\sqrt{Lg}$, (see e.g., Figure 4.33). If $Fn < \approx 0.15$, the wave resistance in deep water is negligible.

As an example, let us consider a ship length $L = 100$ m. The previous discussion then says that

Table 4.3. *Values of Froude numbers for different water depth–to–ship length ratios h/L that correspond to depth Froude numbers equal to 1, that is, critical Froude number*

h/L	$Fn = U/\sqrt{Lg}$
0.050	0.22
0.125	0.35
0.167	0.41
0.208	0.46
0.250	0.50
0.333	0.58

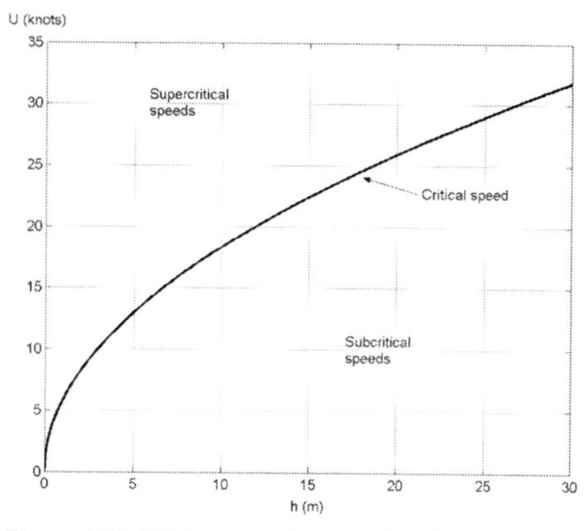

Figure 4.33. Critical speed as a function of water depth h.

$h < 20\,\text{m}$ should be of concern. What the corresponding critical speeds are follows from Figure 4.33. This means the ship speeds are less than 27 knots. However, there is a minimum ship speed that is of concern. We must require that $Fn \geq 0.15$. $Fn = 0.15$ corresponds to a ship speed of 9 knots. However, Figure 4.32 shows that a critical speed of 9 knots corresponds to an unrealistically small water depth for a 100 m–long vessel with realistic drafts in which to operate.

4.4.1 Wave patterns

We start by examining the mathematical representation of the linear far-field waves, as in section 4.2 for deep-water waves. This means we start out with a harmonic long-crested wave in an Earth-fixed coordinate system (see eq. (4.4)). The dispersion relationship for regular waves in finite water depth is (see Table 3.1)

$$\frac{\omega^2}{g} = k \tanh kh. \tag{4.71}$$

The group velocity is given by eq. (3.22). By now, using a ship-fixed coordinate system and requiring the waves to be steady (time independent) relative to the ship-fixed coordinate system, we find, as for deep-water waves, that $\omega = kU\cos\theta$. This means that eq. (4.71) can be written as

$$kU^2\cos^2\theta/g = \tanh kh. \tag{4.72}$$

By introducing $k = 2\pi/\lambda$, we can express this equation for $\theta = 0$ as $Fn^2 = \lambda_T/(2\pi L)\tanh((2\pi L/\lambda_T)h/L)$, where λ_T means the transverse wavelength along the ship's track. Certain values of λ_T/L, independent of the depth, are responsible for the humps and hollows in the wave resistance. Because $\tanh((2\pi L/\lambda_T)h/L) \to 1$ when $(h/L) \to \infty$ and generally $\tanh((2\pi L/\lambda_T)h/L) \leq 1$, it means that when transverse waves exist, the humps and hollows for finite depth correspond to smaller Fn than those for infinite depth. Later we discuss the condition for transverse waves to exist in finite water depth.

It follows from eq. (4.72) that

$$Fn_h^2 kh\cos^2\theta = \tanh kh, \tag{4.73}$$

where we have introduced the depth Froude number given by eq. (4.70). As already described, this plays an important role in classifying shallow-water steady ship waves. By using the expression for ω, we can write the group velocity as

$$V_g = \left(\frac{1}{2} + \frac{kh}{\sinh 2kh}\right)U\cos\theta. \tag{4.74}$$

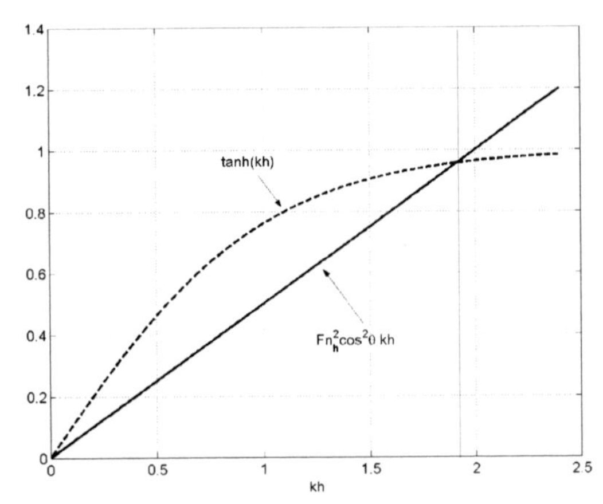

Figure 4.34. Illustration of how we can solve eq. (4.73) to find kh for given wave propagation direction θ and depth Froude number Fn_h.

Eq. (4.73) has to be solved numerically for kh. We can geometrically find the solution as the intersection point between the straight line $Fn_h^2\cos^2\theta \cdot kh$ and tanh kh as a function of kh (Figure 4.34). Because tanh kh and $d(\tanh(kh))/d(kh)$ are monotonically increasing and decreasing with kh, respectively, the slope of tanh kh at $kh = 0$ is important for the existence of a solution of eq. (4.73). We can write

$$\left.\frac{d\tanh(x)}{dx}\right|_{x=0} = \left.\frac{1}{\cosh^2(x)}\right|_{x=0} = 1.$$

This means that it is necessary that

$$Fn_h|\cos\theta| \leq 1 \tag{4.75}$$

for a solution to exist. If $Fn_h < 1$ (subcritical Froude numbers), then solutions exist for all θ. If $Fn_h > 1$ (supercritical Froude numbers), then there exists a minimum $|\theta_{\min}| = \cos^{-1}(1/Fn_h)$ (Figure 4.35) so that solutions exist only for $|\theta_{\min}| < |\theta| < \pi/2$. This means, for instance, that transverse waves along the ship's track ($\theta = 0$) cannot exist for supercritical flow.

In order to show the far-field wave patterns, we could use the "method of stationary phase" as for deep-water waves. We only outline the procedure. We start by representing the waves as in eq. (4.17),

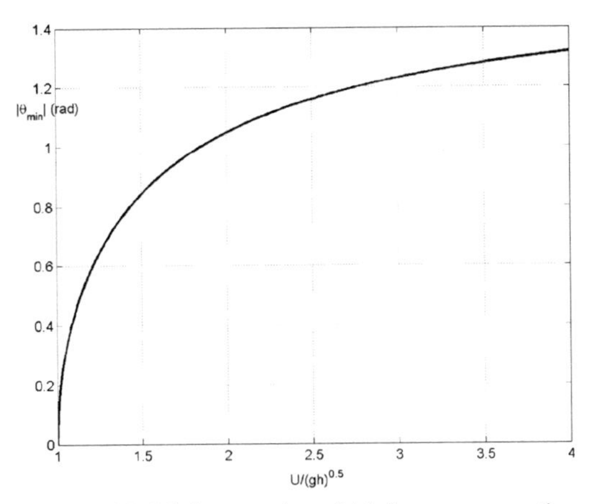

Figure 4.35. Minimum value of $|\theta|$ (wave propagation direction) for steady waves to exist for supercritical-depth Froude numbers.

giving

$$\zeta = \mathrm{Re} \int_{-\pi/2}^{\pi/2} A(\theta) \exp[irG(\theta)]d\theta, \quad (4.76)$$

where

$$G(\theta) = k(\theta)\cos(\theta - \alpha). \quad (4.77)$$

Here $k(\theta)$ is determined by eq. (4.73). Points of stationary phase follow from solutions of $dG/d\theta = 0$. This procedure cannot be done analytically, as was done for infinite depth. The calculations show that for $Fn_h < 1$, both transverse and divergent waves exist in a similar way to those for deep water. However, the Kelvin angle, which is $19°28'$ for deep water (see Figure 4.3), is influenced by Fn_h (Figure 4.36). When $Fn_h < 0.5 - 0.6$, the value is practically the same as for deep water. We will explain this by considering the longest ship waves in deep water, that is, the transverse waves along the ship's track. The wavelength is $\lambda_T = 2\pi U^2/g$ (see eq. (4.11)). This means $\lambda_T/h = 2\pi Fn_h^2$, that is, $\lambda_T/h = 2.3$ for $Fn_h = 0.6$. Because of the exponential decay of the fluid motion with depth, the waves hardly feel the bottom. A rapid increase in the Kelvin angle occurs for $Fn_h > 0.9$, and the angle is $90°$ for $Fn_h = 1$.

We explain the Kelvin angle of $90°$ at $Fn_h = 1$ similarly to the way we explained, by means of Figure 4.7, the Kelvin angle of $19°28'$ for deep water. First we consider eq. (4.73), which determines kh as a function of θ. When $Fn_h = 1$, the solution of kh is small for small θ. So we Taylor expand tanh kh and get

$$\cos^2\theta \approx 1 - \frac{1}{3}(kh)^2 \text{ for small } kh. \quad (4.78)$$

Using this expression, we can express the group velocity given by eq. (4.74). This gives

$$V_g = \left(0.5 + \frac{\sqrt{3}|\sin\theta|}{\sinh(2\sqrt{3}|\sin\theta|)}\right) U\cos\theta \quad (4.79)$$

for small θ.

At time $t_0 - t$, the ship is at position B and generates waves in different directions. After time t, the ship has moved to the new position, A. Now the time is t_0. We consider the position of the wave energy that was created by the ship at the previous time $t_0 - t$. This is given by $V_g t$ and is a function of the propagation direction θ of the waves. This is presented in Figure 4.37 by using eq. (4.79), which is valid for small θ. In order to predict $V_g t$ for other θ-values, eq. (4.73) has to be solved numerically. An exception is when θ is close to $\pi/2$. Then the solution will correspond to large kh so that tanh $kh \approx 1$ and we get the deep water solution $k = g/\left(U^2\cos^2\theta\right)$. The group velocity is then $0.5U\cos\theta$. We have also shown that curve in Figure 4.37.

Because the front of the waves generated at the arbitrary previous time $t_0 - t$ touches the position

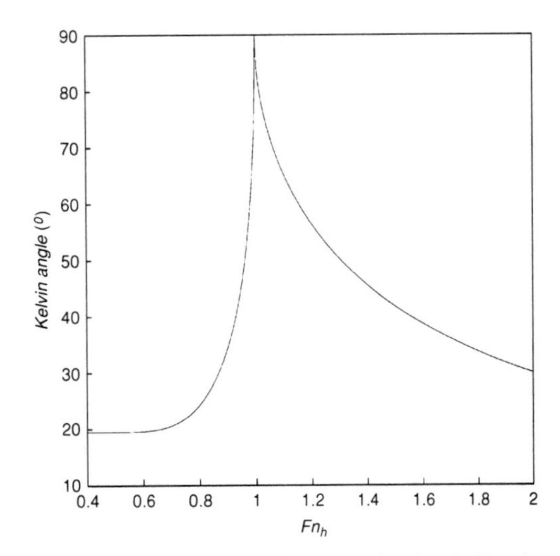

Figure 4.36. The relation between the depth Froude number Fn_h and the Kelvin angle behind the ship in finite water depth. The Kelvin angle is the angle between the boundary of the wave system and the ship course. This curve is based on linear theory in water with an infinite free-surface extent (Yang 2002).

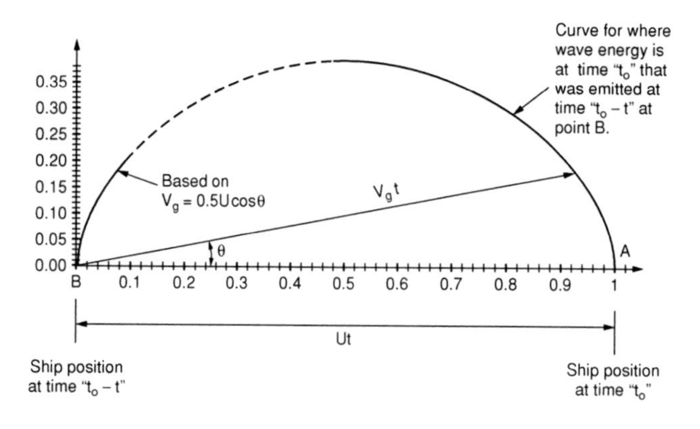

Figure 4.37. Illustration of propagation of waves generated by a ship at critical speed.

A of the ship at time t_0, the Kelvin angle is 90°. We see this if we draw a line from point A tangent to the curve represented by $V_g t$. This is similar to the procedure for Figure 4.7 to define the Kelvin angle for deep water. In the present critical Froude number case, we get a vertical line through A, which means the Kelvin angle is 90°.

Let us now study the case $\theta = 0$ in Figure 4.37. We notice that the group velocity V_g is then equal to U. As the ship moves ahead, the ship continuously generates waves and all the waves with $\theta = 0$ will have an energy propagation velocity equal to the ship's speed. This results is an accumulation of wave energy along $\theta = 0$. This energy cannot escape according to our linear theory. It is constrained to follow the ship's speed at critical speed, according to linear theory. However, there is a limit to how steep waves can be. They will finally break, but nonlinear effects matter before that. This may imply that steady conditions relative to the ship are not obtained. Let us illustrate that for shallow-water waves in which the wave energy propagation velocity V_g is $\sqrt{g(h+A)}$, with A as the wave amplitude. When A increases, V_g becomes larger and larger relative to the ship's speed. This means steady conditions are not obtained. This effect is clearly seen for a ship in a shallow-water channel. Large-amplitude waves will then exist upstream of the ship around critical speed.

The wave amplitude at critical speed depends strongly on h/L. This can be seen indirectly from Figures 4.30 and 4.31. It is a small h/L that causes large waves at critical speed. Yang (2002) presented wave patterns around a Wigley hull at subcritical and supercritical speeds by using thin ship theory.

4.5 Ship in shallow water

We assume shallow-water conditions, that is, $kh \to 0$, as Tuck (1966) did in his analysis of linear shallow-water ship waves. This is a simplification, which does not fully account for eq. (4.73). This will be demonstrated by comparing the results with thin ship results by Yang (2002) for finite but small depth.

The flow analysis is divided into far-field and near-field descriptions. In the far-field description, we look upon the flow at a distance from the ship that is of the order of the length of the ship. We do not see the details of the flow around the cross sections of the ship from that perspective. We need a near-field solution to do that. The flow cannot be completely determined by either the far-field or near-field solution. In order to do that, we match the two solutions.

4.5.1 Near-field description

The variation of the flow in the near field is stronger in the transverse cross-sectional plane than in the longitudinal direction. We divide the velocity potential into two parts, as in eq. (3.4), and want to find the velocity potential φ caused by the ship. In the near field this can be assumed to satisfy the 2D Laplace equation, that is,

$$\frac{\partial^2 \varphi}{\partial y^2} + \frac{\partial^2 \varphi}{\partial z^2} = 0. \tag{4.80}$$

The body boundary condition follows from the requirement that there be no flow through the hull surface. This means

$$\frac{\partial \varphi}{\partial n} = -Un_1 \quad \text{on the hull surface.}$$

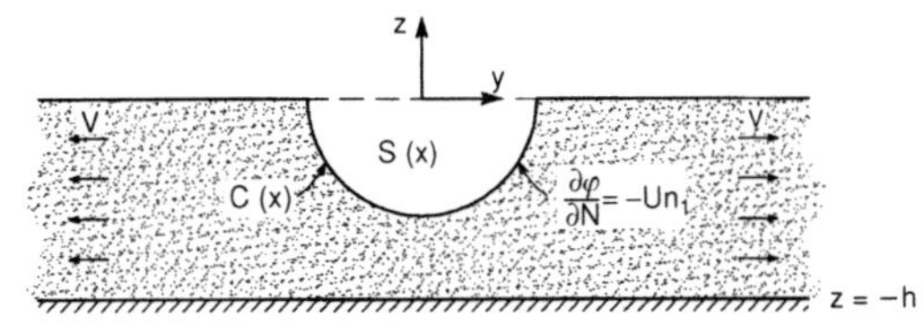

Figure 4.38. Near-field description of the flow.

Here $\mathbf{n} = (n_1, n_2, n_3)$ is the normal vector to the hull surface. The positive normal direction is into the fluid domain. We now assume a slender ship and use the fact that $n_1\partial/\partial x$ is much smaller than $n_2\partial/\partial y$ and $n_3\partial/\partial z$. This means we can approximate $\partial\varphi/\partial n$ as

$$\begin{aligned} n_2\frac{\partial\varphi}{\partial y} + n_3\frac{\partial\varphi}{\partial z} &\equiv \frac{\partial\varphi}{\partial N} \\ &= -Un_1 \quad \text{on the hull surface,} \end{aligned} \tag{4.81}$$

where $\mathbf{N} = (n_2, n_3)$. We must also satisfy the condition that $\partial\varphi/\partial z = 0$ on $z = -h$. The free-surface condition is simplified in Tuck's analysis by using a rigid wall condition, that is, $\partial\varphi/\partial z = 0$ on $z = 0$. At a distance from the cross section, the flow will be depth independent with a velocity V as shown in Figure 4.38. V can be determined by conservation of fluid mass, that is,

$$V2h = \int_{C(x)} \frac{\partial\varphi}{\partial N} dl = -U \int_{C(x)} n_1 \, dl. \tag{4.82}$$

The integral can be rewritten by means of Figure 4.39, in which we consider a strip of the ship of length Δx. Noting the direction of the hull

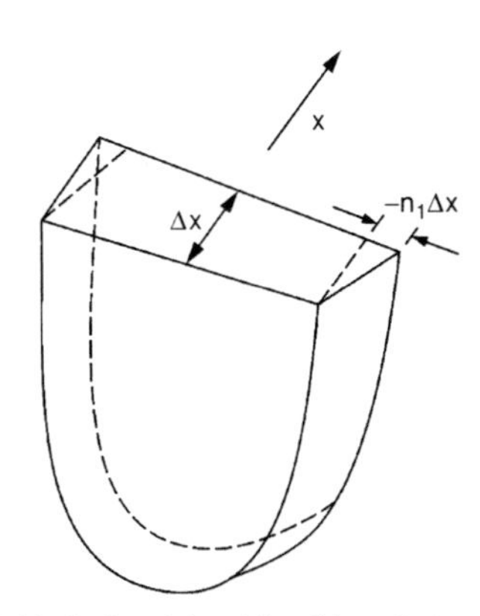

Figure 4.39. Strip of the ship of length Δx. n_1 is the x-component of the unit normal vector to the hull surface pointing outward and into the fluid.

surface normal vector, we then see from geometry that

$$\Delta x \int_{C(x)} n_1 \, dl = -\left(S(x + \Delta x) - S(x)\right).$$

Here S is the submerged cross-sectional area of the ship. In the limit $\Delta x \to 0$, we have

$$\int_{c(x)} n_1 \, dl = -\frac{dS}{dx}. \tag{4.83}$$

It follows from eq. (4.82) that

$$V = \frac{U}{2h}\frac{dS}{dx}. \tag{4.84}$$

So if $S(x)$ increases, V is an outward-going velocity.

4.5.2 Far-field equations

The field equation is now the 3D Laplace equation for the velocity potential φ, that is,

$$\frac{\partial^2\varphi}{\partial x^2} + \frac{\partial^2\varphi}{\partial y^2} + \frac{\partial^2\varphi}{\partial z^2} = 0. \tag{4.85}$$

The free-surface conditions follow from eqs. (3.7) and (3.9) by setting p_0 and $\partial/\partial t$ equal to zero. The bottom condition is $\partial\varphi/\partial z = 0$ on $z = -h$. It follows by integrating eq. (4.85) in the z-direction that

$$\frac{1}{h}\int_{-h}^{0}\left[\frac{\partial^2\varphi}{\partial x^2} + \frac{\partial^2\varphi}{\partial y^2}\right]dz + \frac{1}{h}\frac{\partial\varphi}{\partial z}\bigg|_{z=0} - \frac{1}{h}\frac{\partial\varphi}{\partial z}\bigg|_{z=-h} = 0.$$

The depth-averaged velocity potential

$$\bar{\varphi}(x, y) = \frac{1}{h}\int_{-h}^{0} \varphi dz \tag{4.86}$$

is now introduced in combination with using the kinematic free-surface condition (see eq. (3.9)) and the bottom condition. This gives

$$\frac{\partial^2\bar{\varphi}}{\partial x^2} + \frac{\partial^2\bar{\varphi}}{\partial y^2} + \frac{U}{h}\frac{\partial\zeta}{\partial x} = 0.$$

Using the dynamic free-surface condition (see eq. (3.7)) and approximating φ by $\bar{\varphi}$ gives

$$\frac{\partial^2\bar{\varphi}}{\partial x^2} + \frac{\partial^2\bar{\varphi}}{\partial y^2} - Fn_h^2\frac{\partial^2\bar{\varphi}}{\partial x^2} = 0. \tag{4.87}$$

4.5.3 Far-field description for supercritical speed

We can formally write a solution of eq. (4.87) for $Fn_h > 1$ as

$$\bar{\varphi} = F(u), \tag{4.88}$$

where

$$u = x \pm \left(Fn_h^2 - 1\right)^{1/2} y. \tag{4.89}$$

This can easily be shown by substituting $\bar{\varphi}$ into eq. (4.87). The function F is undetermined at this stage, but the dependence of F on x and y expressed by eq. (4.89) gives us valuable information that will be explored later. In order to determine F, the ship must be introduced into the analysis. This means that F must have a symmetric dependence on y. Further, we require that the flow caused by the ship appear downstream of the ship. We can then write

$$\bar{\varphi} = F\left(x - \left(Fn_h^2 - 1\right)^{1/2} |y|\right). \tag{4.90}$$

The function F must be found by matching with the near-field solution, that is, the inner expansion of eq. (4.90) near the ship must be consistent with eq. (4.84). We assume $y > 0$ without loss of generality. Eq. (4.90) gives horizontal velocity

$$\frac{\partial \bar{\varphi}}{\partial y} = -\left(Fn_h^2 - 1\right)^{1/2} \times F'\left(x - \left(Fn_h^2 - 1\right)^{1/2} y\right) \quad \text{for } y > 0.$$

If we let $y \to 0$, this must be equal to V given by eq. (4.84), that is,

$$-\left(Fn_h^2 - 1\right)^{1/2} F'(x) = \frac{U}{2h}\frac{dS}{dx}.$$

This means

$$\bar{\varphi} = -\frac{U}{2h}\left(Fn_h^2 - 1\right)^{-1/2} S\left[x - \left(Fn_h^2 - 1\right)^{1/2} |y|\right]. \tag{4.91}$$

The free-surface elevation can then be expressed by the dynamic free-surface condition as

$$\zeta = 0.5 Fn_h^2 (Fn_h^2 - 1)^{-1/2} dS(u)/du, \tag{4.92}$$

where

$$u = x - (Fn_h^2 - 1)^{1/2} |y|.$$

It means that ζ is constant along the lines

$$x - \left(Fn_h^2 - 1\right)^{1/2} |y| = \text{const.} \tag{4.93}$$

The lines must originate from the ship, otherwise dS/dx is zero. We can then write the constant lines as

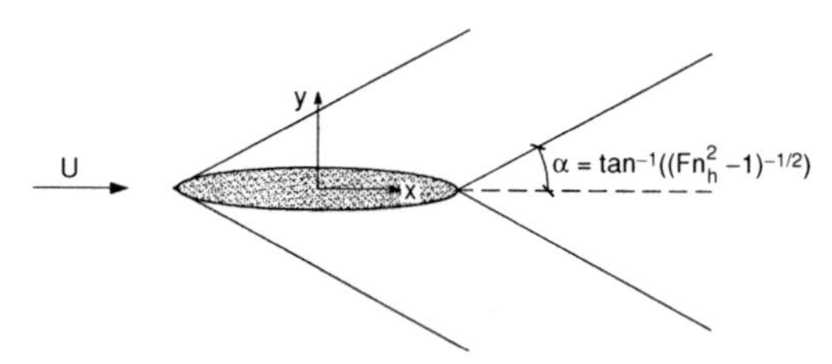

Figure 4.40. Boundaries of the wave system for supercritical flow according to Tuck's (1966) shallow-water theory.

$$|y| = \frac{1}{(Fn_h^2 - 1)^{1/2}}(x - x_0), \tag{4.94}$$

where $-L/2 \le x_0 < L/2$. We have depicted the lines originating from the bow and the stern in Figure 4.40. The angle α of the lines as shown in Figure 4.40 can be related to the wave propagation direction as

$$\theta = \alpha - \frac{\pi}{2}.$$

This means

$$\cos\theta = \sin\left(\tan^{-1}\left(\left(Fn_h^2 - 1\right)^{-1/2}\right)\right) = 1/Fn_h. \tag{4.95}$$

This is the same as the minimum angle $|\theta_{\min}|$ for waves to exist, as presented in Figure 4.35. The wave system shown in Figure 4.40 has a delta-like formation. The higher the speed, the smaller the α.

Eq. (4.92) shows that the wave amplitude does not decay at all with lateral distance. As pointed out in the introduction to this chapter, this has important consequences for wash on the seashore.

Figure 4.41 shows comparisons among Tuck's (1966) shallow-water theory, thin ship theory, and linear and nonlinear Boussinesq-type equations by Yang (2002). The wave elevation is predicted at the two longitudinal cuts $y = L$ and $y = 1.5L$. Fn_h is 1.2 and $h/L = 0.1$. The angle α by Tuck's theory is 56.4°. The wave elevation according to Tuck is a sawtooth function. When $y = L$, the wave elevation ζ is zero upstream from $x/L = -0.66$ and downstream from $x/L = -1.66$. ζ jumps to the value $0.0181L$ at $x/L = -0.66$ and jumps from the value $-0.0181L$ to zero at $x/L = -1.66$. Similar behavior occurs at $y = 1.5L$. Then the wave elevation is zero upstream of $x/L = -0.99$ and downstream of $x/L = -1.99$. The other calculation methods have a smooth behavior in the free

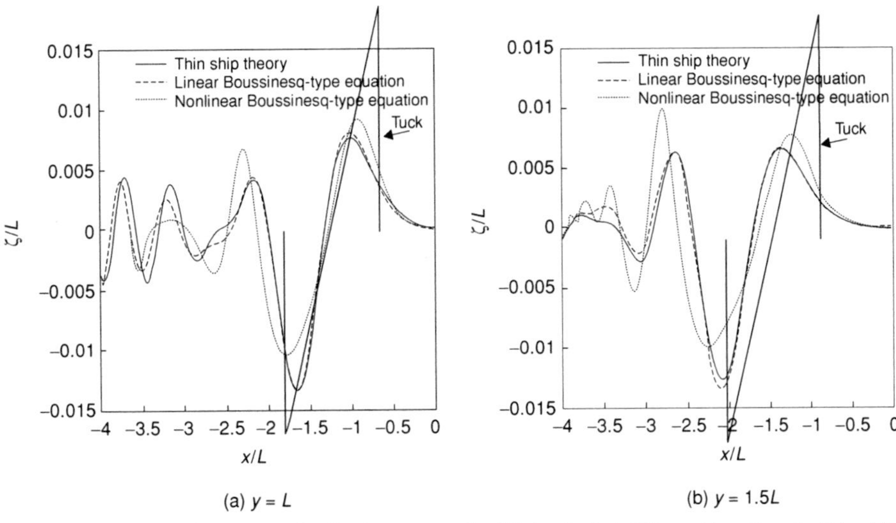

Figure 4.41. Comparison of longitudinal wave cuts for Wigley hull at different locations by thin ship theory and linear and nonlinear Boussinesq-type equations (Yang 2002) as well as Tuck's shallow-water theory. The input of Boussinesq-type equations is obtained by thin ship theory at $y = 0.5L$. The ship is located at $-1 < x/L < 0$, $y/L = 0$. $Fn_h = 1.2$. $h/L = 0.1$. $B/L = 0.1$, $D/L = 0.0625$. $\zeta =$ wave surface elevation.

surface and do not predict zero wave elevation outside a strip, as does Tuck's method. One reason for the difference is that Tuck's shallow-water theory is nondispersive whereas the other methods account for dispersion. The results also show that nonlinearities matter.

By Fourier expanding the sawtooth function, we see that many Fourier components are needed to describe the function. This means that a broad, discrete spectrum of wave numbers is of importance. This is not in contradiction with the shallow-water approximation of eq. (4.73), which simply says that $Fn_h^2 \cos^2\theta = 1$, that is, all wave numbers satisfy this equation. However, this leads to a contradiction if we insert the wave numbers following from the Fourier analysis into eq. (4.73). This means that all the wavelengths corresponding to the necessary wave numbers are not sufficiently long for the shallow-water approximation to be valid.

Later we see that Tuck's shallow-water theory gives a better description of forces and moments, but before doing that, we analyze subcritical speed.

4.5.4 Far-field description for subcritical speed

Eq. (4.87) also applies for subcritical speed, but the solution is not in the form of eqs. (4.88) and (4.89). We see that u, given by eq. (4.89), is imaginary when $Fn_h < 1$. This has no physical meaning. So we proceed differently for subcritical speed. The first step is to introduce the new variable

$$y' = y\left(1 - Fn_h^2\right)^{1/2}. \tag{4.96}$$

This transformation means that eq. (4.87) can be written as

$$\frac{\partial^2 \bar{\varphi}}{\partial x^2} + \frac{\partial^2 \bar{\varphi}}{\partial y'^2} = 0. \tag{4.97}$$

We do not see the details of the ship in the far field. The ship appears as a line from $x = -L/2$ to $x = L/2$. In several previous examples, we have seen that the effect of the ship on the flow is to push the water out in the front part and attract the flow in the stern part. This means we have used a distribution of sources (sinks) to represent the effect of the ship. In those cases, the 3D Laplace equation had to be satisfied in the fluid. In our approximation, we do not have a 3D Laplace equation but a 2D Laplace equation given by eq. (4.97). This means that the sources (sinks) must satisfy this equation outside the source points. This leads to writing the velocity potential in terms of the following line distribution of 2D sources and sinks:

$$\bar{\varphi} = \frac{1}{2\pi} \int_{-L/2}^{L/2} q(\xi) \ln \sqrt{(x-\xi)^2 + y'^2} d\xi. \quad (4.98)$$

Here the source density $q(x)$ is an unknown that will be determined by matching with eq. (4.84). This means we evaluate $\partial\bar{\varphi}/\partial y$ when $y \to 0$. Before doing that, we need to study $\partial\bar{\varphi}/\partial y'$ when $y' \to 0$. We can go through an argument similar to the one in the 3D case in which we used Figure 4.12. The result is

$$\left. \frac{\partial \bar{\varphi}}{\partial y'} \right|_{y'=0+} = \frac{1}{2} q(x).$$

Because it follows from eq. (4.96) that

$$\frac{\partial \bar{\varphi}}{\partial y} = \left(1 - Fn_h^2\right)^{1/2} \frac{\partial \bar{\varphi}}{\partial y'},$$

the matching gives

$$\left(1 - Fn_h^2\right)^{1/2} \frac{1}{2} q(x) = \frac{U}{2h} \frac{dS}{dx}.$$

This means

$$\bar{\varphi} = \frac{U}{2\pi h} \left(1 - Fn_h^2\right)^{-1/2} \times \int_{-L/2}^{L/2} d\xi\, S'(\xi) \ln \sqrt{(x-\xi)^2 + y^2\left(1 - Fn_h^2\right)}. \quad (4.99)$$

We note that this is not a solution containing waves. It was a surprise for some when Tuck first presented these physically based simple analytical models showing that waves are generated in the supercritical flow regime only for a ship at moderate-length Froude numbers in shallow water. Some expected the opposite. However, Tuck's theory gives a correct description.

4.5.5 Forces and moments

We derive formulas for forces and moments based on Tuck's theory and start with his expressions for supercritical speed. An approximation of the free-surface elevation along the hull is obtained by setting $|y| = 0$ in eq. (4.92). We now use that the pressure is hydrostatic relative to the instantaneous free-surface elevation. This can be shown by using the finite-depth expressions in Table 3.1 and letting $kh \to 0$. We separate out the pressure part that is hydrostatic relative to the mean free-surface level $z = 0$. That part gives buoyancy forces. This means that the dynamic pressure along the hull is approximated as

$$p = \rho U^2 0.5 h^{-1} \left(Fn_h^2 - 1\right)^{-1/2} \frac{dS}{dx}, \quad -\frac{L}{2} < x < \frac{L}{2} \quad (4.100)$$

for $Fn_h > 1$. We note that the pressure is independent of y and z for a given x. Further, the pressure is positive in the forward part of the ship where dS/dx is increasing, whereas it is negative in the aft part of the ship where dS/dx is decreasing. We can then intuitively understand that this pressure causes a resistance and a pitch moment that forces the bow up. Further, the net vertical force is zero for a ship with fore-and-aft symmetry. The expression for the wave resistance can be derived by formally writing

$$R_W = -\iint_{S_B} p n_1 \, dS. \quad (4.101)$$

Here S_B is the mean wetted body surface and n_1 is the x-component of the normal vector $\mathbf{n} = (n_1, n_2, n_3)$ to the body surface. The positive normal direction is into the fluid domain. Eq. (4.101) can be simplified by using the fact that p is independent of y and z, together with eq. (4.83). This means the wave resistance for supercritical flow can be expressed as

$$R_W = \frac{\rho U^2}{2h\left(Fn_h^2 - 1\right)^{1/2}} \int_{-L/2}^{L/2} [S'(x)]^2 dx. \quad (4.102)$$

The vertical force follows by exchanging n_1 in eq. (4.101) with n_3. We note from Figure 4.42 that

$$-\int n_3 \, dl = \int_{-b/2}^{b/2} dy = b(x). \quad (4.103)$$

The left-hand integral is along the cross-sectional surface. Further, $b(x)$ is the local beam. We then see by using eqs. (4.103) and (4.100) that the vertical force can be expressed as

$$F_3 = \frac{\rho U^2}{2h\left(Fn_h^2 - 1\right)^{1/2}} \int_{-L/2}^{L/2} b(x) S'(x) \, dx \quad (4.104)$$

for supercritical speed. If the ship has fore-and-aft symmetry, F_3 is zero.

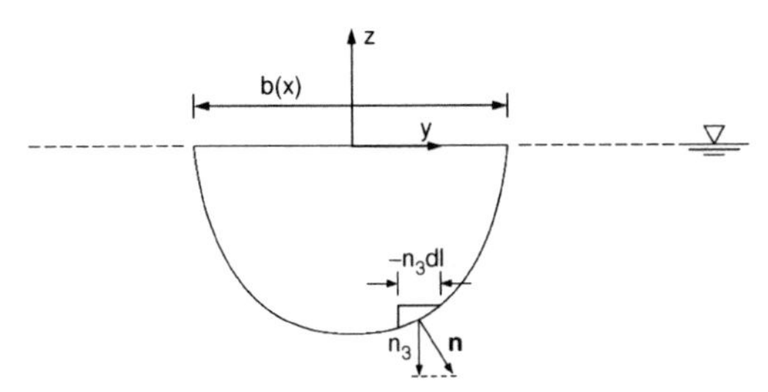

Figure 4.42. Cross section of the ship. n_3 is the z-component of the normal unit vector to the hull surface, pointing outward.

We can now write up the pitch moment as

$$F_5 = -\frac{\rho U^2}{2h(Fn_h^2 - 1)^{1/2}} \int_{-L/2}^{L/2} x b(x) S'(x) dx \quad (4.105)$$

for supercritical flow.

Because the ship according to Tuck's analysis does not generate waves for subcritical speed, the wave resistance is zero. Further, it turns out that the pitch moment is zero for a ship with fore-and-aft symmetry, and relatively small otherwise. However, the vertical force is important. This can be expressed as

$$F_3 = \frac{\rho U^2}{2\pi h(1 - Fn_h^2)^{1/2}} \times \int_{-L/2}^{L/2} b'(x) \int_{-L/2}^{L/2} S'(\xi) \ln|\xi - x| d\xi \, dx. \quad (4.106)$$

The ξ-integral is connected with the pressure. This follows from eq. (4.99) by setting $y = 0$ and using the pressure part $-\rho U \partial\bar{\varphi}/\partial x$ from eq. (3.6). Actually, the ξ-integral, as expressed in eq. (4.106), is connected with the velocity potential and a manipulation of the integral has been done. We will show this by first noting that $-\rho U \partial\bar{\varphi}/\partial x$ on the hull is independent of y and z according to the analysis. This means we can write

$$F_3 = -\rho U \int_{-L/2}^{L/2} b(x) \frac{\partial \bar{\varphi}}{\partial x} dx.$$

By integration by parts and noting that b is zero at the ship ends, we get

$$F_3 = \rho U \int_{-L/2}^{L/2} b'(x) \bar{\varphi} \, dx.$$

Then we can directly identify the terms in eq. (4.106). We recall that in the case with supercritical speed, we expressed the pressure in terms of the wave elevation ζ. However, ζ is connected to $-\rho U \partial\varphi/\partial x$ by the dynamic free-surface condition given by eq. (3.7).

Yang (2002) compared Tuck's theory for forces and moments with his own calculations based on thin ship theory for finite water depth. The water depth–to–ship length ratio was $h/L = 0.1$, and a Wigley hull with $B/L = 0.1$ and $D/L = 0.0625$ was used. The comparison is shown in Figure 4.43. Generally speaking, the agreement is better than that shown for the wave elevation in Figure 4.41.

Figure 4.43a shows the comparison of wave resistance coefficient $C_W = R_W/(0.5\rho U^2 S)$ by the two theories. Here R_W is the wave resistance and S is the mean wetted surface area. As already stated, the wave resistance coefficient is zero at subcritical speeds according to Tuck's (1966) slender body theory for shallow water, whereas it is a small value by the thin ship theory. At critical speed, the result of Tuck's (1966) slender body theory for shallow water goes to infinity, whereas thin ship theory gives finite, large results. At supercritical speeds, the two theories show similarities. Figure 4.43b shows the comparison of vertical force F_3 from the two theories. The two theories agree well at subcritical speeds. At supercritical speeds, Tuck's (1966) shallow-water slender body theory predicts zero result but the thin ship theory predicts nonzero small values. The reason that Tuck's (1966) shallow-water slender body theory predicts zero results is that the Wigley hull has for-and-aft symmetry. This is also the reason Tuck's (1966) slender body theory in shallow water predicts zero pitch moment at subcritical speeds. Figure 4.43c shows the comparison of pitch moment F_5 from the two theories. The two theories agree well at supercritical speeds. At subcritical speeds, Tuck's (1966) shallow-water slender body theory predicts zero whereas the thin ship theory predicts nonzero results. We must note that Tuck's (1966) shallow-water slender body theory is very fast from a computational point of view whereas the thin ship theory is only fast when calculating the wave resistance but takes long CPU time when calculating the vertical force F_3 and the pitch moment F_5. Tuck's (1966) shallow-water

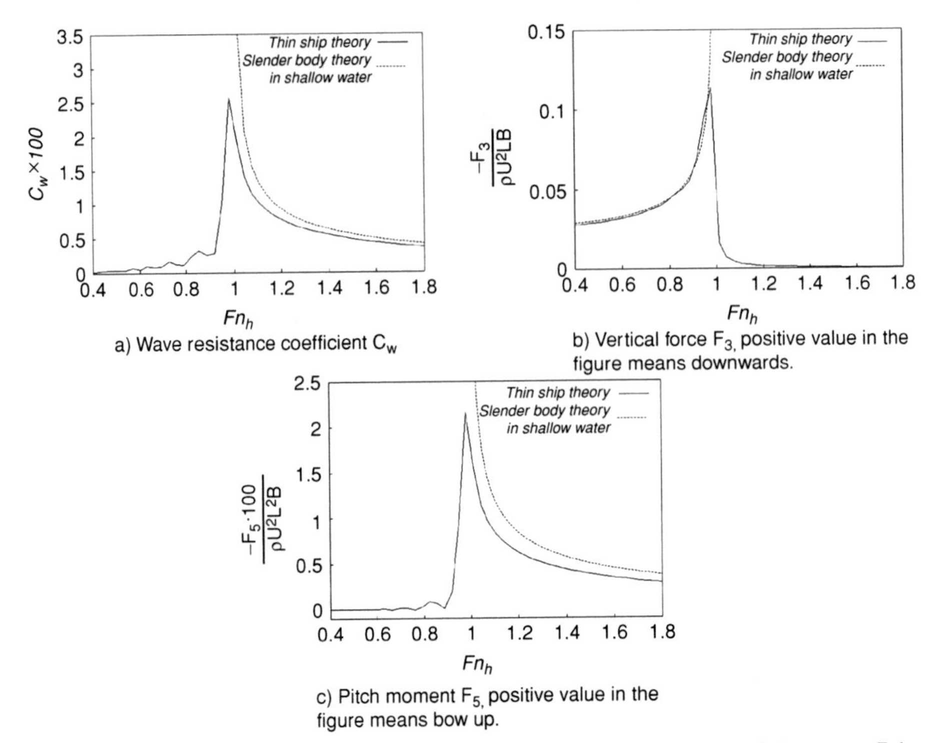

Figure 4.43. Comparison of wave resistance coefficient C_W, vertical force F_3, and pitch moment F_5 by thin ship theory and Tuck's (1966) slender body theory in shallow water. Wigley hull. $B/L = 0.1$, $D/L = 0.0625$, $h/L = 0.1$ (Yang 2002 and Yang et al. 2005).

slender body theory is, however, limited to small water depth, whereas the thin ship theory can be applied to any water depth.

4.5.6 Trim and sinkage

We now use Tuck's shallow-water theory to predict trim and sinkage. We start with sinkage in which only subcritical speed is of interest. Static balance of vertical force requires the hydrodynamic force given by eq. (4.106) to balance the weight and the buoyancy force due to hydrostatic pressure $-\rho g z$. When $U = 0$, the weight and the buoyancy force are in equilibrium. The ship then has a submerged cross-sectional area $S_0(x)$. This area is increased because of sinkage η_s, which we define as positive downward. We assume that the ship is wall-sided at the mean free surface and η_s is small relative to the draft. The submerged cross-sectional area can then be expressed as

$$S(x) = S_0(x) + b(x)\eta_s. \tag{4.107}$$

Because the trim is negligible for subcritical speed, we can assume that η_s is x-independent. We can now set up the following vertical force balance:

$$\frac{Fn_h^2}{2\pi\left(1 - Fn_h^2\right)^{1/2}} \cdot \int_{-L/2}^{L/2} dx b'(x) \times \left\{ \int_{-L/2}^{L/2} S_0'(\xi) \ln|\xi - x| d\xi + \eta_s \int_{-L/2}^{L/2} b'(\xi) \ln|\xi - x| d\xi \right\} + A_W \eta_s = 0. \tag{4.108}$$

The last term is associated with the added buoyancy due to sinkage. This means A_W is the water-plane area of the vessel. The equation is linear in η_s and can easily be solved. The results for the Wigley hull used for Figure 4.43 are presented in Figure 4.44 in terms of η_s/L as a function of Fn_h. We have avoided presenting results too close to $Fn_h = 1$ where the theory is invalid. The nondimensional results are independent of h/L, but of

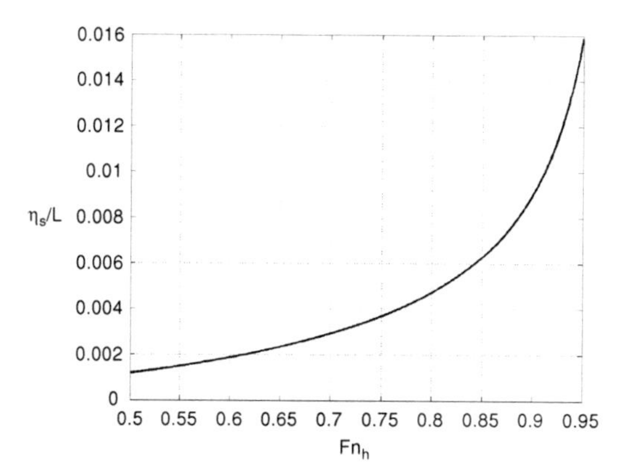

Figure 4.44. Sinkage η_s of a Wigley hull with $B/L = 0.1$ and $D/L = 0.0625$ based on Tuck's shallow-water theory. Zero trim.

course h/L is assumed small. The results show that η_s/L can be of the order of 0.01, which means 1 m sinkage for a 100 m–long vessel. This could be of concern from a grounding point of view, and it also increases the ship's resistance.

The speed dependence of sinkage presented in Figure 4.44 should be emphasized. If you are operating a ship in shallow water at subcritical speed with a risk of grounding due to squat, you should slow down.

We now study supercritical speed, in which the main effect is the result of trim. We express the vertical motion along the ship due to trim as $-x\eta_5$, which means that positive trim (pitch) angle η_5 causes bow up. We consider now the balance of pitch moments in the same way as we did for vertical forces. The hydrodynamic pitch moment is given by eq. (4.105) and we get

$$-\frac{Fn_h^2}{2\left(Fn_h^2-1\right)^{1/2}}\cdot\left[\int_{-L/2}^{L/2} xb(x)\,S_0'(x)\,dx - \eta_5\int_{-L/2}^{L/2} xb(x)\,(b(x)\,x)'\,dx\right]. \quad (4.109)$$

$$-\eta_5\int_{-L/2}^{L/2} x^2 b(x)\,dx = 0$$

Strictly speaking, we should have chosen $x = 0$ to coincide with the longitudinal position of center of gravity (COG). The effect of the vertical distance between COG and center of buoyancy should in principle have been included in the contribution from the hydrostatic pressure. This is important for transverse metacentric height for a monohull but is often neglected for the longitudinal metacentric height. Results for the Wigley hull are presented in Figure 4.45, both in terms of trim angle and resulting local sinkage at the stern due to trim. It indicates that this local sinkage due to trim for supercritical speed may be of the order of twice the sinkage for subcritical speed.

4.6 Exercises

4.6.1 Thin ship theory

a) Michell's thin ship theory for wave resistance is described in the main text. Let φ be the velocity potential due to the ship and approximate the dynamic free-surface condition with $\varphi = 0$ (relevant for very high speed and in the bow region of

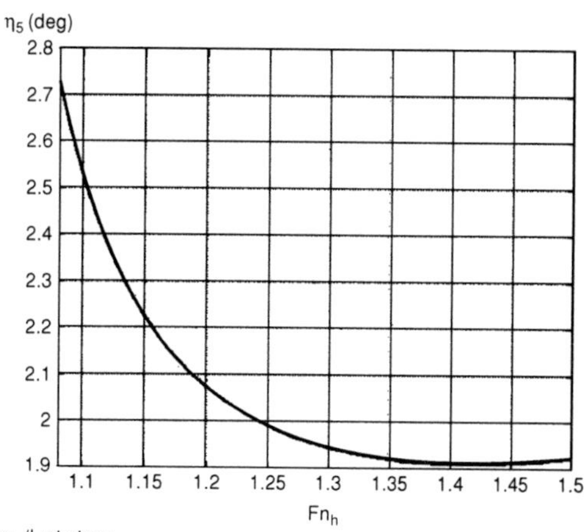

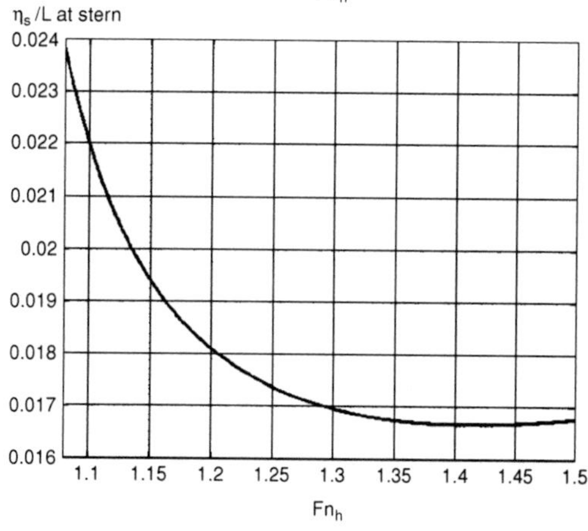

Figure 4.45. Trim (η_5) and local sinkage (η_s) at the stern due to trim. Wigley hull. $B/L = 0.1$, $D/L = 0.0625$. Based on Tuck's shallow-water theory.

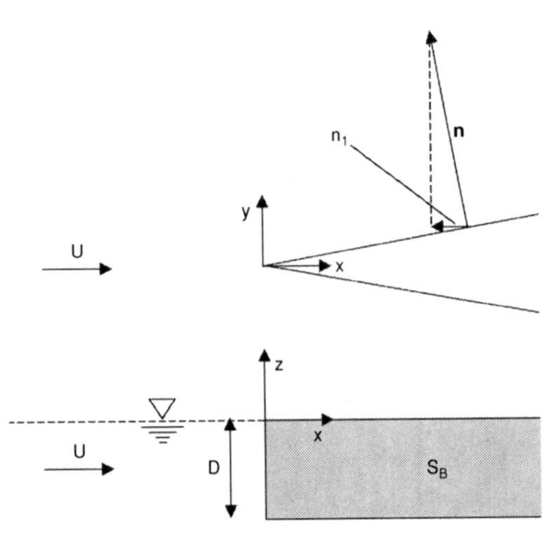

Figure 4.46. Wedge-shaped bow.

slender ships). Show that φ can be written as

$$\varphi = \frac{U}{2\pi} \iint_{cp} d\xi \, d\zeta \, n_1(\xi, \zeta) \Big[\left((x-\xi)^2 + y^2 + (z-\zeta)^2 \right)^{-1/2} - \left((x-\xi)^2 + y^2 + (z+\zeta)^2 \right)^{-1/2} \Big] \quad (4.110)$$

Here cp is the center plane of the ship, n_1 is the x-component of the normal vector to the hull surface, and U is the ship's speed. The coordinate system is shown in Figure 4.46.

b) Assume the ship is wedgelike (Figure 4.46) and semi-infinite in extent. (Physically, one can imagine that one is studying the flow locally around a wedgelike bow.) Show that φ can be written as

$$\varphi = \frac{U}{2\pi} n_1 \int_{-D}^{0} d\zeta \times \ln \left[\frac{-x + \sqrt{x^2 + y^2 + (z+\zeta)^2}}{-x + \sqrt{x^2 + y^2 + (z-\zeta)^2}} \right]. \quad (4.111)$$

c) Assume $x \to -\infty$. Show that $\varphi \approx 0$. Express briefly what it means.

d) Use the kinematic free-surface condition $U\partial\zeta/\partial x = \partial\varphi/\partial z$ on $z = 0$, and show that the wave elevation at the bow ($x = 0$, $y = 0$) is

$$\zeta_0 = \frac{\alpha D}{\pi}, \quad (4.112)$$

where α is half the wedge angle. (Hint: Integrate $\partial\zeta/\partial x$ from far upstream.)

e) Assume x is positive and much larger than y and z. Show that

$$\varphi \approx -\frac{U}{\pi} n_1 \int_{-D}^{0} d\zeta \ln \left[\frac{\sqrt{y^2 + (z-\zeta)^2}}{\sqrt{y^2 + (z+\zeta)^2}} \right] \quad (4.113)$$

For what boundary-value problem is eq. (4.113) the solution? What does this express qualitatively about the flow in the transverse plane relative to the longitudinal direction?

f) Use eq. (4.113) from the bow $x = 0$ of the ship and assume $\zeta = 0$ at $x = 0$. Find an expression for the wave elevation ζ along the ship hull by first showing that

$$\frac{\partial\zeta}{\partial x} = -\frac{n_1}{\pi} \ln \left[\frac{n_1^2 x^2 + D^2}{n_1^2 x^2} \right]. \quad (4.114)$$

Discuss the dependence on Froude number ($U/\sqrt{Lg}$) and ship form.

g) According to eq. (4.114), ζ will increase with increasing x-value. If we think in terms of a realistic ship, what will cause the wave elevation to decrease with increasing x-value?

4.6.2 Two struts in tandem

Consider two parabolic struts in tandem with a constant forward speed U (Figure 4.47). Use thin ship theory (see eq. (4.42)) to express the wave amplitude function $A(\theta)$.

Consider transverse waves for $\theta = 0$ and, as in Figure 4.47, define L to be the length of each strut and L_t to be the distance between the centers of each strut.

Derive an expression for $A(0)$, and use this to discuss when there is amplification and cancellation of the transverse waves generated by each strut.

4.6.3 Steady ship waves in a towing tank

Consider a ship model in a towing tank (Figure 4.48) with breadth b and water depth h. The towing speed U is constant. A Cartesian ship-fixed

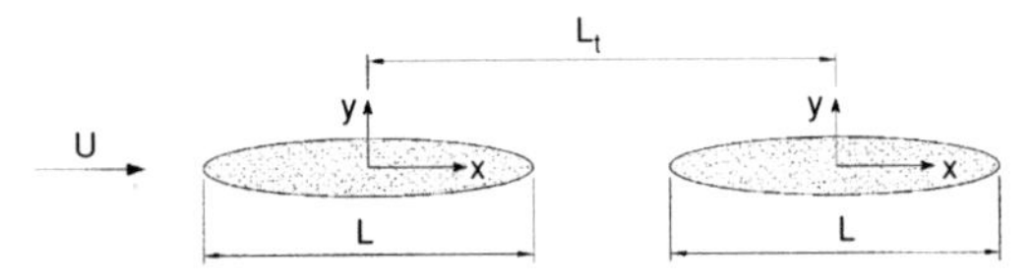

Figure 4.47. Two struts in tandem.

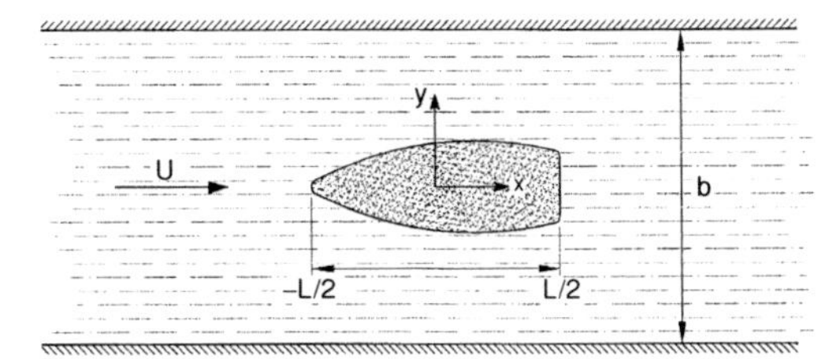

Figure 4.48. Ship model in a towing tank with breadth b and water depth h.

coordinate system (x, y, z) is defined as in Figure 4.48. Consider linear steady ship-generated waves.

a) At what distance behind the ship model will the ship-generated waves start to be reflected from the tank walls?

b) Assume that we are aft of the ship model and at a sufficient distance behind the ship model for tank wall reflection to occur. Show that the wave amplitude can be expressed as

$$\zeta = \sum_{n=0}^{\infty} A_n \cos(w_n x + \varepsilon_n) \cos\left(u_n\left(\frac{b}{2} - y\right)\right), \tag{4.115}$$

where

$$u_n = n\pi/b, n = 0, 1, 2, 3 \ldots$$

$$w_n^2 = k_n^2 - u_n^2$$

$$\frac{g}{U^2} k_n \tanh k_n h = w_n^2.$$

(Hint: Use the Laplace equation for the velocity potential and the boundary conditions of the tank wall and bottom, as well as the free-surface condition given by eq. (4.27).)

c) Express the wave system in an Earth-fixed coordinate system.

4.6.4 Wash

Feasibility studies of submerged floating tunnels crossing deep Norwegian fjords have been made. Figure 4.49 shows one concept using tension-leg mooring. The tunnel length is on the order of 1 km. Consider a horizontal circular cylinder with diameter $D = 12$ m as part of such a tunnel. The vertical distance from the top of the cylinder to the mean free surface is 20 m.

Consider a scenario of a 70 m–long monohull vessel passing over one of the moorings with the

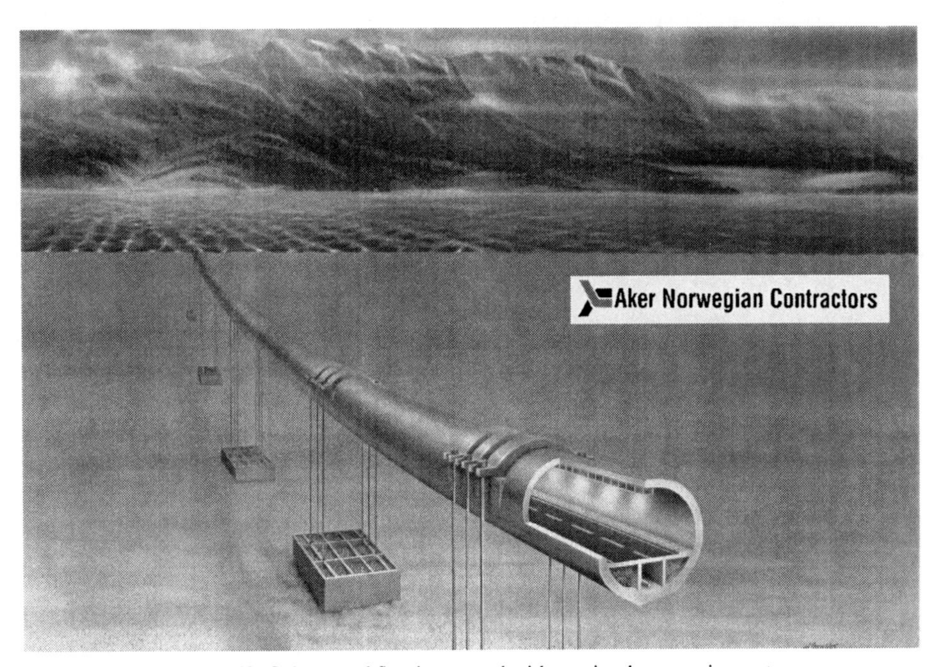

Figure 4.49. Submerged floating tunnel with tension-leg mooring system.

track perpendicular to the tunnel axis. The Froude number is 0.5 and there are deep-water conditions.

Evaluate the vertical wave force on the tunnel due to the passing ship by using the mass term in Morison's equation. This means we write the vertical wave force per unit length as

$$F_z = \rho\pi\frac{D^2}{2}a_z, \tag{4.116}$$

where a_z is vertical incident fluid acceleration at the tunnel axis.

Consider only the transverse waves perpendicular to the ship's track and account for the Kelvin angle and wave decay. Consider first the time instant when there is longitudinally one ship length's distance between the ship stern and the tunnel axis. Further, assume that there is a wave amplitude of 0.5 m at the tunnel position. Use expressions for long-crested regular waves, as presented in Table 3.1, to calculate a_z. Relate the calculations to the requirement that there should be no slack in the tension legs shown in Figure 4.49. Neglect dynamic motions of the tunnel. Assume that the distance between each tension-leg mooring is 200 m and the structural weight of the tunnel is 2.5% lower than the buoyancy of the tunnel. Examine different distances between the ship and the tunnel to find the most severe situation.

4.6.5 Wave patterns for a ship on a circular course

Consider a ship in deep water with constant speed on a circular course. Figure 4.50 illustrates the steady divergent and transverse wave system predicted by linear theory. Explain, as we did in Figure 4.7 for a ship on straight course, the outer boundaries of the waves by considering curves for where wave energy was emitted at a previous time. We have in Figure 4.50 helped this process by drawing two circles that are tangential to the outer boundaries of the wave system.

a) Draw additional circles like this and show a larger part of the outer boundaries of the wave system. Explain what you are doing.

b) Consider now a ship with a constant speed on a circular course in shallow water at critical-depth Froude number. Use arguments similar to those in Figure 4.37 to sketch the outer boundaries of the wave system.

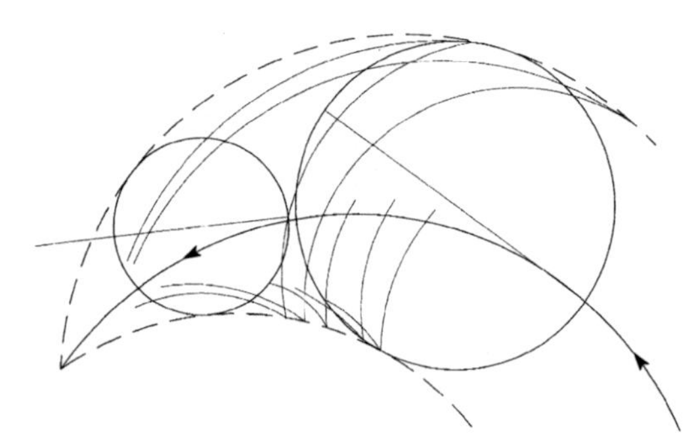

Figure 4.50. Wave crests for a ship with constant speed on a circular course in deep water. Based on a figure in Stoker (1958). Two circles that are tangential to the outer boundaries are added. These follow from wave energy propagation considerations, as in Figure 4.7.

4.6.6 Internal waves

The scenario is shown in Figure 4.51. There are two layers of fluid with different mass densities. There are gravity waves occurring both on the upper surface facing air and on the interface between the two fluids. The mass density ρ' of the upper layer is smaller than the density ρ for the lower layer. The upper and lower layer have mean heights h' and h, respectively. It is convenient for further derivation to use a coordinate system with the xy-plane in the mean interface of the two fluids (Figure 4.51).

a) Consider two-dimensional linear harmonic waves propagating both on the upper surface and on the interface. We assume that the lower layer has infinite depth and that there is no mean flow. The following derivation follows Landau and Lifshitz (1959). The velocity potentials for the lower

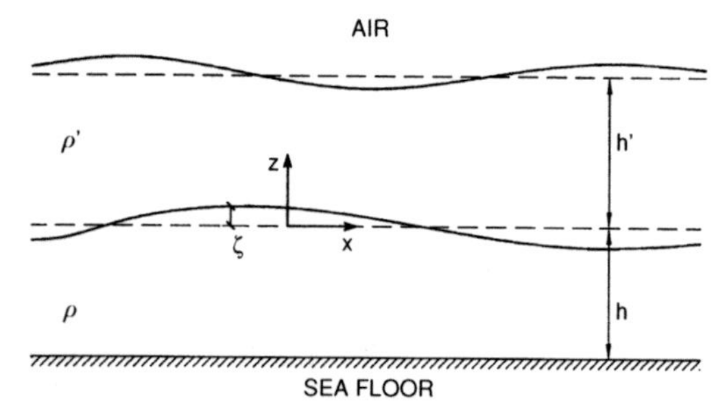

Figure 4.51. Free-surface waves and internal waves. The upper layer has a mean depth h', mass density of fluid ρ'. Corresponding notation for the lower layer is without superscript.

and upper layers are expressed as

$$\varphi = Ae^{kz}\cos(kx - \omega t) \qquad (4.117)$$

$$\varphi' = \left(Be^{-kz} + Ce^{kz}\right)\cos(kx - \omega t). \qquad (4.118)$$

The free-surface condition, as in eq. (3.11), applies on the upper surface for $z = h'$. Linearized conditions for continuity of vertical velocity and pressure on the interface are used.

Two solutions of the circular frequency as a function of the wave number k are as follows:

$$\omega^2 = kg\frac{(\rho - \rho')\left(1 - e^{-2kh'}\right)}{(\rho + \rho') + (\rho - \rho')\,e^{-2kh'}} \qquad (4.119)$$

$$\omega^2 = kg. \qquad (4.120)$$

We recognize that eq. (4.120) is the dispersion relationship for deep-water waves without an interface. Eq. (4.119) is related to the internal waves.

What are the phase velocity and group velocity of the internal waves?

b) The interface between the two fluids is a thin free shear layer. Actually, we have set the thickness equal to zero, which is similar to what we do with the boundary layer of a ship when we find a potential flow solution.

Express the integrated vorticity $\boldsymbol{\omega} = \nabla \times \mathbf{u}$ ($\mathbf{u}$ is fluid velocity) across the interface by using the potential flow solutions given by eqs. (4.117) and (4.118).

c) Consider now $h' \to \infty$ and assume that there is a way that internal waves have been generated, for instance, by an object moving near the interface. Express the solution form of the velocity potential in the upper and lower layers. Formulate the boundary conditions at the interface and find the dispersion relationship.

d) Consider now steady ship waves by transferring the results for the internal waves in problem a) to a ship-fixed coordinate system in a way similar to the one we used in section 4.2. Show that this gives the following equation for the wave number:

$$F^2\cos^2\theta \cdot kh' = \frac{\left(1 - e^{-2kh'}\right)}{2 - \frac{\Delta\rho}{\rho}\left(1 - e^{-2kh'}\right)}. \qquad (4.121)$$

Here $\Delta\rho/\rho = (\rho - \rho')/\rho$, θ is the wave propagation direction as defined by eqs. (4.8) and (4.9) together with Figure 4.6. Further,

$$F = \frac{U}{\left(\frac{\Delta\rho}{\rho}gh'\right)^{1/2}}. \qquad (4.122)$$

F is called the densimetric Froude number. There is a similarity between this problem and ship waves in finite water depth. This means there is a critical speed corresponding to $F = 1$.

Show that divergent and transverse internal waves occur at subcritical speed and that there are only divergent waves for supercritical speed.

What is the minimum $|\theta|$ for the divergent waves as a function of F for supercritical speed?

e) The phenomenon discussed in d) is called "dead water." There is a clear hump in the internal wave resistance at $F = 1$ for small h'/L. The ship speed corresponding to $F = 1$ is very low. Let us consider an example with $h' = 5$ m and $\Delta\rho/\rho = 3 \cdot 10^{-2}$ corresponding to freshwater as the upper layer and saltwater as the lower layer. This may be a scenario in a Norwegian fjord. Calculate the critical speed.

Consider the transverse internal waves along the ship's track for $F = 0.8$ and the divergent waves for $F = 4$ corresponding to minimum$|\theta|$. Assume that the corresponding wave amplitude at the interface is not negligible. Use the solution form given by eq. (4.118) and k determined by eq. (4.121) to discuss, based on your own criterion, if the waves are visible by eye from above in the air.

Use the solution form given by eq. (4.117) to discuss, based on your own criterion, what water depth is needed for these two internal waves to be negligibly influenced by the presence of the sea floor.

f) The ship speeds that we have discussed above are low. Because the ship must have a draft that is not too small relative to h' to cause internal waves, we can generally expect that the length Froude number is low. This means that the ship would cause negligible free-surface waves when there is no interface in the water. Using a rigid wall as the upper surface, then, is a good approximation. Landau and Lifshitz (1959) have presented linearized solutions that account for both finite h and h' in the unsteady flow case. The solution form of the velocity potential in the lower and upper layers are, respectively,

$$\varphi = A\cosh k(z+h)\cos(kx-\omega t) \quad (4.123)$$

$$\varphi' = B\cosh k(z-h')\cos(kx-\omega t) \quad (4.124)$$

By using the boundary conditions at the interface, it follows that

$$\omega^2 = \frac{kg(\rho-\rho')}{\rho\coth kh + \rho'\coth kh'} \quad (4.125)$$

We could now have transferred the solutions to a ship-fixed coordinate system as we did in d). We will instead consider a different application. The problem is a tank that is completely filled by two fluids, and we want to express the natural frequencies for sloshing caused by internal waves.

We must then re-express φ and φ' so that they represent standing waves with zero horizontal velocity at the vertical wall. This can be obtained, for instance, by adding propagating waves in the negative x-direction to φ and φ' given by eqs. (4.123) and (4.124). The zero normal velocity condition on the walls, then, determines k. This implies infinite numbers of k-values. We concentrate on the lowest k-value, that is, the lowest natural mode for sloshing in the tank.

Consider now an example with a tank that has a length of 0.42 m and in which $h = 0.025$ m and $h' = 0.045$ m. We can use eq. (4.125) with wave numbers k that are consistent with the body boundary conditions to determine the natural frequencies of sloshing in the tank. The highest natural period has been measured to be 12 s. Argue by assuming that kh and $k'h$ are small that you can approximate eq. (4.125) by letting kh and kh' tend to zero. Use that expression to find $(\rho-\rho')/\rho$.

5 Surface Effect Ships

5.1 Introduction

Figure 5.1 shows an example of a surface effect ship (SES). Figure 1.8 gives a fish-eye view. The vessel is supported by an air cushion that is bounded by flexible seal systems at the bow and stern and by two side hulls. The aft seal is usually a flexible bag consisting of a loop of flexible material open against the side hulls, with one or two internal webs restraining the aft face of the loop into a two- or three-loop configuration. In equilibrium position, there is a very small gap between the bottom of the bag and the water surface. An example is a gap height of 3 cm. The bow seal (skirt) is usually a finger seal consisting of a row of vertical loops of flexible material. Details are shown in Figures 5.2 and 5.3. The seal material is rubber.

There is an air fan system (Figure 5.4) that provides the excess pressure in the air cushion and lifts the SES up, thereby reducing the water resistance. The excess pressure in the air cushion causes a water level inside the cushion lower than the level outside. Typically, the air cushion carries 80% of the weight of the vessel. The buoyancy of the side hulls carries the rest of the weight at zero speed. When the vessel speed increases, the vertical side hull forces due to the water are caused by both hydrostatic (buoyancy) and hydrodynamic pressures.

An SES on cushion has a lower resistance than a similarly sized catamaran, can achieve a higher speed with less total power, and has better seakeeping characteristics in head sea conditions in moderate sea states. Calm water resistance components of an SES are discussed in Chapters 2 and 4. The relative importance of resistance components is exemplified in Figure 4.2. Section 4.3.6 analyzes the wave resistance due to the air cushion. In this chapter, we discuss added resistance and speed loss in waves, which can be quite severe.

Problem areas for an SES are:

- Wear of skirts
- Power peaks and wear and tear of propulsion/machinery systems caused by ventilation and cavitation
- Speed loss in waves
- Cobblestone oscillations

Cobblestone oscillations cause unpleasant vertical accelerations in small sea states and are the result of resonant compressible flow effects in the air cushion. They are called cobblestone oscillations to highlight the resemblance to driving a car on roughly layed cobblestones. Ride-control systems are used to dampen some of the "cobblestone" effect.

An important consideration is steering and maneuvering of an SES. Figure 5.5 presents results from full-scale trial results with SES "Agnes 200." In the cushionborne mode, a remarkable speed loss while turning was experienced at the speed related to the hump (maximum) of the air cushion drag. This results from high drift angles and associated feeding problems of the waterjets. Turning radii are difficult to measure under these conditions, and vehicle speed becomes fairly unstable, particularly at higher wind speeds. Berthing of an SES at high wind speeds may be difficult. Maneuvering of an SES is discussed further in Chapter 10.

A general problem area for all types of high-speed vessels is the conflict between small weight and sufficient strength. The effect of impact wave loading (slamming) is important for all high-speed vessels. Because air leakage from the air cushion in high sea states can cause an SES to be off-cushion and lower the wetdeck inside the cushion, wetdeck slamming on an SES is of concern. Global wave loads are important for catamarans, monohulls, and SES when the vessel length is larger than, say, 50 m. Global wave loads and slamming are discussed in Chapters 7 and 8, respectively.

5.2 Water level inside the air cushion

We show by an example what the water level typically is inside the air cushion of a 200-tonne SES in calm water and zero speed and in static equilibrium. Let us say that the fans are able to provide an excess pressure p_0 in the cushion corresponding to 0.05 times the standard atmospheric pressure

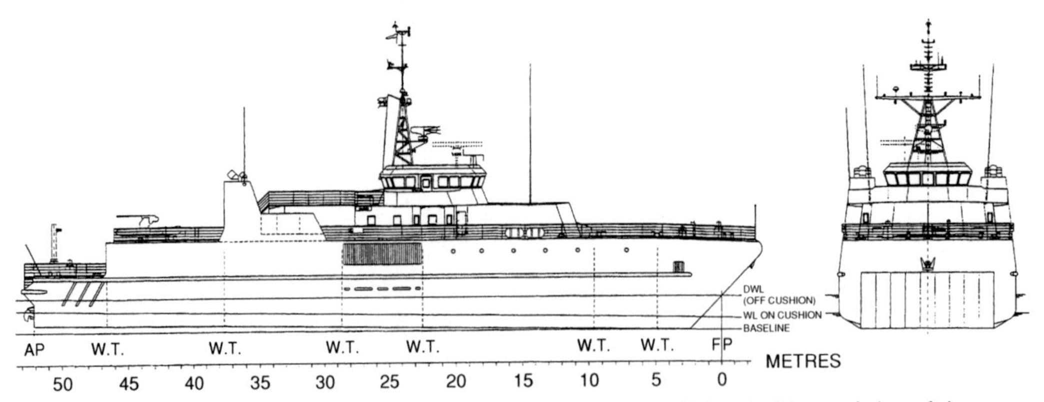

Figure 5.1. SES mine countermeasure vessel by UMOE Mandal. (Printed with permission of the copyrightholder: Royal Norwegian Navy.)

(Figures 5.6 and 5.7). The pressure in the water is simply

$$p = p_a - \rho g z, \tag{5.1}$$

where p_a is the atmospheric pressure. The coordinate system is defined in Figure 5.7, with the z-axis pointing vertically upward and $z = 0$ corresponding to the mean free-surface level outside the cushion. By applying eq. (5.1) on the water surface inside the cushion, we find

$$p_0 + p_a = p_a + \rho g h,$$

that is,

$$h = \frac{p_0}{\rho g}.$$

The *standard* atmospheric pressure at sea level is 1.01×10^5 Pa. This gives $h = 0.5$ m. As we see in Figure 5.6, it is common to use the unit *millimeters water column* (mm Wc) for the excess pressure. This can be translated into pascals (Pa) by noting that 1 mm Wc corresponds to a pressure $10^{-3}\rho g$ in pascals. Here the mass density of water, ρ is in kilogram per meter cube (kgm^{-3}) and $g = 9.81$ ms^{-2}. For instance, using $\rho = 1025$ kgm^{-3} gives that 1 mm Wc is equal to 10.055 Pa.

We can find the buoyant volume of the side hulls of the SES (Figure 5.7) by balancing the weight, cushion pressure forces, and buoyancy forces, that is,

$$p_0 A_b + \rho g V_b = Mg, \tag{5.2}$$

where A_b is the cushion area, V_b is the hull volume below $z = 0$, and M is the mass of the SES. Let us say $Mg = 2 \cdot 10^6$ N, $\rho g = 10^4$ Nm^{-3}, $A_b = 320$ m^2 and h = 0.5 m. This means $V_b = 40$ m^3. This is consistent with the fact that the air cushion carries typically 80% of the weight of the vessel.

We have in eq. (5.2) set the buoyancy force equal to $\rho g V_b$. If we blindly had followed Archimedes' law, this would say that the buoyancy force is equal to ρg times the submerged volume, but what is the submerged volume in this case when there are different free-surface levels inside the cushion and outside the SES? This dilemma can be solved by instead studying the vertical force on

Figure 5.2. Details of the bow seal (skirt) of an SES. (Photo: Hans Olav Midtun.)

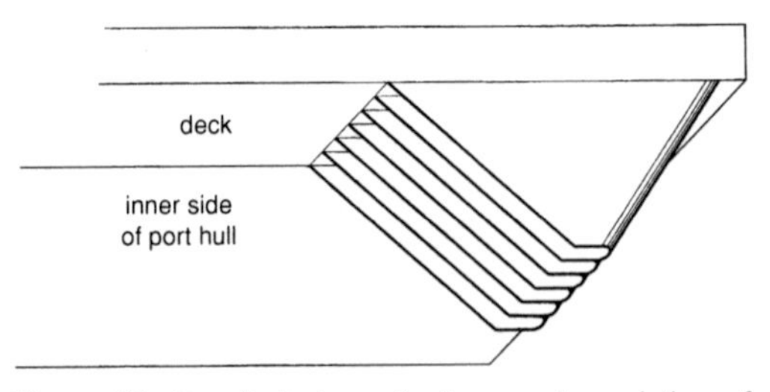

Figure 5.3. Detailed view of a bow seal consisting of individual fingers (Moulijn 2000).

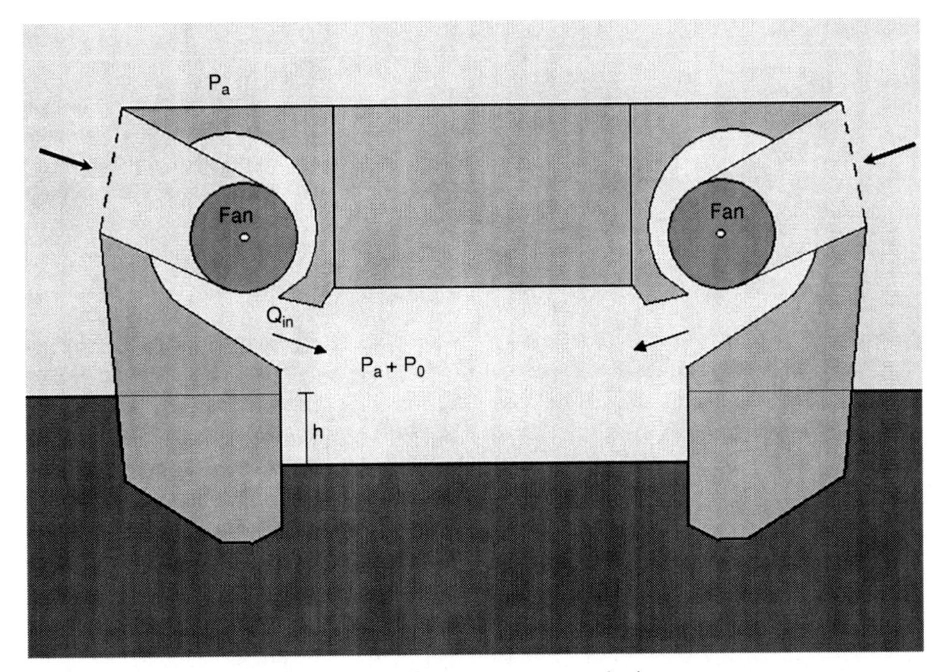

Figure 5.4. Air fan system. Q_{in} = volume flow rate, p_a = atmospheric pressure, p_0 = excess pressure in the air cushion.

the side hulls due to hydrostatic pressure $-\rho g z$. We can then solve the problem by first pretending that the free-surface level inside the cushion is the same as outside the SES. Then we have a situation in which Archimedes' law applies, and the submerged volume is well defined. Then we can take away the water from $z = 0$ to $z = -h$ inside the cushion. However, the hydrostatic pressure in this virtual fluid mass causes only a horizontal force on the vertical side hulls shown in Figure 5.7. So when V_b is as shown in Figure 5.7, our formulation of buoyancy force in eq. (5.2) is correct.

5.3 Effect of air cushion on the metacentric height in roll

The air cushion has a destabilizing effect on the heel (roll)-restoring moment of an SES. We show that by studying the heel-restoring moment about an axis through the center of gravity (COG) of the SES when the SES has a small heel angle η_4 (Figure 5.8).

It is then appropriate to define a body-fixed coordinate system (x', y', z') with the origin in the center of gravity (COG). One part of the heel

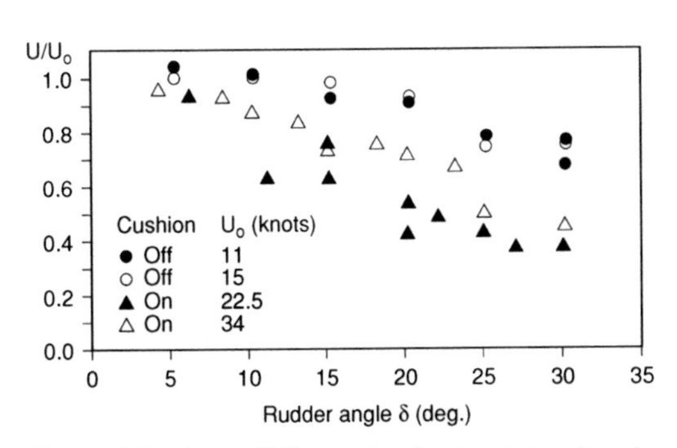

Figure 5.5. "Agnes 200" speed reduction in turning circle. U_0 is maximum speed (Skorupka et al. 1992).

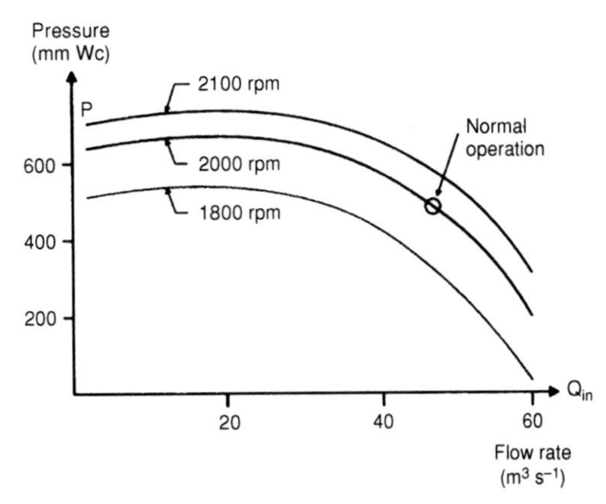

Figure 5.6. An example of fan characteristics for an SES.

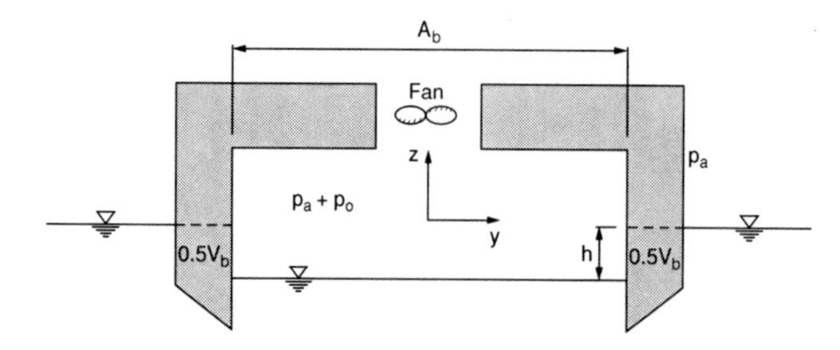

Figure 5.7. Water level inside the air cushion (p_a = atmospheric pressure, p_0 = excess pressure in the cushion). No forward speed. Simplistic drawing of the fan.

moment caused by the excess pressure is a result of the fact that the cushion area A_1 disappears and the cushion area A_2 appears (Figure 5.8). We assume constant cross section along the length of the SES and can write

$$A_2 = 0.5b\ell\eta_4,$$

where b is defined in Figure 5.8 and ℓ is the cushion length. There is an excess pressure p_0 acting perpendicular to A_2. This pressure load causes a heel moment $h_{FG}p_0 0.5b\ell\eta_4$ about the COG. Here h_{FG} is the lever arm for the moment (Figure 5.8). We get a similar contribution from the fact that A_1 disappears. Because $A_1 = A_2$, the total contribution is $h_{FG}p_0 A_b\eta_4$. Here $A_b = bl$ is the waterplane area inside the cushion. This means a destabilizing moment contribution.

In the same way as is known from conventional ships, we get contributions from the hydrostatic pressure. We can divide this effect into two parts. We first calculate as if the water level is the same inside and outside the cushion. Then we get similar expressions as for catamarans. Further, we have to make a correction because the water level is lower inside the cushion. We can write this correction term for the heel moment as

$$F_{4HC} = -\rho g \int_{\Sigma} z(y'n_3' - z'n_2')\, ds,$$

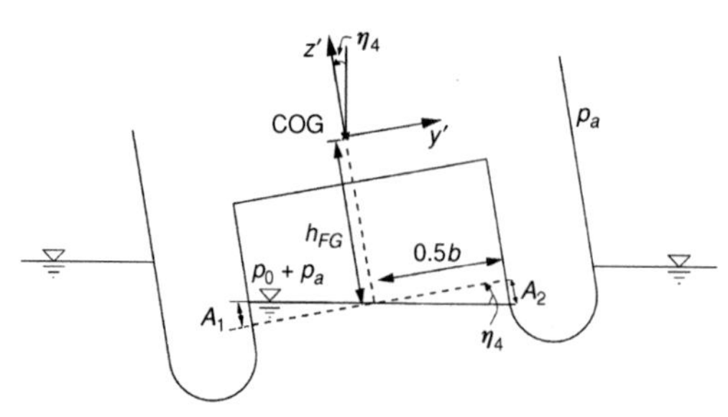

Figure 5.8. Definitions used in calculating metacentric height of an SES on cushion at zero forward speed.

where Σ is the hull surface that we have incorrectly included by assuming the same water level inside and outside the SES. Further, $\boldsymbol{n}' = (n_1', n_2', n_3')$ is the hull surface normal in the body-fixed coordinate system. The positive normal direction is outward from the hull. If we assume wall-sided body surface at the free surface, then $n_3' = 0$ and $n_2' = \pm 1$ for positive and negative y'-values, respectively. We note that both z and z' appear in the formula for F_{4HC}. The origin of the z-term is the hydrostatic pressure, which is defined relative to the Earth-fixed coordinate system. We can write

$$z' \approx z - z_G - y'\eta_4,$$

where z_G is the z-coordinate of the center of gravity of the vessel. This implies that we can write

$$F_{4HC} = \rho g \left[\int_{\Sigma} z^2 n_2'\, ds - z_G \int_{\Sigma} z n_2'\, ds - \eta_4 \int_{\Sigma} y' z n_2\, ds \right].$$

The contribution from the two first terms are zero. This means

$$F_{4HC} = \rho g \eta_4\, lb \int_{-\frac{p_0}{\rho g}}^{0} z\, dz$$

$$= -\rho g A_b 0.5 \left(\frac{p_0}{\rho g}\right)^2 \eta_4 = -p_0 A_b 0.5 h \eta_4,$$

where we have used the fact that $h = p_0/\rho g$.

We now sum up the results. We refer to Figure 5.9 and recall the following definitions:

p_0 = static excess pressure in the cushion

(x', y', z') = body-fixed coordinate system with origin in the center of gravity

(x, y, z) = Earth-fixed coordinate system, where $z = 0$ corresponds to mean free-surface level outside the cushion

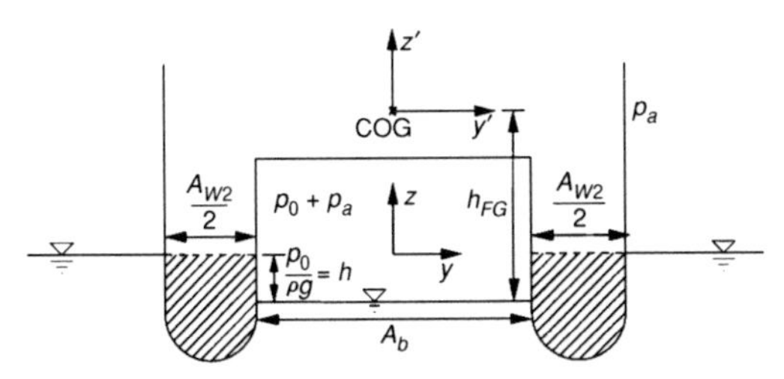

Figure 5.9. Definitions used in calculating metacentric height of an SES on cushion at zero forward speed.

A_{w2} = the total waterplane area inside the hulls
A_b = waterplane area of the cushion
V_b = total volume of the hulls below $z = 0$ (shaded areas in Figure 5.9)
z'_B = z'-coordinate of center of buoyancy of V_b.

The heel (roll) moment F_4 about an x-axis through the center of gravity when the SES has a small heel (roll) angle η_4 can be written

$$F_4 = -\left[-p_0 A_b (h_{FG} - 0.5h) + \rho g \int_{A_{w2}} y^2 \, ds + \rho g z'_B V_b\right] \eta_4. \quad (5.3)$$

The transverse metacentric height $\overline{GM}$ is defined by

$$F_4 = -Mg\overline{GM}\,\eta_4. \quad (5.4)$$

Because about 80% of the weight of the SES is carried by the excess pressure in the air cushion, it makes a big difference if we use $\rho g V_b$ instead of Mg in eq. (5.4). As an example of a typical $\overline{GM}$-value, we can mention a $\overline{GM}$ of 10 m for a 37 m–long SES. Eq. (5.3) illustrates that the static excess pressure p_0 in the cushion gives a negative contribution to the metacentric height. This follows by noting that the center of gravity is in reality higher than the wetdeck and that $h_{FG} - 0.5h$ is positive. Further, z'_B is lower when the SES is on-cushion than when it is off-cushion.

If we want to express the static restoring moment F_4 for finite η_4, we write $F_4 = -Mg\overline{GZ}\,(\eta_4)$. This can be obtained numerically by evaluating the hydrostatic pressure $-\rho g z$ on the wetted hull together with the static excess pressure p_0 on the hull surface in the air cushion for a given η_4. Then we evaluate the heel moment due to this pressure distribution. Leakage under a side hull starts to occur for a certain roll (heel) angle, and the $\overline{GZ}$-value then becomes the same as for a catamaran.

5.4 Characteristics of aft seal air bags

A flexible-bag aft seal is commonly used on an SES. It was originally developed for the hovercraft (ACV). A three-loop bag is shown in Figure 5.10, in which there are two internal webs restraining the aft face of the bag. There are holes in the webs to equalize the pressure in the bag. The bag is open against the side hulls. The bag material is usually reinforced rubber. Dedicated booster fans provide air with a pressure inside the bag that is typically 15% higher than in the air cushion. Figure 5.10 shows a system with booster fans from the air cushion through ducts in the wetdeck. The air is flowing out between the bag and the side hulls and through drain holes in the lower part of the bag and in the part facing the open air. There is a small gap between the lower part of the bag and the free surface in calm water conditions. Advantages of the flexible-bag aft seal are low weight, no water resistance in calm water conditions, and small leakage of air through the gap below the bag in heavy sea conditions. However, the impact between water and the bag in heavy sea causes wear of the material.

Figure 5.11 shows the static shape and the four lowest dynamic modes of a two-loop flexible-bag

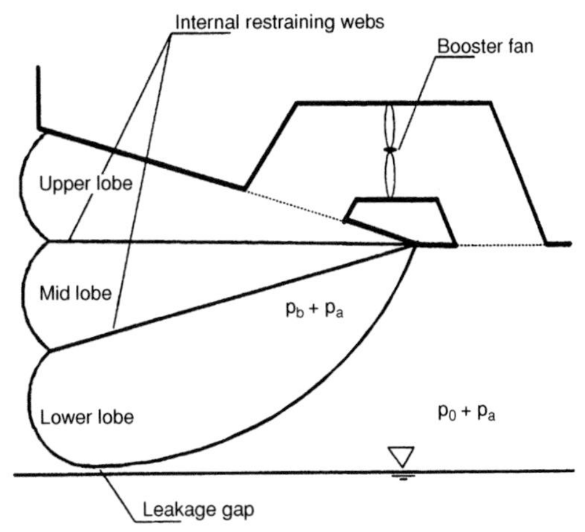

Figure 5.10. Side-view cross section of a flexible bag in three-loop configuration. p_a = atmospheric pressure. p_b and p_0 are static excess pressures in the air bag and in the air cushion, respectively (Steen 1993).

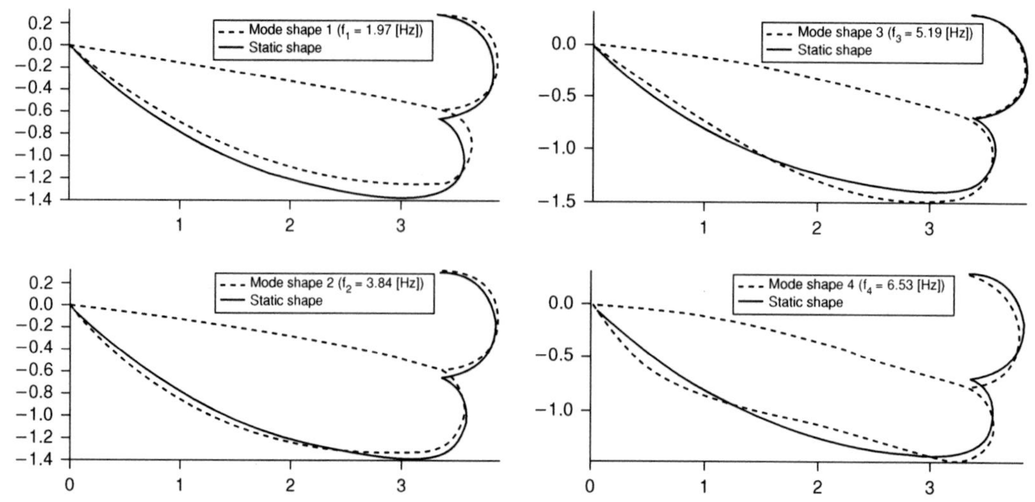

Figure 5.11. Example of the static configuration and the first four mode shapes of a flexible stern seal bag. For this configuration, the static difference pressures ΔP_{01} and ΔP_{02} are 500 Pa and 5500 Pa relative to the air cushion and the atmosphere, respectively. Length dimensions are in meters. The axial and bending stiffnesses of the bag structure are equal to $EA = 0.6 \times 10^6$ N and $EI = 4.0$ Nm^2 and the structural mass per unit length of the bag segment is $M_m = 4.3$ kgm^{-1} (Ulstein 1995).

aft seal. The calculations are based on setting the air gap below the bag equal to zero, thereby neglecting the airflow below the air bag. This airflow will cause a suction pressure on the bag and influence the shape of the bag. It is assumed in Figure 5.11 that there is a constant pressure difference $\Delta P_{01} = p_b - p_0 = 500$ Pa between the pressure inside the bag and the air cushion. Further, there is a constant pressure difference $\Delta P_{02} = p_b = 5500$ Pa between the pressure inside the bag and the atmospheric pressure. This means ΔP_{01} and ΔP_{02} are pressure loadings on the bag structure facing the cushion and the air outside the SES, respectively.

Let us illustrate how the static shape of the bag can be calculated by just considering one loop (Figure 5.12). Only tension (membrane) forces in the structure are considered. The membrane structure is divided into two segments, OA and AB, where OA faces the cushion. It is assumed that the membrane structure has a continuous derivative at the lowest point, A. The tension, T_0, is then constant along the structure. We can set up the following equilibrium conditions for segments 1 and 2:

$$\Delta P_{01} R_1 = T_0 \tag{5.5}$$

$$\Delta P_{02} R_2 = T_0. \tag{5.6}$$

Here R_1 and R_2 are the radii of curvature of the two segments as illustrated in Figure 5.12. So eqs. (5.5) and (5.6) contain three unknowns: R_1, R_2, and T_0. However, we can write

$$H_b = R_1(1 - \cos\theta_1) \tag{5.7}$$

$$L_b = R_1 \sin\theta_1. \tag{5.8}$$

Here the length L_b and the height H_b are known. The angle θ_1 is defined in Figure 5.12. Eliminating

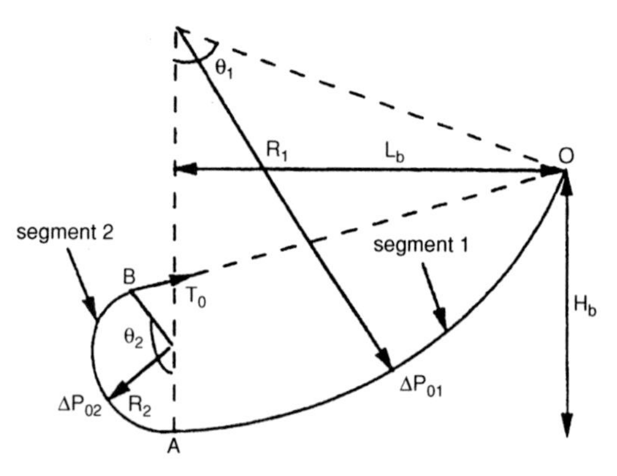

Figure 5.12. Static geometry of a one-loop flexible-bag seal. It consists of two weightless cable segments with constant radii of curvature (R_1 and R_2). The tension in the two segments is constant and equal (Ulstein 1995).

θ_1 between eqs. (5.7) and (5.8) gives

$$H_b = R_1\left(1 - \left(1 - \left(\frac{L_b}{R_1}\right)^2\right)^{0.5}\right). \quad (5.9)$$

The unknown R_1 can now be found numerically from eq. (5.9). Eqs. (5.5) and (5.6) then determine R_2 and T_0. θ_1 follows from either eq. (5.7) or eq. (5.8).

We can now proceed to find the angle θ_2 (Figure 5.12) that defines the end point B of segment 2. The following relationship can be set up:

$$\theta_2 = \pi - \tan^{-1}\left[\frac{H_b - R_2(1 - \cos\theta_2)}{L_b + R_2 \sin\theta_2}\right]. \quad (5.10)$$

This must then be numerically solved to find θ_2.

If we now proceed to a two-loop bag like the one in Figure 5.11, one more segment with unknown radius of curvature R_3 and tension T_{03} has to be introduced. We can set up a relationship like the one in eq. (5.6) by replacing R_2 and T_0 with R_3 and T_{03}. The presence of the internal restraining web implies that T_{03} differs from the tension in the lower part of the structure. We refer to Ulstein (1995) for further details about the analysis. This also includes the linear dynamic analysis leading to the mode shapes shown in Figure 5.11. Second-order differential equations coupling the motions in the transverse (η_n) and longitudinal (η_ℓ) directions of the bag structure are formulated with necessary boundary conditions. The reason for coupling between η_n and η_l is the radius of curvature R of the bag structure. The differential equations can be formulated as

$$M_m \frac{d^2\eta_n}{dt^2} = T_0 \frac{d^2\eta_n}{ds^2} + \frac{T_0 + EA}{R}\frac{d\eta_\ell}{ds} - \frac{EA}{R^2}\eta_n + \Delta P \quad (5.11)$$

$$M_m \frac{d^2\eta_\ell}{dt^2} = EA\frac{d^2\eta_\ell}{ds^2} - \frac{EA}{R}\frac{d\eta_n}{ds}. \quad (5.12)$$

Here M_m is structural mass per unit length, EA is axial stiffness, T_0 is static tension, s is the coordinate along the membrane structure, and ΔP is the dynamic pressure difference across the bag structure. The analysis in Figure 5.11 is based on setting $\Delta P = 0$, and a small bending stiffness term $EI\, d^4\eta_n/ds^4$ was added on the left-hand side of eq. (5.11). General methods for analysis of tensioned structures are described by Leonard (1988).

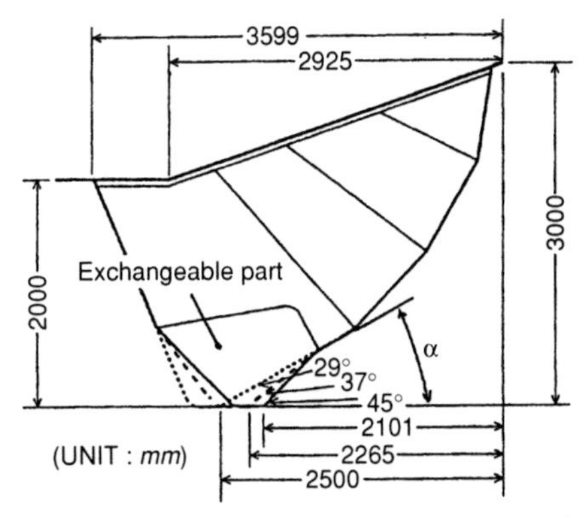

Figure 5.13. Different bow seal systems used by Yamakita and Itoh (1998) during sea trials with the SES test craft Meguro-2. Side view of Series B test pieces.

5.5 Characteristics of bow seal fingers

Bow seal fingers behave as a flag flapping in the wind, and their wear rate is proportional to a high power of the vessel speed, that is, U^4. Yamakita and Itoh (1998) made a broad investigation of the wear characteristics of the bow seal fingers on the SES test craft Meguro-2 by means of sea trials. Detailed views of bow seal fingers have been given in Figures 5.2 and 5.3. The lower part of the bow seal was exchangeable, and different materials and angles α (Figure 5.13) between the bow seal and the mean free surface were investigated. An optimum angle α from a wear-rate point of view was found to be about 40°.

The measured accelerations of the finger tips were extremely high. The largest value was 7450 times the acceleration of gravity. Figure 5.14 shows

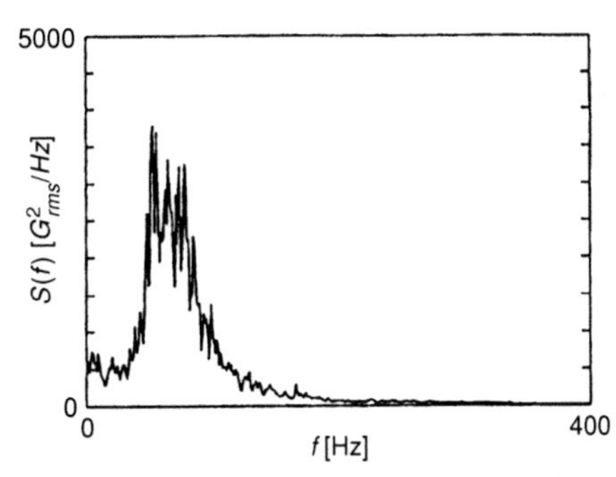

Figure 5.14. Power spectral density of accelerations due to finger vibration on the SES test craft Megura-2. Measurements are 100 mm from the tip. RMS (root mean square) value of accelerations is 418 times the acceleration of gravity g. Ship speed $U = 38$ knots (Yamakita and Itoh 1998).

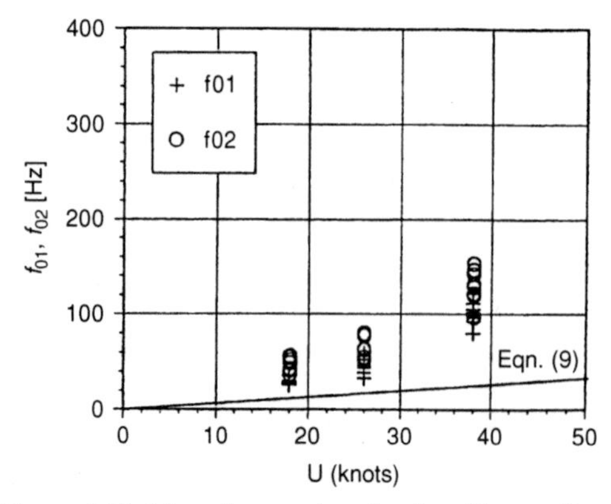

Figure 5.15. Mean frequencies, f_{01}, f_{02} of finger vibrations during sea trals with the SES Megura-2. Eq. (9) refers to theoretical estimate by Yamakita and Itoh (1998).

an example of the measured power spectral density $S(f)$ of vertical accelerations as a function of frequency f in hertz. It is more common in marine hydrodynamics to present a spectrum as a function of $\omega = 2\pi f$ in *rad/s*. The relationship between $S(f)$ and $S(\omega)$ is simply that $S(f)\,df = S(\omega)\,d\omega$. Another way of saying this is that the area under the spectrum should be the same using $S(f)$ and $S(\omega)$. We can define mean frequencies

$$f_{01} = \frac{m_1}{m_0}, \quad f_{02} = \sqrt{\frac{m_2}{m_0}}, \tag{5.13}$$

where

$$m_n = \int_0^\infty f^n S(f)\,df. \tag{5.14}$$

These frequencies are presented in Figure 5.15, together with an estimate by Yamakita and Itoh (1998). We will do a similar analysis based on their ideas. The formulation of their model is based on Figure 5.16. There is a rigid flat plate of length ℓ in the free surface that can rotate with angle θ about

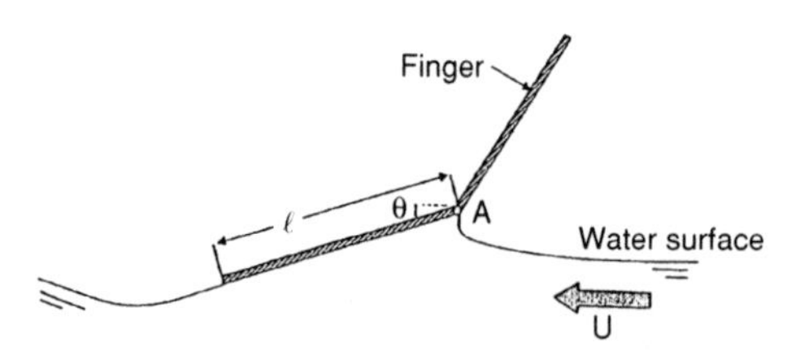

Figure 5.16. A simplified model of finger vibrations proposed by Yamakita and Itoh (1998). Note that the elastic properties are not accounted for.

a point A. One should note that assuming a rigid plate is a simplification. The bow seal is elastic. If θ is fixed, a steady hydrodynamic force acts on the plate that is a function of the forward speed U. Because the Froude number $U/(\ell g)^{0.5}$ of the plate is high and the submergence is low, hydrostatic pressure loads on the plate can be neglected. The difference between our analysis and Yamakita and Itoh's analysis is how this steady hydrodynamic force on the plate is handled. However, our analysis is of a qualitative nature.

A two-dimensional hydrodynamic flow past the rigid flat plate is assumed. The static (steady) equilibrium position of the plate follows by considering the effect of the static excess pressure p_0 in the cushion, the hydrodynamic loads on the plate, and the weight distribution. If the moment about the point A is taken, these three load contributions must cause balance.

The steady hydrodynamic load on the plate is a function of the angle of attack θ shown in Figure 5.16. If θ increases, the steady hydrodynamic loads on the plate will increase. If θ decreases, the steady hydrodynamic loads decrease. If we consider this in combination with the excess pressure and the weight of the plate, we find that the steady hydrodynamic loads cause a restoring (spring) effect.

Let us try to exemplify this restoring effect by using linear potential flow theory for the flow past the plate. A more accurate description of the hydrodynamic loads on a high-aspect–ratio planing surface is given in section 9.2.4. Because the Froude number $U/(g\ell)^{0.5}$ of the plate is very high, we simplify the free-surface condition as $\varphi = 0$, where φ is the velocity potential due to the plate. Further, because linear theory is assumed, this free-surface condition is imposed on the mean free-surface position. If we now take the image of the plate above the mean free surface, the problem is the same as the linear steady flow past a thin flat plate in infinite fluid (see Chapter 6). So we use that solution by noting that, in our case, the hydrodynamic pressure loads act only on the lower part of the plate whereas in the infinite fluid problem, fluid forces act on both the lower and upper parts of the plate. So we divide the infinite fluid loads by two. The lift coefficient on the plate in the infinite fluid is $2\pi\theta$, and the force acts $\ell/4$ from the leading edge (i.e., point A in Figure 5.16). This gives the following pitch moment F_5 per unit length about A:

$$F_5 = \frac{\rho}{8}\pi U^2 \ell^2 \theta. \quad (5.15)$$

We divide θ into a static part and small time-dependent pitch angle η_5 and write the equation for η_5 as

$$(I_{55} + A_{55})\frac{d^2\eta_5}{dt^2} + \frac{\rho}{8}\pi U^2 \ell^2 \eta_5 = 0. \quad (5.16)$$

Here I_{55} and A_{55} are, respectively, the structural mass and added mass moment of inertia in pitch about point A. Yamakita and Itoh (1998) base their estimate of A_{55} on the infinite frequency results by Newman (1977). This means $A_{55} = 9\pi\rho\ell^4/256$. I_{55} can be neglected relative to A_{55}. Assuming a time dependence $\exp(i\omega_n t)$ with i as the complex unit in eq. (5.16) gives a natural frequency:

$$\omega_n = \frac{4}{3}\sqrt{2}U/\ell. \quad (5.17)$$

This value is 27% higher than what Yamakita and Itoh (1998) estimated. The calculations in Figure 5.15 are based on $\ell = 0.18$ m. Eq. (5.17) shows that the frequency increases with speed, that is, similarly to the experimental results in Figure 5.15. However, we cannot quantitatively predict the experimental frequencies with this simple formula.

Another matter is that the finger vibrations are probably caused by instabilities. This means that in our simplified model, we need a negative damping term.

In order to improve our model, we may need to introduce nonlinear free-surface effects for 2D planing (see section 9.2.4). Further, Malakhoff and Davis (1981) demonstrated that the bow seal accelerations are a result of 3D folding and dynamic instability (flutter) of the finger material.

5.6 "Cobblestone" oscillations

An important effect of the air cushion pressure for an SES in small sea states is the so-called cobblestone effect. It is caused by resonance effects in the air cushion. Figure 5.17 shows full-scale measurements of the phenomena on a 35 m–long SES. The power spectrum of vertical accelerations at the bow shows large energy for about 2 Hz and 5 Hz. At the resonance frequency of 2 Hz, the dynamic part of the excess pressure in the cushion is oscillating with nearly the same amplitude all over the cushion. Compressibility effects of the air in the cushion are essential. The oscillations are excited because the waves change the enclosed air-cushion volume. It is the vertical vessel accelerations that are of concern. The vertical vessel motions associated with the cobblestone oscillations are small. This follows simply by noting that the frequency of oscillation ω is high and that the motion amplitudes are equal to acceleration amplitudes divided by ω^2.

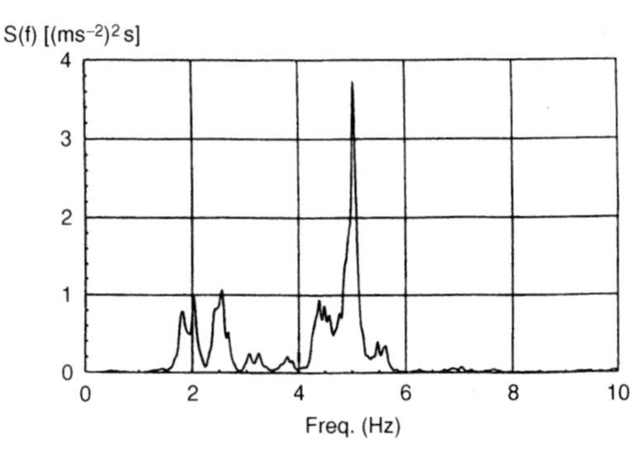

Figure 5.17. Full-scale measured power spectrum $S(f)$ of vertical accelerations at the bow of a 35 m–long SES with flexible-bag aft seal, running at 45 knots in head sea waves with significant wave height estimated to be $H_{1/3} = 0.3$ to 0.4 m, f = frequency in hertz (Steen 1993).

Figure 5.18 gives an overview of the physical effects that matter in describing the cobblestone oscillations (Ulstein 1995). We will not consider all these aspects in detail. The 1D wave equation referred to in the figure means that spatially varying one-dimensional standing acoustic waves and spatially uniform dynamic air-cushion pressures are studied. These two aspects are handled separately in the following text. However, we do not discuss in detail the effect of the dynamic pressure in the air bag, the fact that the water waves impact on the bag, and elastic vibrations of the bag. The vibrations of the bag are like a wave maker for the acoustic wave motions in the air cushion. The figure also mentions spatially varying air pressure in the vicinity of the air bag. Because this occurs on a length scale that is short relative to the important acoustic wavelength, it can be analyzed by assuming incompressible fluid. Because of the continuity of fluid mass, the escaping airflow under the air bag must have a mean velocity that is dependent on the

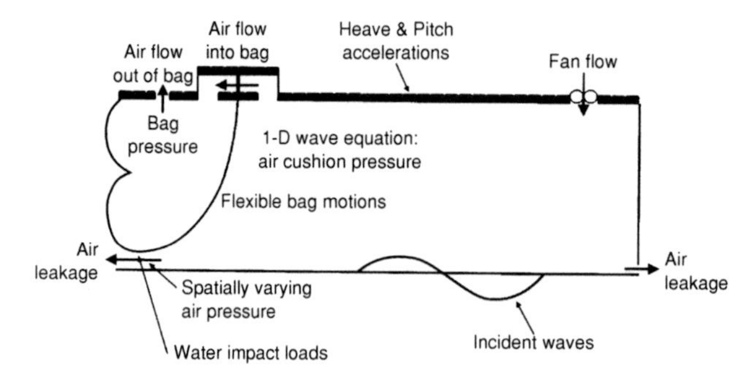

Figure 5.18. Physical effects influencing cobblestone oscillations of an SES (Ulstein 1995).

local height between the air bag and the water surface. Because high velocity implies small pressure, the escaping airflow causes a suction force on the air bag. This influences the mean escape area of the air from the air cushion. Later we see that this influences the damping level of the cobblestone oscillations.

The following text considers uniform pressure and acoustic wave resonance separately. Because the frequency range of these two phenomena are very different and the coupling between the air motion modes associated with uniform pressure and acoustic wave resonance is weak in practice, it is appropriate to consider these two sub-problems separately.

5.6.1 Uniform pressure resonance in the air cushion

Let us present the main principles in the analysis of uniform pressure resonance that shows how a 35 m SES typically has a resonance frequency of 2 Hz. We consider head sea deep-water waves. Before presenting the details, we must introduce the frequency of encounter in regular head sea waves. Let us concentrate on one point P on the vessel and consider the time T_e it takes for two successive wave crests to pass point P. This will obviously be less than the wave period T_0. From Figure 5.19, we can deduce that

$$UT_e + cT_e = \lambda, \qquad (5.18)$$

where U is the ship speed, c is the phase speed of the waves (i.e., the propagation speed of the wave profile), and λ is the wavelength. We can write $c = \omega_0/k$, where $\omega_0 = 2\pi/T_0$ and $k = \omega_0^2/g = 2\pi/\lambda$ (see Table 3.1). This means that the circular frequency of encounter $\omega_e = 2\pi/T_e$ between the ship and the waves for head sea can be written as

$$\omega_e = \omega_0 + \omega_0^2 U/g. \qquad (5.19)$$

In section 7.2, this is derived for a general heading angle.

Equations of heave and dynamic cushion pressure and density

The pressure is assumed to be constant in the whole air cushion. If the longitudinal position of the center of gravity is in the middle of the air cushion and we neglect the effect of the side hulls, there will not be any effect due to pitch. The following analysis is based on Kaplan et al. (1981) and neglects pitch. There are three unknown variables: the heave η_3 at the center of gravity of the SES, the dynamic variations of the pressure, and the mass density of the air in the air cushion. These variables require three equations:

1. Continuity equation for the air mass in the cushion
2. Relationship between pressure and mass density
3. Newton's law

The continuity equation for the air mass in the cushion can formally be written as

$$\rho_a Q_{in} - \rho_a Q_{out} = \frac{d}{dt}(\rho_c \Omega). \qquad (5.20)$$

Here

ρ_a = Air mass density at equilibrium pressure $p_0 + p_a$
$\rho_a Q_{in}$ = Air mass flow per unit time through the fans
$\rho_a Q_{out}$ = Air mass flow per unit time due to leakage

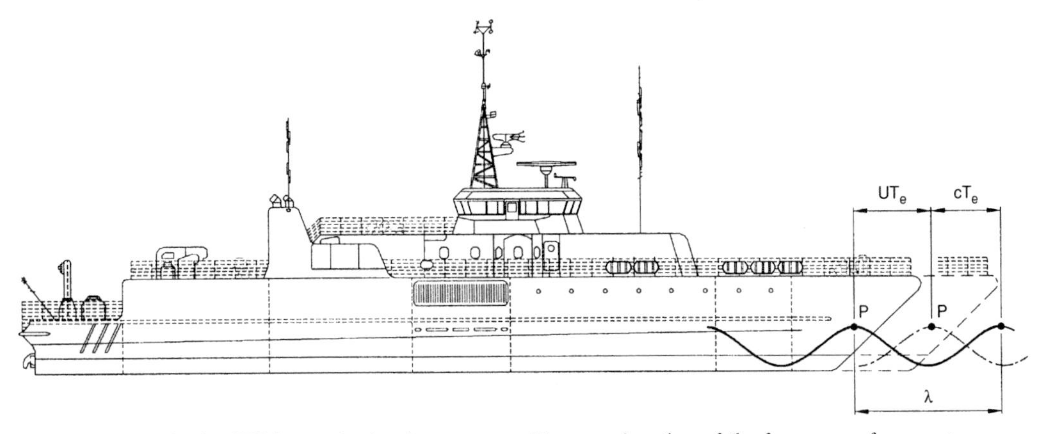

Figure 5.19. An SES in regular head sea waves with an explanation of the frequency of encounter effect (U = forward speed of the SES, T_e = encounter period between the SES and the waves, c = phase speed of the waves, λ = wavelength). (Based on Figure 5.7, copyrightholder: Royal Norwegian Navy.)

ρ_c = Dynamic air mass density of the pressurized air cushion
Ω = Air-cushion volume

We have neglected the influence of the air bag. The relationship between pressure and mass density is based on adiabatic conditions, that is,

$$\frac{p}{p_0 + p_a} = \left(\frac{\rho_c}{\rho_a}\right)^{\gamma}, \tag{5.21}$$

where γ is the ratio of specific heat for air. $\gamma = 1.4$ for air. Further,

$$p = p_0 + p_a + \mu(t)p_0$$

is the total pressure in the air cushion.

Finally Newton's law gives

$$M\frac{d^2\eta_3}{dt^2} = p_0\mu(t)A_b. \tag{5.22}$$

Here $p_0\mu(t)$ is the dynamic part of the pressure in the air cushion and A_b is the cushion area. Eq. (5.22) implies that the hydrodynamic forces on the hull are neglected. The justification is that typically 80% of the weight of the SES is carried by the air cushion. This tells indirectly about the relative importance of hydrodynamic forces and forces due to the dynamic excess pressure in the air cushion.

The next step in the analysis is to linearize the equations. Let us start with the right-hand side of eq. (5.20), which can be expressed as

$$\frac{d\rho_c}{dt}\Omega + \rho_c\frac{d\Omega}{dt} \approx \frac{d\rho_c}{dt}A_b h_b + \rho_a\frac{d\Omega}{dt}, \tag{5.23}$$

where (see Figure 5.20)

A_b = cushion area
h_b = cushion plenum height

We can express $d\rho_c/dt$ by the pressure variations by using eq. (5.21). The first step is to linearize eq. (5.21). We can write

$$\frac{\rho_c}{\rho_a} = \left(1 + \frac{\mu(t)p_0}{p_0 + p_a}\right)^{\frac{1}{\gamma}} \approx \left(1 + \frac{1}{\gamma}\frac{\mu(t)p_0}{p_0 + p_a}\right).$$

This means

$$\frac{d\rho_c}{dt} = \frac{\rho_a}{\gamma}\frac{p_0}{p_0 + p_a}\frac{d\mu}{dt}. \tag{5.24}$$

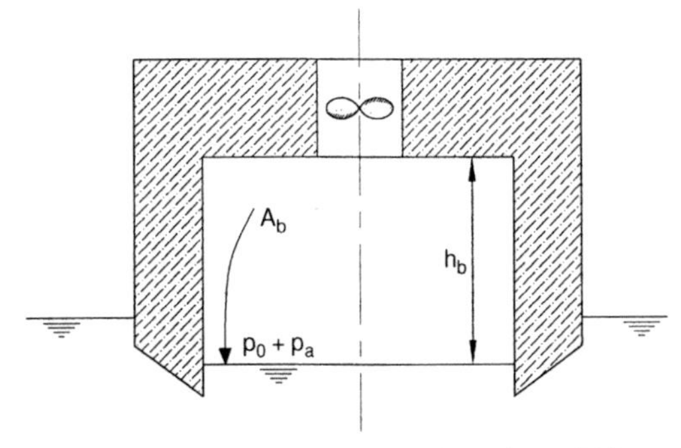

Figure 5.20. Transverse cross section of an SES on-cushion.

The air-cushion volume will change with time because of the wave motion in the cushion area and the heave motion. We can write

$$\frac{d\Omega}{dt} = A_b \frac{d\eta_3}{dt} + \frac{dV_W}{dt}, \tag{5.25}$$

where dV_W/dt is associated with the wave motion and is denoted in the literature as the wave volume pumping.

Let us express dV_W/dt by assuming that this can be described by the incident waves only. This is not appropriate for lower frequencies when $\omega_e^2 b/g$ is of order one (Ogilvie 1969b). Here b is the breadth of the air cushion. The pressure oscillations in the air cushion will then cause free-surface waves.

We can express the incident waves in an Earth-fixed coordinate system as

$$\zeta = \zeta_a \sin(\omega_0 t - kX) \tag{5.26}$$

(see Table 3.1). We then introduce a coordinate system (x, y, z) moving with the ship's forward speed, that is, $X = x - Ut$. This gives

$$\zeta = \zeta_a \sin(\omega_e t - kx). \tag{5.27}$$

We assume an air cushion of constant breadth b and length ℓ. Further, the skirt and air bag have x-coordinates -0.5ℓ and 0.5ℓ, respectively. Then

$$\begin{aligned}\frac{dV_W}{dt} &= -b \int_{-0.5\ell}^{0.5\ell} \frac{d\zeta}{dt} dx \\ &= -b\zeta_a\omega_e \left[\int_{-0.5\ell}^{0.5\ell} [\cos(\omega_e t)\cos(kx) \right. \\ &\qquad \left. + \sin(\omega_e t)\sin(kx)]\, dx \right].\end{aligned}$$

Hence,

$$\frac{dV_W}{dt} = -\frac{2b\zeta_a\omega_e}{k}\sin(0.5k\ell)\cos(\omega_e t)$$

or

$$V_W = V_{Wa}\sin\omega_e t = -\frac{2b\zeta_a}{k}\sin(0.5k\ell)\sin\omega_e t. \tag{5.28}$$

We now linearize the effect of Q_{in} and Q_{out} in eq. (5.20). Linearizing the fan charactistic (see Figure 5.6) gives

$$Q_{in} = Q_0 + \left(\frac{\partial Q}{\partial p}\right)_0 p_0\mu, \tag{5.29}$$

where Q_0 is the mean flow rate due to the fans. This means $(\partial Q/\partial p)_0$ is found by a quasi-steady approach by using the fan characteristics, as illustrated in Figure 5.6.

Q_{out} through small leakage areas A_L are estimated by the formula

$$Q_{out} = 0.61 A_L \left[\frac{2(p_0 + \mu p_0)}{\rho_a}\right]^{1/2}. \tag{5.30}$$

A contraction coefficient of 0.61 for the escaping jet flow has been used here. In reality, this will depend on the local details of the structure at the leakage area. Leakage will, for instance, occur under the air bag (see Figure 5.10). A_L will then be influenced by the relative vertical motions at the leakage area. However, this generally has a smaller influence on the dynamic variations of Q_{out} than the effect of dynamic pressure variations. It implies that A_L is considered a constant in the following analysis. We now Taylor expand eq. (5.30) about $\mu = 0$, that is,

$$Q_{out} \approx Q_{out}\Big|_{\mu=0} + \mu \frac{\partial Q_{out}}{\partial \mu}\Big|_{\mu=0}.$$

By noting that $Q_{out}|_{\mu=0}$ must be equal to Q_0, we find that

$$Q_{out} = Q_0\left(1 + \frac{1}{2}\mu\right). \tag{5.31}$$

It now follows that the linearized version of eq. (5.20) can be written as

$$\begin{aligned}&\frac{A_b h_b}{\gamma(1 + p_a/p_0)}\frac{d\mu}{dt} + A_b\frac{d\eta_3}{dt} \\ &+ \left(0.5Q_0 - \frac{\partial Q}{\partial p}\Big|_0 p_0\right)\mu = -\frac{dV_W}{dt}.\end{aligned} \tag{5.32}$$

We can eliminate μ from eq. (5.32) by using eq. (5.22). This gives a third-order differential equation in η_3. If we integrate this equation once with respect to time, we get

$$\begin{aligned}&\frac{Mh_b}{\gamma(p_0 + p_a)}\frac{d^2\eta_3}{dt^2} + \frac{M}{p_0 A_b}\left[0.5Q_0 - \frac{\partial Q}{\partial p}\Big|_0 p_0\right]\frac{d\eta_3}{dt} \\ &+ A_b\eta_3 = -V_W.\end{aligned} \tag{5.33}$$

This assumes a steady-state solution. The structure of this equation is now in a form that is familiar to us from dynamics of mechanical systems. This means we can make the analogy to a mass term associated with the acceleration $d^2\eta_3/dt^2$, a damping term with the velocity $d\eta_3/dt$, a spring term with the motion η_3, and an excitation on the right-hand side of the equation. This means that the inflow and leakage from the cushion act as damping and the wave motion in the cushion area as excitation. Because $\partial Q/\partial p\,|_0$ is negative in normal operations (see Figure 5.6), the heave damping will always be positive. The air cushion acts like a spring.

The undamped natural circular frequency ω_n is found by setting damping and excitation equal to zero and assuming η_3 is proportional to $\exp(i\omega_n t)$. This gives

$$\omega_n = \left[\frac{A_b\gamma(p_0 + p_a)}{Mh_b}\right]^{\frac{1}{2}}. \tag{5.34}$$

The critical damping of the system given by eq. (5.33) is two times ω_n times the mass term.

Eq. (5.33), therefore, gives a critical damping:

$$b_{cr} = 2\frac{Mh_b}{\gamma(p_0 + p_a)}\omega_n.$$

The ratio ξ between the damping and the critical damping is

$$\xi = \frac{\left[0.5Q_0 - \left.\frac{\partial Q}{\partial p}\right|_0 p_0\right]\gamma\left(1 + \frac{p_a}{p_0}\right)}{2h_b A_b\omega_n}. \tag{5.35}$$

This is an important parameter characterizing the response amplitude at resonance (see section 7.1.4).

Let us write eq. (5.33) as

$$m_{ac}\frac{d^2\eta_3}{dt^2} + b_{ac}\frac{d\eta_3}{dt} + c_{ac}\eta_3 = -V_{Wa}\sin\omega_e t, \tag{5.36}$$

where V_{Wa} is given by eq. (5.28). After finding the steady-state solution of η_3 from eq. (5.36) (see section 7.1.4), we can obtain the corresponding acceleration by differentiating η_3 twice with respect to the time. The amplitude a_{3a} of the vertical acceleration at the center of the gravity of the SES is then

$$a_{3a} = \frac{\omega_e^2\,|V_{Wa}|}{\left[m_{ac}^2\left(\omega_n^2 - \omega_e^2\right)^2 + \omega_e^2 b_{ac}^2\right]^{0.5}}. \tag{5.37}$$

By using eq. (5.22) we can then find the dynamic pressure in the air cushion.

Example: Natural frequency, damping, and vertical accelerations

Consider an SES with mass $M = 2\cdot 10^5$ kg, a cushion length of 40 m, a cushion beam of 10 m, and a cushion plenum height of 2 m. Assume $p_0 = 400$ mm Wc ($= 4.022\cdot 10^3$ Pa when $\rho = 1025$ kgm^{-3}), $Q_0 = 50\,\text{m}^3\text{s}^{-1}$, and $\partial Q/\partial p\,|_0 = -4.5\cdot 10^{-3}\,\text{m}^3\text{s}^{-1}\text{Pa}^{-1}$. This corresponds to realistic values and gives $\omega_n = 12.1$ rad/s, $f_n = \omega_n/(2\pi) = 1.9$ Hz, and $\xi = 8\cdot 10^{-2}$.

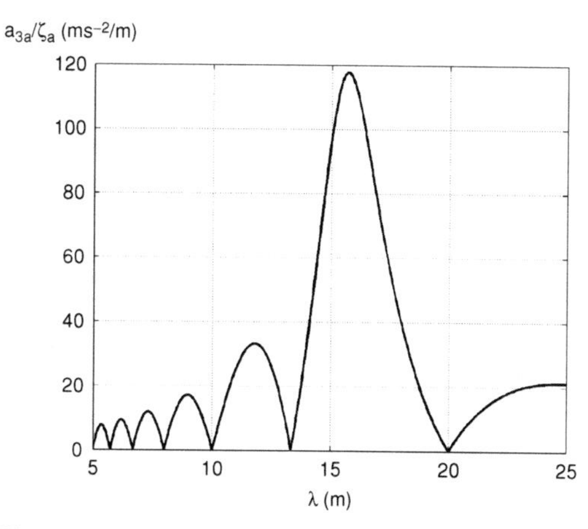

Figure 5.21. Example of calculated amplitude a_{3a} of vertical accelerations at the center of gravity of a 200-ton SES at forward speed 25 ms^{-1} head sea regular waves of wavelength λ and amplitude ζ_a.

We then assume a forward speed $U = 25$ ms^{-1}. The amplitude a_{3a} of the vertical acceleration of the center of gravity of the SES in head sea regular waves with wavelength λ and amplitude ζ_a is presented in Figure 5.21. The natural frequency ω_n corresponds to $\lambda = 15.5$ m. This can be obtained by setting $\omega_e = \omega_n$ in eq. (5.19) and solving this equation with respect to ω_0. This gives

$$\omega_0 = \frac{-\frac{g}{U} + \left[\left(\frac{g}{U}\right)^2 + \left(\frac{4\omega_n g}{U}\right)\right]^{0.5}}{2}.$$

The corresponding wavelength is obtained by assuming that $\omega_0^2/g = k = 2\pi/\lambda$. We see that the largest response occurs in the vicinity of $\lambda = 15.5$ m. The reason a_{3a} is zero for several λ-values in Figure 5.21 is the $\sin(0.5k\ell)$-term in V_W (see eq. (5.28)). Because we have approximated V_W by saying that the air cushion oscillations do not cause free-surface waves, the cancellation of V_W for certain wavelengths is an approximation.

In a sea state, there are many regular wave components (see Chapter 3). The significant wave height $H_{1/3}$ is between 0.3 and 0.4 m for the data presented in Figure 5.17. A wavelength of 15.5 m is realistic in such a sea state. This shows that the resonance oscillations with 1.9 Hz natural frequency can be excited in realistic sea states. How to deal with the response in irregular seas is described in section 7.4.

Scaling

By approximating Mg as $p_0 A_b$ (this gives about 80% of Mg), we can write eq. (5.34) as

$$\omega_n \approx \left[\frac{\gamma g(1 + p_a/p_0)}{h_b}\right]^{0.5}.$$

Further, p_a/p_0 is large. For instance, p_a/p_0 is 25 in the previously presented example. This means

$$\omega_n \approx \left[\frac{\gamma g p_a}{p_0 h_b}\right]^{0.5} = \left[\frac{\gamma p_a}{\rho h h_b}\right]^{0.5}. \quad (5.38)$$

Here ρ is the mass density of the water and h is the vertical distance between mean free-surface levels outside and inside the cushion (see Figure 5.7).

Let us use eq. (5.38) to discuss scaling from model tests to full scale. By using superscripts (m) and (p) to indicate model and full scales, we can write

$$\frac{\omega_n^{(m)}}{\omega_n^{(p)}} = \left[\frac{h^{(p)} h_b^{(p)}}{h^{(m)} h_b^{(m)}}\right]^{0.5}. \quad (5.39)$$

It is then assumed that ρ and p_a are the same in model and full scales. Let us introduce L_m and L_p as the length of the SES in model and full scales. The weight and A_b of the SES scale are Λ_L^3 and Λ_L^2, respectively. Here $\Lambda_L = L_p/L_m$. Because $p_0 A_b = \rho g h A_b \approx Mg$, h scales as Λ_L. The consequence is that h_b also scales as Λ_L. Eq. (5.39) then gives

$$\frac{\omega_n^{(m)}}{\omega_n^{(p)}} = \frac{L_p}{L_m}. \quad (5.40)$$

However, model tests are based on Froude scaling. This means $\omega (L/g)^{0.5}$ should be the same in model and full scales. Eq. (5.40) does not satisfy this. According to eq. (5.40), the natural frequency will be too high in model scale to detect cobblestone oscillations.

So what can we do in order for ω_n to satisfy Froude scaling ? If we go back to eq. (5.38), we could change the atmospheric pressure p_a. However, depressurized towing tanks with wave makers are certainly not common, if they exist at all. Let us say we had been able to Froude scale ω_n. We must also consider the scaling of the damping. However, we will not deal with the details of scaling of damping. It is also difficult in model tests to generate quality waves with the small wavelengths needed to study cobblestone oscillations. Anyway, a common procedure is not to do model tests with cobblestone oscillations. In model tests, one considers only sea states in which the dominant frequency is sufficiently low for the air cushion dynamics to be insignificant. Cobblestone oscillations are instead studied numerically and by full-scale tests. In the next section, we proceed with analyzing cobblestone oscillations due to acoustic wave resonance.

5.6.2 Acoustic wave resonance in the air cushion

The resonance frequency of about 5 Hz illustrated in Figure 5.17 is the result of acoustic standing waves. The interaction with the flexible-bag system is important (Steen 1993). An irrotational linear acoustic flow under adiabatic conditions can be described by the wave equation (Landau and Lifshitz 1959)

$$\frac{\partial^2 \varphi}{\partial x^2} + \frac{\partial^2 \varphi}{\partial y^2} + \frac{\partial^2 \varphi}{\partial z^2} - \frac{1}{c^2}\frac{\partial^2 \varphi}{\partial t^2} = 0 \quad (5.41)$$

and proper boundary and initial conditions. Here φ is the velocity potential for the flow in the pressurized air cushion. The fluid velocity $\mathbf{u}$ is given by $\nabla\varphi$, and c means the speed of sound, which for gases can be expressed as

$$c = (\partial p/\partial \rho)_s^{0.5}. \quad (5.42)$$

The subscript s means constant entropy, that is, we can use eq. (5.21) to describe the relationship between pressure and density. This is the same as an adiabatic relationship. Using eq. (5.21) with mean pressure equal to atmospheric pressure and mean mass density 1.226 kgm^{-3} gives $c = 339$ ms^{-1}. We must emphasize that eq. (5.21) is valid for gases and not for a liquid. Because the speed of sound in water is about 1500 m/s, we see that using eqs. (5.21) and (5.42) does not lead to the correct answer. We must for a liquid introduce the "elasticity of the volume" $\kappa = \rho\, dp/d\rho$. The value of κ for a given liquid varies with the temperature, but only slightly with the pressure. The speed of sound can be expressed as $c^2 = \kappa/\rho_0 = (dp/d\rho)_{\rho_0}$, where ρ_0 is the equilibrium mass density of the liquid.

We should note that eq. (5.41) gives us a Laplace equation when $c \to \infty$. A consequence of this is as follows. Consider a body in infinite fluid. If we assume incompressible fluid, the flow disturbances caused by the body will propagate with infinite velocity, corresponding to $c = \infty$. This means that if a body is set into motion, we immediately see a fluid motion everywhere in the fluid. Consider, for instance, irrotational flow past a circular cylinder in an infinite incompressible fluid. We then know the analytic solution. Whatever value of the coordinate we use in this expression, the velocity potential will have a value. This is true even far away. If we had accounted for compressibility, signals would have taken a finite time to propagate.

We will explain the resonance frequency by a simplified analysis. We assume one-dimensional acoustic waves in the longitudinal direction of the air cushion. We can write the velocity potential of a standing wave system as

$$\varphi = \alpha \cos(\omega t) \cos(\omega x/c). \tag{5.43}$$

We can easily verify that eq. (5.43) is a solution of eq. (5.41). The flow velocity can be written as

$$u = \frac{\partial \varphi}{\partial x} = -\alpha \left(\frac{\omega}{c}\right) \cos(\omega t)\ \sin(\omega x/c) \tag{5.44}$$

and the linearized dynamic pressure as

$$p = -\rho_a \frac{\partial \varphi}{\partial t} = \rho_a \omega \alpha \sin \omega t\ \cos(\omega x/c). \tag{5.45}$$

Here ρ_a is air mass density at static equilibrium. We see that $u = 0$ at $x = 0$, $\pi c/\omega$, $2\pi c/\omega$, and so on. This means if the two ends of the cushion correspond to $x = 0$ and $x = n\pi c/\omega$, $(n = 1, 2, \ldots)$, we have a resonance condition in the air cushion. Let us concentrate on $n = 1$ because this represents the most important case. It means that

$$\frac{\pi c}{\omega} = \ell,$$

where ℓ is the length of the cushion. Let us say that $\ell = 29$ m, $c = 340$ ms^{-1}. We get that the resonance frequency is

$$f = \omega/2\pi = 5.9 \text{ Hz}.$$

Figure 5.17 indicates that the resonance frequency should be lower, that is, about 5 Hz. The reason is interaction with the bag system (Steen 1993). We note that the spatially varying pressures have their maximum absolute values at the ends of the cushion. They are out of phase at the two ends of the cushion. That means the standing acoustic pressure system creates a pitching moment on the wetdeck of the SES, that is, the acoustic resonant waves cause pitching accelerations, resulting in vertical accelerations of the vessel that are largest at one of the ends of the vessel. We should recall that the uniform pressure resonance causes a heave acceleration of the vessel.

Simplified response model

We present a simplified response model for pitch accelerations at acoustic pressure resonance. There are two unknowns in the following equation set: the velocity potential φ for the motion of the air in the cushion and the pitch angle η_5 as defined in Figure 7.11. The two equations are

1. Balance of moments following from Newton's law
2. The wave equation (5.41) with boundary conditions

The wave equation plays the same role as the equations for continuity of air mass and the relationship between pressure and mass density in the uniform pressure resonance problem. However, the wave equation allows for airflow inside the cushion, which is needed in the acoustic resonance problem.

The following assumptions are made:

- Head sea regular wave excitation with frequency of encounter in the vicinity of the lowest natural frequency for standing sound waves in the longitudinal direction
- Hydrodynamic loads on the side hulls are negligible
- The effect of skirt and bag is represented as rigid walls (Figure 5.22).
- Air leakage is through louvers at the deck (Figure 5.22)

We choose a coordinate system (x, y, z), as shown in Figure 5.22, with an origin at the mean free surface inside the cushion. The center of gravity has zero x-coordinate. The x-coordinate of the centers of the louvers and the lift fans are, respectively, x_L and x_F. The closure of the air cushion at the front and aft end is at $x = \pm 0.5\ell$. The mean position of the air cushion has rectangular cross sections with a breadth b, height h_b, and length ℓ.

The velocity potential φ for the flow in the compressible cushion can be described by eq. (5.41). We will integrate this equation over a volume Ω with length Δx in the air cushion (Figure 5.23). The boundary of Ω includes part of the free surface and the wetdeck. We study first

$$\frac{\partial^2 \varphi}{\partial x^2} + \frac{\partial^2 \varphi}{\partial y^2} + \frac{\partial^2 \varphi}{\partial z^2} = \nabla \cdot \nabla \varphi. \tag{5.46}$$

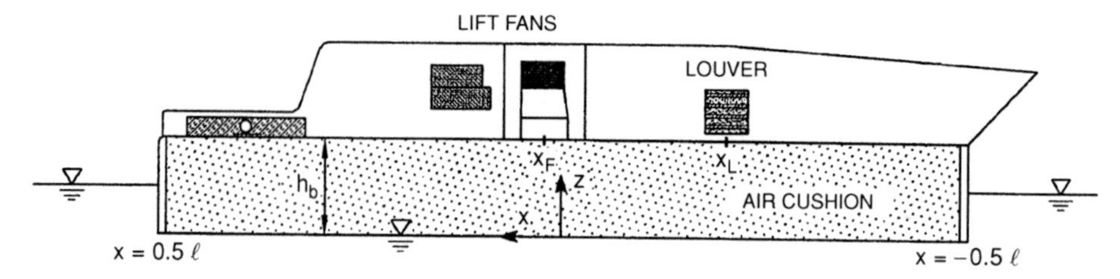

Figure 5.22. Simplified model of an air cushion used in the analysis of acoustic wave resonance.

It follows by the divergence theorem (see eq. (2.205)) that

$$\iiint_{\Omega} \nabla \cdot \nabla \varphi \, d\tau = \oint_{S} \nabla \varphi \cdot \mathbf{n} \, ds, \quad (5.47)$$

where the surface S encloses the volume Ω and $\mathbf{n}$ is the surface normal with positive direction out of Ω. We study first $\partial\varphi/\partial n$ on the free surface S_F. Figure 5.23 illustrates that the incident steady flow velocity U has a component $U\partial\zeta/\partial x$ in the normal direction. Here $z = \zeta$ represents the free-surface shape. Further, the horizontal and vertical velocities $\partial\varphi_0/\partial x$ and $\partial\varphi_0/\partial z$ due to the incident waves must be decomposed in the normal direction. However, only $\partial\phi_0/\partial z$ contributes according to linear theory. The result is

$$\left.\frac{\partial\varphi}{\partial n}\right|_{S_F} = U\frac{\partial\zeta}{\partial x} - \left.\frac{\partial\varphi_0}{\partial z}\right|_{z=0}. \quad (5.48)$$

Expressing ζ as $\zeta_a \sin(\omega_e t - kx)$ and $\partial\varphi_0/\partial z$ as $\omega_0\zeta_a \cos(\omega_e t - kx)$ gives

$$\begin{aligned}\frac{\partial\varphi}{\partial n} &= -\,\omega_e\zeta_a \cos(\omega_e t - kx) \\ &= -\,\omega_e\zeta_a \left[\cos(\omega_e t)\cos(kx)\right. \\ &\quad \left. + \sin(\omega_e t)\sin(kx)\right] \quad \text{on } S_F.\end{aligned} \quad (5.49)$$

This means that we neglect the effect of the air cushion on the water flow and use the incident waves to represent $\partial\varphi/\partial n$ at the interface between the air cushion and the water.

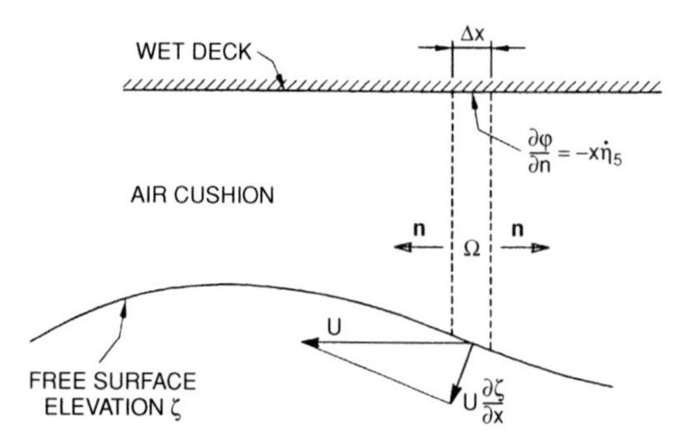

Figure 5.23. Control volume Ω with length Δx in the air cushion of an SES. The boundary of Ω includes part of the free surface and the wetdeck.

At the rigid part of the top of the cushion, that is, outside the outlet of air from the lift fan and inlet of air to the louvers, we can write

$$\frac{\partial\varphi}{\partial n} = \frac{\partial\varphi}{\partial z} = -x\frac{d\eta_5}{dt} \quad \text{on } z = h_b. \quad (5.50)$$

An average $\partial\varphi/\partial z$ at the louvers follows by considering the linearized dynamic part Q^d_{out} of the air mass flux Q_{out} (see eq. (5.30)). We write $A_L = A_{L0} + \Delta A_L$ and perform a series expansion and omit the mean part. This gives

$$Q^d_{out} = 0.61 A_{LO}\left[\frac{2p_0}{\rho_a}\right]^{1/2}\left(\frac{1}{2}\mu_L + \frac{\Delta A_L}{A_{LO}}\right). \quad (5.51)$$

Here A_{LO} is the mean leakage area and ΔA_L is a controlled leakage area, which we discuss in section 5.6.3. Further, $\mu_L p_0$ is the spatially varying dynamic pressure at the louver. The air mass flow rate at the fan lift inlet is described by eq. (5.29). The linearized dynamic part Q^d_{in} is

$$Q^d_{in} = \left(\frac{\partial Q}{\partial p}\right)_0 p_0\mu_F, \quad (5.52)$$

where $p_0\mu_F$ is the dynamic pressure at the lift fan.

We consider now the contribution to the right-hand side of eq. (5.47) from integration along the vertical surfaces at x and $x + \Delta x$. These vertical surfaces extend from $z = 0$ to h_b (see Figure 5.22) and from $y = -0.5b$ to $0.5b$. This means that we study

$$I_1 = \int_{-0.5b}^{0.5b} dy \int_0^{h_b} dz \left[\left.\frac{\partial\varphi}{\partial x}\right|_{x+\Delta x} - \left.\frac{\partial\varphi}{\partial x}\right|_{x}\right].$$

A Taylor expansion in x of the integrand gives

$$I_1 = \int_{-0.5b}^{0.5b} dy \int_0^{h_b} dz \frac{\partial^2\varphi}{\partial x^2}\Delta x. \quad (5.53)$$

We assume now that the variations of φ in the y- and z-directions are much smaller than in the x-direction. We therefore introduce

$$\bar{\varphi} = \frac{1}{h_b b} \int_{-0.5b}^{0.5b} dy \int_{0}^{h_b} \varphi \, dz. \tag{5.54}$$

Eq. (5.53) can then be approximated as

$$I_1 = h_b b \frac{\partial^2 \bar{\varphi}}{\partial x^2} \Delta x.$$

Returning now to eq. (5.41), using the previous approximations and letting $\Delta x \to 0$ gives

$$\frac{\partial^2 \bar{\varphi}}{\partial x^2} - \frac{1}{c^2}\frac{\partial^2 \bar{\varphi}}{\partial t^2} + \frac{1}{h_b b} \int_{-0.5b}^{0.5b} dy \left[\left.\frac{\partial \varphi}{\partial z}\right|_{z=h_b} + \left.\frac{\partial \varphi}{\partial n}\right|_{S_F} \right] = 0. \tag{5.55}$$

We now express the solution of eq. (5.55) as

$$\bar{\varphi} = \beta(t) \sin\left(\frac{\pi}{\ell} x\right). \tag{5.56}$$

Here $\beta(t)$ is an unknown that will follow from solving the problem. We note that eq. (5.56) satisfies a rigid wall condition, that is, $\partial\bar{\varphi}/\partial x = 0$ at $x = \pm 0.5\ell$. The dynamic pressure can be expressed by linearized Bernoulli equation as

$$p = -\rho_a \frac{\partial \bar{\varphi}}{\partial t} = -\rho_a \dot{\beta}(t) \sin\left(\frac{\pi}{\ell} x\right), \tag{5.57}$$

where dot is the same as time derivative. This means that μ_L in eq. (5.51) is

$$\mu_L = -\rho_a \dot{\beta}(t) \sin\left(\frac{\pi}{\ell} x_L\right) \Big/ p_0. \tag{5.58}$$

Similarly, we can find an expression for the lift fan coefficient μ_F. However, we will set $x_F = 0$ in the following mathematical expressions. This gives that μ_F is zero. Another matter is that it is beneficial from a cobblestone oscillation point of view to have both x_F and x_L close to the ends of the cushion. This will be more evident later. We substitute eq. (5.56) into eq. (5.55), multiply the resulting expression by $\sin(\pi x/\ell)$, and integrate over the cushion length. Eq. (5.50) is assumed to be mathematically valid over the whole cushion length. This gives a small difference from that obtained by allowing for the louvers and the fan lift. However, physically, we must incorporate the effect of eqs. (5.51) and (5.52) in addition.

This process involves the integrals

$$\int_{-0.5\ell}^{0.5\ell} \sin^2(\pi x/\ell)\, dx = 0.5\ell \tag{5.59}$$

$$\int_{-0.5\ell}^{0.5\ell} x \sin(\pi x/\ell)\, dx = 2(\ell/\pi)^2 \tag{5.60}$$

$$\int_{-0.5\ell}^{0.5\ell} \cos(kx) \sin(\pi x/\ell)\, dx = 0 \tag{5.61}$$

$$\int_{-0.5\ell}^{0.5\ell} \sin(kx) \sin(\pi x/\ell)\, dx = \frac{2k \cos(0.5k\ell)}{(\pi/\ell)^2 - k^2}. \tag{5.62}$$

This leads to the following equation:

$$-\frac{1}{c^2}\ddot{\beta}(t)\left(\frac{\ell}{2}\right) - K_1 \dot{\beta}(t) - \left(\frac{\pi}{\ell}\right)^2 \frac{\ell}{2}\beta(t) - \frac{1}{h_b} 2\left(\frac{\ell}{\pi}\right)^2 \dot{\eta}_5(t) = K_2 \sin\omega_e t, \tag{5.63}$$

where

$$K_1 = \frac{A_{LO}}{h_b b} 0.61 \left[\frac{2\rho_a}{p_0}\right]^{1/2} 0.5 \sin^2\left(\frac{\pi x_L}{\ell}\right) \tag{5.64}$$

$$K_2 = \omega_e \zeta_a \frac{2k \cos(0.5k\ell)}{h_b((\pi/\ell)^2 - k^2)}. \tag{5.65}$$

When $k\ell = \pi$, we note that $\cos(0.5k\ell)/(\pi - k\ell) = 0.5$.

Then we need to satisfy the balance of moments following from Newton's law. The dynamic air-cushion pressure given by eq. (5.57) gives the following pitch moment on the SES:

$$F_5 = \rho_a \dot{\beta}(t) b \int_{-\ell/2}^{\ell/2} x \sin\left(\frac{\pi}{L} x\right) dx = \rho_a \dot{\beta}(t) 2(\ell/\pi)^2 b.$$

By neglecting the hydrodynamic loads on the side hulls, we then have

$$I_{55}\ddot{\eta}_5 = \rho_a 2(\ell/\pi)^2 b \dot{\beta}(t). \tag{5.66}$$

Here I_{55} is the mass moment of inertia of the SES in pitch. By integrating eq. (5.66) with respect to time and implicitly assuming a steady-state solution with frequency equal to the excitation frequency and by substituting into eq. (5.63) we obtain

$$\frac{\ell}{2c^2}\ddot{\beta}(t) + K_1 \dot{\beta}(t) + K_3 \beta(t) = -K_2 \sin\omega_e t, \tag{5.67}$$

where

$$K_3 = \left(\frac{\pi}{\ell}\right)^2 \frac{\ell}{2} + \frac{4b}{h_b}\left(\frac{\ell}{\pi}\right)^4 \frac{\rho_a}{I_{55}}. \quad (5.68)$$

So we once more have the structure of a mass-spring system with damping that is excited by incident wave volume pumping of the air cushion. Further, the damping source is the air leakage, as it was for the uniform pressure resonance. If the x-coordinates of the lifting fans were different from zero, there would be an additional term in K_1. We note from the damping coefficient K_1 given by eq. (5.64) that maximum damping effect is created by placing the louvers at either the front or aft end of the air cushion. The same applies for the fan outlet. For the uniform pressure resonance, there was a damping effect from the lift fans. Because the lift fans are placed at $x = 0$, there is no effect on the studied acoustic pressure resonance problem.

More complete treatments of acoustic pressure resonance coupled with uniform pressure resonance for realistic air cushions and air bags are found in Steen (1993), Sørensen (1993), and Ulstein (1995).

5.6.3 Automatic control

In reality, one would use an automatic control system to damp out some of the cobblestone effect. This is done by controlling the airflow out from the cushion in such a way that it effectively acts as a damping on the system. In order to do that properly, one needs a simplified but rational mathematical method that accounts for the dynamic pressure variations in the air cushion in combination with the global heave and pitch accelerations of the vessel. We will illustrate a possible automatic control system proposed by Sørensen (1993) (Figure 5.24). It is common jargon to call an automatic control system such as this a ride control system.

Sørensen used a louver system consisting of two vent valves in the front of the air cushion. The opening and closing of the vent valves control the airflow from the air cushion so that one gets a damping effect on the system. There are three pressure sensors in the air cushion and one accelerometer on the vessel as part of the ride control system. By properly filtering the signals from the measurement units and using a mathematical model for the system behavior, the control system can give the correct signals to the louver system. The placement of the louver system is essential. For instance, if the louver system is placed at midships, it will have a negligible effect on the acoustic resonance mentioned above. The reason is simply that the acoustic pressure component, see eq. (5.57), has a node, that is, no amplitude, at midships, whereas it has its maximum amplitude at the ends of the cushion.

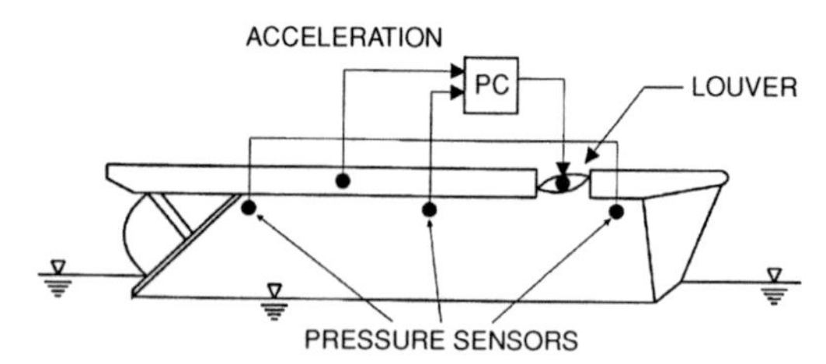

Figure 5.24. Automatic control system for an SES (Sørensen 1993).

Let us illustrate some details of a mathematical model for the ride control in the case of uniform pressure resonance. We express the linear dynamic air mass flux out of the louver as in eq. (5.51). We want to control the time-varying leakage area ΔA_L of the louver. The proposed feedback controller may be written as

$$\frac{\Delta A_L}{A_{LO}} = k_1\mu + k_2\frac{d\mu}{dt}. \quad (5.69)$$

This causes a controlled air mass flux out of the louver that is

$$Q^c = 0.61 A_{LO}\left(\frac{2p_0}{\rho_a}\right)^{1/2}\left(k_1\mu + k_2\frac{d\mu}{dt}\right). \quad (5.70)$$

The constants $k_1 \geq 0$ and $k_2 \geq 0$ have to be determined to maximize damping with due consideration of constraints such as available power, limitations of maximum and minimum leakage area, rate of change of opening with time, and computer capacity.

We now introduce eq. (5.70) into the linearized equations for uniform pressure resonance. Going through the steps that led to eq. (5.33), we see that it is the term associated with k_1 that leads to damping. So we set $k_2 = 0$ and get an additional damping term,

$$\frac{M}{p_0 A_b} 0.61 A_{LO} k_1 \left[\frac{2p_0}{\rho_a}\right]^{1/2} \frac{d\eta_3}{dt}, \quad (5.71)$$

on the left-hand side of eq. (5.33). The ratio ξ_{RC} between this damping coefficient and the critical damping can, as for eq. (5.35), be expressed as

$$\xi_{RC} = \frac{0.61 A_{LO} k_1 \left[\frac{2p_0}{\rho_a}\right]^{1/2} \gamma \left(1 + \frac{p_a}{p_0}\right)}{2h_b A_b \omega_n}. \quad (5.72)$$

By using knowledge about the response of a mass-spring system with damping (see section 7.1.4), this damping ratio shows how much the response can be reduced at resonance. Let us exemplify by using the same example as in section 5.6.1 and set $A_{LO} = 1\,\text{m}^2$. If we, for instance, want $\xi_{RC} = 0.1$, then eq. (5.72) gives $k_1 = 31$. However, we must have the constraints of the system in mind. For instance, using eq. (5.69) determines the maximum leakage area. This must obviously be beyond a certain limiting value. The oscillation amplitudes of ΔA_L, the rate of change with time of ΔA_L, and saturation elements need to be included in the controller to force the command-control signals to be within the physical limitations of the louver system.

One should notice that for the dynamic uniform pressure resonance, the longitudinal location of the vent values is of no concern. However, this matters for acoustic wave resonance (see section 5.6.2). We have left it as exercise 5.9.6 to show how to automatically control the acoustic wave resonance in the air cushion.

5.7 Added resistance and speed loss in waves

The main cause of the added resistance of an SES in waves is not the mean second-order nonlinear wave loads on the side hulls, as described in Chapter 7 for semi-displacement vessels. There are additional effects for an SES. For instance, there is also an added resistance due to the oscillatory cushion pressure. This has been discussed theoretically by Doctors (1978). Both previous effects are caused by wave radiation. However, Moulijn (2000) has shown experimentally that these effects are small relative to the effect of sinkage. The latter is associated with air leakage from the cushion in waves (Faltinsen et al. 1991a, Moulijn 2000). The leakage occurs because of relative vertical motions between the vessel and the waves and causes the SES to sink. The increased sinkage of the SES increases the wetted area of the side hulls and hence changes the still-water viscous resistance on the hulls. A similar effect occurs for the bow seals. Because of the change in the excess pressure in the cushion, there also occurs a change in the still-water wave resistance due to the cushion pressure. The net effect is an increased resistance if the speed is constant.

We present the procedure by Faltinsen et al. (1991a) in some detail. To calculate the air leakage they used the formula

$$Q_{out} = 0.61 A_L \left[\frac{2p_0}{\rho_a}\right]^{1/2}, \quad (5.73)$$

where Q_{out} = volume flux out of the cushion, A_L = leakage area, p_0 = excess pressure in the cushion, and ρ_a is the mass density of the air in the cushion at pressure $p_0 + p_a$. The first step is to find the effect of the waves and wave-induced motions on the leakage area. Faltinsen et al. (1991a) assumed that this increase occurs only at the skirt. For simplicity, we consider only the effect of heave and pitch motions and that the skirt is rigidly connected to the SES. It is assumed that the leakage occurs when the relative vertical motion is larger than $d > 0$, where d depends on the skirt configuration and the mean sinkage of the vessel.

Let us first analyze this problem in regular head sea waves. The relative vertical motion between the vessel and the water at the bow is expressed as $\eta_{Ra} \sin \omega_e t$. We have exemplified this in Figure 5.25 by plotting $\eta_{Ra} \sin(\omega_e t)/d$. Leakage occurs when

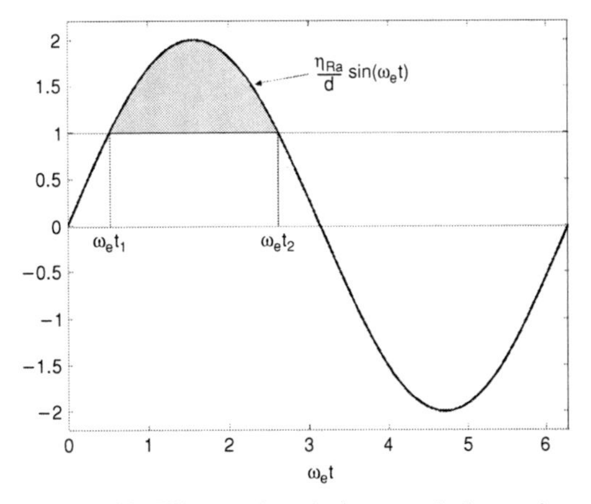

Figure 5.25. Effect of relative vertical motion $\eta_{Ra} \sin \omega_e t$ at the bow on the leakage area under the bow seal of an SES in incident regular head sea waves with frequency of encounter ω_e. d = vertical distance needed for the the bow seal to be out of the water by statically lifting the bow of the vessel. The shaded area is the basis for estimating mean value of leakage area.

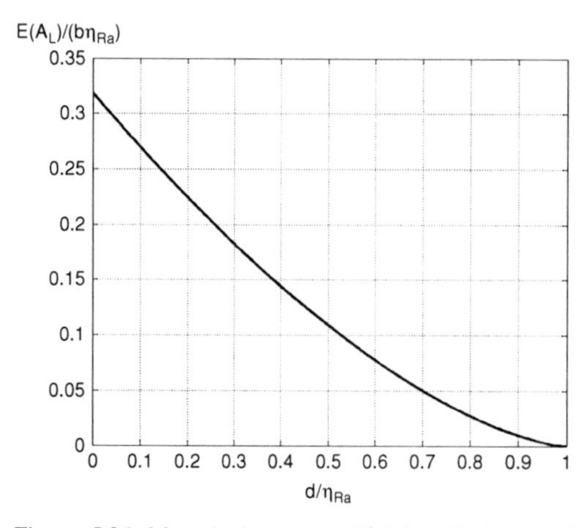

Figure 5.26. Mean leakage area $E(A_L)$ at the bow seal of an SES over one period of oscillation in regular head sea waves with frequency of encounter ω_e. $\eta_{Ra}\sin\omega_e t$ is the relative vertical motion in the bow. d = vertical distance for the bow to be out of the water by statically lifting the bow of the vessel, b = breadth of the bow seal.

$\omega_e t_1 < \omega_e t < \omega_e t_2$, where

$$\omega_e t_1 = \sin^{-1}\left(\frac{d}{\eta_{Ra}}\right) \quad \text{and} \quad \omega_e t_2 = \pi - \omega_e t_1. \tag{5.74}$$

Here d/η_{Ra} is ≤ 1. We can then write the mean leakage area $E(A_L)$ over one period of oscillation as

$$\begin{aligned} E(A_L) &= \frac{\eta_{Ra} b}{2\pi}\int_{\omega_e t_1}^{\omega_e t_2}\left(\sin(\omega_e t) - \frac{d}{\eta_{Ra}}\right) d(\omega_e t) \\ &= \frac{\eta_{Ra} b}{2\pi}\left[2\sqrt{1-(d/\eta_{Ra})^2} + 2\left(\frac{d}{\eta_{Ra}}\right)\right. \\ &\quad \left.\times \sin^{-1}\left(\frac{d}{\eta_{Ra}}\right) - \left(\frac{d}{\eta_{Ra}}\right)\pi\right]. \end{aligned} \tag{5.75}$$

Here b is the breadth of the bow seal. Eq. (5.75) is plotted in Figure 5.26 when $0 \leq d/\eta_{Ra} \leq 1$. There is obviously no leakage when $d/\eta_{Ra} \geq 1$.

Let us then consider irregular head sea waves. We then need a stochastic description of the relative vertical motions. We can write the expected value of the dynamic change in the leakage area at the skirt as

$$\begin{aligned} E[A_L] = b\left\{\frac{\sigma_R}{\sqrt{2\pi}}\exp(-0.5d^2/\sigma_R^2)\right. \\ \left. - 0.5d + 0.5d\Phi\left(\frac{d}{\sqrt{2}\sigma_R}\right)\right\}, \end{aligned} \tag{5.76}$$

where Φ ( ) is the probability integral (Gradshteyn and Ryzhik 1965); see Faltinsen et al. 1991a. σ_R is the standard deviation of the relative vertical motion. Eq. (5.76) was derived by assuming a long-crested irregular sea state and that the amplitudes of the relative motions are Rayleigh distributed. It can be generalized to include the effect of other motion modes and leakage areas. By using the characteristics for the cushion fans in combination with eqs. (5.73) and (5.76), one can find the expected value for the drop in pressure and the volume flux for constant RPM (revolutions per minute) of the fans. Constant RPM is assumed in the following presented examples. However, it is not necessary to assume this. In practice, one should aim at keeping the same excess pressure in the air cushion as in calm water. This depends on available lifting fan power.

When the pressure drop in the cushion has been found, one can estimate the sinkage by balancing the weight of the SES with the vertical forces caused by the excess pressure in the cushion and the buoyancy forces on the hulls. The increased sinkage of the SES changes the still-water viscous and wave resistance on the side hulls. The change in the excess pressure in the cushion will also lead to a change in the still-water wave resistance due to the cushion pressure. All these changes can be interpreted as contributions to the added resistance in waves and are included in the predictions. Figure 4.2 illustrates how the still-water resistance components change with significant wave height.

The SES can in practice experience a significant involuntary speed loss in waves. We distinguish between involuntary and voluntary speed loss. Voluntary speed reduction means that the shipmaster reduces the ship's speed, for instance, because of heavy slamming, water on deck, or large accelerations. Involuntary speed reduction is the result of added resistance of the vessel due to waves, wind and maneuvering and changes in the efficiency of the propulsion system in waves. We can illustrate the speed reduction of an SES by comparing the involuntary speed loss of a 40-m SES with a 40-m catamaran in waves. These results are presented in Figure 5.27 as a function of significant wave height $H_{1/3}$ for head sea long-crested waves. To each value of $H_{1/3}$ there corresponds several realistic values of the mean wave period. The speed loss will be a function of the mean wave period. That is the reason for the shaded areas in the figure. Figure 5.27 illustrates that an SES has a more significant speed loss in waves than does a

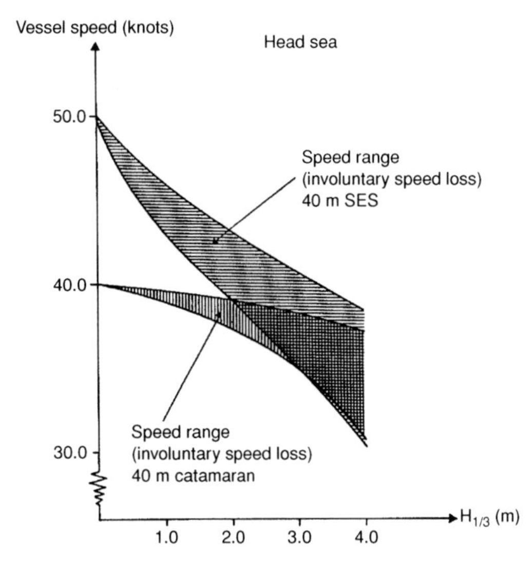

Figure 5.27. Vessel speed range of a 40-m SES and a 40-m catamaran as a function of significant wave height in head sea waves. Only involuntary speed loss effects are shown in these numerical results.

catamaran. As already stated, the important reason for this is the air leakage from the cushion that occurs in waves. It should be realized that the total shaft powers for the catamaran and the SES are, respectively, 8300 kW and 5500 kW. Even allowing for a 20% to 25% increase in power due to the fans of the SES, it is seen that the SES uses less power and keeps a higher speed than does the catamaran for nearly all sea states of practical interest.

5.8 Seakeeping characteristics

Figure 1.12 shows that in a comparative study between a similarly sized catamaran and an SES, the SES has generally the best behavior from a vertical acceleration point of view in head seas, except for small sea states, in which cobblestone oscillation causes problems for the SES.

Table 5.1. *Main particulars of an SES*

Length overall	40.0 m
Length of waterline	34.8 m
Trim angle	0.82°
Longitudinal center of gravity	15.0 m forward stern
Weight	189 tonnes
Cushion support, still water	83.3%
Cushion excess pressure, still water	0.508 m water height
Position of bow seal	33.0 m forward stern
Pitch radius of gyration	10.4 m

We present additional numerical seakeeping results reported by Faltinsen et al. (1991a) for an SES. The characteristics of the SES are given in Table 5.1. Waterjet propulsion is used. The skirt of the SES is assumed to touch the free surface in still water. No "ride control" for the cobblestone oscillations is accounted for. Only head sea waves and wind have been considered. A two-parameter JONSWAP spectrum, as recommended by ITTC, is used. The wind field is assumed to be uniform in space. The wind velocity V_W is estimated from the formula $V_W = (H_{1/3}g/0.21)^{1/2}$, where $H_{1/3}$ is the significant wave height.

Table 5.2 presents seakeeping data for the 40 m–long SES in head seas. For each combination of $H_{1/3}$ and T_1, involuntary speed reduction, standard deviations σ_a of vertical accelerations at the center of gravity, and standard deviations σ_R of relative vertical motions at the waterjet inlet have been calculated. We note the high involuntary speed reductions. For instance, for $H_{1/3} = 2$ m, $T_1 = 5$ s, the speed has dropped 11.4 knots because of air leakage and sinkage. The cushion supports 50.8% of the vessel weight in this sea state compared with 83.3% in still water.

By comparing σ_R in Table 5.2 with the submergence d in Table 5.2 of the waterjet inlet relative to the steady free surface, the exposure of the waterjet inlet to the free air can be evaluated. The submergence d was estimated by accounting for the change in sinkage of the vessel and the steady free-surface deformation inside the cushion due to increased air leakage from the cushion in a seaway. The steady free-surface elevation due to waves was not accounted for. By requiring that $d > 4\sigma_R$, we see that voluntary speed reduction is necessary for all sea states presented in Table 5.2. Even if this may be a too-strict requirement, the data in Table 5.2 indicate that exposure of the waterjet inlet to the free air is a possible reason for the significant engine load fluctuations reported by Meek-Hansen (1991); see Figure 1.13. These calculations indicate that something has to be done with the submergence of the waterjet inlet. The best way may be to use a scoop inlet to the waterjet. The consequence of this is increased resistance. For instance, if the inlet is 1 m below the keel, it means typically a 10% to 15% increase in power.

Table 5.2. *Seakeeping data of a 40 m–long SES (Table 5.1) in head sea long-crested waves described by a two-parameter JONSWAP spectrum*

	T_1 (s)						
$H_{1/3}$ (m)	3.3	5.0	6.7	8.3	10.0	11.7	
	44.8	43.0	43.4	44.7	45.5	46.0	Speed (kn)
	0.42	0.57	0.52	0.41	0.35	0.30	d (m)
1.0	0.19 g	0.15 g	0.13 g	0.10 g	0.08 g	0.06 g	σ_a
	0.22	0.38	0.31	0.21	0.16	0.12	σ_R (m)
	41.5	38.6	38.9	40.8	42.3	43.3	Speed (kn)
	0.64	0.93	0.84	0.64	0.52	0.44	d (m)
2.0	0.31 g	0.26 g	0.23 g	0.18 g	0.14 g	0.11 g	σ_a
	0.44	0.80	0.66	0.45	0.32	0.25	σ_R (m)
	38.9	35.2	35.1	37.5	39.4	40.8	Speed (kn)
	0.83	1.22	1.10	0.86	0.69	0.57	d (m)
3.0	0.39 g	0.36 g	0.31 g	0.24 g	0.19 g	0.15 g	σ_a
	0.66	1.22	0.96	0.67	0.50	0.38	σ_R (m)
		30.9	30.8	34.4	36.7	38.4	Speed (kn)
		1.45	1.29	1.04	0.85	0.69	d (m)
4.0		0.43 g	0.37 g	0.30 g	0.24 g	0.19 g	σ_a
		1.61	1.16	0.88	0.66	0.51	σ_R (m)

$H_{1/3}$ = significant wave height, T_1 = mean wave period. The shaft power is 2 × 2750 kW in all sea states. Waterjet propulsion (flush type), diameter outlet: 0.5 m. Design speed, still water: 50 knots. Total efficiency, still water 0.60. First line: Speed in knots including involuntary speed loss. Second line: Submergence of waterjet inlet relative to the steady free-surface in the cushion. Third line: Standard deviation of the vertical accelerations at COG. Fourth line: Standard deviation of relative vertical motions at the waterjet inlet. (Faltinsen et al. 1991a).

Another possibility is an automatic control system for the propulsion-engine system that accounts for the possibility of air coming into the waterjet system and can minimize the engine power fluctuation. However, it is common to have possibilities to move fluids (water, fuel) in the longitudinal direction to be able to trim the vessel to avoid air ingestion in the waterjets.

An SES may experience bow-dive in waves resulting in an abrupt stop of the vessel. Air lift fans ought to be placed in the bow region to counteract the bow dive. Further, the bow design may matter.

Operational problems with wetdeck slamming may occur both in "on-cushion" and "off-cushion" conditions. Wetdeck slamming on a catamaran occurs most often in the bow part, whereas slamming on an SES may happen in the aft part of the wetdeck.

We can use data such as those in Table 5.2 in an operational study, that is, find out how much of the time the vessel can operate in a given operational area. We then need a scatter diagram showing the occurrence of different combinations of significant wave height and mean wave period. Table 3.4 is an example on such a diagram. This corresponds to long international services, which is obviously questionable for a 40 m long vessel. We should ideally know the probability of different wave headings and the typical course of the ship.

We also need criteria for operational limits, such as those presented in Chapter 1. Then we just go through the different combinations of sea states relative to the ship. We can see for each sea state if the vessel would meet the criteria or not. Each sea state has a probability of occurrence. For instance, Table 3.4 shows that probability of occurrence of 1.5 m $< H_{1/3} <$ 2.5 m and 3.5 s $< T_2 <$ 4.5 s is 2/995. This includes all wave headings. If we do not have the information about the probability of occurrence of different wave headings relative to the ship, we must make our own judgments. It is, for instance, usual in open sea conditions to say that all wave headings have equal probability.

Then we collect the sea states in which the ship satisfies the given operational limits and add their

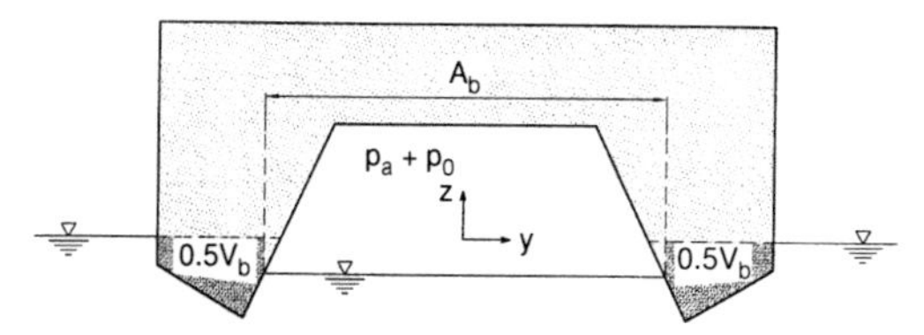

Figure 5.28. Definitions of area A_b and volume A_b to be used in eq. (5.2) when the SES has nonvertical surfaces of the side hulls inside the cushion.

probabilities. This gives an estimated fraction of the total time the ship can operate.

5.9 Exercises

5.9.1 Cushion support at zero speed

Eq. (5.2) relating the weight of the SES, buoyancy of the side hulls, and cushion support was based on the cross section of the SES shown in Figure 5.7 with vertical surfaces of the side-hulls inside the cushion.

Generalize eq. (5.2) to allow for nonvertical surfaces of the side hulls inside the cushion. Assume zero forward speed.

The answer is that eq. (5.2) is still valid, but with different definitions of V_b and A_b. If a 2D situation is considered, as in Figure 5.28, then V_b is defined by first drawing vertical lines through the intersection points between the side hulls and the free surface inside the cushion. If we consider the left side hulls, then half of V_b is obtained by considering the displaced volume below $z = 0$ and to the left of the vertical line. Similarly, we obtain the other half of V_b by considering the right side hull. A_b is the area of the free surface inside the cushion. You should explain this.

5.9.2 Steady airflow under an aft-seal air bag

Consider the situation in Figure 5.29 under steady conditions. There is no wave motion on the free surface. The higher pressure inside the cushion than outside the SES causes a flow under the air bag.

Assume:

- The shape of the air bag is known
- The air is incompressible and inviscid
- The flow can be approximated as one dimensional
- At $x_b = 0$ (see Figure 5.29), there is no flow

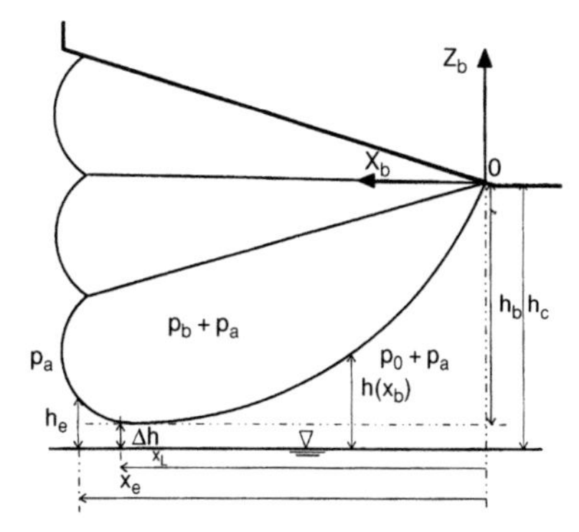

Figure 5.29. Airflow under an aft-seal air bag of an SES (Steen 1993).

- The flow separates at longitudinal coordinate x_e
- There is no contraction of the jet flow escaping from the air cushion, and the pressure is atmospheric at $x = x_e$

a) By using the steady Bernoulli equation without gravity and the conservation of fluid mass, show that the velocity u and pressure p at x_b is

$$u(x_b) = \frac{h_e}{h(x_b)}\left(\frac{2p_0}{\rho}\right)^{0.5} \tag{5.77}$$

$$p(x_b) = p_a + p_0 - \left(\frac{h_e}{h(x_b)}\right)^2 p_0. \tag{5.78}$$

Here p_0 is the excess pressure in the air cushion and $h(x_b)$ (see Figure 5.29) is the vertical distance between the free surface and the bag at x_b. Note that $h_e/h(0)$ is implicitly assumed to be zero.

b) Discuss the fact that the flow causes a suction force on the bag.

c) Suppose that the flow separates at the lowest point of the bag. Use representative values for p_0 and ρ and calculate the ratio between the escape velocity of the air and speed of sound in air.

d) Discuss whether we should have considered the air to be compressible and viscous.

(Hint: Use qualitative arguments based on boundary-layer thickness and air gap for the effect of viscosity.)

5.9.3 Damping of cobblestone oscillations by T-foils

We want to estimate what influence a T-foil, as shown in Figure 7.3, has on the uniform pressure resonance in the air cushion. We express the damping coefficient due to the foil as

$$B_{33}^{L} = 0.5\rho U \frac{dC_L}{d\alpha} A. \tag{5.79}$$

Here $dC_L/d\alpha$ is the steady lift slope of the foil, which is 2π for a foil with infinite aspect ratio. We set $dC_L/d\alpha = 3$ and the planform area $A = 1.5\,\text{m}^2$.

a) Introduce this damping effect into the equations that we derived in section 5.6.1 for the uniform pressure resonance.

b) Use the data for the SES presented in the example in section 5.6.1 and calculate the ratio of the damping due to the T-foil and the critical damping.

5.9.4 Wave equation

Derive the wave equation given by eq. (5.41) by starting out with the equation of continuity

$$\frac{\partial \rho}{\partial t} + \nabla \cdot (\rho\, \mathbf{u}) = 0 \tag{5.80}$$

and Euler's equation without gravity effects

$$\frac{\partial \mathbf{u}}{\partial t} + \mathbf{u} \cdot \nabla \mathbf{u} = -\frac{1}{\rho} \nabla p. \tag{5.81}$$

Further, the flow is irrotational so that the fluid velocity $\mathbf{u} = \nabla \phi$. The relationship between small changes in the pressure p' and small changes in the mass density ρ' is given by the adiabatic relationship

$$p' = (\partial p/\partial \rho)_s\, \rho'. \tag{5.82}$$

Here the subscript s means constant entropy.

By linearization of the equations, show that Euler's equation gives the linear Bernoulli equation.

5.9.5 Speed of sound

Show how the speed of sound is a function of the static excess pressure in the air cushion of an SES.

5.9.6 Cobblestone oscillations with acoustic resonance

Consider an SES with total mass $M = 2 \cdot 10^5$ kg, cushion length $\ell = 40\,\text{m}$, cushion beam $10\,\text{m}$, cushion plenum height $2\,\text{m}$, and $p_0 = 4.022 \cdot 10^3$ Pa. Set the speed of sound in air equal to 340 ms^{-1} and mass density of air $\rho_a = 1.23\,\text{kgm}^{-3}$. Set pitch radius of gyration to be $r_{55} = 0.25\ell$ and note $I_{55} = Mr_{55}^2$.

Use the mathematical model for head sea regular waves described in section 5.6.2, with a louver placed in the front end of the cushion. The fan lifts are at midships.

a) What is the lowest resonance frequency for acoustic wave resonance in the air cushion?

b) Set the mean leakage area of the louver equal to $1\,\text{m}^2$. Assume that the ride control system is not operating.

What is the ratio between damping and critical damping?

Assume a ship speed $U = 25\,\text{ms}^{-1}$. Calculate the amplitude of vertical acceleration at the front end of the cushion divided by incident wave amplitude as a function of incident wavelength, that is, a curve similar to the one presented in Figure 5.21 for uniform pressure resonance.

c) Introduce a ride control model, as in eq. (5.69), in the acoustic model. Express the new equation for β, that is, modify eq. (5.67).

d) Consider the resonant condition and assume an incident wave amplitude equal to 0.2 m, and limit the maximum value of $\Delta A_L/A_{LO}$ in eq. (5.69) to 0.5. How much is it possible to reduce the pitch accelerations at resonance by the ride control system?

6 Hydrofoil Vessels and Foil Theory

6.1 Introduction

Hydrofoil vessels in foilborne conditions generally have good seakeeping characteristics, create small wash, and have small speed loss due to incident waves. This is particularly true for fully submerged foil systems. Foils are normally designed for subcavitating conditions. However, the possibility of cavitation is then an important issue. Our discussion assumes subcavitating foils.

Johnston (1985) pointed out that important aspects when selecting foil and strut configurations of fully submerged hydrofoils are:

- Maintenance of directional and roll stability
- Stable recovery when a foil comes out of the water (broaches)
- Graceful deterioration of performance in severe seas
- Safety

The designer tries to maximize the foil's lift-to-drag ratio and the speed for cavitation inception. Further, the weight of the strut-foil system must be minimized with due consideration of structural strength.

Abramson (1974) discussed relevant structural loads for monohull hydrofoil vessels. In this context, important aspects are slamming, hull-bending moments in foilborne conditions, and bending of the forward foil and strut during recovery from a forward foil broach. Slamming on the side hulls of a foil catamaran is not considered a problem. The reason is large deadrise angles. Because a monohull hydrofoil vessel typically uses a planing hull with relatively small deadrise angles, slamming loads matter. If a foil catamaran is hullborne in bad weather, wetdeck slamming must be considered. The possibility of grounding and hitting of objects like logs against the strut-foil system must also be considered.

Flutter of foils and struts could cause catastrophic failure, but this has never occurred. The classical flutter scenario is dynamic instability of combined bending-torsion vibrations of a strut-foil system. Henry et al. (1959), Besch and Liu (1972), and Abramson (1974) present theoretical and experimental studies of flutter of hydrofoil vessels. Flutter of aircraft is discussed by Bisplinghoff et al. (1996). The mass ratio, that is, the ratio of a typical density of structural material to the density of the fluid is an important parameter. Henry et al. (1959) expressed this as $\mu = m/(\pi \rho b^2)$, where m is the mass per unit spanwise dimension of the lifting surface, $2b$ is a representative chord length, and ρ is the fluid density that for water is the order of 1000 times the density of air. Representative values of μ for hydrofoils are 0.5 and less, but 50 and higher for airplane wings. This difference creates a clear advantage for a hydrofoil vessel relative to an aircraft when it comes to flutter (Henry et al. 1959).

In the following text, we describe the main aspects and important physical features of hydrofoil vessels. Then a detailed description of foil theory will follow. This is a necessary basis to numerically predict the steady and unsteady behaviors of a hydrofoil vessel in waves and during take-off and maneuvering. The description starts by introducing a boundary element method (BEM) based on source and dipole distributions that can account for nonlinearities, 3D flow, and interaction between foils and struts, as well as free-surface effects. The linear theory is presented afterward. The advantage of a linear theory is that we can show more easily how the angle of attack, foil camber, foil flaps, and three-dimensionality of the flow influence lift and drag of the foil. It is also discussed how the free surface and foil interaction affect the steady lift and drag of a foil. The analysis is supported by experimental results. Unsteady flow conditions due to incident waves are also handled. These are used to calculate heave and pitch motions of a foilborne hydrofoil vessel in head sea and following waves.

Foil theory is also relevant in other marine applications than hydrofoil vessels. Propellers, rudders, and trim tabs are examples. The keel and sails on sailboats are lifting surfaces that can be described by foil theory. Lefandeux (1999) has used surface-piercing foils in combination with a multihull

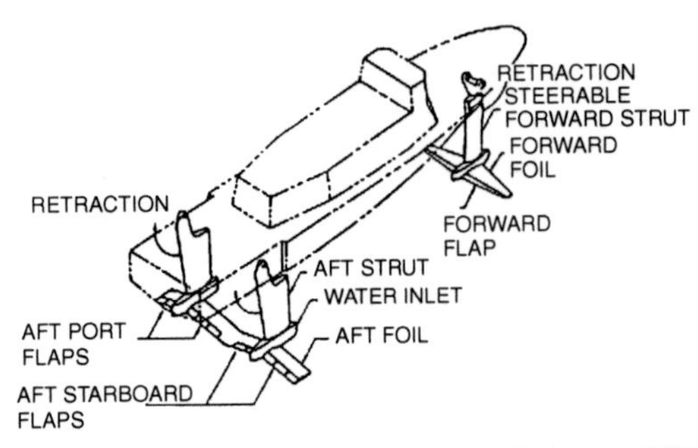

Figure 6.1. Fully submerged foil system (Johnston 1985).

sailboat. Inukai et al. (2001) have considered a sailing catamaran with submerged foils.

6.2 Main particulars of hydrofoil vessels

There are two main categories of hydrofoil vessels. They have either fully submerged or free-surface piercing foil systems (see Figures 1.6 and 1.7). An example of a fully submerged foil system with flaps is also shown in Figure 6.1. The forward strut is used for steering and waterjet propulsion is incorporated in the aft foil arrangement. The waterjet has a ram inlet. The internal ducting goes through the strut, and the water comes out in the air aft of the vessel (see Figure 2.54). Many of the existing hydrofoil craft are equipped with flaps. They are used for the control of the trim and heel and to counteract wave excitation in a seaway.

Figure 6.2 shows different types of foil configurations. If 65% or more of the vessel's weight is supported at maximum speed by the forward or by the aft foils, the foil system is called conventional or canard, respectively. The canard system is the most common type for hydrofoil vessels. A conventional system resembles the foil arrangement of a commercial airplane. Examples of nomenclature used for foils are presented in Figure 6.3. NACA profiles (Abbott and von Doenhoff 1959) are commonly used as foil sections.

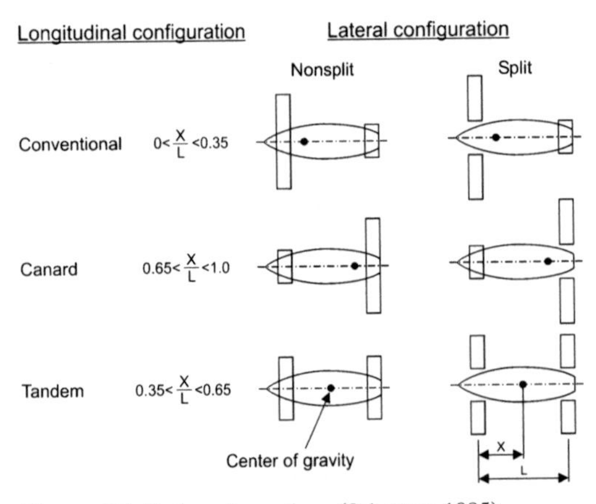

Figure 6.2. Foil configurations (Johnston 1985).

The main dimensions of free surface–piercing and fully submerged hydrofoil craft of monohull type are presented in Tables 6.1 and 6.2. The Series 65 hull form is typically used (Holling and Hubble 1974).

Foil catamarans (Figure 6.4) have been built in Norway and Japan. With a speed of approximately 50 knots, the "Foil Cat 2900" (Svenneby and Minsaas 1992 and Minsaas 1993) has a couple of fully submerged inverted T-foils forward and a full-width foil at the stern. *Z* propeller drives (Meyer and Wilkins 1992) are incorporated in the struts of the aft foil. "Super Shuttle 400" (Rainbow) represents a Japanese foil catamaran. It has full-width fully submerged foils, at both the bow and the stern, and it is equipped with waterjet propulsion. The Japanese foil catamaran is of somewhat lower speed but higher passenger capacity than the Norwegian one. Both vehicles are run by diesel engines, and the catamaran hulls are lifted out of the water completely at operating speeds.

The Foil Cat 2900 has an overall length of 29.25 m, total breadth of 8.36 m, draft of 3.7 m, maximum draft reduction when lifting of 1.9 m, span of aft foil of 7.79 m, span of forward foil of 2.50 m, weight of 112 to 120 tonnes, main engine output of 2 × 2000 kW, and propeller diameter of 1.25 m and can carry up to 160 passengers. Details about the foil system are shown in Figures 6.4 and 6.12. The aft foil carries about 60% of the weight in foilborne conditions. The strut of the forward foils act as rudders and can be turned ±25°. Each forward foil is equipped with a flap, and the aft foil with three flaps.

Minsaas (1993) also describes the 190-tonne Kværner Fjellstrand design FC40, which is run by gas turbines and uses waterjet propulsion. The vessel can carry more than 400 passengers.

6.3 Physical features

6.3.1 Static equilibrium in foilborne condition

The weight of the vessel is balanced by the steady lift force from the foil system in foilborne

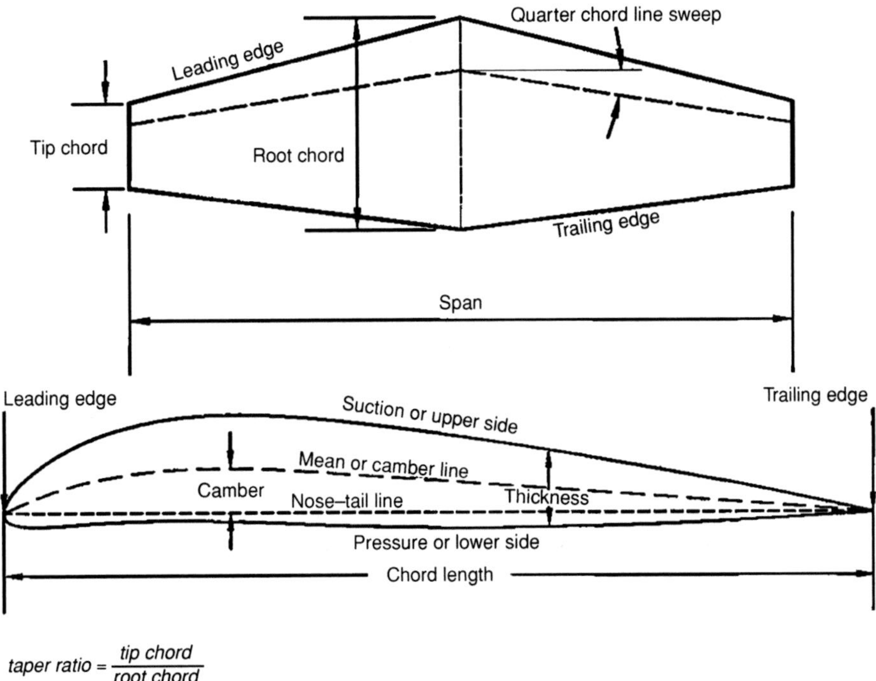

Figure 6.3. Foil geometry. The angle of attack of the incident flow velocity is defined relative to the nose-tail line (see Figure 2.16).

condition. We can write for a submerged horizontal foil system

$$Mg = \frac{\rho}{2} C_L U^2 A. \tag{6.1}$$

Here M is the mass of the vessel, C_L is the steady lift coefficient, and A is the planform area of the foil system. The planform area is defined as the projected area of the foil in the direction of the lift force for zero angle of attack. The lift coefficient C_L is defined by eq. (2.88). A maximum

Table 6.1. *Main dimensions of free surface–piercing hydrofoil craft of the monohull type (van Walree 1999)*

Length (hull)	9–40 m
Beam (hull)	3–7 m
Beam (foils)	3–16 m
Displacement	4–200 tonnes
Speed (foilborne)	28–40 knots
Foil system aspect ratio	6–10

Table 6.2. *Main dimensions for fully submerged hydrofoil craft of the monohull type (van Walree 1999)*

Length (hull)	11–40 m
Beam (hull)	3.5–6.0 m
Beam (foils)	4.0–6.5 m
Displacement	6–250 tonnes
Speed (foilborne)	36–50 knots
Foil system aspect ratio	4–10

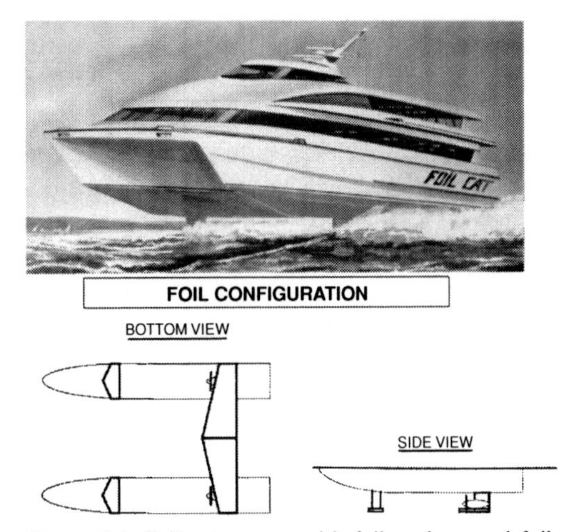

Figure 6.4. Foil catamaran with fully submerged foils and tractor propellers (see Figure 2.48).

achievable C_L for a foil is about 1.0, but 20% to 30% is reserved for control forces in connection with, for instance, trim and maneuvering (Meyer and Wilkins 1992). Let us set $\rho = 1025\,\mathrm{kgm}^{-3}$ and consider a vessel with weight 2 MN and $A = 35\,\mathrm{m}^2$. If the takeoff speed is $U = 12\,\mathrm{ms}^{-1}$, eq. (6.1) shows that C_L must be larger than 0.77 to lift the hull out of the water. If the cruising speed is $20\,\mathrm{ms}^{-1}$, then $C_L = 0.28$.

Eq. (6.1) can be used to illustrate how the planform area increases with ship size. Let us choose a hydrofoil vessel with length L_1, mass M_1, and planform area A_1 as a basis. We then consider a hydrofoil vessel with a geometrically similar hull and length $L_2 > L_1$. The mass of the vessel must balance the displaced fluid mass at zero forward speed. The mass of the foil and strut system will typically be 10% of the total mass of the vessel. Even if the foil and strut systems are not geometrically similar for the vessels of length L_1 and L_2, as a first approximation, we may set the mass M_2 of the larger vessel equal to $M_1(L_2/L_1)^3$ because the displaced fluid mass of the hull at zero speed scales like this. Because cavitation must be avoided, U and C_L must be limited. Increased length means increased submergence of the foils and higher ambient pressure at the foils. The higher the ambient pressure, the higher U can be to avoid cavitation. However, because the hydrostatic pressure part at the foils is clearly smaller than the atmospheric pressure for practical foil arrangements, we could roughly say that an upper limit for U to avoid cavitation is approximately 50 knots for the vessels of lengths L_1 and L_2. The limitation of C_L depends on the foil system. Anyway, we must assume as a first approximation that maximum U and C_L do not depend on the ship's length. Eq. (6.1) then gives that the planform area must scale like the mass of the vessel, that is, the planform area A_2 of the vessel with length L_2 is $A_1(L_2/L_1)^3$. If the aspect ratios of the foils are not changed with ship length, the span of a foil increases as $(L_2/L_1)^{3/2}$. This means the ratio between the foil span and the hull beam increases as $(L_2/L_1)^{1/2}$, that is, the foils outgrow the hull. A potential danger is that the increase of the planform area with ship length increases the ratio between the strut-foil weight and the vessel weight. It will generally increase viscous resistance as well. The increased weight has negative consequences for payload, that is, maximum number of passengers on commercial vessels. However, it is possible to design a foil-strut system from a structural strength point of view so that its weight relative to vessel weight increases only slightly with the vessel weight (Meyer and Wilkins 1992).

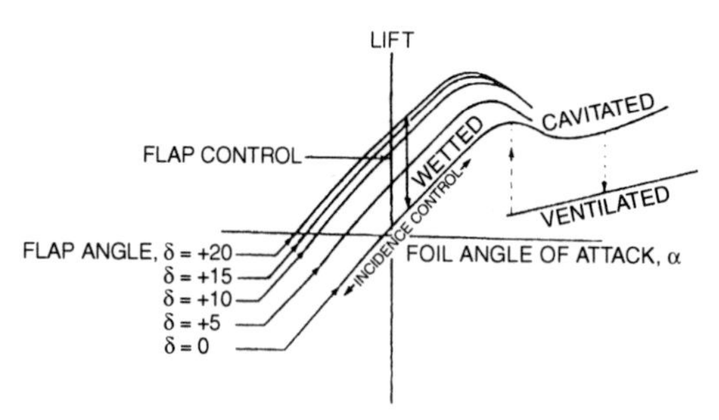

Figure 6.5. Typical foil lift curves (Johnston 1985).

In order to determine the trim angle of the vessel, we need to consider the lift coefficients of the fore and aft foils separately. C_L for an individual foil depends on many parameters, such as

- Angle of attack α of the incident flow
- Flap angle δ
- Camber
- Thickness-to-chord ratio
- Aspect ratio
- Ratio between foil submergence, h, and maximum chord length, c
- Submergence Froude number $Fn_h = U/(gh)^{0.5}$
- Interaction from upstream foils
- Cavitation number
- Reynolds number

Other geometrical details of the foil surface also matter. Further, the fluid-depth Froude number matters in shallow water (see Chapter 4 and Breslin 1994).

Figure 6.5 shows schematically how the steady lift force depends on α and δ. When α and δ are small, the lift is linearly dependent on α and δ. If the foil has a camber, the lift is non-zero when α and δ are zero. When α and/or δ are large, cavitation and ventilation occur, depending on speed and submergence. Figure 6.5 shows a substantial decrease in lift as a consequence of ventilation. The magnitude of the lift force with cavitation depends on the cavitation number. The suction side of the foil may be partially or fully cavitating. Partial cavitation may lead to unsteady lift forces. The flow may also separate from the leading edge

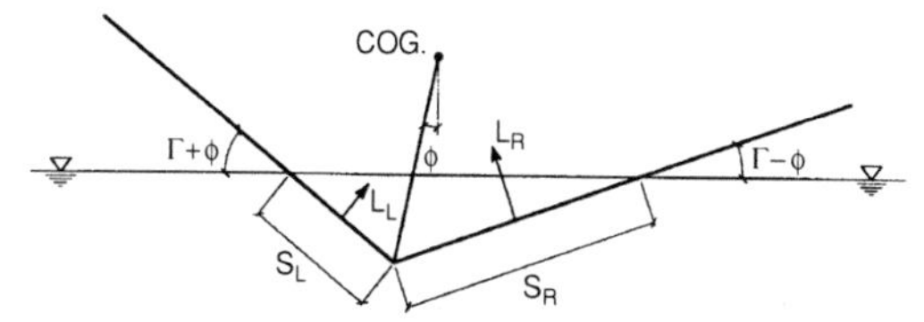

Figure 6.6. Free surface–piercing forward foil of a hydrofoil vessel in foilborne condition on a straight course in calm water. The vessel has a constant heel angle ϕ. Lift forces L_L and L_R act on, respectively, the left- and right-hand sides of the foil. COG = center of gravity; s_L and s_R are submerged spans of, respectively, the left- and right-hands side of the foil; Γ is called the dihedral angle of the foil.

area when α and/or δ are large. This also causes a reduction in the lift force and an increase in drag.

6.3.2 Active control system

A surface-piercing hydrofoil system in foilborne condition stabilizes the vessel in heave, roll, and pitch. This can be understood by means of a quasi-steady analysis. Consider, for instance, that the heave motion increases. Here heave is positive upward. This causes a reduction of the wetted foil area. Because the lift is proportional to the wetted area, the lift due to the foils decreases. The weight of the vessel balances the lift in the equilibrium position. The increased heave implies that the vessel weight will force the vessel downward. Another way of saying this is that there is a restoring force in heave bringing the vessel back to the equilibrium position.

Let us then consider a static roll (heel) angle (Figure 6.6). As a consequence, the wetted area of a surface-piercing foil increases on one side and decreases on the other. The lift force distribution causes then a roll moment that counteracts the roll and forces the vessel back to the upright position. Detailed evaluation of this is given by Hamamoto et al. (1993). The same type of balance can be seen by giving the vessel a static pitch angle.

An active control system is commonly used for a vessel equipped with a fully submerged foil system to stabilize the heave, roll, and pitch in calm water. The system is also used in connection with maneuvering and to minimize wave-induced vessel accelerations and relative vertical motions between the vessel and the waves. Sensors are used to measure the position of the vessel. Change of position is counteracted by the foil flaps. A computer program that describes the vessel behavior is needed as a part of the active control system (ride control system, see Saito et al. 1991 and van Walree 1999).

Platforming and contouring modes are used in connection with an active control system (Figure 6.7). The contouring mode is used in longer waves to minimize relative vertical motion between the vessel and the waves and to avoid ventilation and broaching of the foils. The platform mode is used to minimize vertical accelerations of the vessel in relatively short waves.

6.3.3 Cavitation

Cavitation on foils designed for subcavitating conditions limits the vessel speed to the order of 50 knots. Cavitation appears when the pressure in the water is equal to the vapor pressure, that is, close to zero. The consequence of cavitation is that the material of the foil (or a propeller) can be quickly destroyed and the lift capabilities of the foils may be significantly reduced. Another consequence is that the drag on the foils increases. Because cavitation is accompanied by noise generation, one can hear onboard the vessel when there is the possibility of damage due to cavitation. If the vessel speed is increased substantially beyond 50 knots, supercavitating foils must be used to avoid cavitation damage. They are characterized by lift-to-drag ratios and lift coefficients much lower than those of subcavitating foils. Our discussion is based on subcavitating foils.

We illustrate the possibility of cavitation on a foil by studying the calculated pressure distribution around a two-dimensional foil with a flap (Figure 6.8) used in a preliminary design of a foil catamaran. Camber and thickness distribution are of the same type as published by Shen and Eppler

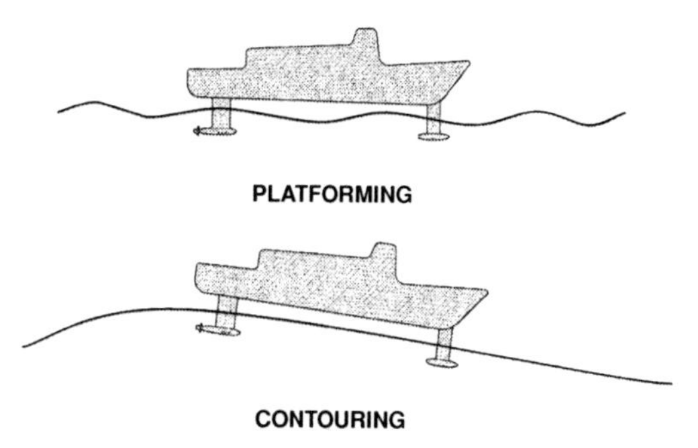

Figure 6.7. Platforming and contouring modes used in connection with an active control system.

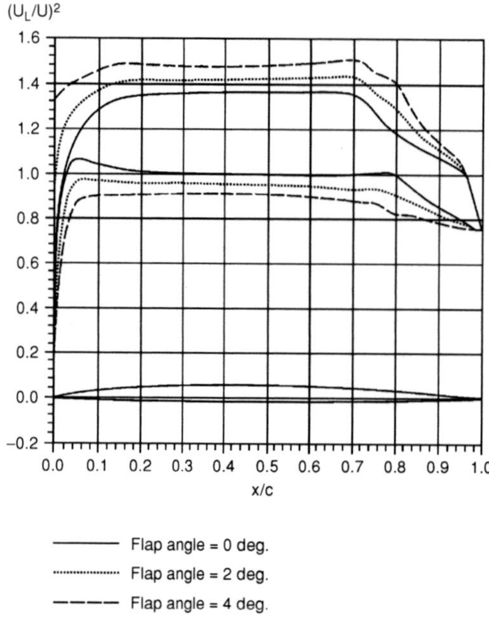

Figure 6.8. Calculated velocity distribution on a two-dimensional foil with flap. U = incident flow velocity, U_L = local flow velocity. Angle of attack of incident flow is zero. The foil is sketched in the lower part of the figure. Flap length–to–chord length ratio = 0.2 (Minsaas, unpublished).

(1979). The thickness-to-chord ratio is 0.075, and the flap length–to-chord ratio and the camber-to-chord ratio are 0.2 and 0.02, respectively. The foil is shown in the lower part of the figure. The angle of attack of the free stream relative to the x-axis is zero, and velocity results are shown for the two sides of the foil in the case of flap angles 0°, 2° and 4°. Taking U as the incident flow velocity and U_L as the local flow velocity in Figure 6.8, the total pressure can be written as

$$p = p_a + \rho g h + \frac{\rho}{2}U^2\left(1 - \left(\frac{U_L}{U}\right)^2\right). \quad (6.2)$$

Here p_a is the atmospheric pressure and h is the submergence of the foil relative to the free surface in calm water (Figure 6.9). The effect of the free surface on U_L is not included in Figure 6.8. Let us say that $U = 25\ \text{ms}^{-1}$, that is, $0.5\rho U^2 \approx 3.2$ times atmospheric pressure, and consider a flap angle of 4°, where $(U_L/U)^2$ may be as large as 1.5 on the suction side of the foil. This means that the foil should be submerged deeper than approximately 6 m to avoid cavitation. The foils are not so deeply submerged on the Foil Cat 2900. This illustrates that the flap angle has to be quite limited and care has to be taken when designing the foils.

The pressure distribution along the foil should be relatively flat in order to minimize the possibility of cavitation, that is, there must not be pronounced local pressure minima (suction peaks). The NACA 16 and 64 series without flaps, for instance, have this characteristic. A badly designed foil with a flap will result in a pronounced suction peak at the flap hinge at moderate flap angles. This limits the operability. A NACA 16 section with a flap is an example (Shen and Eppler 1979). The pressure increase aft of a suction peak, that is, adverse pressure gradient, may cause flow separation with decreased lift-to-drag ratio and flap effectiveness in providing lift. Shen and Eppler (1979) used an inverse numerical design procedure for foil sections with a flap to maximize the critical cavitation speed and the lift-to-drag ratio. The desired pressure distribution was specified, and the resulting foil shape was found by a potential flow code. A boundary-layer calculation that accounts for laminar flow, transition to turbulence, and turbulent flow was used to detect flow separation. The thickness-to-chord and flap length–to-chord ratios were constrained to, respectively, 0.09 and 0.2.

Cavitation-free buckets for a given foil are used to identify the possibility of cavitation. Let us use the example in Figure 6.8 to illustrate how this can be calculated. We introduce the pressure coefficient

$$C_p = \frac{p - p_0}{0.5\rho U^2}, \quad (6.3)$$

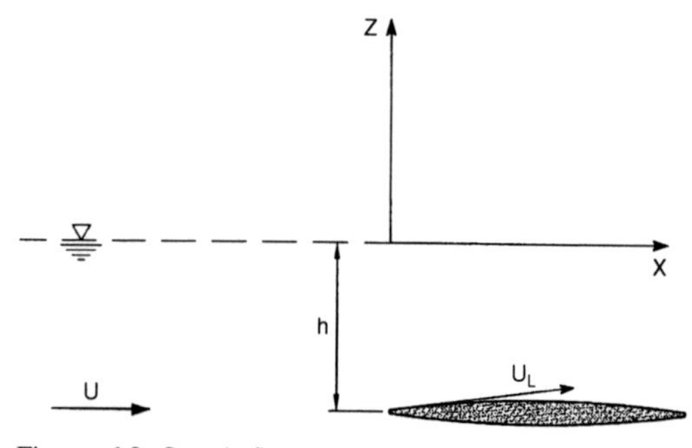

Figure 6.9. Steady flow past a 2D foil at a depth h below the mean free surface. The incident flow velocity is U and the local tangential flow velocity at the foil surface is U_L.

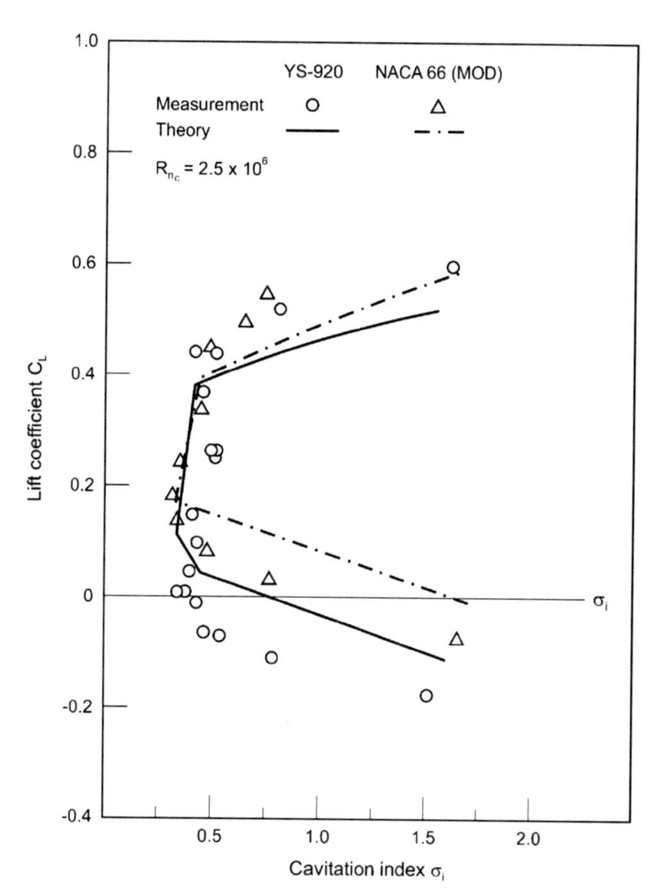

Figure 6.10. Cavitation-free buckets of YS-920 and NACA 66 (MOD) sections (surface smooth) (Shen 1985).

where $p - p_0$ is $0.5\rho U^2(1 - (U_L/U)^2)$. The ambient pressure p_0 was set equal to $p_a + \rho g h$ in eq. (6.2). There corresponds a lift coefficient C_L and a minimum pressure coefficient $C_{p_{\min}}$ for a given flap angle. Cavitation will occur at one point on the foil if the minimum pressure is equal to or lower than the vapor pressure p_v. We can assume that the predicted pressure distribution on the noncavitating foil is then still valid. If the minimum pressure is equal to p_v, eq. (6.3) gives

$$C_{p_{\min}} = \frac{p_v - p_0}{0.5\rho U^2}. \tag{6.4}$$

The cavitation number is defined as

$$\sigma = \frac{p_0 - p_v}{0.5\rho U^2}. \tag{6.5}$$

This is the same as minus $C_{p_{\min}}$ given by eq. (6.4). We can then define a cavitation inception index σ_i as

$$\sigma_i = -C_{p_{\min}}. \tag{6.6}$$

By calculating corresponding values between σ_i and C_L for different flap angles, a boundary curve can be obtained delimiting the region where cavitation occurs.

Theoretical curves are presented together with experimental results in Figure 6.10 for the YS-920 and NACA 66 (MOD) foil sections. These foils do not have flaps, so it is the angle of attack that has been varied. When $\sigma < \approx 0.4$, Figure 6.10 shows that cavitation will occur whatever the angle of attack. Because σ decreases with increasing U, there is a maximum vessel speed beyond which we cannot avoid foil cavitation, whatever the loading on the foil is. Shen and Eppler (1979) have presented numerically calculated cavitation-free buckets for foils with a flap. Using curves like those in Figure 6.10 as a basis, we can define boundaries for vessel speed and steady foil loading that the vessel must operate within to avoid cavitation. Because curves like that are based on steady flow analysis, we must have a safety margin that accounts for unsteady wave effects. Figure 6.11

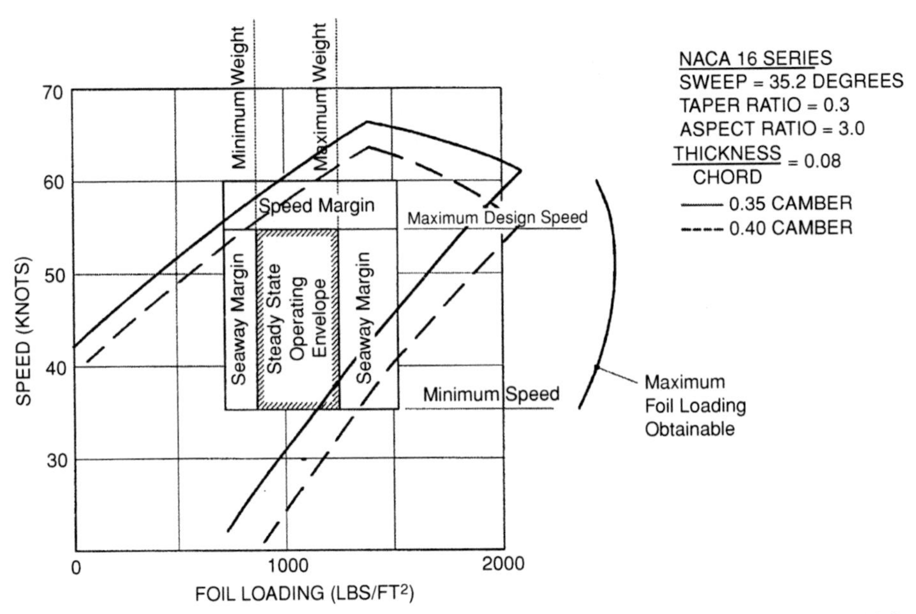

Figure 6.11. Foil loading-speed envelope. Foil parameters are defined in Figure 6.3 (Johnston 1985).

shows cavitation boundary plots for two different foil sections. The foil with 0.35 camber will cavitate slightly for the minimum speed and maximum weight, particularly in a seaway. Using 0.4 camber improves the performance from a cavitation point of view. The foil loading is presented in the unit lbs/ft^2; 2116.2 lbs/ft^2 is equal to 101.32 kPa.

However, one should note that a limited extent of leading-edge cavitation can and must usually be tolerated. This type of cavitation has no unfavorable effects. It may even reduce the response of the foil due to wave-induced angle of attack variations (van Walree, unpublished). Cavitation originating at the mid-chord position does have more unfavorable effects and should be avoided.

Proper design to delay cavitation on the aft foil system requires evaluations of the downwash due to the forward foil system on the aft one. The downwash is influenced both by trailing vorticity (Figure 6.12) and by generation of free-surface waves (Mørch 1992). Numerical calculations of downwash are illustrated in Figure 6.13. The calculations were done with a boundary element method as described in section 6.4. It accounts for the roll-up of the vortex sheet and free-surface wave generation. We note that both positive and negative vertical inflow velocities occur along the span of the aft foil as a result of the two forward foils. Sidewash on the struts connected to the aft foil is also generated. Figure 6.13 shows only the angle of attack α_L due to the downwash on the aft foil. There would then be a resulting local lift

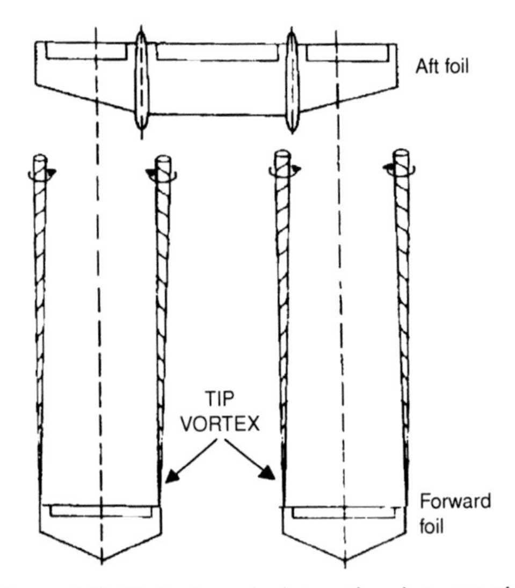

Figure 6.12. Hydrodynamic interaction between the foils of a foil catamaran. The wake (shear layer) including roll-up of tip vortices generated by the two upstream low-aspect–ratio lifting surfaces affects the angle of attack of the flow at the downstream (aft) foil. The waves generated by the forward foils also affect the aft foil.

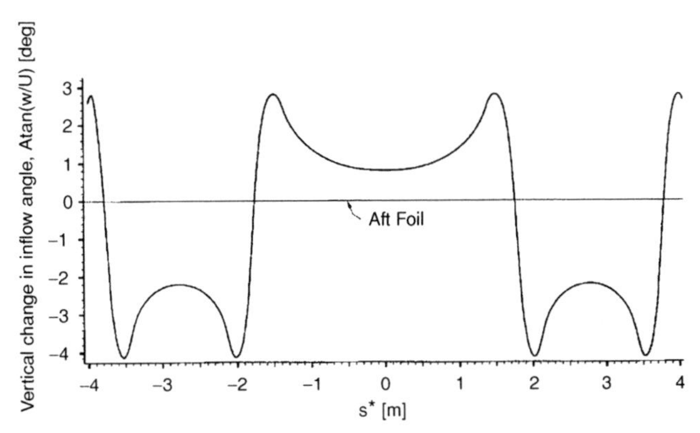

Figure 6.13. Calculated downwash along the aft foil (AF) due to two small-aspect–ratio forward foils (FF) for a foil catamaran (see Figure 6.12) at 50 knots. Displacement = 102 tonnes; 60% of lift generated by aft foil (AF); span AF = 7.8 m; average chord AF = 1.3 m; span FF = 2.5 m; average chord FF = 0.95 m; foil immersion = 1.9 m; distance between midspan of forward foils = 5.45 m; distance from FF to AF = 17.1 m. Effects of roll-up of the vortex sheet and generation of the free-surface waves are included (Mørch 1992; see also Mørch and Minsaas 1991).

contribution on the aft foil that is proportional to α_L. The lift is mainly caused by pressure loads, and as we have seen, the primary concern for cavitation is the pressure on the suction side of the foil. We can then qualitatively understand why downwash is important for cavitation. The results in Figure 6.13 suggest that the aft foil should be twisted so that the angle of attack on the aft foil does not have such large spanwise variations and extreme values as those shown in Figure 6.13.

The downwash effect on the aft foil is most pronounced if the forward foil system contains a foil or foils with smaller span than the aft foil, as illustrated in Figure 6.12 for a foil catamaran. In this case, two small-aspect–ratio foils were selected instead of one high-aspect–ratio foil with a span equal to the distance between the centerplanes of the two hulls. The choice is connected with the possibility of ventilation along one of the two forward struts during maneuvering. When two foils are used, a probability exists that only one of the foils is ventilated and loses lift. In contrast, if a single extended foil is considered, the whole foil will be ventilated because of ventilation along one of the struts. Therefore, the use of only one foil means a much larger loss of lift. Using two small-aspect–ratio foils implies a clear change in downwash along the span of the aft foil. Using winglets on a low-aspect–ratio lifting surface of the front foil system (see Figure 2.20) reduces the magnitude of the downwash on the aft foil system.

Care must also be shown in the design of local details at the junction between a strut and a foil to avoid vortex formation and resultant cavitation.

6.3.4 From hullborne to foilborne condition

An important design consideration is sufficient power and efficiency of the propulsor system to lift the vessel to foilborne condition. This is of special concern when waterjet propulsion is used.

Figures 6.14 and 6.15 show typical examples of the behavior of trim, draft, resistance, and

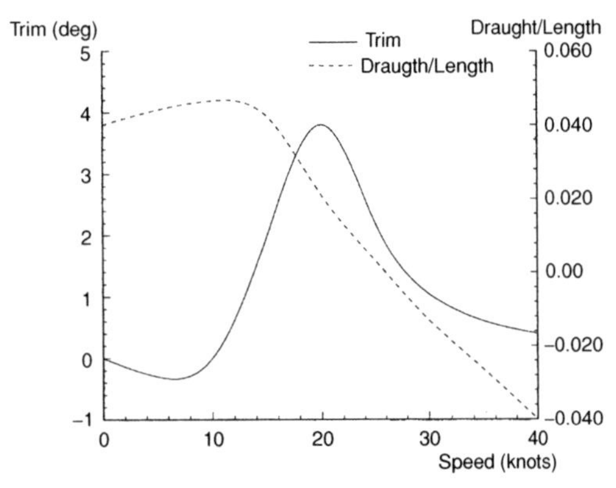

Figure 6.14. Typical example of draft (excluding foils) and trim versus speed for monohull hydrofoil vessels (van Walree 1999).

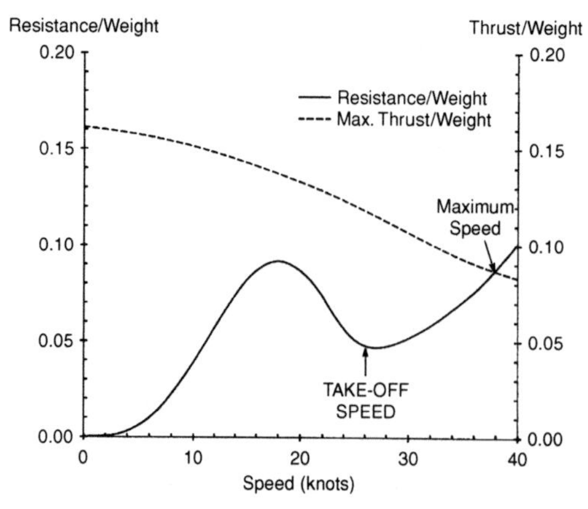

Figure 6.15. Typical example of resistance and thrust versus speed for monohull hydrofoil vessels (van Walree 1999).

available thrust as a function of vessel speed for monohull hydrofoil vessels. Positive trim corresponds to bow up. Draft means hull draft, that is, D in Figure 6.16. This means that the hydrofoil vessel illustrated in Figure 6.14 is foilborne around 26 knots. The trim angle increases from around 0° at 10 knots to about 4° at 20 knots and then decreases. High trim angle, speed, and flap angles are beneficial from a foil-lifting point of view as long as cavitation is avoided. The propulsor thrust generally has a vertical and longitudinal component (see Figure 6.16), with the main component in the longitudinal direction. The difference between thrust and resistance in Figure 6.15 is therefore approximately equal to $(M + A_{11})a_x$, where M is the vessel mass, A_{11} is added mass in surge, and a_x is the forward acceleration of the vessel. A_{11} is small relative to M.

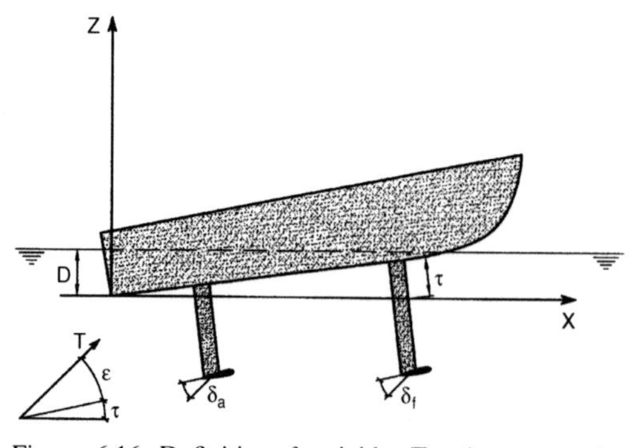

Figure 6.16. Definition of variables. T = thrust, τ = trim angle. δ_a and δ_f are flap angles of aft and forward foils (van Walree 1999).

The propeller thrust line in Figure 6.15 corresponds to a given power setting. It shows that the propeller thrust decreases as the speed increases. Both conventional propellers (section 2.10) and waterjet propulsion (section 2.11) are used. Conventional propellers can be arranged as either tractor propellers on the aft foil (see Figure 6.4) or as propellers at the stern connected to the engine with a straight inclined shaft having a long wetted part in the water similar to the one in Figure 2.1a. Z-drives (Meyer and Wilkins 1992) are used to transmit power from the engine to waterjets and tractor propellers.

We note that there is a resistance hump around 20 knots. The hull is then still in the water. The resistance on the hull can be estimated as described in Chapter 2. In addition, we have the drag on the struts, foils, appendages, and propulsion system. The resistance hump in calm water is important in determining the propulsion power. One must allow for 20% to 25% margin due to a maximum increased resistance in operational sea states. The minimum resistance shown in Figure 6.15 occurs typically after takeoff.

The results in Figures 6.14 and 6.15 are not representative of the Norwegian-designed foil catamarans. Their trim angles are very small, and there is no resistance hump when the vessels go from hullborne to foilborne condition. How it is possible to avoid a resistance hump is illustrated by Figures 6.17 and 6.18 for a 128.5-tonne foil catamaran, which is similar to an existing foil catamaran. Figure 6.17 shows the propulsive power as a function of vessel speed for different drafts. The propulsive power means resistance times speed. The resistance was obtained by scaling model test results. The foil catamaran with the superstructure, the demihulls, and the struts, but without the foils, was tested in the towing tank at MARINTEK. Because the resistance of the foils may be affected by the cavitation number, the foils were separately tested in the depressurized circulating flow channel with free surface at the Technical University of Berlin. The foil system was equipped with flaps that provided the needed lift to balance the weight of the hydrofoil vessel at each speed and submergence. A correct trim moment also had to be ensured. The total resistance of the foil catamaran was simply obtained by adding the results from the two model tanks.

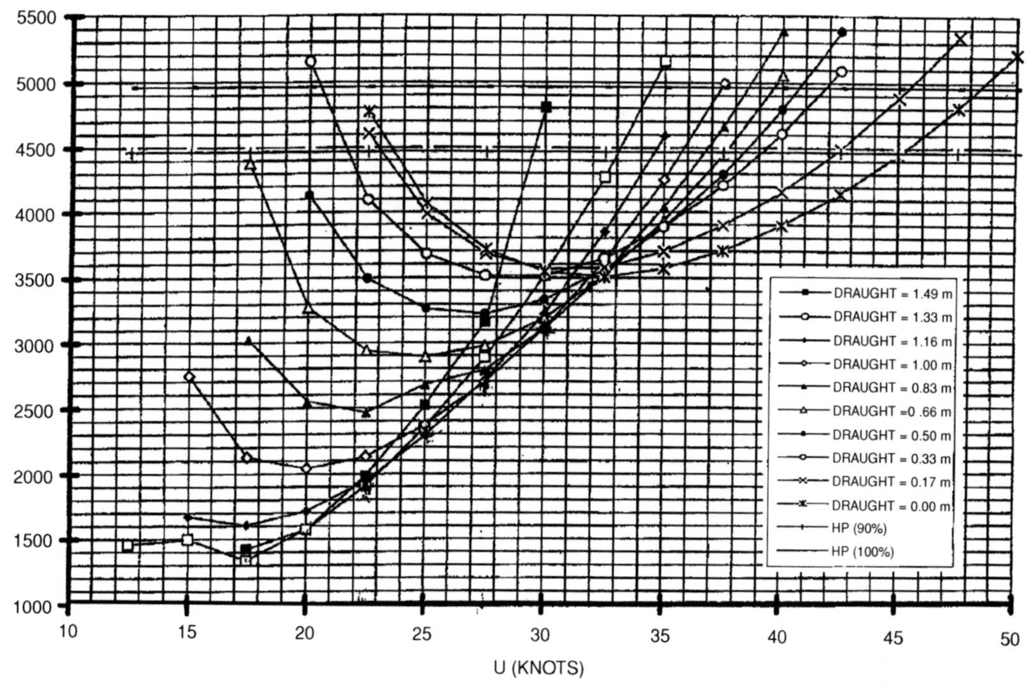

Figure 6.17. Propulsive power (HP) as a function of vessel speed U of a 128.5-tonne foil catamaran at different hull draft. The results are obtained by model tests (Minsaas, unpublished).

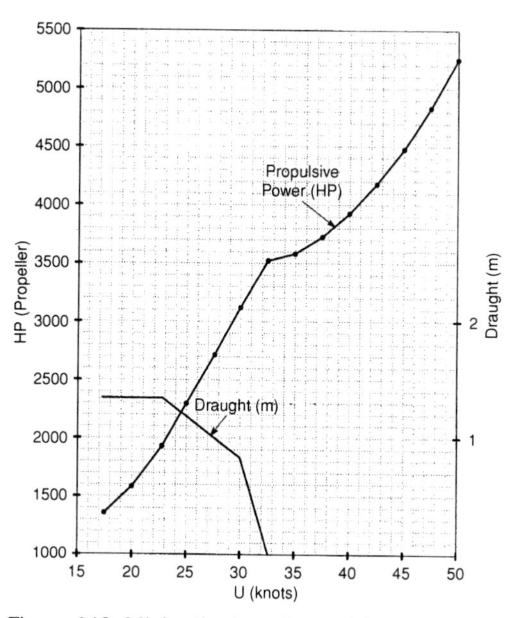

Figure 6.18. Minimalization of propulsive power of a 128.5-tonne foil catamaran going from hullborne to foilborne condition. Based on the experimental data in Figure 6.17 (Minsaas, unpublished).

The results in Figure 6.18 were obtained as follows. For a given speed, a draft that gave the smallest or nearly the smallest propulsive power was selected. Because the draft obviously should decrease or at least not increase with increasing speed, the propulsive power was not the smallest for all speeds. The final results in Figure 6.18 showing the propulsive power and draft as a function of speed demonstrate that it is possible to follow a strategy in which the foil catamaran does not experience a resistance hump going from hullborne to foilborne condition.

One reason a foil catamaran and a monohull hydrofoil vessel show a different behavior before takeoff is that a foil catamaran uses slender hulls whereas the monohull versions use hulls typical for planing vessels. The trim angles are relatively high because of these nonslender hulls. This causes a resistance hump but also delivers more dynamic lift after the hump, just before takeoff.

The resistance components in foilborne condition consist of:

- *Viscous resistance on foils and struts.*

  This can be calculated by eqs. (2.89) and (2.90) for a smooth surface. Nearly zero lift is assumed.

The effect of surface roughness for turbulent boundary-layer flow is given by eq. (2.85).

- *Induced drag.*
  The induced drag due to the trailing vortex sheet of a high-aspect–ratio foil with zero camber and elliptical loading in infinite fluid is given by eq. (2.98). The wake effect due to an upstream foil or strut must be considered.
- *Wave resistance of foils and struts.*
  This is caused both by the lift and by the foil thickness. The thickness effect of struts in deep water is discussed in section 4.3.3 (see Figure 4.16). This was based on linear theory and has a small effect when the Froude number Fn_c based on the chord length is larger than 3. The wave resistance due to thickness and lift on a single foil is handled in section 6.8 for deep water. Koushan (1997) has considered theoretically and experimentally the effect of tank walls and finite water depth on the wave resistance of a foil. The wave effect due to an upstream foil or strut must also be considered.
- *Spray resistance on struts.*
  Spray resistance was discussed for monohull vessels in section 2.4. The spray resistance is caused both by potential flow and viscous flow effects. Hoerner (1965) simply sets the spray drag force equal to

$$R_S = 0.12\rho U^2 t^2 \qquad (6.7)$$

when $Fn_c = U/\sqrt{gc} > 3$. Here t is the thickness of the strut and c is the chord length.

There are, in addition, drag forces on appendages (rudders, propeller nacelles, waterjet intakes, pods) and propeller shafts (van Walree 1999). The rudders can be handled similarly to struts. Because the angle between the incident flow and the propeller shaft axis is small, the cross-flow principle cannot be used to obtain drag forces. This is discussed in section 10.6 in the context of ship maneuvering. We must also consider the air resistance on the vessel (see section 2.3).

The lift-induced drag on the foils will decrease with increasing speed in foilborne condition. This can be explained by first noting that the associated drag coefficient C_D is proportional to the square of the lift coefficient, that is, C_L^2 (see eqs. (2.88) and (2.89) for definitions of C_L and C_D). Eq. (6.1) gives the balance between the weight and the total lift. It means C_L must be reduced proportional to U^{-2} with increasing speed. This implies that the drag coefficient and the drag associated with the lift are proportional to U^{-4} and U^{-2}, respectively.

The minimum resistance occurring after takeoff is caused by the lift-induced drag. When the hydrofoil vessel is at cruising speed, a major resistance component is caused by viscous resistance on the foils and on the struts. Because this is affected by surface roughness, frequent maintenance procedures to clean the foil and strut surfaces are needed.

Figure 6.19 shows recommended speed/height curve for the Foil Cat 2900 as a function of speed. By height, we mean the minimum vertical distance from hull bottom to mean free surface. From the figure, we note that a limit in lifting height exists beyond which the passive stabilization in heave, pitch, and roll is insufficient. The ride control system then regulates the flap angles at the foils in such a way that the vessel is stable. Figure 6.19 also shows limitations in lifting height as a function of significant wave height ($H_{1/3}$). For instance, the vessel can achieve the maximum lifting height at $H_{1/3} < 0.75$ m. When $H_{1/3} > 4.0$ m, one should seek sheltered water.

6.3.5 Maneuvering

It is common to let the hydrofoil vessel heel (bank) in a turn (Figure 6.20). This creates the smallest transverse forces on the struts. It is common to use the forward strut as a rudder. However, the angle of attack of the rudder must be limited, that is, less than 5° to 6°, to avoid ventilation along the strut.

Figure 6.21 illustrates forces acting in a coordinated turn with a heel angle ϕ. Because the transverse forces on the struts are relatively small, they are disregarded in the figure. Further, the lift forces L on the foils are assumed to be acting through the center of gravity of the vessel. The vessel is assumed to have a constant speed U in calm water and perform a turn with constant radius R. A centrifugal force $F_c = MU^2/R$ acts on the vessel. Here M is the vessel mass. The balance of vertical and transverse forces gives

$$L\cos\phi = Mg \qquad (6.8)$$

$$L\sin\phi = \frac{MU^2}{R}. \qquad (6.9)$$

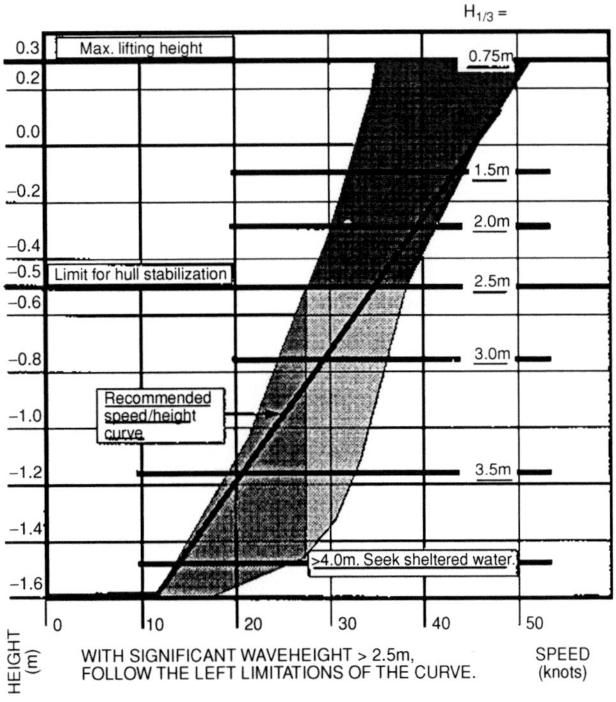

Figure 6.19. Speed and possible lifting height of the foil catamaran "Foil Cat 2900," with limitations given by the significant wave height $H_{1/3}$ (Svenneby and Minsaas 1992).

Further,

$$R\Omega = U, \tag{6.10}$$

where Ω is the turn rate in radians per second. The combination of eqs. (6.8) and (6.9) gives

$$\tan\phi = \frac{U^2}{Rg}. \tag{6.11}$$

Figure 6.20. The PHM hydrofoil vessel by Boeing (Johnston 1985).

Eliminating R between eqs. (6.10) and (6.11) gives

$$\Omega = \frac{g\tan\phi}{U}. \tag{6.12}$$

An object onboard will not be accelerated relative to the vessel during a coordinated turn. Let us consider a cup of coffee with mass M_c on a table and decompose the weight and the centrifugal force along the table. This gives components $M_c g\sin\phi$ and $M_c U^2\cos\phi/R$ acting in opposite directions along the table. Eq. (6.11) states that these two components balance each other, that is, the cup of coffee cannot be accelerated along the table. Further, the components of the weight and of the centrifugal force perpendicular to the table are in the same direction and prevent the cup of coffee from accelerating perpendicularly to the table. There is an analogy between hydrofoil vessel and airplane banking in a turn. The latter is commonly experienced by passengers.

In contrast, a hydrofoil vessel in a flat turn has $\phi = 0$. The centrifugal force is then likely to overcome the friction between the cup and the table and accelerate the cup. Further, a flat turn requires a transverse force on the struts to balance the centrifugal force on the vessel. If the angle of attack of the incident flow on the struts becomes more

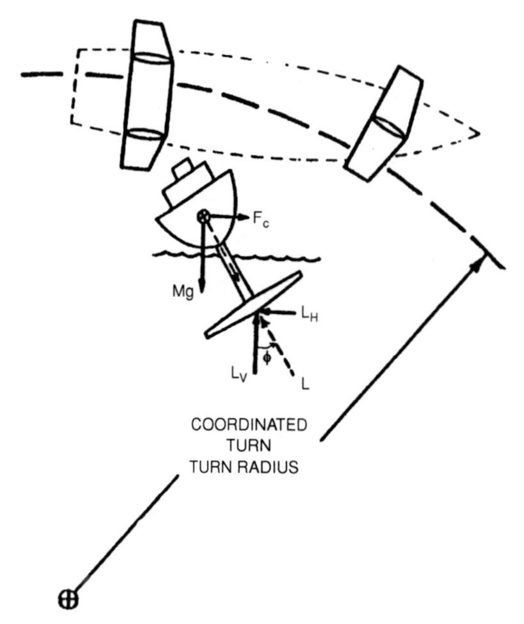

Figure 6.21. Hydrofoil coordinated turn with bank (heel) angle ϕ (Johnston 1985).

than 5° to 6°, ventilation is likely to occur, with a subsequent reduction in the transverse force. A worst-case scenario is that the ventilation penetrates to the foil with a significant reduction in the lift force. Because the transverse force on the struts is proportional to the angle of attack when ventilation does not occur, a larger turning radius must be used in practice for a flat turn than for a coordinated turn. This is needed to decrease the centrifugal force. Ventilation of struts is discussed in section 6.10.

A nonlinear simulation model in six degrees of rigid-body motion will be discussed for a hydrofoil vessel in section 10.9.

6.3.6 Seakeeping characteristics

Hydrofoil vessels in general have good seakeeping behavior, particularly for a fully submerged hydrofoil craft. This is exemplified for the foil catamaran Foil Cat 2900 in Figure 6.22, in which full-scale measurements of vertical accelerations in different sea states are evaluated according to the ISO criteria (see section 1.1). Comparisons are also made with a conventional catamaran of similar length. The operation limit in terms of significant wave height is much higher for the foil catamaran than for the conventional catamaran. Passengers have compared the behavior of a foil catamaran with a train ride, that is, one can experience horizontal accelerations similar to those on a train. In section 6.12, we demonstrate the possibility of large relative vertical motions and pitch in following sea. If these are not counteracted by an active control system, they may lead to foil ventilation and broaching.

6.4 Nonlinear hydrofoil theory

The following description is limited to potential flow. An alternative is to use CFD (computational fluid dynamics) and solve the Navier-Stokes equations.

6.4.1 2D flow

Two-dimensional flow past a foil in infinite fluid is first considered (Figure 6.23). This is a good approximation for the flow past most parts of a hydrofoil when the following conditions are satisfied:

- The aspect ratio Λ (see Figure 6.3) is high
- The ratio between foil submergence h and chord length c is larger than ≈ 4
- There is no wake or downwash because of an upstream foil that is varying strongly in the spanwise direction
- There is no cavitation or ventilation

A Cartesian coordinate system (x, y) moving with the steady forward velocity U of the foil is used. The forward speed appears, then, as an incident flow with velocity U along the x-axis. The velocity vector of the foil in this reference frame is denoted $\boldsymbol{V_B}$. Vorticity is shed from the trailing edge of the foil and is present in the thin free shear layer S_V. The vorticity comes from the thin boundary layer around the foil. Both the boundary layer and the free shear layer have zero thickness in the figure. There is potential flow outside the boundary layer and the free shear layer. If the boundary layer is thin, a separate boundary layer calculation based on the potential flow pressure distribution can be made. This can be used, for instance, to detect flow separation (Shen and Eppler 1979). An important aspect is that the potential flow leaves the trailing edge parallel to the foil surface. This is called the Kutta condition and is consistent with visual flow observation (Kutta 1910). This is essential in

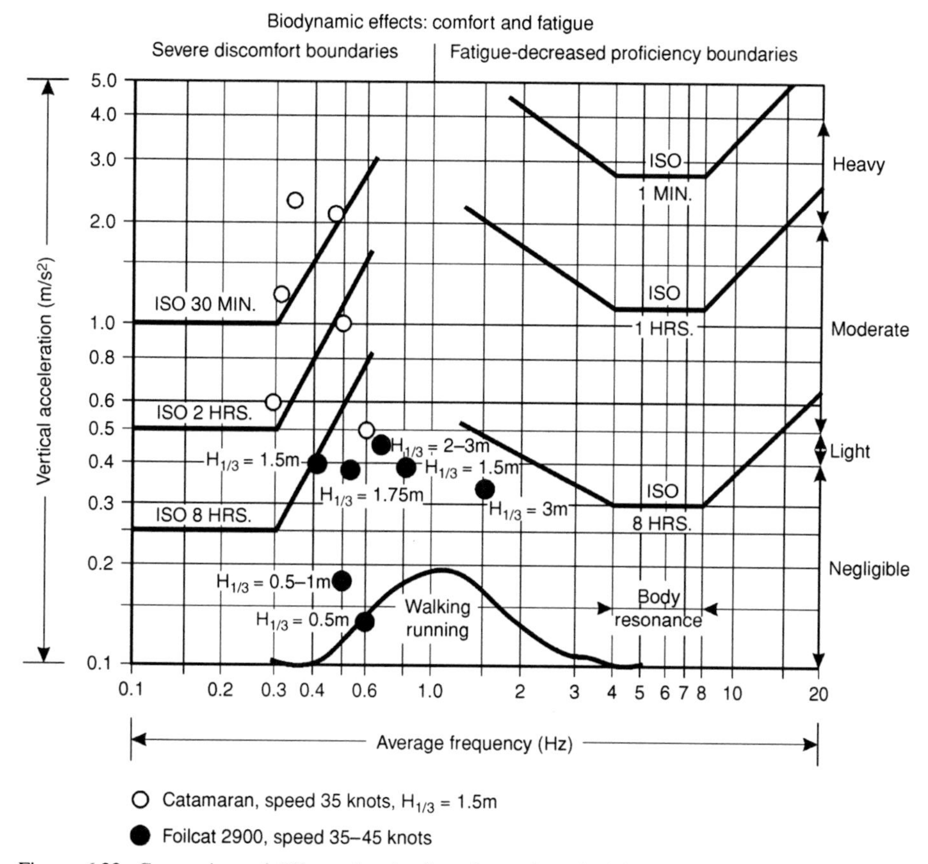

Figure 6.22. Comparison of different levels of comfort onboard a foil catamaran (Foil Cat 2900) with overall length of 29.25 m and onboard a conventional catamaran of similar length at different sea states. Active control is used for the foil catamaran (Minsaas 1993).

developing lift on the foil. Because the interior angle of the foil surface at the trailing edge is finite in Figure 6.23, an unsteady flow has to leave parallel to either the upper or lower foil surface. The argument for this will come later. Which side is determined by the local flow at the trailing edge.

There is a jump $(\partial\Phi^+/\partial s - \partial\Phi^-/\partial s)$ in the tangential velocity across the free shear layer S_V. Here Φ is the velocity potential and + and – refer to the two sides of the free shear layer (Figure 6.23). This jump in tangential velocity is equal to the vorticity across the free shear layer (see discussion after eqs. (2.95) and (2.96)). Because $\partial\Phi/\partial s$ is discontinuous across S_V, Φ is also discontinuous across S_V. We can relate $\Phi^+ - \Phi^-$ to the circulation Γ, which is defined as

$$\Gamma = \oint_C \mathbf{u} \cdot d\mathbf{s}. \tag{6.13}$$

Here $\mathbf{u}$ is the fluid velocity vector and integration is along a closed curve C. The direction followed when we integrate along C matters. Let us illustrate this by applying Stokes's theorem. At the same time, this will permit us to show how Γ is related to the vorticity $\omega = \nabla \times \mathbf{u}$. Stokes's theorem says that

$$\oint_C \mathbf{u} \cdot d\mathbf{s} = \iint_s (\nabla \times \mathbf{u}) \cdot \mathbf{n}\, ds. \tag{6.14}$$

Here S is a surface bounded by C and $\mathbf{n}$ is the normal vector to S. If we walk along C in the

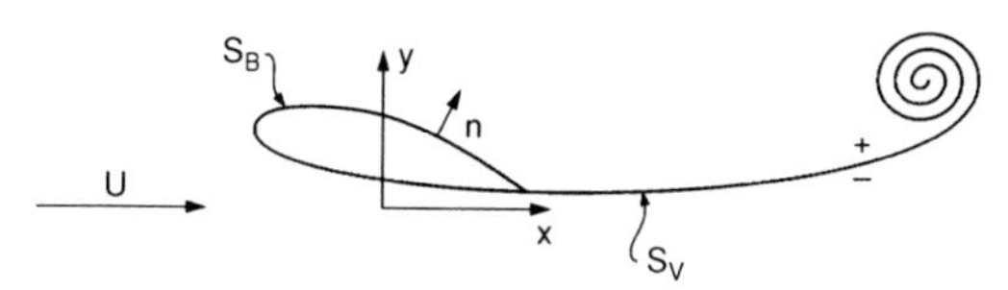

Figure 6.23. 2D flow past a foil in infinite fluid. S_B = body surface, S_V = surface-enclosing free shear layer (the subscript V denotes vorticity).

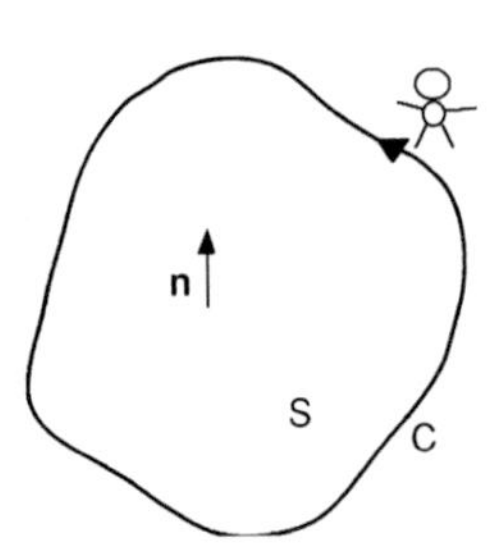

Figure 6.24. Integration direction along C when applying Stokes's theorem.

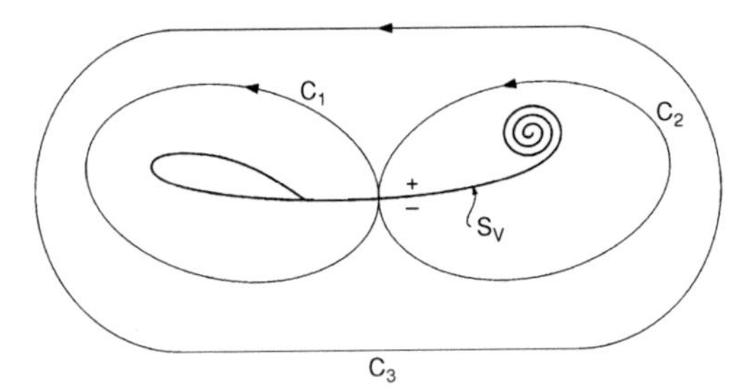

Figure 6.25. Curves C_i, i = 1,2,3 used to calculate circulation. The arrows indicate that the integration is in the counterclockwise direction.

integration direction with the head pointing in the direction of the positive normal to S, the surface should be on the left-hand side (see Figure 6.24).

Let us exemplify how Γ is related to $\Phi^+ - \Phi^-$ by considering three curves C_1, C_2, and C_3 as illustrated in Figure 6.25. The integration is in the counterclockwise direction. We start with curve C_1, which encloses the foil and part of the free shear layer S_V. We can write

$$\Gamma_1 = \int_{C_1} \frac{\partial \Phi}{\partial s} ds = \Phi^- - \Phi^+. \tag{6.15}$$

Further

$$\Gamma_2 = \int_{C_2} \frac{\partial \Phi}{\partial s} ds = \Phi^+ - \Phi^- \tag{6.16}$$

and

$$\Gamma_3 = \int_{C_3} \frac{\partial \Phi}{\partial s} ds = 0. \tag{6.17}$$

So there is zero circulation around a curve enclosing both the foil and the complete free shear layer. This is also a consequence of Kelvin's theorem (Newman 1977) stating that the circulation around a closed curve C moving with the fluid remains constant according to potential flow theory.

Similar to the thin boundary layer of Chapter 2, there is no jump in the pressure across the free shear layer. Bernoulli's equation can be used to express the pressure outside the free shear layer. Zero pressure jump across the free shear layer can then be expressed as

$$\frac{\partial}{\partial t}(\Phi^+ - \Phi^-) + \frac{1}{2}\left[\left(\frac{\partial \Phi^+}{\partial x}\right)^2 - \left(\frac{\partial \Phi^-}{\partial x}\right)^2 + \left(\frac{\partial \Phi^+}{\partial y}\right)^2 - \left(\frac{\partial \Phi^-}{\partial y}\right)^2\right] = 0.$$

This can be rewritten as

$$\frac{\partial}{\partial t}(\Phi^+ - \Phi^-) + \frac{1}{2}\left(\frac{\partial \Phi^+}{\partial x} + \frac{\partial \Phi^-}{\partial x}\right)\frac{\partial}{\partial x}(\Phi^+ - \Phi^-) + \frac{1}{2}\left(\frac{\partial \Phi^+}{\partial y} + \frac{\partial \Phi^-}{\partial y}\right)\frac{\partial}{\partial y}(\Phi^+ - \Phi^-) = 0. \tag{6.18}$$

Eq. (6.18) shows that $\Phi^+ - \Phi^-$ does not change with time when we move with the velocity

$$\mathbf{u}_C = \frac{1}{2}\left[\frac{\partial \Phi^+}{\partial x} + \frac{\partial \Phi^-}{\partial x}\right]\mathbf{i} + \frac{1}{2}\left[\frac{\partial \Phi^+}{\partial y} + \frac{\partial \Phi^-}{\partial y}\right]\mathbf{j}. \tag{6.19}$$

Here $\mathbf{i}$ and $\mathbf{j}$ are unit vectors along the x- and y-axes. Eq. (6.19) implies that $\Phi^+ - \Phi^-$ is convected (advected) with the mean value of the velocity on the two sides of the free shear layer. We should note that only the tangential velocity is discontinuous across the free shear layer.

Let us apply eq. (6.18) at the trailing edge ($T.E.$). The details of the flow at the trailing edge are illustrated in Figure 6.26. The apex angle α of the foil at $T.E.$ is finite. The following analysis is based on Maskell (1972). As we have already said, the free shear layer is parallel to either the upper or lower part of the foil surface at $T.E.$, depending on the sign of the shed velocity. The free shear layer is parallel to the lower part of the foil surface at $T.E.$ in Figure 6.26, and the tangential velocity is U_T at $T.E.$ As a matter of simplicity, we assume that the

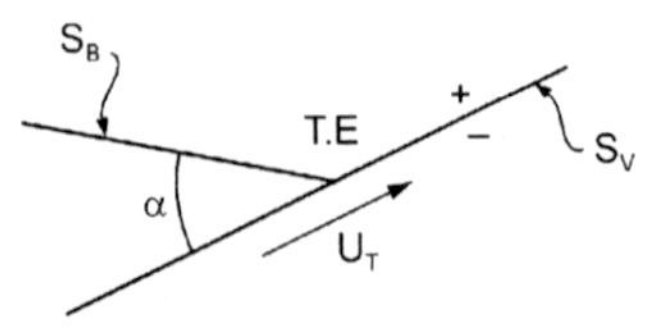

Figure 6.26. Details of the unsteady potential flow at the trailing edge ($T.E.$) of a foil. S_V is the free shear layer. Finite apex angle α of the foil at the trailing edge.

foil does not move. This means the fluid has no normal velocity at *T.E.*

There is a corner flow at the upper side of the free shear layer at *T.E.* Because the corner is a stagnation point, that is, the velocity is zero at the corner, eq. (6.18) applied at *T.E.* gives

$$\frac{\partial}{\partial t}(\Phi^+ - \Phi^-) - \frac{1}{2}\left(U_T^-\right)^2 = 0.$$

Here U_T^- is the tangential velocity at the lower part of the foil surface at *T.E.* Using eq. (6.15) and defining Γ as the circulation around the foil, we have

$$\frac{d\Gamma(t)}{dt} + \frac{1}{2}\left(U_T^-\right)^2 = 0. \qquad (6.20)$$

If the free shear layer had been parallel to the upper part of the foil surface at *T.E.*, eq. (6.20) would be exchanged with

$$\frac{d\Gamma(t)}{dt} - \frac{1}{2}\left(U_T^+\right)^2 = 0. \qquad (6.21)$$

Here U_T^+ is the tangential velocity at the upper part of the foil surface at *T.E.* If the apex angle of the foil at *T.E.* is zero, we get

$$\frac{d\Gamma(t)}{dt} + \frac{1}{2}\left(U_T^-\right)^2 - \frac{1}{2}\left(U_T^+\right)^2 = 0. \qquad (6.22)$$

Let us return to eq. (6.20). What does $0.5(U_T^-)^2$represent physically? Well, $-U_T^-$ is the vorticity across the boundary layer at *T.E.* (see eq. (2.96)). Further, $0.5U_T^-$ is the convection velocity of the vorticity. This means that $-0.5\left(U_T^-\right)^2$ is the vorticity flux in the boundary layer at *T.E.* This states that the vorticity in the free shear layer originates from the boundary layer along the foil surface. If the free shear layer had not been parallel to either the upper or lower side of the foil surface at *T.E.*, there would not have been shed any vorticity according to the previous analysis.

There are different ways to represent a potential flow. One way is to start out with Green's second identity. We will show this approach. Green's second identity in three dimensions states

$$\iiint_{\Omega}(\varphi\nabla^2\psi - \psi\nabla^2\varphi)d\tau = \iint_{S}\left(\psi\frac{\partial\varphi}{\partial n} - \varphi\frac{\partial\psi}{\partial n}\right)ds, \qquad (6.23)$$

where S is the surface enclosing the fluid volume Ω. It is necessary that φ and ψ have continuous derivatives of first and second order in Ω. The normal direction n is into the fluid region. The 2D

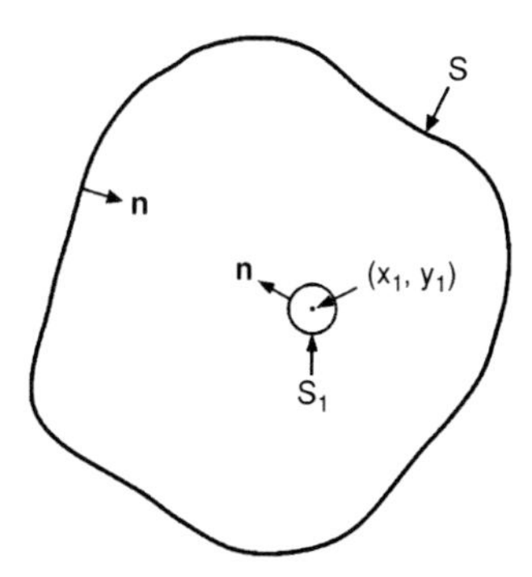

Figure 6.27. Integration surfaces used in applying Green's second identity. (x_1, y_1) = singular point of eq. (6.25).

version of eq. (6.23) is considered in the following text.

Example 1

If $\nabla^2\varphi = 0$ and $\nabla^2\psi = 0$ everywhere in the fluid domain, it follows from eq. (6.23) that

$$\int_{S}\left(\psi\frac{\partial\varphi}{\partial n} - \varphi\frac{\partial\psi}{\partial n}\right)ds = 0. \qquad (6.24)$$

Example 2

If $\nabla^2\varphi = 0$ everywhere in the fluid domain,

$$\psi(x, y; x_1, y_1) = \ln[(x - x_1)^2 + (y - y_1)^2]^{1/2} \qquad (6.25)$$

and the point (x_1, y_1) is inside the fluid volume (Figure 6.27), we have to be careful in how we handle the singular point (x_1, y_1) in eq. (6.25). We note that ψ is the velocity potential for a 2D source of strength 2π with the source at point (x_1, y_1). Outside the singular point, ψ satisfies the Laplace equation. Eq. (6.24) therefore applies if we integrate over $S + S_l$, that is,

$$\int_{S+S_1}\left(\psi\frac{\partial\varphi}{\partial n} - \varphi\frac{\partial\psi}{\partial n}\right)ds = 0,$$

where S_1 is the surface of a circle with small radius r enclosing the point (x_1, y_1). Along S_1 we can write $\frac{\partial\psi}{\partial n} = \frac{\partial\psi}{\partial r} = 1/r$ where

$$r = ((x - x_1)^2 + (y - y_1)^2)^{1/2}. \qquad (6.26)$$

As $r \to 0$, this gives

$$\int_{S_1}\varphi\frac{\partial\psi}{\partial n}ds = \varphi(x_1, y_1)\int_0^{2\pi}\frac{1}{r}rd\theta = 2\pi\varphi(x_1, y_1) \qquad (6.27)$$

and

$$\int_{S_1} \frac{\partial \varphi}{\partial n} \psi \, ds = 0. \tag{6.28}$$

As a result, we can write the velocity potential for 2D flows as

$$2\pi\varphi(x_1, y_1) = \int_S \left(\frac{\partial \varphi(s)}{\partial n} \ln r - \varphi(s) \frac{\partial}{\partial n} \ln r \right) ds(x, y). \tag{6.29}$$

Here S has to be a closed surface. The expression $\partial(\ln r)/\partial n$ in eq. (6.29) is a dipole in the normal direction of S. Eq. (6.29) states that we can represent the velocity potential in a fluid domain by means of a combined source and dipole distribution along the surface S enclosing the fluid domain. However, we need to know the values of $\partial\varphi/\partial n$ and φ along S in order to determine the velocity potential φ at (x_1, y_1) in the fluid domain. Later we explain how this can be done for our foil problem. A numerical method based on representing the fluid flow through a distribution of singularities (source, dipoles, vortices) along boundary surfaces is called boundary element method (BEM).

We now apply eq. (6.29) to our foil problem and write the total velocity potential as

$$\Phi = Ux + \varphi, \tag{6.30}$$

where eq. (6.29) is used to express φ. S consists then of the foil surface S_B, a control surface S_∞ at infinity, and S_V, which contains both sides of the free shear layer. The contribution from integrating over S_∞ can be shown to be zero (see exercise 6.13.2).

We will rewrite eq. (6.29) by using the body boundary condition

$$\frac{\partial \varphi}{\partial n} = \mathbf{V}_B \cdot \mathbf{n} - Un_1, \tag{6.31}$$

where $\mathbf{V}_B$ is the foil velocity and $\mathbf{n} = (n_1, n_2)$. Further, by using the condition that the fluid velocity perpendicular to S_V is the same on the two sides of S_V, we have

$$\begin{aligned}\varphi(x_1, y_1) = &-\frac{1}{2\pi} \int_{S_B} \varphi \frac{\partial}{\partial n} \ln r ds \\ &+ \frac{1}{2\pi} \int_{S_B} (\mathbf{V}_B \cdot \mathbf{n} - Un_1) \ln r ds \\ &- \frac{1}{2\pi} \int_{S_V^+} [\varphi^+ - \varphi^-] \frac{\partial}{\partial n^+} \ln r dS.\end{aligned} \tag{6.32}$$

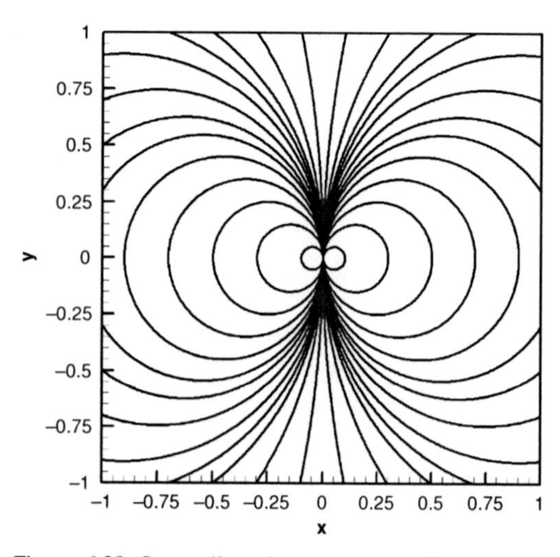

Figure 6.28. Streamlines due to a vertical 2D dipole in infinite fluid.

Here + and − refer to the two sides of the free shear layer. Eq. (6.32) states that the velocity potential can be written as a distribution of sources and dipoles over the body surface and a distribution of dipoles over the free shear layer. This is not the most common way of expressing the flow when vorticity is involved. It is more common to use potential flow vortices than dipoles. However, by integrating the dipole expression in eq. (6.32) by parts, it is possible to replace the dipole distribution by a vortex distribution.

Using potential flow vortices instead of normal dipoles to represent the flow at the free shear layer gives a more directly correct physical picture of the flow. Let us explain this by examining the flow associated with a dipole. Streamlines due to a vertical 2D dipole are shown in Figure 6.28. Imagine that the vertical direction in Figure 6.28 corresponds to the normal direction of S_V^+. This flow picture does not highlight any jump in the tangential velocity across S_V at the singular position of the dipole.

Let us then consider a potential flow vortex. The velocity potential can be expressed as

$$\frac{\Gamma}{2\pi} \tan^{-1}\left(\frac{y_1 - y}{x_1 - x}\right), \tag{6.33}$$

where (x, y) are the coordinates of the vortex center and Γ is the circulation of the vortex. Integrating eq. (6.33) in the counterclockwise direction corresponds to positive Γ. Polar coordinates (r, θ) are introduced so that $x_1 - x = r\cos\theta$ and

$y_1 - y = r \sin\theta$, that is,

$$\theta = \tan^{-1}\left(\frac{y_1 - y}{x_1 - x}\right). \qquad (6.34)$$

Eq. (6.33) leads only to a fluid velocity V_θ in the θ-direction. This can be expressed as

$$V_\theta = \frac{\Gamma}{2\pi r}. \qquad (6.35)$$

If now (x, y) is a point on S_V, eq. (6.35) gives the difference in the velocity direction on the two sides of S_V at the singularity point of the vortex. This is more consistent with the condition that there is a jump in tangential velocity across S_V.

In order to show that the normal dipole distribution can be replaced by a vortex distribution in eq. (6.32), we must note first that

$$\frac{\partial}{\partial y}\ln r = -\frac{\partial\theta}{\partial x}. \qquad (6.36)$$

Further, we consider as a preliminary example the following normal dipole distribution along the x-axis from $x = a$ to b:

$$\int_a^b \frac{\partial}{\partial y}\ln\sqrt{(x - x_1)^2 + (y - y_1)^2}\Big|_{y=0} \times [\varphi^+(x) - \varphi^-(x)]dx. \qquad (6.37)$$

Introducing eq. (6.36) into eq. (6.37) and using partial integration, we obtain

$$-[\varphi^+ - \varphi^-]\theta|_a^b + \int_a^b \theta\left[\frac{\partial\varphi^+}{\partial x} - \frac{\partial\varphi^-}{\partial x}\right]dx. \qquad (6.38)$$

This expression consists of a vortex at $x = a$ and one at b plus a vortex distribution from $x = a$ to b along the x-axis. So we have illustrated the equivalence between a normal dipole distribution and a vortex distribution.

If we want to show that the normal dipole distribution in eq. (6.32) can be replaced by a vortex distribution, we must replace the coordinate system (x, y) with the coordinate system (n, s) and then follow a procedure using partial integration similar to that applied above.

Let us now return to how we determine the velocity potential by using eq. (6.32). The problem is solved as an initial-value problem. The integral over S_V^+ as well as the position of S_V^+ at each time instant is determined by using the following:

a) $\Phi^+ - \Phi^-$ at the free shear layer S_V is convected with the velocity $\mathbf{u}_C$ given by eq. (6.19). The expressions for the velocities on the two sides of the free shear layer are obtained by differentiating eq. (6.32) for $\varphi(x_1, y_1)$ with respect to x_1 and y_1 and adding the incident flow velocity.
b) $\Phi^+ - \Phi^-$ is continuous at the trailing edge.
c) S_V is parallel with either the upper or lower part of S_B at the trailing edge. This depends on the local flow at the trailing edge.

We then let (x_1, y_1) coincide with point (x, y) on S_B in eq. (6.32). This gives an integral equation that determines φ on S_B. Then we know φ everywhere in the fluid by means of eq. (6.32). The pressure follows by Bernoulli's equation. The force on the foil can either be found by integrating pressure loads or by an alternative formula presented by Faltinsen and Pettersen (F&P; 1983).

The solution procedure requires a numerical approximation. In the F&P case, S_V and S_B are divided into a number of straight-line segments. $\varphi^+ - \varphi^-$ is assumed to be linearly varying over each straight-line segment of S_V^+. Further, φ and $(\mathbf{V}_B \cdot \mathbf{n} - Un_1)$ are assumed constant over each straight-line segment of S_B. When setting up the equation system for φ on S_B, (x_1, y_1) in eq. (6.32) is taken equal to the midpoint of each straight-line segment on S_B. However, it is necessary to enforce the Kutta condition at the trailing edge. For doing this, a linear condition is set up following from continuity of $\Phi^+ - \Phi^-$ at the trailing edge. In order to have N equations with N unknown values for the velocity potential associated with each segment on S_B, the body boundary condition was not enforced at the midpoint of one specific segment. For this purpose, an element close to the trailing edge was selected. The resulting linear equation system is solved at each time step by inverting a coefficient matrix. For the time integration, an Euler method is used, showing good stability. The reader is referred to F&P for further details.

We will present some of the numerical results by F&P. The first case is the so-called Wagner problem. This can be analytically solved (Wagner 1925 and Newman 1977). At time $t = 0$, a small angle of attack α is given to a flat plate and kept constant with time. Because the procedure outlined above does not work for a plate with zero thickness, this was circumvented by introducing a small foil thickness. The maximum value was only 1% of the foil chord length c. Results for the lift

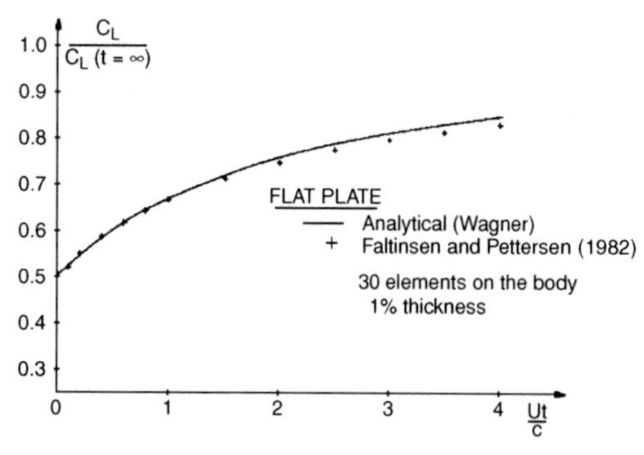

Figure 6.29. Comparison of the lift coefficient C_L between the analytical Wagner solution and the numerical method by Faltinsen and Pettersen (1983).

coefficient $C_L = L/(0.5\rho U^2 c)$, where L is the 2D lift force, are presented in Figure 6.29 as a function of nondimensional time Ut/c. The value of C_L at $t \to \infty$ is $C_L(t = \infty) = 2\pi\alpha$. Figure 6.29 shows that even after the foil has traveled four times the chord length, the lift is less than 90% of the steady lift at $t \to \infty$. The difference is because during the transient phase, the vorticity in the free shear layer causes a downwash at the foil. This reduces the effective angle of attack. The vorticity shed from the foil decreases with time and finally becomes negligible. The shed vorticity during this time is later convected far away from the foil. It is at this stage that the lift on the foil is close to the steady value. The Wagner solution is linearly dependent on α. If α is not small, there is in reality, a nonlinear relation between the lift and α. Further, the flow may also separate from the area around the leading edge for large α. This is associated with stalling, that is, reduction in mean lift with increasing α.

If a heaving foil is considered, the heave velocity divided by the incident flow velocity U is similar to an angle of attack. F&P studied the nonlinear behavior of a harmonically heaving foil with forward velocity. This problem was earlier studied by Giesing (1968), who found good qualitative agreement with experimental pictures of the wake. Giesing used discrete vortices to represent the free shear layer, and in one extreme case, the wake has a "mushroom" configuration. F&P's results for the same situations Giesing studied are considered in the following. The airfoil is a NACA 0015 (Abbott and von Doenhoff 1959). F&P used 30 elements to represent the foil numerically. The chord length of the foil and the incident flow along the longitudinal axis of the airfoil were chosen as 10 m and 1 ms^{-1}, respectively.

The case presented in Figure 6.30 corresponds to circular frequency of heave oscillation $\omega = 1.7$ rad/s. The maximum angle of attack was 17.8°. The time step was 0.05 s. The picture of the wake after 175 time steps is presented in Figure 6.30. It appears as if there is a discontinuity in the slope of the free shear layer at the trailing edge (*T.E.*). However, if a detailed view of the flow in the vicinity of *T.E.* had been shown, it would be evident that the flow is consistent with Figure 6.26. We see very clearly a mushroom configuration in the wake, and the free shear layer details are very complicated. The result in Figure 6.30 agrees qualitatively with Giesing's experimental pictures of the wake.

6.4.2 3D flow

The theoretical and numerical procedure outlined above can be generalized to 3D flows. Let us limit ourselves to infinite fluid. We start with eq. (6.23) and choose

$$\psi = \frac{1}{R}, \tag{6.39}$$

where

$$R = ((x_1 - x)^2 + (y_1 - y)^2 + (z_1 - z)^2)^{1/2} \tag{6.40}$$

Here (x_1, y_1, z_1) is inside the fluid volume. This means a 3D source in infinite fluid is used instead of a 2D source. Generalizing the derivation for 2D flow gives

$$\varphi(x_1, y_1, z_1) = \frac{1}{4\pi}\iint_{S_B}\left(\varphi\frac{\partial}{\partial n}\left(\frac{1}{R}\right) - \frac{1}{R}\frac{\partial\varphi}{\partial n}\right)dS + \frac{1}{4\pi}\iint_{S_V^+}(\varphi^+ - \varphi^-)\frac{\partial}{\partial n^+}\left(\frac{1}{R}\right)dS. \tag{6.41}$$

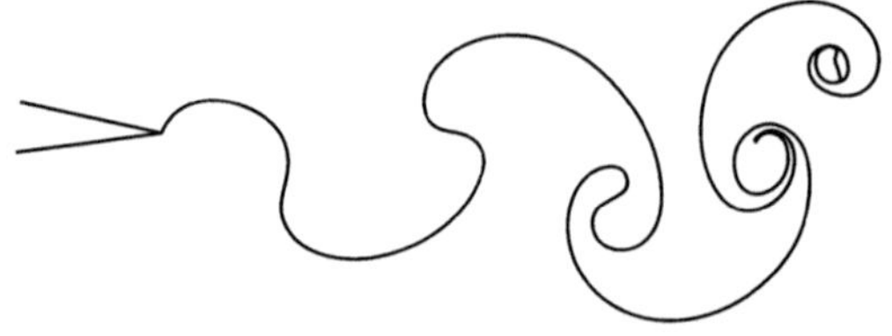

Figure 6.30. Wake profile for a harmonically heaving NACA 0015 airfoil. $\omega c/U = 17$. (Faltinsen and Pettersen 1983).

Eq. (6.18) can be easily generalized to 3D flow showing that $\varphi^+ - \varphi^-$ is convected with the mean value of the velocity on the two sides of the free shear layer. A Kutta condition is also needed to solve the problem. A method like this was used by Skomedal (1985) for a 3D foil in infinite fluid. It is commonly applied in the analysis of propulsors and is capable of handling arbitrary blade, hub, and duct geometries; general inflows; and the presence of blade sheet cavitation (Kinnas 1996).

Cavitation and free-surface effects can be included by letting the bounding surface for the fluid volume include the cavity surface and the free surface (Bal et al. 2001). Boundary conditions on the free and cavity surfaces are, of course, needed to solve the problem. Because a source in infinite fluid is used, a method like this is denoted by a Rankine source method. The free surface has to be truncated to avoid an infinite number of unknowns. This may be done by either a hybrid method in which an analytical solution form is used in the far field of the foil or by a numerical beach, which has an effect similar to that of a wave beach in a model tank. The purpose is to avoid unphysical reflected waves from the outer boundary of the truncated free surface.

It was easy to show in the 2D case how a dipole description of a vortex sheet could be expressed in terms of a vortex distribution. It is more cumbersome to show the equivalence in the 3D case. We start out with a more general representation of the velocity due to the vorticity ω in a volume Ω. If the fluid is incompressible, the velocity $\mathbf{u}$ outside Ω can be expressed as (Batchelor 1967)

$$\mathbf{u}(x_1, y_1, z_1) = \nabla \times \frac{1}{4\pi} \iiint_{\Omega} \frac{\omega(x, y, z)}{R}\, d\Omega. \tag{6.42}$$

Here $\nabla = \mathbf{i}\frac{\partial}{\partial x_1} + \mathbf{j}\frac{\partial}{\partial y_1} + \mathbf{k}\frac{\partial}{\partial z_1}$ and $\boldsymbol{i}, \boldsymbol{j}, \boldsymbol{k}$ are unit vectors along the x_1-, y_1-, and z_1-axes, respectively. We can express ω in eq. (6.42) as

$$\omega = \nabla \times \mathbf{u} = \mathbf{i}\left(\frac{\partial w}{\partial y} - \frac{\partial v}{\partial z}\right) + \mathbf{j}\left(\frac{\partial u}{\partial z} - \frac{\partial w}{\partial x}\right) + \mathbf{k}\left(\frac{\partial v}{\partial x} - \frac{\partial u}{\partial y}\right), \tag{6.43}$$

where $\mathbf{u} = (u, v, w)$. Let us now consider a thin free shear layer in the xy-plane. Consistent with a thin free shear layer (see similar eq. (2.95) of Chapter 2), the vorticity ω can be approximated as

$$\omega = -\mathbf{i}\frac{\partial v}{\partial z} + \mathbf{j}\frac{\partial u}{\partial z}. \tag{6.44}$$

Integrating ω across the free shear layer gives

$$\int_{0^-}^{0^+} \omega\, dz = -\mathbf{i}(v^+ - v^-) + \mathbf{j}(u^+ - u^-).$$
$$= -\mathbf{i}\frac{\partial}{\partial y}(\varphi^+ - \varphi^-) + \mathbf{j}\frac{\partial}{\partial x}(\varphi^+ - \varphi^-) \tag{6.45}$$

This means

$$\mathbf{u} = \nabla \times \frac{1}{4\pi} \iint_{S_V} \frac{1}{R}\left(-\mathbf{i}\frac{\partial}{\partial y}(\varphi^+ - \varphi^-) + \mathbf{j}\frac{\partial}{\partial x}(\varphi^+ - \varphi^-)\right) dx\, dy, \tag{6.46}$$

where

$$\nabla \times \frac{\mathbf{i}}{R} = \mathbf{j}\frac{\partial}{\partial z_1}\left(\frac{1}{R}\right) - \mathbf{k}\frac{\partial}{\partial y_1}\left(\frac{1}{R}\right) \tag{6.47}$$

$$\nabla \times \frac{\mathbf{j}}{R} = -\mathbf{i}\frac{\partial}{\partial z_1}\left(\frac{1}{R}\right) + \mathbf{k}\frac{\partial}{\partial x_1}\left(\frac{1}{R}\right) \tag{6.48}$$

$$R = \left((x_1 - x)^2 + (y_1 - y)^2 + z_1^2\right)^{1/2}. \tag{6.49}$$

If eq. (6.41) is used to represent this vortex sheet, the velocity potential due to the vortex sheet can be expressed as

$$\varphi(x_1, y_1, z_1) = \frac{1}{4\pi} \iint_{S_v} (\varphi^+ - \varphi^-)\frac{\partial}{\partial z}\left(\frac{1}{R}\right) dx\, dy. \tag{6.50}$$

We will not show the details of the equivalence between eqs. (6.46) and (6.50), and refer instead to exercise 6.13.4, in which one has to derive the details in a special case. Eq. (6.42) is instead applied to a vortex line. This leads to Biot-Savart's law.

We consider first a vortex tube as in Figure 6.31 and let the cross-sectional area δa go to zero. The vorticity is a vector along the unit vector $\mathbf{s}$ tangent to the vortex line. The magnitude ω of the vorticity times δa is equal to the circulation Γ around the vortex line (see eq. (6.13)). Γ is a constant along a vortex line. This can be shown by considering a closed curve consisting of C_1, C_2, C_3, C_4 (see

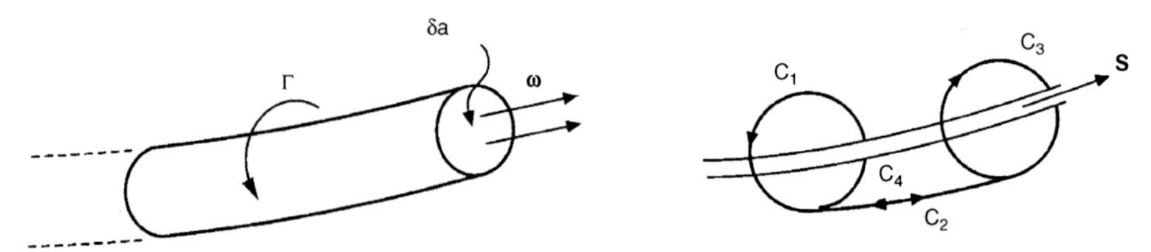

Figure 6.31. The left figure shows a vortex tube with circulation Γ. The right drawing shows a closed curve $C_1 + C_2 + C_3 + C_4$ used in proving that the circulation Γ is constant along a vortex line.

Figure 6.31). Here C_1 and C_3 are in cross-sectional planes of the vortex line. This closed curve does not enclose vorticity. The circulation around the closed curve is therefore zero. Because C_2 and C_4 coincide and the velocity is integrated in opposite directions along C_2 and C_4, the contributions to the circulation from C_2 and C_4 cancel each other. This means the contributions from C_1 and C_3 must also cancel each other. By noting that we integrate along C_1 and C_3 in opposite directions, it follows that the circulation Γ is a constant along a vortex line. A consequence is that a vortex line cannot end in the fluid.

From eq. (6.42), it follows that the velocity induced by a vortex line is

$$\mathbf{u}(x_1, y_1, z_1) = \frac{\Gamma}{4\pi} \oint \nabla \times \left(\frac{\mathbf{s}}{R}\right) dl, \quad (6.51)$$

where the ∇-operator involves differentiation with respect to x_1, y_1, and z_1. The detailed differentiation gives

$$\mathbf{u}(x_1, y_1, z_1) = \frac{\Gamma}{4\pi} \oint \frac{\mathbf{s} \times \mathbf{R}}{R^3} dl, \quad (6.52)$$

where

$$\mathbf{R} = (x_1 - x)\,\mathbf{i} + (y_1 - y)\,\mathbf{j} + (z_1 - z)\,\mathbf{k}. \quad (6.53)$$

Eq. (6.52) is Biot-Savart's law.

Consider as a special case an infinitely long straight vortex line along the x-axis. This means $\mathbf{s} = \mathbf{i}$, $y = 0$, and $z = 0$. We choose $x_1 = 0$, $z_1 = 0$, that is, $\mathbf{R} = -x\mathbf{i} + y_1\mathbf{j}$ (Figure 6.32). This gives $\mathbf{s} \times \mathbf{R} = y_1\mathbf{k}$ or that there is only a velocity component w along the z-axis. This can be expressed as

$$w = \frac{\Gamma}{4\pi} \int_{-\infty}^{\infty} \frac{y_1 dx}{\left(x^2 + y_1^2\right)^{3/2}} = \frac{\Gamma}{4\pi y_1} \left[\frac{x}{\left(x^2 + y_1^2\right)^{1/2}}\right]_{-\infty}^{\infty} = \frac{\Gamma}{2\pi y_1},$$

which is the same velocity as that induced by a 2D vortex. This means a 2D vortex is the same as an infinitely long straight 3D vortex line.

Biot-Savart's law is, for instance, used in connection with vortex-lattice methods for propellers (Kerwin and Lee 1978) and for linear steady flow past 3D foils. The principle is illustrated in Figure 6.33. The foil is divided into panels. A horseshoe vortex with strength Γ_n is associated with each panel. The vortex consists of three connected straight lines. Two of the lines (the trailing vortices) go to infinity downstream and are parallel to the incident flow. The third line connects the two

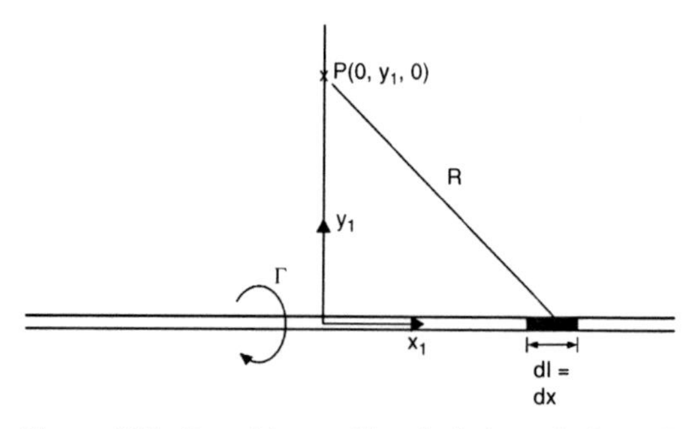

Figure 6.32. Quantities used in calculating velocity at P induced by an infinitely long straight vortex line along the x_1-axis.

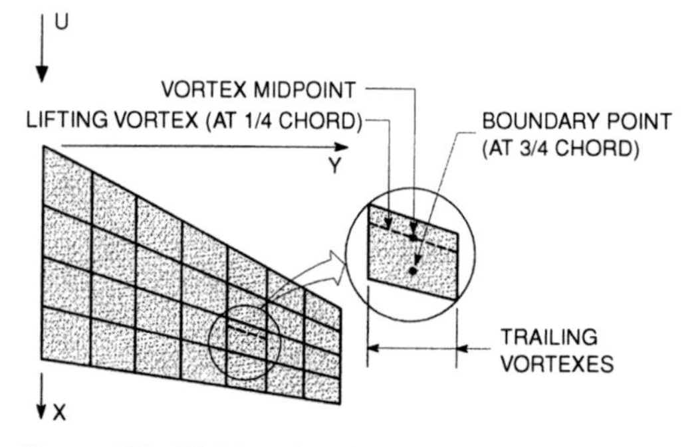

Figure 6.33. Division of surface into panels and location of vortices and control points in a vortex-lattice method (Feifel 1981).

trailing vortices at a position that is $\ell/4$ from the front of the panel. Here ℓ is the length of the panel in the incident flow direction. The horseshoe vortices from each panel are superimposed, and the body boundary condition is satisfied on each panel at a control point $3\ell/4$ from the front of the panel. The induced velocity from each horseshoe vortex at the control points is calculated by Biot-Savart's law. By satisfying the boundary conditions at the foil, we get an equation system that determines the unknown Γ_n. Further details and generalizations in connection with hydrofoil vessels are presented by van Walree (1999).

6.5 2D steady flow past a foil in infinite fluid. Forces

Two-dimensional steady flow past a foil in infinite fluid is considered. This means the free shear layer S_V in Figure 6.23 has been convected to infinity relative to the foil during a transient time period. We use a coordinate system (x, y) as in Figure 6.23 so that the forward speed U of the foil appears as an incident flow along the x-axis. Because the flow is steady, the foil does not move relative to the xy-coordinate system.

Expressions for lift and drag forces on the foil are derived based on potential flow theory and by using conservation of fluid momentum (see section 2.12.4). The fluid volume enclosed by the foil surface S_B and the surface S_R of a circle with the center inside the foil and with large radius R is considered (Figure 6.34).

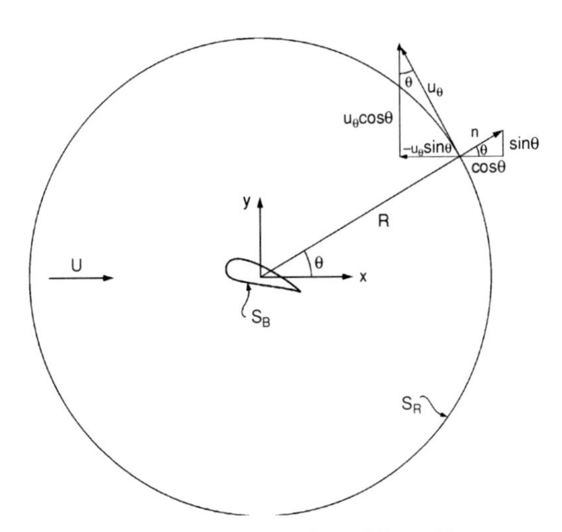

Figure 6.34. Control surfaces S_B and S_R used in expressing lift and drag forces on a 2D foil in infinite fluid by conservation of fluid momentum.

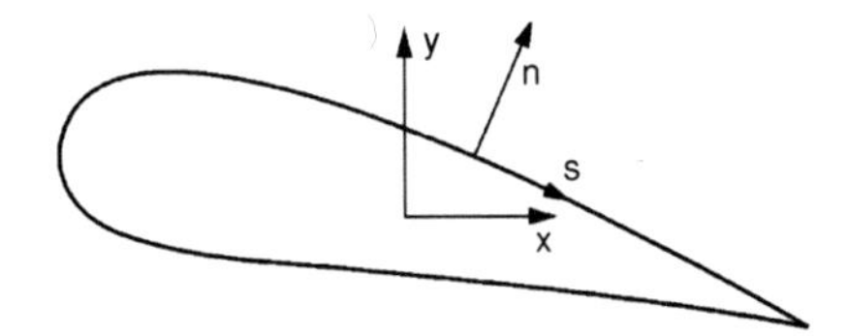

Figure 6.35. Coordinate systems.

If we are close to the foil, the flow caused by the foil can be represented by a distribution of sources, sinks, and vortices over the body surface S_B as described in the previous section. However, if we are far away from the foil, we do not see the details of the foil. As a first approximation, the flow due to the foil can then be represented in terms of a vortex with the center inside the foil. This means that the velocity potential at S_R is

$$\Phi = Ux + \varphi = Ux + \frac{\Gamma}{2\pi}\theta + O\left(\frac{1}{R}\right). \quad (6.54)$$

Here $O(\ )$ means the order of magnitude and Γ is the circulation around the foil. The angle θ is defined in Figure 6.34. We will show that there cannot be a source (sink) term $(Q/2\pi)\log R$ in eq. (6.54). The source strength Q represents the fluid mass flux through S_R caused by the foil. This can be expressed as

$$Q = \int_{S_B} \frac{\partial \varphi}{\partial n} ds \quad (6.55)$$

by using conservation of fluid mass. Here φ is the velocity potential due to the body. The normal derivative of φ follows from the body boundary condition, that is, $\partial\varphi/\partial n = -Un_1$, where n_1 is the x-component of the outward normal vector to S_B. If we introduce a curvilinear coordinate s along the foil surface (Figure 6.35) and express the x and y coordinates of the foil surface as a function of s, we can express n_1 as $-dy/ds$. This means

$$Q = U\int_{S_B} \frac{dy}{ds} ds = U\int_{S_B} dy. \quad (6.56)$$

Because the integration is along a closed curve, that is, the start and end values of y are the same, Q is zero.

We start by examining conservation of fluid momentum in the y-direction. Because first approximations of the vertical fluid velocity v and of the normal velocity are, respectively, $\Gamma \cos\theta/(2\pi R)$ and $U \cos\theta$ at S_R, the y-component of the momentum flux at S_R is

$$J_y = \rho \int_0^{2\pi} \frac{\Gamma}{2\pi R} \cos\theta U \cos\theta R\, d\theta + O\left(\frac{1}{R}\right). \quad (6.57)$$

The y-component of the hydrodynamic force acting on the control volume at boundary S_R is

$$F_y = -\int_0^{2\pi} p \sin\theta R\, d\theta, \quad (6.58)$$

where the pressure p follows from Bernoulli's equation. Expressing the horizontal fluid velocity as $U + u$ and neglecting gravity, we have

$$p = -\frac{\rho}{2}\left[(U+u)^2 + v^2\right] + \frac{\rho}{2}U^2 + p_a$$
$$= -\rho U u - \frac{\rho}{2}(u^2 + v^2) + p_a. \quad (6.59)$$

Here p_a is the ambient pressure. A first approximation of u at S_R is $-\Gamma \sin\theta/(2\pi R)$. This means

$$F_y = \rho U \int_0^{2\pi} \left(-\frac{\Gamma}{2\pi R}\sin\theta\right) \sin\theta R\, d\theta + O\left(\frac{1}{R}\right). \quad (6.60)$$

There is no momentum flux through S_B. Further, the y-component of the force acting on the fluid at S_B is minus the lift force L acting on the foil. This gives

$$J_y = F_y - L,$$

that is,

$$-L = \rho U \frac{\Gamma}{2\pi} \int_0^{2\pi} (\cos^2\theta + \sin^2\theta) d\theta + O\left(\frac{1}{R}\right). \quad (6.61)$$

Letting $R \to \infty$ results in the well-known Kutta-Joukowski formula

$$L = -\rho U \Gamma. \quad (6.62)$$

This formula and that the drag is zero are much more easily derived mathematically using complex potentials and the Blasius theorem (Milne-Thomson 1996). We should note that positive circulation Γ is in the counterclockwise direction. Γ is negative and the lift force positive for the flow situation in Figure 6.34. Another way of seeing this qualitatively is that negative Γ implies an increase in fluid velocity on the top of the foil relative to the bottom of the foil. Because increasing fluid velocity means decreasing pressure, the lift force is positive.

The drag force D on the foil follows from conservation of fluid momentum in the x-direction. A first approximation of the x-component of the fluid velocity at S_R is $U - \Gamma \sin\theta/(2\pi R)$. The x-component of the momentum flux at S_R is then

$$J_x = \rho \int_0^{2\pi} \left(U - \frac{\Gamma}{2\pi R}\sin\theta\right) U \cos\theta R\, d\theta + O\left(\frac{1}{R}\right).$$
$$= O\left(\frac{1}{R}\right)$$

The x-component of the hydrodynamic force acting on S_R is

$$F_x = \rho \int_0^{2\pi} \left(-\frac{\Gamma}{2\pi R}\sin\theta\right) U \cos\theta R\, d\theta + O\left(\frac{1}{R}\right).$$
$$= O\left(\frac{1}{R}\right)$$

Conservation of fluid momentum implies

$$J_x = F_x - D.$$

It follows that the drag force D is zero by letting $R \to \infty$. The hydrodynamic force acting on a 2D foil in steady flow and in infinite fluid is therefore perpendicular to the incident flow velocity according to potential flow theory. If a flat plate with an angle of attack relative to the incident flow is considered, this may sound contradictory. Because pressure loads act perpendicularly to a body surface, one may be tempted to say that the force acts perpendicularly to the flat plate instead of perpendicularly to the incident flow velocity. However, detailed analysis shows that the very large pressure force acting on a very small area at the leading edge causes a finite suction force. The magnitude of this suction force explains why the total hydrodynamic force is perpendicular to the incident flow velocity.

If viscous effects are considered, there is a drag force. This can, for instance, be derived by conservation of fluid momentum similar to what was done in section 2.9 for the viscous flow past a 2D strut with zero angle of angle.

6.6 2D linear steady flow past a foil in infinite fluid

Linear steady flow past a foil in infinite fluid is considered. Linearity means that the fluid velocity $|\nabla\varphi|$ due to the foil is much smaller than U and that only linear terms in φ are kept in the formulation of the boundary-value problem. Linearization implies that an analytical solution can be found for general hydrofoil shapes in an infinite fluid.

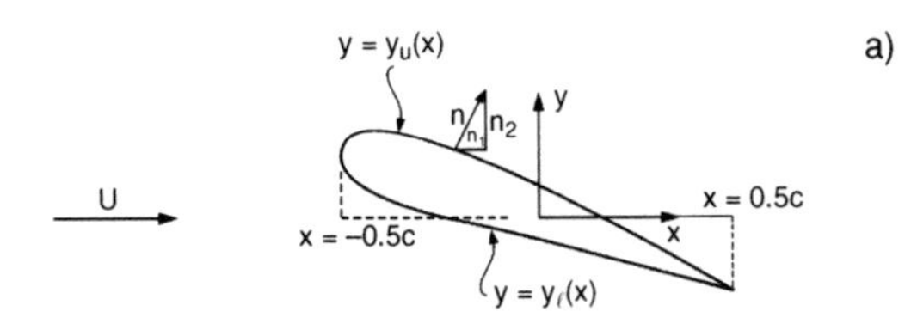

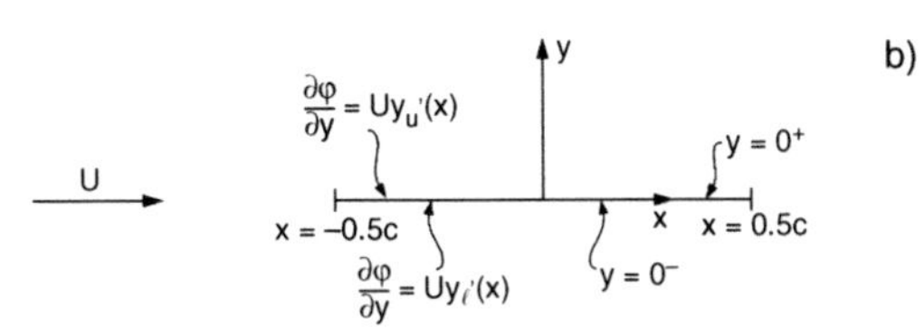

Figure 6.36. Linearization of boundary value problem of 2D steady flow past a foil in infinite fluid. (a) Defines the foil geometry; (b) illustrates how body boundary conditions on the foil through linearization are transferred to a cut along the x-axis from $x = -c/2$ to $x = c/2$. c = chord length.

The lift slope $dL/d\alpha$, where L is the lift and α is the angle of attack, is constant according to linear theory. To how large an α that linear theory applies depends, for instance, on Reynolds number, cavitation number, ventilation, and the foil shape. The linear theory may be applicable up to 13° to 17° for a 2D foil with turbulent boundary layers in infinite fluid (see Figure 2.17). However, this depends on when separation at the leading edge occurs. A laminar boundary layer flow separates more easily than a turbulent boundary layer flow. Leading-edge flow separation causes the lift to decrease with increasing α (stalling). A linear theory cannot describe cavitation inception. However, linear theory is used in describing flow due to cavitating foils (Newman 1977). We assume a noncavitating foil.

We start by linearizing the body boundary condition on the foil. The upper and lower parts of the foil surface are denoted $y = y_u(x)$ and $y = y_l(x)$, respectively (see Figure 6.36a).

The exact body boundary condition on $y = y_u(x)$ is

$$\frac{\partial \varphi}{\partial n} = -Un_1, \tag{6.63}$$

where n_1 is the x-component of the outward normal vector $\mathbf{n} = (n_1, n_2)$ to the foil surface. We can write $\partial\varphi/\partial n = n_1 \partial\varphi/\partial x + n_2 \partial\varphi/\partial y$. Because $n_1 \ll n_2$ on the most part of the foil surface, we write

$$\frac{\partial \varphi}{\partial n} \approx \frac{\partial \varphi}{\partial y} \quad \text{on } y = y_u(x).$$

This is not a good approximation at the nose of the foil. The consequence is a singularity in the analytical solution at the nose of the foil. Because $n_1 \approx -\partial y_u/\partial x$, it follows that

$$\frac{\partial \varphi}{\partial y} = U y_u{}'(x) \quad \text{on } y = y_u(x).$$

Here the symbol $'$ means *derivative*.

The next step is to Taylor expand this boundary condition about $y = 0^+$, that is,

$$\frac{\partial \varphi}{\partial y} = \left.\frac{\partial \varphi}{\partial y}\right|_{y=0^+} + y_u \left.\frac{\partial^2 \varphi}{\partial y^2}\right|_{y=0^+} + \cdots,$$
$$-c/2 < x < c/2.$$

Because y_u is small, we get the following linearized body boundary condition for the upper part of the foil surface:

$$\frac{\partial \varphi}{\partial y} = U y_u{}'(x) \quad \text{on } y = 0^+, \; -c/2 < x < c/2. \tag{6.64}$$

Similarly, for the lower part of the foil surface, we get

$$\frac{\partial \varphi}{\partial y} = U y_l{}'(x) \quad \text{on } y = 0^-, \; -c/2 < x < c/2. \tag{6.65}$$

These linearized body boundary conditions are illustrated in Figure 6.36b.

The next step is to divide the flow into two parts that are, respectively, antisymmetric and symmetric about the x-axis. We then define the mean-camber line of the foil

$$\eta(x) = 0.5[y_u(x) + y_l(x)] \tag{6.66}$$

and express y_u as $\eta(x) + a(x)$ and y_l as $\eta(x) - a(x)$. This means eqs. (6.64) and (6.65) become

$$\frac{\partial \varphi}{\partial y} = U[\eta'(x) + a'(x)] \quad \text{on } y = 0^+,$$
$$-c/2 < x < c/2 \tag{6.67}$$

$$\frac{\partial \varphi}{\partial y} = U[\eta'(x) - a'(x)] \quad \text{on } y = 0^-,$$
$$-c/2 < x < c/2, \tag{6.68}$$

respectively. We introduce then the decomposition $\varphi = \varphi_0 + \varphi_e$ so that

$$\frac{\partial \varphi_0}{\partial y} = U\eta'(x) \quad \text{on } y = 0, \; -c/2 < x < c/2 \tag{6.69}$$

$$\frac{\partial \varphi_e}{\partial y} = \begin{cases} +Ua'(x) & \text{on } y = 0^+, \; -c/2 < x < c/2 \\ -Ua'(x) & \text{on } y = 0^-, \; -c/2 < x < c/2 \end{cases}. \tag{6.70}$$

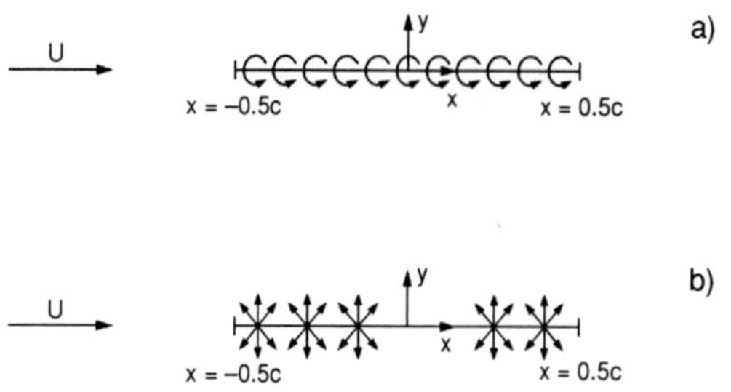

Figure 6.37. The linearized boundary value problem for 2D steady flow past a foil in infinite fluid can be represented as the sum of vortex distribution (a) and source distribution (b) along a cut of the x-axis from $x = -c/2$ to $x = c/2$. c = chord length.

We note that this is consistent with eqs. (6.67) and (6.68). Eq. (6.69) says that the vertical fluid velocity is continuous across the cut $-c/2 < x < c/2$. A distribution of vortices along the cut will satisfy this properly (Figure 6.37a). This flow is antisymmetric about the x-axis. Eq. (6.70) says that the vertical flow velocity has the same magnitude, but opposite directions on the two sides of the cut. This expresses that the flow is either pushed outward or attracted inward by the foil. A source distribution along the cut can be used to represent this part of the flow (Figure 6.37b). Actually, there must be a distribution of sources and sinks so that the net source strength Q is zero as shown in section 6.5.

The flow associated with the source distribution is symmetric about the x-axis and does not cause a vertical force. It is the vortex distribution that causes a lift force on the foil. We will concentrate on this part of the flow. The velocity potential can be expressed as

$$\varphi_0 = \frac{1}{2\pi} \int_{-0.5c}^{0.5c} \gamma(\xi) \tan^{-1} \frac{y}{x - \xi} d\xi. \quad (6.71)$$

Here $\gamma(\xi)\, d\xi$ is the circulation of a vortex with center in $y = 0$ and $x = \xi$. The x- and y-components of the fluid velocity due to the foil are then

$$u = \frac{\partial \varphi_0}{\partial x} = -\frac{1}{2\pi} \int_{-0.5c}^{0.5c} \gamma(\xi) \frac{y}{(x - \xi)^2 + y^2} d\xi \quad (6.72)$$

$$v = \frac{\partial \varphi_0}{\partial y} = \frac{1}{2\pi} \int_{-0.5c}^{0.5c} \gamma(\xi) \frac{x - \xi}{(x - \xi)^2 + y^2} d\xi. \quad (6.73)$$

By using eqs. (6.69) and (6.73), it now follows that

$$\frac{1}{2\pi} PV \int_{-0.5c}^{0.5c} \frac{\gamma(\xi)}{x - \xi} d\xi = U\eta'(x) \qquad -c/2 < x < c/2. \quad (6.74)$$

Here $PV \int$ means principal value integral. This is defined by the limiting process

$$\lim_{\varepsilon \to 0} \left[\int_{-0.5c}^{x-\varepsilon} + \int_{x+\varepsilon}^{0.5c} \right] \frac{\gamma(\xi)}{x - \xi} d\xi.$$

It is necessary to impose the Kutta condition at the trailing edge in order for eq. (6.74) to have a unique solution. This was discussed in the unsteady case in section 6.4 (see eq. (6.22)). However, there is no vorticity shed into a free shear layer in the steady case. This means $U_T^- = U_T^+$ in eq. (6.22). Here $U_T^{\pm} = U + u_T^{\pm}$, where $u_T^{\pm}$ follows by evaluating eq. (6.72) at $y = 0^{\pm}$ at the trailing edge.

Eq. (6.72) gives

$$u = \mp \frac{\gamma(x)}{2} \quad \text{on } y = 0^{\pm}, \ -c/2 < x < c/2. \quad (6.75)$$

So it is only the local vortex strength that contributes to the horizontal velocity at the cut. This can be understood by the flow picture caused by the vortex distribution in Figure 6.37. The vortices with centers different from x will only cause a vertical velocity at x. This means that we can write

$$u(x, \pm\varepsilon) = \mp \frac{\gamma(x)}{2\pi} \int_{-0.5c}^{0.5c} \frac{\varepsilon d\xi}{\left[(x - \xi)^2 + \varepsilon^2\right]}$$
$$= \mp \frac{\gamma(x)}{2\pi} \tan^{-1} \left(\frac{\xi - x}{\varepsilon} \right) \Bigg|_{-0.5c}^{0.5c},$$

where ε is small. Letting $\varepsilon \to 0$ gives eq. (6.75). Because we require that $U_T^- = U_T^+$ at the trailing edge and $U_T^{\pm} = U + u_T^{\pm}$, eq. (6.75) gives

$$\gamma(c/2) = 0. \quad (6.76)$$

This is the mathematical formulation of the Kutta condition for this problem.

Newman (1977) and Breslin and Andersen (1994) showed how we generally can solve the singular integral equation for γ given by eq. (6.74). Because $C((c/2)^2 - x^2)^{-0.5}$, where C is any constant, satisfies eq. (6.74) with the right-hand–side of the eq. (6.74) equal to zero, it is not sufficient to find a particular solution for eq. (6.74). The Kutta condition is needed to make the solution unique.

An approach in which the solution form of γ is guessed follows below. There is some rationality behind the representation of γ. However, knowing the answer certainly helps. The first step is to make the substitution

$$x = -(c/2)\cos \chi, \xi = -(c/2)\cos\theta, \quad (6.77)$$

where χ and θ vary between 0 and π. The assumed representation of γ is

$$\gamma(\xi) = 2U\left\{a_0\frac{1+\cos\theta}{\sin\theta} + \sum_{n=1}^{\infty} a_n \sin n\theta\right\}. \quad (6.78)$$

This vortex density is consistent with the Kutta condition, that is, γ is zero at $\theta = \pi$ or $\xi = c/2$. If $a_0 \neq 0$, then γ is infinite at the leading edge, that is, at $\theta = 0$ or $\xi = -c/2$. It will turn out that the singular term in eq. (6.78) represents the complete solution for a flat plate with an angle of attack. Eq. (6.74) can now be expressed as

$$\frac{1}{\pi} PV \int_0^{\pi} \frac{\left\{a_0(1+\cos\theta) + \sum_{n=1}^{\infty} a_n \sin n\theta \cdot \sin\theta\right\}}{\cos\theta - \cos\chi} d\theta = \eta'(x). \quad (6.79)$$

The trigonometric identity

$$\sin n\theta \cdot \sin\theta = \frac{1}{2}\left[\cos(n-1)\theta - \cos(n+1)\theta\right] \quad (6.80)$$

is then introduced in eq. (6.79). This will in eq. (6.79) require the evaluation of the Glauert integrals (Newman 1977), that is, that

$$PV \int_0^{\pi} \frac{\cos n\theta \, d\theta}{\cos\theta - \cos\chi} = \pi \frac{\sin n\chi}{\sin\chi}. \quad (6.81)$$

It means by using trigonometric identities that

$$\begin{aligned} PV \int_0^{\pi} \frac{\sin n\theta \sin\theta}{\cos\theta - \cos\chi} d\theta &= \frac{1}{2}\left[PV \int_0^{\pi} \frac{\cos(n-1)\theta}{\cos\theta - \cos\chi} d\theta - PV \int_0^{\pi} \frac{\cos(n+1)\theta}{\cos\theta - \cos\chi} d\theta\right] \\ &= \frac{\pi}{2}\left[\frac{\sin(n-1)\chi - \sin(n+1)\chi}{\sin\chi}\right] \\ &= -\pi \cos n\chi \end{aligned} \quad (6.82)$$

Eq. (6.79) then becomes

$$a_0 - \sum_{n=1}^{\infty} a_n \cos n\chi = \eta'(x). \quad (6.83)$$

This is just a Fourier cosine series representation of $\eta'(x)$. Because we can always represent $\eta'(x)$ like that, the assumed solution form of γ is consistent with both the Kutta condition and a Fourier series representation of $\eta'(x)$. It follows by multiplying eq. (6.83) with $\cos n\chi$, $n = 0, \ldots$, and integrating from 0 to π that

$$\begin{aligned} a_0 &= \frac{1}{\pi}\int_0^{\pi} \eta'(x) d\chi, \\ a_n &= -\frac{2}{\pi}\int_0^{\pi} \eta'(x)\cos n\chi \, d\chi \end{aligned} \quad (6.84)$$

where x is $-0.5c\cos\chi$.

When $\gamma(x)$ is obtained, we can find the linearized pressure p on the foil. This follows from eq. (6.59) by noting that $-0.5\rho(u^2+v^2)$ is nonlinear. Because the ambient pressure p_a does not contribute to integrated loads on the foil, we disregard p_a. This gives $p = -\rho U u$ or that

$$p^{\pm} = \pm\rho U \frac{\gamma(x)}{2}. \quad (6.85)$$

The lift force on the foil is then

$$L = -\rho U \int_{-c/2}^{c/2} \gamma(\xi)\, d\xi \equiv -\rho U \Gamma. \quad (6.86)$$

This is consistent with eq. (6.62). Using eqs. (6.77), (6.78), and (6.84) gives

$$\begin{aligned} L &= -\rho U \int_0^{\pi} \gamma 0.5c \sin\theta \, d\theta, \\ &= -2\rho U^2 \pi (a_0 + 0.5a_1) 0.5c \end{aligned} \quad (6.87)$$

that is,

$$L = -2\rho U^2 \left(\frac{c}{2}\right) \int_0^{\pi} \frac{d\eta}{d\xi}(1-\cos\theta)\, d\theta.$$

We now use eq. (6.77) once more and that $d\xi = c/2 \sin\theta d\theta = \sqrt{(c/2)^2 - \xi^2} d\theta$. This implies

$$L = -2\rho U^2 \int_{-c/2}^{c/2} \frac{d\eta}{d\xi}\left[\frac{c/2+\xi}{c/2-\xi}\right]^{1/2} d\xi. \quad (6.88)$$

The hydrodynamic moment on the foil about $x = 0$ can be expressed as

$$M = -\rho U \int_{-c/2}^{c/2} \gamma(\xi)\xi \, d\xi \qquad (6.89)$$

by means of eq. (6.85). This can be rewritten in a similar way as for the lift, that is,

$$M = \rho U \left(c/2\right)^2 \int_0^{\pi} \gamma \cos\theta \sin\theta \, d\theta$$

$$= \rho 2U^2 \left(\frac{c}{2}\right)^2 \cdot \frac{\pi}{2}\left(a_0 + \frac{1}{2}a_2\right). \qquad (6.90)$$

Introducing eq. (6.84) now gives

$$M = 2\rho U^2 \int_{-c/2}^{c/2} \frac{d\eta}{d\xi}[(c/2)^2 - \xi^2]^{1/2} \, d\xi. \qquad (6.91)$$

The *center of pressure*, x_{cp} is the ratio $x_{cp} = M/L$, that is, the ratio between the hydrodynamic moment M about $x = 0$ and the lift force L.

6.6.1 Flat plate

A flat plate with angle of attack α is considered, that is, $d\eta/d\xi = -\alpha$ in eqs. (6.88) and (6.91). The integrals are evaluated by introducing a new variable θ so that

$$\xi = 0.5c \sin\theta. \qquad (6.92)$$

This gives

$$L = 2\rho U^2 \alpha \int_{-\pi/2}^{\pi/2} \frac{c/2 + (c/2)\sin\theta}{[(c/2)^2 - (c/2)^2 \sin^2\theta]^{1/2}} \times (c/2)\cos\theta \, d\theta$$

or

$$L = \rho U^2 c \pi \alpha. \qquad (6.93)$$

We could, of course, also use the transformation given by eq. (6.92) to facilitate the integration. The lift coefficient is

$$C_L = \frac{L}{0.5\rho U^2 c} = 2\pi\alpha. \qquad (6.94)$$

The moment M can be expressed as

$$M = -2\rho U^2 \alpha \left(0.5c\right)^2 \int_{-0.5\pi}^{0.5\pi} \cos^2\theta \, d\theta$$

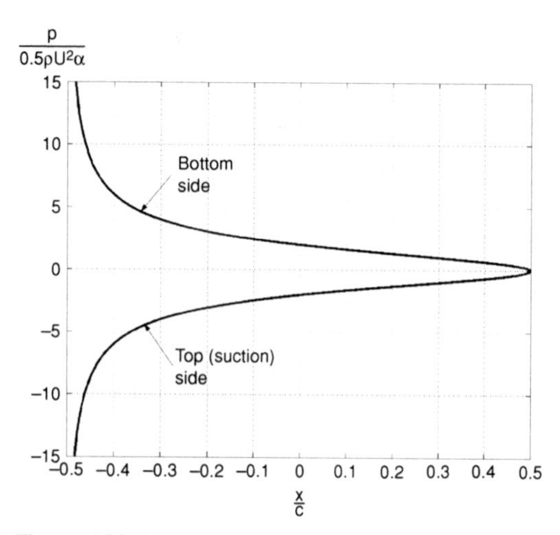

Figure 6.38. Linear pressure distribution on the top (suction) and bottom (pressure) sides of a flat plate with a positive angle of attack α. c = chord length. Ambient pressure is not included.

or

$$M = -0.25\rho U^2 c^2 \pi\alpha. \qquad (6.95)$$

The moment coefficient is then

$$C_M = \frac{M}{0.5\rho U^2 c^2} = -\frac{\pi}{2}\alpha. \qquad (6.96)$$

This gives that the center of pressure is $0.75c$ from the trailing edge.

The vortex distribution can be found by eqs. (6.78) and (6.84). It follows from eq. (6.84) that $a_0 = -\alpha$ and $a_n = 0$ for $n \geq 1$. Eq. (6.78) with x as a variable instead of ξ gives then that

$$\gamma = -2U\alpha\frac{1 + \cos\chi}{\sin\chi} = -2U\alpha\frac{0.5c - x}{\sqrt{(0.5c)^2 - x^2}},$$

that is,

$$\gamma(x) = -2\alpha U \left[\frac{0.5c - x}{0.5c + x}\right]^{1/2}. \qquad (6.97)$$

We can then use eq. (6.85) to express the pressure on the top and bottom of the foil, that is,

$$\frac{p^{\pm}}{0.5\rho U^2 \alpha} = \mp 2\left[\frac{0.5c - x}{0.5c + x}\right]^{1/2}. \qquad (6.98)$$

This is illustrated in Figure 6.38. The pressure is infinite at the leading edge. However, the linearized solution is a good approximation of the pressure distribution away from the close vicinity of the leading edge. The flow can be corrected and made finite at the leading edge (Lighthill 1951).

The pressure given by eq. (6.98) is integrably singular at $x = -c/2$, so the forces and moments are finite as we have already seen by eqs. (6.93) and (6.95).

6.6.2 Foil with angle of attack and camber

The camber $\eta_f(x)$ of a foil section is measured relative to the nose-tail line (see Figure 6.3). The maximum f of $\eta_f(x)$ is called the physical camber. If we represent the camber by a parabola and the foil has an angle of attack α relative to the nose-tail line, we can express the mean line of the foil as

$$\eta = -\alpha x + f \cdot \left(1 - \left(\frac{x}{0.5c}\right)^2\right). \tag{6.99}$$

Because the problem is linear, the effect of the angle of attack α and of the camber can be handled separately. We therefore set α equal to zero first. Using eq. (6.88) gives the lift force as

$$L = \frac{16\rho U^2 f}{c^2} \int_{-0.5c}^{0.5c} \xi \left[\frac{0.5c + \xi}{0.5c - \xi}\right]^{1/2} d\xi.$$

Introducing the θ-variable by eq. (6.92), we have

$$L = 4\rho U^2 f \int_{-0.5\pi}^{0.5\pi} \sin\theta(1 + \sin\theta)\, d\theta$$
$$= 2\rho U^2 f\pi.$$

The lift coefficient is then

$$C_L = \frac{L}{0.5\rho U^2 c} = 4\pi \frac{f}{c}. \tag{6.100}$$

This means, for instance, that a small camber ratio $f/c = 0.2/4\pi = 0.0159$ gives a lift coefficient as large as 0.2 when the angle of attack is zero. Accounting for both the angle of attack α and camber gives then

$$L = \rho U^2 c\pi\,(\alpha + 2f/c), \tag{6.101}$$

where eq. (6.93) has been used for the angle of attack effect. This means zero lift occurs when $\alpha = \alpha_0 = -2f/c$. A parabolic camber will not contribute to the moment. Further, the pressure due to the parabolic camber will be finite everywhere along the foil.

6.6.3 Ideal angle of attack and angle of attack with zero lift

An angle of attack is called *ideal* if the linear theory predicts finite pressure at the leading edge. This is also called *shockless entry*. The constant a_0 in eq. (6.84) is zero at the ideal angle of attack α_i. This leads to a vortex distribution (see eq. (6.78)) and a pressure distribution (see eq. (6.85)) that are finite everywhere. Because large negative pressure implies the possibility of cavitation, the ideal angle of attack is important in minimizing the risk for cavitation inception. If we express the mean line of the foil as $\eta = -\alpha x + \eta_f(x)$, then eq. (6.84) gives

$$\alpha_i = \frac{1}{\pi}\int_0^{\pi} \eta'_f(x)\,d\chi \tag{6.102}$$

for a foil in infinite fluid. Here $x = -0.5c\cos\chi$. The ideal angle of attack is therefore zero for a foil with a camber that is even in x. The previous case with a parabolic camber is an example of that.

The angle of attack α_0 that causes zero lift follows from eqs. (6.87) and (6.84), that is,

$$\alpha_0 = \frac{1}{\pi}\int_0^{\pi} \eta'_f(x)\,(1 - \cos\chi)\,d\chi. \tag{6.103}$$

The linear lift coefficient can then be expressed as

$$C_L = 2\pi(\alpha - \alpha_0). \tag{6.104}$$

We saw in Section 6.6.2 that $\alpha_0 = -2f/c$ for a foil with parabolic camber.

Let us consider as another example the mean line of the NACA 08 (modified) section (Abbott and von Doenhoff 1959). This is commonly used in the design of conventional propellers. The number *08* means that the vortex density γ is constant from the leading edge downstream over a distance 0.8 times the chord length. Using eqs. (6.102) and (6.103) gives $\alpha_i = 1.40^\circ$ and $\alpha_0 = -7.72^\circ$. The lift coefficient at the ideal angle of attack is $2\pi\,(1.4 + 7.72)\,(\pi/180) = 1$.

6.6.4 Weissinger's "quarter-three-quarter-chord" approximation

According to Weissinger's (1942) theory, the circulation Γ around a 2D foil in steady motion in infinite fluid can be determined by placing a vortex with strength Γ at one quarter of a chord length from the leading edge and satisfying the boundary

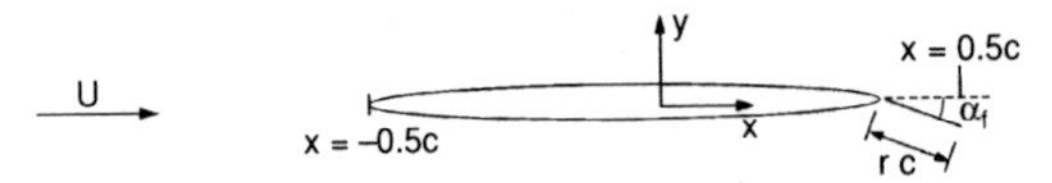

Figure 6.39. Foil with a flap of length rc and angle α_f.

condition at three quarters of a chord length from the leading edge. If this is applied to a foil with angle of attack and camber as described by eq. (6.99), it means that

$$\frac{\Gamma}{2\pi 0.5c} = -U(\alpha + 2f/c).$$

Using the Kutta-Joukowski formula for the lift, that is, eq. (6.62), gives the same result as eq. (6.101).

6.6.5 Foil with flap

The foil with flap illustrated in Figure 6.39 is considered. The camber is zero. The flap has an angle of attack α_f and a length rc, where c is the chord length including the flap. The other part of the foil has a zero angle of attack. Using eq. (6.88) to calculate the lift gives

$$L = 2\rho U^2 \alpha_f \int_{0.5c-rc}^{0.5c} \left[\frac{0.5c+\xi}{0.5c-\xi}\right]^{1/2} d\xi. \quad (6.105)$$

We introduce θ as an integration variable by means of eq. (6.92). When $\xi = 0.5c - rc$, the corresponding angle θ is $0.5\pi - \theta_f$, that is, $0.5c - rc = 0.5c\sin(0.5\pi - \theta_f)$ or

$$r = 0.5(1 - \cos\theta_f). \quad (6.106)$$

Eq. (6.105) can then be expressed as

$$L = \rho U^2 \alpha_f c \int_{0.5\pi-\theta_f}^{0.5\pi} (1 + \sin\theta)\, d\theta$$

$$= \rho U^2 \alpha_f c\, (\theta_f + \sin\theta_f). \quad (6.107)$$

It follows from eq. (6.106) that $r \approx 0.25\theta_f^2$ for small r. Eq. (6.107) gives then

$$L \approx 4\rho U^2 c r^{1/2} \alpha_f. \quad (6.108)$$

We can introduce an efficiency coefficient η_f that expresses the foil capability to provide lift relative to a flat plate with an angle of attack α_f. Using eqs. (6.93) and (6.107) gives

$$\eta_f = \frac{\rho U^2 \alpha_f c\,(\theta_f + \sin\theta_f)}{\rho U^2 \alpha_f c\pi} = \frac{\theta_f + \sin\theta_f}{\pi}. \quad (6.109)$$

Applying the asymptotic formula for small r, that is, eq. (6.108), we obtain

$$\eta_f = \frac{4}{\pi} r^{1/2} \quad \text{for small } r. \quad (6.110)$$

Eqs. (6.109) and (6.110) are plotted in Figure 6.40 showing that eq. (6.110) is a reasonable approximation for $r < \approx 0.2$. Further, using, for instance, $r = 0.15$ gives, respectively, $\eta_f = 0.48$ and 0.49 for the exact and asymptotic formula. This illustrates that a flap is an efficient means of generating lift. It is not possible to use Weissinger's "quarter-three-quarter-chord" approximation to describe the effect of a flap. This would lead, for instance, to zero lift force when the flap length is less than one quarter of the chord length.

Using eq. (6.91) to calculate the hydrodynamic moment about $x = 0$ gives

$$M = -2\rho U^2 \alpha_f \int_{0.5c-rc}^{0.5c} ((0.5c)^2 - \xi^2)^{1/2} d\xi.$$

Once more introducing θ as an integration variable, we get

$$M = -0.5\rho U^2 \alpha_f c^2 \int_{0.5\pi-\theta_f}^{0.5\pi} \cos^2\theta\, d\theta \quad (6.111)$$

$$= -0.25\rho U^2 \alpha_f c^2 (\theta_f - 0.5\sin 2\theta_f).$$

Assuming θ_f is small and relating θ_f to r gives

$$\theta_f - 0.5\sin 2\theta_f \approx \theta_f - 0.5\left(2\theta_f - \frac{(2\theta_f)^3}{6}\right)$$

$$= \frac{4^2}{3} r^{3/2},$$

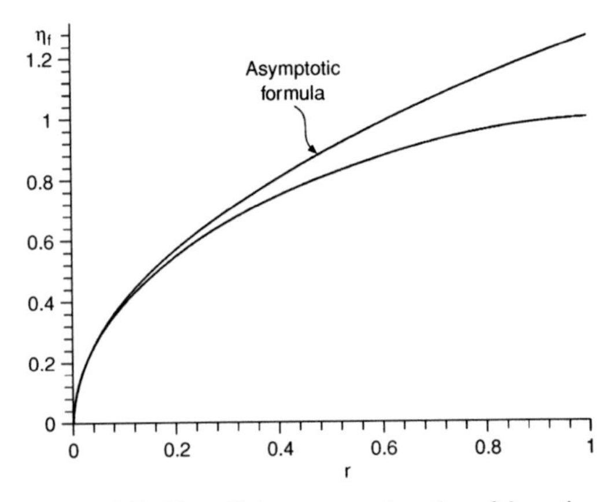

Figure 6.40. Flap efficiency η_f as a function of the ratio r between flap length and chord length.

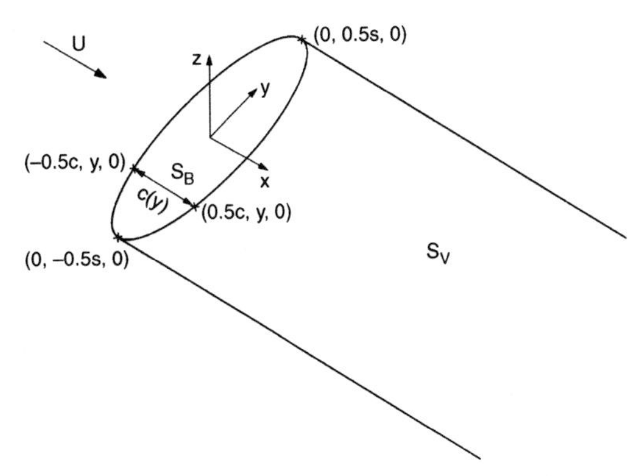

Figure 6.41. Projections of the foil surface S_B and vortex sheet S_V on the xy-plane. These are used in the solution of the 3D linearized problem.

that is,

$$M = -\frac{4}{3}\rho U^2 c^2 r^{3/2} \alpha_f. \tag{6.112}$$

The center of pressure x_{CP} is then

$$x_{CP} = \frac{M}{L} = -\frac{rc}{3}. \tag{6.113}$$

Because rc is assumed small, x_{CP} is close to $x = 0$. It implies that we cannot simply handle the flap as an independent appended foil to the rest of the foil. The center of pressure would then be $0.75rc$ from the trailing edge of an uncambered flap.

6.7 3D linear steady flow past a foil in infinite fluid

Eq. (6.41) is a representation of the velocity potential φ for 3D nonlinear flow past a foil in infinite fluid. The problem will be linearized similarly to that in the 2D case. This is illustrated in Figure 6.41, in which the vortex sheet S_V and the foil surface S_B have been projected on the xy-plane.

The foil has a span s and a local chord length $c(y)$. The incident flow velocity U is along the positive x-axis. We will change the notation in eq. (6.41) so that the integration variable (x, y, z) is replaced by (ξ, η, ζ) and the field point (x_1, y_1, z_1) is exchanged with (x, y, z). We call I_W the linearized integral over S_V^+ in eq. (6.41). This can be written as

$$I_W = \frac{z}{4\pi}\int_{-0.5s}^{0.5s} d\eta \int_{0.5c}^{x_E} \frac{[\varphi^+ - \varphi^-]d\xi}{[(x-\xi)^2 + (y-\eta)^2 + z^2]^{3/2}}. \tag{6.114}$$

Here x_E corresponds to the x-coordinate of the longitudinal extent of the vortex sheet in the downstream direction. This is assumed independent of y.

The linearized zero pressure jump condition over the vortex sheet follows from eq. (6.18) by neglecting the time-dependent part, setting the convection velocity $\boldsymbol{u}_c$ given by eq. (6.19) equal to $U\boldsymbol{i}$, and transferring eq. (6.18) to $z = 0$. This means

$$\frac{\partial}{\partial x}(\varphi^+ - \varphi^-) = 0 \quad \text{on } z = 0 \quad \text{on } S_V. \tag{6.115}$$

$[\varphi^+ - \varphi^-]$ in eq. (6.114) can then be replaced with its value at the trailing edge *(T.E.)*. Further, in steady conditions, the vortex sheet extends to $x_E = \infty$. This gives

$$I_W = \frac{z}{4\pi}\int_{-0.5s}^{0.5s} d\eta [\varphi^+ - \varphi^-]_{T.E.} \times \int_{0.5c}^{\infty} \frac{d\xi}{[(x-\xi)^2 + (y-\eta)^2 + z^2]^{3/2}}$$
$$= \frac{z}{4\pi}\int_{-0.5s}^{0.5s} d\eta \frac{[\varphi^+ - \varphi^-]_{T.E.}}{[(y-\eta)^2 + z^2]} \times \left\{1 + \frac{x - 0.5c(\eta)}{[(x - 0.5c(\eta))^2 + (y-\eta)^2 + z^2]^{1/2}}\right\}. \tag{6.116}$$

6.7.1 Prandtl's lifting line theory

We now consider Prandtl's lifting line theory. This has been presented, for instance, by Newman (1977) but a different approach will be followed here. Prandtl's lifting line theory assumes a high-aspect–ratio foil, that is, the span length is much larger than the maximum chord length. Lifting line theory is also used in the analysis of propellers (Breslin and Andersen 1994). However, because the shed vorticity behind a propeller is not located in a flat plane, as in the following analysis, the procedure becomes more complicated. However, the principle behind the analysis for a propeller is the same as described below.

We introduce far-field and near-field descriptions of the flow around the foil. The details of the transverse foil dimensions are not seen in the far-field view. The foil appears as a straight line along the y-axis between $y = -0.5s$ and $0.5s$ (Figure 6.42). The consequence is that in the far-field description of I_W, we can set $c(\eta) = 0$ in eq. (6.116).

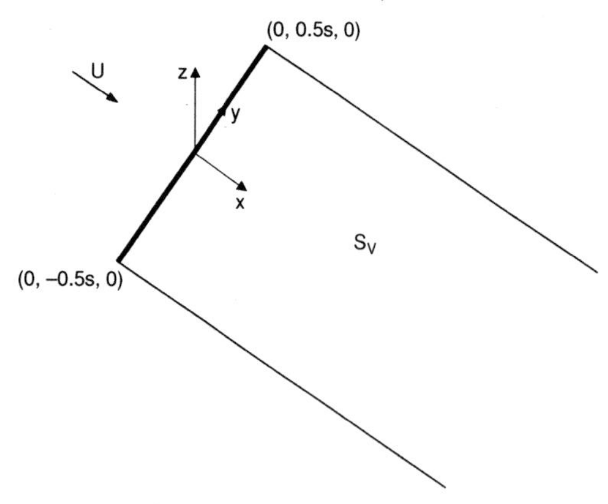

Figure 6.42. Far-field view of a high-aspect–ratio foil. This means we do not see the details of the transverse dimensions of the foil. Otherwise, the explanations are as in Figure 6.41. The foil appears to be a straight line in the y-axis between $y = -0.5\,s$ and $0.5\,s$.

The details of the transverse dimensions of the foil are seen in the near-field description. The flow is locally two-dimensional in the transverse cross-sectional xz-plane (Figure 6.43). The effect of the far-field description is a vertical inflow velocity w_i that changes the angle of attack of the incident flow. One should note the fact that the calculated w_i is negative; therefore, the induced velocity is downward in Figure 6.43. This reduces the angle of attack. The far-field version of eq. (6.116) is used to calculate w_i. This means both x and c are set equal to zero in eq. (6.116). I_W can be re-expressed by using partial integration and noting that

$$\frac{d}{d\eta}\tan^{-1}\left(\frac{z}{y-\eta}\right) = \frac{z}{(y-\eta)^2 + z^2}. \qquad (6.117)$$

Using condition $\varphi^+ - \varphi^- = 0$ for $y = \pm 0.5s$, we have

$$I_W = -\frac{1}{4\pi}\int_{-0.5s}^{0.5s} \frac{d}{d\eta}[\varphi^+ - \varphi^-]_{T.E.}\tan^{-1}\frac{z}{y-\eta}\,d\eta \quad \text{for } x = c = 0. \qquad (6.118)$$

The induced vertical velocity w_i is then obtained by differentiating eq. (6.118) with respect to z and setting $z = 0$. Further, $[\varphi^+ - \varphi^-]_{T.E.}$ can be related to the circulation $\Gamma(y)$ around the foil in the 2D flow description in Figure 6.43 (see eq. (6.15) and Figure 6.25). This gives

$$w_i = \frac{1}{4\pi} PV \int_{-0.5s}^{0.5s} \frac{d\Gamma}{d\eta}\frac{d\eta}{y-\eta}. \qquad (6.119)$$

There is a circulation Γ_{2D} around the foil when w_i is zero. However, w_i causes an angle of attack $\alpha_i = w_i/U$ of the incident flow, and this changes the circulation around the foil. Using eqs. (6.62) and (6.93) gives the change in circulation due to α_i as $-Uc\pi\alpha_i$. This gives the following equation:

$$\Gamma(y) = \Gamma_{2D}(y) - \pi c(y) w_i;$$

that is, noting signs, we write

$$\Gamma(y) = \Gamma_{2D}(y) + \frac{1}{4}c(y)PV \times \int_{-0.5s}^{0.5s} \frac{d\Gamma(\eta)}{d\eta}\frac{d\eta}{\eta - y}, \quad -0.5s < y < 0.5s. \qquad (6.120)$$

Eq. (6.120) can be rewritten by introducing the new variables θ and θ' defined as

$$y = \frac{s}{2}\cos\theta \qquad (6.121)$$

and

$$\eta = \frac{s}{2}\cos\theta'. \qquad (6.122)$$

This means that $y = \pm\frac{s}{2}$ corresponds to $\theta = 0$ and $\theta = \pi$, respectively. Γ can be expressed as the following Fourier series:

$$\Gamma = 2Us\sum_{n=1}^{\infty} a_n \sin n\theta. \qquad (6.123)$$

Because Γ is zero for $\theta = 0$ and π, Fourier series terms with $\cos n\theta$ cannot be included. Introducing the Fourier series representation gives

$$PV\int_{-0.5s}^{0.5s}\frac{d\Gamma(\eta)}{d\eta}\frac{d\eta}{\eta - y} = -4U\sum_{n=1}^{\infty} n a_n PV\int_0^{\pi}\frac{\cos n\theta' d\theta'}{\cos\theta' - \cos\theta}. \qquad (6.124)$$

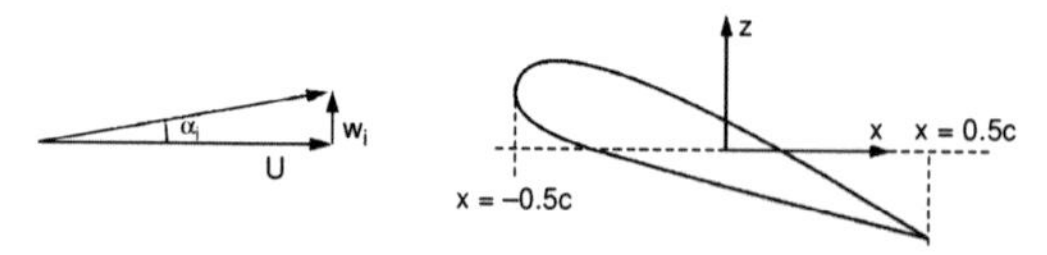

Figure 6.43. Near-field view of the steady flow past a high-aspect–ratio foil in infinite fluid. The far-field 3D flow described by Figure 6.42 causes a vertical inflow velocity w_i that changes the angle of attack of the incident flow. Note that the calculated w_i is negative.

The Glauert integrals appearing in eq. (6.124) can be evaluated by eq. (6.81). Eq. (6.120) can then be expressed as

$$\sum_{n=1}^{\infty} a_n \sin n\theta = \frac{1}{2Us}\Gamma_{2D}\left(\frac{1}{2}s\cos\theta\right)$$

$$-\frac{\pi}{2s}c\left(\frac{1}{2}s\cos\theta\right)\sum_{n=1}^{\infty} na_n\frac{\sin n\theta}{\sin\theta}, 0 < \theta < \pi. \quad (6.125)$$

A linear equation system with N unknowns a_n follows by satisfying eq. (6.125) for N different θ-values. However, an analytical solution can be found if an uncambered lifting surface with elliptical planform is considered. In this case,

$$c = c_0 \sin\theta, \quad (6.126)$$

where c_0 is the chord at midspan. Further,

$$\Gamma_{2D} = -Uc\pi\alpha, \quad (6.127)$$

where α is the geometrical angle of attack. Using eq. (6.125) then gives

$$a_1 = -\frac{\pi\alpha c_0}{2s}\left(1+\frac{\pi c_0}{2s}\right)^{-1}. \quad (6.128)$$

This can be re-expressed in terms of the aspect ratio $\Lambda = s^2/A$, where $A = 0.25\pi c_0 s$ is the projected area of the elliptical foil on the xy-plane. This gives $a_1 = -2\alpha/(\Lambda+2)$. The lift force L on the foil is then

$$L = -\rho U \int_{-s/2}^{s/2} \Gamma(y)dy$$

$$= \rho U 2Us\frac{2\alpha}{\Lambda+2}\cdot\frac{s}{2}\int_0^{\pi}\sin^2\theta\, d\theta. \quad (6.129)$$

$$= \rho U^2 s^2\frac{\alpha\pi}{\Lambda+2}$$

The lift coefficient C_L is defined as $L/(0.5\rho U^2 A)$, where $A = s^2/\Lambda$, that is,

$$C_L = \frac{2\pi\alpha}{(1+2/\Lambda)}, \quad (6.130)$$

where $\Lambda = 4s/\pi c_0$. Eq. (6.130) is consistent with the 2D result $2\pi\alpha$ when $\Lambda \to \infty$. However, when $\Lambda \to 0$, the result $\pi\Lambda\alpha$ is inconsistent with low-aspect–ratio theory, the value predicted by eq. (6.130) being twice the correct value. However, generally speaking, eq. (6.130) gives a good indication of 3D effects. Lifting line theory gives conservative estimates, and by using it, the relative error is 5% at $\Lambda = 8$, 10% at $\Lambda = 4$, and 20% at $\Lambda = 2$.

In 1942, the German aerodynamicist H. B. Helmbold modified eq. (6.130) to give improved predictions for low-aspect–ratio straight wings (Anderson 2001). The expression is

$$C_L = \Lambda\frac{2\pi\alpha}{2+\sqrt{\Lambda^2+4}}. \quad (6.131)$$

When $\Lambda \to 0$, we get $C_L = 0.5\pi\Lambda\alpha$. This is consistent with low-aspect–ratio theory (see section 10.3.1). Søding (1982) presented the formula

$$C_L = \frac{\Lambda(\Lambda+1)}{(\Lambda+2)^2}2\pi\alpha, \quad (6.132)$$

which agrees better with the 3D results than Helmbold's formula does.

Kuchemann (1978) presents the following modification to Helmbold's equation for swept wings:

$$C_L = \Lambda\frac{2\pi\alpha\cos\gamma}{\sqrt{\Lambda^2+4\cos^2\gamma}+2\cos\gamma}. \quad (6.133)$$

Here γ means the sweep angle of the wing at the half–chord line (see Figure 6.3, in which the sweep angle is defined at the quarter–chord line).

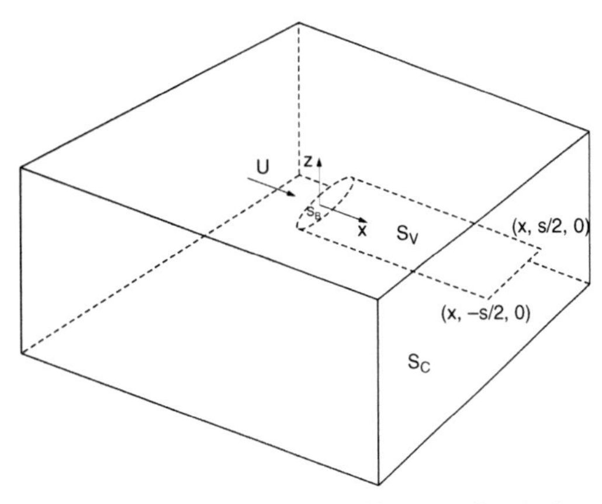

Figure 6.44. Control volume used in expressing the drag on a foil by conservation of fluid momentum.

6.7.2 Drag force

Drag force on the foil is derived by using conservation of fluid momentum. A fluid volume Ω exterior to the foil surface and the trailing vortex sheet is considered. Ω is bounded far away from the foil by the surfaces of a box with sides that are parallel to either the xy-, xz-, or yz-plane (Figure 6.44).

Eq. (6.116) will be used to evaluate the fluid velocity due to the vortex sheet. At the control surface S_c that is perpendicular to the vortex sheet far downstream of the foil, we can write

$$I_W = \frac{z}{2\pi} \int_{-0.5s}^{0.5s} d\eta \frac{[\varphi^+ - \varphi^-]_{T.E.}}{[(y-\eta)^2 + z^2]}$$
$$\text{for } x - c/2 \to +\infty.$$

This can be rewritten in a way similar to that used for eq. (6.119), that is,

$$I_W = \frac{1}{2\pi} \int_{-0.5s}^{0.5s} \frac{d\Gamma}{d\eta} \tan^{-1} \frac{z}{y-\eta} d\eta$$
$$\text{for } x - c/2 \to +\infty. \tag{6.134}$$

Here we have used $\Gamma = (\varphi^- - \varphi^+)_{T.E.}$. This shows that the flow due to the vortex sheet is two-dimensional in the yz-plane at S_c, which means that the longitudinal velocity is U at S_c. The momentum flux through the surface enclosing the control volume is therefore zero. The longitudinal pressure force acting on S_c is

$$\frac{\rho}{2} \iint_{S_c} \left[\left(\frac{\partial \varphi}{\partial y} \right)^2 + \left(\frac{\partial \varphi}{\partial z} \right)^2 \right] dy\, dz,$$

where φ is the same as that used in the expression of I_W. The only additional longitudinal force acting on the control volume is the force opposite the drag force D on the foil. This means

$$D = \frac{\rho}{2} \iint_{S_c} \left[\left(\frac{\partial \varphi}{\partial y} \right)^2 + \left(\frac{\partial \varphi}{\partial z} \right)^2 \right] dy\, dz. \tag{6.135}$$

This can be rewritten using the divergence theorem and by noting that $\nabla\varphi \cdot \nabla\varphi = \nabla \cdot (\varphi \nabla \varphi)$. By using the fact that $\partial\varphi/\partial z = \pm\partial\varphi/\partial n$ is continuous across the vortex sheet, the result is

$$D = \frac{\rho}{2} \oint \varphi \frac{\partial \varphi}{\partial n} dl = -\frac{1}{2}\rho \int_{-0.5s}^{0.5s} (\varphi^+ - \varphi^-) \frac{\partial \varphi}{\partial z} dy. \tag{6.136}$$

The vertical velocity $\partial\varphi/\partial z$ in eq. (6.136) follows by differentiating I_W given by eq. (6.134) with respect to z and then setting $z = 0$. This gives

$$D = \frac{\rho}{4\pi} \int_{-0.5s}^{0.5s} dy \Gamma(y) PV \int_{-0.5s}^{0.5s} \frac{d\Gamma}{d\eta} \frac{d\eta}{y-\eta}. \tag{6.137}$$

An uncambered lifting surface with elliptical planform is now considered. This means $\Gamma = 2Usa_1 \sin\theta$, where a_1 is given by eq. (6.128). Further, θ is defined by eq. (6.121). Using eqs. (6.81) and (6.124) gives then

$$D = \frac{\rho}{4\pi} \int_{-0.5s}^{0.5s} dy \Gamma(y) \pi 4Ua_1$$
$$= \rho U a_1 \int_0^{\pi} 2Usa_1 \sin\theta \frac{s}{2} \sin\theta\, d\theta$$
$$= \rho U^2 a_1^2 s^2 \frac{\pi}{2}$$

The corresponding drag coefficient is

$$C_D = \frac{D}{0.5\rho U^2 A} = \pi \Lambda a_1^2.$$

a_1 can be expressed as $-2\alpha/(\Lambda + 2)$, so

$$C_D = \frac{4\pi\alpha^2\Lambda}{(\Lambda+2)^2} = \frac{C_L^2}{\pi\Lambda} \tag{6.138}$$

for an uncambered lifting surface with elliptical planform. Letting $\Lambda \to \infty$, this gives $C_D = 0$, which is consistent with 2D results for a foil in infinite fluid.

The same result for the drag force can be obtained by using the Kutta-Joukowski formula for 2D flow. We consider, then, 2D flow as in Figure 6.43 and calculate $\alpha_i = w_i/U$, where w_i is given by eq. (6.119). Here $\Gamma = 2Usa_1 \sin\theta'$, where $a_1 = -2\alpha/(\Lambda + 2)$ and θ' is defined by eq. (6.122). This leads to $\alpha_i = -2\alpha/(\Lambda + 2)$, which is constant along the span of the foil. There is then an incident velocity with an angle α_i relative to the x-axis. The magnitude of the incident velocity can be approximated as U. According to the Kutta-Joukowski formula, the force acts perpendicularly to the inflow direction. The force magnitude can be approximated with the lift force L. The force acting along the x-axis is $-L\alpha_i$. Because α_i is negative, this is a drag force D. If α_i had been positive, it would have led to a thrust force. Using eq. (6.129) for the lift force gives

$$D = \rho U^2 s^2 \frac{\alpha\pi}{\Lambda+2} \frac{2\alpha}{\Lambda+2}$$

and

$$C_D = \frac{4\pi\alpha^2\Lambda}{(\Lambda+2)^2}.$$

This is the same result as eq. (6.138).

These results can be generalized to include the foil camber and a flap. The 2D results with parabolic camber showed that $C_L = 2\pi[\alpha - \alpha_0 + \eta_f \alpha_f]$. Here $\alpha_0 = -2f/c$ is the camber effect when the mean line of the foil is expressed

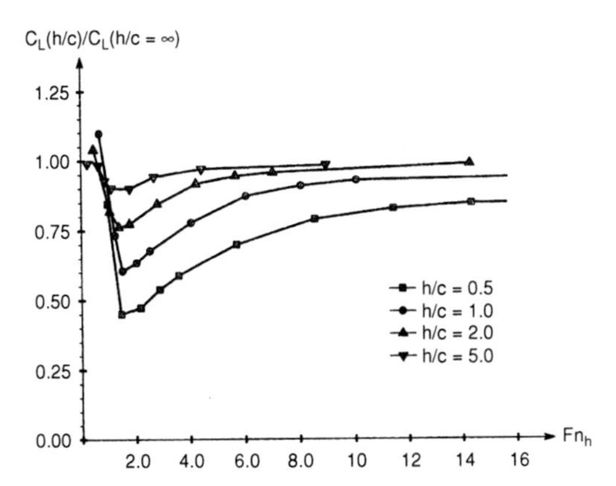

Figure 6.45. Free-surface effects on steady lift for an uncambered 2D thin foil at an angle of attack. C_L = lift coefficient, h = foil submergence, c = chord length, Fn_h = Froude number with h as length parameter (Hough and Moran 1969).

as in eq. (6.99). Further, α_f is the flap angle (see Figure 6.39) and η_f is the flap efficiency (see eq. (6.109)). The two-dimensional circulation Γ_{2D} in eq. (6.120) is then $-Uc\pi(\alpha - \alpha_0 + \eta_f \alpha_f)$. If $\alpha - \alpha_0 + \eta_f\alpha_f$ does not vary along the foil span, eqs. (6.130) and (6.138) for C_L and C_D can be generalized by replacing α with $\alpha - \alpha_0 + \eta_f\alpha_f$.

6.8 Steady free-surface effects on a foil

6.8.1 2D flow

We assume 2D steady flow and consider a thin flat foil at a submergence h below the mean free surface. Infinite water depth is assumed. The linearized lift force can be expressed as in eq. (6.86). However, the circulation Γ is influenced by the presence of the free surface. Figure 6.45 shows numerically predicted lift coefficient C_L as a function of submergence Froude number

$$Fn_h = \frac{U}{\sqrt{gh}} \tag{6.139}$$

for different values of h/c. These calculations are based on a linear body boundary condition and the linear free-surface condition

$$U^2\frac{\partial^2\varphi}{\partial x^2} + g\frac{\partial\varphi}{\partial z} = 0 \quad \text{on } z = 0,$$

where φ is the velocity potential due to the foil. This is a free-surface condition similar to the one used for the Green function G in eq. (4.27). The presence of the free surface causes generally lower C_L than for infinite fluid. An exception is small Froude numbers. The interesting Froude numbers will, in practice, be high. Consider, for instance, $U = 15\,\mathrm{ms}^{-1}$, $c = 1\,\mathrm{m}$ and $h = 2\,\mathrm{m}$. This gives $Fn_h = 3.4$ and about a 12% reduction in lift due to free-surface effects. However, Figure 6.45 shows that the influence of the free surface can be even higher for smaller submergences. The results are partly influenced by foil-generated free-surface waves. According to linear theory, these waves have a wavelength (see Chapter 4)

$$\lambda = \frac{2\pi}{g}U^2. \tag{6.140}$$

We know from linear wave theory that there is a negligible flow at a depth λ from the free surface. We can therefore say that an object at a larger depth than λ from the free surface causes negligible free-surface waves. This means that free-surface waves are generated when

$$Fn_h > \approx \frac{1}{\sqrt{2\pi}} \approx 0.4. \tag{6.141}$$

Consequently, Figure 6.45 shows large variations in C_L with increasing values of Fn_h up to approximately $10/\sqrt{h/c}$ or $U/\sqrt{gc} \approx 10$. When $Fn_h > 10/\sqrt{h/c}$, C_L is nearly independent of Fn_h. The same results can then be predicted by neglecting gravity and using the dynamic free-surface condition $\varphi = 0$, where φ is the velocity potential due to the foil. This is often referred to as the biplane approximation and implies that the fluid velocity due to the foil is vertical at the free surface. The case is discussed more, later in the text.

When the Froude number is very small, the free surface acts like a rigid wall. The problem is then similar to a lifting wing close to the ground. The wing-in-ground (WIG) effect causes the lift to increase with decreasing distance of the foil from the ground. The results in Figure 6.45 show this. A WIG vehicle, as presented in Figure 1.11, takes advantage of this effect.

The behavior when $Fn_h \to 0$ or $Fn_h > 10/\sqrt{h/c}$ follows by a simple analysis based on Weissinger's quarter-three-quarter-chord approximation (see section 6.6.4). We start out with $Fn_h \to 0$ and represent the flow due to the foil by two vortices with opposite circulations Γ. One vortex has a center in the foil at a distance $c/4$ from the leading edge. The other vortex has a center at the image point about the mean free surface (see Figure 6.46). Note that we have used the opposite

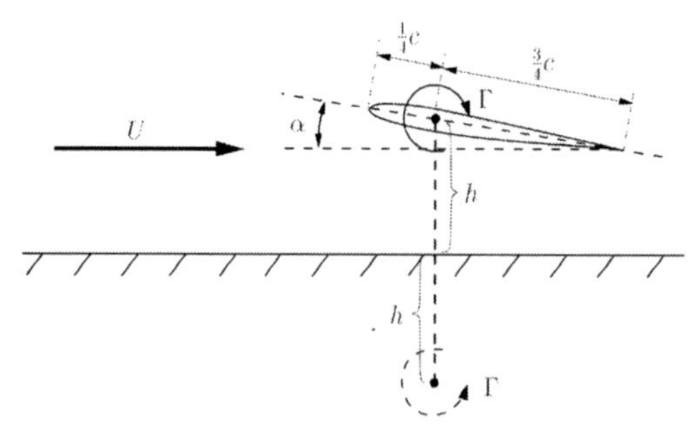

Figure 6.46. Approximate flow around a foil near the ground. Also representative for submerged foil near the free surface when the Froude number is asymptotically small.

circulation direction relative to the previous definition. The reason is that the sign of the circulation used in Figure 6.46 gives a direct illustration that the flow velocity increases on the suction side of the foil. Because the two vortices have opposite signs, the rigid free-surface condition is satisfied. According to the Weissinger approximation, the body boundary condition has to be satisfied only at one point, that is, at a distance 3c/4 from the leading edge. The image vortex effectively causes an increase in the angle of attack at this point. This means a higher C_L-value. The closer the image vortex, the higher C_L. We have earlier shown that the Weissinger approximation is appropriate for steady flow around a 2D foil with angle of attack and parabolic camber in infinite fluid. However, it does not apply to a foil with a flap. Let us examine whether it can account for the free-surface effect when $Fn_h \to 0$ for a foil with angle of attack and camber. The body boundary condition gives

$$\frac{\Gamma}{\pi c} - \frac{0.5c\Gamma}{2\pi(4h^2 + 0.25c^2)} = U(\alpha + 2f/c), \quad (6.142)$$

where α and f are defined by eq. (6.99). Solving eq. (6.142) for Γ and using the Kutta-Joukowski formula (see eq. (6.62)), we can evaluate the lift on the foil. The result of the analysis is that the lift coefficient can be written as

$$C_L\left(\frac{h}{c}\right) = C_L\left(\frac{h}{c} = \infty\right)\left[1 + \frac{1}{16}\left(\frac{c}{h}\right)^2\right]$$
$$\text{when } Fn_h \to 0. \quad (6.143)$$

Eq. (6.143) shows the correct trend that the lift increases strongly for small h/c-values, but the infinite predicted lift for $h/c = 0$ must obviously be wrong. The ratio between C_L and its infinite fluid value in the case of the lowest Fn_h-value in Figure 6.45 is, respectively, 1.1 and 1.05 for $h/c = 1.0$ and 2.0. The corresponding values by using eq. (6.143) are 1.06 and 1.02.

When $Fn_h > 10/\sqrt{h/c}$, the only difference in the analysis is that the two vortices have the same sign (Figure 6.47). This ensures that the dynamic free-surface condition $\varphi = 0$ is satisfied. The analysis gives

$$C_L\left(\frac{h}{c}\right) = C_L\left(\frac{h}{c} = \infty\right) \cdot \left[\frac{1 + 16(h/c)^2}{2 + 16(h/c)^2}\right]$$
$$\text{when } Fn_h > 10/\sqrt{h/c}. \quad (6.144)$$

This illustrates that C_L decreases with decreasing distance of the foil from the free surface. Eq. (6.144) predicts $C_L/C_L(h/c = \infty)$ equal to 0.83 and 0.94 for, respectively, $h/c = 0.5$ and 1.0. This agrees reasonably with Figure 6.45. When $h/c = 0$, eq. (6.144) gives $C_L/C_L(h/c = \infty) = 0.5$. This is consistent with linear theory for a planing foil on the free surface. In reality, cavitation and ventilation may happen to a foil at high Froude number very close to the free surface (see discussion associated with eq. (6.2)). This may cause a significant reduction of the lift. C_L of a supercavitating thin flat foil in infinite fluid is $0.5\pi\alpha$, whereas the corresponding value with no cavitation and ventilation is $2\pi\alpha$.

Kochin et al. (1964) have given expressions for the wave resistance. By using linear theory, we can divide the flow into a thickness and lifting problem, which can be solved by using, respectively, source (sink) and vortex distributions along the foil (see Figure 6.37). Both effects cause wave resistance. We consider first the lifting problem. The wave resistance can be expressed as

$$R_{W\Gamma} = \frac{\rho g \Gamma^2}{U^2} \exp(-2/Fn_h^2). \quad (6.145)$$

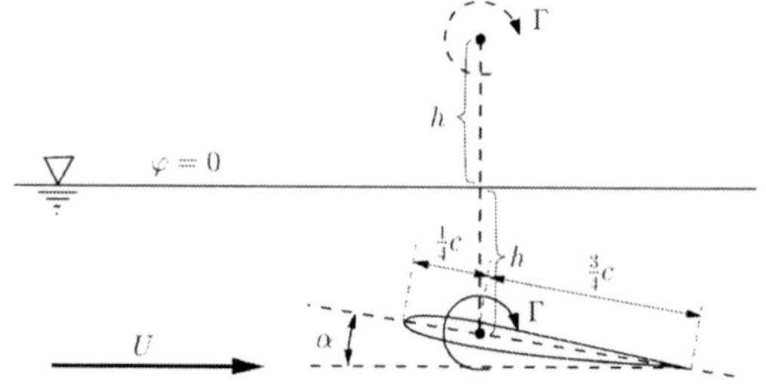

Figure 6.47. Approximate flow around a foil below the free surface when the Froude number is very high.

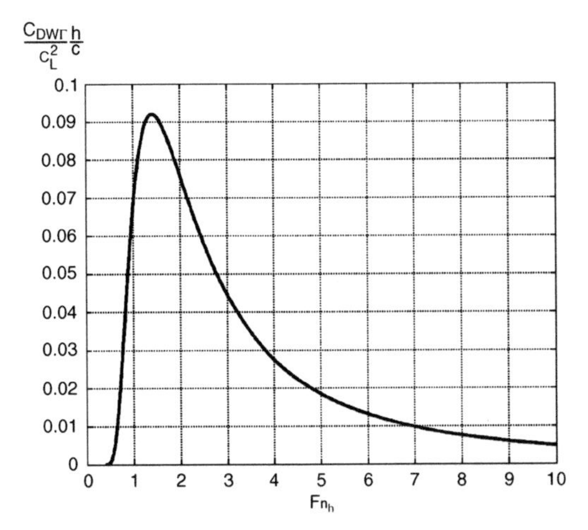

Figure 6.48. Wave resistance due to lift effects on a 2D foil with submergence h and chord length c, expressed in terms of drag coefficient $C_{DW\Gamma}$ as a function of submergence Froude number Fn_h. C_L is the lift coefficient for the foil.

This means

$$C_{DW\Gamma} = \frac{R_{W\Gamma}}{0.5\rho U^2 c} = \frac{g\Gamma^2}{0.5U^4 c}\exp(-2/Fn_h^2). \tag{6.146}$$

We introduce now the lift coefficient $C_L = L/\left(0.5\rho U^2 c\right)$ and evaluate the lift force L by eq. (6.86). This means

$$C_L^2 = \frac{\Gamma^2}{0.25U^2c^2}. \tag{6.147}$$

Combining eqs. (6.146) and (6.147) gives

$$\frac{C_{DW\Gamma}}{C_L^2} = \frac{0.5}{Fn_h^2}\left(\frac{c}{h}\right)\exp\left(-2/Fn_h^2\right). \tag{6.148}$$

This expression is plotted in Figure 6.48. The maximum value of $C_{DW\Gamma}/C_L^2$ occurs at $Fn_h = \sqrt{2}$ and is equal to $0.25e^{-1}c/h = 0.092c/h$.

The thickness of a foil also causes a wave resistance component R_{WS}. This is similar to what we described in Chapter 4 for ships. Kochin et al. (1964) have presented a formula for a 2D foil with elliptical cross section and without lift. The formula is

$$R_{WS} = \pi^2\rho g t^2 \frac{c+t}{c-t} e^{-2/Fn_h^2} J_1^2 \times \left[\frac{c}{2h}\frac{1}{Fn_h^2}\left(1-\left(\frac{t}{c}\right)^2\right)^{1/2}\right]. \tag{6.149}$$

Here, t is the thickness of the foil and J_1 is the Bessel function of the first kind and order one (Abramowitz and Stegun 1964). Eq. (6.149) is presented in Figure 6.49 as a function of submergence Froude number Fn_h for difference ratios between the submergence h and the chord length c. The calculations are done for $t/c = 0.075$, but $R_{WS}/\rho g t^2$ is not very sensitive to t/c.

In order to quantify the importance of R_{WS}, we have compared it with the viscous resistance R_V obtained by using eq. (2.90). The chord length and the submergence were both chosen as 1 m, that is, $h/c = 1$. In the calculations, the foil thickness is 0.075 m and the kinematic viscosity coefficient is $1.35 \times 10^{-6}\,\text{m}^2\text{s}^{-1}$. The ratio $(R_{WS} + R_V)/R_V$ is 1.026 for $U = 5\,\text{ms}^{-1}$ and decreases rapidly with increasing U. This indicates that wave resistance due to thickness effect is small. The same is not true for wave resistance due to lift effects. It is not rigorous to separate the wave resistance due thickness and lifting effects even if we can separate the flow due to thickness and lifting effects. The reason is that the wave resistance can be related to the square of the downstream wave amplitude generated by the foil. This gives interaction terms. However, in the following text, we will assume that the wave resistance due to lifting effects dominates and neglect the thickness effect.

A consequence of wave resistance R_W is that the foil generates steady regular waves with amplitude A far downstream of the foil. By using energy conservation (Newman 1977), it can be shown that

$$R_W = \frac{\rho g}{4}A^2. \tag{6.150}$$

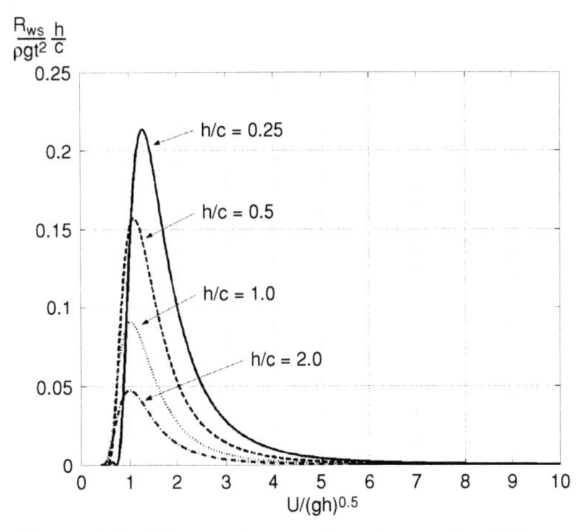

Figure 6.49. Wave resistance R_{WS} due to the thickness effect of a 2D foil with elliptical cross section at submergence h. c = chord length, t = foil thickness, t/c = 0.075.

6.8.2 3D flow

The free surface will affect the 3D steady flow around a foil. The free-surface effect on the lift has been presented earlier for 2D steady flow (see Figure 6.45).

Eqs. (6.143) and (6.144) can be combined with the lifting line theory to account for the free surface when $Fn_h \to 0$ and $Fn_h \to \infty$. If high Froude number is considered, Γ_{2D} given by eq. (6.127) can be generalized to $-Uc\pi\alpha(1+16(h/c)^2)/(2+16(h/c)^2)$. This implies that C_L of an elliptical planform also satisfies eq. (6.144). This gives that $C_L/C_L(h/c=4)$ is 0.67, 0.84, and 0.95 for, respectively, $h/c = 0.25, 0.5,$ and1.0. This is in reasonable agreement with the experimental and theoretical results by van Walree (1999) for large Fn_h. He studied a rectangular foil with aspect ratio 6; however, his small Fn_h results do not agree so well with the asymptotic results for an elliptical planform when $Fn_h \to 0$.

The formula for the drag coefficient C_D of a horizontal foil with elliptic loading based on potential flow has been presented by Breslin (1957). Linearized free-surface conditions are assumed. When the Froude number is asymptotically large, we can write

$$\frac{C_D}{C_L^2} = \frac{1}{\pi\Lambda} + \frac{\sigma(\lambda)}{\pi\Lambda}, \quad Fn_h \to \infty, \quad (6.151)$$

where

$$\sigma(\lambda) = 1 - \frac{4}{\pi}\lambda\sqrt{1+\lambda^2}\left[K\left(\frac{1}{\sqrt{1+\lambda^2}}\right) - E\left(\frac{1}{\sqrt{1+\lambda^2}}\right)\right] \quad (6.152)$$

and $\lambda = 2h/s$, with s as the foil span. Here K and E are complete elliptic integrals of the first and second kind. The aspect ratio Λ for an elliptical foil can be expressed as

$$\Lambda = \frac{4}{\pi}\frac{s}{c_o}, \quad (6.153)$$

where c_o is maximum chord length. We can derive eq. (6.151) for the drag force by conservation of fluid momentum and by following a procedure similar to the one in section 6.7.2 for a foil in infinite fluid. The upper surface of the control volume will coincide with the mean free surface. The drag force D can be expressed as in eq. (6.135). The difference now is that the vertical control surfaces extend only up to the mean free surface. By using the divergence theorem as in eq. (6.136), it follows that

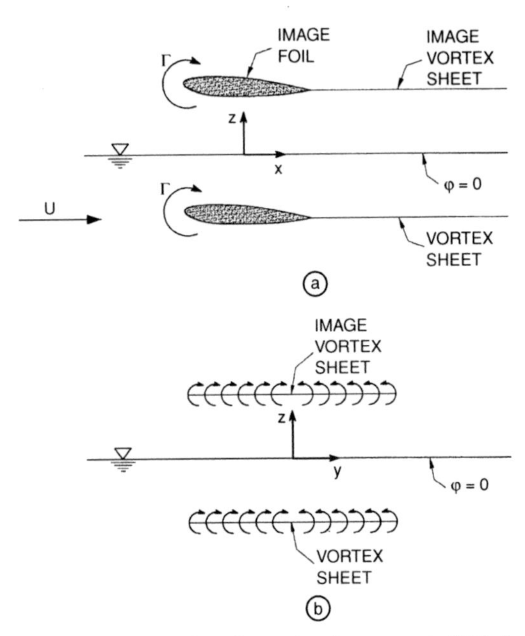

Figure 6.50. High Froude number flow around a 3D foil. (a) The solution is represented in terms of an image flow above the free surface. (b) Representation of the vortex sheet and the image vortex sheet in the yz-plane far downstream the foil.

$$D = -\frac{1}{2}\rho\int_{-0.5s}^{0.5s}(\varphi^+ - \varphi^-)\frac{\partial\varphi}{\partial z}dy, \quad (6.154)$$

where $\partial\varphi/\partial z$ is the vertical fluid velocity at the intersection between S_C and the vortex sheet behind the foil. The difference is now in the way $\partial\varphi/\partial z$ is evaluated. We can express a solution for $\partial\varphi/\partial z$ by imagining that the flow around the foil at high Froude number is the same as the flow around the foil and an image foil above the free surface. The image foil causes a vortex sheet with the same vorticity as in the vortex sheet behind the foil (see Figure 6.50). This image system ensures that the high Froude number free-surface condition $\varphi = 0$ on the mean free surface is satisfied. Here φ is the velocity potential caused by the foil. This is similar to the 2D case shown in Figure 6.47.

$\partial\varphi/\partial z$ in eq. (6.154) consists of two parts. The first part is the result of the vortex sheet behind the foil. The expression is the same as that used in eq. (6.137). The second contribution comes from the image vortex sheet above the mean free surface. This can be expressed as the vertical velocity induced by a two-dimensional distribution of vortices along the image vortex sheet, that is, similar to the evaluation of $\partial\varphi/\partial z$ from eq. (6.134). We get

$$w_{i_2} = \frac{1}{2\pi}\int_{-0.5s}^{0.5s} \frac{d\Gamma}{d\eta}\frac{y-\eta}{(y-\eta)^2+4h^2}d\eta. \quad (6.155)$$

We introduce now an elliptical circulation distribution along the span of the foil, that is,

$$\Gamma = \Gamma_0\left(1-\left(\frac{2\eta}{s}\right)^2\right)^{1/2}. \quad (6.156)$$

If a high-aspect–ratio plane foil is considered, Prandtl's lifting theory (see section 6.7.1) shows that an elliptical planform implies an elliptical circulation distribution. However, eq. (6.156) is also possible to achieve with other planforms. By integrating eq. (6.155) by parts we find that

$$w_{i_2} = -\frac{\Gamma_0}{2\pi}\int_{-0.5s}^{0.5s}\left(1-\left(\frac{2\eta}{s}\right)^2\right)^{1/2} \times\frac{(y-\eta)^2-4h^2}{[(y-\eta)^2+4h^2]^2}d\eta. \quad (6.157)$$

Introducing $y = y_1\frac{s}{2}$ and $\eta = y_2\frac{s}{2}$ gives

$$w_{i_2} = -\frac{\Gamma_0}{\pi s}\int_{-1}^{1}\left(1-y_2^2\right)^{1/2}\frac{\left[(y_1-y_2)^2-\frac{16h^2}{s^2}\right]}{\left[(y_1-y_2)^2+\frac{16h^2}{s^2}\right]^2}dy_2. \quad (6.158)$$

By inserting $\partial\varphi/\partial z$ into eq. (6.154), we have

$$D = \frac{\rho}{4\pi}\int_{-0.5s}^{0.5s} dy\Gamma(y)PV\int_{-0.5s}^{0.5s}\frac{d\Gamma}{d\eta}\frac{d\eta}{y-\eta} + D_2 \quad (6.159)$$

The first term leads to the expression for the drag in infinite fluid (see eq. (6.138)) and is the first term on the right-hand side of eq. (6.151). The second term, D_2, is the result of the image vortex sheet and can be expressed as

$$D_2 = \frac{\rho}{2}\int_{-0.5s}^{0.5s}\Gamma(y)w_{i_2}(y)\,dy$$

$$= \frac{\rho s}{4}\Gamma_0\int_{-1}^{1}\left(1-y_1^2\right)^{1/2}w_{i_2}(y_1)\,dy_1 \quad (6.160)$$

$$= \rho\Gamma_0^2\frac{\pi}{8}\sigma,$$

where

$$\sigma = -\frac{2}{\pi^2}\int_{-1}^{1}dy_1\left(1-y_1^2\right)^{1/2} \times\int_{-1}^{1}\frac{\left[(y_1-y_2)^2-\frac{16h^2}{s^2}\right]\left(1-y_2^2\right)^{1/2}}{\left[(y_1-y_2)^2+\frac{16h^2}{s^2}\right]^2}dy_2. \quad (6.161)$$

D_2 can be expressed in terms of the lift, which is

$$L = -\rho U\Gamma_0\int_{-s/2}^{s/2}\left(1-\left(\frac{2y}{s}\right)^2\right)^{1/2}dy = -\rho U\Gamma_0\frac{\pi s}{4} \quad (6.162)$$

for elliptical loading. The lift coefficient is

$$C_L = \frac{L}{0.5\rho U^2 A} = -\frac{\pi}{2}\frac{\Gamma_0 s}{UA}. \quad (6.163)$$

The drag coefficient associated with D_2 is

$$C_{D_2} = \frac{D_2}{0.5\rho U^2 A} = \frac{\rho\Gamma_0^2\frac{\pi}{8}\sigma}{0.5\rho U^2 A}. \quad (6.164)$$

Using eq. (6.163) to express Γ_0/U gives

$$\frac{C_{D_2}}{C_L^2} = \frac{\sigma}{\pi\Lambda}. \quad (6.165)$$

This means the same as the second term in eq. (6.151) if σ given by eq. (6.161) is the same as eq. (6.152). To show this equivalence analytically is not trivial. However, numerical calculations give the same answer from the two expressions.

When the Froude number goes to zero, we can follow a similar analysis as for infinite Froude number. However, the circulation around the image foil and the vorticity in the image vortex sheet must now be opposite the circulation around the foil and to the vorticity in the vortex sheet, respectively, in order to satisfy the rigid free-surface condition. The analogue case for 2D flow was illustrated in Figure 6.48. The analysis gives that

$$\frac{C_D}{C_L^2} = \frac{1}{\pi\Lambda} - \frac{\sigma(\lambda)}{\pi\Lambda}, \quad Fn_h \to 0 \quad (6.166)$$

when the Froude number is asymptotically small.

Eqs. (6.151) and (6.166) are presented in Figure 6.51 as a function of h/c when the aspect ratio is 6. The asymptotic value for $h/c \to \infty$ is 0.053. We note that decreasing h/c increases C_D in the infinite Froude number case. The opposite trend happens in the low Froude number case. The results for infinite Froude number are in good agreement

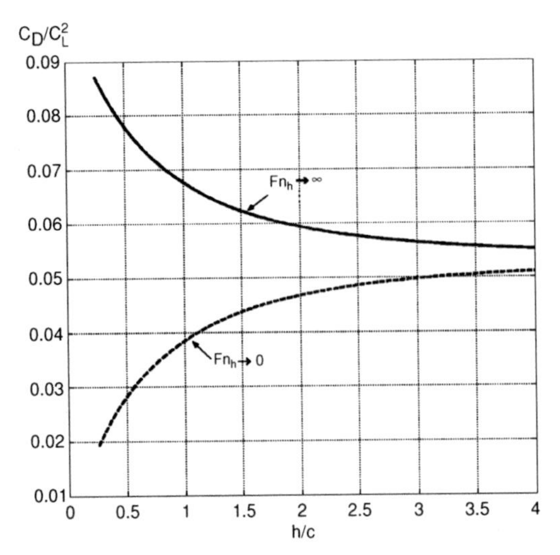

Figure 6.51. Theoretical drag coefficient C_D for a horizontal foil with elliptical loading as a function of ratio between submergence h and maximum chord length c at infinite and zero Froude numbers. The aspect ratio is 6.

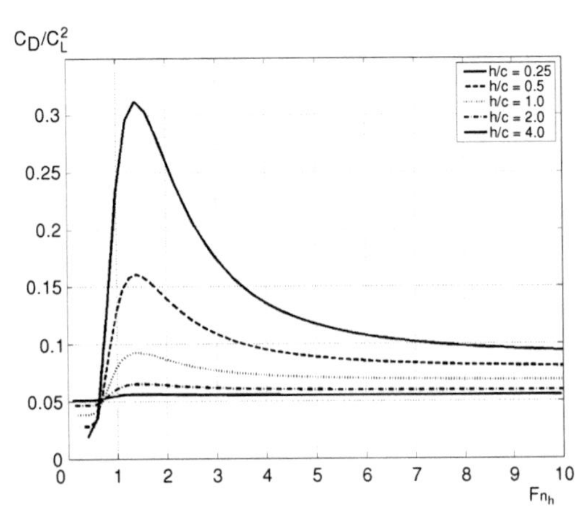

Figure 6.52. Drag coefficient C_D of a horizontal foil with elliptical loading, submergence h, and maximum chord length c presented as a function of submergence Froude number Fn_h. Aspect ratio = 6.

with the results by van Walree (1999) even if those results are for a rectangular foil.

Breslin's formula for C_D for any Froude number is

$$\frac{C_D}{C_L^2} = \frac{1}{\pi\Lambda} - \frac{\sigma(\lambda)}{\pi\Lambda} + \frac{8}{\pi\Lambda} \times \int_0^{\pi/2} \frac{J_1^2\left(0.5\nu s \sec^2\theta \sin\theta\right) e^{-2\nu h \sec^2\theta}}{\sin^2\theta\cos\theta} d\theta. \tag{6.167}$$

Here J_1 is a Bessel function of the first kind and $\nu = g/U^2$. Eq. (6.167) accounts for both transverse and divergent waves and is based on using the linear free-surface condition given by eq. (4.27) for the flow caused by foil.

Results for aspect ratio 6 and different h/c-values are presented in Figure 6.52. These results are in reasonable agreement with the experimental and numerical results for a rectangular foil with the same aspect ratio as the one presented by van Walree (1999). One exception is for the largest C_D/C_L^2-values occurring at $h/c = 0.25$. Because the foil is then close to the free surface, nonlinear free-surface effects may matter. An important contribution to the peak values of C_D/C_L^2 presented in Figure 6.52 is the result of transverse waves. This is illustrated in Figure 6.53 for the elliptical foil at $h/c = 1.0$ by separating the contribution from transverse waves. This is estimated by integrating from 0 to $\sin^{-1}(1/\sqrt{3})$ in the integral in eq. (6.167). We note that the effect of the transverse waves is small when $Fn_h > \approx 4.0$. The effect of the divergent waves will then contribute. However, the results are quite close to the drag at infinite Froude number when $Fn_h > \approx 4.0$. The latter does not include wave effects. The 2D results presented in Figure 6.48 contain only the effect of transverse waves and, to a large extent, can explain

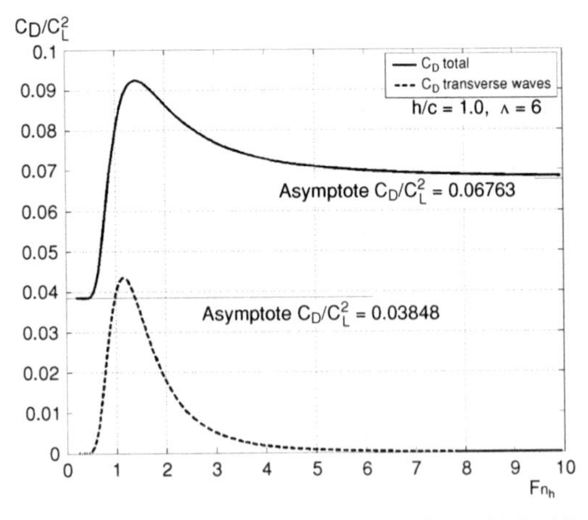

Figure 6.53. Drag coefficient C_D of a horizontal foil with elliptical loading at submergence (h) to maximum chord length (c) ratio 1 presented as a function of submergence Froude number Fn_h. Aspect ratio = 6. The contribution to C_D from transverse waves is illustrated.

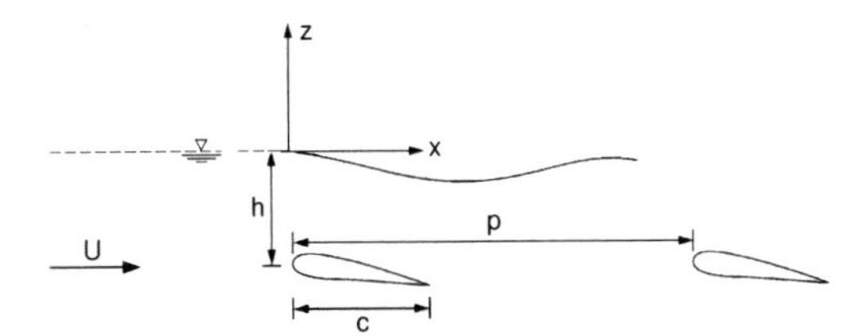

Figure 6.54. Tandem arrangement of two foils with equal submergence h.

how C_D/C_L^2 depends on Fn_h and h/c in the vicinity of maximum C_D/C_L^2.

6.9 Foil interaction

The wake and the free-surface waves generated by a foil affect the inflow and thereby the lift and drag on a downstream foil. This was studied numerically by, for instance, Andrewartha and Doctors (2001). Mørch (1992) studied foil interaction experimentally and numerically for a hydrofoil catamaran. His numerical studies included the combined effect of divergent waves and roll-up of far-field vortex sheets behind upstream foils at high Froude numbers. A nonlinear 2.5D analysis was used. Because the 2.5D method assumes that the flow variations are smaller in the inflow direction than in a plane perpendicular to the inflow velocity, the method can be applied only to the wake at a distance that is larger than at least the maximum chord length. Mørch (1992) assumed an elliptical circulation distribution as a starting condition. This is a function of the loading on the forward foil.

Nakatake et al. (2003) examined two identical rectangular hydrofoils in a tandem arrangement (Figure 6.54). The foil cross section is a NACA 0012 profile (Abbott and von Doenhoff 1959). The aspect ratio Λ is 5 and the angle of attack of the two foils is 5°. The ratios between the stagger p and the chord length c were $p/c = 2.5, 6.0$, and 10. The submergence h of each foil was equal to c. Experimental and theoretical results for lift coefficients were presented as a function of Froude number $Fn = U/\sqrt{gc}$ for the fore and aft foil. Nakatake et al. (2003) also presented drag results. However, the experimental chord length was only 0.06 m. Using a kinematic viscosity coefficient $\nu = 1.35 \cdot 10^{-6}\,\mathrm{m}^2 s^{-1}$ and a Froude number range from 0.5 to 5 means that the Reynolds number $Rn = Uc/\nu$ varies from $1.7 \cdot 10^4$ to $1.7 \cdot 10^5$. The boundary-layer flow is then likely to be laminar in the whole Reynolds number range for the fore foil. However, the inflow to the aft foil is turbulent because of the wake of the fore foil, affecting the transition to turbulence. This means eqs. (2.90) and (2.4) for turbulent flow cannot be used to estimate the viscous drag force. The Blasius solution given by eq. (2.5) is a better basis to find viscous drag force on the foil. An additional uncertainty is that drag forces on the struts were also included in the measurements. In principle, there is also an influence of the viscosity on the lift; however, in this case, the effect is of secondary importance. One should be very careful with model tests at such low Reynolds numbers. Laminar separation may lead to an irregular lift coefficient as function of incidence (van Walree and Luth 2000).

C_L of the aft foil shows a large influence of p/c. There is a strong variation with Froude number for Fn in the vicinity of 1. This is caused by the waves generated by the upstream foil that change the angle of attack and thereby the lift force on the aft foil. We analyze this problem as a 2D flow. This means only transverse waves are considered. Let us take a vortex with circulation Γ at $x = 0$ and $z = -h$. The wave effect caused by this vortex is an idealization of the lifting effect. We disregard the thickness effect of the foil on the wave field.

According to Kochin et al. (1964), the steady downstream wave elevation in the far field can be expressed as

$$\zeta = \frac{2\Gamma}{U} e^{-\nu h} \sin \nu x, \tag{6.168}$$

where $\nu = g/U^2$. By using eq. (6.150), we see that this is consistent with eq. (6.145). According to our definition of positive Γ, the linear lift force L is equal to $-\rho U \Gamma$. Using the fact that the lift coefficient C_L is equal to $L/\left(0.5\rho U^2 c\right)$ shows eq. (6.168) can be expressed as

$$\zeta = -C_L c e^{-gh/U^2} \sin(gx/U^2). \tag{6.169}$$

The linearized kinematic free-surface condition states that (see eq. (4.52b))

$$\left.\frac{\partial \varphi}{\partial z}\right|_{z=0} = U\frac{\partial \zeta}{\partial x} = -UC_L c \frac{g}{U^2} e^{-gh/U^2} \cos(gx/U^2). \tag{6.170}$$

Using the exponential decay of $\partial\varphi/\partial z$ with z as described in Table 3.1 and using $\nu = g/U^2$ instead of $k = \omega^2/g$, we obtain

$$\left.\frac{\partial\varphi}{\partial z}\right|_{z=-h} = \left.\frac{\partial\varphi}{\partial z}\right|_{z=0} e^{-gh/U^2}, \text{ that is,}$$

$$\left.\frac{\partial\varphi}{\partial z}\right|_{z=-h} = -UC_L Fn^{-2} e^{-2/Fn_h^2} \cos\left(\frac{1}{Fn^2}\frac{x}{c}\right), \tag{6.171}$$

where $Fn = U/(gc)^{0.5}$ and $Fn_h = U/(gh)^{0.5}$. This means Fn will be equal to Fn_h in our case. The effect of the transverse waves generated by the upstream foil on the angle of attack α_i for the downstream foil can then be approximated as

$$\alpha_i = \frac{\partial\varphi/\partial z|_{z=-h}}{U} = -C_L Fn^{-2} e^{-2/Fn_h^2} \cos\left(\frac{1}{Fn^2}\frac{x}{c}\right). \tag{6.172}$$

The cosine function in eq. (6.172) causes an oscillatory dependence on x. For instance, if the foils at p/c equal to 2.5 and 6.0 are considered, then the cosine functions are $\cos(2.5/Fn^2)$ and $\cos(6.0/Fn^2)$. If the difference between $6.0/Fn^2$ and $2.5/Fn^2$ is π, then α_i has the opposite sign for these two foils. This means that α_i for the foils at p/c equal to 2.5 and 6.0 is 180° out of phase when $\pi Fn^2 = 3.5$ or $Fn = 1.06$. Further, α_i on the foils at p/c equal to 2.5 and 10.0 is 180° out of phase when $\pi Fn^2 = 7.5$ or $Fn = 1.55$. When Fn increases, the phase difference in α_i between the aft foils becomes smaller. We have estimated lift coefficient C_{L_2} on the aft foil by setting the lift coefficient C_L for the forward foil equal to 0.35 for all Froude numbers and setting for the aft foil

$$C_{L_2} = 0.35 + 2\pi\alpha_i.$$

The results are presented in Figure 6.55 as a function of Fn for p/c equal to 2.0, 6.0, and 10. The theoretical estimates and the experiments by Nakatake et al. (2003) are in qualitative agreement. However, the experimental results show larger differences among the different p/c values for large Froude numbers. Important contributions are caused by 3D flow effects that are partly associated with divergent waves. This is indirectly illustrated by Figure 6.53. Anyway, the results in Figure 6.55 help in explaining qualitatively the oscillatory behavior of the lift forces on the aft foils as a function of Fn as well as the phase difference in lift between foils located as

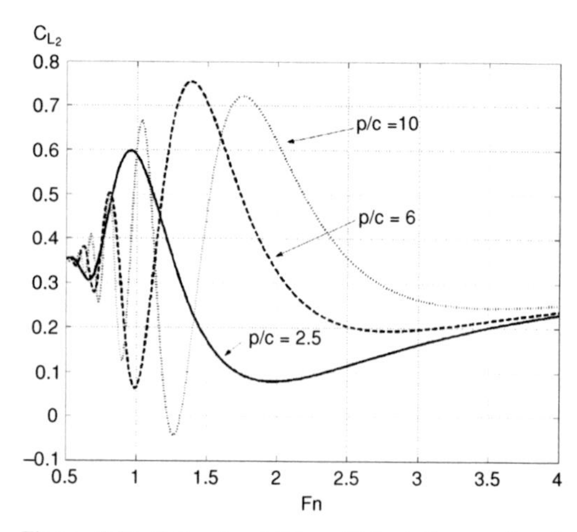

Figure 6.55. Estimate of lift coefficient C_{L_2} on the aft foil as a function of Froude number Fn at different stagger-to–chord length ratios p/c (see Figure 6.54). The effect of transverse waves generated by the upstream foil at the position of the aft foil is accounted for.

p/c equal to 2.5, 6, and 10. Because part of the drag coefficient C_D is related to wave resistance and part is the result of lift effect and is related to C_L, there is a similarity in the behavior of C_L and C_D.

3D flow effects in the wake

When $Fn > \approx 4$, the previously described results by Nakatake et al. (2003) show that C_L of the aft foil is clearly lower than that of the fore foil at high Froude numbers. On the aft foil, an important reason is the 3D effect of the wake from the fore foil. This effect is evaluated for an elliptical planform by first neglecting the effect of the free surface. If we are far downstream of the fore foil, eq. (6.116) gives eq. (6.134). Using that $w_i = \partial I_w/\partial z$ leads to a vertical velocity,

$$w_i = \frac{1}{2\pi} PV \int_{-0.5s}^{0.5s} \frac{d\Gamma}{d\eta}\frac{d\eta}{y-\eta} = -\frac{4U\alpha}{\Lambda + 2} \tag{6.173}$$

for a point on free shear layer S_V^+. We have here used that $\Gamma = 2Usa_1 \sin\theta'$, where $a_1 = -2\alpha/(\Lambda+2)$ and θ' is defined by eq. (6.122). The detailed calculations are similar as those for eq. (6.119). Eq. (6.173) means that $w_i/U = -4\alpha/7$ for an aspect ratio $\Lambda = 5$. Let us consider two foils with the same dimensions and angles of attack, and the aft foil is in the far field of the free shear layer of the fore foil. The lift force on the aft foil is only 3/7 of the lift force on the fore foil. This

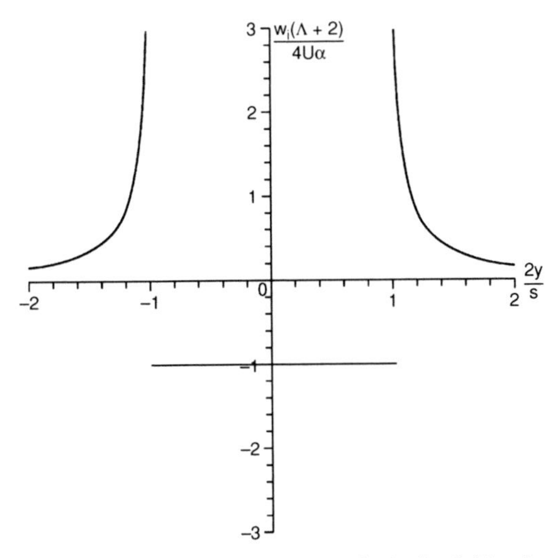

Figure 6.56. Downwash velocity w_i in the far-field wake of an elliptical planform with aspect ratio Λ, angle of attack α, and span s in infinite fluid. The foil is located between $y = -s/2$ and $y = s/2$.

result is independent of p/c. However, the results by Nakatake et al. (2003) show the dependence on p/c for large Fn. An investigation of this requires that the finite cross-sectional dimensions of the foil and the full three-dimensional dependence of w_i are accounted for.

In general, an aft foil does not have the same dimensions as a fore foil (see Figures 6.2 and 6.12). Let us assume that the aft foil has a larger span than the fore foil. We need then to calculate w_i outside the wake of the fore foil. If the submergence of the two foils is the same, for $|y| > s/2$ we can write

$$\frac{w_i}{U} = \frac{4\alpha}{\pi(\Lambda+2)}\int_0^{\pi}\frac{\cos\theta}{2y/s - \cos\theta}d\theta = \frac{4\alpha}{\Lambda+2}\left[\frac{|y^*|}{\sqrt{y^{*2}-1}} - 1\right], \quad (6.174)$$

where $y^* = 2y/s$ and s is the span of the upstream foil. The derivation of eq. (6.174) is similar as for eq. (6.173). However, because $|y| > s/2$, we do not have a principal value integral and the Glauert integrals given by eq. (6.81) are not needed. The integral in eq. (6.174) can be found using integral tables.

Eqs. (6.173) and (6.174) are illustrated in Figure 6.56, which shows that w_i is singular at $|y| = s/2$ and is positive and nonconstant for $|y| > s/2$. The singular nature does not occur in reality. The vortex sheet rolls up in the vicinity of $|y| = s/2$ and tip vortices like those in Figure 6.12 occur. If a strip theory approach for the aft foil is used, according to eq. (6.174), a cross section with $|y| > s/2$ will get an increased lift because of the wake of the fore foil. Because the increased lift may lead to cavitation, this is not necessarily beneficial. In order to minimize the magnitude and spanwise variation of the angle of attack, it is beneficial to use a twisted aft foil that is adapted to a typical spanwise inflow. The upward wake velocity predicted by eq. (6.174) is the reason nature has taught geese to fly in V-form.

The free-surface effect on the vertical velocity in the far-field wake at high Froude number can be accounted for by adding eq. (6.158) to the results in Figure 6.56. We set $\Gamma_0 = -c_0\pi\alpha U$, where c_0 is the maximum chord length, into eq. (6.158). This means only the angle of attack α is accounted for in calculating the circulation of the upstream foil. Using also that the aspect ratio Λ of an elliptical foil is $s/(0.25\pi c_0)$ gives the vertical induced velocity due to the image vortex sheet presented in Figure 6.57. We should note that the normalization of w_i is different in Figures 6.56 and 6.57.

Roll-up of vortex sheets

The previous analysis neglects the roll-up of vortex sheets, which has been numerically studied by Krasny (1987) in the far-field wake of an elliptically loaded foil in infinite fluid. The velocity potential due to the vortex sheet satisfies, then, the 2D Laplace equation in planes perpendicular to

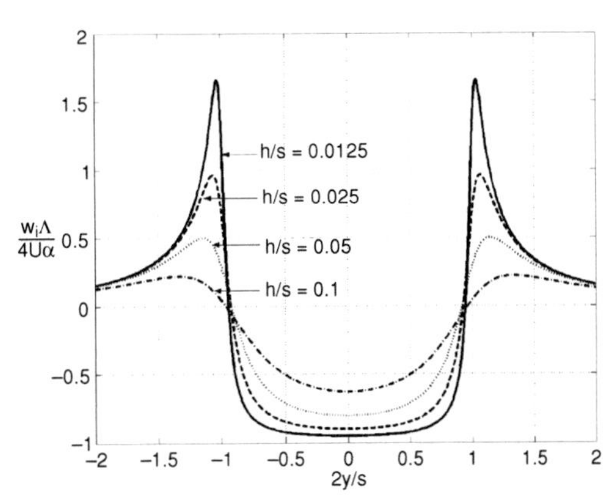

Figure 6.57. Vertical induced velocity w_i due to an image vortex sheet in the far-field wake of an elliptical planform with submergence h, aspect ratio Λ, angle of attack α, and span s. The foil is located between $y = -s/2$ and $y = s/2$. Adding these results together with the results in Figure 6.56 gives a total description of the vertical velocity in the far-field wake at a high Froude number.

the incident flow direction. A similar assumption was made in the drag force analysis in section 6.7.2. We can then use a 2.5D (2D+t) method to analyze the roll-up of the vortex sheet. Mørch (1992) further developed the analysis to include free-surface effects, which was the basis of the results presented in Figure 6.13. All these effects change the vertical position of the free shear layer downstream.

We have up until now assumed that the trailing vortex sheet is infinitely thin. However, turbulence in the free shear layer causes diffusion of the vorticity. The trailing vortex sheet then has a finite thickness that increases with downstream distance from the foil. This requires a numerical analysis based on the Navier-Stokes equations. However, we will not pursue such analysis here.

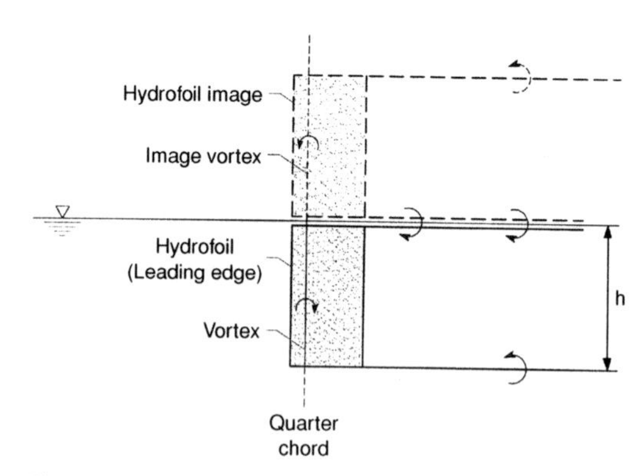

Figure 6.59. Horseshoe vortex system that satisfies the infinite–Froude number, free-surface condition for a surface-piercing strut (Wadlin 1958).

6.10 Ventilation and steady free-surface effects on a strut

The steady free-surface effects on a strut without lifting effects were discussed in section 4.3.3. The effect of the angle of attack on the side force coefficient is illustrated numerically and experimentally for a strut with aspect ratio 0.5 in Figure 6.58. The angle of attack is 4.6°, and no ventilation occurred in the experiments. A more representative aspect ratio of struts on hydrofoil vessels is 1. Further, the strut is connected with a foil, which influences the flow. As an example, if we consider a speed of 20 ms^{-1} and a chord length of 1.5 m, the length Froude number is 5.2. Figure 6.58 suggests then that an infinite Froude number free-surface condition can be used, that is, there is a small effect of free-surface wave generation on the side force.

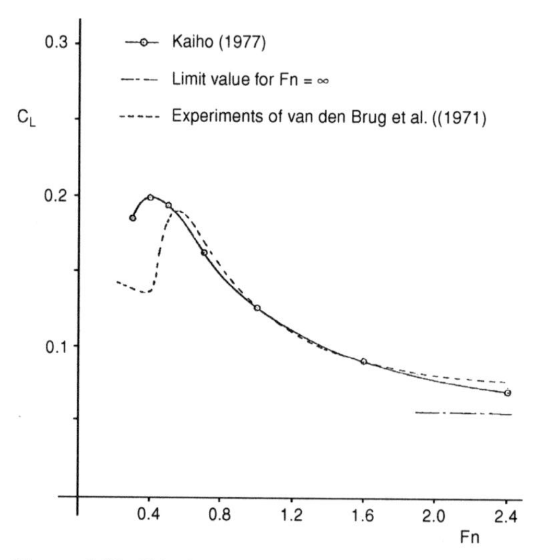

Figure 6.58. Side-force coefficient on yawed strut as a function of length Froude number Fn. Aspect ratio = 0.5; $\tan \alpha = 0.08$, where α is the angle of attack (Ogilvie 1978).

Ventilation means that air enters from the atmosphere to low-pressure areas on the strut. It occurs typically at angles of attack ψ higher than 4° to 6° for a strut on a hydrofoil vessel in foilborne condition. However, this depends on the Froude number. Ventilation causes a significant drop in the transverse force on the strut and in the lift on a foil attached to the strut. Hysteresis is associated with ventilation, which means that ψ may have to be lowered significantly to restore nonventilated conditions. Examples on factors influencing ventilation are

- Flow separation from the leading-edge area
- Cavitation
- Aeration of the trailing vortex developed at the lower tip of the strut (Figure 6.59)

Flow separation from the leading-edge area depends on the leading-edge curvature and the boundary-layer condition upstream of the separation line. A laminar boundary layer separates more easily than a turbulent boundary layer. This means model tests may suffer from scale effects if turbulence stimulators are not used. Further, the model tests must be done at the same cavitation and Froude numbers as those in full scale.

Based on experimental results for struts, Breslin (1958) proposed

$$C_L \geq \frac{5}{Fn_h^2} \tag{6.175}$$

as a condition for full ventilation when the submergence Froude number $Fn_h = U/(gh)^{0.5} \geq 3$. Here $C_L = F_2/0.5\rho U^2 A$, where F_2 is the transverse force and A is the projected area of the strut. Eq. (6.175) illustrates that ventilation is strongly dependent on Fn_h, that is, the smaller Fn_h is, the larger the angle of attack can be to avoid ventilation.

Eq. (6.175) states that the upper bound F_2/A for nonventilated flow is $2.5\rho gh$ when $Fn_h > 3$. Here F_2/A is the average pressure difference between the pressure and suction sides of the strut. It is natural that the level of the suction pressure influences the occurrence of ventilation. The detailed pressure distribution must matter. If the suction pressure has a pronounced minimum at the leading edge, the ventilation is expected to start at the leading edge.

Figure 6.60 shows model test results for a front strut-foil system (see Figure 2.20) used as a rudder on a foil catamaran. Similar results are presented by Minsaas (1993). The side force coefficient C_L is presented as a function of the yaw angle ψ both at atmospheric air condition and at the cavitation number $\sigma = 0.349$ corresponding to the full-scale condition. The cavitation number is defined as $\sigma = (p_a + \rho gh - p_v)/(0.5\rho U^2)$. Here p_a and p_v are, respectively, atmospheric and vapor pressures. The pitch angle is 2.7° and the submergence Froude number Fn_h is 5.96, corresponding to a full-scale speed of 50 knots. The drop in the absolute value of C_L with either increasing positive ψ-values or decreasing negative ψ-values is an indication of the start of ventilation. This occurs, for instance, at $\psi \approx 4°$ when $\sigma = 0.349$ and at $\psi = 6°$ at atmospheric air conditions. If we use eq. (6.175), this gives $C_L = 0.14$ as the upper bound for nonventilated conditions. This agrees with the results for $\sigma = 0.349$, but it is a conservative estimate for ventilation during model tests at atmospheric conditions.

A consequence of the ventilation at the strut is that air penetrates to the foil and causes a significant drop in the lift force on the foil. This drop occurred at ψ approximately equal to ±6°. The lift force on the ventilated foil was the order of 50% of the nonventilated lift.

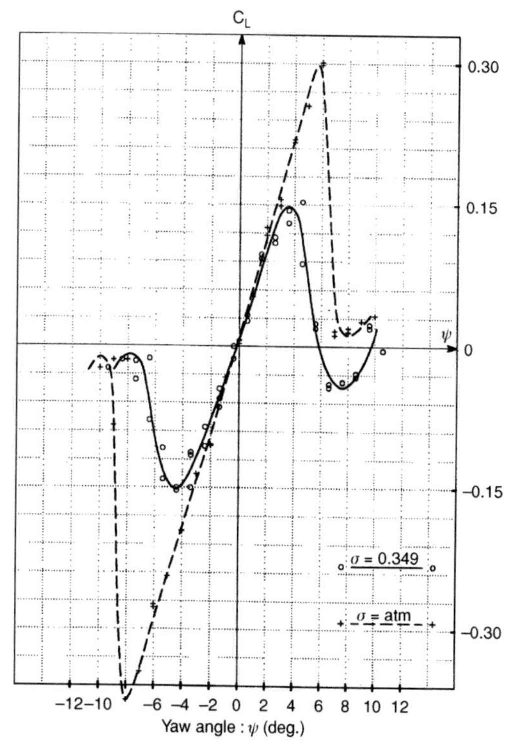

Figure 6.60. Model test results of a front strut-foil system (see Figure 2.20) used as a rudder on a foil catamaran. The side force coefficient C_L is presented as a function of the yaw angle ψ, both at atmospheric air condition (+) and at the cavitation number $\sigma = 0.349$ corresponding to full-scale conditions (o). The submergence Froude number Fn_h is 5.96, corresponding to a full-scale speed of 50 knots. The pitch angle is 2.7° (Minsaas, unpublished).

6.11 Unsteady linear flow past a foil in infinite fluid

6.11.1 2D flow

We consider a 2D foil in infinite fluid and use a coordinate system (x, y) following the constant forward speed U of the foil. The forward speed appears, then, as an incident flow along the positive x-axis. The foil has an unsteady translatory and rotational velocity. In addition, there can be an incident (ambient) unsteady velocity, for instance, due to ocean waves.

The problem is linearized, and only the lifting problem is considered. It has already been shown how to linearize the body boundary conditions in the steady-flow case which led to eq. (6.69). However, this condition must be modified by accounting for the relative vertical velocity $V_R(x,t)$ between the foil and the incident flow. The linearized body boundary condition for the unsteady lifting problem becomes

$$\frac{\partial \varphi}{\partial y} = V_R(x,t) + U\eta'(x,t) \quad \text{on} \quad y = 0, \; -c/2 < x < c/2. \qquad (6.176)$$

The total velocity potential is $Ux + \varphi$.

Boundary conditions on the free shear layer downstream from the trailing edge must also be accounted for. This was discussed in the nonlinear case in section 6.4. Requiring a zero pressure jump across the free shear layer leads to eq. (6.18). Linearization implies that linear terms in φ are kept in Bernoulli's equation and that the boundary conditions are transferred to $y = 0$ for $x > c/2$. The trailing edge corresponds to $x = c/2$ in the linearized formulation of the boundary-value problem. Eq. (6.18) can then be simplified as

$$\frac{\partial}{\partial t}(\varphi^+ - \varphi^-) + U\frac{\partial}{\partial x}(\varphi^+ - \varphi^-) = 0 \quad \text{on } y = 0 \quad \text{for } c/2 < x < c/2 + Ut. \qquad (6.177)$$

Here $t = 0$ corresponds to initial time. The effect of the free shear layer can be represented in terms of a vortex distribution with density $\gamma(x,t)$. Differentiating eq. (6.177) with respect to x and using eq. (6.75) give

$$\frac{\partial \gamma}{\partial t} + U\frac{\partial \gamma}{\partial x} = 0 \quad \text{on } y = 0, \quad c/2 < x < c/2 + Ut, \qquad (6.178)$$

that is,

$$\gamma(x,t) = \gamma(x - Ut). \qquad (6.179)$$

This can be confirmed by substituting eq. (6.179) into eq. (6.178). The body boundary condition expressed by eq. (6.74) for the steady-flow case can now be generalized to

$$\frac{1}{2\pi} PV \int_{-c/2}^{c/2} \frac{\gamma(\xi,t)\,d\xi}{x - \xi} + \frac{1}{2\pi}\int_{c/2}^{c/2+Ut} \frac{\gamma(\xi - Ut)\,d\xi}{x - \xi} = V_R(x,t) + U\eta'(x,t) \qquad (6.180)$$

on $y = 0, \; -c/2 < x < c/2$. The Kutta condition on $x = c/2$ requiring that γ is finite is needed to solve this equation (see Newman 1977 and Bisplinghoff et al. 1996 for further details).

When γ is found, the linear vertical force L on the foil follows from Bernoulli's equation, that is,

$$L = \rho \int_{-0.5c}^{0.5c} \left(\frac{\partial}{\partial t}(\varphi^+ - \varphi^-) + U\frac{\partial}{\partial x}(\varphi^+ - \varphi^-) \right) dx, \qquad (6.181)$$

where the superscripts + and − refer to $y = 0^+$ and $y = 0^-$, respectively. Eq. (6.181) can be rewritten in terms of γ by using eq. (6.75) and noting that

$$\int_{-c/2}^{c/2} \frac{\partial}{\partial t}(\varphi^+ - \varphi^-)\,dx = (x - c/2)\frac{\partial}{\partial t}(\varphi^+ - \varphi^-)\Big|_{-c/2}^{c/2} - \int_{-c/2}^{c/2} (x - c/2)\frac{\partial}{\partial t}\left(\frac{\partial \varphi^+}{\partial x} - \frac{\partial \varphi^-}{\partial x}\right) dx = \int_{-c/2}^{c/2} (x - c/2)\frac{\partial \gamma}{\partial t}\,dx$$

Here we have used $\varphi^+ = \varphi^-$ at $x = -c/2$. This means

$$L = \rho \int_{-c/2}^{c/2} \left[(x - c/2)\frac{\partial \gamma}{\partial t} - U\gamma \right] dx. \qquad (6.182)$$

The Wagner (1925) solution results presented in Figure 6.29 follow from an analysis such as this, which assumes a transient flow. We will now consider steady-state conditions.

6.11.2 2D flat foil oscillating harmonically in heave and pitch

The steady-state solution for a flat foil that is harmonically oscillating in heave and pitch is considered. The vertical motion (heave) of the center of the foil is expressed as

$$h(t) = \mathrm{Re}\{h_0 e^{i\omega t}\}. \qquad (6.183)$$

Further, the pitch angle δ is given as

$$\delta(t) = \mathrm{Re}\{\delta_0 e^{i\omega t}\}. \qquad (6.184)$$

Here h_o and δ_0 are complex numbers and i is the complex unit. The vertical motion of the mean-camber line is then expressed as $\eta = h - \delta x$. Eq. (6.176) can be expressed as

$$\frac{\partial \varphi}{\partial y} = \dot{h} - \dot{\delta}x - U\delta \quad \text{on } y = 0, \; -c/2 < x < c/2. \qquad (6.185)$$

Here dot means time derivative. From eq. (6.179) and because of the harmonic time dependence of the flow, it follows that the vortex density in the wake has the mathematical form

$$\gamma = \mathrm{Re}\left\{\gamma_0 e^{i\left(\omega t - \frac{\omega}{U}x\right)}\right\}. \qquad (6.186)$$

So the vortex density propagates as a sinusoidal wave with phase speed U along the x-axis. The wave number is ω/U, which gives a wavelength of $2\pi U/\omega$.

A complete solution was found by Theodorsen (1935). The details are also given by Bisplinghoff et al. (1996) and Newman (1977). It follows that

$$L = -\rho 0.25\pi c^2(\ddot{h} - U\dot{\delta}) - \rho\pi UcC(k_f) \times (\dot{h} - U\delta - 0.25c\dot{\delta}), \qquad (6.187)$$

where

$$k_f = \frac{\omega c}{2U} \qquad (6.188)$$

is the reduced frequency. Further,

$$C(k_f) = F(k_f) + iG(k_f) = \frac{H_1^{(2)}(k_f)}{H_1^{(2)}(k_f) + iH_0^{(2)}(k_f)}. \qquad (6.189)$$

Here F and G are the real and imaginary parts of $C(k_f)$. $H_n^{(2)}$ are Hankel functions (Abramowitz and Stegun 1964), which can be expressed in terms of Bessel functions of the first and second kinds, that is,

$$H_n^{(2)} = J_n - iY_n. \qquad (6.190)$$

$C(k_f)$ is called the Theodorsen function in memory of the person who first derived the lift force expression. When ω is zero, $C(k_f)$ is equal to one. Eq. (6.187) gives then that the lift force is equal to $\rho\pi cU^2\delta$, which is consistent with the linear steady result for a 2D flat plate in infinite fluid (see eq. (6.93)).

Both h and δ in eq. (6.187) should be expressed as complex quantities. It is the real part of the expression for L that has physical meaning. Let us elaborate more on that and consider heave only. We express $h = h_0 \cos\omega t$, where h_0 is a real amplitude. This means $\dot{h} = -\omega h_0 \sin\omega t$. The real part of the term $C(k_f)\dot{h}$ in eq. (6.187) becomes

$$\begin{aligned} &\mathrm{Re}\left\{(F(k_f) + iG(k_f))\, i\omega h_0 (\cos\omega t + i\sin\omega t)\right\} \\ &= -F(k_f)\omega h_0 \sin\omega t - G(k_f)\omega h_0 \cos\omega t \\ &= F(k_f)\dot{h} - \omega G(k_f)h \end{aligned}$$

Figure 6.61. Quasi-steady analysis of a heaving foil. $\dot{h}$ is the heave velocity.

This means that the lift force with $\delta = 0$ is

$$L = -\rho 0.25\pi c^2\ddot{h} + \rho\pi Uc\omega G(k_f)h - \rho\pi UcF(k)\dot{h}. \qquad (6.191)$$

When $\omega \to 0$, $L \to -\rho Uc\pi\dot{h}$. This can be explained by a quasi-steady analysis of a heaving horizontal flat, thin foil that moves with a forward speed. There is an incident flow velocity U parallel to the foil and an incident flow velocity $-\dot{h}$ perpendicular to the foil (Figure 6.61). This means that there is an ambient flow velocity

$$V = (U^2 + (\dot{h})^2)^{\frac{1}{2}} \approx U$$

with an instantaneous angle of attack

$$\alpha = -\dot{h}/U \qquad (6.192)$$

relative to the foil. Here we have implicitly assumed small α-values. If a quasi-steady approach is used, eq. (6.93) can be directly applied. This gives a two-dimensional vertical force:

$$L = -\rho Uc\pi\dot{h}. \qquad (6.193)$$

When $\omega \to \infty$, the lift force is dominated by the acceleration term $-\rho 0.25\pi c^2\ddot{h}$ of eq. (6.191). The term in phase with the velocity $\dot{h}$ approaches $-0.5\rho U\pi c\dot{h}$ when $\omega \to \infty$.

An interesting consequence of the quasi-steady analysis is illustrated by Figure 6.62. It was shown in section 6.5 that the steady lift force L is perpendicular to the incident flow velocity. Because the incident flow direction changes, the direction of the lift force changes. However, there will always be a component of L with the same sign in the longitudinal direction of the foil. This force component acts as a thrust (T) on the foil. Horizontal foils on a ship can therefore act as a propulsion unit in waves. We must, of course, have in mind

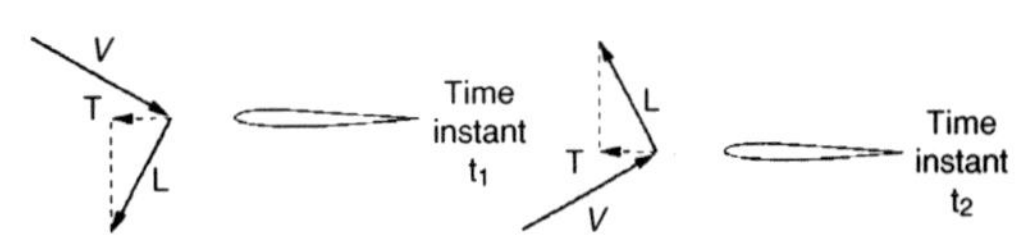

Figure 6.62. Thrust force T on a foil occurring as a consequence of quasi-steady 2D potential flow analysis of a heaving foil.

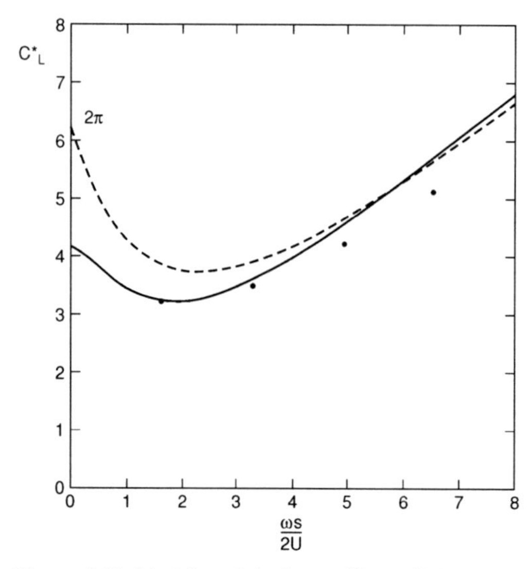

Figure 6.63. Modulus of the heave lift-coefficient defined in eq. (6.194) for a wing of elliptical planform of aspect ratio $\Lambda = 4$ as a function of the reduced frequency made nondimensional by the wing span s. (- - - - - -) strip theory, (——) unified theory, (•) numerical solution of Lee (1977) using a vortex-lattice method (Sclavounos 1987).

the presence of viscous drag forces, that potential flow effects may cause drag due to the free vortex systems in 3D flow, and free-surface effects.

6.11.3 3D flow

When it comes to unsteady flow effects due to continuously shed vorticity, the general tendency is that the higher the frequency of oscillation, the smaller the 3D effect (Sclavounos 1987). We present some of Sclavounos's results and then define a lift coefficient

$$C_L^* = \frac{L_a}{0.5\rho U A \dot{h}_a}, \tag{6.194}$$

where L_a and $\dot{h}_a$ mean amplitudes of L and $\dot{h}$, respectively. Further, A is the projected area of the foil on the plane of the incident flow. The results are presented in Figure 6.63 together with numerical results by Lee (1977) and strip theory based on eq. (6.191). Sclavounos's (1987) results are referred to as unified theory, which assumes a high-aspect–ratio foil, that is, similar to Prandtl's lifting line theory. When $\omega \to 0$, the results converge to Prandtl's results. This can be seen by first re-expressing eq. (6.130) as

$$C_L^* = \frac{2\pi\Lambda}{\Lambda + 2}. \tag{6.195}$$

We have then used that α in eq. (6.130) is $\dot{h}_a/U$ for our problem and the fact that C_L used in eq. (6.130) is related to C_L^* as $C_L = C_L^* \dot{h}_a/U$. This gives $C_L^* = 4.19$ for $\Lambda = 4$ and is in agreement with Figure 6.63 for $\omega = 0$. The results in Figure 6.63 agree with strip theory results when $0.5\omega s/U > \approx 4$. Here s means the wingspan. We can relate ω to a wavelength λ of the vorticity waves propagating along the vortex sheet. This can be found from eq. (6.186) and expressed as $\lambda = 2\pi U/\omega$, which means $0.5\omega s/U > \approx 4$ corresponds to $\lambda/s < \approx \pi/4$. There is a parallel to these results in ship motion theory in waves in which strip theory is a high-frequency (small-wavelength) theory.

6.12 Wave-induced motions in foilborne conditions

A hydrofoil vessel in foilborne conditions generally has good seakeeping characteristics compared with semi-displacement vessels. This is illustrated in Figure 6.22 by comparing vertical accelerations of a hydrofoil catamaran and a conventional catamaran. We try to explain this by considering a monohull hydrofoil vessel with a fully submerged foil system in regular head sea waves in deep water. The vessel in foilborne conditions is illustrated in Figure 6.64. The forward and aft foils are assumed to have the same rectangular planform area and high-aspect ratios. A coordinate system (x, y, z) moving with the forward speed U of the vessel is used. The x-coordinate of the center of gravity (COG) is zero, and $z = 0$ corresponds to the mean free-surface level. z is positive upward. The geometrical centers of the two foils have coordinates $(\pm 0.5L_f, 0, -h)$.

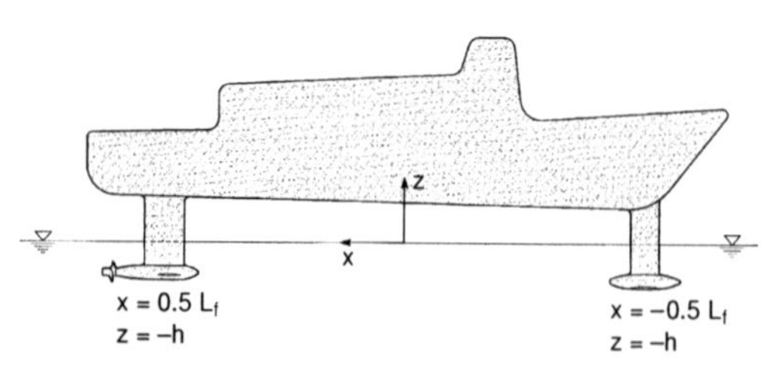

Figure 6.64. A hydrofoil vessel with submerged foils that have geometrical centers at $(\pm 0.5L_f, 0, -h)$, shown together with pods and struts.

A complete analysis would require use of a numerical method as described in section 6.4. However, the analysis is simplified by neglecting the hydrodynamic interaction between the foils. We saw in section 6.9 that this matters for the steady hydrodynamic loads on the aft foil. Further, the interaction with the free surface will not be complete in the analysis. A strip theory approach using the 2D unsteady analysis in section 6.11.2 is followed. This implies that the 2D results are multiplied with the foil span s to obtain 3D results. The hydrodynamic loads on the pods shown in Figure 6.64 are neglected.

The linear heave motion η_3 of the COG and the linear pitch motion η_5 of the vessel are analyzed. Linear response means that η_3 and η_5 are proportional to the incident wave amplitude ζ_a. The vertical motion of the vessel is then expressed as $\eta_3 - x\eta_5$.

A basis for describing η_3 and η_5 is Newton's second law. This means

$$M\frac{d^2\eta_3}{dt^2} = \sum F_{3j}. \qquad (6.196)$$

Here M is the vessel mass and the right-hand side is the sum of vertical linear unsteady hydrodynamic forces acting on the hydrofoil vessel in foilborne condition.

In addition, by considering pitch moments about the COG, it follows that

$$I_{55}\frac{d^2\eta_5}{dt^2} = \sum F_{5j}, \qquad (6.197)$$

where $I_{55} = Mr_{55}^2$ is the moment of inertia in pitch about the COG and r_{55} is the radius of gyration in pitch. The right-hand side of eq. (6.197) is the sum of linear unsteady hydrodynamic pitch moments about the COG.

Steady-state oscillations are assumed. That means the hydrodynamic loads oscillate with the frequency of encounter of the incident waves.

We start by describing the incident waves by transforming the results presented in Table 3.1 to the coordinate system moving with the forward velocity U of the vessel. We show how this can be done in Chapter 5 (see eqs. (5.19) and (5.27)). This means that there is a vertical incident fluid velocity that can be expressed as

$$w = \omega_0\zeta_a e^{kz}\cos(\omega_e t - kx), \qquad (6.198)$$

where the circular frequency of encounter ω_e is related to the circular frequency of the waves ω_0 by

$$\omega_e = \omega_0 + \omega_0^2 U/g. \qquad (6.199)$$

Further, $k = \omega_0^2/g = 2\pi/\lambda$, where λ is the wavelength. We now introduce complex notation and express eq. (6.198) as

$$w = \omega_0\zeta_a e^{kz}e^{-ikx+i\omega_e t}. \qquad (6.200)$$

In all complex expressions such as this, it is implicitly understood that it is the real part that has physical meaning. The incident waves cause two types of hydrodynamic forces on the foil. The first is called Froude-Kriloff force, which represents the effect of integrating the loads due to the linear hydrodynamic pressure p_D in the incident waves, that is, due to

$$p_D = \rho g\zeta_a e^{kz}\sin(\omega_e t - kx). \qquad (6.201)$$

If the incident wavelength is long relative to the chord length c of a foil, the resulting vertical force on the foil with geometric center at $x = \pm L/2$ and $z = -h$ is

$$F_3^{FK} = -\rho\Omega\omega_0^2\zeta_a e^{-kh}\sin(\omega_e t \mp kL/2). \qquad (6.202)$$

Here Ω is the displaced volume of the foil. Eq. (6.202) can be proven by means of Gauss' theorem. We start out formally expressing

$$F_3^{FK} = -\iint_{S_B} p_D n_3 dS, \qquad (6.203)$$

where S_B is the body surface of the considered foil and n_3 is the z-component of the normal vector n to S_B. The positive normal direction is into the fluid domain. Gauss' theorem (see eq. (2.205)) implies that

$$\iint_{S_B} p_D n_3 dS = \iiint_{\Omega} \frac{\partial p_D}{\partial z}d\tau. \qquad (6.204)$$

The derivative $\partial p_D/\partial z$ can be approximated as a constant in Ω by using the long-wavelength assumption. Then eq. (6.202) now follows by setting $\partial p_D/\partial z$ outside the integral in eq. (6.204) and using eq. (6.201) to obtain $\partial p_D/\partial z$ in the geometrical center of the foil. Having now obtained F_3^{FK}, we will neglect it. The reason is that this term is small relative to other terms to be derived below. A more physical way to see that the Froude-Kriloff force is not important is to note that p_D is practically the same on the upper/lower side of the foil.

The second effect associated with the incident waves is brought about because the normal component of the incident wave velocity has to be counteracted at the foil surface. This causes a pressure field and flow in the water. The analysis is done by introducing a diffraction potential φ_D satisfying the Laplace equation so that

$$\frac{\partial \varphi_D}{\partial n} = -\frac{\partial \varphi_0}{\partial n} \quad \text{on } S. \tag{6.205}$$

Here φ_0 means the incident wave potential. By using the same linearization procedure as the one we followed in section 6.6 and by considering only the lifting problem, we can replace eq. (6.205) by

$$\frac{\partial \varphi_D}{\partial z} = -w \quad \text{on} - 0.5c \pm 0.5L_f \le x \le 0.5c \pm 0.5L_f, \quad z = -h. \tag{6.206}$$

Here $\pm$ corresponds to which foil we are considering. Further, c is the chord length, which will be constant along the span in the case of a rectangular planform. Once more, we use the fact that the wavelength λ of the incident waves is much longer than the chord length. This means w in eq. (6.206) is evaluated at $z = -h$ and $x = \pm 0.5L_f$. An indication of how large the wavelength λ ought to be relative to the chord length can be found by averaging w over the chord length. If, for instance, $\lambda/c = 5$, this averaged w is 0.94 times the value obtained by evaluating w at the midpoint of the foil. There is a parallel to this in the calculations of wave loads on offshore structures. If we consider as an example a vertical circular cylinder with diameter D, it is generally agreed that a similar long-wavelength approximation of the wave load calculations can be used if $\lambda/D > 5$ (Faltinsen 1990).

Assuming 2D flow in the xz-plane implies that the φ_D problem with the long-wavelength approximation is the same as the heaving problem considered in section 6.11.2. Another coordinate system was used then, that is, the y-coordinate was vertically upward. The consequence is that we can account for the effect of the diffraction potential on the 2D unsteady vertical hydrodynamic forces by expressing $\dot{h}$ in eq. (6.187) as

$$\dot{h} = i\omega_e\bar{\eta}_3 e^{i\omega_e t} - xi\omega_e\bar{\eta}_5 e^{i\omega_e t} - \omega_0\zeta_a e^{-kh} \times (\cos kx - i\sin kx)\, e^{i\omega_e t}, \tag{6.207}$$

where $x = \pm 0.5L_f$, depending on which foil we consider. Further, $\bar{\eta}_3$ and $\bar{\eta}_5$ are the complex amplitudes of heave and pitch, respectively. The angle δ in eq. (6.187) is the same as η_5. Adding together the vertical force on the two foils then gives the following vertical hydrodynamic force:

$$\begin{aligned} F_3^{HD} &= -\rho 0.5\pi c^2 s\big(-\omega_e^2\bar{\eta}_3 - Ui\omega_e\bar{\eta}_5 - i\omega_0\omega_e\zeta_a e^{-kh} \\ &\quad \times \cos(0.5kL_f)\big)\, e^{i\omega_e t} - 2\rho\pi UcC(k_f) \\ &\quad \times s\big(i\omega_e\bar{\eta}_3 - U\bar{\eta}_5 - 0.25ci\omega_e\bar{\eta}_5 - \omega_0\zeta_a e^{-kh} \\ &\quad \times \cos(0.5kL_f)\big)\, e^{i\omega_e t}, \end{aligned} \tag{6.208}$$

where $k_f = 0.5\omega_e c/U$.

Through a similar analysis, we can derive the linear unsteady hydrodynamic pitch moment on the vessel. The contribution from one foil is approximated as $-xLs$, where $x = \pm 0.5L_f$. Here L is given by eq. (6.187). The corresponding total pitch moment is then

$$\begin{aligned} F_5^{HD} &= -\rho 0.5\pi c^2 s\big[-0.25L_f^2\omega_e^2\bar{\eta}_5 \\ &\quad + 0.5L_f\omega_e\omega_0\zeta_a e^{-kh}\sin(0.5kL_f)\big]\, e^{i\omega_e t} \\ &\quad - 2\rho\pi UcC(k_f)\, s\big[0.25L_f^2 i\omega_e\bar{\eta}_5 \\ &\quad - 0.5L_f\omega_0\zeta_a e^{-kh} i\sin(0.5kL_f)\big]\, e^{i\omega_e t}. \end{aligned} \tag{6.209}$$

This approach does not account for a complete interaction with the free surface. Let us now return to eqs. (6.196) and (6.197). We set F_3^{HD} and F_5^{HD} equal to $\sum F_{3j}$ and $\sum F_{5j}$, respectively. Dividing by the common factor $\exp(i\omega_e t)$ and rearranging the terms so that the unknowns appear on the left-hand side of the two equations, gives

$$\begin{aligned} &[-\omega_e^2[M + \rho 0.5\pi c^2 s] + 2\rho\pi UcC(k_f)si\omega_e]\bar{\eta}_3 \\ &- [\rho 0.5\pi c^2 sUi\omega_e + 2\rho\pi UcC(k_f) \\ &\times s\,[U + 0.25ci\omega_e]]\,\bar{\eta}_5 \\ &= [\rho 0.5\pi c^2 si\omega_e + 2\rho\pi UcC(k_f)s]\,\omega_0\zeta_a e^{-kh} \\ &\times \cos(0.5kL_f) \end{aligned} \tag{6.210}$$

$$\begin{aligned} &[-\omega_e^2\left(I_{55} + \rho 0.5\pi c^2 s\, 0.25L_f^2\right) \\ &+ 2\rho\pi UcC(k_f)s0.25L_f^2 i\omega_e]\,\bar{\eta}_5 \\ &= -\left[\rho 0.5\pi c^2 s\omega_e - 2\rho\pi UcC(k_f)si\right] \\ &\times 0.5L_f\omega_0\zeta_a e^{-kh}\sin(0.5kL_f) \end{aligned} \tag{6.211}$$

From eq. (6.211), we note that pitch is uncoupled from heave. This is a consequence of the symmetry of the foil system. However, eq. (6.210) shows that heave is influenced by pitch. We can then first solve eq. (6.211) for pitch and use eq. (6.210) to find heave afterward.

An asymptotic solution for small ω_e can be derived. Acceleration terms will be neglected. This means that the mass terms on the left-hand sides of eqs. (6.196) and (6.197) are neglected. Further, hydrodynamic terms proportional to heave, pitch, and incident wave accelerations are disregarded. An analysis of this provides simplified explanations of what causes the hydrofoil vessel to oscillate in heave and pitch in incident waves with long wavelengths. It also leads to simple expressions for heave and pitch as a function of the incident waves. Only hydrodynamic forces due to lift effects are considered. Because ω_e is small, a quasi-steady approach is followed. This means the linear unsteady lift force on one foil is expressed as

$$L = 0.5\rho U^2 A \frac{dC_L}{d\alpha}\alpha, \qquad (6.212)$$

where A is the planform area of the foil, $dC_L/d\alpha$ is the slope of the steady lift coefficient C_L for the foil, and α is the unsteady angle of attack. If we consider a 2D flat and thin foil in infinite fluid, $dC_L/d\alpha$ is 2π. The angle of attack α can be expressed as

$$\alpha = \eta_5 - \frac{\dot{\eta}_3}{U} + x\frac{\dot{\eta}_5}{U} + \frac{\omega_0\zeta_a}{U}e^{-kh} \times (\cos kx - i\sin kx)\, e^{i\omega_e t}, \qquad (6.213)$$

where $x = \pm 0.5L_f$. This is consistent with the quasi-steady analysis of a foil illustrated in Figure 6.61. We note that the pitch angle and the vertical velocities due to heave, pitch, and incident waves contribute to the unsteady angle of attack α. One should recognize that this quasi-steady approach does not include the steady lift forces that balance the weight of the vessel and cause zero pitch moment about the center of gravity.

Assuming that $dC_L/d\alpha$ and A are the same for the two foils and requiring that the total quasi-steady lift force on the foils is zero, we get

$$U\bar{\eta}_5 - i\omega_e\bar{\eta}_3 + \omega_0\zeta_a e^{-kh}\cos(0.5kL_f) = 0. \qquad (6.214)$$

Here $\bar{\eta}_j$ means the complex amplitude, that is, $\eta_j = \bar{\eta}_j e^{i\omega_e t}$, $j = 3$, and 5. Requiring also that the total quasi-steady pitch moment on the foils is zero results in

$$\bar{\eta}_5 = \frac{2}{L_f}\frac{\omega_0}{\omega_e}\zeta_a e^{-kh}\sin(0.5kL_f). \qquad (6.215)$$

Substituting this into eq. (6.214) gives

$$\bar{\eta}_3 = -i\frac{\omega_0}{\omega_e}\zeta_a e^{-kh} \left[\frac{2U}{\omega_e L_f}\sin(0.5kL_f) + \cos(0.5kL_f)\right]. \qquad (6.216)$$

We note that both $\bar{\eta}_3$ and $\bar{\eta}_5$ are independent of $dC_L/d\alpha$. The first term in the brackets in eq. (6.216) is the result of pitch. The equation illustrates that coupling with pitch increases the heave amplitude. An important reason heave and pitch are relatively small in head seas is the factor ω_0/ω_e in eqs. (6.215) and (6.216), which is less than one at forward speed. Eqs. (6.215) and (6.216) show that heave and pitch are 90° out of phase when the frequency of encounter is small.

An important response variable is the wave-induced vertical acceleration of the vessel whose amplitude at a longitudinal coordinate x can be expressed as

$$a_3 = \omega_e^2\,|\bar{\eta}_3 - x\bar{\eta}_5|\,. \qquad (6.217)$$

However, a practical evaluation of the vertical accelerations requires that an irregular sea state is considered. If long-crested waves are assumed, a wave spectrum $S(\omega_0)$ will describe the energy content of the waves as a function of ω_0 for a given significant wave height $H_{1/3}$ and mean wave period T_1. This is explained in section 3.3. How we can obtain the standard deviation σ_{a3} of the vertical acceleration in long-crested head sea waves in a short-term sea state is described in section 7.4.1. The result is simply

$$\sigma_{a3}^2 = \int_0^\infty \left(\frac{\omega_e^2\,|\bar{\eta}_3 - x\bar{\eta}_5|}{\zeta_a}\right)^2 S(\omega_0)d\omega_0. \qquad (6.218)$$

Another important response variable is the relative vertical motion between the waves and the vessel. This variable measures the possibility of foil ventilation, broaching, and slamming. We now consider regular waves and start by expressing the incident wave elevation ζ. This must be done with a phasing that is consistent with the vertical incident velocity given by eq. (6.198). This means

$$\zeta = \zeta_a \sin(\omega_e t - kx)\,. \qquad (6.219)$$

The complex representation is

$$\zeta = -i\zeta_a e^{-ikx}e^{i\omega_e t}. \qquad (6.220)$$

The amplitude η_{Ra} of the relative vertical motions at a longitudinal position x is then

$$\eta_{Ra} = \left|\bar{\eta}_3 - x\bar{\eta}_5 + i\zeta_a e^{-ikx}\right|. \quad (6.221)$$

The standard deviation σ_{η_R} of relative vertical motion in long-crested head sea waves in a sea state can be expressed as

$$\sigma_{\eta R}^2 = \int_0^\infty \left(\frac{\left|\bar{\eta}_3 - x\bar{\eta}_5 + i\zeta_a e^{-ikx}\right|}{\zeta_a}\right)^2 S(\omega_0)d\omega_0. \quad (6.222)$$

Table 6.3. *Main parameters used in seakeeping analysis of a hydrofoil vessel with a fully submerged and symmetric foil system in foilborne condition (see Figure 6.64)*

	Symbol	
Vessel mass/mass density of water	M/ρ	$185.0\,\text{m}^3$
Foil span	s	12.0 m
Chord length of foil	c	1.5 m
Foil submergence	h	1.5 m
Distance between fore and aft foil	L_f	25.0 m
Radius of gyration in pitch	r_{55}	$0.25\,L_f$

Following sea

We will demonstrate that heave, pitch, and relative vertical motions may be large in following seas when the frequency of encounter ω_e is small and no active control is used. We consider regular incident waves and generalize the quasi-steady approach for head seas that leads to eqs. (6.215) and (6.216) for the complex amplitudes of the pitch and heave.

The incident wave elevation ζ and vertical incident fluid velocity w in following seas can be expressed as

$$\zeta = -i\zeta_a\left(\cos kx + i\sin kx\right)e^{i\omega_e t} \quad (6.223)$$

$$w = \omega_0\zeta_a e^{-kh}\left(\cos kx + i\sin kx\right)e^{i\omega_e t}, \quad (6.224)$$

where

$$\omega_e = \omega_0 - \omega_0^2 U/g. \quad (6.225)$$

The expression for $\bar{\eta}_3$ and $\bar{\eta}_5$ can be obtained by noting that a difference in w for head and following sea is a result of the term $\cos kx - i\sin kx$ for head seas (see eq. (6.200)) and $\cos kx + i\sin kx$ for following seas. So we just change the sign of the $\sin(0.5kL_f)$-terms in eqs. (6.215) and (6.216) to obtain the following expressions

$$\bar{\eta}_5 = -\frac{2}{L_f}\frac{\omega_0}{\omega_e}\zeta_a e^{-kh}\sin(0.5kL_f) \quad (6.226)$$

$$\bar{\eta}_3 = -i\frac{\omega_0}{\omega_e}\zeta_a e^{-kh} \times \left[-\frac{2U}{\omega_e L_f}\sin(0.5kL_f) + \cos(0.5kL_f)\right] \quad (6.227)$$

for $\bar{\eta}_5$ and $\bar{\eta}_3$ in following sea. The amplitude of vertical acceleration can be obtained by eq. (6.217), whereas the amplitude of relative vertical motions is

$$\eta_{Ra} = \left|\bar{\eta}_3 - x\bar{\eta}_5 + i\zeta_a e^{ikx}\right|. \quad (6.228)$$

Standard deviation of vertical accelerations and relative vertical motions in a sea state can be obtained similarly to that for head seas.

Eqs. (6.226) and (6.227) show that heave and pitch are proportional to ω_0/ω_e. Because eq. (6.225) shows that ω_e may be zero for following seas, infinite vertical motions theoretically may occur. However, this has to be counteracted by an active control system. If we had been able to record the vertical incident wave velocities at the two foils, by using independently controlled flaps at the front and aft foil, we could create lift forces on the two foils that counteract the angle-of-attack effect of the vertical incident wave velocities and minimize the wave excitation. However, a procedure like that is not practical. The flap commands are instead based on measurements of vertical accelerations, pitch, and relative vertical motions by means of accelerometers, gyros, and height sensors. A contouring mode, as illustrated in Figure 6.7 is selected for long wavelengths.

6.12.1 Case study of vertical motions and accelerations in head and following waves

Let us consider a hydrofoil vessel in a foilborne condition with main parameters as presented in Table 6.3. The vessel speed is $25\,\text{ms}^{-1}$.

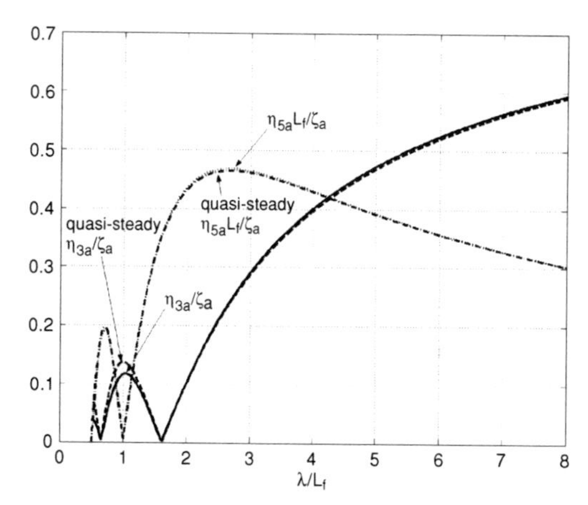

Figure 6.65. Heave amplitude η_{3a} and pitch amplitude η_{5a} in regular head sea waves of a hydrofoil vessel with a fully submerged foil system in foilborne condition. ζ_a = incident wave amplitude, λ = incident wavelength, L_f = longitudinal distance between the geometric centers of the fore and aft foils. Vessel speed = 25 ms^{-1}. No active control is applied. Vessel data are given in Table 6.3.

The transfer function of heave, that is, the heave amplitude $\eta_{3a} = |\bar{\eta}_3|$ divided by the incident wave amplitude ζ_a is presented together with $\eta_{5a}L_f/\zeta_a$ in Figure 6.65 as a function of the ratio between the incident wavelength λ and L_f. Here η_{5a} means $|\bar{\eta}_5|$. We note that the quasi-steady approximation given by eqs. (6.215) and (6.216) is a good approximation in the considered wavelength range. The values are small compared with the corresponding values for a semi-displacement vessel (see Figures 7.29 and 7.30). This confirms the good seakeeping behavior in head sea of a hydrofoil vessel with a submerged foil system in foilborne condition. There are two important reasons why a semi-displacement vessel has worse seakeeping behavior than a hydrofoil vessel in head seas. One reason is that resonant vertical motions with relatively small damping are excited for a semi-displacement vessel. This does not occur for a hydrofoil vessel. Further, the wave excitation heave force and pitch moments are larger for a semi-displacement vessel than for a hydrofoil vessel. An important reason is the significance of the Froude-Kriloff loads.

The numerical and experimental results by Falck (1991) for a foil catamaran without active control show a behavior similar to that of the heave and pitch results in Figure 6.65. Falck's numerical method is a generalization of Sclavounos's (1987) unsteady lifting line theory. Strut-foil interactions and free-surface effects are accounted for. The free-surface condition is the high–Froude number, linear free-surface condition $\varphi = 0$ on $z = 0$ (see Figure 6.50), where φ is the velocity potential due to the hydrofoil vessel.

When λ/L_f is smaller than the values considered in Figure 6.65, the quasi-steady approximation of the equations of motion is no longer adequate to describe the seakeeping behavior of the vessel. This matters, for instance, in calculating the vertical accelerations of the vessel in realistic sea states. The mass inertia terms of the vessel must then be considered, whereas hydrodynamic loads proportional to the acceleration of the vessel are still of secondary importance.

Figure 6.66 shows calculated standard deviation σ_{a3} of vertical accelerations of the vessel at $x = 0$ and $\pm 0.5L_f$ in irregular long-crested head sea waves for mean wave periods T_2 between 3 and 10 s. However, the approximation of $\dot{h}$ (see eq. 6.207) by evaluating the vertical incident wave velocity w at the mean position of a foil becomes inaccurate for small wave periods or wavelengths, that is, when the wavelength is smaller than approximately five times the chord length. The effect of the incident waves should then be evaluated by the Sears function (Newman 1977). The inadequacy of our theoretical model for the

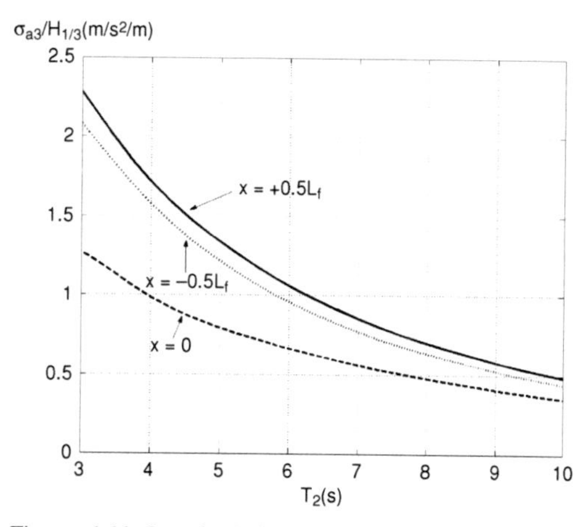

Figure 6.66. Standard deviation σ_{a3} of vertical acceleration of a hydrofoil vessel in foilborne condition at $x = 0, \pm 0.5L_f$ in long-crested irregular head sea waves described by a Pierson-Moskowitz wave spectrum. $H_{1/3}$ = significant wave height, T_2 = mean wave period. Vessel speed = 25 ms^{-1}. No active control is applied. Vessel data are given in Table 6.3.

smaller wavelengths is the reason we do not present results for mean wave periods smaller than $T_2 = 3\,\text{s}$ in Figure 6.66. However, because waves below 3 s are associated with small significant wave heights, they are of little practical significance.

The results in Figure 6.66 are in qualitative agreement with numerical and experimental results by Falck (1991) for a foil catamaran without active control in irregular head sea waves. However, because the vessels are not the same and the mean wave periods and wave spectra were not specified, quantitative comparisons are not possible. Eq. (6.218) with a Pierson-Moskowitz wave spectrum (see eq. (3.55)) has been used in our calculations. The calculated values should be related to operability-limiting criteria for vertical accelerations, as, for instance, presented in Table 1.1. A commonly used criterion is $\sigma_{a3} = 0.2\,\text{g}$ at the center of gravity. There are, of course, other wave headings and response variables, such as relative vertical motion at the foils, that must be considered. If we consider only head sea and vertical accelerations, then Figure 6.66 directly determines what the maximum operational $H_{1/3}$ is for a given T_2 in head sea when the vessel speed is $25\,\text{ms}^{-1}$. By having information about the occurrence of different combinations of $H_{1/3}$ and T_2, we can then directly calculate the percentage of time that the ship can operate. However, we should realize that the results are dependent on the use of active control. For instance, the operational limit for the Jetfoil by Kawasaki is larger than $H_{1/3} = 3.5$ m. The longer the mean wave period is, the higher the limiting value for $H_{1/3}$. The results presented in Figure 6.66 lead to lower operational limits. For instance, if we use $\sigma_{a3} = 2\,\text{ms}^{-2}$ at $x = 0$ as a criterion for operational limits, this shows that the vessel can operate up to $H_{1/3} = 2$ m for $T_2 = 4\,\text{s}$. The results in Figure 6.66 also show that the operational limit increases with T_2.

The full-scale values of vertical accelerations presented in Figure 6.22 are lower than the calculated values in Figure 6.66. However, because the ISO criteria (see Chapter 1) have been used, the presented values in Figure 6.22 have to be lower than the standard deviation. Further, active control has been used for the foil catamaran whereas this is neglected in our theoretical calculations. This has an important effect. Another matter that influences the results is that Figure 6.22 does not indicate what the wave headings are.

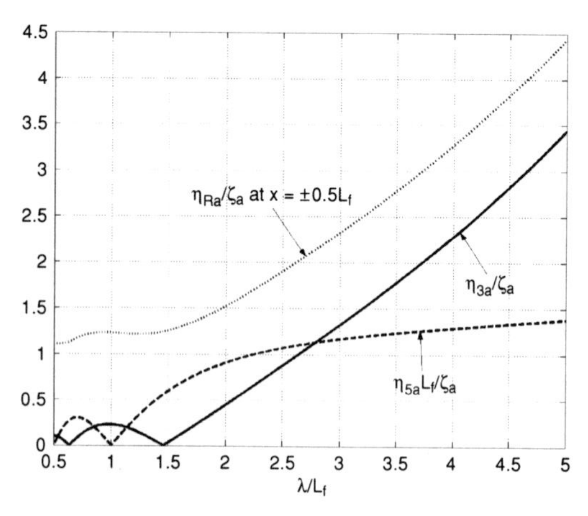

Figure 6.67. Heave amplitude η_{3a}, pitch amplitude η_{5a}, and amplitude η_{Ra} of relative vertical motions at $x = \pm L_f$ as a function of ratio between incident wavelength λ and longitudinal distance L_f between fore and aft foils in regular following waves. ζ_a = incident wave amplitude. The hydrofoil is in foilborne condition with a speed of $25\,\text{ms}^{-1}$. No active control is applied. Vessel data are given in Table 6.3.

Saito et al. (1991) theoretically studied the effect of a control system on the vertical accelerations and relative vertical motions at the bow of the Jetfoil-115 in regular waves. The results show in general a clear difference between platforming and contouring modes. For instance, the platforming mode clearly gives smaller vertical accelerations in head sea waves. However, no result without active control was presented. If we compare vertical accelerations in regular head sea waves for our hydrofoil vessel without active control with their results, our results are clearly larger. This is an indirect way of saying that active control matters.

Figure 6.67 shows calculated heave amplitude (η_{3a}), pitch amplitude (η_{5a}), and relative vertical motion amplitude (η_{Ra}) at $x = \pm 0.5L_f$ in regular following waves. The low-frequency formulas given by eqs. (6.226), (6.227), and (6.228) have been used. Comparing Figure 6.65 and Figure 6.67 shows that vertical motions are clearly larger in following waves than in head sea waves. If results had been presented for higher λ/L_f-values than those in Figure 6.67, the vertical motion would continue to increase because of decreasing frequency of encounter ω_e until $\lambda/L_f = 16.01$ when $\omega_e = 0$. The theoretical model predicts, then, infinite heave and pitch amplitudes. However,

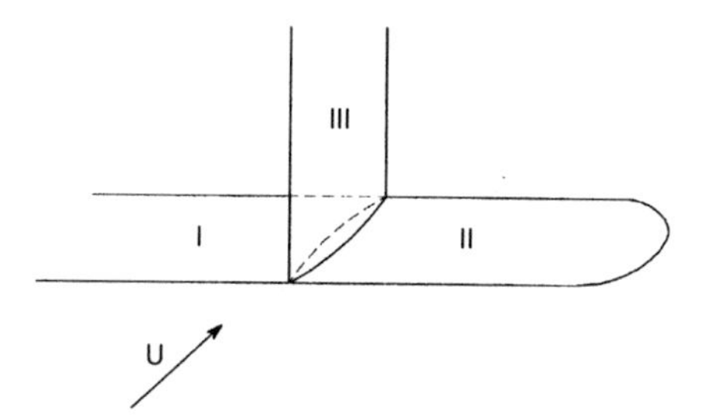

Figure 6.68. Foil-strut intersection.

this long wavelength may not be of practical concern.

Automatic control has to be used in practice when ω_e is small to avoid ventilation and broaching of the foils. By using numerical calculations with the Jetfoil in following regular waves, Saito et al. (1991) showed that active control is effective.

6.13 Exercises

6.13.1 Foil-strut intersection

Part of a foil-strut system is shown in Figure 6.68. Set up the relationship between circulations Γ_1 for the foil part I, Γ_2 for the foil part II, and Γ_3 for the strut III and the intersection between the foil and the strut.

(Hint: Introduce a closed curve C that includes the curves defining $\Gamma_i, i = 1, 3$ so that the circulation along C is zero.)

6.13.2 Green's second identity

a) Green's second identity is used to represent the velocity potential due to a 2D foil as a distribution of sources (sinks) and normal dipoles over the foil surface S_B and normal dipoles over the free shear layer S_V (see eq. (6.32)). It is then necessary that the contribution from a control surface S_∞ at infinity is zero. Show that this is true.

(Hint: Represent the velocity potential at S_∞ as in eq. (6.54)).

b) Use eq. (6.32) to derive the dominant term for the velocity potential due to the foil at large distances from the foil and free shear layer.

(Hint: Taylor expand $\ln r$ and $\partial(\ln r)/\partial n$ about $x = 0$, $y = 0$).

6.13.3 Linearized 2D flow

a) Derive the pressure distribution on a foil due to parabolic camber

b) Assume the mean line of the foil is described by

$$\eta = -\alpha x + C_1 x^3. \tag{6.229}$$

What is the ideal angle of attack α_i?

6.13.4 Far-field description of a high-aspect–ratio foil

Consider the far-field linear steady flow past a high-aspect–ratio foil in infinite fluid as it is illustrated in Figure 6.42. Assume a foil with an elliptic planform. Use eq. (6.46) to show that the vertical fluid velocity can be expressed as

$$w(x_1, y_1, z_1) = \frac{2\alpha U}{\pi(\Lambda + 2)}$$

$$\int_0^\pi d\theta \frac{\cos\theta\,(y_1^* - \cos\theta)}{\left[(y_1^* - \cos\theta)^2 + z_1^{*2}\right]} \tag{6.230}$$

$$\times \left[1 + \frac{x_1^*}{\left[x_1^{*2} + (y_1^* - \cos\theta)^2 + z_1^{*2}\right]^{1/2}}\right],$$

where α is the angle of attack, U is the forward speed, Λ is the aspect ratio, $x_1^* = 2x_1/s$, $y_1^* = 2y_1/s$, and $z_1^* = 2z_1/s$.

Derive also eq. (6.230) by starting out with a dipole representation of the vortex sheet, that is, eq. (6.116) for the velocity potential. It is then useful to note that

$$\frac{d}{dy}\left(\tan^{-1}\left[\frac{z_1}{x_1}\frac{\left(x_1^2 + (y_1 - y)^2 + z_1^2\right)^{1/2}}{(y_1 - y)}\right]\right)$$

$$= \frac{z_1 x_1}{\left((y_1 - y)^2 + z_1^2\right)\left(x_1^2 + (y_1 - y)^2 + z_1^2\right)^{1/2}}.$$

6.13.5 Roll-up of vortices

In the far-field wake, the linear steady flow around a 3D foil can be represented by 2D flow in the cross-sectional plane S_C shown in Figure 6.44. We use this flow picture as a start condition for roll-up of the vortices. This means the start condition is as shown in Figure 6.69.

a) Represent the flow field due to this initial description of the vortex sheet in two different ways by using either a vortex distribution or a

Figure 6.69. 2D representation of the vortex sheet in the far-field linear steady wake of a 3D foil.

dipole distribution. Use Prandtl's lifting line theory for an uncambered lifting surface with elliptical planform to represent the vortex and dipole densities.

b) Now let the vortex sheet be free to move in the cross-sectional plane. Sketch how one can numerically calculate the motion of the vortex sheet, and discuss qualitatively why this causes roll-up of the vortex sheet.

6.13.6 Vertical wave-induced motions in regular waves

a) Eqs. (6.210) and (6.211) are the coupled equations of heave and pitch motions in head sea regular waves of a hydrofoil vessel with a symmetric submerged foil system in foilborne condition. We will concentrate on the hydrodynamic terms associated with $\bar{\eta}_3$ in eq. (6.210). Using seakeeping nomenclature, these can be expressed as

$$A_{33}\frac{d^2\eta_3}{dt^2} + B_{33}\frac{d\eta_3}{dt},$$

where A_{33} and B_{33} are called, respectively, added mass and damping coefficients in heave. Show that

$$A_{33} = \rho 0.5\pi c^2 s + 2\rho\pi Ucs \,\mathrm{Im}\left(C\left(k_f\right)\right)/\omega_e \tag{6.231}$$

$$B_{33} = 2\rho\pi Ucs \,\mathrm{Re}\left(C\left(k_f\right)\right). \tag{6.232}$$

Show that the right-hand side of eq. (6.210) can be expressed as

$$\left(i\omega_e A_{33} + B_{33}\right)\omega_0\zeta_a e^{-kh}\cos\left(0.5kL_f\right).$$

b) A_{33} and B_{33} given by eqs. (6.231) and (6.232) neglect influence of the free surface. We will study a quasi-steady influence of the free-surface by using the asymptotic lift force formula for large Froude numbers given by eq. (6.144). Only heave will be considered. Show that this leads to the following linear unsteady vertical force on the hydrofoil vessel

$$F_3^{RF} = -C_{33}\eta_3, \tag{6.233}$$

where

$$C_{33} = \frac{32Mg\left(\frac{h}{c}\right)\frac{1}{c}}{(1+16(h/c)^2)(2+16(h/c)^2)}. \tag{6.234}$$

(Hints: Start with eq. (6.144) and write $h = h_m - \eta_3$, where h_m is the average vertical position of the foil. Use Taylor expansion about $h = h_m$ and the fact that at $h = h_m$, the lift force on the two foils balances the weight Mg of the hydrofoil vessel. h in eq. (6.234) corresponds to h_m.)

c) Use the main parameters given in Table 6.3 and a ship speed of $U = 25\,\mathrm{ms}^{-1}$. Consider a range of incident wavelengths to L_f-ratios between 0.25 and 8 in head sea.

Discuss the relative importance between

- A_{33} and M
- $-\omega_e^2\left(M + A_{33}\right)$ and C_{33}
- $-\omega_e^2\left(M + A_{33}\right) + C_{33}$ and $\omega_e B_{33}$

7 Semi-displacement Vessels

7.1 Introduction

Monohulls and catamarans, often equipped with foils, trim tabs, and/or interceptors that control the trim angle and minimize wave-induced motions, are nowadays the most established concepts for high-speed vessels. Interceptors are relatively new concepts and are illustrated in Figure 7.5. The vessels have transom sterns, that is, there is a flat part of the aft ship below the mean free surface that is perpendicular to the centerplane. Catamaran designs include the wave-piercing and semi-SWATH (small waterplane area twin hull)-style hulls. The length of high-speed catamarans used for passenger transportation in coastal water is typically 30 to 40 m. Both monohulls and catamarans longer than 100 m have been built. Trimarans and pentamarans are new types of multihull vessels that are considered. They consist of a long center hull with smaller outrigger hulls. The outrigger hulls are important for static heeling stability. The larger vessels are typically ro-pax ferries, which means they carry passengers and allow roll-on/roll-off payloads, most often cars.

Calm water resistance of semi-displacement vessels is dealt with in Chapters 2 and 4. This chapter concentrates on linear wave-induced motions and loads; however, added resistance in waves is also handled. Common statistical procedures for calculating short- and long-term responses based on linear results in regular waves are also shown. One important load aspect is slamming, which is relevant for all high-speed vessels. This is dealt with in detail in Chapter 8. Maneuvering is considered in Chapter 10.

7.1.1 Main characteristics of monohull vessels

Table 7.1 lists examples of main characteristics of large high-speed monohulls. The data are partly based on *Jane's 2002–2003 High-Speed Marine Transportation*. These vessels have very low draft and a high beam-to-draft ratio relative to conventional ships. Steel or aluminium is used as the hull material. All the vessels use waterjet propulsion. The use of motion control systems is also listed. L_{PP}, L_{WL}, C_B, and H_b are the length between perpendiculars, the length of the designer's load waterline, the block coefficient, and the height of the bow above calm water, respectively.

7.1.2 Main characteristics of catamarans

The side hulls of a catamaran have main dimensions different from those of a monohull. We will describe typical main characteristics of high-speed catamarans with a length between perpendiculars L_{PP} varying from 40 to 90 m. The ratio $L_{PP}/\nabla^{1/3}$ may vary from 6 to 7.5. Here ∇ is the displaced volume. The center of buoyancy is typically 40% to 48% of L_{PP} from the stern. When it comes to the cross-sectional form of each hull, the aft part often has a constant local beam (see Figure 7.53). U-frames are typical for the midship section. In order to reduce base drag, the draft at the transom is smaller than that for the midship section. There are either concave frames or V-frames in the forward part of the hull, with nearly constant local draft over most of the fore part. A picture of the waterline is shown in Figure 7.1. b_1 should be larger than b_2 (see Figure 7.1). The ratios $b_1/b_2 = 1.5$ and $b_2/L_{PP} = 1/12$ are typical for 40-m catamarans, whereas b_1/b_2 from 2 to 2.25, b_2/L_{PP} from $1/14$ to $1/15$, and b_1/L_{PP} of 1/7 are representative for 80 to 90 m–long catamarans. The minimum value of b_2 is determined by the engine and is about 3 m for a 40 m–long catamaran. The distance $2p = b_1 + b_2$ between the centerplanes of the two hulls may vary between four and six times the draft midships. The trim angle is often 0° at zero forward speed. An example of the wetdeck for a 40 m–long catamaran is illustrated in Figure 7.2. The longitudinal cross section at the centerplane is illustrated. The transverse cross section of the wetdeck may be wedge-formed or horizontal (see Figure 7.53). The wetdeck of a "wave-piercing" catamaran is very different (see Figure 1.3). The center of gravity is a small distance above the wetdeck. The radii of gyration in pitch and roll are typically 26% of L_{PP} and 50% of the beam of the vessel, respectively. Figure 7.53 and Table 7.5 show

Table 7.1. *Examples of high-speed monohulls with lengths larger than 50 m (Jullumstrø, unpublished)*

Type/ Manufacturer	L_{PP} (m)	Beam B (m)	Draft D (m)	L_{WL}/B	B/D	Bow height H_b (m)	H_b/L_{WL}	C_B	U (knots)	Material	Op. limit $H_{1/3}$ (m)	$\frac{H_b+D}{L_{WL}}$	Propulsion power (kW)	Motion control
Corsair 6000/ Alstom	58.0	10.9	2.0	5.32	5.45	3.8	0.065		34			0.1	7200	4 roll stabilization fins
Corsair 10000/ Alstom	95.5	13.9	2.6	6.87	5.34				36	Steel + aluminum superstructure	4.5		24,300	Stern flaps, fins, bow T-foil
Corsair 11500/ Alstom	100.0	15.7	2.6	6.36	6.03	6.0	0.060		35	HTS + aluminum superstructure	3.1	0.086	28,320	Stabilization fins, transom flaps
Corsair 12000/ Alstom	105.0	15.7	2.6	6.69	6.03				36	Steel + aluminum superstructure	4.5		32,400	Stern flaps, fins, bow T-foil
Corsair 14000/ Alstom	126.0	21.8	3.8	5.78	5.74				40	Steel + aluminum superstructure	6.0		66,200	Stern flaps, fins, bow T-foil
Lursen 70 m	62.5	10.4	2.0	6.00	5.20	5.45	0.087		38	Aluminum		0.129		
MDV 1200 Pegasus/ Fincantieri	82.0	16.5	2.7	4.96	6.11	7.70	0.094		37	HTS		0.127	24,000	
MDV 1200 Superseacat/ Fincantieri	88.0	17.1	3.0	5.14	5.70	8.30	0.094		38	Aluminum		0.128	27,500	

TMV 50/ Rodriguez CN	43.0	9.2	1.35	4.67	6.81	3.90	0.091	0.34	30	Aluminum	0.122	4700	
TMV 70/ Rodriguez CN	64.0	12.4	2.45	5.16	5.06	5.40	0.084	0.32	32	HTS + aluminum superstructure	0.122	9400	
TMV 103/ Rodriguez CN	85.3	14.5	2.1	5.88	6.09			0.41	43	HTS +aluminum superstructure		24,065	
TMV 114/ Rodriguez CN	96.2	16.5	2.5	5.83	6.6				45	HTS + aluminum superstructure		36,000	
TMV 115/ Rodriguez CN	96.2	17.0	2.5	5.66	6.8			0.37	40	Aluminum		28,800	4 roll stabilization fins, 2 T-foils, interceptor
Kattegat/ Mjellem & Karlsen	86.5	17.4	3.7	4.97	4.70	7.40	0.085		38	Aluminum	0.128	23,200	4 roll stabilization fins
Alhambra DF110/Izar	96.2	17.0	2.6	5.65	6.54	8.10	0.084		38	Aluminum	0.111	28,800	2 active fins forward, 2 active transom flaps
Alhambra DF125/Izar	110.0	18.7	2.5	5.88	7.48				40			33,900	2 active fins forward, 2 active transom flaps
Albozyn/Izar	84.0	14.6	2.08	5.75	7.02	7.10	0.085		39	Aluminum	0.109	22,600	Stabilization fins
Mean values:				5.68	6.04						0.115		

HTS = high-tensile steel, L_{WL} = Length of the designers load waterline.

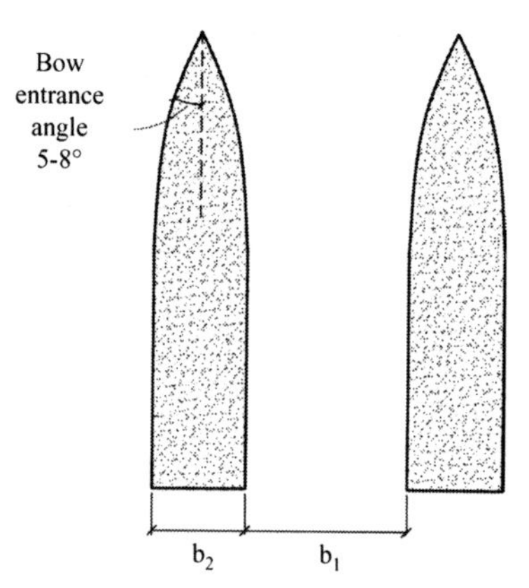

Figure 7.1. Typical waterline of a high-speed catamaran.

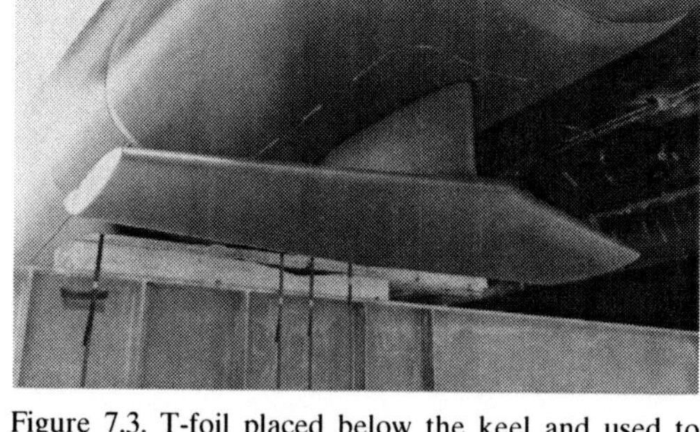

Figure 7.3. T-foil placed below the keel and used to damp vertical wave-induced ship motions at high speed (Seastate).

an example of the body plan and main particulars of a catamaran to be used later in discussing global wave loads.

7.1.3 Motion control

Wave-induced vertical accelerations may cause seasickness and represent an important factor for the limited operability of high-speed vessels (see section 1.1 for criteria). However, motion control may be effective in reducing the heave and pitch of a high-speed vessel. T-foils, trim tabs (flaps), and interceptors are used as parts of control systems for monohull and multihull semi-displacement vessels. Trim tabs and interceptors are situated at the transom stern, whereas T-foils are in the forward part of the vessel. Because the vertical ship motions are largest in the bow, it is an advantage that a heave and pitch damping device is placed close to the bow. The damping effect of T-foils, trim tabs, and interceptors increases with speed and would not be efficient for conventional ships operating at moderate speeds.

Figure 7.3 shows how the T-foil is placed below the keel. A deeply submerged foil is an advantage in avoiding slamming, cavitation, and ventilation. Cavitation is a function of the ambient pressure (including the atmospheric pressure) at the foil position. The higher the ambient pressure, the smaller the probability of cavitation and ventilation. However, the onset of cavitation and ventilation also depends on the local flow around the foil, which is affected by the foil design, the angle of attack of the incident flow to the foil, and the foil motion. The higher the ship speed, the larger the probability of cavitation and ventilation. Cavitation and ventilation on a foil operating at speeds higher than approximately 50 knots are difficult to avoid. Because T-foils add drag to the vessel, it is an advantage that they can be retracted in calm water conditions. Retractable T-foils are obviously also an advantage during operation in shallow water.

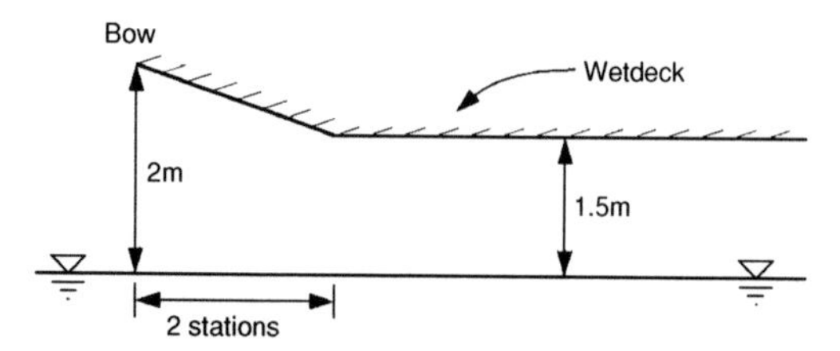

Figure 7.2. The wetdeck of a high-speed catamaran at the centerplane. $L_{PP} = 40$ m. The measure "2 stations" means 10% of the length between perpendiculars (L_{PP}).

The phasing of the flap angles can be controlled relative to the vertical velocities to increase the heave and pitch damping. However, a T-foil also works well as a passive damping device. This will be discussed further later in this chapter.

Figure 7.4 shows a trim tab (flap) installation. It is a flat plate that is hinged from the hull and actuated by a hydraulic cylinder. A transom flap may be either mounted aft of the transom or recessed into the hull forward of the transom. We could say that the trim tab is like a horizontal rudder integrated with the hull and it is only wetted on the lower side when it is in operation at high speed. By changing the angle of the trim tab, we change the trim moment on the vessel so that the trim angle is either increased or reduced. This also can be done as a part of an automatic control system. Further, the roll can also be controlled by having independently operating flaps.

Figure 7.4. Trim tab installation (Seastate). The trim tab is an elongation of the hull bottom. By changing the angle of the trim tab by means of an hydraulic actuator, the trim moment on the ship is changed so that the trim angle is either increased or reduced.

An interceptor is illustrated in Figure 7.5 and has clearly less weight than a trim tab. Minimization of the weight is important for high-speed vessels. Figure 7.6 illustrates how an interceptor with height h works in steady 2D flow. The height h is adjustable. An incident boundary-layer flow is shown. The interceptor is typically inside the boundary layer of the vessel. When the Froude number is sufficiently high, let us say $Fn > 0.4$, the flow separates from the interceptor and leaves a hollow behind the vessel. This is similar to what is discussed in Chapter 4 for transom stern flow (see Figure 4.18). The presence of the interceptor causes a pressure distribution on the hull, illustrated by a hydrodynamic pressure coefficient C_p in Figure 7.6. There is a pressure maximum at the intersection between the hull and the interceptor. If we had potential flow, this hydrodynamic pressure maximum would be $0.5\rho U^2$, where U is the inflow velocity. If the interceptor were not there, the pressure would, in a small area of the hull bottom, decrease to atmospheric pressure at the transom stern.

The pressure distribution illustrated in Figure 7.6 causes a trim moment on the vessel that reduces the trim angle. (Positive trim angle corresponds to bow up.) This is similar to how a trim tab works. Brizzolara (2003) has numerically studied the steady 2D flow situation in Figure 7.6 and how C_p depends on the relative longitudinal coordinate x/h. The interceptor causes a drag force on the vessel, but this is of small significance for typical interceptor heights. Numerical studies of how the interceptor works in incident waves are also needed. If the interceptors are placed on the hull side at the transom stern instead of on the bottom, they can be used for steering control. This may also be possible by trim tabs.

T-foils are most efficient in damping vertical motions. They may also be advantagous in following sea when there is a tendency toward dive-in of the bow. This is a quasi-steady flow situation, and proper change of the flap angles may help in lifting the bow up.

An important effect of roll on catamarans is that roll causes vertical motions of the side hulls. Everything that has been said for heave and pitch damping previously in the text is also relevant for roll damping of catamarans. By independently controlling the vertical motions of the side hulls by T-foils, trim tabs, and/or interceptors, we can

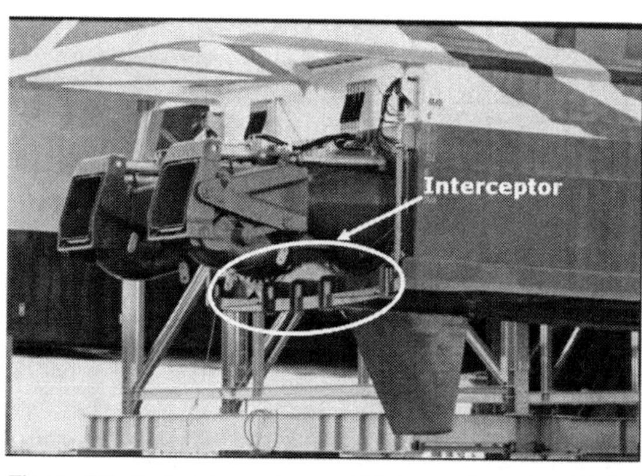

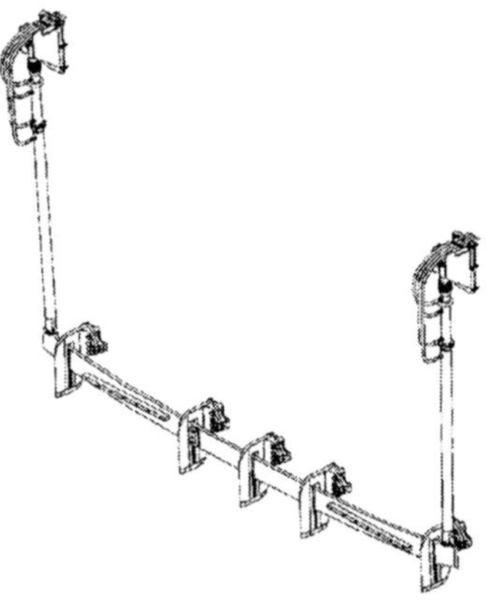

Figure 7.5. Left, interceptor and high-speed rudder. Right, the Seastate interceptor assembly. The horizontal part is the interceptor (Seastate).

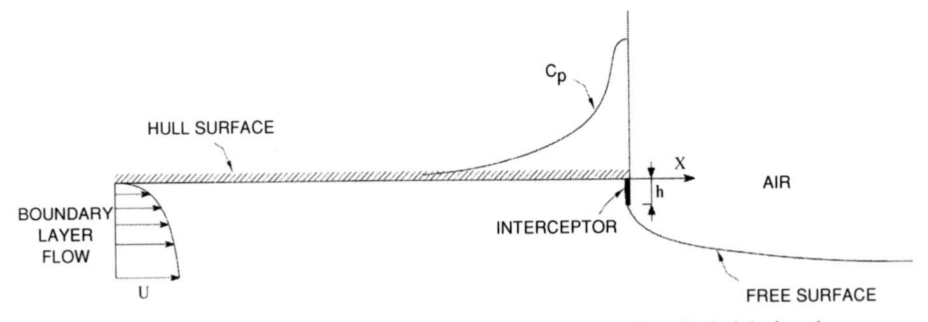

Figure 7.6. 2D steady boundary-layer flow past an interceptor with adjustable height h at the transom stern. C_p = hydrodynamic pressure coefficient on the hull surface.

control the roll motions of a multihull vessel. Roll may represent a problem for monohull vessels. A rudder (see Figure 7.5) provides roll damping at high speeds.

7.1.4 Single-degree mass-spring system with damping

The theoretical description of wave-induced motions and loads on monohull and multihull vessels is, to a large extent, based on a mass-spring type of system with damping. Coupling between the different modes of motions – for example, coupling between heave and pitch or among sway, roll, and yaw – matters. However, to exemplify essential features of the dynamic system, we will limit ourselves to a single degree of freedom system, even if this may be oversimplified for quantitative predictions. This means we study the second-order differential equation

$$m\ddot{y}(t) + b\dot{y}(t) + cy(t) = f(t). \quad (7.1)$$

Here $y(t)$ is the response variable, which may be the heave motion. m is the mass term, which in the case of heave motion, includes the vessel mass and the added mass in heave. b is the damping coefficient. Heave damping is, for instance, caused by wave radiation due to heave oscillations. Other damping contributions, such as hull-lift damping, foil-damping, and viscous damping, are discussed later in this chapter. c is the restoring (spring) coefficient. For instance, changes in the buoyancy force due to heave motion cause a restoring force. There are also restoring components from a foil's angle of attack. $f(t)$ is the excitation force. In the main text of this chaper, we deal with continuous wave excitation. Transient excitation due to slamming (wave impact) is discussed in Chapter 8. Actually, eq. (7.1) does not apply in the time domain for a ship in waves. This is discussed in section 7.3. The problem looks like the one in eq. (7.1) only in steady-state monochromatic waves in which the frequency-dependent added mass and damping are defined.

If the right-hand side $f(t)$ of eq. (7.1) is zero, eq. (7.1), together with the initial conditions, describes the free vibrations. This can be used in experimental studies to obtain the damping coefficient and natural frequency of oscillations (free decay tests).

If $f(t)$ is nontransient, as in continuous wave loading, we are normally interested in a steady-state solution. This means the effect of the initial conditions has vanished. However, if the damping is zero, the effect of the initial conditions does not vanish.

Let us now study the different cases mathematically.

Free vibrations

We set $f(t) = 0$ in eq. (7.1) and specify initial conditions at $t = 0$ for $y(t)$ and $\dot{y}(t)$. Possible solution forms are obtained by substituting $y(t) = \exp(\lambda t)$ into eq. (7.1). This gives

$$\lambda^2 + \frac{b}{m}\lambda + \frac{c}{m} = 0 \quad (7.2)$$

or

$$\lambda_{1,2} = -\frac{b}{2m} \pm \frac{1}{2m}\sqrt{b^2 - 4mc}. \quad (7.3)$$

Here λ_1 and λ_2 are the two possible solutions. They are generally complex. The sign of the discriminant $b^2 - 4mc$ defines whether the solution has an imaginary part. This can be used to classify the

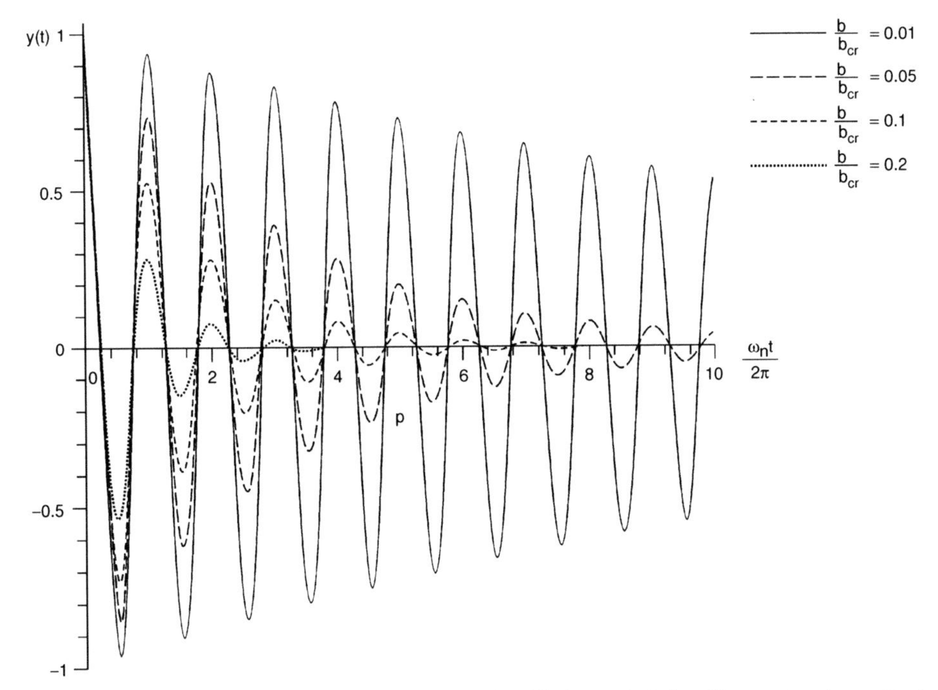

Figure 7.7. Free vibrations as a function of number of oscillation periods for different damping–critical damping ratios $\xi = b/b_{cr} = 0.5b/(mc)^{0.5}$.

solution into three classes:

$$\begin{array}{llll} i) & b^2 > 4mc, & \text{Overdamped} & \\ ii) & b^2 = 4mc, & \text{Critically damped} & (7.4) \\ iii) & b^2 < 4mc, & \text{Underdamped} & \end{array}$$

Overdamping means that the solution has no imaginary part, that is, the solution decays exponentially without oscillating. This is not a practical case for the dynamic systems that we will study. In our case, the systems will be underdamped. The critical damping $b_{cr} = 2\sqrt{mc}$ is often used as a relative measure of the damping level. We concentrate now on an underdamped system and introduce

$$\omega_n = \frac{1}{2m}\sqrt{4mc - b^2} \qquad (7.5)$$

$$\alpha = \frac{b}{2m}. \qquad (7.6)$$

Here ω_n is the damped natural frequency. The general solution of eq. (7.1) can then be expressed as

$$y(t) = e^{-\alpha t}\{A\cos\omega_n t + B\sin\omega_n t\}, \qquad (7.7)$$

where A and B are determined by the initial conditions. Eq. (7.7) can be rewritten as

$$y(t) = \exp\left(-\frac{\xi}{(1-\xi^2)^{1/2}}\omega_n t\right) \times \{A\cos\omega_n t + B\sin\omega_n t\}, \qquad (7.8)$$

where ξ is the ratio between the damping b and the critical damping b_{cr}, that is,

$$\xi = \frac{b}{b_{cr}} \equiv \frac{b}{2\sqrt{mc}}. \qquad (7.9)$$

Eq. (7.8) is exemplified in Figure 7.7 by $A = 1$, $B = 0$, and $\xi = 0.01, 0.05, 0.1, 0.2$ as a function of $\omega_n t/2\pi$. This illustrates the decay rate as a function of the damping ratio ξ.

If m and c are known, we can use an experimental free-decay curve like the one in Figure 7.7 to obtain the damping. Let us exemplify this for small ξ so that $\xi/(1-\xi^2)^{0.5} \approx \xi$. We consider then the two values y_i and y_{i+n} that are recorded at time t_i and $t_i + n\,2\pi/\omega_n$. Here n is an integer. Eq. (7.8) gives then

$$\frac{y_i}{y_{i+n}} = \exp(2\pi n\xi),$$

which is solved for the damping coefficient b:

$$b = \sqrt{mc}\ \ln(y_i/y_{i+n})/\pi n. \qquad (7.10)$$

Forced harmonic oscillations

We express now $f(t)$ in eq. (7.1) as $F_0 \cos \omega t$. The general solution of eq. (7.1) is the sum of the homogeneous solution y_h and the particular solution y_p. Here y_h is the same as eq. (7.8), whereas the particular solution solves the equation when the right-hand side of eq. (7.1) is different from zero. It follows that

$$y_p = \frac{F_0}{(c - m\omega^2)^2 + \omega^2 b^2} \times [(c - m\omega^2)\cos \omega t + \omega b \sin \omega t]. \tag{7.11}$$

So we see that the two parts of the solution, that is, y_h and y_p, oscillate with different frequencies ω_n and ω. This causes a "beating" effect in the time series until y_h is damped out. We will later be interested in the steady-state solution. This means the transient effects represented by y_h have vanished and the solution is steady state and given by eq. (7.11).

The dynamic amplification ratio D is defined as the ratio between the amplitude of y_p and the amplitude $|y_{st}|$ of the quasi-static response. y_{st} is obtained by setting m and b equal to zero in eq. (7.11). It then follows that

$$D = \frac{|y_p|}{|y_{st}|} = \left(\left(1 - \left(\frac{\omega}{\omega_{n0}}\right)^2\right)^2 + 4\left(\frac{\omega}{\omega_{n0}}\right)^2 \xi^2\right)^{-1/2} \tag{7.12}$$

where $\omega_{n0} = (c/m)^{0.5}$ is the undamped natural frequency and ξ is the damping ratio defined by eq. (7.9). Eq. (7.12) is plotted in Figure 7.8 for various damping ratios ξ. Eq. (7.12) shows that $D = 0.5/\xi$, when $\omega = \omega_{n0}$. If ξ is small, maximum response occurs at $\omega = \omega_{n0}$.

These results are of importance later when we study the response of a ship in regular waves. However, m, b, and F_0 will be frequency dependent in that case. Further, coupling between motion modes will matter. This means the response amplitude is only qualitatively similar to that presented in Figure 7.8.

When ω/ω_{n0} is very small, the response is in phase with the excitation. The response is 180° out of phase with the excitation when ω/ω_{n0} is very large. A rapid change in phase occurs when

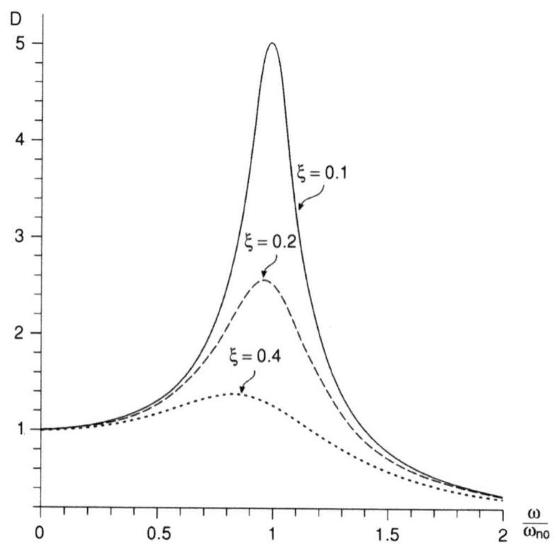

Figure 7.8. Dynamic amplification factor D as a function of ratio ω/ω_{n0} between forcing frequency ω and undamped natural frequency ω_{n0}. Note that when the ratio ξ between damping and critical damping is high, the maximum response occurs at a frequency clearly different from ω_{n0}.

ω/ω_{n0} is near 1. The smaller ξ is, the more rapid the change. When $\omega/\omega_{n0} = 1$, the phase of the response is 90° out of phase with the excitation.

Response to impulsive loads

We now allow $f(t)$ to be any excitation function. The general solution of eq. (7.1) is the sum of the homogeneous solution y_h and the particular solution y_p. Here y_h is the same as eq. (7.8). The particular solution is

$$y_p(t) = \frac{1}{m\omega_n} \int_0^t f(\tau) \sin[\omega_n(t - \tau)] \times \exp[-\xi\omega_{n0}(t - \tau)] d\tau, t > 0, \tag{7.13}$$

where ω_n and ξ are given by eqs. (7.5) and (7.9). Further, $\omega_{n0} = (c/m)^{0.5}$ is the undamped natural frequency.

A slamming load is an example of impulsive loads in which the excitation force $f(t)$ in eq. (7.1) has a limited duration T_d. The character of the response is dependent on the ratio T_d/T_n, where $T_n = 2\pi/\omega_n$ is the natural period. For short-duration loads, that is, $T_d/T_n < \approx 0.25$, the force impulse written as

$$I = \int_0^{T_d} f(t)dt \tag{7.14}$$

determines the maximum response. We can qualitatively understand this from eq. (7.13) by assuming the maximum response occurs on the time scale of T_n. The integral can then be approximated as

$$y_p(t) = \frac{1}{m\omega_n} \sin(\omega_n t) \exp(-\xi \omega_{n0} t) \int_0^{T_d} f(\tau) d\tau. \tag{7.15}$$

This shows that the maximum response is proportional to the force impulse.

For long-duration loading, that is, $T_d / T_n > \approx 1$, the dynamic magnification factor D is between 1 and 2. The tendency is that the longer the rise time of the loading to its maximum value, the lower D is. Detailed descriptions of the response to sine-wave, rectangular, and triangular loading are given by Clough and Penzien (1993).

7.2 Linear wave-induced motions in regular waves

Linear theory can, to a large extent, describe the wave-induced motions of a semi-displacement ship in operational conditions. However, nonlinear effects matter in severe sea states.

Consider a ship in incident regular waves of amplitude ζ_a. The wave steepness is small, that is, the waves are far from breaking. Linear theory implies that the wave-induced motion amplitudes are linearly proportional to ζ_a.

A useful consequence of linear theory is that we can obtain results in irregular waves by adding together results from regular waves of different amplitudes, phases, wavelengths, and propagation directions. This means it is sufficient from a hydrodynamic point of view to analyze a ship in incident regular sinusoidal waves of small wave steepness; this is done in the following text. We assume a steady-state condition, meaning there are present no transient effects due to initial conditions. It implies that the linear dynamic motions and loads on the ship are harmonically oscillating with the same frequency as the wave loads that excite the ship. The hydrodynamic problem in regular waves is normally dealt with as two sub-problems, namely:

1. The forces and moments on the ship when the body is restrained from oscillating and there are incident regular waves. The hydrodynamic loads are called wave excitation loads and are composed of the so-called Froude-Kriloff and diffraction forces and moments. Froude-Kriloff loads are caused by the pressure field in incident waves, which are undisturbed by the ship. Newman (1977) names what we call the diffraction problem the scattering problem. In his nomenclature, the diffraction loads are the sum of the Froude-Kriloff and scattering loads.
2. The forces and moments on the body when the structure is forced to oscillate in calm water with the wave excitation frequency in any rigid-body motion mode. There are no incident waves, but the oscillating body causes radiating waves. The hydrodynamic loads are identified as added mass, damping, and restoring forces and moments. This sub-problem is often termed the radiation problem.

Because of linearity, the forces obtained in items 1 and 2 can be added to give the total hydrodynamic force. One cannot separate the diffraction and radiation problems in a nonlinear theory. Before we go into detail and describe the different hydrodynamic loads, we define coordinate systems and the rigid-body motion modes. A right-handed coordinate system (x, y, z) fixed with respect to the mean oscillatory position of the ship is used, with positive z vertically upward through the center of gravity of the ship, and the origin in the plane of the undisturbed free surface. If the ship moves with a mean forward speed, the coordinate system moves with the same speed. In addition, we define a body-fixed coordinate system $(\bar{x}, \bar{y}, \bar{z})$ that coincides with the (x, y, z) when the ship does not oscillate (Figure 7.9). We show the connection between these two coordinate systems by considering either head or following sea. The ship will surge, heave, and pitch.

We define η_1 (surge) and η_3 (heave) as the translatory motions of the origin of the $(\bar{x}, \bar{y}, \bar{z})$ system along the x- and z-axes, respectively. Positive rotational angle η_5 (pitch) about the y- or $\bar{y}$-axis corresponds to bow up. We consider then a fixed point P on the ship (Figure 7.10) with coordinates $(\bar{x}, \bar{y}, \bar{z})$. The corresponding x and z-coordinates can be derived as illustrated in Figure 7.10, that is,

$$x = \bar{x} \cos \eta_5 + \eta_1 + \bar{z} \sin \eta_5 \tag{7.16}$$

$$z = \bar{z} \cos \eta_5 + \eta_3 - \bar{x} \sin \eta_5. \tag{7.17}$$

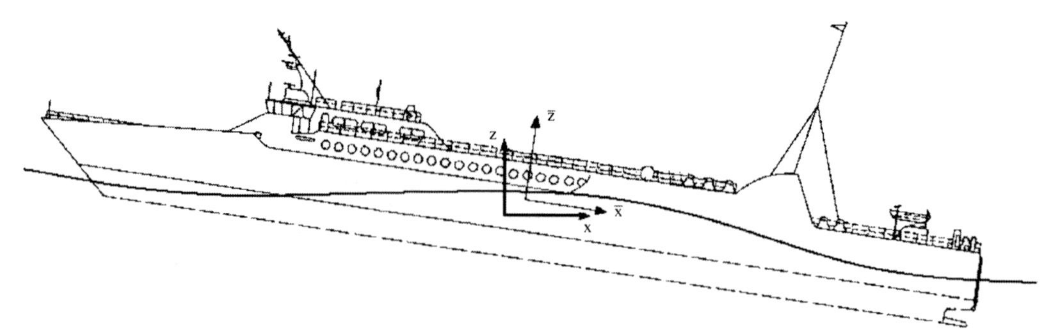

Figure 7.9. Inertial system (x, y, z) moving with the constant ship speed U. Body-fixed coordinate system $(\bar{x}, \bar{y}, \bar{z})$.

Because linear theory is considered, we keep only linear terms in η_i. This gives as a first approximation, $x = \bar{x}$ and $z = \bar{z}$. A second approximation of eqs. (7.16) and (7.17) are

$$x = \bar{x} + \eta_1 + z\eta_5 \tag{7.18}$$

$$z = \bar{z} + \eta_3 - x\eta_5. \tag{7.19}$$

The longitudinal and vertical motions of point P on the ship can therefore be expressed in the (x, y, z)-system as, respectively, $\eta_1 + z\eta_5$ and $\eta_3 - x\eta_5$. This means we do not need the body-fixed coordinate system in describing the linear motions. Because the (x, y, z) coordinate system is an inertial system, we can directly apply Newton's second law and Bernoulli's equation in this system. If we had used the body-fixed coordinate system, we would have had to modify these equations. The body-fixed coordinate system would be natural to use if the complete nonlinear ship-wave interaction problem were to be solved.

Let us now return to a more general formulation of the linear motions in combination with the (x, y, z) coordinate system. Let the translatory displacements in the x-, y-, and z-directions with respect to the origin be η_1, η_2, and η_3 respectively, so that η_1 is the surge, η_2 is the sway, and η_3 is the heave displacement. Furthermore, let the angular displacements of the rotational motions about the x-, y-, and z-axes be η_4, η_5, and η_6, respectively, so that η_4 is the roll, η_5 is the pitch, and η_6 is the yaw angle. The coordinate system and the translatory and angular displacement conventions are shown in Figure 7.11.

The motion of any point on the ship can be written as

$$\mathbf{s} = \eta_1\mathbf{i} + \eta_2\mathbf{j} + \eta_3\mathbf{k} + \omega \times \mathbf{r},$$

where "×" denotes vector product and

$$\omega = \eta_4\mathbf{i} + \eta_5\mathbf{j} + \eta_6\mathbf{k}, \quad \mathbf{r} = x\mathbf{i} + y\mathbf{j} + z\mathbf{k},$$

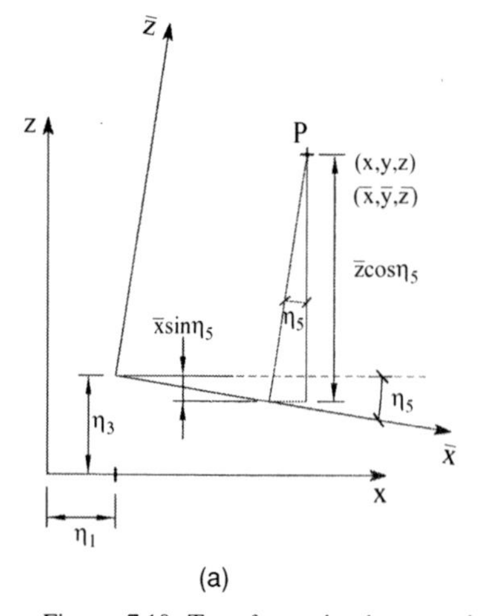

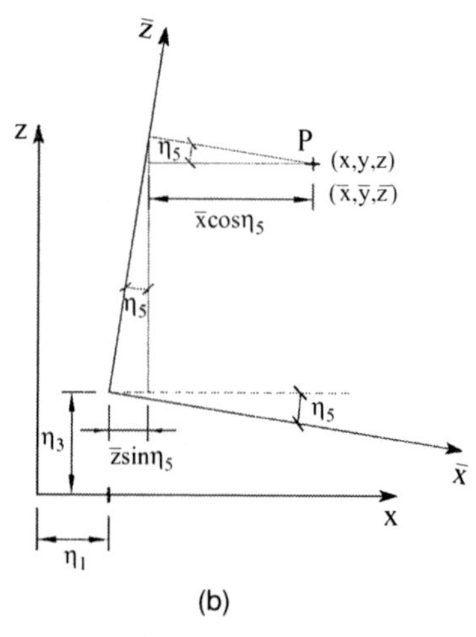

Figure 7.10. Transformation between body-fixed $(\bar{x}, \bar{y}, \bar{z})$ and inertial (x, y, z) coordinate systems.

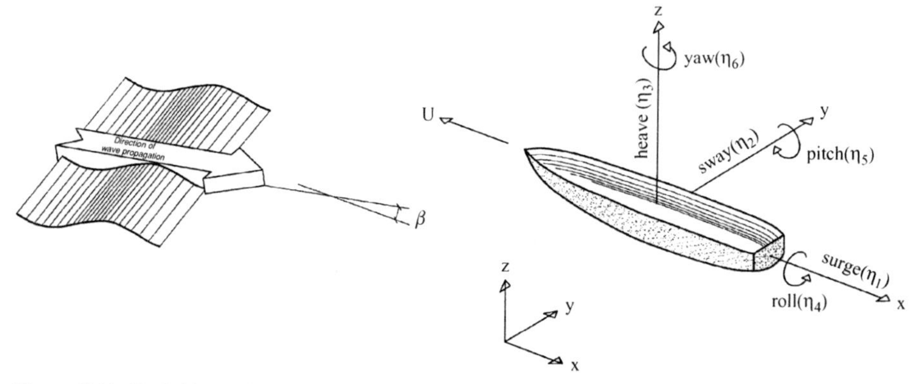

Figure 7.11. Definitions of coordinate system, rigid-body motion modes, and wave propagation direction. U is the forward speed of the ship. The coordinate system moves with the forward speed of the vessel but does not oscillate with the ship. The origin is in the mean free surface. The z-axis goes through the center of gravity of the vessel when the vessel does not oscillate.

and **i**, **j**, **k** are unit vectors along the x-, y-, and z-axes, respectively. This means

$$\mathbf{s} = (\eta_1 + z\eta_5 - y\eta_6)\,\mathbf{i} + (\eta_2 - z\eta_4 + x\eta_6)\,\mathbf{j} + (\eta_3 + y\eta_4 - x\eta_5)\,\mathbf{k}. \tag{7.20}$$

We will now express η_j in the case of steady-state harmonic oscillations in regular incident waves. We then need an expression for the incident waves to relate to. The wave elevation in an Earth-fixed coordinate system XYZ (see Figure 4.6) can be expressed as in eq. (4.4), that is,

$$\zeta = \zeta_a \cos(kX\cos\beta + kY\sin\beta - \omega_0 t - \varepsilon),$$

where we have replaced ω with ω_0 to express the frequency of the waves in the XYZ system. We are free to select the phase angle ε. It is just a question of what we define as $t = 0$. We choose $\varepsilon = -\pi/2$. Then a coordinate transformation to the xyz system (see Figure 4.6), $X = x - Ut$, $Y = y$, gives

$$\zeta = \zeta_a \sin((\omega_0 + kU\cos\beta)t - kx\cos\beta - ky\sin\beta). \tag{7.21}$$

This means the wave elevation oscillates in the xyz-system with the frequency $\omega_0 + kU\cos\beta$. This defines the frequency of encounter, ω_e, that is,

$$\omega_e = \omega_0 + kU\cos\beta \quad \text{and} \quad k = \frac{\omega_0^2}{g} = \frac{2\pi}{\lambda}. \tag{7.22}$$

Here $\beta = 0, 90^\circ, 180^\circ$ corresponds to head sea, beam sea, and following sea, respectively. The linear steady-state motion η_j in six degrees of freedom can now be expressed as

$$\eta_j = |\eta_j|\sin(\omega_e t + \varepsilon_j), \quad j = 1, \ldots, 6. \tag{7.23}$$

Positive ε_j means a phase lead relative to the wave elevation at $x = 0$ and $y = 0$. (Note that the literature and computer programs may have different definitions of phases, but as long as we know the definitions, we can transform one definition of phase angle into another.) The amplitude $|\eta_j|$ is proportional to ζ_a in linear theory. The ratio $|\eta_j|/\zeta_a$ is called a transfer function (or response amplitude operator, RAO) for motion mode j. It is a function of ω_e, U and β and has to be found by either experiments or numerical calculations. Assuming that $|\eta_j|$ and ε_j are known, let us illustrate how other response variables can be derived.

Vertical accelerations in the bow

We consider as the first example head sea, and we want to express the linear wave-induced vertical acceleration at the bow. Using eq. (7.20) and differentiating it twice with respect to time gives

$$a_3 = -\omega_e^2\left[|\eta_3|\sin(\omega_e t + \varepsilon_3) + \frac{L}{2}|\eta_5|\sin(\omega_e t + \varepsilon_5)\right]. \tag{7.24}$$

Here $x = -L/2$ is used for the x-coordinate of the bow. If we want to find the amplitude of the vertical acceleration at the bow, we have to collect $\sin\omega_e t$ and $\cos\omega_e t$ terms separately in eq. (7.24).

This means we write eq. (7.24) as

$$- \omega_e^2 \left\{ \sin \omega_e t \left[|\eta_3| \cos \varepsilon_3 + \frac{L}{2} |\eta_5| \cos \varepsilon_5 \right] + \cos \omega_e t \left[|\eta_3| \sin \varepsilon_3 + \frac{L}{2} |\eta_5| \sin \varepsilon_5 \right] \right\}.$$

The acceleration amplitude then becomes $a_{3a} = \omega_e^2 \sqrt{A^2 + B^2}$, where $A = |\eta_3| \cos \varepsilon_3 + 0.5L|\eta_5| \cos \varepsilon_5$ and $B = |\eta_3| \sin \varepsilon_3 + 0.5L|\eta_5| \sin \varepsilon_5$. The time-dependent vertical acceleration at the bow is

$$a_3(t) = a_{3a} \sin(\omega_e t + \varepsilon_a),$$

where

$$\cos \varepsilon_a = -\frac{A}{\sqrt{A^2 + B^2}} \quad \text{and} \quad \sin \varepsilon_a = -\frac{B}{\sqrt{A^2 + B^2}}.$$

Sway and roll

It is common to say that a vessel rolls about a certain axis (often called roll axis). We now show that this is generally not possible and consider a situation in which yaw is negligible, as it may be in a beam sea condition. By using eqs. (7.20) and (7.23), we can write the linear transverse translatory motion of a point on the vessel as

$$\begin{aligned} s_2(z;t) &= |\eta_2| \sin(\omega_e t + \varepsilon_2) - z|\eta_4| \sin(\omega_e t + \varepsilon_4) \\ &= (|\eta_2| \cos \varepsilon_2 - z|\eta_4| \cos \varepsilon_4) \sin \omega_e t \\ &\quad + (|\eta_2| \sin \varepsilon_2 - z|\eta_4| \sin \varepsilon_4) \cos \omega_e t. \end{aligned}$$

If the vessel is to roll about an axis, we must require there to be a z-value in which the transverse translatory motion is always zero, that is,

$$|\eta_2| \cos \varepsilon_2 - z|\eta_4| \cos \varepsilon_4 = 0 \qquad (7.25)$$

$$|\eta_2| \sin \varepsilon_2 - z|\eta_4| \sin \varepsilon_4 = 0. \qquad (7.26)$$

Solving eq. (7.25) for $z|\eta_4|$ and introducing this into the left-hand side of eq. (7.26) gives

$$|\eta_2| (\sin \varepsilon_2 - \cos \varepsilon_2 \tan \varepsilon_4).$$

Because the sway amplitude is different from zero, in order to satisfy eq. (7.26), we must require that $\tan \varepsilon_2 = \tan \varepsilon_4$. ε_2 and ε_4 depend on ω_e and U and generally do not satisfy this relationship. In the same way, we may show that the ship in general will not pitch about a certain axis in waves. This is the same as saying that the vertical motion amplitude is non-zero everywhere along the ship.

Complex expressions of response variables

It is common to use complex variables to express the linear response variables. This means the motions are written as

$$\eta_j = \bar{\eta}_j e^{i\omega_e t}, \qquad (7.27)$$

where i is the complex unit, $\bar{\eta}_j$ is the complex amplitude, and it is understood in all expressions that it is the real part of the total complex expression, for instance, $\mathrm{Re}\left(\bar{\eta}_j \exp(i\omega_e t)\right)$, that has physical meaning. This is allowed as long as we assume a linear system. Let us illustrate what we mean by using eq. (7.27). The real and imaginary parts of $\bar{\eta}_j$ are called η_{Rj} and η_{Ij}, respectively. This means the physical part of eq. (7.27) is

$$\begin{aligned} &\mathrm{Re}\left\{(\eta_{Rj} + i\eta_{Ij}) e^{i\omega_e t}\right\} \\ &= \eta_{Rj} \cos \omega_e t - \eta_{Ij} \sin \omega_e t, \end{aligned} \qquad (7.28)$$

where we have used $e^{i\omega_e t} = \cos \omega_e t + i \sin \omega_e t$. We expand eq. (7.23) into $\cos \omega_e t$ and $\sin \omega_e t$ terms and compare this with eq. (7.28), giving

$$\eta_{Rj} = |\eta_j| \sin \varepsilon_j$$

$$\eta_{Ij} = -|\eta_j| \cos \varepsilon_j$$

or

$$\varepsilon_j = \tan^{-1}\left(\frac{\eta_{Rj}}{-\eta_{Ij}}\right).$$

It is more convenient to use complex variables when we want to combine linear response variables. Having obtained the final answer for the complex amplitude, we multiply it by $e^{i\omega_e t}$ as in eq. (7.28) and take the real part of the resulting expression to get the physical variable.

Wave-induced accelerations of cargo and equipment

We want to stress that the coordinate system xyz in Figure 7.11 is not fixed relative to the instantaneous position of the ship. This is, for instance, important when we want to study the effect of wave-induced ship oscillations on objects (cargo or equipment) on the deck of the vessel. We might want to find out when the object loses grip or to design foundations or other lashing devices. Consider a head sea condition in which the vessel is oscillating in surge, heave, and pitch. There is then a linear component $g\eta_5$ along the ship-fixed $\bar{x}$-axis. Here g is gravitational acceleration.

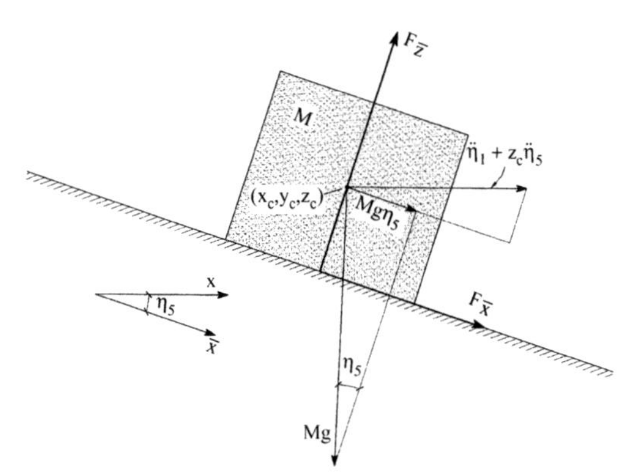

Figure 7.12. Object with mass M on the deck in head sea conditions. Center of gravity coordinates (x_c, y_c, z_c) in the global (x, y, z) coordinate system. The $(\bar{x}, \bar{y}, \bar{z})$ coordinate system is body fixed. A force with components $F_{\bar{x}}$ and $F_{\bar{z}}$ acts from the deck on the object.

Let us consider the consequences of that. We consider an object with mass M on the deck (Figure 7.12). The deck is assumed horizontal in the mean oscillatory position of the vessel. The center of gravity of the object has coordinates (x_c, y_c, z_c) in the (x, y, z) coordinate system defined in Figure 7.11. The longitudinal wave-induced acceleration component at the center of gravity of the object is consistent with linear theory equal to $\ddot{\eta}_1 + z_c\ddot{\eta}_5$ if we refer to either the x- or $\bar{x}$-axis. Here dot means time derivative. We now consider equilibrium conditions of the object along the body-fixed coordinate axes $\bar{x}$ and $\bar{z}$. Consistent with linear theory and using Newton's second law in the $\bar{x}$-direction, we have

$$M(\ddot{\eta}_1 + z_c\ddot{\eta}_5) = Mg\eta_5 + F_{\bar{x}}. \qquad (7.29)$$

Here $F_{\bar{x}}$ is the force component in the $\bar{x}$-direction acting *on the object* as a result of sea fastening and/or friction forces from the deck. If the object is not fastened, we write

$$F_{\bar{x}} = -\mu F_{\bar{z}}, \qquad (7.30)$$

where μ is a friction coefficient and $F_{\bar{z}}$ is the force component in the $\bar{z}$-direction acting from the deck on the object. Consistent with linear theory and by using Newton's second law in the z-direction gives

$$M(\ddot{\eta}_3 - x_c\ddot{\eta}_5) = -Mg + F_{\bar{z}}. \qquad (7.31)$$

We see from eq. (7.31) that it is necessary that $F_{\bar{z}}$ is positive. Otherwise, the object will leave the deck.

Eq. (7.29) shows that the derived response variable (relative acceleration)

$$a_{\bar{x}} = \ddot{\eta}_1 + z_c\ddot{\eta}_5 - g\eta_5 \qquad (7.32)$$

is important in evaluating whether an object on the deck will lose grip or in designing foundations or other lashing devices for the object.

If we consider oblique sea, eq. (7.32) can be generalized to

$$a_{\bar{x}} = \ddot{\eta}_1 + z_c\ddot{\eta}_5 - y_c\ddot{\eta}_6 - g\eta_5. \qquad (7.33)$$

There is, then, also a relative acceleration component

$$a_{\bar{y}} = \ddot{\eta}_2 - z_c\ddot{\eta}_4 + x_c\ddot{\eta}_6 + g\eta_4 \qquad (7.34)$$

along the $\bar{y}$-axis to be considered. We note that the signs of the g-terms in eqs. (7.33) and (7.34) are different. We have already explained the sign in eq. (7.33). If we look at Figure 7.11, we see that positive roll η_4 means a gravity acceleration component $g\eta_4$ along the negative $\bar{y}$-axis.

We also emphasize that eqs. (7.33) and (7.34) are the accelerations needed to estimate the dynamic forces on the object while the forces on the deck/seafastening have opposite directions.

7.2.1 The equations of motions

When the hydrodynamic forces have been found, we can set up the equations of rigid-body motions. This follows by using the equations of linear and angular momentum. For steady-state sinusoidal motions, we may write

$$\sum_{k=1}^{6} [(M_{jk} + A_{jk})\ddot{\eta}_k + B_{jk}\dot{\eta}_k + C_{jk}\eta_k] = F_j e^{i\omega_e t} \quad (j = 1, \ldots, 6), \qquad (7.35)$$

where M_{jk}, A_{jk}, B_{jk}, and C_{jk} are, respectively, the components of the generalized mass, added mass, damping, and restoring matrices of the ship. For example, the subscripts in $A_{jk}\ddot{\eta}_k$ refer to the force (moment) component in j-direction because of motion in k-direction. F_j are the complex amplitudes of the exciting force and moment components. Obtaining the hydrodynamic forces is by no means trivial.

The equations for $j = 1, 2, 3$ follows from Newton's second law, which assumes an inertial system like the (x, y, z) system. For instance, let us consider $j = 1$. For a structure that has lateral symmetry (symmetric about the xz-plane) and with center of gravity at $(0, 0, z_G)$ in its static equilibrium position, we can write the linearized acceleration of the center of gravity in the x-direction as

$$\frac{d^2\eta_1}{dt^2} + z_G\frac{d^2\eta_5}{dt^2}.$$

From this, the components of the mass matrix M_{jk} follow as

$$M_{11} = M, \quad M_{12} = 0, \quad M_{13} = 0$$
$$M_{14} = 0, \quad M_{15} = Mz_G, \quad M_{16} = 0.$$

Here M is the ship mass. We have similar results for the other translatory directions, that is, $j = 2, 3$. For $j = 4, 5, 6$, we have to use the equations derived from the angular momentum. We can then set up the following mass matrix

$$M_{jk} = \begin{bmatrix} M & 0 & 0 & 0 & Mz_G & 0 \\ 0 & M & 0 & -Mz_G & 0 & 0 \\ 0 & 0 & M & 0 & 0 & 0 \\ 0 & -Mz_G & 0 & I_{44} & 0 & -I_{46} \\ Mz_G & 0 & 0 & 0 & I_{55} & 0 \\ 0 & 0 & 0 & -I_{46} & 0 & I_{66} \end{bmatrix}, \tag{7.36}$$

where I_{jj} is the moment of inertia in the jth mode and I_{jk} is the product of inertia with respect to the coordinate system (x, y, z). Explicitly:

$$I_{44} = \int (y^2 + z^2)dM, \quad I_{55} = \int (x^2 + z^2)dM$$
$$I_{66} = \int (x^2 + y^2)dM, \quad I_{46} = \int xz\, dM. \tag{7.37}$$

Here dM is the mass of an infinitesimally small structural element located at (x, y, z). The integration in eq. (7.37) is over the whole structure. This is done in practice by a summation. I_{46} can often be neglected. Further, it is common to express I_{jj} as Mr_{jj}^2, where the radius of gyration r_{jj} corresponding to pitch is typically 0.25 times the ship length.

One may wonder why we did not choose the origin of the coordinate system in the center of gravity of the vessel. That would be natural from a ship mass point of view. However, it is more convenient to use the chosen coordinate system with origin in the mean free surface when hydrodynamic sub-problems are considered. Then we can solve the hydrodynamic problem without considering where the vertical position of the center of gravity is.

The added mass and damping loads are steady-state hydrodynamic forces and moments due to forced harmonic rigid-body motions. There are no incident waves; however, the forced motion of the structure generates outgoing waves. The forced motion results in oscillating fluid pressure on the body surface. If the linearized pressure is expressed as in eq. (3.6), and no interaction with local steady flow is considered, then it is the pressure part

$$p_1 = -\rho\frac{\partial\varphi}{\partial t} - \rho U\frac{\partial\varphi}{\partial x} \tag{7.38}$$

that is considered in the equation of added mass and damping loads. The velocity potential φ is linearly dependent on the forced motion amplitude and is harmonically oscillating with the forcing frequency. Integration of these pressure loads over the mean position of the ship's surface gives resulting forces and moments on the ship. By defining the force components in the x-, y-, and z-directions as F_1, F_2, and F_3 and the moment components along the same axes as F_4, F_5, and F_6, we can formally write the hydrodynamic added mass and damping loads due to harmonic motion mode η_j as

$$F_k = -A_{kj}\frac{d^2\eta_j}{dt^2} - B_{kj}\frac{d\eta_j}{dt}. \tag{7.39}$$

What we have implicitly said is that added mass has nothing to do with a finite mass of the fluid that is oscillating. The latter is a common misunderstanding. Later we actually see an example of a catamaran in which heave-added mass is negative in a certain frequency domain.

Similarly, if we integrate the pressure loads due to the hydrostatic pressure $-\rho gz$, it results in restoring forces and moments. It is then necessary to integrate over the instantaneous position of the ship. Because the COG of the ship is not chosen as the origin of the coordinate system, one must also consider moments due to the ship's weight acting through the COG. We may write the force and moment components as

$$F_k = -C_{kj}\eta_j. \tag{7.40}$$

The only non-zero restoring coefficients for a ship in intact condition, that is, the xz-plane is the symmetry plane for the submerged volume, are

$$C_{33} = \rho g A_W$$
$$C_{35} = C_{53} = -\rho g \iint_{A_W} x \, ds \tag{7.41}$$

$$C_{44} = \rho g \nabla (z_B - z_G) + \rho g \iint_{A_W} y^2 \, ds = \rho g \nabla \overline{GM}$$

$$C_{55} = \rho g \nabla (z_B - z_G) + \rho g \iint_{A_W} x^2 \, ds = \rho g \nabla \overline{GM_L}.$$

Here A_W is the waterplane area; ∇ is the displaced volume of water; z_G and z_B are the z-coordinates of the center of gravity and center of buoyancy, respectively; $\overline{GM}$ is the transverse metacentric height; and $\overline{GM_L}$ is the longitudinal metacentric height. We can, for instance, deduce C_{33} by considering forced heave motion and analyzing the additional buoyancy forces due to hydrostatic pressure $-\rho g z$. This can be linearly approximated as $-\rho g A_W \eta_3$. From this, C_{33} follows from eq. (7.40). The analysis of C_{33} and C_{55} is similar to that used when deriving eqs. (4.108) and (4.109) for sinkage and trim estimates of a ship in calm water. If the ship is equipped with T-foils, one can also associate restoring terms with the foil angle of attack.

If the ship has partially filled tanks, the fluid motion in the tanks will influence the dynamics of the vessel. If the behavior of the fluid in the tanks is assumed quasi-steady, this causes reductions in the longitudinal and transverse metacentric heights. If the period of oscillation is high relative to the highest natural period of the fluid motion in the tank, this is a good approximation. Because resonant fluid motions (sloshing) in a tank may occur, changing the metacentric heights would be wrong in general. An analysis of the coupling between the dynamic fluid motions in the tank and the ship motions as described by Rognebakke and Faltinsen (2003) is necessary. This would account for the proper phasing of the forces and moments caused by sloshing.

For a ship with lateral (port-starboard) symmetry, the six coupled equations of motions reduce to two sets of equations, one set of three coupled equations for surge, heave, and pitch and another set of three coupled equations for sway, roll, and yaw. Thus for a ship with lateral symmetry, surge, heave, and pitch are not coupled with sway, roll, and yaw.

The equations of motions (7.35) can be solved by substituting $\eta_k = \bar{\eta}_k e^{i\omega_e t}$ into the left-hand side. Here $\bar{\eta}_k$ is the complex amplitude of the motion mode k. Dividing by the factor $e^{i\omega_e t}$, the resulting equations can be separated into real and imaginary parts. This leads to six coupled algebraic equations for the real and imaginary parts of the complex amplitudes for surge, heave, and pitch. This is exemplified for coupled heave and pitch in section 9.5.2. A similar algebraic equation system can be set up for sway, roll, and yaw. These matrix equations can be solved by standard methods. When the motions are found, the wave loads can be obtained by using the expressions we discussed previously for hydrodynamic forces .

It should be stressed that equation (7.35) is only generally valid for steady-state sinusoidal motions. For instance, in a transient free-surface problem, the hydrodynamic forces include memory effects and do not depend only on the instantaneous values of body velocity and acceleration (Cummins 1962 and Ogilvie 1964). This is discussed further in section 7.3.

There are different ways to calculate the added mass, damping, and wave excitation loads that appear in the equations of motions in the frequency domain. Before showing this in some more detail, we discuss in the following sections what causes large wave-induced ship motions.

Heave, pitch, and roll are response variables in which the resonance frequencies play an important role. We will show that the wavelength causing resonant heave (and pitch) increases with the Froude number in head sea. The consequence is that the wave excitation loads per unit of wave amplitude causing resonant vertical motion increase with the Froude number in head sea. The resonant response amplitude is obviously also dependent on damping. There are four main sources of damping for a rigid ship. They can be categorized as

- Wave radiation damping
- Hull-lift damping
- Foil-lift damping
- Viscous damping

As long as the flow does not separate and vortices are created, viscous damping will be small and will not be considered here.

In section 7.2.3, we analyze heave motion in beam sea of a monohull at zero speed. This demonstrates that the wave radiation damping decreases, and hence the resonant heave motion per unit of wave amplitude increases with decreasing beam-to-draft ratio B/D. This illustrates that small B/D-values for a monohull and demihulls of a catamaran are not beneficial from a seakeeping point of view. Because wave radiation damping is a function of ship-generated waves, in section 7.2.4 we discuss how these waves depend on ship speed and frequency of oscillation. Wave trapping may occur between the hulls of a multihull vessel. The consequence is small-wave radiation, which is discussed in section 7.2.5 for vertical motions and section 7.2.11 for roll. The effect is most pronounced for smaller ship speeds.

Hull-lift damping in heave and pitch is connected with flow separation from the transom stern at high Froude numbers, which leaves the transom stern dry. This is explained in section 7.2.7 by making a simplifying high-frequency assumption. Because a semi-displacement vessel at maximum operating speed may have large heave and pitch motions, passive and active control by means of hydrofoils is important. This is discussed in sections 7.2.8, 7.2.9, and 7.2.10.

Two common numerical methods are then presented in section 7.2.12: a 3D Rankine panel method and a 2.5D (2D+t) method. Comparisons are made with model tests, and it is shown that interaction between unsteady flow and local steady flow matters in heave and pitch predictions.

7.2.2 Simplified heave analysis in head sea for monohull at forward speed

We now demonstrate how increasing speed in a head sea condition causes higher wave excitation loads for resonant heave motions. Several simplifications are made. The first is that we decouple the heave motion from other motions, that is, we write the equation of motion based on eq. (7.35) as

$$(M + A_{33})\frac{d^2\eta_3}{dt^2} + B_{33}\frac{d\eta_3}{dt} + C_{33}\eta_3 = F_3 e^{i\omega_e t}. \tag{7.42}$$

In practice, one must include the coupling with pitch, but it is common to neglect the coupling with surge in strip theories and in a 2.5D theory. The argument is that a slender hull causes small hydrodynamic forces and moments due to forced surge relative to forced heave and pitch motion. A typical added mass in surge is the order of 5% of the vessel mass.

The next simplification is the wavelength dependence of the vertical excitation loads. We obtain that by considering the vertical Froude-Kriloff force on a box-shaped body of length L and beam B in head sea waves in deep water. We find the corresponding pressure on the bottom of the body by using Table 3.1, that is,

$$p = \rho g \zeta_a e^{-kD} \sin(\omega_e t - kx).$$

Here D is the draft. This results in the following vertical force

$$F_3 = \rho g \zeta_a e^{-kD} B \int_{-L/2}^{L/2} \sin(\omega_e t - kx)\,dx = \rho g \zeta_a e^{-kD} B \frac{2}{k} \sin\left(\frac{kL}{2}\right) \sin \omega_e t.$$

This means

$$F_3^N = \frac{2}{kL}\left|\sin\left(\frac{kL}{2}\right)\right| \tag{7.43}$$

can be used to discuss the wavelength dependence of the heave excitation force. Here $k = 2\pi/\lambda$ is the wave number and L is the ship length. Eq. (7.43) qualitatively expresses the integrated effect of the phase differences of the excitation loads along the ship. For instance, if $\lambda \to \infty$, the sectional excitation loads along the ship are in phase and $F_3^N \to 1$. If $\lambda = L, F_3^N = 0$. This is a consequence of the 180° phase difference between the vertical force from FP to midships and from midships to AP. You can see this by drawing a picture of the instantaneous incident wave along the ship. The relationship between the frequency of encounter ω_e and the wave frequency ω_0 for head sea (see eq. (7.22)) is

$$\omega_e = \omega_0 + \frac{\omega_0^2}{g} U, \tag{7.44}$$

where U is the ship speed. We are interested in studying when ω_e is equal to the undamped natural frequency ω_{n3} in heave. ω_{n3} follows by setting B_{33} and F_3 equal to zero in eq. (7.42) and looking for the nontrivial solutions that are oscillating as $\exp(i\omega_{n3}t)$. This gives that

$$\omega_{n3}^2 = \frac{C_{33}}{M + A_{33}} = \frac{\rho g A_W}{M + A_{33}}.$$

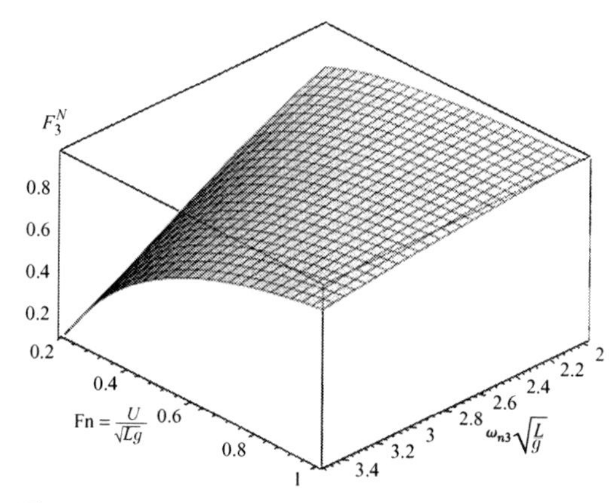

Figure 7.13. Qualitative estimation of how the heave excitation force at natural heave frequency ω_{n3} depends on the ship speed U. This is expressed by the function F_3^N (see Eq. (7.43)). The larger F_3^N is, the higher the heave excitation per unit of wave amplitude. The effect of natural frequency is most pronounced at low speeds.

This can be rewritten as

$$\omega_{n3}\sqrt{\frac{L}{g}} = \sqrt{\frac{L}{D}\frac{C_{WP}}{C_B}\frac{1}{(1 + A_{33}/M)}}. \qquad (7.45)$$

Here $C_{WP} = A_W/(L \cdot B)$ and C_B are the waterplane area coefficient and block coefficient, respectively. By using eq. (7.44) and that $\omega_0^2/g = k$ and $\omega_{n3} = \omega_e$, we can also write

$$\omega_{n3}\sqrt{\frac{L}{g}} = \sqrt{kL} + kLFn.$$

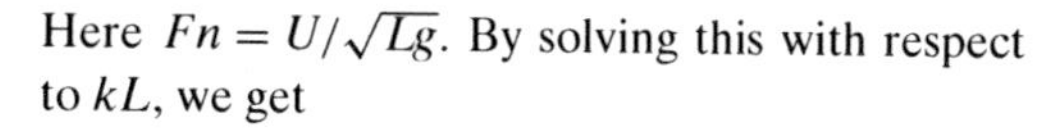

Here $Fn = U/\sqrt{Lg}$. By solving this with respect to kL, we get

$$\sqrt{kL} = \frac{-1 + \sqrt{1 + 4\omega_{n3}\,(L/g)^{1/2}\,Fn}}{2Fn}. \qquad (7.46)$$

This represents λ/L values that for a given speed, give resonance. By using eqs. (7.46) and (7.43), we can get a qualitative picture of how the heave excitation force at heave resonance increases with speed. The results are graphically presented in Figure 7.13 for various values of $\omega_{n3}\sqrt{L/g}$. The range of $\omega_{n3}\sqrt{L/g}$ is found by eq. (7.45).

7.2.3 Heave motion in beam seas of a monohull at zero speed

We now examine how the resonant heave motion of a monohull in beam sea is influenced by the beam-to-draft ratio at zero forward speed. We start by discussing how two-dimensional added mass and damping coefficient vary with beam, draft, and frequency. The damping is caused by the wave radiation. The smaller the ship speed and the higher the frequency of oscillation, the more relevant are these data. We can then use a strip theory approach (Salvesen et al. 1970). This means the flow at different transverse cross sections of the ship are assumed independent of one another, and 2D results are used as building blocks.

Figure 7.14 shows two-dimensional added mass and damping in heave for a rectangular cross section for various beam-draft ratios B/D. The highest B/D value is 8, which means that the B/D

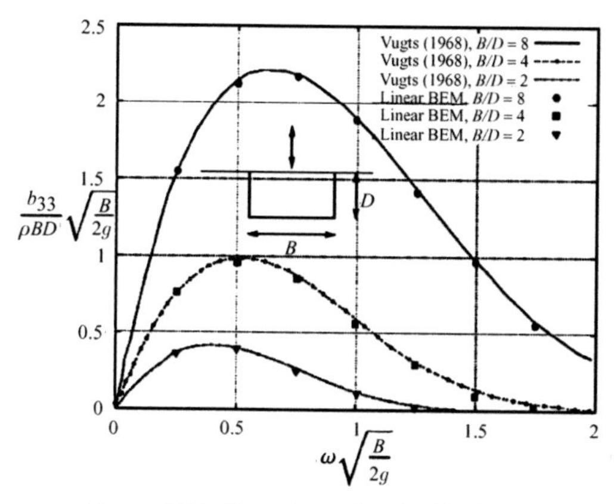

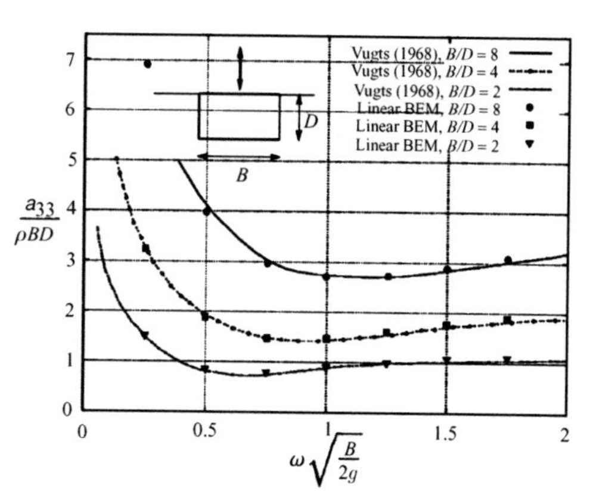

Figure 7.14. Two-dimensional added mass and damping in heave for a rectangular cylinder oscillating on the free surface, for different B/D ratios. B is the beam of the cylinder and D is the draft. Infinite water depth is used. BEM (Baarholm 2001) and results given by Vugts (1968) are shown. a_{33} = 2D added mass in heave, b_{33} = 2D damping in heave, ω = circular frequency of oscillation.

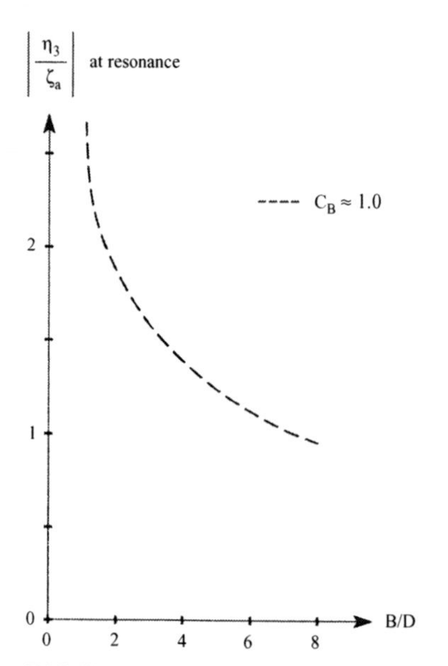

Figure 7.15. Heave amplitude $|\eta_3|$ at resonance of a ship in regular beam sea waves as a function of beam-to-draft ratio. The ship has a constant cross section along its length. ζ_a = incident wave amplitude, B = beam, D = draft. Infinite water depth (Faltinsen 1990).

variations include realistic values for high-speed monohulls (see Table 7.1). The wave radiation damping goes to zero when $\omega \to 0$ and ∞. This implies that the body is then a poor wave generator. The added mass increases strongly when $\omega \to 0$. Kotik and Mangulis (1962) have shown that the added mass in heave for a 2D surface-piercing body goes logarithmically to infinity when $\omega \to 0$. We note that the damping clearly increases with B/D. We illustrate the consequence of this for wave-induced heave motion by considering a two-dimensional rectangular cross section in beam sea regular waves. The heave equation is given by eq. (7.42). For an infinitely long cylinder in beam sea, Newman (1962) expressed the exciting force amplitude $|F_j|$ per unit length as

$$|F_j| = \zeta_a \left(\frac{\rho g^2}{\omega} b_{jj} \right)^{1/2}, \quad j = 2, 3, 4, \tag{7.47}$$

where b_{jj} is the two-dimensional damping coefficient in mode j. Using eq. (7.42), it follows that

$$\eta_3 = \frac{|F_3|\, e^{i(\omega t + \varepsilon)}}{-\omega^2 (m + a_{33}) + b_{33} i\omega + c_{33}}. \tag{7.48}$$

Here m and c_{33} are the structural mass- and heave-restoring coefficients per unit of length. Further, ε is the phase angle of the heave excitation force. The largest heave response occurs approximately when $-\omega^2(m + a_{33}) + c_{33} = 0$, that is, at the undamped natural frequency ω_{n3} for heave motion. This means that resonant heave amplitude is

$$|\eta_3| = \zeta_a \frac{g}{\omega_{n3}^{3/2}} \left[\frac{\rho}{b_{33}} \right]^{1/2}. \tag{7.49}$$

The results are presented in Figure 7.15, which shows that the transfer function for heave $|\eta_3|/\zeta_a$ at the resonance clearly decreases with increasing B/D values. This means that it is beneficial to have the large B/D values that high-speed monohulls have. In contrast, we see from Figure 7.15 that choosing very small B/D values has a negative consequence.

We should, of course, realize that the transverse cross sections of high-speed monohulls are far from rectangular and that C_B can be less than 0.5 (see Table 7.1). However, this does not change the general trend that it is beneficial to have large B/D values also at other headings and at forward speed.

We now start discussing the influence of forward speed by considering the wave system that a ship generates at forward speed.

7.2.4 Ship-generated unsteady waves

A ship generates unsteady waves both when it oscillates and when there are incident waves and

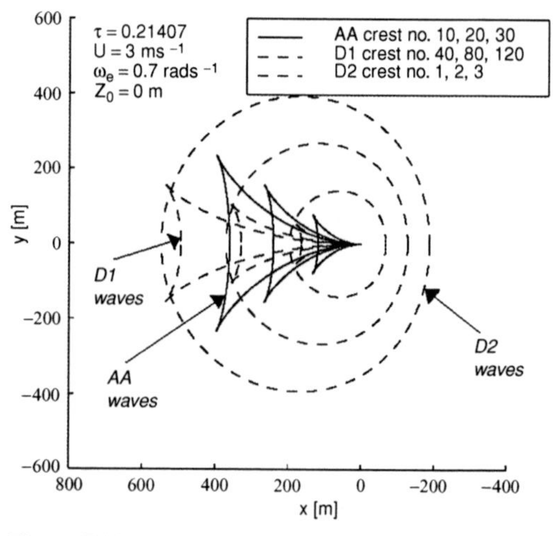

Figure 7.16. Wave crests due to source in (0,0,0) for $\tau < 1/4$; $U = 3.0$ ms^{-1} and $\omega_e = 0.7$rad/s, giving $\tau = 0.21$. Positive x is in the downstream direction (Ronæss 2002).

the ship is restrained from oscillating. Both transverse and divergent waves occur. Later we see that these waves depend on both the frequency and the forward speed. When the frequency is zero and the ship has a forward speed, the wave system is the same as the one we discussed in Chapter 4. The frequency and forward speed dependence of the waves generated by the oscillating ship causes the added mass A_{jk} and damping coefficients B_{jk} to be frequency and forward-speed dependent. Forward-speed dependence is also caused by the body boundary conditions and the pressure expressed by eq. (7.38).

The parameter

$$\tau = \frac{\omega_e U}{g}$$

is important for the characteristics of the waves. When $\tau < 1/4$, deep-water waves may be generated upstream of the ship. This is not true for $\tau > 1/4$. Later we explain why this is so.

In order to find the wave systems and their characteristics, we can proceed with an analysis similar to the one in Chapter 4 on steady waves. However, more wave systems are generated because of a harmonically oscillating source at forward speed. This is illustrated in Figures 7.16 and 7.17 for $\tau < 1/4$ and $\tau > 1/4$, respectively. When $\tau < 1/4$, there are circular waves (D2) propagating outward and two sets of divergent and transverse waves (D1 and AA). The ring waves D2 have their shortest wavelength upstream of the source. When $U = 0$, only ring waves are present. The wavelength is then the same in all directions. The D1 waves have propagation direction with a positive x-component, whereas the x-component is negative for the AA waves. When $\tau > 1/4$, the circular D2 waves degenerate into circular and divergent parts and the transverse waves of system AA disappear. The lines shown in Figures 7.16 and 7.17 are lines with equal phase. They will represent the crests at a specific time instant. Because some of the wave systems have small wavelengths relative to other wave systems, not all wave "crests" are shown for all the wave systems. This is defined in the figure captions.

We illustrate how the wave systems depend on ω_e and U by considering the transverse waves along the ship's track. In order for the Laplace equation to be satisfied, the velocity potential of

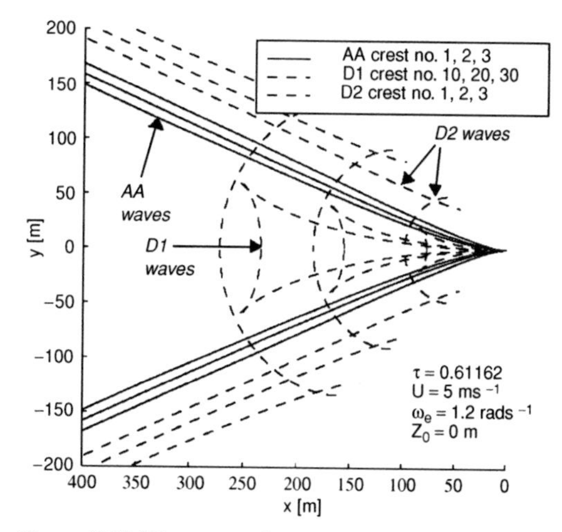

Figure 7.17. Wave crests due to source in (0, 0, 0) for $\tau > 1/4$; $U = 5.0\ \mathrm{ms}^{-1}$ and $\omega_e = 1.2$rad/s, giving $\tau = 0.61$. Positive x is in the downstream direction (Ronæss 2002)

these wave systems has to be of the form

$$\varphi_T e^{-i\omega_e t} = A e^{kz \pm ikx - i\omega_e t}, \qquad (7.50)$$

where A is a constant. The k-value is real and positive in order for the flow to vanish deep in the fluid and for the velocity potential to represent waves at large positive or negative x-values. When ω_e and U are given, the wave number k follows from satisfying the free-surface condition (see eq. (3.10) with $p_0 = 0$), that is,

$$\left[\left(-i\omega_e + U\frac{\partial}{\partial x}\right)^2 + g\frac{\partial}{\partial z}\right]\varphi_T = 0 \quad \text{at } z = 0. \qquad (7.51)$$

We first consider waves propagating along the negative x-axis, that is, in the ship's forward direction. This means the combination $\exp(-ikx - i\omega_e t)$ in eq. (7.50). This gives $-\omega_e^2 - 2k\omega_e U - U^2k^2 + kg = 0$. The corresponding solutions of k are

$$k_{1,2} = \frac{g}{2U^2}\left[1 - 2\tau \pm \sqrt{1 - 4\tau}\right]. \qquad (7.52)$$

Because real solutions are required, τ must be less than one fourth.

Waves propagating along the positive x-axis means the combination $\exp(ikx - i\omega_e t)$ in eq. (7.50). This gives the following two wave numbers:

$$k_{3,4} = \frac{g}{2U^2}\left[1 + 2\tau \pm \sqrt{1 + 4\tau}\right]. \qquad (7.53)$$

This is possible for all τ. In order to find out if the waves corresponding to the various k-values are upstream or downstream of the source (ship), we must evaluate the group velocity. We should then consider the problem from a relative frame of reference system $x' = x - Ut$. In this coordinate system, the ship moves with a velocity U in the negative x'-direction. The various wave systems must be considered in this coordinate system. The group velocity (energy propagation velocity) of the different wave systems can be written as

$$C_g = 0.5\sqrt{g/k}. \tag{7.54}$$

We then find

$$\begin{aligned} C_g^{1,2} &= 0.5\sqrt{g/k_{1,2}} \\ &= \frac{\sqrt{2}}{2}U\left(1 - 2\tau \pm (1 - 4\tau)^{1/2}\right)^{-1/2}. \end{aligned} \tag{7.55}$$

This means that $C_g^1 < U$ and $C_g^2 > U$. Because the phase velocity is in the negative x-direction, the waves corresponding to k_2 are upstream of the source (ship). However, both k_1 and k_2 require $\tau < 1/4$. We get similarly

$$C_g^{3,4} = \frac{\sqrt{2}}{2}U\left(1 + 2\tau \pm (1 + 4\tau)^{1/2}\right)^{-1/2}. \tag{7.56}$$

This means $C_g^3 < U$ and $C_g^4 > U$. However, both these wave systems propagate in the positive x-direction and cannot appear upstream of the source.

Let us calculate the wavelengths corresponding to k_i for the conditions presented in Figures 7.16 and 7.17. It means that the ring waves D2 in Figure 7.16 have wavelengths $\lambda_2 = 59.8$ m upstream and $\lambda_4 = 175.5$ m downstream along the ship's track. The AA and D1 waves have wavelengths $\lambda_1 = 12.1$ m and $\lambda_3 = 4.1$ m. The AA waves have a propagation direction along the negative x-axis but appear downstream of the ship because the group velocity is less than the ship speed.

The case for $\tau > 1/4$ presented in Figure 7.17 has only two transverse wave systems along the ship's track. They both appear downstream of the source (ship). The D1 and D2 waves have wavelengths $\lambda_3 = 7.9$ m and $\lambda_4 = 87.3$ m.

The various wave systems are not equally important. For instance, if the wavelength is of the order of the ship draft, the corresponding wave system is unimportant. Figure 7.18 presents a case by Ronæss (2002) in which only the D2 waves matter. The figure shows the wave pattern for a modified Wigley hull that is harmonically oscillating in heave at $Fn = 0.2$ and $\omega_e\sqrt{(L/g)} = 3$. The results are presented as contour plots of the real and imaginary parts of the complex amplitude. The divergent waves dominate.

A simplified picture of the wave system can be obtained by a strip theory approach. This means we consider first a two-dimensional cross section of a ship that is forced to oscillate with the frequency ω_e. Two-dimensional waves propagating away from the cross section are created. The wave front will move with the group velocity $0.5g/\omega_e$. Let us now introduce the forward speed U and the complete ship hull. The wave front generated initially at the bow part moves a distance $0.5gt/\omega_e$ outward, whereas the ship moves a distance Ut forward. The angle α between the ship and the outer border of the wave system can then be approximated as

$$\tan\alpha = \frac{0.5gt/\omega_e}{Ut} = \frac{g}{2U\omega_e} = \frac{1}{2\tau}. \tag{7.57}$$

This is illustrated in Figure 7.19. τ cannot be too small for this to be valid; for instance, it is not true for $\tau < 1/4$. However, this angle agrees well with the results in Figure 7.18. Further, for a high-speed vessel, the divergent waves are the ones that dominate. This was shown analytically by Ohkusu and Faltinsen (1990). This is also the basis for the 2.5D theory to be considered in section 7.2.12. The wave angle α is consistent with the 2.5D theory.

7.2.5 Hydrodynamic hull interaction

There will be wave interference between the waves generated by each hull of a multihull vessel. *Wave interference* means that the waves generated from each hull are superimposed without accounting for the fact that the waves generated by one hull will be modified because of the presence of another hull (see section 4.3.5). Another matter is wave interaction, that is, that the waves generated by one hull become incident to another hull and wave diffraction occurs as a consequence.

A first check on whether there will be any wave interaction between the two side hulls of a catamaran can be assessed by first assuming no hydrodynamic hull interaction and then considering the wave angle α for one hull given by eq. (7.57) to see if the waves inside the wave angle α become incident to the other hull. By using a procedure

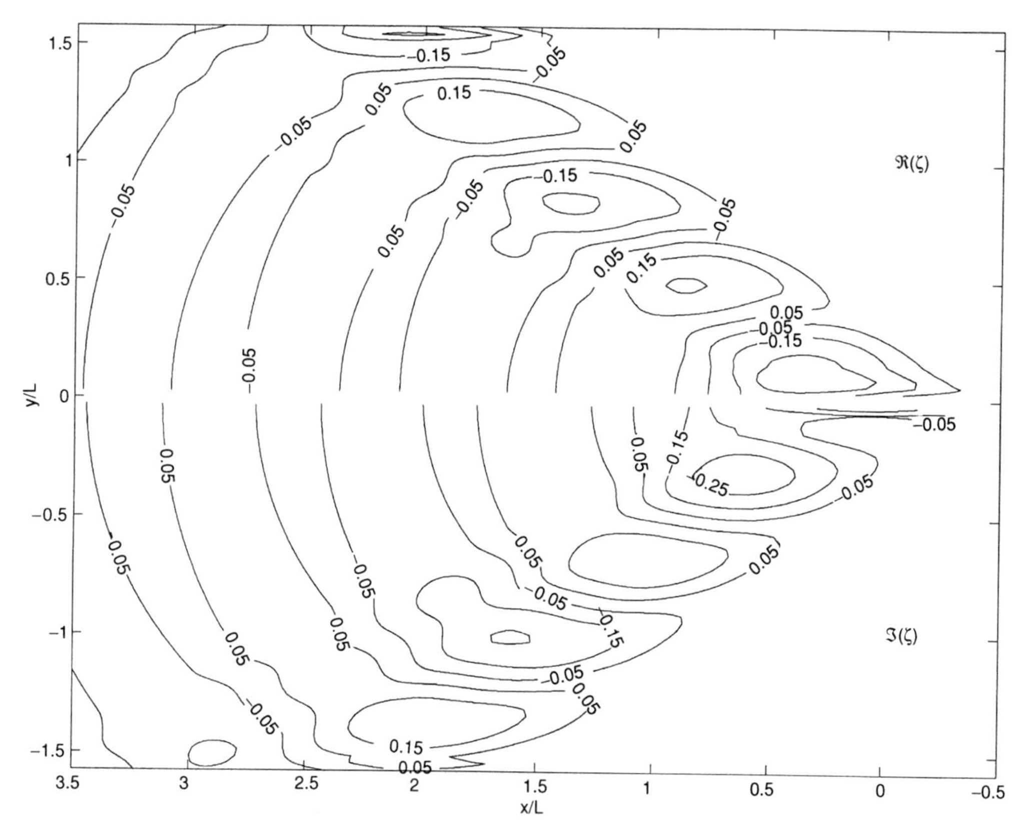

Figure 7.18. Contour plot of the real and imaginary parts of the free-surface elevation due to a modified Wigley hull advancing at $Fn = 0.2$ while oscillating in pure heave at frequency $\omega_e(L/g)^{0.5} = 3$. The upper part is real, the lower part is imaginary. The elevation is made dimensionless with respect to forced heave amplitude. Only D2 waves are included (Ronæss 2002).

similar to the one in Figure 4.26, the length L_I of the aft part of the side hull that is affected by the other hull can be expressed as

$$L_I = L - (b_1 + 0.5b_2) / \tan\alpha.$$

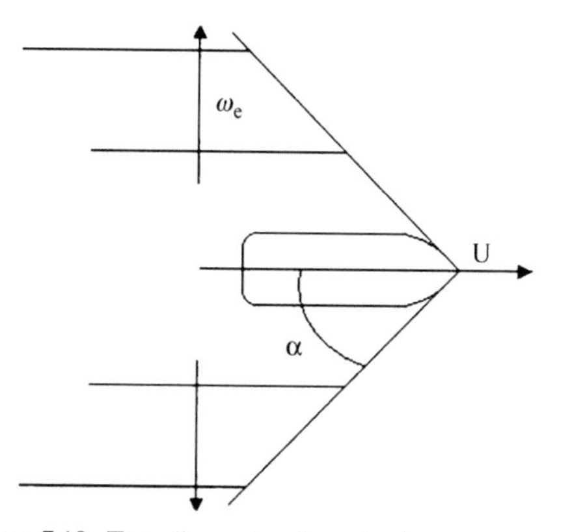

Figure 7.19. Two-dimensional unsteady wave pattern with forward-speed effect (Ronæss 2002).

This means

$$\frac{L_I}{L} = 1 - \left(\frac{b_1 + 0.5b_2}{L}\right) 2\frac{\omega_e U}{g}. \tag{7.58}$$

Eq. (7.58) shows that for a given ω_e, the wave interaction decreases with increasing speed.

Wave trapping due to vertical motions

Wave interaction between the two side hulls becomes particularly strong when resonant wave motion occurs between the two hulls. This problem will be discussed for vertical motions. Section 7.2.11 considers the wave trapping for roll motion. We start by showing two-dimensional results, that is, there is no effect of the forward speed. The features associated with resonance between the two hulls will then be exaggerated. Both 3D flow and forward speed will reduce the effect. However, even if we consider a high-speed ship, it also has to be analyzed for zero-speed

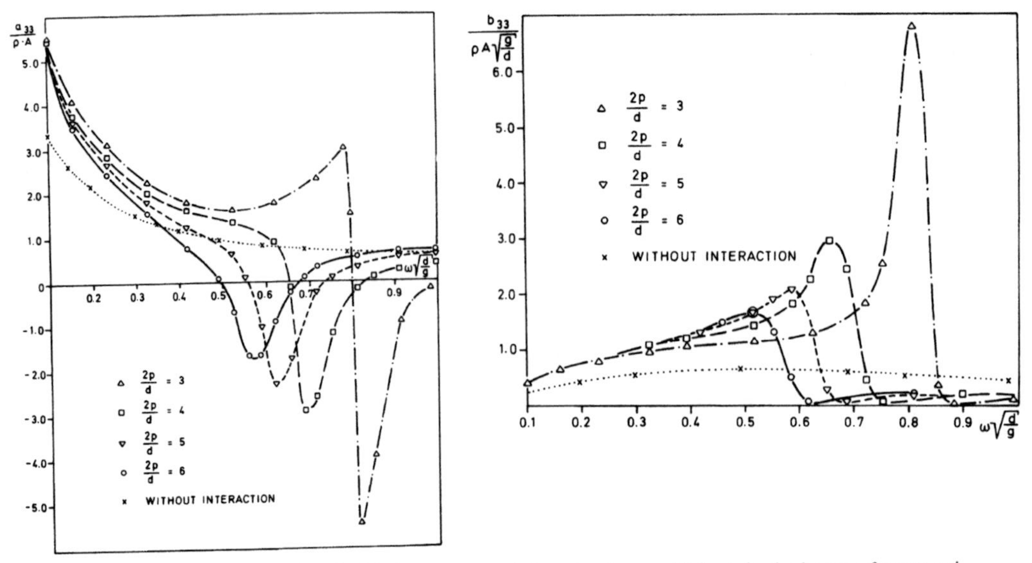

Figure 7.20. Two-dimensional added mass a_{33} and damping b_{33} coefficients in the heave of two semi-submerged circular cylinders with axes in mean free surface. $2p$ = distance between cylinder axis, A = cross-sectional submerged area of the two cylinders, d = draft (Nordenstrøm et al. 1971).

conditions. The reason can be engine failure in bad weather.

Figure 7.20 shows two-dimensional added mass a_{33} and damping b_{33} in the heave of two semi-submerged circular cylinders with axis in the mean free surface, presented by Nordenstrøm et al. (1971). The results are in agreement with theoretical and experimental results by Ohkusu (1969). There is also a curve for added mass and damping coefficients without accounting for hydrodynamic hull interaction. We note a pronounced effect of interaction in the whole frequency domain presented, as well as an even stronger effect in a limited frequency domain. For instance, the added mass becomes negative and the damping coefficient has both a peak value and is close to zero in this limited frequency domain. No waves are radiated when the damping coefficient is zero. Ohkusu (1969) commented that a high standing wave occurred between the maximum and minimum damping coefficients. As an example, he mentioned that the amplitude of the standing wave was 16 times the heaving amplitude for $2p/d = 3$. Here $2p$ is the distance between the centerplanes of the two hulls and d is the draft. Figures 7.21 and 7.22 show similar numerical results for two nearly rectangular-shaped cross sections. Here the ratio $\bar{A}_3$ between the radiated wave amplitude and the forced heave amplitude is presented as function of the square of nondimensionalized frequency. By conservation of energy, $\bar{A}_3$ can be related to the 2D heave damping coefficient b_{33} by noting that the work done in one period T by forcing the body in heave is $b_{33}\omega^2|\eta_3|^2 0.5T$ and that the energy flux due to the generated waves over any period is $2 \cdot 0.5\rho g \bar{A}_3^2 \cdot |\eta_3|^2 \cdot (g/2\omega)T$. This gives

$$b_{33} = \frac{\rho g^2}{\omega^3}\bar{A}_3^2, \tag{7.59}$$

where ω is the forcing frequency. The beam-to-draft ratio and the block coefficient for each hull used in Figures 7.21 and 7.22 are 2 and 0.984, respectively.

Piston mode resonance

The results shown above clearly indicate a resonance effect between the two hulls. We will try to relate these results to what Molin (1999) calls piston mode resonance. His concept is illustrated in Figure 7.23. There is a one-dimensional resonant fluid motion between the two hulls causing an oscillating mass flux with large amplitude at the lower end of the gap between the two hulls.

In the analysis, we use a coordinate system yz, where the y-axis is the plane of the lower horizontal parts of the two rectangular hulls (Figure 7.24). The origin is in the centerplane of the two hulls,

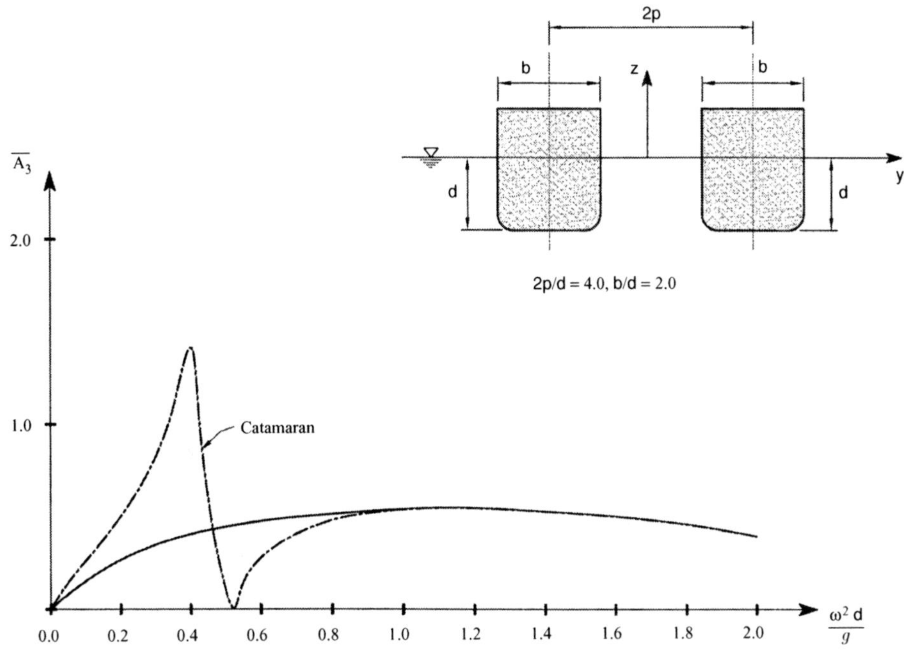

Figure 7.21. Amplitude $\bar{A}_3$ of heave-induced radiation waves per unit of heave amplitude. The solid line shows results when only one of the side hulls is present (Okhusu 1996).

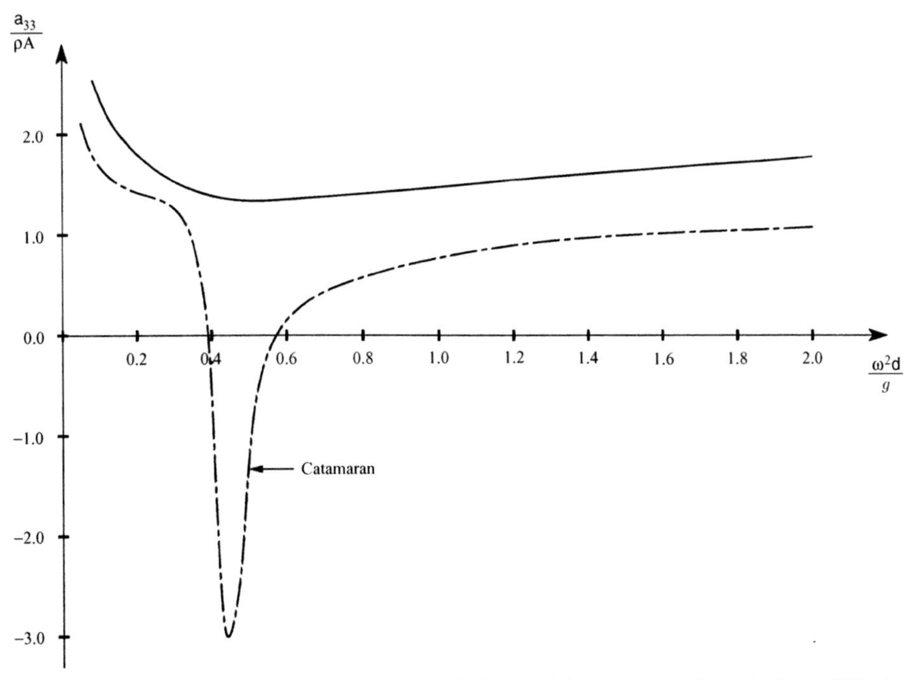

Figure 7.22. Two-dimensional added mass a_{33} in the heave of the catamaran shown in Figure 7.21. $A =$ cross-sectional submerged area of the two hulls, $d =$ draft. The solid line shows results when only one of the side hulls is present (Ohkusu 1996).

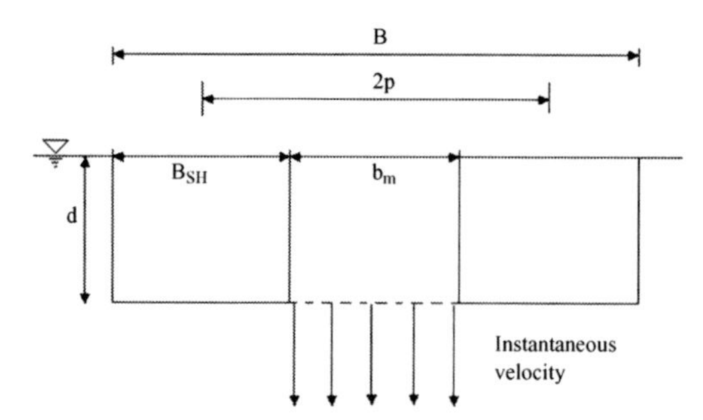

Figure 7.23. Piston mode resonance between the two hulls, illustrated by instantaneous fluid velocity vectors indicating large fluid mass flux. The velocity is harmonically oscillating with the natural frequency of the resonance. As a first approximation, the velocity is constant across the gap between the two hulls.

and the mean free surface is at $z = d$, where d is the draft. In an eigenvalue analysis, we are looking for the nontrivial solutions when there is no forcing. This means the body is restrained from moving.

We assume a one-dimensional fluid motion in the gap between the two hulls, that is, the velocity potential is approximated as

$$\varphi = A_0 + B_0 \frac{z}{d}, \tag{7.60}$$

where A_0 and B_0 are independent of y and z. The later analysis will determine a relationship between A_0 and B_0. Because it is an eigenvalue problem, we cannot determine the values of both A_0 and B_0. Molin represents the flow for negative z-value by a source distribution along the y-axis. If we now make an analogy to thin ship theory in the previous chapter (see Figure 4.12 and accompanying discussion), then the source density is expressed by the vertical velocity along the y-axis. The case in Figure 4.12 was a 3D problem, but the same procedure applies to a 2D problem. The difference is only in the source expression. A two-dimensional source in infinite fluid is $(Q/2\pi)\ln r$,

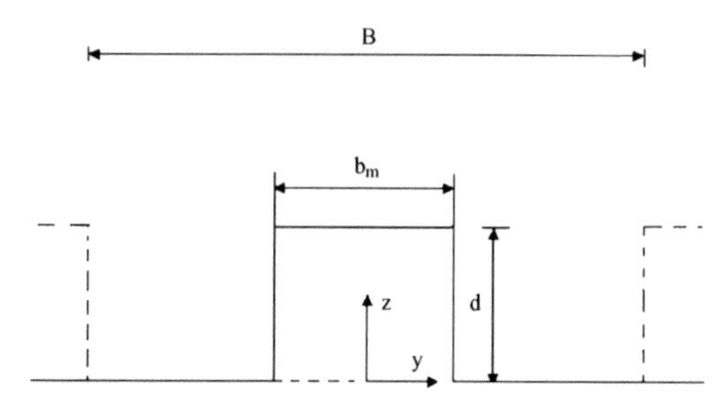

Figure 7.24. Definitions in Molin's (1999) model for piston mode resonance.

where r is the radial distance between the source point and the field point. This representation of the flow implies a source density in the gap expressed by the vertical fluid velocity $\partial\varphi/\partial z$ at $z = 0$, that is, for y between $-0.5b_m$ and $0.5b_m$, where b_m is the breadth of the gap. Because $\partial\varphi/\partial z = 0$ on the bottom of each hull, the source density is zero there.

So far, no approximations have been made in solving the problem. The difficulty is to represent the vertical velocity at $z = 0$ outside the two hulls. We do not really know that before we solve the complete problem for the fluid motion in the whole fluid. It is not sufficient to use only a source distribution for y between $-0.5b_m$ and $0.5b_m$. This will cause infinite pressure at infinity. It is at this stage that Molin makes a big simplification to ensure that the flow at infinity is not source-like, by placing two sinks at $y = \pm B/2$. Here B is the total beam of the catamaran. When we have used the terms *source* and *sinks*, we have not been completely precise. The reason is that $\partial\varphi/\partial z$ at the gap is harmonically oscillating and changing between causing a source and a sink effect. The flow seen from infinity must look like a flow with a non-zero source (sink) strength. Molin's procedure ensures that. We can then represent the velocity potential φ at $z = 0$ in the gap as

$$\varphi(y, 0, t) = -\frac{1}{\pi} \int_{-0.5b_m}^{0.5b_m} \frac{\partial\varphi}{\partial z}(\eta, 0, t) \times \left[\ln|y-\eta| - \frac{1}{2}\ln\left(\frac{B}{2} - y\right) - \frac{1}{2}\ln\left(\frac{B}{2} + y\right) \right] d\eta. \tag{7.61}$$

Here the field point and the source (sink) point coordinates are at $(y, 0)$ and $(\eta, 0)$, respectively. The integration is in the η-direction, which is the same as the y-direction in Figure 7.24. By using the fact that $|y| \le 0.5b_m \ll 0.5B$, eq. (7.61) can be further approximated as

$$\varphi(y, 0, t) = -\frac{1}{\pi} \int_{-0.5b_m}^{0.5b_m} \frac{\partial\varphi}{\partial z}(\eta, 0, t) \ln\frac{|y-\eta|}{0.5B} d\eta. \tag{7.62}$$

Now we equate the equations (7.60) and (7.62). We cannot satisfy this relationship exactly, but we can do it in an average way. We first integrate the right-hand side of eq. (7.62) by assuming $\partial\varphi/\partial z = B_0/d$. This expression will depend on y. This is

Table 7.2. *2D piston mode resonance frequency for two semi-submerged circular cylinders with axes in the mean free surface and horizontal distance 2p between the cylinder axes*

$2p/R$	$\omega\sqrt{R/g}$ at max(b_{33})	$\omega\sqrt{R/g}$ at min(b_{33})	$\omega_n\sqrt{R/g}$
3	0.81	0.88	0.75
4	0.66	0.75	0.67
5	0.59	0.69	0.62
6	0.51	0.61	0.59

R = cylinder radius. $\omega_n\sqrt{R/g}$ is according to Molin's formula (see eq. 7.64). $\omega\sqrt{R/g}$ at max(b_{33}) and min(b_{33}) refer to maximum and minimum values of the two-dimensional damping coefficients presented in Figure 7.20 in the vicinity of $\omega_n\sqrt{R/g}$.

inconsistent with the left-hand side of eq. (7.62), which, by using eq. (7.60), says that $\varphi(y, 0, t) = A_0$. Then we do the averaging. This means we integrate both the left- and right-hand sides from $y = -0.5b_m$ to $0.5b_m$ and divide by b_m. This gives

$$A_0 = \frac{1}{\pi}\frac{b_m}{d}B_0\left[\frac{3}{2} + \ln\frac{B}{2b_m}\right]. \quad (7.63)$$

We have found the relationship between A_0 and B_0, but not the natural frequency. We assume A_0 and B_0 to be harmonically oscillating as $\exp(i\omega_n t)$, where ω_n is the natural frequency, and use the free-surface condition $-\omega_n^2\varphi + g\,\partial\varphi/\partial z = 0$ for $z = d$ and y between $-0.5b_m$ and $0.5b_m$. Eq. (7.60) gives then

$$-\omega_n^2(A_0 + B_0) + g\,B_0/d = 0$$

or

$$\omega_n^2 = \frac{g}{d(1 + A_0/B_0)}.$$

Using eq. (7.63) gives Molin's formula for piston mode resonance frequency, that is,

$$\omega_n\sqrt{\frac{d}{g}} = \sqrt{\frac{1}{1 + \frac{b_m}{\pi d}\left(1.5 + \ln\frac{B}{2b_m}\right)}}. \quad (7.64)$$

It is noted that the right-hand side approaches the value 1 for small b_m and is then in accordance with page 99 in Faltinsen (1990) for moon pools.

We will show that eq. (7.64) gives good estimates for the natural frequency even for the case with two semi-submerged circular cylinders presented in Figure 7.20. Table 7.2 presents the frequency corresponding to the maximum and minimum values of b_{33} for two circular cylinders with axis in the mean free surface based on the calculated results in Figure 7.20. According to Ohkusu's experiments, the resonance condition occurred between the frequencies corresponding to maximum b_{33} and minimum b_{33}. In the same table, we present the resonance frequency ω_n for piston mode resonance according to Molin's formula, even though he assumed rectangular hull sections. We note that ω_n is between the frequencies for maximum and minimum b_{33} when $2p/R$ is 4, 5, and 6. However, when the distance $2p$ between the centerplane of the cylinders is equal to three times the radius R of a cylinder, ω_n is at a slightly lower frequency than the frequency corresponding to maximum b_{33}.

The two nearly rectangular hull sections that are used in the calculated results presented in Figure 7.21 are more in accordance with the assumed hull form in Molin's formula. This case corresponds to $2p/d = 4$. Molin's formula gives $\omega_n^2 d/g = 0.45$, whereas maximum value of $\bar{A}_3$ (the amplitude of radiated wave per unit of heave amplitude) corresponds to $\omega_n^2 d/g = 0.42$. The local minimum value of $\bar{A}_3$ in the vicinity of maximum $\bar{A}_3$ occurs when $\omega_n^2 d/g = 0.53$, thus the average $(0.42 + 0.53)/2 = 0.475$ is in good agreement with Molin's formula.

Ronæss (2002) pointed out that Molin's formula was also useful in indicating piston mode resonance for a catamaran at forward speed. The experimental studies by Kashiwagi (1993) for a Lewis form catamaran at $Fn = 0.15$ were used to validate Ronæss' numerical calculations based on the unified theory concept by Newman (1978) and Newman and Sclavounos (1980). The unified theory is appropriate for all frequencies but implicitly assumes a moderate Froude number. In the analysis, the demihulls were assumed to be in the far field of each other. There is a clear difference in the behavior of the results by Ronæss (2002) relative to the previous 2D results. The added mass in heave now becomes only slightly negative in a certain frequency domain. Further, the damping coefficient in heave does not become zero except when $\omega \to 0$ and $\omega \to \infty$, but that is only a consequence of the fact that a ship is a very bad wave generator for very small and very high frequencies.

However, this very strong interaction effect for this catamaran did not have an important effect on predicted heave and pitch motions by Ronæss (2002). The reason is a clear difference in the piston mode resonance frequency and the heave and pitch resonance frequencies. This conclusion does not need to be so for all catamarans and Froude numbers.

7.2.6 Summary and concluding remarks on wave radiation damping

We have shown that the wave system generated by an oscillating ship is dependent on the frequency of oscillation ω and the forward speed U. The wave radiation damping is therefore also dependent on ω and U. The hull form and hull interaction will matter. A general tendency for heave and pitch wave radiation damping is that the smaller the beam-to-draft ratio of a monohull or the demihull of a catamaran, the smaller the damping relative to the critical damping.

Because both added mass and damping arise from integrating hydrodynamic pressures due to forced oscillations in different modes of motion, added mass will also depend on ω and U. Because the ship also generates waves when it is restrained from oscillating and there are incident waves, the wave excitation loads due to diffraction are also dependent on ω and U.

We must either rely on experiments or numerical methods to estimate the effect of wave generation on hydrodynamic coefficients. When the Froude number is high, this can, for instance, be achieved by a 2.5D theory, to be described later in the text.

All the discussions are based on linear theory and neglect of interaction between local steady flow and unsteady flow.

We now look at two other sources of damping: hull-lift and foil-lift damping. Hull-lift damping is an integrated part of calculations of damping coefficients for the hull. However, we treat it in the next section as a separate effect without any wave generation in order to illustrate the physics.

7.2.7 Hull-lift damping

To illustrate what we mean by hull-lift damping, some simplifications will be made. We neglect free-surface wave generation and use the dynamic free-surface condition $\varphi_B = 0$, where φ_B is the velocity potential due to the vessel. This can be considered a first approximation for a high-speed vessel. The dynamic free-surface condition $\varphi_B = 0$ is also consistent with a high-frequency assumption. The fact that φ_B is assumed to be a constant (i.e., zero) on the free surface implies that the fluid velocity is vertical on the free surface. The next simplification we will make is to use strip theory. As an example, we consider forced heave motion of the vessel. We can formally write the velocity potential as

$$\varphi = \varphi_B + Ux = \varphi_3 \dot{\eta}_3 + Ux, \qquad (7.65)$$

where φ_3 is independent of heave velocity and will be estimated by a strip theory approach. Strip theory means that there is no hydrodynamic interaction between the flows at the various cross sections of the ship. We have here used a coordinate system that moves with the forward speed U of the vessel. The x-direction is in the longitudinal direction of the vessel and is positive aftward. Eq. (7.65) assumes that steady flow around the ship is simply represented by Ux, that is, the steady wave pattern and local flow effects are neglected. We then consider the hydrodynamic pressure that follows from Bernoulli's equation. The Bernoulli equation can be written as

$$p - p_a = -\rho \frac{\partial \varphi}{\partial t} - \frac{\rho}{2} |\nabla \varphi|^2 - \rho g z + \frac{\rho}{2} U^2. \qquad (7.66)$$

Here p_a is the atmospheric pressure and $-\rho g z$ represents the hydrostatic pressure. $z = 0$ is at the mean free surface, and z is positive upward. We now linearize eq. (7.66) by inserting eq. (7.65). This means we keep only the linear terms in η_3. Further, we do not consider the hydrostatic pressure in this context. The hydrostatic pressure is accounted for in the restoring coefficients in the equations of motions. This means we will consider the linearized pressure term

$$-\rho \left(\frac{\partial}{\partial t} + U \frac{\partial}{\partial x} \right) \varphi_3 \dot{\eta}_3 \qquad (7.67)$$

following from eq. (7.66). The last term in eq. (7.67) follows from noting that $|\nabla \varphi|^2 = |\nabla \varphi_3|^2 \dot{\eta}_3^2 + 2U \frac{\partial \varphi_3 \dot{\eta}_3}{\partial x} + U^2$. Let us first consider the pressure term $-\rho \partial \varphi / \partial t$. This gives rise to a vertical hydrodynamic force

$$F_3' = \rho \iint_S \varphi_3 \ddot{\eta}_3 n_3 \, ds = \int_L dx \left[\rho \int_{c(x)} \varphi_3 n_3 \, ds \right] \ddot{\eta}_3, \qquad (7.68)$$

where S is the mean wetted surface and n_3 is the z-component of the normal vector $\boldsymbol{n}$ of the body surface. The positive direction of $\boldsymbol{n}$ is into the fluid. $c(x)$ means the average submerged cross-sectional boundary curve. As a matter of definition, we can also write eq. (7.68) as $-A_{33}\ddot{\eta}_3$, where A_{33} is the added mass in heave. A_{33} can, by a strip theory approach, be written as $\int_L a_{33}(x)dx$, where $a_{33}(x)$ is the two-dimensional added mass in heave for the cross section at longitudinal coordinate x. By comparing this with eq. (7.68) we obtain

$$a_{33}(x) = -\rho \int_{c(x)} \varphi_3 n_3 \, ds. \qquad (7.69)$$

We will use eq. (7.69) when considering the vertical force F_3'' due to the pressure term $-\rho U \dot{\eta}_3 \partial\varphi_3/\partial x$. We can write

$$F_3'' = \rho U \int_L dx \int_{c(x)} \frac{\partial \varphi_3}{\partial x} n_3 \, ds \dot{\eta}_3. \qquad (7.70)$$

We perform a partial integration of the $\partial\varphi_3/\partial x$ term along x. It is then important to note that φ_3 is zero at the bow, but non-zero at the stern. It is then assumed that the flow leaves tangentially from the transom stern. The consequence of the partial integration is that eq. (7.70) can be written as

$$F_3'' = \rho U \int_{c(x_T)} \varphi_3 n_3 \, ds \dot{\eta}_3, \qquad (7.71)$$

where x_T is the x-coordinate of the transom stern. If we now use eq. (7.69), we see that

$$F_3'' = -U a_{33}(x_T)\dot{\eta}_3. \qquad (7.72)$$

This is a force part that originally belongs on the right-hand side of the heave equation of motion. If we move it to the left-hand side of the equation, it appears as a damping term:

$$B_{33}^{HL} = U a_{33}(x_T), \qquad (7.73)$$

which we call the hull-lift damping in heave. If we consider a monohull with a half-circular transom stern with beam b_2, then

$$B_{33}^{HL} = \rho \frac{\pi}{8} b_2^2 U, \qquad (7.74)$$

again assuming the high-frequency free-surface condition $\varphi_3 = 0$. This illustrates a strong dependence on U and b_2. If the transom stern has a non-circular form, we can still say that B_{33}^{HL} is approximately proportional to b_2^2. The reason we use the term *hull-lift damping* is as follows. Let us consider the double body, that is, we image the hull reflected in the free surface. Because of the free-surface condition $\varphi_3 = 0$ and the fact that we consider only vertical motions, the flow around the double body gives the correct flow below the free surface. Let us now make a quasi-steady approximation. The vertical velocity $\dot{\eta}_3$ causes an angle-of-attack effect. Observed from the double body, there appears to be an incident flow with approximate velocity U and an angle of attack $-\dot{\eta}_3/U$. The double body can be considered a low-aspect ratio–lifting surface with the trailing edge corresponding to where the transom stern is. If we apply the low-aspect ratio–lifting surface theory (see section 10.3.1), we will get the same result as the one we have already derived.

One should note from the derivation of eq. (7.72), that the force is not acting at the transom stern even though the hull dimensions at the transom stern are what determine the force. If we go back to eq. (7.70), we see that we get a force contribution from where $\partial\varphi_3/\partial x$ is non-zero. So let us consider an idealized ship with constant cross section aft of the bow region. The force is then acting in the bow region. If we had neglected the Kutta condition of the trailing edge or, equivalently, that the flow leaves tangentially from the transom stern, our force would be very different. For instance, if we said that the velocity potential becomes zero just after the ship stern, then we would get zero total force and no hull-lift damping in heave. This type of analysis can also be performed for other modes of motion. The terms that arise are the same as the "end terms" in the strip theory of Salvesen et al. (1970; see also section 8.5.1).

7.2.8 Foil-lift damping

Foils can provide important heave, roll, and pitch damping. The damping increases linearly with forward speed and is therefore more important for high-speed vessels than for displacement vessels. The damping depends on the details of the foil and is, in a quasi-steady approximation, proportional to the lift on the foil.

We assume that the flow is two-dimensional and that there is no effect of boundaries such as the free surface and the vessel. It also implicitly means that cavitation and ventilation are not considered.

Let us now consider a heaving thin foil without camber that moves with a forward speed U. We assume zero mean angle of attack. Relative to the foil, there is an incident flow velocity component U parallel to the foil and an incident flow velocity $-d\eta_3/dt$ orthogonal to the foil (see Figure 6.61).

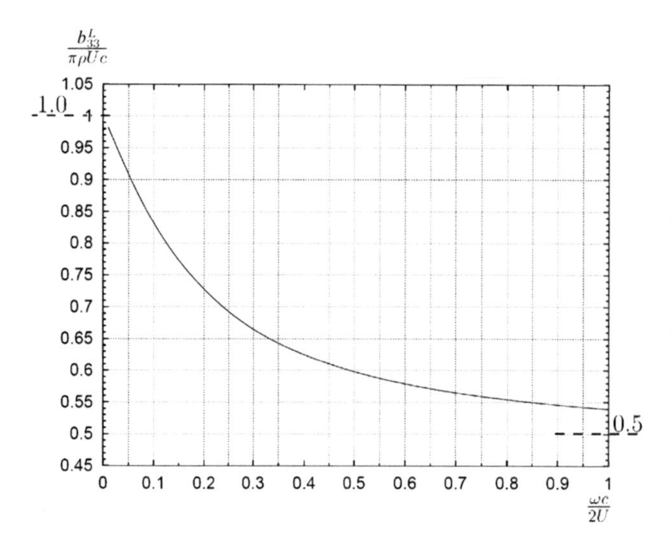

Figure 7.25. Two-dimensional damping coefficient b_{33}^L in heave due to a harmonically heaving flat (uncambered) thin foil in infinite fluid estimated by the Theodorsen function. ω = circular frequency of oscillation, c = chord length, U = forward speed, ρ = mass density of fluid.

Here η_3 means the heave motion of the foil. This means that there is an ambient flow velocity

$$V = \left(U^2 + \left(\frac{d\eta_3}{dt}\right)^2\right)^{\frac{1}{2}} \approx U,$$

with an instantaneous angle of attack

$$\alpha = -\frac{d\eta_3}{dt}/U \tag{7.75}$$

relative to the foil. We have implicitly assumed small α-values. If we use a quasi-steady approach, the lift L per unit length is $0.5\rho U^2 c 2\pi\alpha$ (see eq. (6.93)). This gives a two-dimensional vertical force:

$$L = -\rho U c\pi \frac{d\eta_3}{dt}. \tag{7.76}$$

If this expression is used in the equations of motions, it means that we can interpret

$$b_{33}^L = \rho U c\pi \tag{7.77}$$

as a two-dimensional damping coefficient in heave due to foil lift.

We can generalize this quasi-steady approach as follows, by starting with expressing the steady lift force as $L = 0.5\rho C_L U^2 A$. C_L is the lift coefficient, and A is the planform area of the foil. We then consider C_L as a function of the angle of attack α and assume unsteady variations $-\dot{\eta}_3/U$ in the angle of attack about a mean value α_0. The unsteady part of the lift force can then be expressed as

$$0.5\rho U^2 A \left.\frac{dC_L}{d\alpha}\right|_{\alpha=\alpha_0} (-\dot{\eta}_3/U).$$

If this expression is used in the equations of motion, it gives the following damping coefficient in heave due to foil lift:

$$B_{33}^L = 0.5\rho U A \left.\frac{dC_L}{d\alpha}\right|_{\alpha=\alpha_0}. \tag{7.78}$$

A more accurate determination of foil-lift damping can be obtained by accounting for the continuously shed vorticity from the trailing edge. The continuously shed vorticity influences the effective angle of attack of the flow at the foil and therefore the pressure distribution and the force on the foil. If we assume two-dimensional flow in infinite fluid and consider a thin flat foil that is harmonically oscillating in heave, the lift force can be expressed by the Theodorsen function $C(k)$ (see eq. (6.189)). If ω is the circular frequency of oscillation,

$$k = \frac{\omega c}{2U} \tag{7.79}$$

is the so-called reduced frequency. It is implicitly assumed that the amplitude of the heave oscillation is small and that steady-state conditions have been obtained. It follows from the lift expression presented in Chapter 6 that we can write

$$b_{33}^L = \rho U c\pi \,\mathrm{Re}[C(k)], \tag{7.80}$$

where $\mathrm{Re}[C(k)]$ means the real part of $C(k)$. This formula for b_{33}^L is graphically presented in Figure 7.25 as a function of reduced frequency k. b_{33}^L is monotonically decreasing with increasing values of k. When $k \to 0$, $\mathrm{Re}[C(k)] \to 1$, and when $k \to \infty$, $\mathrm{Re}[C(k)] \to 0.5$. Our previous quasi-steady analysis corresponds to $\omega = 0$. k will in reality be small in practical applications. Consider, for instance, $\omega = 1$ rad/s, $c = 1$ m, and $U = 10$ ms^{-1}. This gives $k = 0.5\,\omega\,c/U = 0.05$. Figure 7.25 shows that $b_{33}^{L} = 0.91\rho\pi Uc$, that is, a 9% reduction relative to quasi-steady predictions. Figure 7.25 shows that unsteady analysis always gives smaller damping than does a quasi-steady analysis, but the difference may not be that significant.

We demonstrated in Chapter 6 that free-surface effects, 3D flow, and possible interaction from an upstream foil ought to be considered. In addition, the interaction with the flow around the hull may matter. However, we will not deal with these effects here.

The discussion of heave damping can be easily generalized to pitch damping by first expressing the local vertical motion of the foil as $\eta_3 - x_F\eta_5$, where x_F is an average x-coordinate of the foil relative to the center of gravity (COG) of the vessel, and η_5 is the pitch angle in radians. Eq. (6.187) giving the lift in terms of the Theodorsen function includes the effect of pitch angle. The pitch damping coefficient and coupled pitch-heave damping coefficients follow by considering the pitch moment of the lift force about the COG of the vessel (see exercise 3.3 in Faltinsen 1990). We can, of course, generalize the procedure and include other modes of motion and other orientations of the lifting surface. A rudder can, for instance, cause important roll damping at high speed.

7.2.9 Example: Importance of hull- and foil-lift heave damping

We illustrate the importance of hull- and foil-lift damping by considering an example of a 560-tonne catamaran. The beam b_2 of each hull at the transom stern is 3.75 m. Each hull is equipped with vertical-motion damping foils. These are T-foils located at 0.17 L_{PP} measured from FP and trim tabs located at AP. The slope of the lift coefficient is $dC_L/d\alpha = 3.0$ for the T-foils and 2.0 for the trim tabs. The projected horizontal foil areas are 2.5 m^2 and 4.0 m^2 for the T-foils and trim tabs, respectively.

We will study heave damping. The critical damping in heave is

$$B_{cr} = 2(M + A_{33})\omega_n, \tag{7.81}$$

where M is the total mass of the vessel, A_{33} is the added mass in the heave of the vessel, and ω_n is the natural circular frequency in heave. We estimate the critical damping in heave to be $2 \cdot 10^6$ N sm^{-1}. The hull-lift damping in heave for each hull is estimated by eq. (7.74). We consider $U = 15$ ms^{-1}. This means that the ratio ξ^{HL} between hull-lift damping and critical heave damping is

$$\xi^{HL} = \frac{2\rho\frac{\pi}{8}b_2^2 U}{2(M + A_{33})\omega_n} = 0.08. \tag{7.82}$$

The foil-lift damping for each foil is calculated by eq. (7.78). This means that the ratio ξ^{FL} between foil-lift damping and critical heave damping is 0.12.

Wave radiation damping will also contribute to the total heave damping, but this has to be assessed by either experiments or a direct numerical method, such as the 2.5D method of Faltinsen and Zhao (1991a). A general tendency is that the more slender the hull – that is, the lower the beam-to-length ratio – the lower the wave radiation damping (see section 7.2.3).

7.2.10 Ride control of vertical motions by T-foils

We consider a T-foil like that in Figure 7.3 with flaps that we will actively control by changing the flap angle to reduce the heave and pitch motions of the vessel. Our analysis will not consider the frequency effect on the lift characteristics, as we did by using the Theodorsen function for a 2D flat plate (thin foil without camber) in infinite fluid. A quasi-steady approach will be followed instead. We consider incident regular head sea waves that have vertical velocity

$$w = \omega_0\zeta_a e^{kz_F}\cos(\omega_e t - kx_F) \tag{7.83}$$

at the mean coordinates $x = x_F$ and $z = z_F$ of the foil. The vertical wave-induced motion of the foil due to heave and pitch is represented as $\eta_3 - x_F\eta_5$. The flap angle is denoted as $\alpha_f(t)$, as in Figure 6.39. The lift coefficient C_L of the foil is a function of the angle of attack α and the flap angle α_f. Similarly, as described in section 7.2.8, there is a linearized dynamic lift on the foil that can be expressed as

$$F_3 = \frac{\rho}{2}\left.\frac{\partial C_L(\alpha,\alpha_f)}{\partial\alpha}\right|_{\substack{\alpha=\alpha_0\\ \alpha_f=\alpha_{f0}}} AU[w - \dot{\eta}_3 + x_F\dot{\eta}_5 + U\eta_5] + \frac{\rho}{2}\left.\frac{\partial C_L(\alpha,\alpha_f)}{\partial\alpha_f}\right|_{\substack{\alpha=\alpha_0\\ \alpha_f=\alpha_{f0}}} AU^2\alpha_f. \tag{7.84}$$

Here $\alpha = \alpha_0$ and $\alpha = \alpha_{f0}$ refer to the mean angle of attack and mean flap angle, respectively, and A is the projected foil area that is equal to the chord length for a 2D foil.

If we consider linear flow past a 2D flat foil with a flap in infinite fluid, we can write $C_L(\alpha,\alpha_f) = 2\pi\alpha + 2\pi\eta_f\alpha_f$, that is,

$$\left.\frac{\partial C_L(\alpha,\alpha_f)}{\partial\alpha}\right|_{\substack{\alpha=\alpha_0\\ \alpha_f=\alpha_{f0}}} = 2\pi$$

$$\left.\frac{\partial C_L(\alpha,\alpha_f)}{\partial\alpha_f}\right|_{\substack{\alpha=\alpha_0\\ \alpha_f=\alpha_{f0}}} = 2\pi\eta_f.$$

Here the flap efficiency coefficient η_f was presented in Figure 6.40 as a function of the ratio between flap length and chord length.

The pitch moment F_5 due to the dynamic lift on the foil can be approximated as

$$F_5 = -x_F F_3. \tag{7.85}$$

Because we focus on damping effects, we disregard added mass forces. Further, we assumed the foil is in the forward part of the ship so that x_F is negative. We now express

$$\alpha_f = -k_1\dot{\eta}_3 - k_2\dot{\eta}_5, \tag{7.86}$$

where the positive gain constants k_1 and k_2 are a consequence of tuning the ride control system with due consideration for the maximum change of α_f, available power, cavitation, and ventilation. In addition, a saturation element maintaining α_f between maximum and minimum limits should be included in the control software. This saturation effect causes nonlinearities. Further, dynamic stability of the system must be ensured. We can now introduce the following damping terms associated with the flaps in the equations of motions:

$$B_{33}^{RC} = \frac{\rho}{2}\left.\frac{\partial C_L}{\partial\alpha_f}\right|_{\substack{\alpha_0\\ \alpha_{f0}}} AU^2k_1 \tag{7.87}$$

$$B_{35}^{RC} = \frac{\rho}{2}\left.\frac{\partial C_L}{\partial\alpha_f}\right|_{\substack{\alpha_0\\ \alpha_{f0}}} AU^2k_2 \tag{7.88}$$

$$B_{53}^{RC} = -\frac{\rho}{2}\left.\frac{\partial C_L}{\partial\alpha_f}\right|_{\substack{\alpha_0\\ \alpha_{f0}}} AU^2k_1x_F \tag{7.89}$$

$$B_{55}^{RC} = -\frac{\rho}{2}\left.\frac{\partial C_L}{\partial\alpha_f}\right|_{\substack{\alpha_0\\ \alpha_{f0}}} AU^2k_2x_F. \tag{7.90}$$

The superscript RC is an abbreviation for ride control. We must, of course, also introduce the other dynamic terms due to the foil in the equations of motion. We can now use the equations of motion to tune k_1 and k_2 to obtain improved performance of the heave and pitch motions. We can get an idea about what damping effect k_1 causes by examining the case in section 7.2.9 and considering the damping ratio for uncoupled heave. In practice, the final tuning of the gain coefficients will be done when the system is installed on the ship.

The situation described above refers to resonant heave and pitch motions. Another scenario is following sea in which deck diving can occur (see Figure 7.39). This is a quasi-steady phenomenon, which means in this case we would express $\alpha_f = -k_3\eta_3 - k_5\eta_5$.

If we consider a multihull vessel, then we can control the roll motion by individual control of the flaps of a T-foil on each hull.

It could be assumed that the controller gains are updated based on the sea state and operational conditions. The most simplified way is to use a set of predefined controller gains, in which the operator or the ride control system itself selects the proper gain settings based on indirect or direct measurements of sea states. Indirect measurements mean that wind, roll, and/or pitch are measured.

7.2.11 Roll motion in beam sea of a catamaran at zero speed

Roll damping of monohulls can be small and result in large resonant roll. On conventional ships, this is counteracted by using bilge keels, antirolling tanks, and fins. Roll damping fins and rudders as those shown in Figures 7.58 and 7.5 are effective at high speed.

We focus on roll motion of a catamaran in beam sea at zero speed. Strong wave interaction may then occur between the two demihulls. Our studies are limited to two-dimensional flow. However, 3D flow as well as forward speed will reduce the wave trapping between the demihulls.

Figure 7.26 shows 2D added mass a_{44} and wave radiation damping b_{44} in roll of two semi-submerged circular cylinders with axis in the mean free surface as a function of nondimensional frequency $\omega(d/g)^{0.5}$ for various $2p/d$-values. Here d is the cylinder radius and $2p$ is the distance

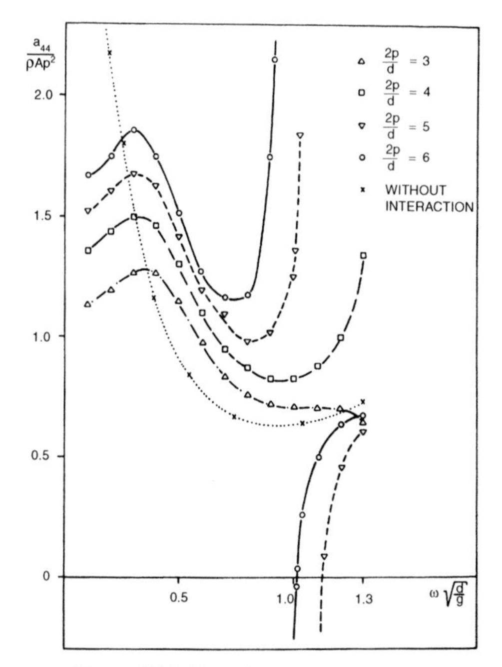

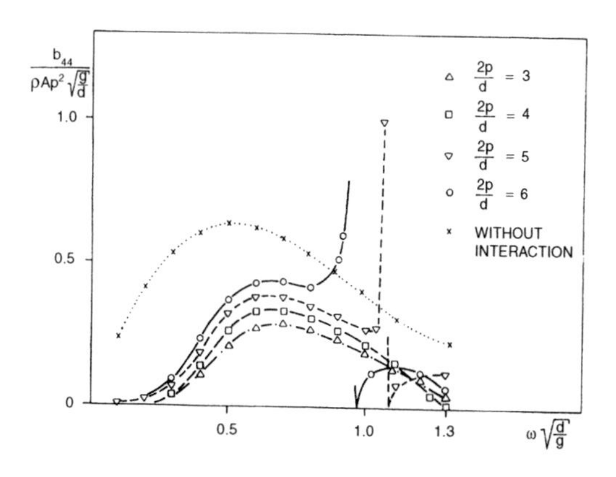

Figure 7.26. Two-dimensional added mass a_{44} and damping b_{44} coefficients in the roll of two semi-submerged circular cylinders with axes in the mean free surface. The coefficients are with respect to an x-axis in the mean free surface, as in Figure 7.49. $2p$ = distance between cylinder axes, A = cross-sectional submerged area of the two cylinders, d = draft (Nordenstrøm et al. 1971).

between the two cylinder axes. When considering a_{44} and b_{44}, we must be precise about what axis we are using. A longitudinal x-axis at the intersection between the centerplane of the catamaran and the mean free surface is the reference axis for the results in Figure 7.26.

Figure 7.26 also shows results calculated by neglecting hull interaction. These are obtained as follows. According to the definition of added mass and damping in section 7.2.1, we start with studying forced roll motion η_4 about the x-axis defined above. This gives, according to eq. (7.20), a vertical motion $p\eta_4$ at the intersection point between the free surface and the centerplane of the demihull with positive y-values. Let us define an x'-axis going through this intersection point and being parallel to the x-axis. The hydrodynamic problem for this demihull can now be decomposed into two parts. One part is the result of forced heave motion $p\eta_4$. The other part is the result of rolling about the x'-axis. However, because the cross section is circular, the pressure distribution due to forced rolling about the x'-axis causes no roll moment about the x'-axis. This means we can focus on the forced heave problem, which gives a vertical hydrodynamic force on one hull:

$$F_3 = -0.5 a_{33} p \frac{d^2\eta_4}{dt^2} - 0.5 b_{33} p \frac{d\eta_4}{dt}. \qquad (7.91)$$

Here $0.5a_{33}$ and $0.5b_{33}$ are the added mass and damping in heave of the demihull. Then we take the moment pF_3 of this vertical force about the x-axis. By making similar assumptions with the other demihull and using the definition of added mass and damping given by eq. (7.39), we find that

$$a_{44} = p^2 a_{33}, \quad b_{44} = p^2 b_{33} \qquad (7.92)$$

when hydrodynamic interaction is neglected.

We note in Figure 7.26 a pronounced interaction effect in the whole frequency domain presented, as well as a very strong effect in a limited frequency domain. For instance, the added mass changes sign and the damping coefficient has both large positive values and zero value in this limited frequency domain. Zero damping means wave trapping between the two demihulls. A similar phenomenon occurs for added mass and damping in heave, as presented in Figure 7.20, in which we

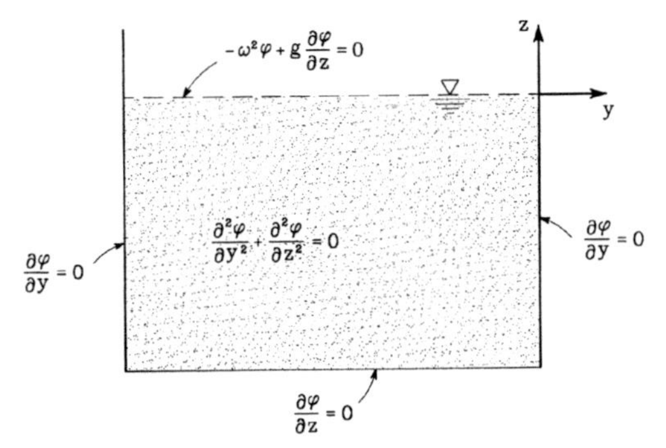

Figure 7.27. The boundary-value problem for determination of natural frequencies and eigenmodes of sloshing in a 2D rectangular tank, used to estimate the natural frequency of wave trapping between the demihulls of a catamaran due to roll motions.

explained this as a piston mode resonance. However, because forced roll motion causes the flow to be antisymmetric at the two demihulls, piston mode resonance cannot be present. We instead explain the wave trapping by relating the resonant motion between the demihulls to sloshing in a ship tank. We consider therefore the eigenvalue problem for sloshing in a rectangular tank, as presented in Figure 7.27. Because there is no obvious depth in the catamaran problem that we can relate to the fluid depth in a tank, we assume infinite depth in the tank. A solution of the boundary-value problem in Figure 7.27 can then be expressed as

$$\varphi = Ae^{kz}\cos(\omega t + \varepsilon)\cos ky, \qquad (7.93)$$

where $\omega^2/g = k$ and $kb = n\pi$, $n = 1, 2, 3\ldots$ and b is the tank breadth. We leave it to the reader to check that this solution satisfies the boundary-value problem in Figure 7.27 for infinite fluid depth. The solution for $n = 1$ gives the lowest natural frequency and corresponds to a half-wavelength between the tank walls. Further, the flow is antisymmetric about the centerplane of the tank. The latter is a necessary requirement for application to our catamaran problem. We will consider the solution for $n = 1$. There are higher antisymmetric modes, but the corresponding natural frequencies are too high to be of interest in this context. We will set $b = 2p - 2d$ corresponding to the breadth of the waterplane area between the demihulls. We then see if b_{44} in Figure 7.26 becomes zero when ω is equal to the natural frequency ω_s for sloshing of the lowest mode. This natural frequency can be expressed as

$$\omega_s\sqrt{\frac{d}{g}} = (\pi/(2p/d - 2))^{0.5}. \qquad (7.94)$$

This agrees reasonably with the results in Figure 7.26. For instance, $\omega_s(d/g)^{0.5}$ are 0.89, 1.02, and 1.25 for, respectively, $2p/d = 6$, 5, and 4. We could not expect exactly the same because, for instance, we are using results for sloshing in a rectangular tank whereas the shape of the demihulls is circular. Further, there is a connection between the flow between the demihulls and what is happening outside this area.

When the incident wave period is equal to both the undamped natural roll period T_n and the trapping period $T_s = 2\pi/\omega_s$, infinite roll response will occur according to potential flow theory and when coupling with sway is neglected.

Let us examine if this is possible by considering uncoupled roll. However, in reality, coupling with sway will matter (Vugts 1968). The undamped natural roll frequency can be expressed as

$$\omega_n = \sqrt{\frac{\rho g A\overline{GM}}{I_{44} + a_{44}}}, \qquad (7.95)$$

where A is submerged volume per unit length. The metacentric height $\overline{GM}$ is defined in eq. (7.41). We set the z-coordinate of the center of gravity equal to d above the mean free surface. This gives

$$\rho g A\overline{GM} = \rho g d^3\left(\left(\frac{2p}{d}\right)^2 - \pi\right). \qquad (7.96)$$

Further, we will set the vessel mass moment of inertia in roll I_{44} and the added mass in roll equal to

$$I_{44} = \rho\pi d^2(p + d)^2, \quad a_{44} = 0.7\rho\pi d^2 p^2. \qquad (7.97)$$

This means that a_{44} does not account for the hull interaction and the frequency dependency. This should of course be done, but it becomes too troublesome to account for the strong frequency dependency of a_{44} in a simple calculation like this. Further, we should, strictly speaking, refer a_{44} to an axis going through the center of gravity. Our estimate now becomes

$$\omega_n\sqrt{\frac{d}{g}} = \sqrt{\frac{(2p/d)^2 - \pi}{\pi[0.25 \cdot 1.7(2p/d)^2 + (2p/d) + 1]}}. \qquad (7.98)$$

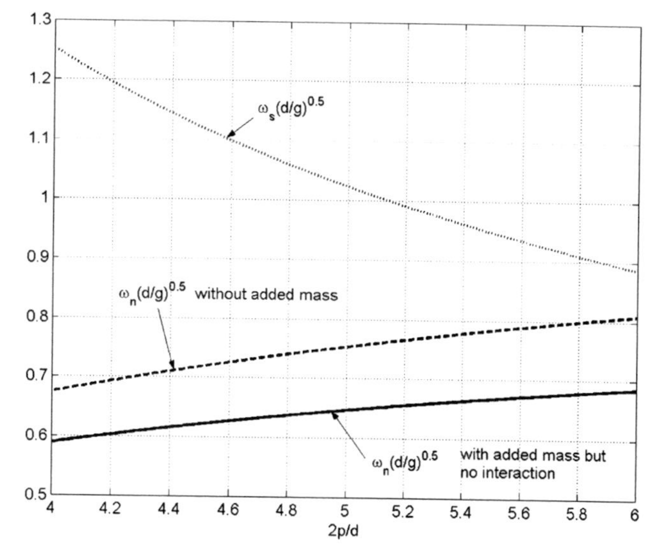

Figure 7.28. Estimate of wave-trapping frequency ω_s for the forced roll motion of two semi-submerged circular cylinders with axes in the mean free surface. ω_n is uncoupled and undamped natural roll frequency, $2p$ = distance between cylinder axes, d = draft.

This is plotted in Figure 7.28 together with $\omega_s(d/g)^{0.5}$ and an estimate $\omega_n(d/g)^{0.5}$ of natural roll frequency obtained by setting $a_{44} = 0$. This gives an indication of the uncertainty caused by our estimate of a_{44}.

The results show that $\omega_n = \omega_s$ is possible when $2p/d$ is close to 6. However, detailed studies of this require that we investigate roll response by using accurate estimates of a_{44} and b_{44} and account for coupling with sway. The trapping frequency ω_s is also important for sway motion and the antisymmetric part of the diffraction potential of the restrained ship.

If we consider uncoupled roll, we can use eq. (7.47) to obtain the wave excitation moment in roll. However, if we consider coupled roll and sway, we need to know the phasing between the wave roll excitation moment and the sway excitation force. Eq. (7.47) does not tell us that. However, there are computer programs that can do calculations like this. Our objective here is to qualitatively discuss the important effects on roll associated with hull interaction.

7.2.12 Numerical predictions of unsteady flow at high speed

There are different ways to calculate the added mass, damping, and wave excitation loads that appear in the equations of motions in the frequency domain. One way is the 2.5D theory (Faltinsen and Zhao 1991a,b). We have already outlined this method for steady flow in section 4.3.4. Differences for unsteady flow are the result of both the body boundary and free-surface conditions. We divide the velocity potential φ due to the body into different parts, that is,

$$\varphi = \sum_{j=1}^{6} \varphi_j \bar{\eta}_j e^{i\omega_e t} + \varphi_7 e^{i\omega_e t}. \qquad (7.99)$$

Here φ_j, $j = 1, \ldots, 6$ is the velocity potential due to unit motion in mode number j when there are no incident waves. φ_7 is the diffraction potential when the ship is restrained from oscillating and there are incident waves. The sum of the velocity potential due to the incident waves and the diffraction potential must satisfy no flow normal to the ship's surface. Addition of the seven subproblems (six radiation and one diffraction) in eq. (7.99) is possible because of linearity. When we solve these hydrodynamic problems, we do not know the motions. They follow from using this procedure to calculate added mass, damping, and excitation forces and moments and by inserting them into the equations of motions given by eq. (7.35).

The free-surface conditions for steady flow were given by eq. (4.52). These must be modified according to eqs. (3.7) and (3.9). This means when interaction with local steady flow is neglected, the dynamic and kinematic free-surface conditions are

$$\frac{\partial \varphi_j}{\partial x} = -\frac{g\zeta_j}{U} - \frac{i\omega_e \varphi_j}{U} \quad \text{on } z = 0$$
$$\frac{\partial \zeta_j}{\partial x} = \frac{1}{U}\frac{\partial \varphi_j}{\partial z} - \frac{i\omega_e \zeta_j}{U} \quad \text{on } z = 0. \tag{7.100}$$

$\zeta_j e^{i\omega_e t}$ in eq. (7.100) is the wave elevation corresponding to $\varphi_j e^{i\omega_e t}$. Each velocity potential part satisfies the two-dimensional Laplace equation in the cross-sectional plane, as in eq. (4.51). The numerical procedure is the same as for the steady flow, that is, we start at the bow and use eq. (7.100) to step the solution downstream. When φ_j are found, we use eq. (7.38) to find the hydrodynamic pressure giving forces and moments on the hull. Using eq. (7.39) defines the added mass and damping coefficients. Together with the pressure loads due to the incident waves (Froude-Kriloff loads), the forces and moments associated with φ_7 give the wave excitation loads on the ship.

Interaction with local steady flow is studied by Faltinsen and Zhao (1991a,b). Comparisons with model tests for added mass and damping show that interaction with local steady flow matters.

We commented in the discussion of the steady flow that this procedure accounts only for divergent waves, requiring a Froude number Fn larger than 0.4 to 0.5 in practice. The same is true for the unsteady flow that is symmetric about the centerplane of the vessel, that is, for the surge, heave, and pitch problem. For sway, roll, and yaw motions, no transverse wave systems are created along the track of the ship. The reason is that the flow has to be antisymmetric about the centerplane. Further, the transverse waves have to be small in the vicinity of the ship's track. This suggests that the method is also applicable for lower Froude numbers in the case of lateral motions. If the hull ends in a transom stern, for the sway, roll, and yaw problems, it is important to assume that there is a vortex sheet leaving from the transom stern in the downstream direction and to consider the ship as a low-aspect ratio–lifting surface.

Figure 7.29 presents an experimental validation of the theoretically predicted heave, pitch, and vertical acceleration of a high-speed monohull in head sea by the 2.5D method by Faltinsen and Zhao (1991a,b). Interaction with local steady flow was neglected. Transom stern effects are important for

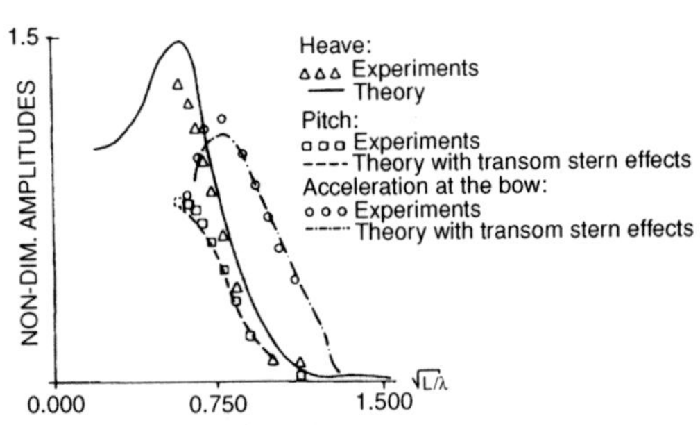

Figure 7.29. Heave, pitch, and vertical amplitudes for a monohull in head sea regular deep-water waves. $Fn = 1.14$. Trim 1.62°. Experiments by Blok and Beukelman (1984). Numerical calculations by 2.5D theory (Faltinsen and Zhao 1991b). Heave: $|\eta_3|/\zeta_a$, pitch: $|\eta_5|/(k\zeta_a)$, accelerations $a_3 L/(50 g\zeta_a)$, L = ship length, ζ_a = incident wave amplitude, k = incident wave number. Length of test waterline = 2 m, beam of test waterline = 0.25 m, draft = 0.0624 m, block coefficient = 0.396.

hull-lift damping (see section 7.2.7). Validation of steady wave elevation, wave resistance, and added mass and damping are also presented. The figure shows that the resonant wavelength for heave corresponds to $\lambda/L = 2.9$. If the ship had zero speed, the corresponding wavelength that excites the resonance oscillations in heave would be much smaller (see discussion in section 7.2.2). The value of heave at resonance is about 1.5 times the incident wave amplitude according to the theory. This is high relative to conventional ships at moderate forward speed, but a high-speed catamaran may have even higher values for the transfer function of heave at resonance. One reason for this is that the wavelength in head sea that causes heave and pitch resonance increases with speed. Increased wavelength tends to increase the wave excitation loads due to smaller phase differences along the hull.

Bertram and Iwashita (1996) and Takaki and Iwashita (1994) have systematically investigated the ability of numerical methods to predict wave-induced motions of semi-displacement monohulls and catamarans. Bertram and Iwashita (1996) conclude that conventional strip methods (e.g., Salvesen et al. 1970) are valid only roughly for $Fn < 0.4$ and that 2.5D theories are at present, that is 1996, the most suitable methods for fast ships, for practical purposes.

We illustrate that state-of-the-art numerical methods do not always give satisfactory predictions of wave-induced heave and pitch motions of semi-displacement vessels. The catamaran model presented in Figure 4.21 is used in the examples. The pitch radius of gyration is 0.26 times the ship length L. A linear 3D Rankine panel method (RPM) similar to the one described in section 4.3.4 is used. The transom stern is assumed wet, and there is therefore no effect of hull-lift damping. The interaction between the unsteady and steady flow is handled in two different ways. One approach assumes that the steady flow can be calculated by a rigid free-surface condition saying that the free surface acts as a wall. This provides interaction between the local steady flow and the unsteady flow. However, this steady free-surface condition is only appropriate for low Froude numbers (Fn), let us say $Fn < 0.2$. Because the steady flow can be calculated by considering a double body in which the submerged hull surface is mirrored about the mean free surface, the model is referred to as a double-body model (DM). The second method assumes that the steady flow can be approximated as a uniform flow with a velocity equal to the ship speed. The free-surface conditions are given by eq. (7.100). In further discussion, the second method is referred to as the Neumann-Kelvin (NK) method.

Head sea waves are considered and response amplitude operators (RAO) of heave and pitch are experimentally and numerically predicted. The RAO of heave refers to steady-state amplitude of heave $|\eta_3|$ divided by the incident wave amplitude ζ_a in regular incident waves. Similarly, the RAO of pitch means $|\eta_5|/\zeta_a$. The RAO can experimentally be obtained either by considering a transient test technique (Colagrossi et al. 2001) or by simply considering incident regular waves. The following results are for $Fn = 0.3, 0.4$, and 0.5 and have been reported by Lugni et al. (2004).

The heave and pitch frequency resonance for $Fn = 0.3, 0.4$, and 0.5 was first identified by the transient test technique. Then, tests in regular incoming waves with different wave amplitudes were performed in the resonance frequency range.

The RAO experimental data are presented in Figure 7.30 together with the predictions by the 3D linear RPM code. The standard deviation (σ) connected with the transient test technique is also given in the plots, showing good reliability of the experiments. For the numerics, both the NK and DM approximations are considered. The numerical results overestimate the pitch motion. For all investigated speeds, the DM linearization shows the best agreement with the experiments. This is consistent with the conclusions in Bertram (1999) documenting an important role of the interaction with the local steady flow in the wave-induced body motion predictions, even at a Froude number around 0.2. A strong amplification of the motions is generally observed near the resonance because of a small damping level. Because each demihull has a small beam-to-draft ratio, that is, $B/D = 1.14$, we should expect this from the discussion in section 7.2.3 (see Figure 7.15). The monohull results presented in Figure 7.29 show smaller numerical and experimental resonant vertical motions than those in Figure 7.30. One reason is that $B/D = 4$ for the monohull, that is, it is clearly larger than the ratio for the demihull of the catamaran. Figure 7.15 suggests, then, smaller resonant vertical motions for the monohull than for the catamaran. The 3D RPM used in connection with Figure 7.30 neglects hull-lift damping. This matters and should be included for $Fn = 0.5$ when the flow separates from the transom. However, it is not correct to include this effect for a wet transom, that is, for $Fn = 0.3$ and 0.4.

The heave and pitch results are not affected by resonant wave trapping between the demihulls. Using eqs. (7.46) and (7.64) gives that the resonant frequency for piston mode resonance corresponds to $\lambda/L = 1.97, 2.31, 2.63$ for $Fn = 0.3, 0.4$, and 0.5, respectively. The results in Figure 7.30 do not show any strong variations due to resonance in the vicinity of these wavelengths.

The small damping in heave and pitch for the catamaran suggests the need for proper active control systems and foils. However, it should also be noted that the values of the wavelength-to-ship length ratio giving resonance in heave and pitch in head sea increase with the Froude number (see section 7.2.2). This implies that the excitation loads along the ship become stronger in phase as Fn increases. The consequence is larger excitation loads.

The experiments show clear nonlinear effects. The regular wave results do not converge to the transient test results as the wave amplitude reduces. One possible error source is a variation of the wave amplitude along the track of the model. This aspect was not investigated. At high Froude numbers, the RAO for the pitch motion shows a double peak behavior, typical

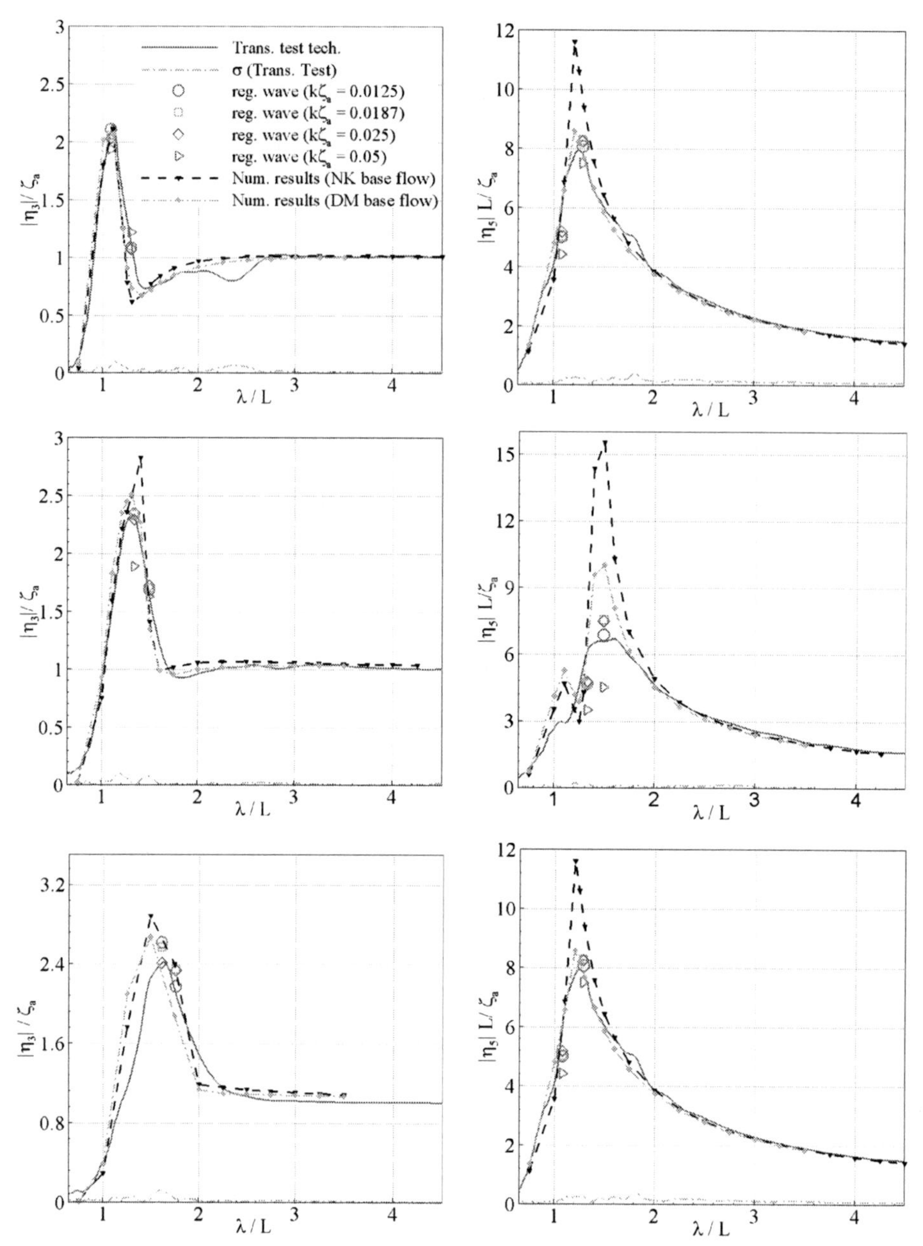

Figure 7.30. Catamaran: heave (η_3, left) and pitch (η_5, right). Response amplitude operators. From top to bottom: $Fn = 0.3, 0.4, 0.5$; ζ_a and $k = 2\pi/\lambda$ are the regular incoming wave amplitude and wave number, respectively; λ is the incoming wavelength. The ship model is presented in Figure 4.21. Lugni et al. (2004).

for the multihull vessels. The results for pitch shows that it is important to account for the hull interaction.

From the experiments, the mean trim and sinkage are not influenced substantially by the incident wave steepness, even at a wave frequency equal to the heave and pitch resonance frequency. It implies that they are dominated by the steady flow. These results are relevant, for instance, for the wetdeck slamming that is sensitive to the trim angle (Ge 2002). An accurate estimate of the relative motions in the impact area matters also in the wetdeck slamming predictions. The presented results suggest that the presented theoretical methods to evaluate wave-induced motions have to be improved for a better prediction of, for instance, wetdeck slamming. Wetdeck slamming is discussed further in section 8.4.

7.3 Linear time-domain response

Section 7.2 deals with wave-induced ship response in the frequency domain by analyzing the steady-state solution in regular incident waves. If we consider the behavior of the ship in irregular sea, there are many excitation frequencies. Because added mass and damping are frequency dependent, we cannot directly use the equation system given by eq. (7.35) in the time domain. However, if we are interested in the steady-state solution, we can circumvent this problem by adding the response to regular wave components. This is what is done in the next section when we consider the statistical description of linear wave-induced response in irregular seas.

However, there are scenarios in which we need the transient response. One example is transient waves generated by a passing ship. Another example is coupling between nonlinear sloshing in a ship tank and ship motions (Rognebakke and Faltinsen 2003). A third example is wetdeck slamming on a catamaran in regular incident waves (see section 8.4). The wetdeck slamming causes a transient vertical force that excites transient response in heave, pitch, and global elastic vibration modes. Let us concentrate on the two-node vertical bending mode that has a natural period on the order of 1 s. There is, in addition, important vessel response at the wave encounter period. This is the order of 10 s. We then have the conflict of which frequency we should use in calculating added mass and damping in eq. (7.35). The added mass and damping for these two periods will be quite different.

These examples illustrate when we cannot use eq. (7.35). We must then formulate the equations of motions in a different way. This was discussed by Cummins (1962) and Ogilvie (1964). If we limit ourselves to heave and pitch motions, we can write the linear equations of motions as

$$
\begin{aligned}
&(M + A_{33}(\infty))\,\ddot{\eta}_3 + B_{33}(\infty)\dot{\eta}_3 + C_{33}\eta_3 \\
&+ \int_0^t h_{33}(\tau)\dot{\eta}_3(t-\tau)d\tau \\
&+ A_{35}(\infty)\ddot{\eta}_5 + B_{35}(\infty)\dot{\eta}_5 + C_{35}\eta_5 \\
&+ \int_0^t h_{35}(\tau)\dot{\eta}_5(t-\tau)d\tau = F_3(t)
\end{aligned}
\tag{7.101}
$$

$$
\begin{aligned}
&A_{53}(\infty)\ddot{\eta}_3 + B_{53}(\infty)\dot{\eta}_3 + C_{53}\eta_3 \\
&+ \int_0^t h_{53}(\tau)\dot{\eta}_3(t-\tau)d\tau \\
&+ (I_{55} + A_{55}(\infty))\,\ddot{\eta}_5 + B_{55}(\infty)\dot{\eta}_5 + C_{55}\eta_5 \\
&+ \int_0^t h_{55}(\tau)\dot{\eta}_5(t-\tau)d\tau = F_5(t).
\end{aligned}
\tag{7.102}
$$

The integrals are often referred to as convolution integrals (often used in connection with Laplace and Fourier transforms) or as Duhamel integrals. Here the vessel mass terms M and I_{55}, as well as the restoring terms C_{jk}, are the same as in eq. (7.35). $A_{jk}(\infty)$ and $B_{jk}(\infty)$ mean infinite-frequency added mass and damping coefficients. $F_3(t)$ and $F_5(t)$ are heave excitation force and pitch excitation moment, respectively. $h_{jk}(t)$ are the retardation functions (also referred to as impulse response functions) that can be evaluated by

$$
\begin{aligned}
h_{jk}(t) &= -\frac{2}{\pi}\int_0^\infty \omega(A_{jk}(\omega) - A_{jk}(\infty))\sin\omega t\, d\omega \\
&= \frac{2}{\pi}\int_0^\infty (B_{jk}(\omega) - B_{jk}(\infty))\cos\omega t\, d\omega.
\end{aligned}
\tag{7.103}
$$

Calculation of $h_{jk}(t)$ requires information on the behavior of either A_{jk} or B_{jk} at all frequencies. It is no problem to calculate A_{jk} and B_{jk} for infinite frequency and for frequencies typical of ship motions. It is more difficult to estimate how A_{jk} and B_{jk} behave asymptotically for large frequencies. Let us illustrate this by referring to a boundary element method. The hull surface is then approximated by panels. The length dimensions of the panels must be small relative to the wavelength. There are, as we saw in section 7.2.4, different wavelengths created by an oscillating ship at forward speed. Let us simplify to illustrate our point. We return then to Table 3.1. The relationship between wavelength λ and ω is $\lambda = 2\pi g/\omega^2$. So increasing ω causes small wavelengths and therefore small panels. There is a practical limit to how small the panels can be because of CPU time. Further, the equation system may be ill-conditioned at high frequencies because of irregular frequencies (Faltinsen 1990). A different strategy is to patch asymptotic expansions of A_{jk}

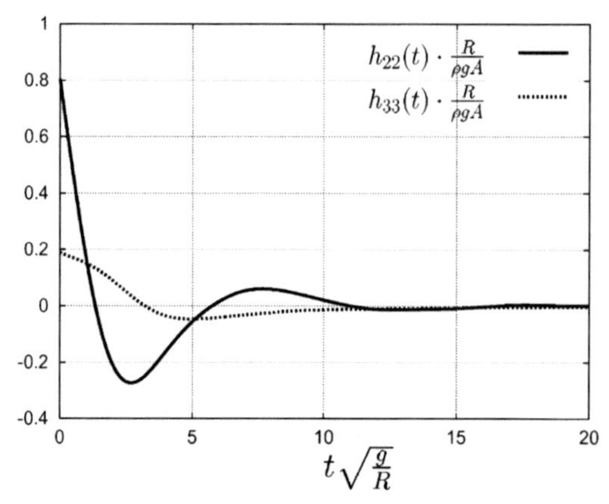

Figure 7.31. Retardation functions h_{jj} *(t)* in sway ($j = 2$) and heave ($j = 3$) for a rectangular cross section with a beam-to-draft ratio of 2.0. R is the half-breadth of the section, and A is the mean submerged area (Rognebakke and Faltinsen 2003).

and B_{jk} for large ω with an ordinary numerical solution for finite ω (Greenhow 1986, Rognebakke and Faltinsen 2003).

The response calculated by eqs. (7.101) and (7.102) is influenced by the high-frequency behavior of A_{jk} and B_{jk}. Adegeest (1995) and Kvålsvold (1994) have, for instance, reported that it is not straightforward to use eqs. (7.101) and (7.102). However, it has become common to use a formulation like eqs. (7.101) and (7.102) in combination with simplified nonlinear hydrodynamic loads on the hull.

Figure 7.31 shows calculated retardation functions h_{22} and h_{33} in sway and heave for a 2D problem with a rectangular cross section. We note that h_{ij} is practically zero when $t(g/R)^{0.5}$ is larger than 12. A similar behavior with h_{ij} being practically non-zero in a finite time is also true for h_{ij} in eqs. (7.101) and (7.102). Let us illustrate the consequence of this by examining the term

$$\int_0^t h_{ij}(\tau)\dot{\eta}_j(t-\tau)d\tau \tag{7.104}$$

in eqs. (7.101) and (7.102). We are going to find $\dot{\eta}_j$ at time t, which corresponds to $\tau = 0$ in eq. (7.104). It is then only necessary to integrate eq. (7.104) over previously found $\dot{\eta}_j$ up to a "cutoff" value $t = t^*$ when h_{ij} is practically zero. The numerical results are not sensitive to t^*. The time-consuming part of the analysis is the calculation of h_{ij}.

We can also solve the hydrodynamic problem and the equations of motions directly in the time domain. Sclavounos and Borgen (2004) used a three-dimensional Rankine panel method that accounts for flow separation at the transom stern. Figure 7.32 illustrates panels distributed over the mean wetted hull surface and a truncated part of the free surface. A numerical beach is used to avoid unphysical wave reflection from the border of the truncated free surface. More details about the Rankine panel method are found in Sclavounos (1996).

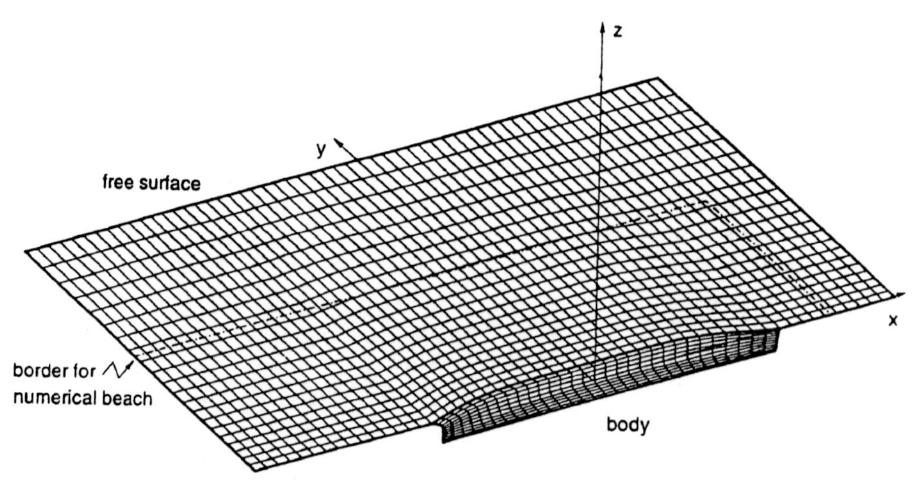

Figure 7.32. Illustration of panels used in the Rankine panel method described by Sclavounos and Borgen (2004).

7.4 Linear response in irregular waves

7.4.1 Short-term sea state response

A short-term sea state refers to wave conditions defined by a given and constant significant wave height $H_{1/3}$ and mean wave period T_2. In addition, we need to specify a mean wave heading, wave energy spreading, and duration. The duration typically used is three or six hours.

A useful consequence of linear theory is that we can obtain results in irregular waves by adding together results from regular waves of different amplitudes, wavelengths, and propagation directions.

The so-called frequency of encounter wave spectrum is sometimes used in connection with calculations of statistical values in an irregular short-term sea. If we measure the incident wave elevation relative to a coordinate system that is moving with the forward speed of the vessel and then evaluate the spectrum, we will obtain the frequency of encounter wave spectrum $S_e(\omega_e)$. However, if we use a standard wave spectrum like JONSWAP or Pierson-Moskowitz, it is more convenient to represent all response variables as a function of the wave frequency ω_0. We will describe how we normally make statistical predictions in a short-term sea. Long-crested seas will be assumed. The following procedure is applicable to all linear wave-induced response values, such as the six degrees of motions, accelerations, global loads and so on. We will see that it is not necessary to evaluate the time-domain solution presented in section 7.3.

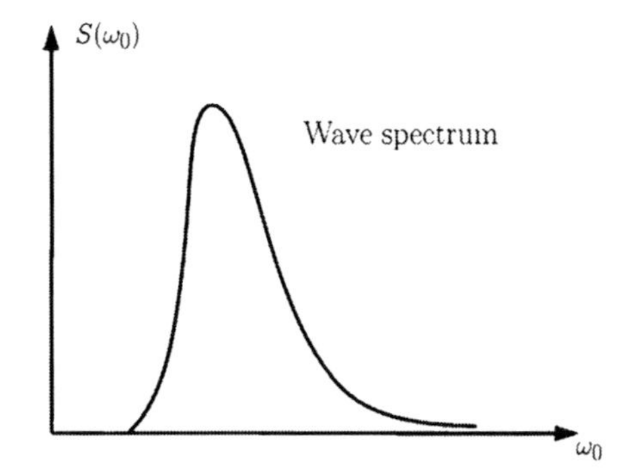

Figure 7.33. Wave spectrum.

1. We start out with a wave spectrum that is representative for the sea state. This could be a measured or an empirical spectrum, for instance, the spectra that ISSC and ITTC have recommended (see Chapter 3). This can be graphically presented as in Figure 7.33.

   We can also express the wave elevation in the time domain as

$$\zeta = \sum_{j=1}^{N} A_j \sin(\omega_{0j}t - k_j X + \varepsilon_j), \quad (7.105)$$

   where $0.5A_j^2 = S(\omega_{0j})\Delta\omega_0$ (see eqs. (3.53) and (3.54) with explanations). We have here assumed that the wave propagation direction is along the X-axis. Further, ω_{0j} is the circular frequency of wave component number j. Ideally, both A_j and ε_j are variables with a probability distribution, but normally only the phasing is treated with a certain likelihood. This might not always be a good assumption (Tucker et al. 1984).

2. We should then combine this wave spectrum with the transfer function of the variable that we study. Let us, for instance, consider heave. We will then calculate the transfer function $|\eta_3|/\zeta_a$ for different regular waves with the same propagation direction as the irregular sea. This can graphically be presented as in Figure 7.34, in which we have also indicated that $|\eta_3|/\zeta_a \rightarrow 1$ when $\omega_0 \rightarrow 0$, that is, for very long wavelengths. The ship behaves, then, like a cork floating on the water. The transfer function depends on the ship speed U, ω_0, and the wave heading β. Because there is a relationship between the frequency of encounter ω_e and U, ω_0, and β, we could also say that the transfer function depends on U, ω_e, and β.

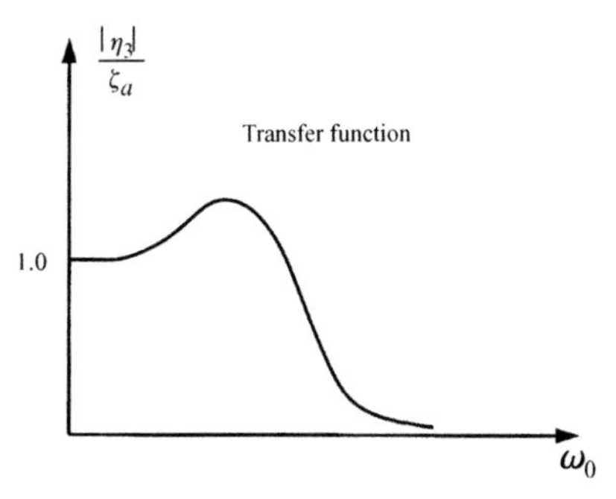

Figure 7.34. A typical transfer function for heave at moderate forward speed in head sea.

3. The transfer function (Figure 7.34) and the wave spectrum (Figure 7.33) can now be combined. We can show this by using a time-domain representation. Formally we can write the steady-state response to one regular wave as

$$A_j |H(\omega_{ej}, U, \beta| \times \sin(\omega_{ej} t + \delta(\omega_{ej}, U, \beta) + \varepsilon_j). \quad (7.106)$$

Here ω_{ej} is the frequency of encounter associated with wave component number j. Further, $|H(\omega_{ej}, U, \beta)|$ is the transfer function, which is the response amplitude per unit of wave amplitude. We note also that there is a phase angle $\delta(\omega_{ej}, U, \beta)$ associated with the response. ε_j is the same phase angle as the one in eq. (7.105). Having obtained the response due to one wave component, we can linearly superpose the response from the different wave components, that is, we can write

$$\sum_{j=1}^{N} A_j |H(\omega_{ej}, U, \beta)| \times \sin(\omega_{ej} t + \delta(\omega_{ej}, U, \beta) + \varepsilon_j). \quad (7.107)$$

We calculate now the variance σ^2 of eq. (7.107)

$$\sigma^2 = \overline{\left\{ \sum_{j=1}^{N} A_j |H(\omega_{ej}, U, \beta)| \sin(\omega_{ej} t + \delta(\omega_{ej}, U, \beta) + \varepsilon_j) \right\}^2}, \quad (7.108)$$

where the line over the expression means time average. Before taking the time average, we multiply out the square expression, that is, the square of eq. (7.107). This gives terms proportional to

$$\cos((\omega_{ej} \pm \omega_{ek})t + \delta(\omega_{ej}, U, \beta) \pm \delta(\omega_{ek}, U, \beta) + \varepsilon_j \pm \varepsilon_k). \quad (7.109)$$

Then we can take the time average of each term. When $\omega_{ej} \neq \omega_{ek}$, the time average of eq. (7.109) is zero. The only contribution occurs when $\omega_{ej} = \omega_{ek}$. The result is

$$\sigma^2 = \sum_{j=1}^{N} 0.5 A_j^2 |H(\omega_{ej}, U, \beta)|^2, \quad (7.110)$$

where we recall that $0.5 A_j^2 = S(\omega_{0j}) \Delta\omega_0$.

4. We now formally let $N \to \infty$ and $\Delta\omega_0 \to 0$. This gives

$$\sigma^2 = \int_0^{\infty} S(\omega_0) \left| H(\omega_e, U, \beta) \right|^2 d\omega_0. \quad (7.111)$$

5. By using the Rayleigh distribution, we can find the probability of exceeding a given value of the heave amplitude. This means we write the probability that the heave amplitude is larger than x as

$$P(\eta_3 > x) = e^{-x^2/2\sigma_3^2}. \quad (7.112)$$

We have used the subscript 3 on the variance to indicate heave. By using eq. (7.112), we can calculate the most probable largest value x_{max} in N oscillations. A good approximation for large N is

$$x_{max} = \sigma_3 \sqrt{2 lnN}. \quad (7.113)$$

Here N can be set equal to t/T_2, where t is the duration of the storm and T_2 is the zero upcrossing period.

7.4.2 Long-term predictions

By combining the Rayleigh distribution with a joint frequency table (scatter diagram) for $H_{1/3}$ and the modal wave period or mean wave period (see Table 3.4), we can obtain the long-term probability distribution of a response. Summing over both period and wave height gives

$$P(R) = 1 - \sum_{j=1}^{M} \sum_{k=1}^{K} \exp\left(-0.5 R^2 / \left(\sigma_r^{jk}\right)^2\right) p_{jk}, \quad (7.114)$$

where $P(R)$ is the long-term probability that the peak value of the response does not exceed R, and σ_r^{jk} is the standard deviation of the response for a mean $H_{1/3}$ and wave period T_2 in the significant wave height interval j and the wave period interval k. Further, p_{jk} is the joint probability for a significant wave height and mean wave period to be in interval-numbers j and k, respectively. For instance, by referring to Table 3.4, the joint probability for the wave period T_2 to be between 9.5 and 10.5 s and $H_{1/3}$ to be between 3.5 and 4.5 m

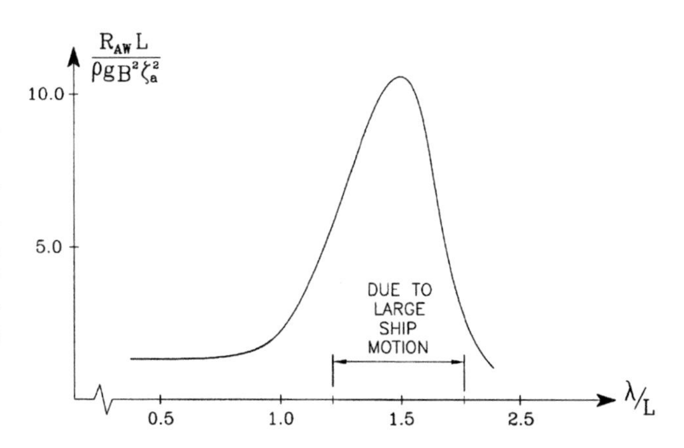

Figure 7.35. Typical wavelength dependency of nondimensional added resistance $R_{AW}L/\rho g B^2 \zeta_a^2$ of a ship at forward speed in regular head sea waves. ζ_a = amplitude of incident waves, λ = wavelength, L = ship length, B = beam of the ship. The general tendency is that the larger the Froude number, the higher λ/L corresponding to maximum nondimensional added resistance.

is 15/995. The values of M and K are, respectively, 13 and 16. However, strictly speaking, the table contains too few observations to be used for extreme value statistics. Typically, one needs on the order of 100,000 observations to create a reliable scatter diagram. However, most high-speed vessels have service restrictions, so extreme sea states may not be that important for these vessels. The probability level $Q = 1 - P(R)$ and the number of response cycles N are related by $Q = 1/N$. A return period of 100 years corresponds approximately to $Q = 10^{-8.7}$, depending on the response variable and its zero upcrossing period. The corresponding response amplitude R can then be found from eq. (7.114). The procedure is often performed by fitting the results to a Weibull distribution (Nordenstrøm 1973). The Weibull distribution is then used for extreme value predictions.

7.5 Added resistance in waves

7.5.1 Added resistance in regular waves

Added resistance R_{AW} in waves is sometimes misunderstood as wave resistance. Wave resistance refers to calm water conditions and is extensively discussed in Chapter 4. Added resistance in waves is caused by interaction between incident waves and the ship. Wave drift forces of importance for stationkeeping of offshore floating platforms are physically the same as added resistance in waves (Faltinsen 1990). An example of the importance of added resistance in waves relative to other resistance components is presented in Figure 4.1.

A typical nondimensional added resistance curve in regular head sea waves is illustrated in Figure 7.35. Added resistance is, as a first approximation, proportional to the square of the incident wave amplitude, that is, ζ_a^2. That is the reason for using ζ_a^2 to nondimensionalize R_{AW} in Figure 7.35.

It can be shown by conservation of fluid momentum and energy (Gerritsma and Beukelman 1972, Maruo 1963) and assuming potential flow that *added resistance in waves is due to the ship's ability to generate unsteady waves*. A ship in the vicinity of resonant heave and pitch conditions will generate the largest waves per unit of wave amplitude. A reason is the large relative vertical motions between the ship and the waves. This condition causes the peak in the nondimensional curve in Figure 7.35.

When the ratio λ/L between the incident wavelength λ and the ship's length L is small (let us say $\lambda/L < 0.5$), the ship will not move much because of the incident waves. However, the incident waves will be reflected from the ship. This is the reason why there is a finite added resistance for small wavelengths in Figure 7.35. When λ/L is very large, the relative motions between the ship and the water goes to zero. It means that the ship does not generate unsteady waves, that is, R_{AW} goes to zero.

Even though the nondimensional added resistance presented in Figure 7.35 is largest at resonant heave and pitch conditions, the finite added resistance at small wavelengths matters. The reason is that small wavelengths are associated with small sea states and small sea states are most frequently encountered by a ship. As an example, let us take the North Atlantic routes shown in Figure 7.36. The average involuntary speed loss over one year was calculated for a 198 m–long container vessel. The target speed was 22 knots and average involuntary speed loss was 1.7 knots westbound and

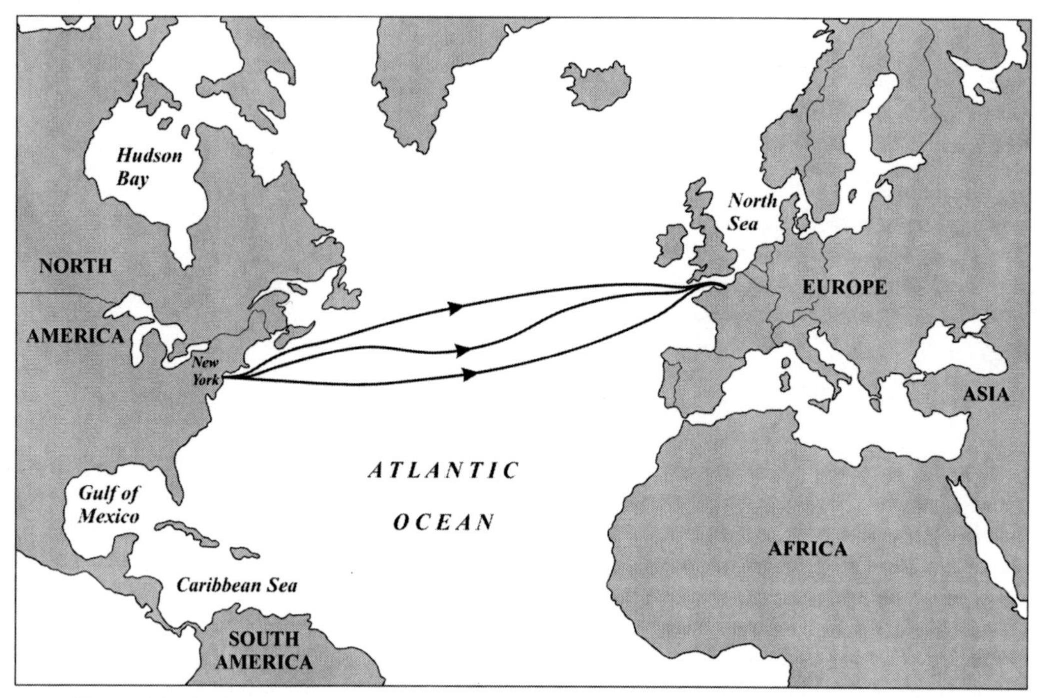

Figure 7.36. Different routes in the North Atlantic of a 198 m–long container vessel with target speed of 22 knots. The calculated average involuntary speed loss over one year is 1.7 knots westbound and 0.9 knots for a roundtrip voyage.

0.9 knots roundtrip voyage. A major contribution was added resistance in small sea states (Faltinsen and Svensen 1990).

An alternative way of using conservation of momentum and energy to estimate added resistance in regular waves is by direct pressure integration (Faltinsen et al. 1980, 1991a). In order to describe this method, we should first recall what is done in linear theory. We then satisfy free-surface conditions at mean free-surface and body boundary conditions at the mean oscillatory position of the ship. The resulting linearized hydrodynamic pressure (this excludes the hydrostatic pressure that must be evaluated at the instantaneous position) is found at the mean oscillatory ship position. Further, the resulting hydrodynamic forces and moments are found by integrating the pressure over the mean wetted surface. The pressure is multiplied by the normal vector of the hull surface in a coordinate system that is not oscillating with the ship. The consequence of the analysis is hydrodynamic forces that are harmonically oscillating in time, that is, the mean value is zero. This means a linear theory will not lead to added resistance in waves.

We must at least include terms that are second order in incident wave amplitude, that is, terms proportional to ζ_a^2 to predict added resistance. So we must go through the whole list of approximations that were done in the linear theory and make corrections to second order in ζ_a. We will demonstrate one such contribution and use Figure 7.37 for illustration. This figure shows a shaded area along the hull that is in and out of the water. This is caused by the relative vertical motions, η_R, between the ship and the water, which can formally be expressed as

$$\eta_R = \eta_{Ra}(x)\cos(\omega_e t + \varepsilon(x)). \qquad (7.115)$$

This means that the wetted area per unit length at longitudinal position x and time t has a difference η_R from the mean wetted area per unit of length at the same position. We will account for this difference in wetted area that linear theory does not account for. We need to know the pressure distribution over this time-dependent wetted area. We can use an argument similar to the one that led to Figure 3.5, that is, the pressure has a "hydrostatic" depth dependence relative to instantaneous water

Figure 7.37. Illustration of hull area (shaded) that is in and out of the water for a ship in regular head sea waves for given incident waves and ship speed. The hydrodynamic pressure on this shaded area contributes to added resistance in waves.

elevation in the close vicinity of the free surface. The resulting longitudinal force is

$$F_1 = -\frac{\rho g}{2}\int_C \eta_R^2 n_1 \, ds, \qquad (7.116)$$

where the integration is with respect to the waterline curve, C, and n_1 is the longitudinal component of the outward normal to C.

Then we need to time average eq. (7.116). Because the time average of $\cos^2(\omega_e t + \delta)$ is 0.5, we get

$$\bar{F}_1 = -\frac{\rho g}{4}\int_C \eta_{Ra}^2(x) n_1 \, dS. \qquad (7.117)$$

Because η_{Ra} is proportional to ζ_a, $\bar{F}_1$ is proportional to ζ_a^2. We want to emphasize that eq. (7.117) is only one of several contributions to added resistance in waves following from direct pressure integration. For instance, integrating the pressure loads due to the quadratic velocity term in Bernoulli's equation over the mean wetted hull surface gives another contribution.

7.5.2 Added resistance in a sea state

We will illustrate how added resistance in a long-crested short-term irregular sea state can be calculated by starting with eq. (7.116). However, we now represent η_R as in eq. (7.107). Then we take the time average of η_R^2. This is the same as was done in eq. (7.108), and this led to eq. (7.110). Now we replace $0.5\,|H|^2$ in eq. (7.110) by $\bar{F}_1/\zeta_a^2$. As for eq. (7.111), this leads to

$$R_{AW} = 2\int_0^\infty S(\omega_0)\frac{\bar{F}_1}{\zeta_a^2}\, d\omega_0. \qquad (7.118)$$

Here $\bar{F}_1/\zeta_a^2$ is a function of ω_0, U, and the ship's heading. Of course, we took only one component of the added resistance in waves when we started out with eq. (7.116). However, doing the same with the other components does not change the fact that we can express R_{AW} in irregular long-crested sea, as in eq. (7.118).

7.6 Seakeeping characteristics

A semi-displacement vessel, hydrofoil vessel, and SES have quite different seakeeping behaviors. We discussed in Chapter 6 that a hydrofoil vessel with a submerged foil system had clearly better seakeeping behavior than a semi-displacement vessel in head sea even without using automatic control. However, this type of hydrofoil vessel needs, in general, a ride control system. Figure 1.12 illustrates the difference for an SES and a catamaran. Calculated operational limits based on vertical accelerations at the center of gravity (COG) of a 40 m–long catamaran without a foil and a 40 m–long SES without a ride control system are shown. An RMS value of 0.2 g is used as a criterion in this example. The vertical accelerations will, of course, depend on the details of the hull. The intention of Figure 1.12 is to illustrate the main features of the acceleration level at COG in head seas. The results in Figure 1.12 are for long-crested waves described by the two-parameter JONSWAP spectrum recommended by ITTC. T_1 is the mean wave period defined by the first moment of the wave spectrum and $H_{1/3}$ is the significant wave height. The vessel speed in calm water is 50 knots for the SES and 40 knots for the catamaran. The SES will have the highest involuntary speed loss in a seaway.

Figure 1.12 shows that the operational limits for small wave periods are clearly lowest for the SES. The reason is the "cobblestone" effect, which is caused by resonances occurring in the air cushion. No ride control was accounted for in the calculations. Use of ride control will increase the operational limits for the SES at lower wave periods.

Figure 1.12 illustrates that the SES has the lowest vertical acceleration level when $T_1 > \approx 5$ s. We illustrate in section 7.2.9 that using foil appendages on a catamaran may improve the seakeeping qualities. However, care must be shown in doing that. We must have in mind that there are different contributions to damping of vertical ship motion. An example will illustrate this for the catamaran. Three hulls were numerically investigated. Hull 1 was a basis hull without foils. Hulls 2 and 3 were modifications of hull 1. The displacement was reduced respectively by 10% and 20% relative to hull 1. This was done by reducing the displacement volume in the aft end of the ship. The hulls had transom sterns, and the reduction in the displaced volume resulted in a decrease in the local beam at the transom stern. The reduction in displacements of hulls 2 and 3 were compensated by lift from foils at the aft end of the hulls. The net result of this was that the vertical accelerations of the three hulls did not differ very much. One reason is probably that the increase in damping by the foils was compensated by a decrease in the damping due to the hull. The latter is the result of both wave radiation damping and dynamic lifting effects. The magnitudes of the lift force and moment on the hull depend on the local beam at the transom stern. However, adding passive foils to a hull without changing the hull will lower the vertical accelerations. The foils are most effective if they are placed in the bow part where the relative vertical motions are largest. However, out-of-water effects of the foils or cavitation will degrade the damping effect of the foils.

There is no doubt that foils improve the seakeeping behavior of a high-speed catamaran. We will quantify this by an example presented by Haugen et al. (1997). A 580-tonne catamaran with 54 m length between perpendiculars was investigated. Long-crested head sea is assumed. According to service regulations by DNV, the vessel can operate in a sea with maximum significant wave height $H_{1/3} = 3.5$ m. Service regulations by DNV specify reductions in speed relative to the significant wave height. The maximum allowable speed is 40 knots for $H_{1/3}$ between 0 and 1 m, 35 knots for $H_{1/3}$ between 1 and 2 m, 30 knots for $H_{1/3}$ between 2 and 2.5 m, and 25 knots for $H_{1/3}$ between 2.5 and 3.5 m. When $H_{1/3}$ is larger than 3.5 m, the vessel should seek shelter at slow speed or it should not leave port.

An alternative way to estimate speed changes in head sea waves is to use operational criteria for slamming and vertical accelerations (see section 1.1). Ochi (1964) has expressed the slamming probability in terms of a critical relative vertical velocity between the hull and the water surface, which is 2.14 ms^{-1} for a ship with $L_{PP} =$ 54 m. This was used in the absence of any other established criteria known to the authors. When the wetdeck slamming probability at FP is 0.03 or larger in a sea state, speed changes are assumed. This depends both on the wave period and the significant wave height. Peak periods between 4 and 12 s were studied. A modified Pierson-Moskowitz wave spectrum was used. The linear transfer functions for vertical relative velocities and motions have been calculated by strip theory (Salvesen et al. 1970) when the Froude number $Fn < 0.4$. When $Fn \geq 0.4$, the 2.5D high-speed theory by Faltinsen and Zhao (1991a) has been used. The two hulls are assumed hydrodynamically independent. The vessel is studied with and without vertical motion damping foils. These are T-foils located at $0.17L_{PP}$ measured from FP and trim tabs located at AP. The slope of the lift coefficient $dC_L/d\alpha = 3.0$ for the T-foils and 2.0 for the trim tabs. The projected horizontal foil areas are 2.5 m^2 and 4.0 m^2 for the T-foils and the trim tabs, respectively. In the calculations, the foils are handled as appendages and no degradation of the foil performance due to possible air ventilation is accounted for. The probability of ventilation increases with sea state and ship speed. Figure 7.38 shows the results of the computations. For a given ship speed, there is a maximum significant wave height $H_{1/3\max}$ that the ship can operate in because of the wetdeck slamming. $H_{1/3\max}$ shown in Figure 7.38 is the minimum value for all the investigated wave periods. There is a limiting maximum value of $H_{1/3}$ that can occur because of the wave breaking for a given wave period. This is not accounted for in the figure. However, wave breaking will not occur for the wave periods that give the $H_{1/3\max}$ value shown in Figure 7.38.

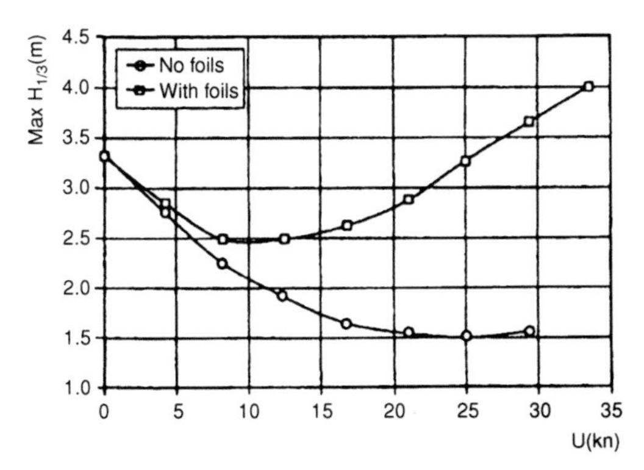

Figure 7.38. Maximum operational significant wave height $H_{1/3}$ due to wetdeck slamming as a function of ship speed U for a 54 m–long catamaran. Head sea long-crested waves (Haugen et al. 1997).

Introducing criteria for speed changes due to vertical accelerations will not significantly change the picture seen in Figure 7.38. It is clear that the catamaran without foils cannot operate in $H_{1/3} =$ 3.5 m in head sea waves; however, the vessel can, of course, change sailing course. When the catamaran is equipped with vertical motion damping foils, it can operate in much higher sea states at high speed. Actually, Figure 7.38 shows that a good strategy to lower the wetdeck slamming probability for $U > \approx 12$ knots is to increase the ship's speed. The reason is that the lift and therefore the damping provided by the foils increase with speed. However, the results in Figure 7.38 are dependent on the slamming criterion by Ochi (1964) and its applicability to wetdeck slamming on a high-speed ship. This needs to be studied further.

When characterizing the seakeeping behavior of a vessel, it is not sufficient to have only figures like Figure 7.38. Other important quantities are vertical accelerations along the length of the ship, rolling, relative vertical motions and velocities between the ship and the waves, and the influence of other wave directions. The relative vertical motions and velocities are closely related to slamming loads. The relative vertical motions also express the danger of green water on the fore deck or of the waterjet inlet being exposed to air. The latter was examined by Faltinsen et al. (1991a) for a 40-m catamaran and a 40-m SES with flush inlets. It was clear that the waterjet inlet on an SES could easily be exposed to air for small sea states. The catamaran has a better behavior. In order to set operational limits for the performance of a waterjet system in a seaway, it is necessary to better understand the physics and how to relate a criterion to the relative vertical motion at the waterjet inlet (Faltinsen et al. 1991a).

We present here more detailed results by Faltinsen et al. (1991a) for the catamaran. The results for the SES are given in section 5.8. The main characteristics of the catamaran are given in Table 7.3. Waterjet propulsion is used. Only head sea waves and wind were considered. A JONSWAP spectrum (see eq. (3.61)) was used. The wind velocity, V_W, was assumed uniform in space and calculated by the formula $V_W = (H_{1/3}g/0.21)^{0.5}$.

The results are presented in Table 7.4. Involuntary speed loss is calculated for each combination of $H_{1/3}$ and T_1, where T_1 is the mean wave period defined by the first moment of the spectrum. We note that the speed is dependent on T_1. This is because of the added resistance in waves, which can be explained by eq. (7.118) together with Figure 7.35 showing typical behavior of added resistance in regular waves. Because of the dominant peak in the added resistance curve for regular waves, the added resistance in irregular sea will be sensitive to the mean wave period (see also Figure 5.27). This example demonstrates that one should not use a sea-state number system in which there is only one mean wave period associated with one significant wave height. As an indication of the relative importance of added resistance, the predicted added resistance is 7.8% of the total resistance when $H_{1/3} = 2$ m, $T_1 = 5$ s. The importance of added resistance increases strongly with increasing significant wave height.

Table 7.4 also presents standard deviations for vertical accelerations at the center of gravity of the catamaran. If we require that $\sigma_a < 0.2\,g$, we

Table 7.3. *Main particulars of the high-speed catamaran*

Length overall	40.00 m
Length of waterline	35.30 m
Beam of each hull	3.53 m
Draft of each hull	1.78 m
Distance between centerplanes of the hulls	8.00 m
Block coefficient of each hull	0.54
Pitch radius of gyration	8.57 m
Longitudinal position of center of gravity from FP	20.73 m

Table 7.4. *Seakeeping data of a 40 m–long high-speed catamaran (see Table 7.3) in head sea long-crested waves with a two-parameter JONSWAP-spectrum*

	T_1(s)						
$H_{1/3}$(m)	3.3	5.0	6.7	8.3	10.0	11.7	
1.0	39.3	39.1	39.0	39.2	39.3	39.3	Speed (kn)
	0.003 g	0.15 g	0.21 g	0.14 g	0.11 g	0.08 g	σ_a
	0.27	0.27	0.25	0.17	0.12	0.09	σ_R(m)
2.0	38.6	37.7	37.4	38.2	38.6	38.8	Speed (kn)
	0.07 g	0.31 g	0.42 g	0.28 g	0.21 g	0.16 g	σ_a
	0.54	0.56	0.48	0.32	0.23	0.18	σ_R(m)
3.0	37.9	35.4	35.0	36.9	37.8	38.2	Speed (kn)
	0.10 g	0.49 g	0.61 g	0.41 g	0.31 g	0.24 g	σ_a
	0.81	0.82	0.67	0.44	0.34	0.27	σ_R(m)
4.0		30.5	31.9	35.0	36.8	37.6	Speed (kn)
		0.73 g	0.74 g	0.52 g	0.40 g	0.32 g	σ_a
		1.05	0.73	0.57	0.45	0.37	σ_R(m)

$H_{1/3}$ = significant wave height, T_1 = mean wave period. Shaft power is 2×4150 kW in all sea states. Waterjet propulsion (flush type), diameter outlet: 0.55 m. Design speed, still water: 40 knots. Total efficiency, still water: 0.603. First line: Speed in knots including involuntary speed loss. Second line: Standard deviation of vertical accelerations at COG. Third line: Standard deviation of relative vertical motions at waterjet inlet.

see that the involuntary speed prediction is too high for one sea state ($T_1 = 6.7$ s) when $H_{1/3} = 1$ m, and for most sea states when $H_{1/3} \geq 2$ m. If we use $\sigma_a = 0.1$ g as a limit, voluntary speed reduction will also occur in most of the sea states when $H_{1/3} = 1$ m. We have also presented the standard deviation σ_R for the relative vertical motions at the waterjet inlet. By requiring that the draft $d = 1.7$ m at the waterjet inlet be larger than four times the standard deviation σ_R of the relative vertical motion at the waterjet inlet, we see that the limiting criterion for voluntary speed reduction is satisfied when $H_{1/3} = 1$ m and for three of six sea states when $H_{1/3} = 2$ m. Here $4\sigma_R$ is close to the most probable largest value in the sea state. One may question whether this is a correct criterion for operational limit.

It is important to investigate different wave headings. For instance, when a catamaran in following regular waves has a speed close to the phase speed of the incident waves, the catamaran can come to a position relative to the waves so that the fore part of the vessel dives into a wave crest (Figure 7.39). If there is not sufficient buoyancy in the fore part of the vessel, a critical situation may occur. The problem can be qualitatively analyzed by approximating the fluid loads by only including hydrostatic and Froude-Kriloff loads.

7.7 Dynamic stability

The dynamic stability of high-speed vessels both in calm water and in waves is, in general, poorly understood. A classification of phenomena that can happen for monohulls is shown in Figure 7.40. We should note that the importance of hydrostatic pressure relative to hydrodynamic pressure decreases with increasing forward speed. One

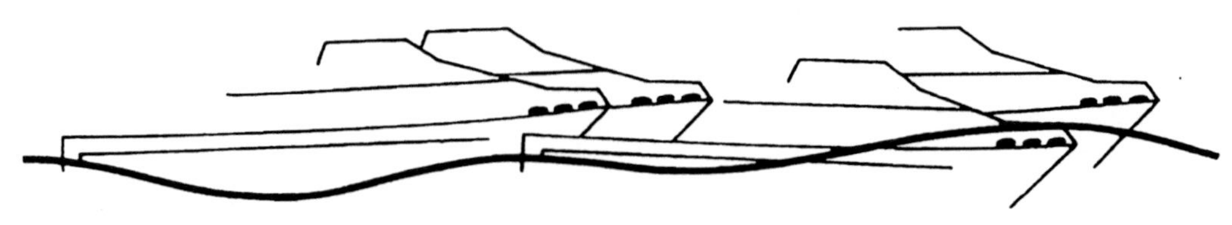

Figure 7.39. Danger of diving of the fore part of a high-speed catamaran in following seas (Werenskiold, unpublished).

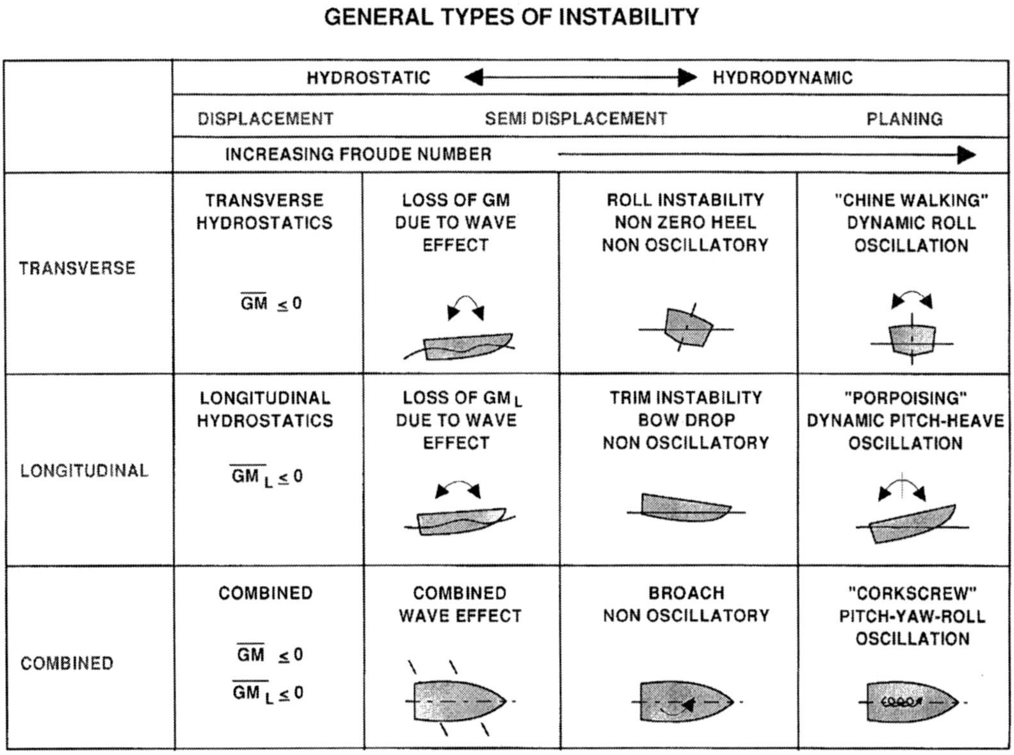

GENERAL TYPES OF INSTABILITY
HYDROSTATIC
HYDRODYNAMIC
DISPLACEMENT
SEMI DISPLACEMENT
PLANING
INCREASING FROUDE NUMBER
TRANSVERSE
TRANSVERSE HYDROSTATICS
$\overline{GM} \leq 0$
LOSS OF GM DUE TO WAVE EFFECT
ROLL INSTABILITY NON ZERO HEEL NON OSCILLATORY
"CHINE WALKING" DYNAMIC ROLL OSCILLATION
LONGITUDINAL
LONGITUDINAL HYDROSTATICS
$\overline{GM}_L \leq 0$
LOSS OF GM_L DUE TO WAVE EFFECT
TRIM INSTABILITY BOW DROP NON OSCILLATORY
"PORPOISING" DYNAMIC PITCH-HEAVE OSCILLATION
COMBINED
COMBINED
$\overline{GM} \leq 0$
$\overline{GM}_L \leq 0$
COMBINED WAVE EFFECT
BROACH NON OSCILLATORY
"CORKSCREW" PITCH-YAW-ROLL OSCILLATION

Figure 7.40. General types of instability for monohulls (Cohen and Blount 1986).

should also note that the rudder, cavitation, and ventilation phenomena will influence the dynamic stability of high-speed vessels. Figure 7.41 is based on static tests of monohulls. A given weight was placed off the centerplane of the vessel. The resulting steady roll moment causes a steady roll angle. The figure illustrates how the constant heel angle changes with increasing forward speed for four different high-speed full-scale craft. The reason is the change in hydrodynamic pressure on the hull, influence of the rudder, and possible cavitation effects in the propeller tunnels.

The loss of steady restoring moment in heel with forward speed can cause a sudden list of a round-bilge monohull to one side. This can at high speed be followed by a violent yaw to one side. The consequence can be capsizing. This "calm water broaching" is the main reason round-bilge hulls should not operate beyond a Froude number of 1.2 (Lavis 1980).

The loss of steady restoring heel moment with speed should be accounted for in the design by having a sufficiently high metacentric height $\overline{GM}$ at zero speed. Recommendations are given by Müller-Graf (1997). Suggested minimum values for round-bilge monohulls are given in Figure 7.42 and are a function of the beam-to-draft ratio and the maximum operating Froude number. The

ROLL ANGLE, CALM WATER
14°
10°
6°
2°
0
0.4
0.8
1.2
1.6
FROUDE NUMBER
CRAFT
A
B
C
D

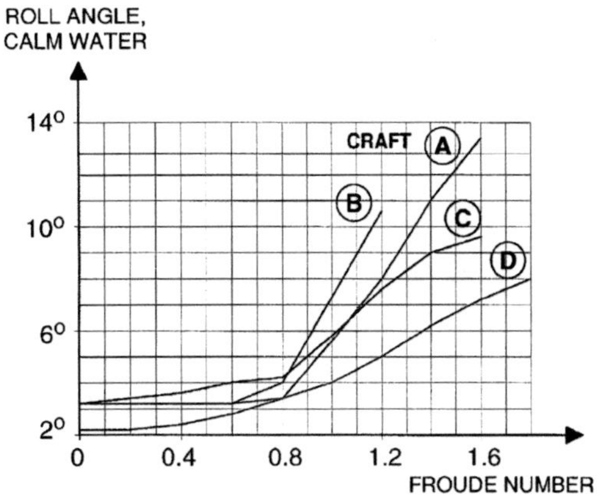

Figure 7.41. Static heel stability of different monohulls in calm water at forward speed. Experimental results. A given weight was placed off the centerplane. The figure illustrates the decreasing importance of hydrostatic pressure with increasing Froude number (Werenskiold 1993).

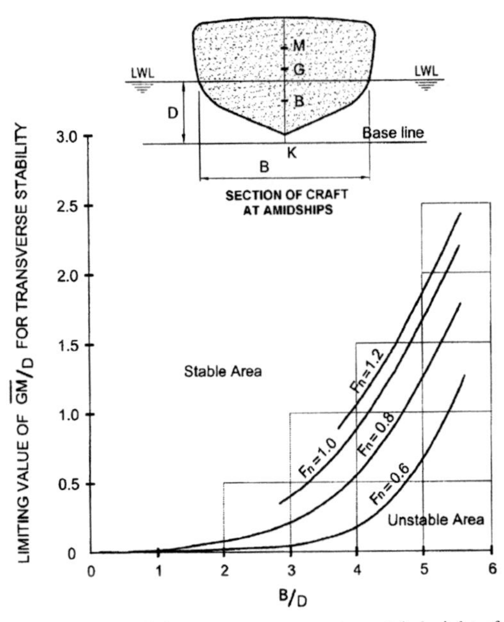

Figure 7.42. Minimum transverse metacentric height of a round-bilge monohull at zero speed as a function of beam-to-draft ratio B/D and maximum operating Froude number Fn (Bailey 1976).

forward speed dependence on the steady heel moment should be considered if Fn is larger than 0.5. There may be a significant effect even at Froude number 0.6. $\overline{GM}$ between 1 and 1.5 m are representative recommended values for a round-bilge monohull of 10 to 30 m length. $\overline{GM}$ less than 0.8 m should be avoided (Müller-Graf 1997). A spray rail system will decrease the loss of steady restoring heel moment at high speed (Müller-Graf and Schmiechen 1982). This can be understood by considering a spray rail a low-aspect–ratio lifting surface. The lift force is proportional to the square of the forward speed of the vessel. The speed dependence of the restoring heel moment of a hard-chine monohull at planing speed is discussed in section 10.9.3.

Because the speed dependence of the the steady restoring heel moment is partly the result of the generation of free-surface waves, shallow water effects matter. This depends on the water depth–to-ship length ratio h/L and the depth Froude number $Fn_h = U/\sqrt{hg}$ (see section 4.4). If $h/L >$ 0.4, there is small water depth influence. When the finite water depth effects matter, the behavior is clearly different at subcritical, critical, and supercritical flows. Even though the discussion in sections 4.4 and 4.5 is for the wave resistance problem, it is also relevant for the speed dependence of the steady heel moment. However, we are not aware of quantitative information on how much shallow water effects can matter.

Because a catamaran has a larger beam than a monohull, the catamaran will with the same position of the center of gravity as the monohull have the clearly highest metacentric height. It means that the loss of steady heel moment with forward speed is a less severe problem for the catamaran.

Let us elaborate more on Figure 7.40. The most classic situation for capsizing is when the vessel rolls with large amplitudes in beam sea. An extreme case is when breaking waves occur. The combination of green water on deck and wind heeling moments increases the probability of capsizing.

Broaching is mentioned in Figure 7.40 as an instability problem for semi-displacement vessels. However, this is also true for ordinary displacement vessels and quite typical for sailboats. It is of particular concern in following seas and occurs in long and steep waves. The wavelength is longer than the ship's length, and the wave shape propagates faster than the ship. If one single wave like this is at the position of the ship, it will tend to move the ship with the same speed as the horizontal fluid velocity in the wave. It means that the relative horizontal velocity between the ship and the water becomes small. The ship then loses its directional stability in the horizontal plane, and the rudder loses its effect. The consequence is a change in the ship's course. This becomes more critical if several (say three or four) similar waves are passing the ship. Each wave will change the ship's course. The result of the change in course is that the ship heels to the leeward side and the waves can come in a beam sea situation relative to the ship. If the ship has a low transverse metacentric height, capsizing may occur. Broaching in waves is described, for instance, by Wahab and Swaan (1964), Nicholson (1974), and Vassalos et al. (2000) (see also section 10.10.3).

Cork-screw instabilities are discussed in section 10.9.3. Porpoising is an instability phenomenon only for planing vessels and is discussed in detail in section 9.4.1.

7.7.1 Mathieu instability

A special type of roll instability that is not explicitly mentioned in Figure 7.40 is the Mathieu-type instability. This is a problem particularly for ships

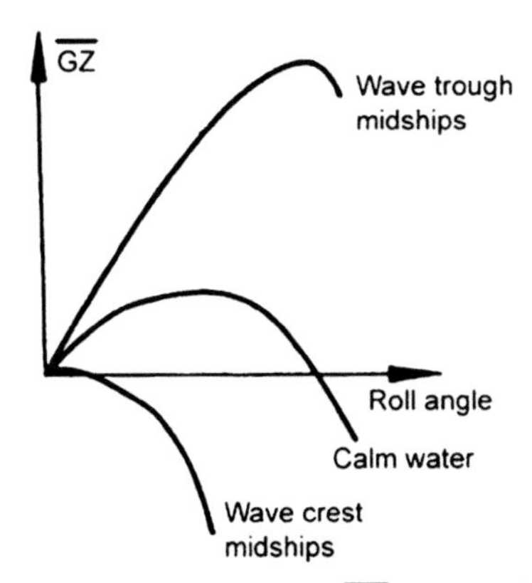

Figure 7.43. Example on how the $\overline{GZ}$-curve is influenced by the wave. Based on quasi-steady analysis. The slope $d\overline{GZ}/d\eta_4$ at zero roll angle η_4 is the metacentric height.

with nonvertical ship sides near the waterline. Changes in the local beam with changes in the local draft are most significant in the bow and stern regions. These changes may cause the transverse $\overline{GM}$ to oscillate with large amplitudes in a seaway. The latter is a necessary requirement for a Mathieu-type instability to occur, but there are other necessary requirements. For instance, the waves must not change much in amplitude or period, which means that it is easy to demonstrate the phenomena in regular waves in a model tank, but difficult to demonstrate in a sea state with many wave components of different frequencies. Further, the ratio between the natural roll frequency and the frequency of oscillation of the ship has to be approximately either 0.5, 1.0, 1.5, 2.0 and so on. A low roll damping is also a necessity for the buildup of large roll amplitudes due to a Mathieu instability.

Let us try to explain the phenomenon theoretically. The first step is to find out how the $\overline{GZ}$-curve is influenced by the heave, pitch, and wave motions. If we know these motions, then we can calculate by a hydrostatic stability program how the $\overline{GZ}$-curve changes at moderate Froude numbers. Let us first illustrate this by considering only the effect of the waves (Figure 7.43). Here only two wave positions are shown. For us, it is the initial metacentric height (the slope of the $\overline{GZ}$-curve at $\eta_4 = 0$) that is most interesting. We see in Figure 7.43 that this is largest when the wave trough is midships and smallest when there is a wave crest midships. One should, of course, remember that the results in Figure 7.43 are for a given wave and a given ship. We should also note that in Figure 7.43 if the wave were stationary relative to the ship (i.e., $\omega_e = 0$) and there were a wave crest midships, this ship could easily capsize as the result of static effects only. If we had drawn the $\overline{GZ}$– curve for other wave positions, we would have found that the uprighting moment for small angles η_4 could be approximated as

$$F_4 \approx -(\rho g \nabla\, \overline{GM}_m + h_w^* \sin(\omega_e t + \alpha))\eta_4, \quad (7.119)$$

where $-\rho g \nabla\, \overline{GM}_m \eta_4$ is the uprighting moment in calm water. We note that the additional term due to waves oscillates with the frequency of encounter ω_e. If we had calculated the GZ-curve for other wave amplitudes ζ_a, we would have found that h_w^* is approximately proportional to ζ_a, that is,

$$h_w^* \approx h_w \zeta_a. \quad (7.120)$$

The magnitude of h_w will strongly depend on the slope of the ship's side near the waterline.

In the same way as for the waves, we can find similar effect of heave and pitch. The coupling due to pitch is generally less than the coupling from heave. We can now write the roll equation approximately as

$$(I_{44} + A_{44})\frac{d^2\eta_4}{dt^2} + B_{44}\frac{d\eta_4}{dt} + \rho g \nabla(\overline{GM}_m + \delta\overline{GM}\sin(\omega_e t + \beta))\eta_4 = 0, \quad (7.121)$$

where the effect of heave, pitch, and wave motions is represented by the term

$$\rho g \nabla \delta \overline{GM} \sin(\omega_e t + \beta)\eta_4. \quad (7.122)$$

Eq. (7.121) is approximate because we have neglected the coupling from sway and yaw and the damping term has been linearized. We have assumed head or following seas, so the roll excitation moment from the waves is zero. However, we may still experience rolling due to the instabilities. This means that a small perturbation from the equilibrium position will result in strongly increased motions with time.

Eq. (7.121) can be rewritten as

$$\frac{d^2\eta_4}{dt^2} + 2\xi\omega_n\frac{d\eta_4}{dt} + \omega_n^2\left(1 + \frac{\delta\overline{GM}}{\overline{GM}_m}\sin(\omega_e t + \beta)\right)\eta_4 = 0. \quad (7.123)$$

Here ω_n is the undamped natural roll frequency, that is,

$$\omega_n = \sqrt{\frac{\rho g \nabla \overline{GM}_m}{I_{44} + A_{44}}} \quad (7.124)$$

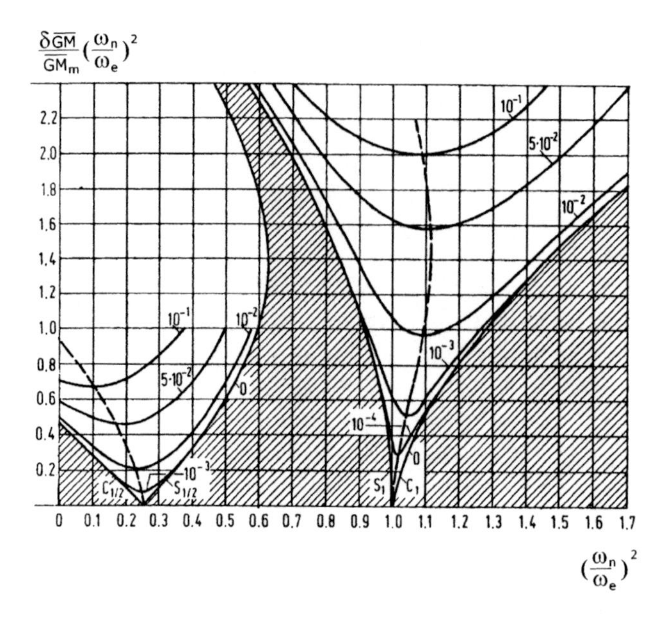

Figure 7.44. Stability diagram for the Mathieu equation applied to ship rolling. ω_n = natural roll frequency, $\delta\overline{GM}$ = amplitude of the harmonically oscillating part of the metacentric height, ω_e = frequency of encounter. Shaded areas represent stable domains when the ratio ξ between roll damping and the critical damping is zero. There are also lines shown with values of ξ (ω_n/ω_e) equal to 10^{-4}, 10^{-3}, 10^{-2}, $5 \cdot 10^{-2}$, and 10^{-1}. These lines are boundaries between stable and unstable domains (Klotter 1978).

and ξ is the ratio between the damping and the critical damping (see eq. (7.9)), that is,

$$\xi = \frac{B_{44}}{2\sqrt{(I_{44} + A_{44})\,\rho g \nabla \overline{GM}_m}}. \qquad (7.125)$$

If $\xi = 0$, eq. (7.123) is the classical Mathieu equation. It can be shown that the stability will depend on ω_n/ω_e, $\delta\overline{GM}/\overline{GM}_m$, and the damping ratio ξ. Klotter (1978) has presented curves that show the instability domains (Figure 7.44). The unshaded areas represent instability domains when $\xi = 0$. The boundaries between the stability and unstability domains are shown for different ξ-values. The domains for dangerous combinations of $\delta\overline{GM}/\overline{GM}_m$ and ω_n/ω_e are located in the vicinity of $\omega_n/\omega_e = 0.5$, 1.0, 1.5, 2.0, and so on, when $\delta\overline{GM}/\overline{GM}_m$ is small. Figure 7.44 concentrates on ω_n/ω_e up to $\sqrt{1.7}$. For instance, $\omega_n/\omega_e = 0.5$ means that the encounter period T_e is half of the natural roll period. Further, larger values of $\delta\overline{GM}/\overline{GM}_m$ are undesirable. This means either small initial stability (i.e., small $\overline{GM}_m$) or that the waves or the heave motion is sufficiently large to cause large values of $\delta\overline{GM}$.

Damping will have a positive influence. The higher the damping, the higher $\delta\overline{GM}/\overline{GM}_m$ has to be for instability to occur. Roll damping tends to be small for monohulls. However, it is an advantage from a wave radiation damping point of view to have the large beam-to-draft ratio that is typical for semi-displacement monohulls. The small roll damping caused by the bare hull is counteracted in conventional ships by, for instance, using bilge keels and antirolling tanks. Antirolling fins and rudders cause a roll damping that increases with the speed. If the vessel is in an area with relatively large waves and is not moving because of some machinery failure, the roll damping fins will be ineffective. The vessel may move broadside to the waves and experience heavy roll in an emergency situation. The roll damping due to wave radiation is frequency dependent, as we have seen in Figure 7.26. If the natural roll period, T_n, is chosen too high, let us say higher than 20 s, then the wave radiation damping tends to be negligible. If the other components of roll damping are also small, it means that, from a Mathieu instability point of view, we should be concerned near the vicinity of $\omega_n/\omega_e = T_e/T_n = 0.5$. Let us say $\xi = 0.02$. Figure 7.44 shows, then, that instabilities may occur for $\delta\overline{GM}/\overline{GM}_m$ slightly larger than 0.8 for $\omega_n/\omega_e = 0.5$.

7.8 Wave loads

There are two different levels at which wave loads may be needed for structural design purposes:

1. Instantaneous local hydrodynamic pressures on the surface of the hull as a result of ship motions and ship-wave interactions. These pressures may be needed over the entire hull surface or only on some portion of it. The

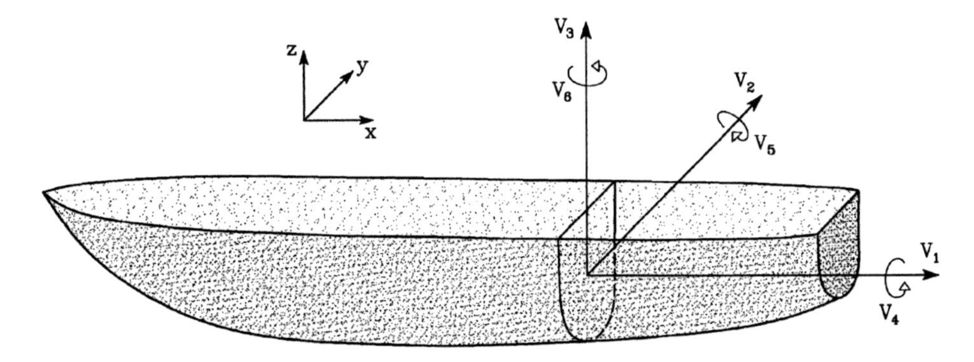

V_1 = compression force
V_2 = horizontal shear force
V_3 = vertical shear force
V_4 = torsional moment
V_5 = vertical bending moment
V_6 = horizontal bending moment

Figure 7.45. Global forces and moments in hull beam.

important case of slamming (water impact) pressures is addressed in Chapter 8.

2. Integrated instantaneous pressures (global wave loads), giving for instance:
   a) Vertical and torsional bending moments and shear forces at midships or other stations (Figure 7.45).
   b) Transverse vertical bending moments, vertical shear forces, and pitch connecting moments on half of a part of a catamaran obtained by intersecting along the centerplane (see Figure 7.48).

Global wave loads are expected to be significant for monohulls and catamarans of lengths larger than 50 m. As stated in the *DNV Rules for High Speed and Light Craft* (part 3, chapter 1, section 1) for craft less than 50 m, the minimum strength standard for hull girder strength is normally satisfied for scantlings obtained from local strength requirements (of plates and stiffeners due to lateral pressure). Often for small ships, but also for ships longer than 50 m, actually the minimum global strength standards are satisfied by a considerable margin, hence there may be possibilities to optimize these vessels' weight slightly more.

7.8.1 Local pressures of non-impact type

The finite element structural analysis techniques have given an impetus to the development of methods for calculating the distribution of instantaneous hydrodynamic pressures over individual sections and hence over the entire surface of a hull oscillating in waves. Earlier in this chapter, we show that the calculation of ship motions requires the determination of integrated unsteady forces and moments due to hydrostatic and hydrodynamic pressures over the surface of the hull. This is done by using integral theorems in the strip theory by Salvesen et al. (1970), but most often it means that the pressure distribution at a particular instant is implicitly known when the motions have been calculated.

When linear theory is used, the hydrodynamic pressure results in the added mass, damping, Froude-Kriloff, and diffraction forces and moments. The hydrodynamic pressure in the linear theory is evaluated at the mean oscillatory position of the ship. It is consistent in linear theory to use the same pressure terms at the instantaneous position of the ship. The hydrostatic pressure term that gives rise to restoring coefficients in the equation of motions must be evaluated from its value at the instantaneous ship position. Some care must be shown at the intersection between the free surface and the hull. This introduces nonlinear effects that a linear theory cannot completely describe.

Knowledge about the pressure distribution close to the waterline is needed in studies of fatigue damage in the side longitudinals in the midship area (Berstad et al. 1997). This type of damage has occurred in oil tankers operating on the western coast of North America and on the eastern coast of South Africa, but they are certainly not the only vessels that experience this type of cracking.

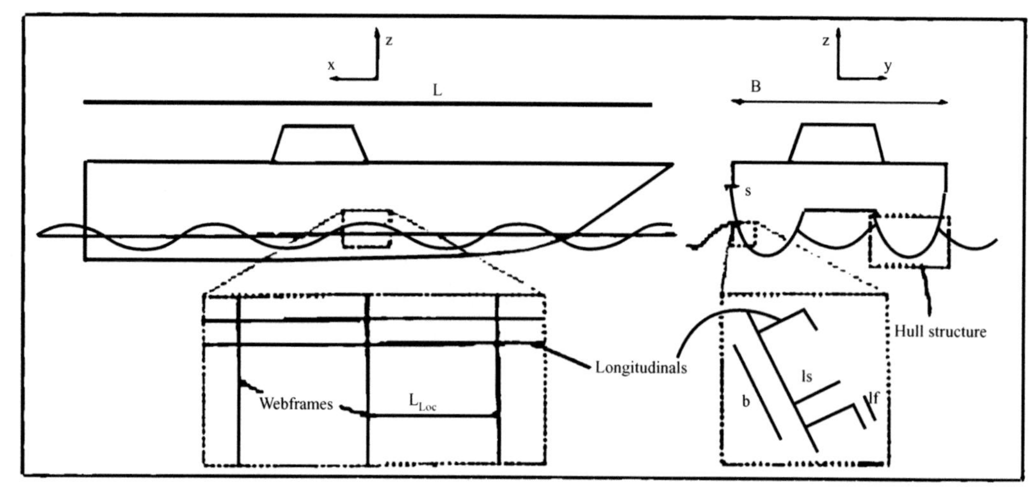

Figure 7.46. Details of the local structure in the waterline zone of a high-speed catamaran used in a fatigue damage investigation by Berstad and Larsen (1997). The stress response in the longitudinals is calculated as the sum of local bending responses from lateral pressure and axial stresses due to global horizontal and vertical bending moments.

The introduction of HT steel into the side structures during the 1980s increased this problem on tankers. This became a big issue after the Exxon Valdez accident and the introduction of OPA-90 (Oil Pollution Act). Cracks propagating through the side shell or contaminating ballast water giving oil spills is a severe problem for the owner during loading/off-loading because oil is easily seen on the water, resulting in a bad reputation and so on. However, this is not a big issue on high-speed vessels from a pollution point of view.

Smaller ships may have transverse instead of longitudinal stiffening. This kind of stiffening will give less redundancy for global buckling and is not often considered with respect to fatigue. However, the method below may also be used for these kinds of details.

The theoretical procedure presented by Berstad et al. (1997) was applied to a high-speed catamaran by Berstad and Larsen (1997). The structural details are shown in Figure 7.46.

We will present the theoretical procedure of Berstad et al. (1997) for finding the pressure in the waterline zone.

The total pressure below the instantaneous free surface is found as

$$p_{tot} = -\rho\left(\frac{\partial}{\partial t} + U\frac{\partial}{\partial x}\right)\varphi_T - \rho g z + p_a, \quad (7.126)$$

where φ_T is the total time-dependent velocity potential. This means φ_T consists of the sum of eq. (7.99) and the velocity potential of the incident waves. Eq. (7.126) assumes no interaction between the local steady flow and the unsteady flow. Because linearity is assumed, linear superposition can be assumed in irregular seas. The atmospheric pressure p_a can be omitted in the analysis. The local wave elevation is

$$\zeta = -\frac{1}{g}\left(\frac{\partial}{\partial t} + U\frac{\partial}{\partial x}\right)\varphi_T. \quad (7.127)$$

where φ_T is evaluated at the mean free surface. When calculating the pressure at a point on the ship's side, one has to account for the vertical motion of the point due to ship motions. The motion is $\eta_3 - x\eta_5 + y\eta_4$. The relative position between point A (Figure 7.47) and the water is then

$$\zeta_T = \zeta - (\eta_3 - x\eta_5 + y\eta_4 + z_A), \quad (7.128)$$

where (x, y, z_A) is the equilibrium coordinates of point A. If the relative water elevation is less than zero, the point is out of water and the pressure is

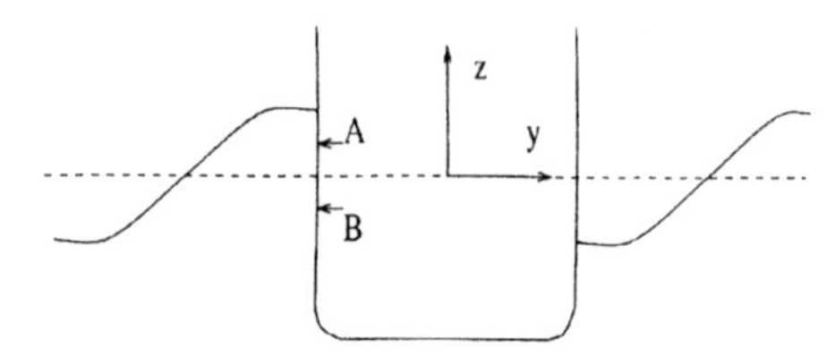

Figure 7.47. Local external fluid pressures in the waterline zone.

zero. Otherwise, the pressure for a point above the mean free surface is calculated as

$$p = \rho g(\zeta - (\eta_3 - x\eta_5 + y\eta_4 + z_A)).$$

This implies that the pressure is "hydrostatic" relative to the instantaneous wave elevation. (see Figure 3.5). Using eq. (7.127) means that

$$p = -\rho\left(\frac{\partial}{\partial t} + U\frac{\partial}{\partial x}\right)\varphi_T\bigg|_{z=0} - \rho g(\eta_3 - x\eta_5 + y\eta_4 + z_A). \qquad (7.129)$$

As seen from eqs. (7.128) and (7.129), the condition that point A is out of water corresponds to the conditions that the pressure calculated from eq. (7.129) is less than zero. Therefore, the pressure is calculated as

$$p = \max\left(-\rho\left(\frac{\partial}{\partial t} + U\frac{\partial}{\partial x}\right)\varphi_T\bigg|_{z=0} - \rho g\left(\eta_3 - x\eta_5 + y\eta_4 + z_A\right), 0\right). \qquad (7.130)$$

For a point B on the ship's side with an equilibrium position below the mean free surface, a Taylor series expansion of the potential φ_T about the point B is performed. Keeping the linear terms, the following expression for the pressure is derived:

$$p = \max\left(-\rho\left(\frac{\partial}{\partial t} + U\frac{\partial}{\partial x}\right)\varphi_T\bigg|_{z=B} - \rho g\left(\eta_3 - x\eta_5 + y\eta_4 + z_B\right), 0\right). \qquad (7.131)$$

Here (x, y, z_B) is the equilibrium coordinates of point B. If this procedure is used to calculate the global loads on the ship, it will lead to nonlinear second-order terms in addition to linear first-order terms. However, it is not a consistent second-order theory. For instance, a second-order potential and other terms in Bernoulli's equation have to be introduced. Further, the local steady flow and its interaction with the unsteady flow have not been considered above.

7.8.2 Global wave loads on catamarans

Important global loads for catamarans are transverse vertical bending moment (often called split moment), vertical shear force, and pitch connecting moment, as illustrated in Figure 7.48. Torsional moments, vertical shear forces, and vertical bending moments at transverse cross sections are also of concern, as they are for monohulls.

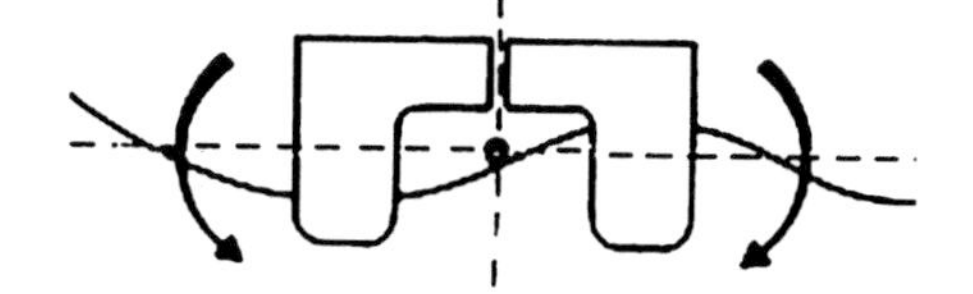

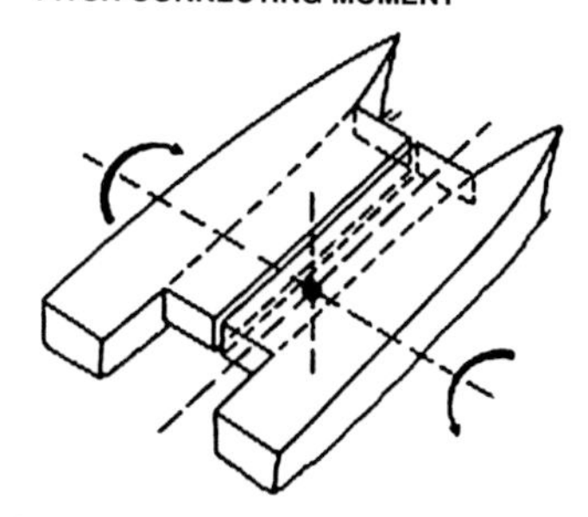

Figure 7.48. Examples of global wave loads on a catamaran.

Both continuous wave loading and slamming will cause global loads. In Chapter 8, we deal with wetdeck slamming–induced global loads, called whipping. The vessel must then be considered elastic. When we consider the effect of continuous wave action, the catamaran can normally be considered rigid in the determination of the global loads. However, it should be ensured that the natural frequencies of the global elastic modes are higher than the encounter frequencies of practical interest. In Chapter 8, we discuss how to account for the global elastic modes during continuous wave loading. This phenomenon is called springing. We discuss unsteady effects; however, there are also steady Froude number effects that must be considered.

Global wave loads in regular waves

Because the numerical method by Faltinsen and Zhao (1991a,b) finds the dynamic pressure distribution on the hull, one can directly integrate the pressure and inertia loads to find the global dynamic loads. In the following, we present a different and simplified method to find the dynamic loads on the half-parts that are obtained by a cut along the centerplane of the catamaran. These global loads are considered to be of main interest in the design of a large catamaran.

Consider a catamaran at a high Froude number in incident regular waves in deep water. A right-handed coordinate system (x, y, z) fixed with

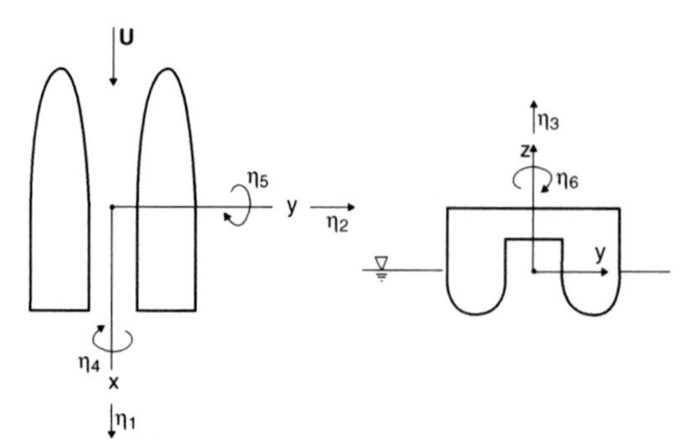

Figure 7.49. Coordinate system and definitions of translatory and angular displacement of a catamaran.

respect to the mean oscillatory position of the ship is used, with positive z vertically upward through the center of gravity of the ship and the origin in the plane of the undisturbed free surface. The catamaran is assumed to have the xz-plane as a plane of symmetry in its mean oscillatory position. Each hull is assumed symmetric about the hull's centerplane below the waterline. Let the translatory displacement in the x-, y-, and z-directions with respect to the origin be η_1, η_2, and η_3, respectively, so that η_1 is the surge, η_2 is the sway, and η_3 is the heave displacement. Furthermore, let the angular displacements of the rotational motions about the x-, y-, and z-axes be η_4, η_5, and η_6, respectively, so that η_4 is the roll, η_5 is the pitch, and η_6 is the yaw angle. The coordinate system and the translatory and angular displacement conventions are shown in Figure 7.49. The incident wave potential is written in complex variables as

$$\varphi_0 \exp(i\omega_e t) = \frac{g\zeta_a}{\omega_0} \exp(kz - i\,kx\cos\beta - i\,ky\sin\beta + i\omega_e t). \quad (7.132)$$

This is consistent with the wave elevation given by eq. (7.21).

The dynamic force vector (V_1, V_2, V_3) on the half-part of the catamaran, obtained by cutting along the centerplane of the catamaran, is the difference between the inertia forces and the sum of external forces acting on the above-mentioned half-part of the catamaran. The external dynamic forces can be divided into added mass and damping forces, restoring forces, and wave excitation forces. Similarly, the moment vector (V_4, V_5, V_6) is equal to the difference between the moment due to the inertia forces and the moment due to the external forces. V_2, V_3, V_4, and V_5 are the important structural loads and will be considered in more detail. In the following text, V_2, V_3, V_4, and V_5 denote horizontal athwartships force, vertical shear force, transverse vertical bending moment, and pitch connecting moment between the two hulls, respectively. By using the symmetry property of each hull and the assumption of no hydrodynamic interaction between the hulls, we can set up simplified expressions.

We go through the details for the vertical shear force V_3 and refer to Figure 7.50. We can use Newton's second law, which implies that:

Mass of right-hand half-part of the catamaran
times
Vertical acceleration of the center of gravity of this half-part
(Vertical structural inertia force)
=
Sum of vertical forces acting on this half-part

These vertical forces consist of

- Vertical shear force V_3
- Vertical added mass, damping, and restoring forces on the half-part
- Vertical wave excitation force on the half-part

The vessel mass of the half-part is $0.5M$, where M is the mass of the catamaran. The vertical acceleration of the center of gravity of the half-part follows by differentiating the z-component of eq. (7.20) twice with respect to time and setting $y = y_A$, which is the y-coordinate of the center of gravity of the half-part of the catamaran. Further, $x = 0$. This means the longitudinal position of the

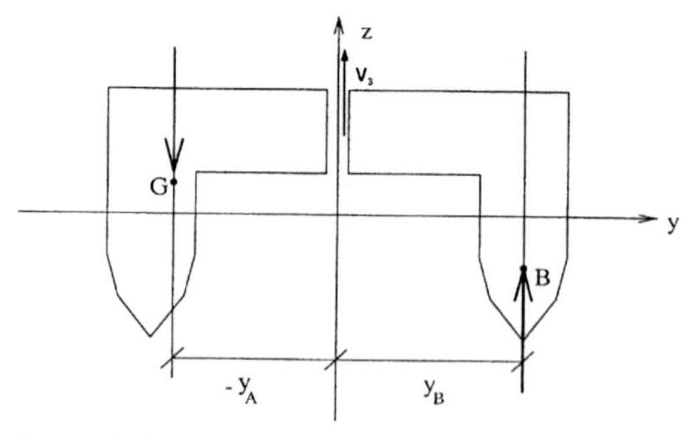

Figure 7.50. V_3 is the vertical shear force acting on one part of the catamaran obtained by cutting the catamaran along the centerplane. y_A is the y-coordinate of the center of gravity for the half-part. y_B is the y-coordinate for the centerplane of the demihull.

center of gravity for the half-part is assumed to be the same as for the whole catamaran. This gives the following vertical structural inertia force for the half-part:

$$0.5M(\ddot{\eta}_3 + y_A\ddot{\eta}_4). \quad (7.133)$$

Here η_3 is heave motion of the center of gravity of the catamaran. Because no fluid interaction is assumed between the demihulls of the catamaran and the underwater geometry of each demihull is symmetric about its own centerplane, the vertical hydrodynamic forces on the half-part act in the centerplane of the demihull. We can then write vertical added mass force on the half-part:

$$-0.5(A_{33}(\ddot{\eta}_3 + y_B\ddot{\eta}_4) + A_{35}\ddot{\eta}_5). \quad (7.134)$$

Here $y = y_B$ is the y-coordinate of the centerplane of the demihull. Further, A_{33} and A_{35} are heave and coupled heave-pitch added mass coefficients for the catamaran. We set a factor 0.5 in eq. (7.134) to account for the fact that we consider only half a part of the catamaran.

The vertical damping and restoring forces can be expressed similarly as eq. (7.134). This gives

$$\begin{aligned}&-0.5\,(B_{33}\,(\dot{\eta}_3 + y_B\dot{\eta}_4) + B_{35}\dot{\eta}_5)\\&-0.5\,(C_{33}\,(\eta_3 + y_B\eta_4) + C_{35}\eta_5)\,.\end{aligned} \quad (7.135)$$

In order to express the vertical wave excitation loads on the half-part we express the incident wave potential given by eq. (7.132) in a local coordinate system (x, y', z) for the half-part of the catamaran. Here

$$y' = y - y_B. \quad (7.136)$$

We can then write

$$\begin{aligned}\varphi_0 = {}&\frac{g\zeta_a}{\omega_0}e^{kz-ikx\cos\beta-iky'\sin\beta}(\cos(ky_B\sin\beta)\\&-i\sin(ky_B\sin\beta))\,.\end{aligned} \quad (7.137)$$

This means the vertical wave excitation force $X_3\exp(i\omega_e t)$ on the half-part is proportional to $\cos(ky_B\sin\beta) - i\sin(ky_B\sin\beta)$ and can formally be written as

$$\begin{aligned}X_3e^{i\omega_e t} = {}&0.5F_3(\cos(ky_B\sin\beta)\\&-i\sin(ky_B\sin\beta))e^{i\omega_e t}.\end{aligned} \quad (7.138)$$

Shortly we will see what F_3 is. If we similarly introduce a local coordinate $y' = y + y_B$ for the other demihull, we find that the vertical wave excitation force on that demihull can be written as

$$0.5F_3(\cos(ky_B\sin\beta) + i\sin(ky_B\sin\beta))e^{i\omega_e t}. \quad (7.139)$$

Adding together eqs. (7.138) and (7.139) gives the total vertical wave excitation. $F_3\cos(ky_B\sin\beta)\exp(i\omega_e t)$ on the catamaran.

Based on Newton's second law and rearrangements of terms, we can now set up the following equation:

$$\begin{aligned}&0.5\left\{(M + A_{33})\,\ddot{\eta}_3 + B_{33}\dot{\eta}_3 + C_{33}\eta_3 + A_{35}\ddot{\eta}_5\right.\\&\left.+B_{35}\dot{\eta}_5 + C_{35}\eta_5 - F_3\cos(ky_B\sin\beta)\,e^{i\omega_e t}\right\}\\&+0.5\,[My_A + A_{33}y_B]\,\ddot{\eta}_4 + 0.5B_{33}y_B\dot{\eta}_4\\&+0.5C_{33}y_B\eta_4 + 0.5F_3 i\sin(ky_B\sin\beta)\,e^{i\omega_e t} = V_3\end{aligned} \quad (7.140)$$

Now we see that the terms inside the { } brackets represent all the terms in the heave equation of motion for the whole catamaran, so this part is equal to zero. So we have found that the vertical shear force V_3 is a function of the roll angle but not a function of heave and pitch.

We can now follow a similar analysis for V_2, V_4, and V_5. This leads as a first step to the following expressions:

$$\begin{aligned}V_2 = 0.5[&(M(\ddot{\eta}_2 - z_G\ddot{\eta}_4) + A_{22}\ddot{\eta}_2 + A_{24}\ddot{\eta}_4\\&+ A_{26}\ddot{\eta}_6 + B_{22}\dot{\eta}_2 + B_{24}\dot{\eta}_4 + B_{26}\dot{\eta}_6\\&- F_2\exp(i\omega_e t - iky_B\sin\beta)]\end{aligned} \quad (7.141)$$

$$\begin{aligned}V_4 = 0.5\Bigg[&M\,(-z_G\ddot{\eta}_2 + y_A\ddot{\eta}_3) + I_{44}\ddot{\eta}_4\\&-2\int_A xydM\ddot{\eta}_5 - I_{46}\ddot{\eta}_6 + A_{42}\ddot{\eta}_2 + y_BA_{33}\ddot{\eta}_3\\&+ A_{44}\ddot{\eta}_4 + y_BA_{35}\ddot{\eta}_5 + A_{46}\ddot{\eta}_6 + B_{42}\dot{\eta}_2\\&+ y_BB_{33}\dot{\eta}_3 + B_{44}\dot{\eta}_4 + y_BB_{35}\dot{\eta}_5 + B_{46}\dot{\eta}_6\\&+ y_BC_{33}\eta_3 + y_BC_{35}\eta_5 + \rho g\nabla\overline{GM}\eta_4\\&-(F_4^L + y_BF_3)\exp(i\omega_e t - i\,ky_B\sin\beta)\Bigg]\end{aligned} \quad (7.142)$$

$$\begin{aligned}V_5 = 0.5\Bigg[&-2\int_A xydM\ddot{\eta}_4 + I_{55}\ddot{\eta}_5 - 2\int_A zydM\ddot{\eta}_6\\&+ A_{53}(\ddot{\eta}_3 + y_B\ddot{\eta}_4) + A_{55}\ddot{\eta}_5 + B_{53}(\dot{\eta}_3 + y_B\dot{\eta}_4)\\&+ B_{55}\dot{\eta}_5 + C_{53}(\eta_3 + y_B\eta_4) + C_{55}\eta_5\\&- F_5\exp(i\omega_e t - iky_B\sin\beta\,)\Bigg]\end{aligned} \quad (7.143)$$

Here A_{jk}, B_{jk}, and C_{jk} are the added mass, damping, and restoring coefficients, respectively, for the catamaran; ∇ is the displacement; $\overline{GM}$ is the roll metacentric height; I_{44} and I_{55} are the moment of inertia with respect to the x- and y-axes, respectively; and I_{46} is the roll-yaw product of inertia for the catamaran. The center of gravity for the catamaran is located at $(0, 0, z_G)$. dM is an infinitesimal mass element located at a point (x, y, z). $\int_A$ means that the integration is over the above-mentioned half-part of the catamaran. Further $F_2 \exp(i\omega_e t) \cos(ky_B \sin\beta)$ is the transverse wave excitation force on the catamaran. The total roll wave excitation moment on the catamaran is $[F_4^L \cos(ky_B \sin\beta) - iy_B F_3 \sin(ky_B \sin\beta)] \exp(i\omega_e t)$ and the total pitch wave excitation moment is $F_5 \cos(ky_B \sin\beta) \exp(i\omega_e t)$. This means F_4^L is proportional to the local roll moment for each hull about an axis coinciding with the intersection between the mean waterplane and the centerplane of the hull.

The expressions for V_j can be simplified by using the equations of motions for the catamaran. We showed this for V_3. We can now write

$$V_2 = 0.5 F_2 i \sin(ky_B \sin\beta) \exp(i\omega_e t) \quad (7.144)$$

$$V_3 = 0.5 \left[(M\, y_A + A_{33} y_B)\ddot{\eta}_4 + y_B B_{33} \dot{\eta}_4 + y_B C_{33} \eta_4 + F_3 i \sin(ky_B \sin\beta) \exp(i\omega_e t) \right] \quad (7.145)$$

$$V_4 = 0.5 \left[F_4^L i \sin(ky_B \sin\beta) \exp(i\omega_e t) + M(y_A - y_B)\ddot{\eta}_3 - 2 \int_A xy\, dM \ddot{\eta}_5 \right]. \quad (7.146)$$

It should be recalled that V_4 is the transverse vertical bending moment about the x-axis. Further,

$$V_5 = 0.5 \left[\left(-2 \int_A yx\, dM + y_B A_{53} \right) \ddot{\eta}_4 + y_B (B_{53} \dot{\eta}_4 + C_{53} \eta_4) - 2 \int_A zy\, dM \ddot{\eta}_6 + F_5 i \sin(ky_B \sin\beta) \exp(i\omega_e t) \right]. \quad (7.147)$$

Eqs. (7.144), (7.145), and (7.147) show that V_2, V_3, and V_5 are zero in head and following seas, that is, wave directions $\beta = 0^\circ$ and 180°. The reason V_2 is zero is that there is no transverse wave forces or motions that appear for $\beta = 0^\circ$ and 180°. Because the vertical hydrodynamic and body inertia force and pitch moments on the two demihulls are equal and in phase when $\beta = 0^\circ$ and 180°, there cannot be any internal vertical force (shear force) and pitch connecting moment in a longitudinal cut along the centerplane.

Eq. (7.146) shows that V_4 is non-zero when $\beta = 0^\circ$ and 180°. The reason is that the vertical hydrodynamic force and structural inertia force on the half-part of the catamaran act through different points with y-coordinates y_B and y_A (see Figure 7.50).

V_2 is zero when the wavelength $\lambda = 2y_B \sin\beta / n (n = 1, 2 \ldots)$. For very long wavelengths, $\sin(ky_B \sin\beta)$, ω_e, η_4, η_5, and η_6 all go to zero. This means that V_2, V_3, V_4, and V_5 go to zero. This is also true for very short wavelengths, because the wave excitation loads and the motions go to zero then.

Eqs. (7.144) to (7.147) show that roll motion is important for vertical shear force and pitch connecting moment, whereas heave and pitch accelerations influence the vertical bending moment. The horizontal athwartships force is independent of the ship motions.

Faltinsen et al. (1992) presented numerical and experimental results of global wave loads on a catamaran at Froude number 0.49. The numerical method is based on the high-speed theory by Faltinsen and Zhao (1991a,b). A simplified rudder model with automatic control was included. Global wave loads were evaluated as described above. The experiments were done with a free-running model (see Figure 7.53 and Table 7.5) in a basin of length 80 m, breadth 50 m, and depth 10 m, that is, at the Ocean Environment Laboratory at the Marine Technology Center in Trondheim. Regular incident waves of different wave headings were used. During the tests, the mean trim angle was between 2.2° and 2.6°. The loads were measured in the deck at $z_c = 0.185$m in a longitudinal cross section parallel to the centerplane at a distance of 0.075 m. Sufficient information on mass distribution is sometimes lacking when model test results are presented. However, the data presented in Table 7.5 are sufficient according to the previously presented theoretical method.

Table 7.5. *Main particulars of the tested catamaran model (see Figure 7.53)*

Designation	Symbol	Unit	Value
Length between perpendiculars	$L = L_{PP}$	m	3.778
Beam at waterline amidships	B	m	0.918
Draft – even keel	D	m	0.235
Displacement	∇	m^3	0.257
Block coefficient	C_B		0.542
Breadth of one hull at waterline amidships	b	m	0.267
Distance between centerplanes of demihulls	$2p$	m	0.652
Transverse metacentric height	$\overline{GM}$	m	0.556
Center of gravity above keel	$\overline{KG}$	m	0.332
Center of gravity aft of amidships	LCG	m	0.296
Pitch radius of gyration with respect to axis through center of gravity	r_{55}	m	0.981
Roll radius of gyration with respect to axis through center of gravity	r_{44}	m	0.334
Yaw radius of gyration with respect to axis through center of gravity	r_{66}	m	1.022
Distance from centerplane of the catamaran to center of gravity of one half-part[a]	y_A	m	0.298
Coupled inertia moment in roll-pitch of one half-part[a]	$-\int_A xy dM$	kgm^2	0.366
Coupled inertia moment in roll-yaw of one half-part[a]	$-\int_A xz dM$	kgm^2	−1.118
Coupled inertia moment in pitch-yaw of one half-part[a]	$-\int_A yz dM$	kgm^2	0.0170

[a] Half-part ($y > 0$) obtained by an intersection along the centerplane of the catamaran.

Error sources in the theory are the result of:

- *Transverse wave systems*
  (The high-speed theory neglects this effect. We saw in Chapter 4 that this is an appropriate approximation when $Fn > \approx 0.4 - 0.5$.)
- *Hull interaction*
  (The theory neglects this effect, but we saw in section 7.2.5 that this effect is present in hydrodynamic coefficients. The effect decreases with increasing forward speed.)
- *Interaction steady/unsteady flow*
  (The theory neglects interaction between local steady and unsteady flow. When the Froude number increases, this interaction get increased importance. We saw that in section 7.2.12 and will see this for planing vessels in Chapter 9.)
- *Rudder model*
- *Nonlinear effects are disregarded*

Experimental error sources are the result of:

- *Time window*
  (Transient effects did not die out because of few oscillation periods. This was particularly true for roll, vertical shear force and pitch connecting moment in beam and following seas.)
- *Heading control*
  (The results may be sensitive to heading.)
- *Non-constant wave condition*
  (The incident wave amplitude may vary up to 10% along the track of the model.)
- *Nonlinearities*
  (The experimental results showed that this was a minor effect. This is because of small incident wave slopes.)

Five wave headings were tested: 0°, 45°, 90°, 135°, and 180°. The pitch connecting moment (RAO) was generally largest for 45°, and the vertical shear force and the vertical bending moment were generally largest for 90°. For short-term statistics, it is not only important how large the maximum value in the transfer function is, but how large a frequency range with large RAO values overlaps frequencies with significant wave energy.

Figure 7.51 shows results for transfer functions (RAOs) of vertical shear force and vertical bending moment for $\beta = 90°$ as a function

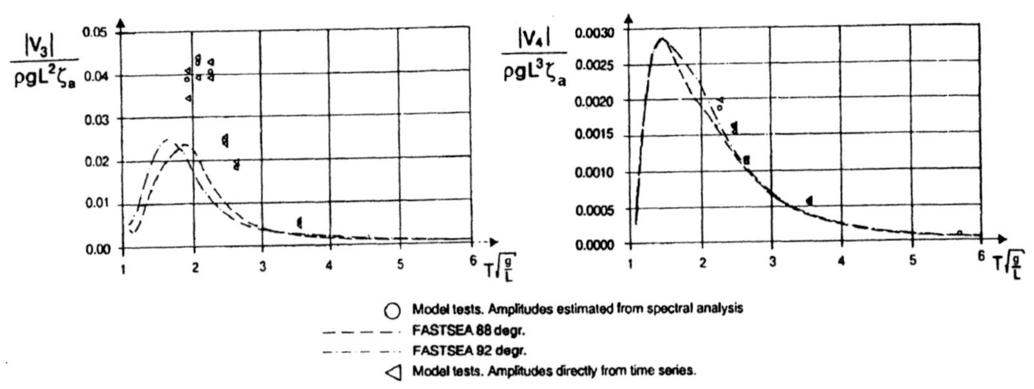

Figure 7.51. Vertical shear force $|V_3|$ and vertical bending moment amplitudes $|V_4|$ in beam sea regular waves for the catamaran model presented in Figure 7.53 and Table 7.5. $Fn = 0.49$.

of nondimensional wave period $T\sqrt{g/L}$ (T = wave period, L = length between perpendiculars). FASTSEA refers to the computer program. The calculations have been done with a trim angle of 2.4°. Calculations with zero trim angle will give different results. For instance, for 90° heading, the maximum roll angle will be 0.2° higher at a trim angle of 2.4° relative to zero trim.

Numerical values for 88° and 92° are presented because the response amplitude may be sensitive to the wave heading and it was not possible to keep the desired heading of 90° during the tests. The error was within 2°. The figures show differences in response amplitudes estimated by a spectral analysis and by inspecting the time series. This gives an indication of the experimental errors and difficulties in performing the tests.

An important reason for the differences between numerically and experimentally predicted vertical shear forces was believed to be the differences in numerically and experimentally predicted roll angles. Figure 7.51 shows quite good agreement between numerical and experimental values for vertical bending moments. The maximum values of $|V_4|$ occur close to the case in which there is half an incident wavelength between the centerplanes of the two demihulls. The latter gives $T(g/L)^{0.5} = 1.47$ and suggests that the wave excitation loads on the two demihulls are 180° out of phase at maximum$|V_4|$.

Results for pitch connecting moment in 45° heading are presented in Figure 7.52. The maximum value occurs for an incident wavelength that is 0.7 times the ship length.

Global wave loads in a short-term sea state

Having obtained the linear transfer functions for V_j as described above we can use eq. (7.111) to

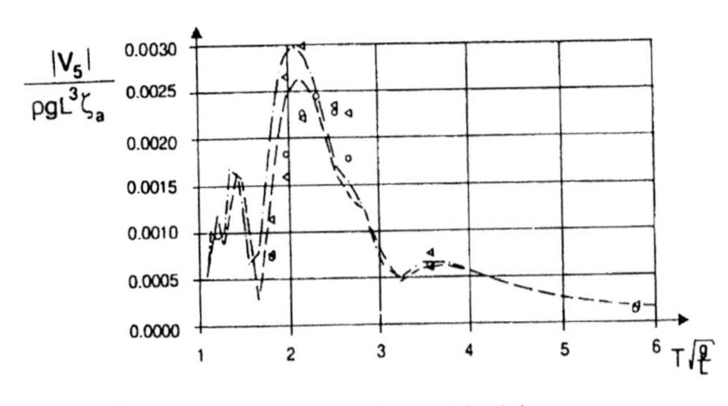

Figure 7.52. Pitch connecting moment amplitude $|V_5|$ in 45° heading for the catamaran model presented in Figure 7.53 and Table 7.5. $Fn = 0.49$.

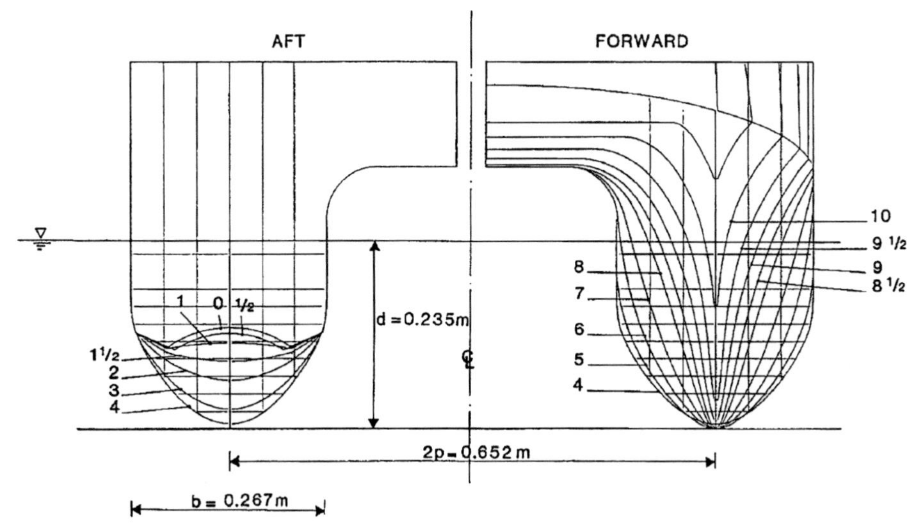

Figure 7.53. Body plan of a catamaran model tested in the Ocean Environment Laboratory of the Marine Technology Center in Trondheim and used as a basis for the calculation of the global wave loads presented in Figure 7.54.

predict the standard deviation σ in a long-crested short-term sea state. Because $S(\omega)$ is proportional to $H_{1/3}^2$, σ is proportional to $H_{1/3}$. Nondimensional values of standard deviations of vertical shear force (σ_3), transverse vertical bending moment (σ_4), and pitch connecting moments (σ_5) in the centerplane of a catamaran in long-crested irregular sea are presented as functions of wave heading and nondimensional mean wave period T_1 in Figure 7.54. The Froude number is 0.7, and the catamaran is described in Figure 7.53 and Table 7.5. The results illustrate that global loads are sensitive to the wave heading. The vertical shear forces and vertical bending moments are generally largest in beam sea, whereas the largest values for pitch connecting moments occur at wave heading 60° when $T_1\,(g/L)^{0.5} > 2$.

Let us consider an example with $H_{1/3} = 5\,\text{m}$ and $L = 70\,\text{m}$. In order to find corresponding values for T_1, we use the scatter diagram presented in Table 3.4. This means $T_2 \in (5.5\,\text{s}, 13.5\,\text{s})$. We now use eq. (3.62), that is, $T_1 = 1.073 T_2$ for a JONSWAP spectrum. This gives the range (2.2, 5.4) for $T_1\,(g/L)^{0.5}$. In this range, we find the following maximum values for the normalized standard deviations for global sectional loads:

$$\max\left\{\frac{\sigma_3}{\rho g L^2 H_{1/3}}\right\} = 0.0032$$

for a heading of 90.0° and $T_2 = 5.5$ s,

$$\max\left\{\frac{\sigma_4}{\rho g L^3 H_{1/3}}\right\} = 0.00035$$

also for a heading of 90.0° and $T_2 = 5.5$ s, and

$$\max\left\{\frac{\sigma_5}{\rho g L^3 H_{1/3}}\right\} = 0.0007$$

for a heading of 60.0° and $T_2 = 5.5$ s. We now want to calculate the most probable largest value. This can be approximated (see eq. (7.113)) as

$$V_j \approx 4\sigma_j, \quad (j = 3, 4, 5).$$

The chosen sea states now give us the following values:

$$V_3 = 4\sigma_3 = \rho g L^2 4 H_{1/3} \cdot 0.0032 = 3.1 \cdot 10^6\,[N]$$
$$V_4 = 4\sigma_4 = \rho g L^3 4 H_{1/3} \cdot 0.00035 = 2.4 \cdot 10^7\,[Nm]$$
$$V_5 = 4\sigma_5 = \rho g L^3 4 H_{1/3} \cdot 0.0007 = 4.8 \cdot 10^7\,[Nm]$$

The results in Figure 7.54 are short-term predictions based on linear theory. However, it should be recognized that nonlinearities may matter in design conditions. Nonlinear corrections are made in practice. However, the state of the art in calculating nonlinear effects from first principles needs to be improved. If the ship has significant flare in the bow part, nonlinearities become particularly important. This can be illustrated by the full-scale measurements on the container vessel CTS Tokyo Express in the northern Atlantic. Short-term statistical representations of the wave-induced

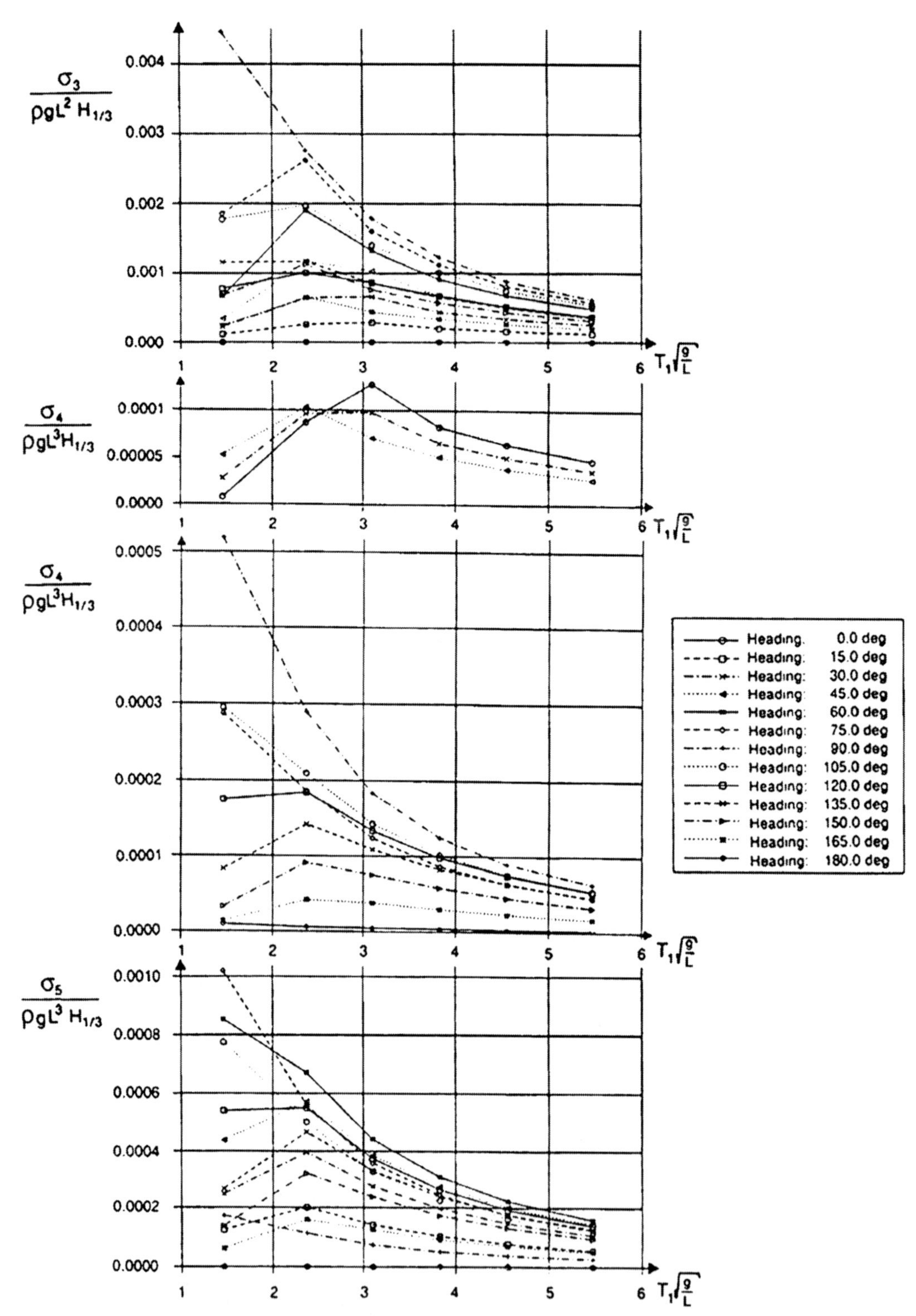

Figure 7.54. Standard deviations of vertical shear forces (σ_3), transverse vertical bending moments (σ_4), and pitch connecting moments (σ_5) in the centerplane of a catamaran in long-crested irregular sea as a function of wave heading and mean wave period T_1. $Fn = 0.7$. The catamaran is described in Figure 7.53 and Table 7.5 (Faltinsen et al. 1992).

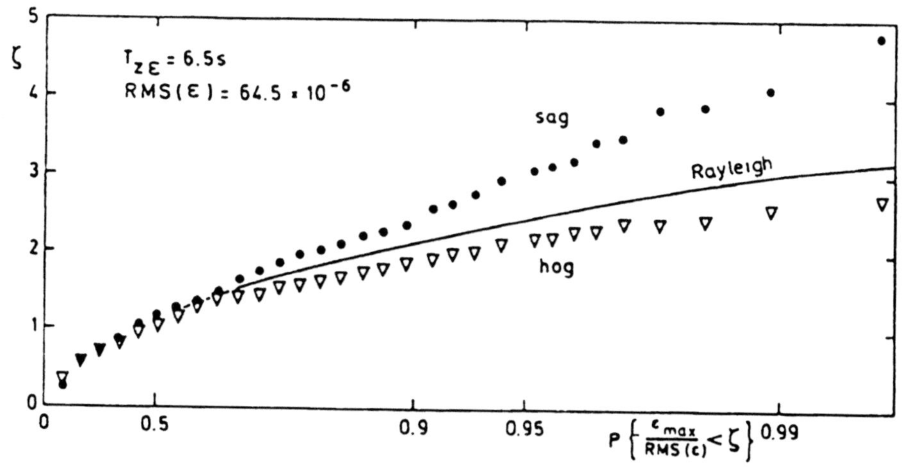

Figure 7.55. Short-term statistical representation of the peaks ε_{max} of wave-induced bending strain ε derived from northern Atlantic measurements on CTS TOKYO EXPRESS (1018 GMT December 27, 1973). A low-pass filter was applied to remove contributions from the two-node vibration taking place at 5 rad/s. (Jensen and Pedersen 1978, Hackmann 1979).

bending strain amplitudes ε_{max} are presented in Figure 7.55. The wave-induced sagging and hogging bending moments are different. Sagging describes a state in which at midships, there are tensile bending stresses in the bottom and compression bending stresses in the upper deck. Hogging is the opposite of sagging, that is, tensile bending stresses occur at the midships upper deck. A linear theory predicts the same values for wave-induced sagging and hogging. These values are denoted as Rayleigh in the figure. The data are obtained during an operational period of 20 minutes. It is difficult to explain completely the nonlinear behavior leading to differences in hogging and sagging, but different nonlinear generalizations of strip theory have been used. They are able to explain a correct trend. A discussion is given by the Load Committee of the 12th ISSC (see also proceedings of the 14th ISSC).

Long-term predictions

The previously described numerical model was combined by Faltinsen et al. (1992) with the procedure for long-term response outlined in section 7.4.2 to predict design values for vertical shear forces, transverse vertical bending moments, and pitch connecting moments on the catamaran model presented in Figure 7.53 and Table 7.5. The full-scale vessel length varied between 50 m and 120 m. The catamaran was assumed to have 2.4° trim. The design speed corresponds to $Fn = 0.7$. The global loads were evaluated in the deck at $z = 0.165$ m (model scale) in the centerplane of the catamaran. An operational area between Korea and Japan was selected. The weather data were in the form of joint frequency tables (scatter diagrams) for significant wave heights and mean wave periods, and included data for all seasons. The data were taken from Takaishi et al. (1980) and represented area E02S in their weather atlas. Each sea state was represented by long-crested sea and the JONSWAP spectrum given by ITTC. For simplicity, it was assumed that each wave heading relative to the vessel had equal probability. This is a normal assumption for conventional vessels in the ocean but not, for example, for ships that have operation related to the offshore oil industry. The calculations were made for 13 wave headings between 0° and 180°. If no rudder effects were accounted for, numerical difficulties occurred at small frequencies of encounter because of high predicted values for transverse motion amplitudes. The results for standard deviation of global loads in a given sea state are presented in Figure 7.54.

Operational limits were accounted for. However, only a criterion related to wetdeck slamming was used. Strictly speaking, criteria due to vertical accelerations and the performance of the propulsion and engine system in a seaway should be included.

The effect of involuntary speed reduction due to added resistance in waves and wind was neglected.

Table 7.6. *Example of long-term statistical values of global loads and operational limits of catamarans of different lengths*

L (m)	Vertical shear force (kN)	Transverse vertical bending moment (kNm)	Pitch connecting moment (kNm)	% Time in harbor because of operational limits
50	172	914	1804	6.7
60	300	1918	3961	5.0
70	485	3796	8560	3.8
80	735	6774	18793	2.7
90	1179	13504	37635	2.0
100	1766	22410	64206	1.7
110	2685	38120	110780	1.4
120	3487	53989	156890	0.7

Probability level 10^{-8}. Design speed $Fn = 0.7$. The catamarans are obtained by scaling the model presented in Figure 7.53 and Table 7.5 (Faltinsen et al. 1992).

If the catamaran did not satisfy the criterion that the slamming probability be less than 0.03 after reducing the ship speed to Froude number 0.45, it was decided to exclude the sea state from the long-term prediction of the response. In practice, operational limits have to be decided in a different way. The most common way is to specify a maximum operating $H_{1/3}$ in combination with allowed speed. This will influence the final results. The long-term predictions were based on section 7.4.2 by also including the effects of wave heading. This means that for each sea state, the proper Rayleigh distribution was multiplied by the probability of occurrence of the sea state. These products were then summed up. It was decided to select design values corresponding to a probability of 10^{-8}. This corresponds to a return period on the order of 20 years. The results in Table 7.6 are for different ship lengths between 50 and 120 m. The time spent in the harbor according to the criteria for operational limits is also shown.The predicted design values for global loads are very much lower than recommended values by DNV. However, if operational limits are omitted, the predictions are in closer agreement with DNV rules.

7.9 Exercises

7.9.1 Mass matrix

Derive the mass matrix elements M_{jk}, $j = 4,6$, $k = 1,6$ in eq. (7.36). You should start by considering an infinitesimal small mass element with mass dM and accelerations following from eq. (7.20). Then you should take the moment about the (x, y, z) axis of the inertia force associated with the mass element. Integrating the resulting moment due to all mass elements of the vessel leads to the answer.

7.9.2 2D heave-added mass and damping

a) Consider the boundary-value problem for determining high-frequency 2D heave-added mass in heave for a semi-submerged circular cylinder shown in Figure 7.56. Show that the velocity potential due to forced heave velocity $\dot{\eta}_3$ can be

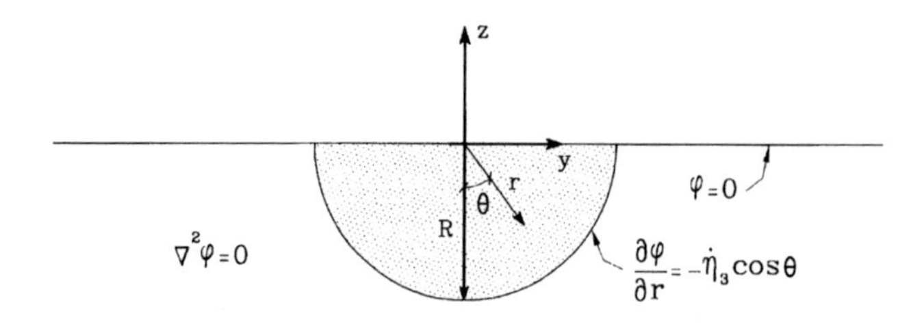

Figure 7.56. The boundary-value problem for high-frequency 2D heave-added mass in heave for a semi-submerged circular cylinder. $\dot{\eta}_3$ = heave velocity.

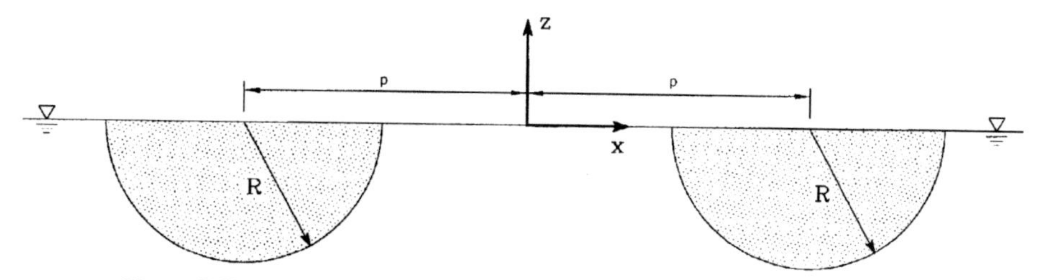

Figure 7.57. 2D catamaran consisting of two semi-submerged circular cylinders.

expressed as

$$\varphi = \dot{\eta}_3 \frac{R^2}{r} \cos\theta. \qquad (7.148)$$

b) Express the vertical force on the cylinder due to the pressure component $-\rho \partial\varphi/\partial t$. Show that this leads to the following 2D heave-added mass and damping coefficients:

$$a_{33} = 0.5\rho\pi R^2, \quad b_{33} = 0.$$

Explain physically why the damping coefficient $b_{33} = 0$.

c) Assume harmonic heave oscillations and consider a time instant when the heave acceleration is maximum. Plot the pressure distribution due to $-\rho\partial\varphi/\partial t$ on the cylinder. What is the fluid velocity at this time instant?

Now consider a time instant when the heave velocity is maximum. Plot the fluid velocity on the free surface. What is the fluid pressure component $-\rho\partial\varphi/\partial t$?

d) Consider a 2D catamaran consisting of two semi-submerged cylinders as shown in Figure 7.57. Assume a high-frequency free-surface condition. Make an estimation of heave-added mass by the following steps:

Step 1: Express the flow field as if the two hulls did not interact with each other.

Step 2: Express an average normal fluid velocity at the position of one of the hulls caused by the other hull, based on the result from step 1.

Step 3: Introduce an additional part to the velocity potential so that the boundary condition on the hull is satisfied. (Hint: Step 2 has caused a flow through the hull surface. This must be counteracted. The solution of the boundary-value problem is similar but not identical to the solution at step 1.)

e) Consider the 2D catamaran in Figure 7.57. Allow for free-surface waves due to forced heave motions. We shall now find an approximation of heave damping by considering interference between the far-field wave systems generated by each hull. By interference, we mean that one considers the waves generated by each hull as if the other hull was not there. The phasing of the waves generated by each hull will then cause either amplification or reduction in the far-field wave amplitude relative to those that arise if we do not consider the phasing. (Hint: The far-field wave elevation caused by the hulls can be expressed as $A_3 \cos(\omega t - k|y \pm p| + \varepsilon)$. Explain!) The far-field wave amplitude A_3 without interaction can be found by means of Figure 7.20 and eq. (7.59). Then you should once more use eq. (7.59) to compare with the damping results, accounting for the interaction in Figure 7.20. Of course, you cannot expect your approximate results to be the same as those in Figure 7.20.

7.9.3 Linear wavemaker solution

A 2D wave tank with semi-infinite length and constant water depth h is considered. When the water is at rest, it occupies the region $0 \leq x \leq \infty$, $0 \leq z \leq -h$. The wavemaker is situated along the z-axis at $x = 0$ and is assumed to oscillate harmonically with small amplitude in the x-direction. A linear steady-state solution for the fluid motion is considered. The horizontal velocity of the wavemaker is described as

$$u_{WM} = -\omega s(z) \sin\omega t \quad \text{at} \quad x = 0. \qquad (7.149)$$

Here $s(z)$ represents horizontal oscillation amplitude of the wavemaker that varies in general with z. If the wavemaker is a piston, then $s(z)$ is a constant. If a paddle that is hinged to the bottom is used to generate waves, then $s(z) = \alpha(z + h)$.

Here α is the angular oscillation amplitude of the paddle.

Billingham and King (2000) have presented a solution to this problem. The velocity potential for the fluid motion is represented as

$$\varphi = A_0 \cosh\left[k_0\left(z+h\right)\right]\cos\left(k_0 x - \omega t\right) + \sum_{n=1}^{\infty} A_n e^{-k_n x}\cos\left[k_n\left(z+h\right)\right]\sin\omega t, \quad (7.150)$$

where k_n for $n \geq 1$ are the positive roots of $\omega^2 = -gk\tan kh$ and k_0 is the unique positive root of $\omega^2 = gk\tanh kh$.

a) Confirm that eq. (7.150) satisfies the Laplace equation, the linearized free-surface condition, the bottom condition at $z = -h$, and a radiation condition ensuring outgoing waves in the far field of the wavemaker.

b) Show by satisfying the body boundary condition on the wavemaker that A_n are determined by

$$A_0 k_0 \int_{-h}^{0} \cosh^2[k_0(z+h)]dz = -\int_{-h}^{0} \omega s(z)\cosh[k_0(z+h)]dz \quad (7.151)$$

$$A_n k_n \int_{-h}^{0} \cos^2[k_n(z+h)]dz = \int_{-h}^{0} \omega s(z)\cos[k_n(z+h)]dz,\ n \geq 1, \quad (7.152)$$

(Hint: Use that

$$\int_{-h}^{0} \cos[k_n(z+h)]\cos[k_m(z+h)]dz = 0 \quad \text{for} \quad n \neq m$$

$$\int_{-h}^{0} \cosh[k_0(z+h)]\cos[k_n(z+h)]dz = 0 \quad \text{for} \quad n = 1, 2, \ldots)$$

c) Consider a piston wavemaker and show that the ratio between the amplitude a of the far-field waves and the amplitude s_0 of the piston's oscillation is

$$\frac{a}{s_0} = \frac{4\sinh^2 k_0 h}{2k_0 h + \sinh 2k_0 h}. \quad (7.153)$$

Figure 7.58. Antiroll damping fin (Seastate).

Discuss this ratio for different water depths and wavelengths.

d) Consider the linearized hydrodynamic force on the piston wavemaker by using Bernoulli's equation. Show that this can be expressed as

$$F_h = -\sum_{n=1}^{\infty} \frac{2\rho\sin^2 k_n h}{k_n^2\left(k_n h + \frac{1}{2}\sin 2k_n h\right)}\ddot{x}_p - \frac{2\rho\omega\sinh^2 k_0 h}{k_0^2\left(k_0 h + \frac{1}{2}\sinh 2k_0 h\right)}\dot{x}_p, \quad (7.154)$$

where $\dot{x}_p$ and $\ddot{x}_p$ are the piston velocity and acceleration, respectively.

Express eq. (7.154) in terms of added mass and damping.

e) Show that the damping term in eq. (7.154) can also be derived by considering conservation of energy.

(Hint: Generalize the derivation of eq. (7.59).)

f) Express the equation of motion of the piston wavemaker.

7.9.4 Foil-lift damping of vertical motions

a) Foil-lift damping in heave was presented in eq. (7.80) based on the Theodorsen function. Generalize the derivation to include a contribution from the foil to the added mass coefficients A_{33}, A_{35}, A_{53}, and A_{55} and the damping coefficients B_{35}, B_{53}, and B_{55}.

b) In order to use the time-domain solution presented in section 7.3, one needs to calculate the retardation function h_{jk}. This means that high-frequency values of either added mass or damping are needed. Derive the infinite frequency values of A_{jk} and B_{jk}.

c) We described in section 7.2.10 automatic control of vertical motions by using controlled flap motion. As in section 7.2.10, assume steady-state heave and pitch in regular head sea waves. Assume the whole foil can be separately pitched in order to automatically control the heave and pitch motions. Generalize the procedure in section 7.2.10, and use the Theodorsen function to calculate vertical hydrodynamic forces on the T-foil.

7.9.5 Roll damping fins

Consider a roll damping fin such as the one in Figure 7.58 on a monohull vessel at forward speed. Base the dynamic lift force on the fin on a quasi-steady analysis, strip theory, and the assumption that the local flow at the ship does not influence the incident flow on the fin. Choose the strips in the chordwise direction, that is, in the longitudinal direction of the ship.

a) Express linear damping terms in sway, roll, and yaw due to the fin.

b) Consider the ship in regular oblique waves. Neglect the lifting effect on the added mass coefficients due to the foil. Express the linear dynamic bending moments about the longitudinal axis at different cross sections of the fin. (Hint: You must account for the fin's mass, added mass, and lifting loads. The effect of the incident waves also must be accounted for.)

c) Assume the fins are equipped with flaps that are automatically controlled to damp the roll motions. Show how this can be incorporated into the equations of motions.

7.9.6 Added mass and damping in roll

The axis to which we refer added mass and damping in roll matters. Let us say that A_{44}, A_{42}, A_{24}, B_{44}, B_{42}, B_{24} have been obtained relative to the coordinate system defined in Figure 7.49. Express A_{44} and B_{44} with respect to center of gravity.

7.9.7 Global wave loads in the deck of a catamaran

a) Derive the mass inertia terms in transverse vertical bending moment V_4 and pitch connecting moment V_5 presented in eqs. (7.142) and (7.143). (Hint: Start with the mass inertia force on a small structural element with mass dM and the acceleration using eq. (7.20).)

b) Derive the expression for pitch connecting moment V_5 given by eq. (7.147).

8 Slamming, Whipping, and Springing

8.1 Introduction

Slamming (water impact) loads are important in the structural design of all high-speed vessels. Further, the occurrence of slamming is an important reason for a shipmaster to reduce the ship's speed. It is also an important effect in calculating operational limits (see section 1.1). The probability of slamming is found by defining a threshold relative impact velocity for slamming to occur (see eq. (8.142)). An often-used criterion is that a typical shipmaster reduces the speed if slams occur more frequently than three out of 100 waves passing the ship. It may be misleading to talk about a threshold velocity. There is no threshold for slamming as a physical process. Further, the conventional way of defining a threshold velocity does not reflect the effect of the structural shape. For instance, for a high-speed vessel with slender lines in the bow, the procedure may say that slamming occurs on the bow part of the hull, although, in reality, it is not a problem. In order to come up with better criteria, it is necessary to study theoretical models for or perform experiments on water impact against wetdecks and hull structures typical for high-speed vessels. This is also necessary in order to develop rational criteria for operational limits due to slamming. The criteria should be related to slamming loads used in the structural design or, ideally, to structural response due to slamming.

Wetdeck slamming (Figure 8.1) is important for multihull vessels. The wetdeck is the lowest part of the cross-structure connecting two adjacent side hulls of a multihull vessel. Figure 8.2 shows an example of a cross section in the forward part of a catamaran, where wetdeck slamming is likely to occur for a vessel with forward speed in head sea conditions. In this case, the wetdeck has a wedge-shaped cross section with deadrise angle β_W. For some vessels, this can be zero or as large as it is for the wave-piercer catamaran shown in Figure 1.3.

When the side hulls come out of the water as a consequence of the relative vertical motions between the vessel and the water surface, subsequent slamming on the side hulls will occur. Because the deadrise angle β in Figure 8.2 is large, the local slamming loads are not expected to be important. However, this also depends on the relative impact velocity V_R. When β is larger than about 5°, the maximum slamming pressure is proportional to V_R^2 for constant V_R.

A more critical situation for the side hulls is shown in Figure 8.3, in which a steep wave impacts on the side hull and the relative angle β_R between the impacting free surface and the hull surface is small. The presence of roll can, as illustrated in the figure, decrease β_R and thereby cause increased slamming loads. The slamming loads are sensitive to β_R, particularly for small angles of β_R.

Another scenario is green water on deck. This is illustrated in Figure 8.4 as a consequence of "dive-in" in following seas, especially when speed is reduced in large waves and the frequency of encounter becomes small, but green water on deck can also happen as a consequence of large relative vertical motions between the vessel and the water. The water can then enter the deck as a plunging breaker, causing slamming loads on the deck. The subsequent fluid motion can have an impact on obstructions such as the wheelhouse in Figure 8.4.

Figure 8.5 shows drop tests at DNV (Det Norske Veritas) of a wedge cross section representing a typical side hull of an SES (surface effect ship) or a catamaran. Hayman et al. (1991) presented results from these drop tests. One model was made of GRP (glass-reinforced plastic) sandwich and had a deadrise angle of 30°; the other was an aluminium model with a deadrise angle of 28.8°. The effect of tilting the models was studied. Experience of vessels with GRP sandwich in severe conditions in service has shown that an important mode of failure is cracking of the core due to shear stresses, generally followed by delamination. Impact pressures and structural strains were measured during the experiments. Negative pressures relative to atmospheric pressure were also seen in the elastic cases. Why this occurs is explained later in the text.

We see in Figure 8.5 a lot of spray as a consequence of the impact. Figure 8.6, taken during drop tests of a wedge-formed cross section in model scale, gives a better view of the spray in combination with the local uprise of the water. Actually,

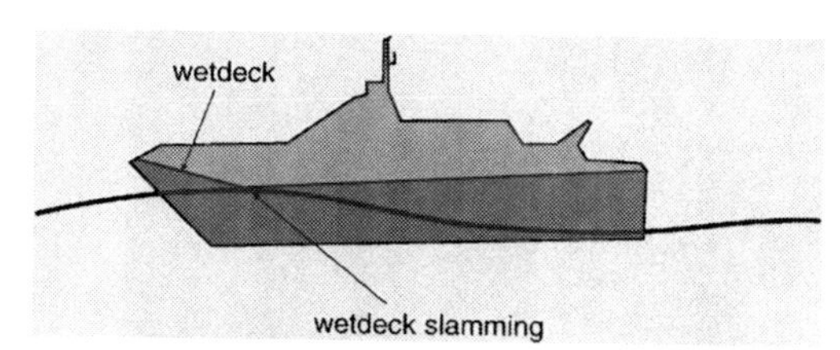

Figure 8.1. Wetdeck slamming.

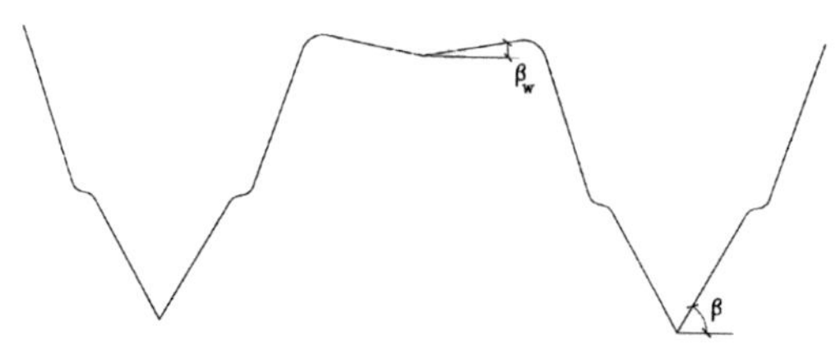

Figure 8.2. Example of a cross section in the forward part of a catamaran. β = deadrise angle of a demi-hull, β_w = deadrise angle of the wetdeck.

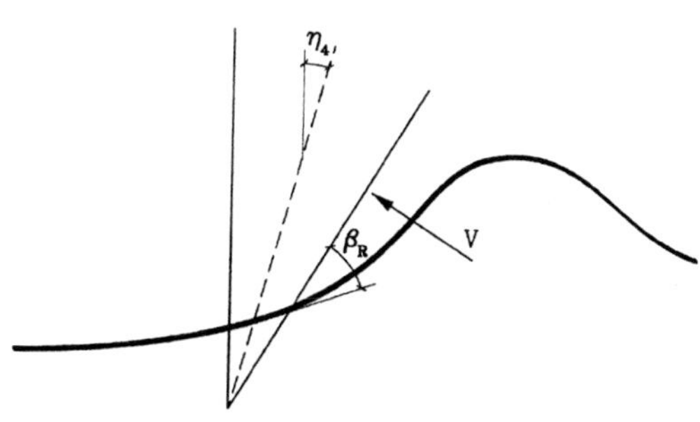

Figure 8.3. Slamming due to a steep wave impacting on the ship's hull. η_4 = roll.

one should not focus too much on the spray. There is close to atmospheric pressure in the spray. The large pressures will typically occur at the "spray root," where there is green water meaning that the flow is not aerated here. This is associated with high free-surface curvature caused by large pressure gradients that accelerate the water into a jet flow with high velocities. This jet flow then changes into spray under the action of surface tension.

In order to translate the results from either numerical or experimental drop tests, we need information about the relative vertical motions and velocities between the ship and the water in the impact area. Figure 8.7 gives calculated relative vertical motions and velocities in irregular long-crested head sea waves in a sea state described by mean wave period T_2 and significant wave height $H_{1/3}$. A two-parameter JONSWAP (Joint North Sea Wave Project) wave spectrum recommended by the ITTC (International Towing Tank Conference) was used. The calculations were done without accounting for the interactions between slamming loads and ship motions. This is the common procedure. However, one should ideally account for this interaction effect.

Slamming causes both local and global effects (Figure 8.8). The global effect is often called whipping. Hydroelasticity may be important for global loads but also matters for local effects in the case of very high slamming pressures of very short duration. Very high pressures may occur when the angle between the impacting free surface and hull surface is small.

Hydroelasticity means that the fluid flow and the structural elastic reaction are considered simultaneously and that we have mutual interaction, that is,

- The elastic vibrations cause a fluid flow with a pressure field

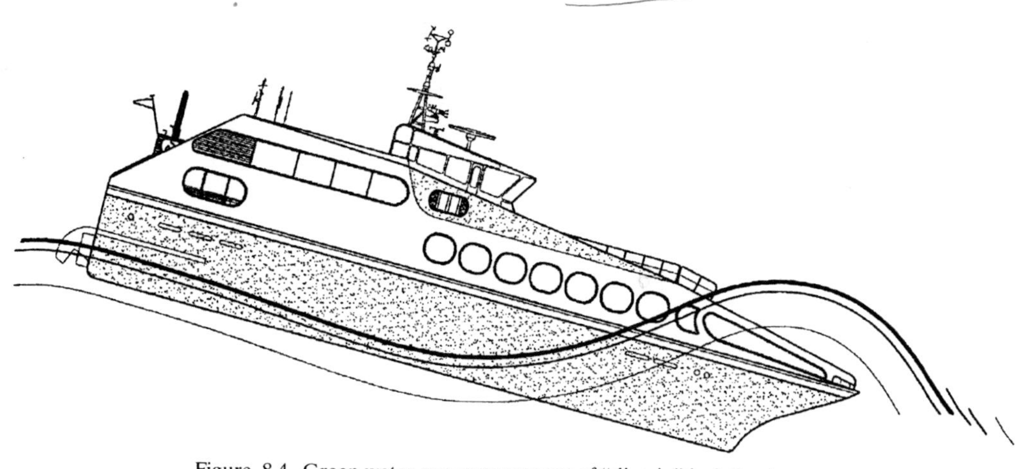

Figure 8.4. Green water as a consequence of "dive-in" in following sea.

Figure 8.5. Drop tests at DNV of a wedge section of a typical sidehull design of an SES or a catamaran. The top picture shows the section at its initial drop height of about 9 m above water level. The lower picture illustrates the large spray created as a consequence of the impact. Note the circular end plates fitted to reduce the three-dimensional effect.

- The hydrodynamic loading affects the structural elastic vibrations

The classical book on hydroelasticity of ships is by Bishop and Price (1979).

A conventional structural analysis without hydroelasticity or dynamic effects considers the hydrodynamic loading by assuming a rigid structure. The loading is then applied in a quasi-steady manner when the resulting static structural elastic and plastic deformations and stresses are calculated. The fluid flow is affected by many physical features, such as compressibility and air cushions. However, it is complicated to solve the complete hydrodynamic problem, and approximations must be made. The guideline in making simplifications

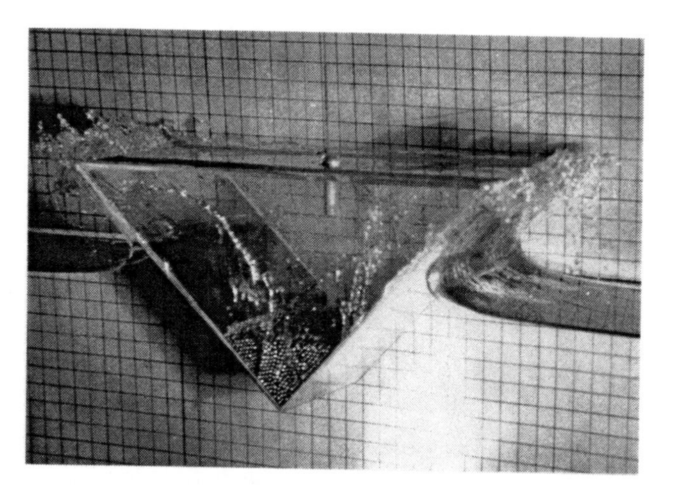

Figure 8.6. Drop test of a wedge (Greenhow and Lin 1983).

is to consider what physical features of the fluid flow have a nonnegligible effect on maximum slamming-induced structural stresses. This implies, in general, that the compressibility of the water can be neglected.

Very high slamming pressures may occur when the angle between the impacting body and the free surface is small. If we want to make a "black-and-white picture," we can say that very high slamming pressures are not important for steel and aluminum structures. The high pressure peaks are localized in time and space, and it is the force impulse that is important for the structural response. We will make it clearer in the main text why this is so.

An important message is that

Slamming must always be analyzed as a combination of hydrodynamics and structural mechanics.

This includes the case in which the impacting body can be considered rigid in the hydrodynamic analysis.

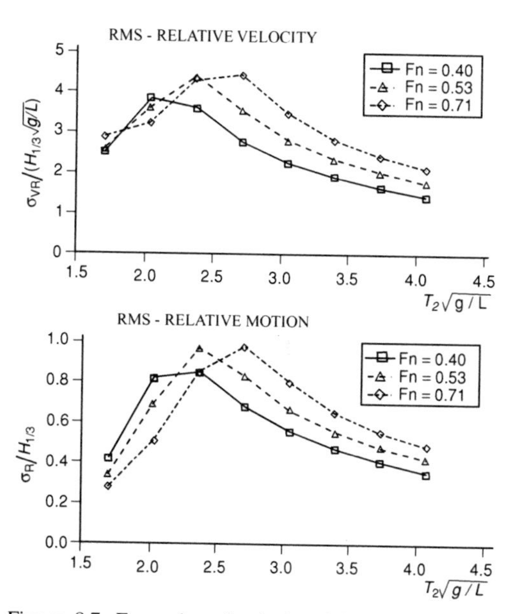

Figure 8.7. Examples of calculated RMS (root mean square) values of relative vertical motions (σ_R) and velocities (σ_{VR}) at FP for a catamaran in head sea long-crested waves. L = ship length, g = acceleration of gravity, Fn = Froude number, T_2 = mean wave period, $H_{1/3}$ = significant wave height.

We start out in the main text of this chapter by discussing local slamming effects by first considering hydroelastic slamming. We then discuss local slamming loads on a rigid body before whipping is studied. Springing is another global effect similar to whipping. This is also dealt with in this chapter. Whereas whipping is a transient vibration caused by slamming, springing is steady-state vibration caused by the oscillating wave forces along the hull. Springing (based on linear theory) is excited by the waves with an encounter frequency equal to the vertical two-node resonant frequency. Springing is a continuous process. The vibration amplitude varies in accordance with the irregular waves. The importance of springing increases with increasing speed, length, and hull girder flexibility. The phenomenon is normally more important for fatigue damage than for extreme design loads. For multihull vessels or open monohull vessels, other vibration modes may have low natural frequencies and may be excited by the oscillating waves. In addition, springing and whipping in irregular sea may occur simultaneously, making it difficult to separate the two phenomena.

Figure 8.8. Artist's impression of bow slamming causing global elastic vibrations (whipping) of the ship's hull. (Artist: Bjarne Stenberg)

8.2 Local hydroelastic slamming effects

Different physical effects occur during slamming. The effects of viscosity and surface tension are, in general, negligible. When the local angle between the water surface and the body surface is small at the impact position, an air cushion may be formed between the body and the water. Compressibility influences the flow of the air in the cushion. The airflow interacts with the water flow. When the air cushion collapses, air bubbles are formed.

The large loads that may occur during impact when the angle between the water surface and body surface is small, can cause important local dynamic hydroelastic effects. The vibrations may lead to subsequent cavitation and ventilation. These physical effects have different time scales. The important time scale from a structural point of view is when maximum stresses occur. This scale is given by the highest wet natural period (T_{n1}) for the local structure.

Compressibility and the formation and collapse of an air cushion are important initially, and normally in a time scale, that is smaller than the time scale for local maximum stresses to occur. Hence, the effect of compressibility on maximum local stress is generally small. However, we cannot exclude the possibility that the shape of the impacting free surface generates an air cushion of sufficiently long duration from a structural reaction point of view (Greco et al. 2003).

Theoretical and experimental studies of wave impact on horizontal elastic plates of steel and aluminium are presented in Kvålsvold (1994), Kvålsvold et al. (1995), Faltinsen (1997), and Faltinsen et al. (1997). The theoretical studies were made assuming 2D beam theory for strips of the plates. Significant dynamic hydroelastic effects were demonstrated. The main parameters for plates used in the drop tests are shown in Table 8.1. The test sections were divided into three

Table 8.1. *Main parameters for plates used in the drop test results presented in Figures 8.9 and 8.10*

Parameter	Plate I (steel)	Plate II (aluminum)
Structural mass per unit length and breadth, M_B	62 kgm^{-2}	21 kgm^{-2}
Modulus of elasticity, E	210 × 10^9 Nm^{-2}	70 × 10^9 Nm^{-2}
Length of plate, L	0.50 m	0.50 m
Breadth of plate, B	0.10 m	0.10 m
Bending stiffness, EI	8960 Nm^{2}m^{-1}	17060 Nm^{2}m^{-1}
Structural mass parameter, $M_B/\rho L$	0.124	0.042
Connecting spring parameter, $k_\theta L/(2EI)$	2.85	1.50
Distance from neutral axis to strain measurements, z_a	0.004 m	0.01375 m

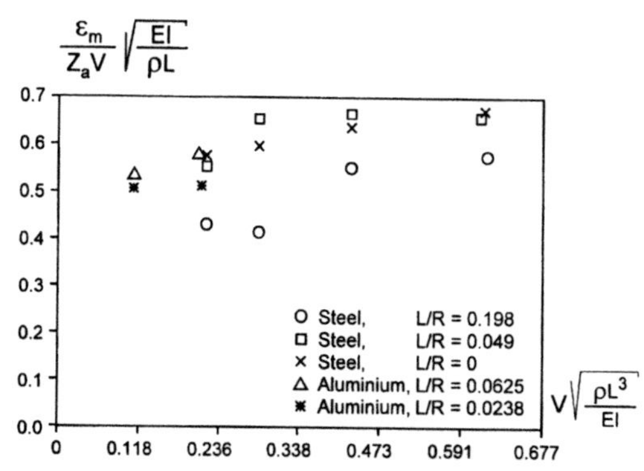

Figure 8.9. Measured maximum nondimensional strain amplitude ε_m in the centers of the horizontal steel and aluminum plates described in Table 8.1 as function of nondimensional water entry velocity. V = water entry velocity, ρ = mass density of water, R = radius of curvature of the waves at the impact position.

parts: one measuring section with a dummy section on each side. The measured nondimensional maximum strains in the middle of the plate are presented in Figure 8.9 as a function of nondimensional impact velocity V. The scaling assumes that maximum strain is proportional to V and that T_{n1} is the time scale. The plates have only vertical velocity, ρ means mass density of the water, and L is the length of the plate. I means the cross-sectional area moment of inertia about the neutral axis divided by the cross-sectional plate breadth B. One may wonder why it is only the mass density of the water that is involved in the nondimensional expression and not also the mass density of the material. Later, it becomes more evident why the mass density of the material does not have a dominant influence.

The impact position and radius R of curvature of the waves at the impact position were varied. A wave crest was intended to initially hit between the plate ends in all cases presented in Figure 8.9. The experimental results decrease slightly with decreasing nondimensional impact velocity. If the largest L/R value is disregarded, the maximum strain shows small influence of L/R. The largest L/R value is normally unrealistic for slamming. Table 8.1 shows that the steel and aluminum plates have different structural mass and connecting spring parameters expressing the end connection of the plates. The difference, however, is not significant (Faltinsen 1997). The asymptotic theory by Faltinsen (1997) agrees well with the experiments (see Figure 8.14). The theory gives the nondimensional maximum strain for a given structural mass and the connecting spring parameter that is independent of impact speed and wave characteristics.

The physics can be explained as follows. Because it takes time to build up elastic deformations w of the plate, the pressure loads from either the water or an air cushion balance the structural inertia force of the plate initially. This is why it is called the *structural inertia phase*. We can formally express the vertical velocity of the plate as $\dot{w} - V$, where V is the rigid-body water entry velocity of the plate. The plate experiences a large force impulse during a small time relative to the highest natural period for the plate vibrations in the structural inertia phase. This causes the space-averaged elastic vibration velocity $\dot{w}$ to be equal to the water entry velocity V at the end of the initial phase. Another way of saying is that the space-averaged velocity of the body is zero. The whole plate is wetted at the end of the structural inertia phase. The plate then starts to vibrate similarly to a free vibration of a wet beam with an initial space-averaged vibration velocity V and zero initial deflection. Maximum strains occur during the *free vibration phase*.

The details of the pressure distribution during the structural inertia phase are not important, but the impulse of the impact is. Very large pressures that are sensitive to small changes in the physical conditions may occur in this initial phase. This can be seen from the collection of measured maximum pressures during the tests (Figure 8.10). They appear to be stochastic in nature.

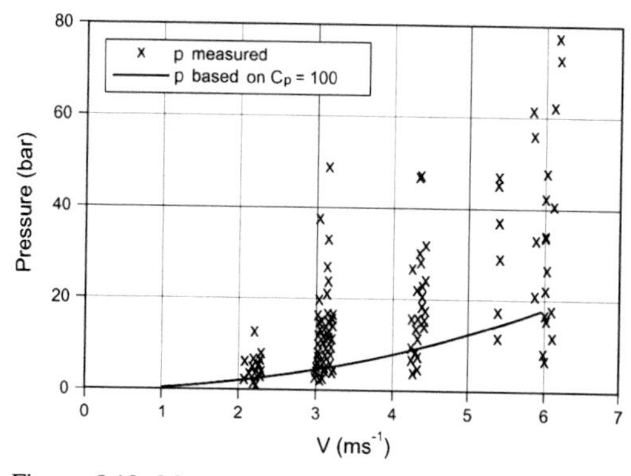

Figure 8.10. Measured maximum pressure from different drop tests of the horizontal plates described in Table 8.1 as a function of the water entry velocity V. $C_p = (p - p_a)/(0.5\rho V^2)$ is the pressure coefficient.

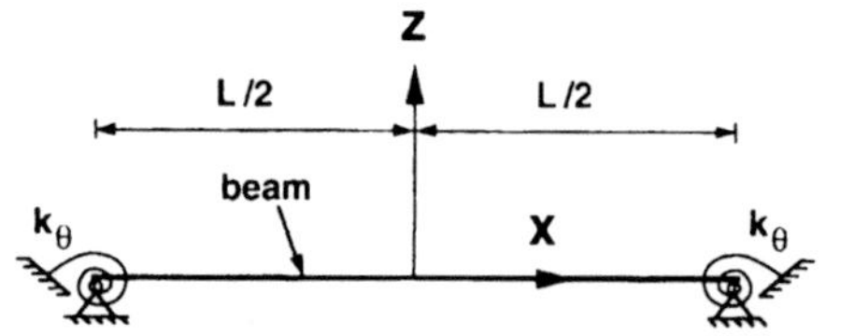

Figure 8.11. Coordinate system used for local hydroelastic analysis of beam of length L. k_θ is the spring stiffness of spiral springs at the beam ends.

The measured maximum strains showed a very small scatter for given impact velocity and plate, even though the maximum pressure varied strongly. The largest measured pressure was approximately 80 bar for V equal to 6 ms^{-1}. This is close to the acoustic pressure $\rho c V \approx 1000 \cdot 1500 \cdot 6 \approx 90$ bar, hence much larger pressures are not expected using smaller pressure gauges. The diameter of each pressure cell was 4 mm. A sampling frequency up to 500 kHz was used. These results document that it can be misleading from a structural point of view to measure the peak pressures for the effect of hydrodynamic impact on aluminum and steel structures.

Free vibration phase of hydroelastic slamming

We will analyze the free vibration phase of hydroelastic slamming. The whole plate is then wetted. The structure is represented by an Euler beam model, that is, the load levels do not cause plastic deformations. The analysis by Kvålsvold and Faltinsen (1995) showed that shear deformation was insignificant. The structure is in the following analysis assumed to be a beam of constant thickness and finite breadth. We later generalize the results to stiffened plates. The beam equation of motion is written as

$$M_B \frac{\partial^2 w}{\partial t^2} + EI \frac{\partial^4 w}{\partial x^4} = p(x, w, t). \quad (8.1)$$

Here w is the beam deflection, t is the time variable, and x is a longitudinal coordinate with $x = 0$ in the middle of the beam (Figure 8.11). Further, p is the hydrodynamic pressure that is a function of the beam deflection. Because we are analyzing the free vibration phase, the slamming pressure is zero. So p is a consequence of the vibrations of the beam. This means that we expect an added mass effect. M_B (mass of plate per length square) and EI (bending stiffness per length width) are assumed constant. Initial conditions are a consequence of the structural inertia phase, as already mentioned. The boundary conditions at the ends of the beam are expressed as

$$w(x, t) = 0 \quad \text{at } x = \pm L/2 \quad (8.2)$$

$$\frac{k_\theta}{EI} \frac{\partial w}{\partial x} \pm \frac{\partial^2 w}{\partial x^2} = 0 \quad \text{at } x = \pm L/2. \quad (8.3)$$

Eq. (8.3) consists of two terms. The first term expresses the effect of a rotational spring at one of the beam ends (Figure 8.11). The restoring moments of the spring at $x = \pm L/2$ can be written as $\mp k_\theta \partial w / \partial x$, where k_θ is the spring stiffness and $\partial w / \partial x$ is the slope of the beam. The second term in eq. (8.3) is proportional to the beam bending moment $-EI \partial^2 w / \partial x^2$. Eq. (8.3) follows from continuity of the bending moment at the rotational springs at the beam ends. When k_θ is zero, we have zero bending moment at the beam ends. This means a hinged-hinged beam model. The case of infinite k_θ corresponds to zero slope of the beam ends, that is, a clamped-clamped beam model. If we think of the beam as a part of a larger structure, the rotational springs are simplifications of the effect of the adjacent structure on the beam. The fact that eq. (8.2) states that the deflection w is zero at the beam ends implies that the adjacent structure to the beam is assumed to be much stiffer than the beam.

The solution is expressed in terms of dry normal modes Ψ_n, that is,

$$w(x, t) = \sum_{n=1}^{\infty} a_n(t) \Psi_n(x). \quad (8.4)$$

The dry normal modes are a good approximation of the wet normal modes when the added mass distribution is similar to the mass distribution. The eigenfunctions Ψ_n are found by first setting $p = 0$ in eq. (8.1) and assuming a solution on the form $\exp(i\omega_n t)\,\Psi_n$, where ω_n are dry natural frequencies associated with the nth eigenmode Ψ_n. This gives

$$-\omega_n^2 M_B \Psi_n + EI \frac{d^4 \Psi_n}{dx^4} = 0. \quad (8.5)$$

We consider only modes that are symmetric about $x = 0$. The reason is that the considered beam loading is symmetric about $x = 0$. This becomes evident later in the text. Solutions of eq. (8.5) can then be expressed as

$$\Psi_n = B_n \cos p_n x + D_n \cosh p_n x, \quad (8.6)$$

where

$$p_n^4 = \frac{M_B \omega_n^2}{EI}. \tag{8.7}$$

We find equations for ω_n, B_n, and D_n by requiring that Ψ_n satisfies the same boundary conditions as w, that is, eqs. (8.2) and (8.3). We cannot determine both B_n and D_n, only how B_n and D_n depend on each other. In order to simplify the following presentation, we assume $k_\theta = 0$, implying the case of a hinged-hinged beam, even though this may not be the most realistic case. The end conditions are then that both w and $\partial^2 w/\partial x^2$ are zero. This gives the mode shapes:

$$\Psi_n = B_n \cos(p_n x) \quad \text{with } p_{n+1} L/2 = \pi/2 + n\pi,$$
$$n = 0, 1, 2, 3, \ldots$$

We can easily control that the end conditions are satisfied. Further, the experiments showed that the first mode shape, that is, $n = 1$, was dominating. This means that the mode shape that we are studying is

$$\Psi_1 = B_1 \cos(p_1 x), \tag{8.8}$$

where

$$p_1 \frac{L}{2} = \frac{\pi}{2}. \tag{8.9}$$

By now combining eqs. (8.7) and (8.9), we have that the lowest dry natural frequency is

$$\omega_1 = \left(\frac{EI}{M_B}\right)^{1/2} \left(\frac{\pi}{L}\right)^2, \tag{8.10}$$

where we see that L is an important parameter. We can normalize Ψ_1 as we want and choose to set $B_1 = 1$. The consequence of these simplifications is that the beam deflection is expressed as

$$w(x, t) = a_1(t) \cos p_1 x. \tag{8.11}$$

We now have to introduce the effect of the pressure p in eq. (8.1). This causes an added mass effect due to the vibrations of the beam. In order to analyze this, we study first the flow due to unit velocity $\dot{a}_1(t)$. This can be described by two-dimensional fluid potential flow theory for an incompressible fluid. The linearized body boundary condition requiring no flow through the beam is

$$\frac{\partial \varphi}{\partial z} = \cos p_1 x, \quad z = 0, \quad -L/2 < x < L/2. \tag{8.12}$$

Because the frequency of oscillations is high, gravity is neglected. The high-frequency free-surface condition is

$$\varphi = 0 \quad \text{on } z = 0, \quad |x| > L/2.$$

An analytical solution to φ can be found in Kvålsvold (1994). However, the solution becomes particularly simple if we average $\cos p_1 x$ over the beam length. This means we replace eq. (8.12) with

$$\frac{\partial \varphi}{\partial z} = \frac{1}{L} \int_{-L/2}^{L/2} \cos p_1 x \, dx = \frac{2}{\pi} \quad \text{on } z = 0,$$
$$-L/2 < x < L/2. \tag{8.13}$$

The problem then becomes the forced heave problem. A solution to this problem may be found in many textbooks (see, e.g., Kochin et al. 1964). It is also discussed in more detail in section 8.3.1. We can write the velocity potential on the body as

$$\varphi = \frac{2}{\pi}((L/2)^2 - x^2)^{1/2}, \quad |x| < L/2, \quad z = 0. \tag{8.14}$$

This means that if the deflection of the body is represented as in eq. (8.11), the corresponding velocity potential is $\dot{a}_1(t)$ times eq. (8.14), that is, $\phi = \dot{a}(t)\varphi(x)$. The considered problem is linear in $\dot{a}(t)$. The corresponding pressure follows from the Bernoulli equation. We only consider pressure terms that are linear in $\dot{a}(t)$. This means that it is sufficient to consider the pressure term $p = -\rho \partial \phi / \partial t$, that is,

$$p = -\rho \ddot{a}_1(t) \frac{2}{\pi}((L/2)^2 - x^2)^{1/2}, \quad |x| < L/2. \tag{8.15}$$

Substituting p and w given by eqs. (8.15) and (8.11) into eq. (8.1) leads to the following equation:

$$M_B \ddot{a}_1(t) \cos p_1 x + EI \cdot p_1^4 a_1(t) \cos p_1 x$$
$$= -\rho \ddot{a}_1(t) \frac{2}{\pi}((L/2)^2 - x^2)^{1/2}.$$

This equation depends on both x and t. In order to find a solution for $a_1(t)$, we follow the standard solution technique when a solution is represented in terms of normal modes (Clough and Penzien 1993). This means that we now multiply the equation above with the lowest mode $\cos p_1 x$ and integrate between $-L/2$ and $L/2$. The final equation can now be written as

$$(M_{11} + A_{11})\frac{d^2 a_1}{dt^2} + C_{11}a_1 = 0. \quad (8.16)$$

Here

$$M_{11} = M_B \int_{-L/2}^{L/2} \cos^2 p_1 x \, dx = 0.5 M_B L. \quad (8.17)$$

This can be interpreted as a generalized structural mass. Further, we find

$$C_{11} = EI p_1^4 \int_{-L/2}^{L/2} \cos^2 p_1 x \, dx = 0.5 \omega_1^2 M_B L. \quad (8.18)$$

This can be interpreted as a generalized restoring (stiffness) term. Finally, the generalized added mass A_{11} can be obtained by

$$A_{11} = \rho \frac{2}{\pi} \int_{-L/2}^{L/2} ((L/2)^2 - x^2)^{1/2} \cos p_1 x \, dx. \quad (8.19)$$

Now we have to solve eq. (8.16) with the initial conditions that followed from the structural inertia phase. A solution that satisfies zero initial deflection is

$$a_1 = C \sin \omega_w t, \quad (8.20)$$

where

$$\omega_w = \left(\frac{C_{11}}{M_{11} + A_{11}} \right)^{1/2} \quad (8.21)$$

is the wet natural frequency of the lowest mode. In order to find C in eq. (8.20), we use the initial condition for the velocity, that is, $\dot{w}\,|_{t=0} = V$, where V is the water entry velocity. This can only be satisfied in an average way, that is,

$$\omega_w C \cos \omega_w t|_{t=0} \int_{-L/2}^{L/2} \cos^2(p_1 x) dx = V \int_{-L/2}^{L/2} \cos p_1 x \, dx.$$

This means

$$C = \frac{4V}{\pi \omega_w}. \quad (8.22)$$

The bending stress σ_b follows from

$$\sigma_b = -E z_a \frac{\partial^2 w}{\partial x^2}, \quad (8.23)$$

where z_a is the distance from the neutral axis to the stress point. This means

$$\sigma_b = E z_a \frac{4V}{\pi \omega_w} \left(\frac{\pi}{L} \right)^2 \cos\left(\pi \frac{x}{L} \right) \sin \omega_w t. \quad (8.24)$$

We note that this equation tells that bending stresses are proportional to the impact velocity V and have a maximum at the middle of the stiffener for a given z_a and simply supported beam ends. Because the slamming pressure is proportional to V^2, a quasi-steady analysis would lead to σ_b being proportional to V^2. Further, the expression shows that the first maximum stress for positive z_a will occur at time $0.5\pi/\omega_w$. The hydrodynamic pressure at that time follows from eq. (8.15) and by noting that

$$\ddot{a}_1(t) = -\frac{4V}{\pi} \omega_w \sin \omega_w t. \quad (8.25)$$

This means that the pressure is maximum at the same time. The hydrodynamic pressure becomes negative after time π/ω_w as a consequence of the $\sin \omega_w t$-dependence. Depending on the magnitude of V and ω_w, eq. (8.15) for the hydrodynamic pressure can at a certain time instant be less than minus atmospheric pressure, or the total pressure is predicted to be less than zero. However, the total pressure cannot be less than the vapor pressure, which is close to zero for normal water temperature. If that happens in our theory, cavitation occurs. Eq. (8.15) tells that this occurs first in the middle of the beam. As time increases, the cavity will spread toward the beam ends. This is illustrated in Figure 8.12 by using $V = 2.94\ \mathrm{ms}^{-1}$, $T_w = 0.0262$ s, and a beam length 0.5 m. Here $T_w = 2\pi/\omega_w$ is the wet natural period. Because the submergence of the beam is low ($Vt = 0.06$ m at $t = 0.02$ s), we have set the total pressure equal to atmospheric pressure for $|x| \geq 0.25$ m (outside the beam) in Figure 8.12. As time increases, there is an increased probability of ventilation, that is, air is sucked in under the beam. We can understand this by noting the increase in the pressure gradient from the air next to the beam ends toward the cavity. When the beam is fully ventilated, it starts to oscillate as if it were in air. We can see these phenomena in Figures 8.13 and 8.14, in which a comparison with experiments is also made with our simplified theory based on one mode only. However, because the spring coefficient k_θ is different from zero (see Table 8.1), we used the more general mode shapes given by eq. (8.6). The experiments confirm that

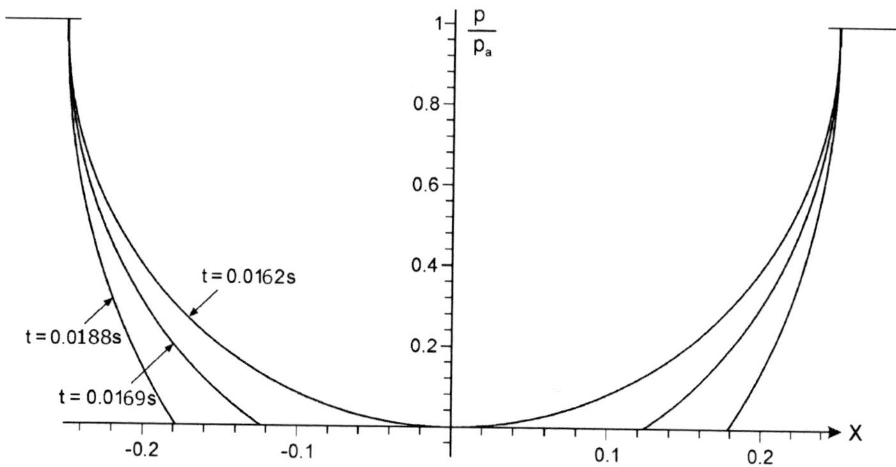

Figure 8.12. Total pressure p during the free vibrations of hydroelastic slamming on a horizontal beam. p_a = atmospheric pressure. Beam end conditions: zero displacement and zero moment. Wet natural period $T_w = 0.0262$ s. Impact velocity $V = 2.94\ \mathrm{ms}^{-1}$. Beam length = 0.5 m. This case shows how the cavitation length increases toward the beam ends as time t increases.

only the lowest mode is important. Further, we see that the strains presented in Figure 8.14 do not react to the very high pressure peaks in the pressure records in Figure 8.13. The details of the elastic test plate used during the drop test results presented in Figure 8.14 are given in Figure 8.15 and Table 8.1.

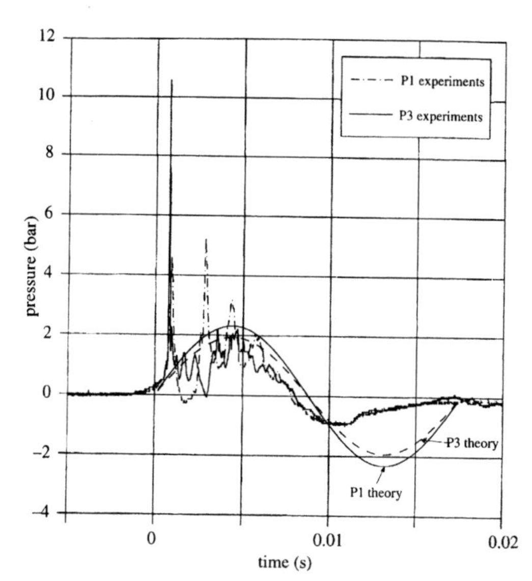

Figure 8.13. Pressure at two positions of the plate as a function of time (see Figure 8.15 and Table 8.1). Comparison between asymptotic theory and drop tests. Drop height is 0.5 m. P1 is located at the middle of the plate and P3 at the quarter length (Faltinsen 1997).

The total pressure can be obtained by adding atmospheric pressure and the pressure shown in Figure 8.13. The total pressure equals vapor pressure, and cavitation starts at approximately 0.01 s after initial impact, according to both theory and experiments. Because the theory does not account for cavitation, it does not give correct predictions after that. However, maximum strains have already occurred. We note that the experiments show that the pressure becomes atmospheric (i.e., ventilation) some time after cavitation has been initiated. Further, Figure 8.14 shows that the strains have then started to oscillate with a

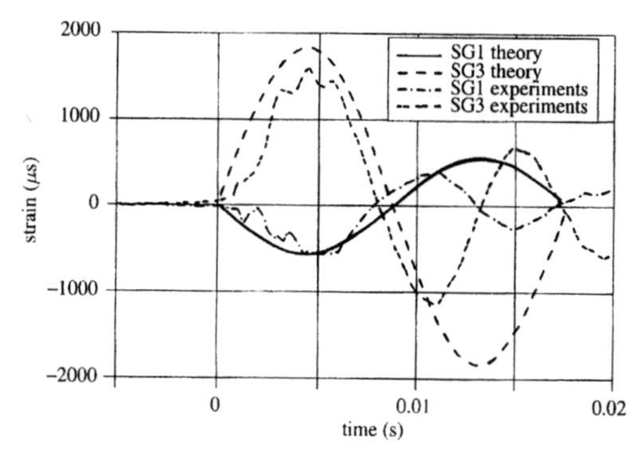

Figure 8.14. The strains at the locations SG1 and SG3 along the beam as a function of the time (see Figure 8.15 and Table 8.1). Comparison between asymptotic theory and drop tests. The drop height is 0.5 m (Faltinsen 1997).

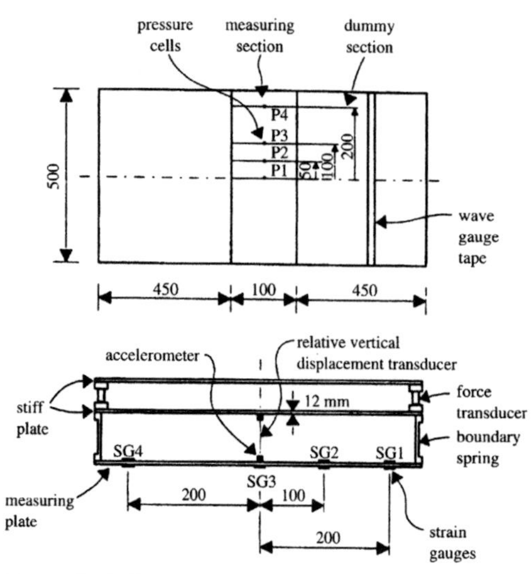

Figure 8.15. Details of the elastic test plate used during drop tests (Kvålsvold et al. 1995).

higher frequency. This is the dry natural frequency of the lowest mode. Then the maximum strain becomes smaller also. This can be understood from eq. (8.24) by exchanging the wet natural frequency ω_w with the corresponding dry natural frequency.

Figure 8.14 shows that the oscillation period for the lowest mode is about 0.018 s. This is also a representative time scale for local hydroelastic slamming effects on the wetdeck of a full-scale catamaran. Phenomena occurring on a much smaller time scale are not important for maximum local slamming-induced strains for a structure like this. An example is the effect of the fluid compressibility. When the structure initially hits the water, signals are sent out with the speed of sound, which is about 1500 ms^{-1} in water without bubbles, depending on temperature and salinity. Actually, what we say when we are assuming an incompressible fluid is that the speed of sound is infinite and that the fluid everywhere is immediately affected by the impact. Of course, this effect decays asymptotically to zero at infinity. Let us then return to the effect of compressibility. We then need a length scale to derive a time scale. A representative length scale is the length of the wetted beam. So we get a time scale that is the order of 10^{-3} s. This means it is correct for us to assume an incompressible fluid in our previous hydroelastic slamming analysis.

Haugen (1999) theoretically and experimentally studied the hydroelastic impact of plates with three beam elements. Each of the beam elements was intended to model a longitudinal stiffener with effective flange between two transverse frames of a wetdeck (see Figure 8.16 for structural arrangement of a wetdeck). The physics in the initial impact phase of three beam elements is somewhat different from that described above for one beam element. The wetting of the plates lasts relatively longer and air cushion effects matter more. Further, there is one dominant high natural period for one beam element whereas there are several for three beam elements. The final results in terms of maximum strains are similar, but somewhat higher for three beams than for one.

Arai and Myanchi (1998) presented a numerical and experimental hydroelastic study of water impact on cylindrical shells. Ulstein and Faltinsen (1996) analyzed the hydroelastic impact between the stern seal bag of an SES and the water surface representative for low sea states, in which cobblestone oscillations may matter. A high forward speed was assumed. The elastic structural behavior is dominated by membrane effects. Examples of mode shapes are shown in Figure 5.11. The hydrodynamic behavior is analogous to transient oscillations of a lifting elastic foil with a time-dependent length.

Although most theoretical and experimental approaches to hydroelasticity have been performed for two-dimensional bodies, Scolan and Korobkin (2003) performed hydroelastic slamming analysis of a three-dimensional cone using Wagner's approach. The results show significant influence of the elasticity compared with the rigid case.

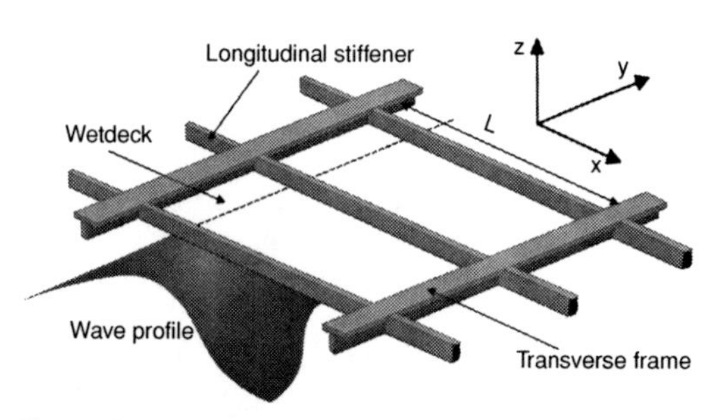

Figure 8.16. A detail of the wetdeck structure of a multihull vessel.

Scaling

When we want to scale the results to other types of materials or to other length scales, it is important to introduce nondimensional variables that reflect the physics of the problem. We show how to do that and start with the wet natural frequency ω_w given by eq. (8.21). We can use eq. (8.10) to rewrite C_{11} given by eq. (8.18) to express how C_{11} depends on L and EI. This gives

$$C_{11} = EI\left(\frac{\pi}{L}\right)^4\left(\frac{L}{2}\right). \tag{8.26}$$

We then use eq. (8.17) for M_{11} and find that we can write eq. (8.21) as

$$\omega_w = \left[\frac{EI\pi^4}{\rho L^5\left(\frac{M_B}{\rho L} + 2\frac{A_{11}}{\rho L^2}\right)}\right]^{1/2}. \tag{8.27}$$

Because both $M_B/(\rho L)$ and $A_{11}/(\rho L^2)$ are nondimensional, it follows from eq. (8.27) that

$$\omega_w\left(\frac{\rho L^5}{EI}\right)^{1/2} \tag{8.28}$$

is a nondimensional wet natural frequency. If we then use the more general mode shape given by eq. (8.6), it will also lead to a nondimensional frequency, as in eq. (8.28). This means eq. (8.28) is a proper way to nondimensionalize ω_w. There are, of course, other ways to nondimensionalize frequency, but the way we do it must be relevant to our problem. For instance, $\omega_w(L/g)^{1/2}$ is a nondimensional frequency following from Froude scaling, but as we have seen from our analysis, the gravitational acceleration does not appear. This means $\omega_w(L/g)^{1/2}$ is not a physically relevant way to nondimensionalize frequency for hydroelastic slamming. However, it is a relevant way to nondimensionalize frequency when we consider the effect of gravity waves on added mass and damping in the equations of motions of the ship.

Let us then continue our work concerning nondimensional variables. Eq. (8.24) shows that

$$\frac{\sigma_{ba}\omega_w L^2}{Ez_a V} \tag{8.29}$$

is nondimensional. The use of eq. (8.28) gives that

$$\frac{\sigma_{ba}}{\left(\frac{z_a}{L}\right)V\left(\frac{\rho L^3 E}{I}\right)^{1/2}} \tag{8.30}$$

is nondimensional. Because $\sigma = E\varepsilon$, where σ is stress and ε is strain and ε is nondimensional, σ_{ba}/E will be nondimensional. Using this in expression eq. (8.30) means that

$$V\left(\frac{\rho L^3}{EI}\right)^{1/2} \tag{8.31}$$

is a relevant nondimensional velocity for our problem. Now we are able to recognize the nondimensional strains and velocities used in Figure 8.9.

Table 8.2 shows by means of the theoretical model how nondimensional bending stress as in eq. (8.30) and nondimensional natural frequencies as in eq. (8.28) are influenced by the nondimensional spring stiffness and structural mass ratio $M_B/\rho L$. Because a realistic range of $M_B/\rho L$ has been used, it means that maximum nondimensional bending stress is more dependent on beam end conditions (i.e., k_θ) than on $M_B/\rho L$.

What we have done in the previous analysis is to use the mathematical model to introduce nondimensional variables. If the mathematical model does not properly describe the phenomena, the scaling would obviously be wrong. For instance, the theory only predicts the initiation of the cavitation, not the detailed behavior leading to ventilation. We must then introduce a cavitation number $\sigma = (p_{amb} - p_v)/(0.5\rho U^2)$, where p_{amb} is the ambient pressure, p_v is the vapor pressure, and U is a characteristic velocity. For our problem, we can set p_{amb} equal to atmospheric pressure and choose U equal to the water entry velocity V. However, because we were interested in maximum strains, and cavitation has not yet occurred, σ is not a parameter in this context.

An alternative to introducing dimensionless parameters is to use the Pi-theorem (Buckingham 1915, see section 2.2.4). This does not require a mathematical model, but does require a good physical understanding of relevant variables.

It is common for engineers to dislike nondimensional variables. One needs to have a quantitative measure of dimensions to understand what they means. However, nondimensional parameters are particularly useful in model testing and scaling up to full-scale. This may be the only approach for certain physical problems for which reliable numerical tools are not available.

Table 8.2. *Generalized added mass A_{11}, lowest wet natural frequency ω_w, and bending stress σ_b as a function of mode shape*

				$\frac{\sigma_{ba}}{(z_a/L)\,V}\sqrt{I/(\rho L^3 E)}$ $x/L=$			
$M_B/(\rho L)$	$\frac{k_\theta L}{2EI}$	$\frac{A_{11}}{\rho L^2}$	$\omega_w\sqrt{\rho L^5/(EI)}$	0.0	0.2	0.4	0.5
0.02	0.0	0.21	14.76	0.85	0.69	0.26	0.0
	0.5	0.23	17.3	0.78	0.59	0.09	−0.21
	1.75	0.24	21.27	0.72	0.48	−0.12	−0.48
	2.85	0.24	23.43	0.70	0.44	−0.22	−0.61
	5.0	0.23	26.04	0.68	0.40	−0.32	−0.75
0.124	0.0	0.21	13.29	0.95	0.76	0.29	0.0
	0.5	0.23	15.57	0.87	0.66	0.10	−0.23
	1.75	0.24	19.15	0.80	0.54	−0.13	−0.53
	2.85	0.24	21.08	0.77	0.49	−0.24	−0.68
	5.0	0.23	23.42	0.76	0.44	−0.36	−0.83
	20.0	0.22	27.97	0.74	0.37	−0.55	−1.09
	10^4	0.21	30.68	0.74	0.33	−0.65	−1.21

$\sigma_b = \sigma_{ba}\sin\omega_w t$, x = coordinate along the beam, beam ends at $x = \pm L/2$, V = water entry velocity, ρ = mass density of water. Other variables are defined in Table 8.1 (Faltinsen 1997).

8.2.1 Example: Local hydroelastic slamming on horizontal wetdeck

Consider the details of the wetdeck structure of a multihull vessel shown in Figure 8.16. The transverse frames can be considered much stiffer than the longitudinal stiffeners. We assume that the wetdeck is horizontal when it hits the water. The main concern here is to find the maximum bending stress in a longitudinal stiffener supported by transverse frames. We then consider one longitudinal stiffener together with its plate flange, as shown in Figure 8.17. The width of the plate flange corresponds to the distance between two longitudinal stiffeners. When we want to apply the results in Figure 8.9, we have to calculate the area moment of inertia about the neutral axis of the cross section shown in Figure 8.17 and divide this by the width of the flange. This gives $I = 11 \cdot 10^{-6}$ m^4/m neglecting the effective flange effect. The distance z_a from neutral axis to maximum strain position is 0.12 m and the length L of the longitudinal stiffener between two transverse frames is 1 m.

The wetdeck material is aluminum, with $E = 70 \cdot 10^9$ Nm^{-2}. Let us assume the impact velocity V is 1 ms^{-1} and set the mass density ρ of water equal to 1000 kgm^{-3}. This gives a nondimensional impact velocity

$$V\sqrt{\frac{\rho L^3}{EI}} = 0.036.$$

We disregard the effect of the radius of curvature R of the impacting free surface and use a value

$$\frac{\varepsilon_m}{z_a V}\sqrt{\frac{EI}{\rho L}} = 0.5$$

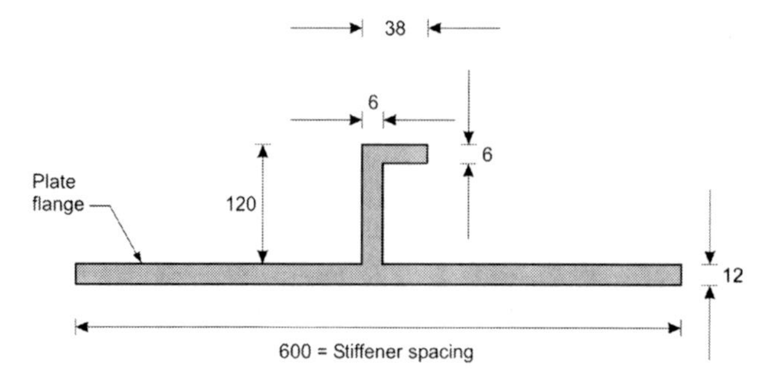

Figure 8.17. The cross-section of a longitudinal stiffener and the plate flange. Dimensions are in millimeters.

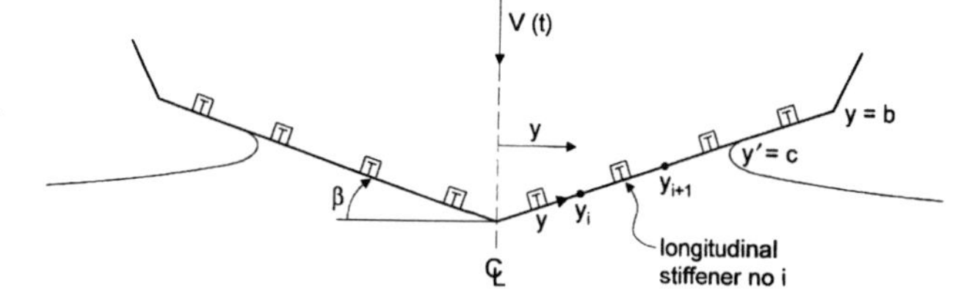

Figure 8.18. Water entry of a wedge-shaped elastic cross section.

based on Figure 8.9. This gives a maximum strain in the longitudinal stiffener midway between two transverse frames that is equal to

$$\varepsilon_m = 2.16 \cdot 10^{-3}.$$

This corresponds to a stress

$$\sigma_m = E\varepsilon_m = 151 \text{ MPa}.$$

This is also the maximum stress anywhere along the longitudinal stiffener for realistic end conditions of the longitudinal stiffener at the transverse frames.

The yield stress of aluminum may vary significantly, for example, from 200 to 300 MPa. It is important to design against the yield strength in the heat-affected zone (HAZ) of the welds, which may be of significant size for aluminum, and not against the yield strength in the weld material or in the base material outside the HAZ. Allowable nominal bending stress on a stiffener according to DNV's rules for direct strength calculations is 142 MPa. So we are not satisfying this requirement. A more complete analysis is needed for slamming-induced buckling. The local analysis must then be combined with a global analysis.

The relative impact velocity may very well be higher than 1 ms^{-1}. This implies that operational restrictions of the vessel are necessary. An alternative is to change the dimensions of the wetdeck. The simple results given by Figure 8.9 tell us how to do that for a given design value of the impact velocity. However, it may be more practical to avoid the wetdeck being horizontal or to increase the wetdeck height above the mean free surface. An alternative is to have a wedge-shaped cross section of the wetdeck. We will present relevant results later in the text.

8.2.2 Relative importance of local hydroelasticity

Faltinsen (1999) studied the relative importance of hydroelasticity for an elastic hull with wedge-shaped cross sections penetrating an initially calm water surface (Figure 8.18). Wagner's theory was generalized to include elastic vibrations. In the following section, it is pointed out that the Wagner theory is approximate for large deadrise angles. However, it is believed that the approximate theory demonstrates the main parameter dependence. Stiffened plating between two rigid transverse frames was examined (Figure 8.19). A hydrodynamic strip theory in combination with orthotropic plate theory was used. The water entry velocity was assumed constant.

The importance of hydroelasticity for the local slamming-induced maximum stresses increased with decreasing deadrise angle β and increasing impact velocity V. The nondimensional parameter $\xi = \tan\beta/[V(\rho L^3/EI)^{1/2}]$ was introduced. L is the length of the analyzed longitudinal stiffener between the two transverse frames. EI is the bending stiffness per width of the longitudinal stiffener including the effective plate flange. The parameter ξ is proportional to the ratio between the wetting time of the rigid wedge and the highest natural period of the longitudinal stiffener. We can see this as follows. If Wagner's theory is used (see eq. (8.52)), the wetting time of a rigid wedge with beam B is $B\tan\beta/(\pi V)$. Because $\omega_w\sqrt{\rho L^5/EI}$ is a constant (see eq. (8.28)) and $T_w = 2\pi/\omega_w$ is the wet natural structural period, T_w is proportional

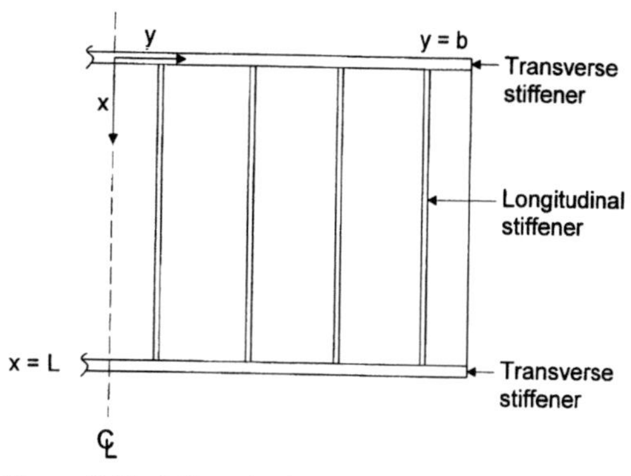

Figure 8.19. Stiffened plating consisting of plate and longitudinal stiffeners.

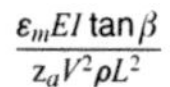

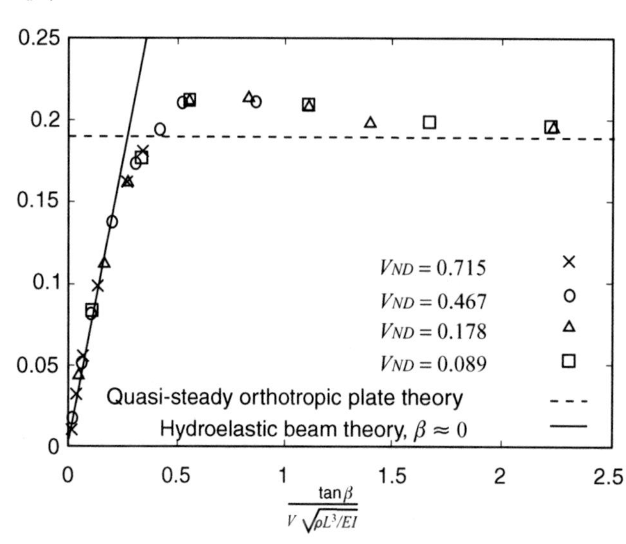

Figure 8.20. Nondimensional maximum strain ε_m in the middle of the second longitudinal stiffener from the keel. The strain is presented as a function of a parameter that is proportional to the ratio between wetting time of a rigid wedge and natural period of the longitudinal stiffener based on beam theory. Different nondimensional constant impact velocities V_{ND} (see eq. (8.32)) are given. β = deadrise angle. Calculations by hydroelastic orthotropic plate theory (Faltinsen 1999) are shown.

to $\sqrt{\rho L^5/EI}$. Assuming B/L is a constant gives the desired result. We can associate the wetting time of the wedge with the duration of the loading. If we make an analogy to a simple mechanical system consisting of a mass and a spring, then we know that the duration of the loading relative to the natural period characterizes the dynamic effects of a transient system (see section 7.1.4). Nondimensional results are presented in Figure 8.20. Also presented are results based on quasi-steady analysis and asymptotic hydroelastic analysis for small deadrise angles β. The quasi-steady analysis assumes the structure is rigid in the hydrodynamic calculations. The pressure is then proportional to V^2. The analysis of the structural deformations due to the water impact gives that $\varepsilon_m EI \tan\beta/(z_a V^2 \rho L^2)$ is independent of the abscissa $\tan\beta/(V\sqrt{\rho L^3/EI})$ in Figure 8.20. The asymptotic hydroelastic analysis is based on writing the ordinate in Figure 8.20 as a function of the abscissa, that is,

$$\varepsilon_m \frac{EI \cdot \tan\beta}{z_a V^2 \rho L^2} = E_{HE} \frac{\tan\beta}{V\sqrt{\rho L^3/EI}},$$

where

$$E_{HE} = \frac{\varepsilon_m}{z_a V}\sqrt{\frac{EI}{\rho L^3}}$$

is estimated by Faltinsen's (1997) hydroelastic analysis for $\beta = 0$. Examples of E_{HE}-values can be found in Table 8.2 by noting that the bending stress σ is equal to $E\varepsilon$, where ε is the strain. A representative value of E_{HE} equal to 0.7 was used in presenting the results in Figure 8.20. This is based on setting $M_B/(\rho L) = 0.015$ and $k_\theta L/(2EI) = 3.0$. This means that the results by the asymptotic hydroelastic analysis appear as a straight line in Figure 8.20.

The particular way of nondimensioning the results gives small explicit dependence on the dimensionless impact velocity

$$V_{ND} = V\sqrt{\frac{\rho L^3}{EI}}. \tag{8.32}$$

Figure 8.20 illustrates that hydroelastic effects are present when $\tan\beta < \approx 1.5\, V(\rho L^3/EI)^{1/2}$ for the studied stiffened plating. The stress from the hydroelastic case may also exceed the stress from the quasi-steady case. A large influence of hydroelasticity occurs when $\tan\beta < \approx 0.25\, V(\rho L^3/EI)^{1/2}$. By independently varying terms in $\xi = \tan\beta/(V\sqrt{\rho L^3/EI})$, we see that small ξ-values are obtained when

- The deadrise angle β is small.
- The water entry velocity V is large.
- $\sqrt{\rho L^3/EI}$ is large. Because $\omega_w L\sqrt{\rho L^3/EI}$ is a constant (see eq. (8.28)), this means that large $\sqrt{\rho L^3/EI}$ corresponds to small values of $\omega_w L$, where ω_w is the lowest wet natural frequency. If L is a constant, large $\sqrt{\rho L^3/EI}$ corresponds to a high wet natural period $2\pi/\omega_w$.

If hydroelasticity is not important, maximum strain ε_m is proportional to V^2. This is a consequence of the fact that the impact pressure is proportional to V^2.

The parameter study assumed constant V during the impact. The relative impact velocity between the vessel velocity and the ambient water velocity, V, may, in reality, vary substantially during the impact of a wedge section with finite β. This happened in the comparative studies with full-scale experiments of local slamming-induced strains in the wetdeck of a 30 m–long catamaran reported by Faltinsen (1999). Fair agreement between theory and experiments was documented, but the predicted strains were sensitive to the time-varying impact velocity. The impact velocity was not measured and is strongly dependent on speed and frequency-dependent nonlinear hull and free-surface effects.

The wetdeck of the 30-m catamaran had a wedge-shaped cross-sectional form with deadrise angle of 14° in the initial impact area. Local hydroelasticity is then insignificant, as in the case of bow flare slamming studied by Kapsenberg and Brizzolara (1999). The measured maximum strains corresponded to about half the yield stress. This occurred in head seas with significant wave height $H_{1/3} = 1.5$ m and a ship speed of 18 knots. The ship was allowed to operate up to $H_{1/3} = 3.5$ m. The classification rules did not predict well that the ship had sufficient height of the wetdeck above sea level to avoid wetdeck slamming. In contrast to computer simulations, the full-scale tests demonstrated that a change in ship course was effective in avoiding heavy wetdeck slamming. In general, proper operational criteria due to wetdeck slamming are lacking.

8.3 Slamming on rigid bodies

When the local angle between the water surface and the body surface is not very small at the impact position, slamming pressures can be used in a static structural response analysis to find local slamming-induced stresses. The body can be assumed rigid in the hydrodynamic calculations. Several approximations can be made in the analysis. The airflow is usually unimportant, and irrotational flow of incompressible water can be assumed. Because the local flow acceleration is large relative to gravitational acceleration when slamming pressures matter, gravity is neglected.

The terms *Wagner method* and *von Karman method* are often mentioned in the following text. A von Karman method neglects the local uprise of the water, whereas a Wagner method accounts for that. However, a Wagner method assumes impact of a blunt body.

Most theoretical studies assume 2D vertical water entry of a symmetric body. An indicator of the importance of 3D flow effects is the ratio $64/\pi^4 \approx 0.66$ between maximum pressures during water entry of a cone and a wedge with constant velocity and small deadrise angles (Faltinsen and Zhao 1998b). A cone represents an extreme case of 3D flow. We cannot say without further investigation whether it is the most extreme case. Scolan and Korobkin (2001) used the Wagner method to study the impact of a three-dimensional body with elliptical contact line on the free surface. By *Wagner method*, we mean that the body boundary condition is transferred to a disc. The free-surface conditions are the same as those Wagner used in the outer flow domain. Chezhian (2003) has investigated the impact of a more general 3D geometry using a generalized Wagner's approach. Comparisons were made with model tests. Beukelman (1991) presented experimental results for three-dimensional models that showed that forward speed has a strong influence on the pressure level when the deadrise angle was lower than $\approx 2°$.

When the exact nonlinear free-surface conditions are used, it is difficult numerically to handle the intersection between the body and the free surface for small local deadrise angles. Small errors in the predicted, very small intersection angle between the free surface and the body may cause large errors in the predictions of the intersection points and destroy the numerical solution. The 2D boundary element method (BEM) by Zhao and Faltinsen (1993) avoided this by introducing a control surface normal to the body surface at the spray root. Because the pressure is approximately atmospheric in the spray, this control surface can be handled similarly to a free surface. This method is applicable to a broad class of body shapes as well as time-varying water entry velocity. General 3D geometry, forward speed with incident waves, and ship-generated steady and unsteady waves further complicate the impact analysis to a situation that does not seem feasible to solve numerically at the moment.

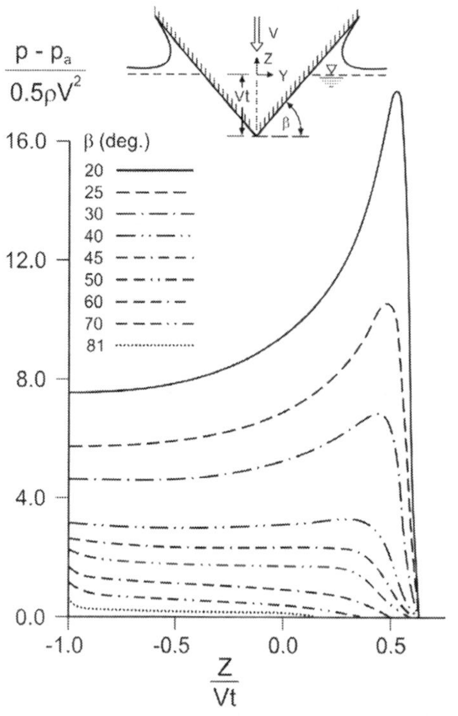

Figure 8.21. Predictions of pressure (p) distribution during water entry of a rigid wedge with constant vertical velocity V. p_a = atmospheric pressure, β = deadrise angle (Zhao and Faltinsen 1993).

Pressure distribution

Numerical results based on the similarity solution by Dobrovol'skaya (1969) for water entry of rigid wedges with constant entry velocity was presented by Zhao and Faltinsen (1993) for $4^\circ < \beta < 81^\circ$. Figure 8.21 shows the predicted pressures for $20^\circ \leq \beta \leq 81^\circ$. The pressure distribution becomes pronouncedly peaked and concentrated close to the spray root when $\beta < \approx 20^\circ$. The smaller β is, the more sensitive slamming loads are to β. A consequence is that large rolling may have an important effect on slamming loads on a bow flare section. The higher the local angle between the water surface and the body, the more uniformly spaced the impact pressure. The maximum pressure occurs at the apex (or keel) when $\beta > 45^\circ$. For larger angles and low impact velocities, other pressure contributions may be as important as the slamming part.

Parameters characterizing slamming on a rigid body with small deadrise angles are the position and value of the maximum pressure, the time duration, and the spatial extent of high slamming pressures. A measure of the spatial extent ΔS_s of the high slamming pressure is explained in Figure 8.22. The results by Zhao and Faltinsen (1993) show that ΔS_s has meaning only when $\beta \leq \approx 20^\circ$. Table 8.3 shows predictions of $C_{p\max}$, $z_{\max}$, and ΔS_s up to $\beta = 40^\circ$ by the similarity solution of Dobrovol'skaya (1969). $z_{\max}$ is defined in Figure 8.22, and $C_{p\max}$ is the maximum value of the pressure coefficient $(p - p_a)/(0.5\rho V^2)$ (ρ = mass density of the water). Results for nondimensional water entry force F_3 are also presented in Table 8.3. A similarity solution implies that the pressure, force, and coordinates can be made nondimensional in such a way that the time is not an explicit parameter. For instance, $C_{p\max}$, $z_{\max}/(Vt)$, and $F_3/(\rho V^3 t)$ presented in Table 8.3 are not a function of time. However, the dimensional variables are obviously a function of time.

The fact that the pressure distribution becomes very peaked and concentrated close to the spray

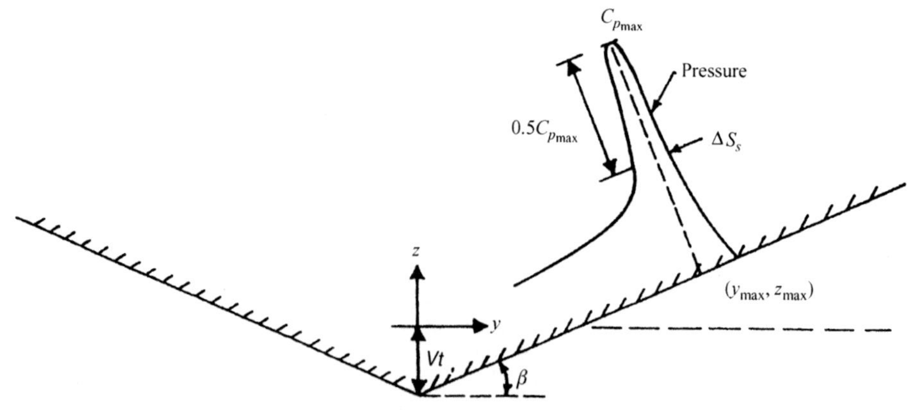

Figure 8.22. Definition of parameters characterizing slamming pressure during water entry of a blunt 2D rigid body. C_p = pressure coefficient = $(p - p_a)/(0.5\rho V^2)$.

Table 8.3. *Calculation of slamming parameters by similarity solution during water entry of a wedge with constant vertical velocity V*

β	$C_{p_{max}}$	$\frac{z_{max}}{Vt}$	$\frac{\Delta S_s}{c}$	$\frac{F_3}{\rho V^3 t}$
4°	503.030	0.5695	0.01499	1503.638
7.5°	140.587	0.5623	0.05129	399.816
10°	77.847	0.5556	0.09088	213.980
15°	33.271	0.5361	0.2136	85.522
20°	17.774	0.5087	0.4418	42.485
25°	10.691	0.4709		23.657
30°	6.927	0.4243		14.139
40°	3.266	0.2866		5.477

β = deadrise angle, $C_{p_{max}}$ = pressure coefficient at maximum pressure, z_{max}= z-coordinate of maximum pressure (see Figure 8.22), ΔS_s = spatial extent of slamming pressure (see Figure 8.22), $c = 0.5\pi\, Vt \cot\beta$, F_3 = vertical hydrodynamic force on the wedge, t = time (Zhao and Faltinsen, 1993).

root at small values of β illustrates that measurement of slamming pressure requires high sampling frequency and "small" pressure gauges. In the literature, several reported experimental values exist for the maximum pressure for wedges, and opinions vary on how well theory for the maximum pressure agrees with experimental results. However, experimental error sources due to the size of the pressure gauge and the change of the body velocity during drop tests are not always taken sufficiently into account. Takemoto (1984) and Yamamoto et al. (1984) did that and showed good agreement with Wagner's (1932) theory for maximum pressure when the deadrise angle was between $\approx 3^\circ$ and 15°. The reason for the disagreement for $\beta < \approx 3^\circ$ is the creation of an air cushion when the wedge enters the water. Another matter is that hydroelasticity should be considered for small deadrise angles.

One should be careful in applying the results for wedges to other cross sections. The local deadrise angle is not the only important body parameter. For instance, the local curvature as well as the time history of the angle and curvature also matter. Further, a time-varying water entry velocity occurs in reality and the water surface is not calm either.

Zhao et al. (1996) presented a generalized Wagner theory that is a simplification of the more exact solution of the water entry problem by Zhao and Faltinsen (1993). The generalized Wagner method is more numerically robust and faster than the original exact solution. It gives satisfactory results and is therefore preferred in engineering practice. Generalized Wagner theory means that the exact body boundary conditions are satisfied. The free-surface conditions are approximated as Wagner (1932) did in the outer flow domain, that is, not for the details at the spray roots. The wetted body surface is found by integrating in time the vertical velocity of the fluid particles on the free surface and determining when the particles intersect with the body surface. This is done by predetermining the intersection points on the body and then determining the time to reach these points in a time-stepping procedure. Because the velocity in the generalized Wagner method is singular at the body-water surface intersection, special care is shown by using a local singular solution. Direct pressure integration is used to predict the water entry force. All terms in Bernoulli's equation are included except the hydrostatic pressure term. If the predicted pressure becomes less than the atmospheric pressure, p_a, the pressure is simply set equal to p_a. This occurs at the spray root and is caused by the velocity-square term in Bernoulli's equation.

Water entry force

Theoretical slamming force results for wedges are presented in Figure 8.23. Constant water entry velocity is assumed. Different methods are used and related to an exact solution of the potential flow incompressible water entry problem without gravity. Wagner's flat plate approximation is only good for small deadrise angles, whereas the generalized Wagner solution can be applied to

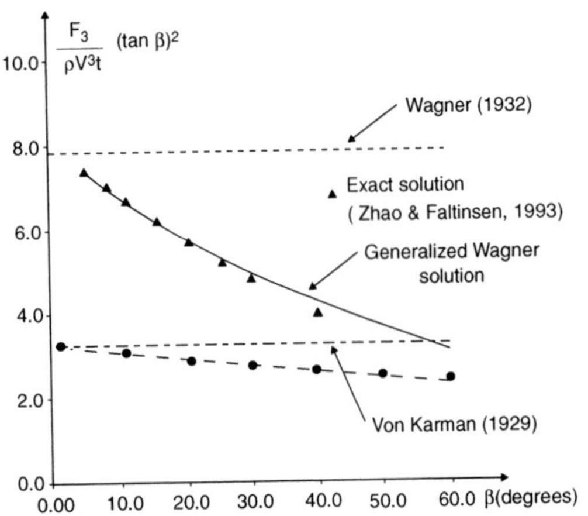

Figure 8.23. The vertical slamming force, F_3, on symmetric wedges during water entry with constant vertical drop (water entry) velocity, V. β = deadrise angle, ρ = mass density of the fluid, t = time variable, Vt = instantaneous draft relative to calm water. Exact vertical slamming force (Zhao and Faltinsen 1993) ▲▲▲▲; generalized Wagner solution——————; Wagner solution---------; von Karman––––––––––––; von Karman-momentum----•--------•-------•------. (Zhao et al. 1996).

large deadrise angles. A von Karman type of solution clearly underpredicts the force for $\beta < \approx 30^\circ$ to 40°.

Separation from knuckles (chines)

Zhao et al. (1996) extended the method by Zhao and Faltinsen (1993) to include separation from knuckles. The hydrodynamic water entry force cannot be neglected after the flow has separated from the knuckles (Figure 8.24). The peak in the vertical force with constant water entry velocity occurs when the spray roots are at the knuckles. If the hydrodynamic vertical water entry force is expressed in terms of the time derivative of infinite-frequency added mass as a function of submergence relative to undisturbed free surface (von Karman method), the force part after flow separation will be negligible. This is common in commercial computer programs for nonlinear wave load analysis. An approach like this will also give a too-low maximum force and a wrong time history of the force. The reason is that the force is proportional to the time rate of change of the wetted area. The local water rise-up, neglected by the von Karman method, implies a larger rate of change of the wetted area.

Because gravity is neglected in the previous water entry studies, the generation of surface waves as well as the Froude-Kriloff and hydrostatic forces are disregarded. The latter two force components can easily be added, which is common in commercial computer programs. The nonlinear Froude-Kriloff and hydrostatic forces on a flared section increase their importance relative to slamming forces with decreasing relative vertical velocity between the ship cross section and the water. The relative importance is also influenced by the local deadrise angle slamming being more important for small angles.

Assuming zero gravity in the present case is similar to assuming zero cavitation number flow in which there is an infinite cavity behind the body in steady flow. Gravity may cause a finite-length air cavity behind an impacting body, which may cause secondary impact and possible entrapped air. The cavity will collapse after some time, and the whole body surface becomes wet.

Asymmetric impact

A hull structure may have asymmetric transverse sections, the hull structure may be tilted, the water surface may be sloping, and/or the structure may have both a horizontal and vertical velocity during an impact. If asymmetric water entry of a wedge is considered, the occurrence of cross-flow at the apex is always expected initially to cause a ventilated area near the apex of the wedge. One side of the wedge could be fully ventilated, depending on

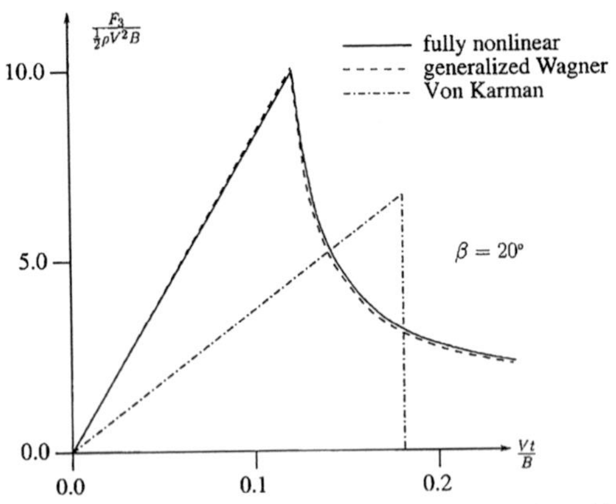

Figure 8.24. Vertical slamming force F_3 on a wedge with knuckles. The deadrise angle is 20°. Constant water entry velocity V. B = maximum wedge breadth. The nondimensional time between the predicted peaks by the different methods is an effect of the uprise of water.

the heel angle, the deadrise angle, and the velocity direction of the body. If partial ventilation occurs only initially, flow separation from the apex associated with viscosity may occur at a later stage.

de Divitiis and Socio (2002) studied the unsymmetric impact of wedges with constant velocity by means of a similarity solution. Irrotational flow of an incompressible fluid was assumed. The symmetry axis of the wedge is vertical, and the water entry velocity has a horizontal component U and a vertical component V. Depending on the deadrise angle β and the direction of the velocity, $\alpha = \tan^{-1} V/U$, the flow can separate from the wedge apex and be fully ventilated on the leeward side of the wedge. If $\beta > 45^\circ$, the critical value α^* of α for separation to occur is very small, whereas $\alpha^* = 60^\circ$ for $\beta = 7.5^\circ$. When the flow separates from the wedge, it is similar to water entry of a flat plate. The latter problem has been studied for small values of β and $\kappa = \pi - \alpha$ by Sedov (1940) and Ulstein and Faltinsen (1996).

8.3.1 Wagner's slamming model

We give a more detailed description of Wagner's (1932) slamming model in this section. Even though this model assumes a local small deadrise angle, it is useful because it provides simple analytical results. These can be used to assess how slamming pressures depend on structural form and time-dependent water entry velocity. Further, this model will be used to show that it is the space-averaged pressure that matters for structural stresses.

Wagner's detailed description of the flow at the intersections between the free surface and the body surface will not be presented. This local flow describes a jet flow that, in practice, ends up as spray. We focus on what is called the outer flow domain. This is located below (outside) the inner and jet domains shown in Figure 8.25. There are

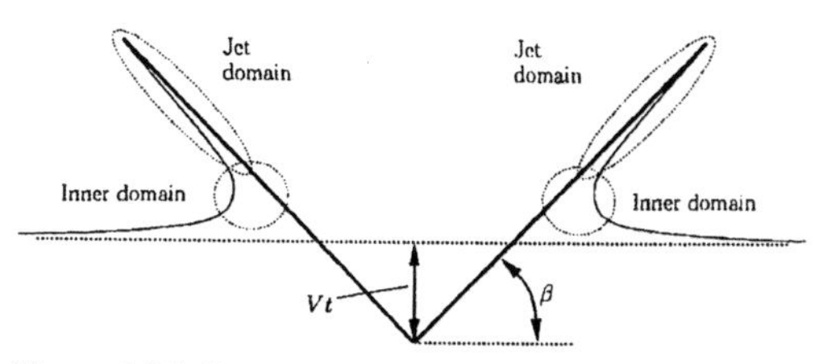

Figure 8.25. Water entry of a wedge with constant velocity V. Definition of inner and jetflow domains.

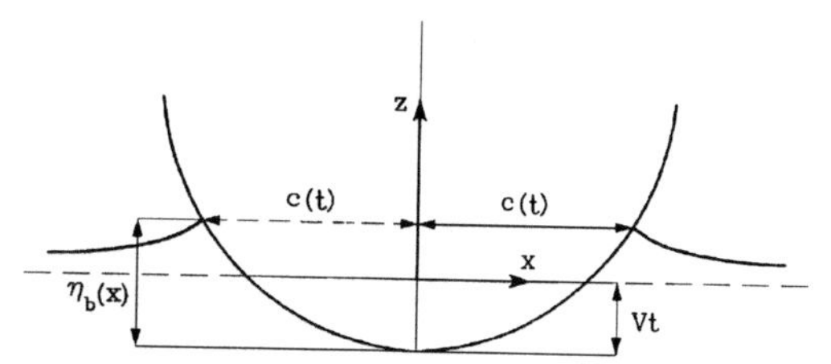

Figure 8.26. Definition of parameters in the analysis of impact forces and pressures on a body by means of Wagner's outer flow domain solution. Constant water entry velocity V is assumed. Vt is the instantaneous draft relative to the undisturbed free surface.

then no details on the spatially rapidly varying flow at the spray roots (inner domain). The predicted intersections between the free surface and the body surface in the outer flow domain model are in a very close vicinity of the spray roots.

Figure 8.26 presents the impacting symmetric body and the free surface in the outer flow domain. The water entry velocity V is constant, and Vt represents the submergence of the lowest point of the body relative to the calm water surface. However, as we see from Figure 8.26, there is an uprise of the water caused by the impact. The volume of the water above $z = 0$ is equal to the volume of water that the body displaces for $z \leq 0$. The difference between the von Karman and Wagner methods is that a von Karman method neglects the local uprise of the water (hence the wetted surface length is smaller).

Figure 8.27 describes the boundary-value problem that must be solved at each time instant. The body boundary condition requiring no flow through the body surface is transferred to a straight line between $x = -c(t)$ and $c(t)$ using Taylor expansion. This can be done because the body is blunt, which means the local deadrise angle is small. This angle is the angle between the x-axis and the tangent to the body surface. The end points $x = \pm c$ correspond to the instantaneous intersections between the outer flow free surface and the body surface (see Figure 8.26). We note in Figure 8.27 that the free-surface condition $\varphi = 0$ on $z = 0$ has been used. This is a consequence of fluid accelerations in the vicinity of the body dominating over gravitational acceleration during impact of a blunt body. Let us express this by first examining Euler's equations, which are a basis when deriving Bernoulli's equation. In an Earth-fixed (inertial) coordinate system with positive z-axis

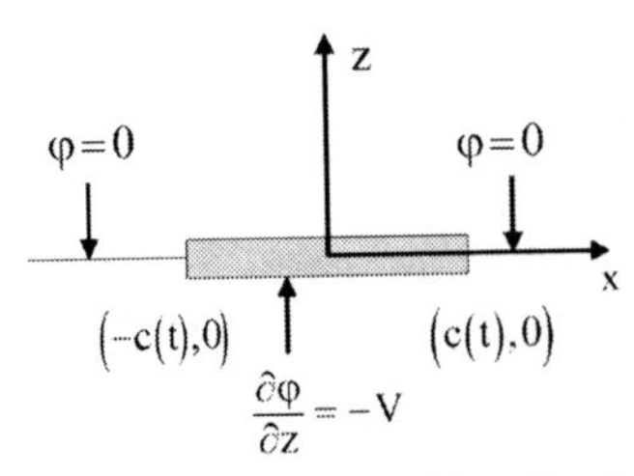

Figure 8.27. Boundary-value problem for the velocity potential φ in a simplified analysis of the impact between a two-dimensional body and the water.

upward, Euler equation states that

$$\frac{\partial \mathbf{u}}{\partial t} + \mathbf{u} \cdot \nabla \mathbf{u} = -\frac{\nabla p}{\rho} - g\mathbf{k}.$$

Here $\mathbf{u}$ is the fluid velocity, p is the pressure, and $\mathbf{k}$ is the unit vector along the z-axis. Saying that fluid accelerations dominate means that both $\mathbf{u} \cdot \nabla \mathbf{u}$ and $g\mathbf{k}$ are small relative to $\partial \mathbf{u}/\partial t$, that is, we can set

$$\rho \frac{\partial \mathbf{u}}{\partial t} = -\nabla p$$

as a first approximation. Substituting $\mathbf{u} = \nabla \varphi$ gives that

$$\nabla \left(\rho \frac{\partial \varphi}{\partial t} + p \right) = 0.$$

This means that $\rho \partial \varphi/\partial t + p$ is a constant. If we assume no surface tension and atmospheric pressure p_a on the free surface, this gives

$$p - p_a = -\rho \frac{\partial \varphi}{\partial t}. \quad (8.33)$$

Because $p = p_a$ on the free surface, we get that $\partial \varphi/\partial t = 0$ on the free surface. If we now follow fluid particles on the free surface, they start at initial time with $\varphi = 0$. Because $\partial \varphi/\partial t = 0$, $\varphi = 0$ remains for all time as a condition on the free surface. However, the free surface moves because $\partial \varphi/\partial n \neq 0$. The final step then is to assume small deviations between φ on $z = 0$ and the free surface and transfer this condition to $z = 0$, again by Taylor expansion. The reason for this is that it simplifies considerably the solution to our problem. The same is true for transferring the body boundary condition to $z = 0$.

The solution to the boundary-value problem shown in Figure 8.27 may be found in many textbooks. Complex variables $Z = x + iz$, in which i is the complex unit, are then introduced. The complex velocity potential can be expressed as (Kochin et al. 1964)

$$\Phi = \varphi + i\psi = iVZ - iV(Z^2 - c^2)^{1/2}, \quad (8.34)$$

where φ is the velocity potential and ψ is the stream function. The complex velocity is

$$\frac{d\Phi}{dZ} = u - iw = iV - iV\frac{Z}{(Z^2 - c^2)^{1/2}}. \quad (8.35)$$

Let us control that the boundary conditions are satisfied. Care must then be shown in evaluating the complex function $(Z^2 - c^2)^{1/2}$ which has a branch cut along the line from $Z = -c$ to c. We introduce $Z - c = r_1 e^{i\theta_1}$ and $Z + c = r_2 e^{i\theta_2}$, where θ_1 and θ_2 vary from $-\pi$ to π (Figure 8.28). This means

$$(Z^2 - c^2)^{1/2} = \sqrt{r_1 r_2}\, e^{i\frac{1}{2}(\theta_1 + \theta_2)}.$$

We can write $\theta_1 = -\pi$ and $\theta_2 = 0$ when $|x| < c$ and $z = 0^-$. This gives

$$(Z^2 - c^2)^{1/2} = -i(c^2 - x^2)^{1/2} \quad \text{for } |x| < c,\ z = 0^-. \quad (8.36)$$

Here $z = 0^-$ corresponds to the underside of the body. When $x > c$ and $z = 0$, both θ_1 and θ_2 are zero, that is,

$$(Z^2 - c^2)^{1/2} = (x^2 - c^2)^{1/2} \quad \text{for } x > c,\ z = 0. \quad (8.37)$$

Further, $x < -c$ and $z = 0$ means that $\theta_1 = \theta_2 = \pi$, that is,

$$(Z^2 - c^2)^{1/2} = -(x^2 - c^2)^{1/2} \quad \text{for } x < -c,\ z = 0. \quad (8.38)$$

Eq. (8.34) gives, then, $\varphi = 0$ for $|x| > c$ on $z = 0$. Further, eq. (8.35) gives

$$\frac{d\Phi}{dZ} = u - iw = iV + V\frac{x}{(c^2 - x^2)^{1/2}} \quad \text{for } |x| < c \quad \text{on } z = 0^-. \quad (8.39)$$

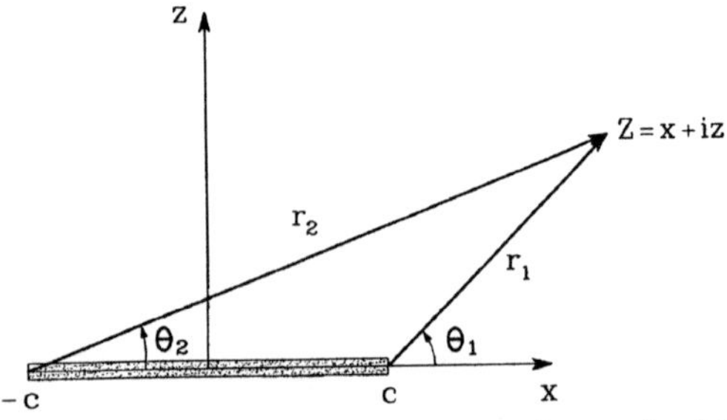

Figure 8.28. Definition of polar coordinates (r_1, θ_1) and (r_2, θ_2) used in evaluating the complex function $(Z^2 - c^2)^{1/2}$. The angles θ_i vary from $-\pi$ to π.

Because $w = \partial\varphi/\partial z$, we see from eq. (8.39) that the body boundary condition is satisfied. Further, eq. (8.35) gives that the fluid velocity goes asymptotically to zero when $|Z| \to \infty$.

Eq. (8.34) gives

$$\Phi = \varphi + i\psi = iVx - V(c^2 - x^2)^{1/2} \quad \text{for } |x| < c(t), \quad z = 0^-. \tag{8.40}$$

We can then write the velocity potential on the body as

$$\varphi = -V(c^2 - x^2)^{1/2}, \quad |x| < c(t). \tag{8.41}$$

Eq. (8.33) gives the hydrodynamic pressure. There is nothing in our derivations so far that prevents us from letting V be time-dependent except that the vertical distance of the lowest point on the body is $\int_0^t V(\tau)\,d\tau$ relative to the calm free surface. This gives

$$p - p_a = \rho V \frac{c}{(c^2 - x^2)^{1/2}} \frac{dc}{dt} + \rho \frac{dV}{dt}(c^2 - x^2)^{1/2}. \tag{8.42}$$

The first term is denoted as the slamming pressure. It is associated with the rate of change of the wetted surface which is approximately $2dc/dt$. Why the second term is called the added mass pressure will be more evident when we later consider the resulting hydrodynamic force. We note that the slamming pressure is infinite at $x = \pm c$. This is unphysical. A detailed analysis near the spray roots (inner domain solution) is needed to find the correct pressure near $x = \pm c$. If V is constant, this gives a maximum pressure of $p - p_a = 0.5\rho\,(dc/dt)^2$. Armand and Cointe (1986) and Howison et al. (1991) showed how to match the inner and outer domain solutions. A composite expression for the pressure that is valid in both domains can then be constructed. Cointe (1991) also studied the details of the solution in the jet domains defined in Figure 8.25. However, in the next section, we see that in practice, we are interested in space-averaged pressures. Because eq. (8.42) is integrable, the singularity appearing in the outer domain solution is not serious.

Let us now derive the two-dimensional vertical force acting on the impacting body. This can be expressed as

$$\begin{aligned} F_3 &= \int_{-c}^{c} p\,dx = \rho V c \frac{dc}{dt} \int_{-c}^{c} \frac{dx}{\sqrt{c^2 - x^2}} \\ &\quad + \rho \frac{dV}{dt} \int_{-c}^{c} \left(c^2 - x^2\right)^{1/2} dx \\ &= \rho\pi V c \frac{dc}{dt} + \rho \frac{\pi}{2} c^2 \frac{dV}{dt}. \end{aligned} \tag{8.43}$$

The term $\rho\pi c^2/2$ appearing in the last term is the two-dimensional added mass in heave a_{33} for the plate shown in Figure 8.27. We can understand this by returning to our definition of added mass in eq. (7.39). We studied forced oscillations of a body, linearized the problem, and defined added mass in terms of resulting linear hydrodynamic forces. The slamming term in eq. (8.43) is nonlinear in V. This will be evident later when we express c. Another way of saying it is that in a linearized problem, we should not allow for a change in the wetted area, that is, dc/dt should be zero. By now noting that V is positive downward, we have by the definition in eq. (7.39) that $\rho\pi c^2/2$ is the added mass in heave. Eq. (7.39) also includes a damping term. Because the damping is caused by wave radiation and we have here the free-surface condition $\varphi = 0$, which implies that no waves can be generated, it is consistent that we find here that the damping is zero. In order to develop waves, we had to include either surface tension or gravity.

We should note that the added mass that we have found is half the heave-added mass of a plate in infinite fluid. We can understand this by studying forced oscillations in the heave of the plate in infinite fluid. Because the velocity potential then is antisymmetric about the x-axis, it implies that $\varphi = 0$ on the x-axis for $|x| > c$. This is the same condition as the one we have used in solving our problem. When we find the resulting hydrodynamic force on the plate in infinite fluid, we have to integrate pressure on both sides of the plate. In our problem, we have only to integrate pressures on the lower side. It then follows that $\rho\pi c^2/2$ is half the heave-added mass of the plate in infinite fluid.

We see from eq. (8.43) that the force can also be expressed as

$$F_3 = \frac{d}{dt}(a_{33}V) = a_{33}\frac{dV}{dt} + V\frac{da_{33}}{dt}, \tag{8.44}$$

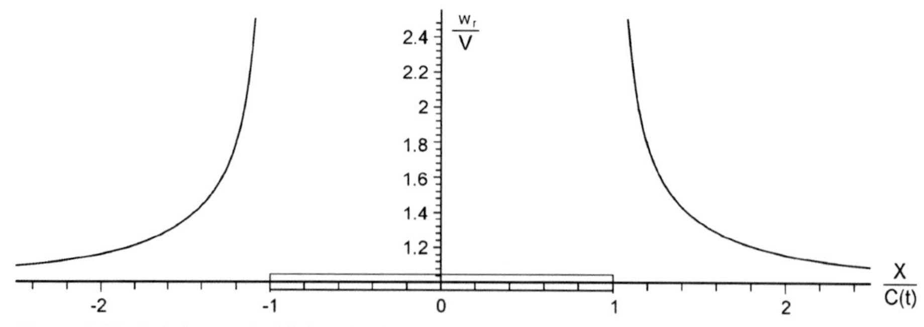

Figure 8.29. Relative vertical fluid velocity w_r between the fluid particles on the free surface and the body due to an impacting blunt symmetric body. V = water entry velocity, $2c(t)$ = instantaneous beam.

where Vda_{33}/dt is the slamming force. This is a common way to express the slamming force in connection with the von Karman method. When this is done, one does not necessarily use a flat plate approximation, as we have done. Further, eq. (8.44) is also applied to a 3D body by replacing the 2D heave-added mass a_{33} with the 3D heave-added mass A_{33}. So what one does is to calculate "infinite frequency" – added mass as a function of submergence. *Infinite frequency* means that the free-surface condition is $\varphi = 0$.

If a flat plate approximation is made, as in Wagner's 2D case, and the instantaneous contact line between the body and the free surface is elliptic, we can express (Scolan and Korobkin 2001) the three-dimensional heave-added mass as

$$A_{33}(t) = \frac{2\pi}{3}\frac{\rho a^2 b}{E(e)}. \tag{8.45}$$

Here $a(t)$ and $b(t)$ are, respectively, the shortest and longest semi-axes of the ellipse. Further, $e = (1 - (a/b)^2)^{0.5}$ is the ellipse eccentricity and E is the complete elliptic integral of the second kind (Abramowitz and Stegun 1964). In the particular case of a circular disk, that is, $a = b$, we get that $A_{33} = 4\rho a^3/3$. Using eqs. (8.44) and (8.45) in combination with a von Karman method is straightforward. This provides a simple way to qualitatively assess the importance of 3D flow effects during impact.

When eq. (8.44) is used in practice, the relative impact velocity between the body velocity and the ambient water velocity is introduced. Further, it is common to add hydrostatic force and forces due to the pressure in the incident waves (Froude-Kriloff forces). Because this has to account for the instantaneous submergence, which is not necessarily small, the Froude-Kriloff and hydrostatic forces are evaluated exactly. We will demonstrate this when we later consider wetdeck slamming loads in the context of a catamaran and bow flare forces.

Prediction of wetted surface

Let us now return to our 2D slamming problem. If we use von Karman's method, $c(t)$ is determined by the geometrical intersection between the undisturbed free surface and body surface. If we use Wagner's method, then we need to follow fluid particles on the free surface and see when they intersect with the body surface. Because $\varphi = 0$ on the free surface, the horizontal velocity $\partial\varphi/\partial x$ is zero on the free surface. We can use eqs. (8.35), (8.37), and (8.38) to express the vertical velocity $w = \partial\varphi/\partial z$ on the free surface. This gives

$$\frac{\partial\varphi}{\partial z} = \frac{V|x|}{\sqrt{x^2 - c^2(t)}} - V \quad \text{on } z = 0, \quad |x| > c(t). \tag{8.46}$$

It should be stressed that this expression does not apply to $z = 0$, $|x| < c(t)$ where $\partial\varphi/\partial z = -V$. We need to know the relative vertical velocities $w_r = \partial\varphi/\partial z + V$ between the fluid particles on the free surface and the body. This relative velocity is presented in Figure 8.29. We now focus on one fluid particle with a given $|x| > c$ and express when this particle intersects the body surface. It has then moved a vertical distance $\eta_b(x)$ (see Figure 8.26) relative to the body. This means

$$\eta_b(x) = \int_0^t \frac{V|x|}{\sqrt{x^2 - c^2(t)}}\, dt. \tag{8.47}$$

Here $t = 0$ corresponds to initial impact and $\eta_b(x)$ is a known function. Eq. (8.47) is an integral equation that determines $c(t)$. We will now derive the details. We consider positive x and introduce c as an integration variable instead of t. c varies from

0 to x. This gives

$$\eta_b(x) = \int_0^x \frac{x\mu(c)\,dc}{\sqrt{x^2 - c^2}}, \qquad (8.48)$$

where

$$\mu(c)\,dc = V\,dt. \qquad (8.49)$$

We do not know $\mu(c)$. Eq. (8.48) is therefore an integral equation that determines $\mu(c)$. When $\mu(c)$ is found, we can use eq. (8.49) to find c as a function of time. We will try to find an approximate solution to eq. (8.48) by guessing that

$$\mu(c) \approx A_0 + A_1 c. \qquad (8.50)$$

Here A_0 and A_1 are unknown constants. By integrating the right-hand side of eq. (8.48), it follows that

$$\eta_b(x) = A_0 \frac{\pi}{2} x + A_1 x^2. \qquad (8.51)$$

If $\eta_b(x)$ is given as a second-order polynomial, we can determine A_0 and A_1 from eq. (8.51). Having determined $\mu(c)$, we can now integrate eq. (8.49) to find c as a function of time. We illustrate this for a wedge and a parabola.

The solution to symmetric impact on a wedge with deadrise angle β, $\eta_b(x) = |x| \tan\beta$ and constant V is

$$c(t) = \frac{\pi V t}{2\tan\beta}. \qquad (8.52)$$

The solution to a parabola with $\eta_b(x) = 0.5x^2/R$ and constant V is

$$c(t) = 2\sqrt{VtR}. \qquad (8.53)$$

If one wants to find $c(t)$ for time varying V as well as a body shape defined by $\eta_b(x) = Ax + Bx^2$, one has to solve the equation

$$\left(\frac{2A}{\pi}c + \frac{B}{2}c^2\right) = \int_0^t V\,dt \qquad (8.54)$$

following from eq. (8.49). Assuming a linearly changing impact velocity, $V(t) = V_0 + V_1 t$, the expression for c becomes

$$c(t) = -\frac{2A}{\pi B} + \frac{\sqrt{(2A/\pi)^2 + 2BV_0 t + BV_1 t^2}}{B}. \qquad (8.55)$$

It requires, of course, that c is real and positive.

Wagner's method does not work for water exit, that is, diminishing wetted surface. One will not find intersection points. This is a consequence of the free-surface condition $\varphi = 0$. It means that fluid accelerations are, for instance, no longer dominant relative to gravitational acceleration. One should, in principle, use the exact free-surface conditions given in section 3.2.1. This requires a numerical method, which by no means is trivial to apply. von Karman's method provides a solution during water exit, but how correct it is depends on the duration T_d of the sum of the water entry and exit phases relative to a characteristic time. If we consider wave impact, this characteristic time is the wave encounter period T_e. This ratio T_d/T_e should be small. We can exemplify what *small* means by referring to two examples. Ge (2002) showed good agreement between a von Karman model and experimental results of vertical forces during wetdeck slamming. The ratio T_d/T_e was less than 0.2. Baarholm (2001) also examined wetdeck slamming loads. However, the time duration of the water exit phase was not satisfactorily predicted by a von Karman model. The ratio T_d/T_e was about 0.65 in this case.

It is common in a von Karman method to neglect the slamming term $V dA_{33}/dt$ during water exit. When we later study global effects due to slamming, both water entry and water exit phases have to be considered.

8.3.2 Design pressure on rigid bodies

When the deadrise angle is small, one should not put too much emphasis on the peak pressures. It is the pressure integrated over a given area that is of interest in structural design as long as hydroelasticity does not matter. When hydroelasticity matters, maximum pressures cannot be used to estimate structural response (see Figures 8.9 and 8.10). In designing experiments on slamming loads, one should have in mind what the results should be used for.

Let us illustrate how we can obtain average pressures appropriate for the design of a local rigid structure. Consider a structural part like those in Figures 8.18 and 8.19 with longitudinal stiffeners and transverse frames and the outside shell plating of the hull. By assuming the transverse frame to be much stiffer than the longitudinal stiffener, the resulting stresses in the longitudinal stiffener are normally more important than those in the transverse frame (the stresses in the transverse frame are, of course, also important depending on its size and length versus the total loading, and sometimes in accidental cases, the frame has been deformed without deforming the stiffeners). If the x-direction means the longitudinal direction of the ship, the instantaneous slamming pressure does not vary much with the position x between two transverse frames. A first approximation of the instantaneous slamming loads of importance for the stresses in the longitudinal stiffener number i is then the space-averaged slamming pressure between y_i and y_{i+1} (see Figure 8.18). This space-averaged pressure varies with time, and it is the largest value that is of prime importance. We will use Wagner's (1932) solution for water entry of a wedge to find the space-averaged pressure. This assumes the deadrise angle to be small. The so-called outer solution is used, which means the details of the spray root are not described. When the water entry velocity V is constant, the intersection point $y = \pm c$ between the free surface and the body surface is given by eq. (8.52). The pressure p on the wedge is given by eq. (8.42). We assume V is constant.

Eq. (8.42) can be integrated analytically to obtain space-averaged pressures p_{av} and the total force. The space-averaged pressure from y_i to y_{i+1} (see Figure 8.18) has a maximum when $c = y_{i+1}$. It follows that the maximum value is

$$p_{av}^{\max} - p_a = 0.5\rho V^2 \frac{\pi}{\tan\beta}\left(\frac{y_{i+1}}{y_{i+1} - y_i}\right) \times \left(\frac{\pi}{2} - \sin^{-1}\left(\frac{y_i}{y_{i+1}}\right)\right). \tag{8.56}$$

It should be noted that eq. (8.42) gives infinite pressure when $y = \pm c$ but this is not true for eq. (8.56).

8.3.3 Example: Local slamming-induced stresses in longitudinal stiffener by quasi-steady beam theory

Consider water entry of a body with a wedge-shaped cross section, as in Figure 8.18, and stiffened platings, as in Figure 8.19. We examine the bending stress in the second longitudinal stiffener from the keel by using steady beam theory. This means the longitudinal stiffener together with the plate flange is considered, that is, similar to Figure 8.17 but with other dimensions. The effective flange should in reality be accounted for in the bending stiffness, whereas the pressure acts on the whole flange. The stiffener is assumed to be independent of the rest of the plating, and secondary stresses, such as plate stresses with Poisson effect, are neglected. The maximum space-averaged impact pressure loading is obtained by eq. (8.56). The resulting stress distribution from the stiffener bending is found by

$$\sigma = \frac{z_a\left(p_{av}^{\max} - p_a\right)}{2I} \times \left[\left(\frac{L}{2}\right)^2 \frac{(1+\alpha/3)}{(1+\alpha)} - (x - 0.5L)^2\right], \tag{8.57}$$

where x is defined in Figure 8.19. Further, z_a is the distance from the neutral axis to where the stress is evaluated and

$$\alpha = 0.5k_\theta L/EI.$$

Here k_θ is a spring stiffness that is related to the restoring beam end moment M_r by $M_r = -k_\theta\theta_b$. θ_b is the rotation angle at the beam end (see also eq. (8.3) and accompanying discussion). We use the following values in eqs. (8.56) and (8.57): $\rho = 1000\,\mathrm{kgm^{-3}}$, $V = 1\,\mathrm{ms^{-1}}$, $\beta = 15^\circ$, $y_i = 0.317$ m, $y_{i+1} = 0.634$ m, $z_a = 0.045$ m, $I = 8.73 \cdot 10^{-7}\,\mathrm{m^4m^{-1}}$, $\alpha = 3$, and $L = 1.25$ m. This gives

$$p_{av}^{\max} - p_a = 12.3\,\mathrm{kNm^{-2}}$$

and the following stress at $x = L/2$:

$$\sigma = 62\,\mathrm{MPa}.$$

This is an acceptable stress level provided V is the design value. The impact velocity V as a design value would be related to a probability level of exceedance. V and β are crucial parameters for the results and should be determined with significant confidence.

8.3.4 Effect of air cushions on slamming

When a body with a horizontal flat bottom or a small deadrise angle hits a horizontal free surface, a compressible air pocket is created between the body and the free surface in an initial phase (Figure 8.30). This has been numerically investigated by Koehler and Kettleborough (1977) for a rigid structure. The air flow causes the water to rise at the edges of the body and encloses an air pocket. The cushioning effect of the air pocket reduces the pressure on the structure.

The pressure in the air cushion will in reality deform both the structure and the free surface. The scenario in Figure 8.30 for an air cushion may have too short a duration for the detailed behavior to influence the maximum

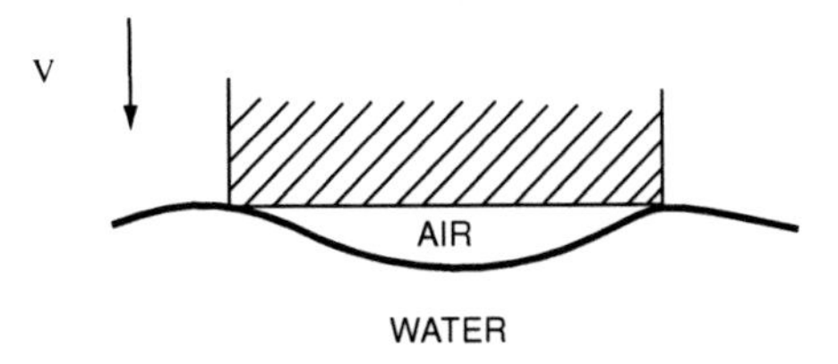

Figure 8.30. Deformation of the free surface and formation of an air pocket during entry of a rigid body with horizontal flat bottom. The thickness of the air layer is exaggerated.

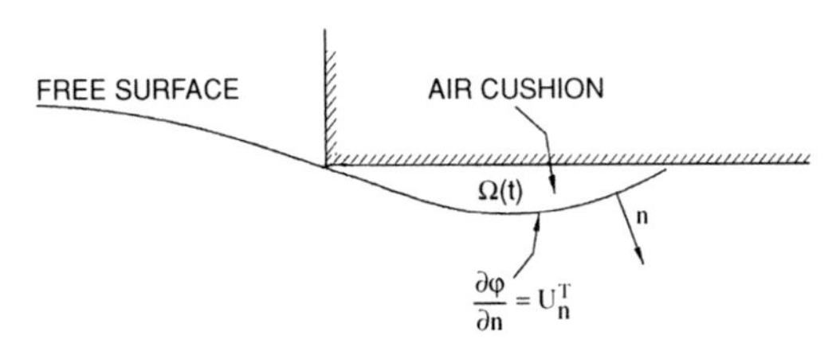

Figure 8.31. Formation of air pocket as a consequence of the shape of the impacting free surface. φ = velocity potential for the water motion, U_n^T = normal velocity of air pocket.

slamming-induced structural stresses. However, air pockets may be created as a consequence of the shape of the impacting free surface. One scenario could be plunging breaking waves against the ship side. This causes an air cushion in a 2D flow situation (Zhang et al. 1996). However, the air has the possibility to escape in a 3D flow situation. Another scenario is in connection with wetdeck slamming (Figure 8.31).

Let us study a situation like the one in Figure 8.31. An impact on the bottom of a semi-infinite long flat plate is examined, and an air cushion with volume $\Omega(t)$ is created. The air cushion dimensions are assumed to be sufficiently large so that surface tension effects can be neglected. Two-dimensional flow in the (x, z) plane is considered with the origin of the coordinate system at the leading edge of the plate (Figure 8.32). The plate is assumed rigid and situated at $x \geq 0$, $z = 0$. The presence of the air cushion influences the flow in the water, which will be described by potential flow of an incompressible fluid. The total velocity potential for the water flow can be divided into several parts. For instance, if this were the linear ship motion problem, the decomposition of the total velocity potential would be given by eq. (7.99).

The boundary-value problem for a velocity potential φ caused by the air pocket is shown in Figure 8.32. A free-surface condition $\varphi = 0$ is imposed on $z = 0$ for $x < 0$. The body boundary condition on the wetted part of the flat plate is $\partial\varphi/\partial z = 0$. On the air cavity surface, we must satisfy $\partial\varphi/\partial n = U_n$, where U_n is the normal velocity of the air cushion surface S_A. The positive normal direction is into the water. In Figure 8.32, this condition is transferred to the plate between $x = a$ and b. In Figures 8.31 and 8.32, we use notations U_n^T and U_n for the normal velocity of the air cushion surface. Here U_n^T means the total normal velocity of the air cushion surface. This includes the effect of the incident waves. However, the situation in Figure 8.32 is similar to an eigenvalue problem. There is no excitation. We just assume that an air cushion is formed and will look for eigenvalues for the oscillation of the air pocket.

On the air cushion surface, we must also require that the pressure in the air cushion be the same as the pressure on the water surface. We assume spatially constant pressure p inside the air cushion. The air must be considered compressible. The continuity equation for the air cushion can be expressed as

$$\rho \frac{d\Omega}{dt} + \frac{d\rho}{dt}\Omega = 0, \tag{8.58}$$

where ρ is the mass density of the air and $\Omega(t)$ is the air cushion volume. We can also write

$$\frac{d\Omega}{dt} = \int_{S_A} U_n \, ds. \tag{8.59}$$

An adiabatic pressure-density relationship

$$\frac{p}{p_a} = \left(\frac{\rho}{\rho_a}\right)^{\gamma} \tag{8.60}$$

with $\gamma = 1.4$ is assumed. Here p_a and ρ_a are values of p and ρ without an air cushion, that is, at the time of the closure of the air pocket. Eqs. (8.58) and (8.60) are similar to the equations used in Chapter 5 to describe cobblestone oscillations of

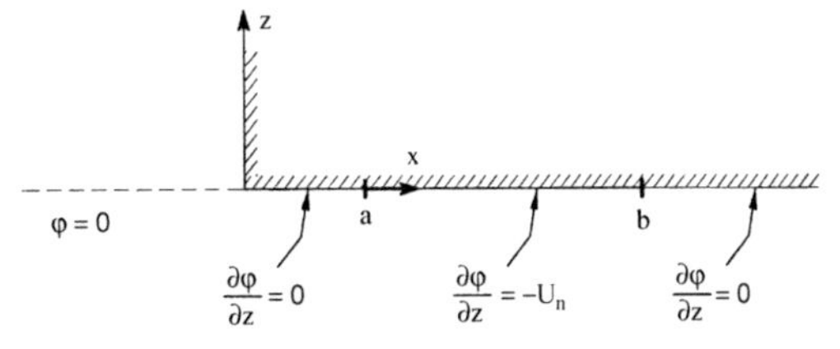

Figure 8.32. Boundary-value problem for the velocity potential φ due to the air cushion between $x = a$ and b. The body is assumed rigid and nonmoving.

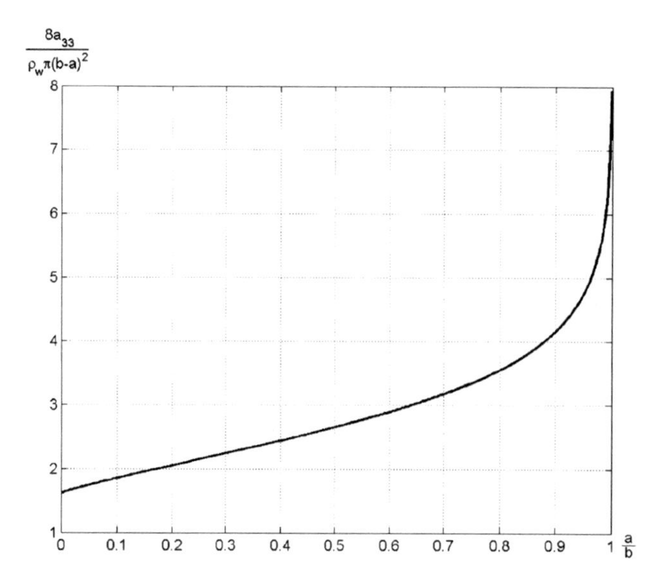

Figure 8.33. Two-dimensional added mass a_{33} due to an oscillating air cushion of length $(b-a)$ on the bottom of a semi-infinitely long plate. a and b with $b > a$ are distances from the leading edge of the plate (see Figure 8.32). The high-frequency free-surface condition is used. ρ_w = mass density of the water.

an SES. However, in the latter case, we accounted for leakage and inflow to the air cushion.

Eqs. (8.58), (8.59), and (8.60) are now linearized by expressing $p = p_a + p_1$, where $p_1/p_a \ll 1$. It follows from eq. (8.60) by first writing $\rho/\rho_a = (p/p_a)^{1/\gamma}$ and then using a Taylor series expansion that

$$\frac{\rho}{\rho_a} \approx 1 + \frac{1}{\gamma}\frac{p_1}{p_a}. \tag{8.61}$$

Further, eq. (8.59) can be approximated as

$$\frac{d\Omega}{dt} \approx U_n(b-a). \tag{8.62}$$

This implies that constant U_n is assumed. The following linearized equation follows from eqs. (8.58), (8.61), and (8.62):

$$\rho_a U_n(b-a) + \Omega_0 \frac{1}{\gamma}\frac{\rho_a}{p_a}\frac{dp_1}{dt} = 0. \tag{8.63}$$

Here Ω_0 is an average air cushion volume and $(b-a)$ an average length of the air cushion. We still have two unknowns, that is, U_n and p_1. However, they can be related as follows. We consider the boundary-value problem in Figure 8.32. We could say this is the same as the problem for a heaving flat plate between $x = a$ and b in combination with a free surface from $x = -\infty$ to 0, and fixed flat plates from $x = 0$ to a and from $x = b$ to ∞. We can solve this problem for unit U_n by a numerical method or conformal mapping. We will not show the details on how to solve the problem, but instead recall from Chapter 7 how added mass is defined. This means that forced oscillation of the plate with velocity U_n will cause a vertical force on the plate that can be expressed as $a_{33}dU_n/dt$, where a_{33} is the two-dimensional added mass in heave. (Note that the sign of the force is consistent with the fact that positive U_n is in the negative z-direction.) This force comes from integrating a dynamic pressure. This pressure is the same as p_1, which is approximated as a uniform pressure from the force, that is,

$$p_1 = \frac{a_{33}}{b-a}\frac{dU_n}{dt}. \tag{8.64}$$

We can express a_{33} in a nondimensional way as follows

$$\frac{a_{33}}{\rho_w(\pi/8)(b-a)^2} = K. \tag{8.65}$$

Here ρ_w means mass density of the water. Calculated values of K based on conformal mapping are presented in Figure 8.33. The added mass is

$$\begin{aligned} a_{33} = \frac{2b}{\pi}\rho_w \int_a^b dx \Big\{ & 0.5\left(\frac{a}{b} - \frac{x}{b}\right) \\ & \times \ln\left[\left(\left(\frac{x}{b}\right)^{0.5} - \left(\frac{a}{b}\right)^{0.5}\right) \Big/ \left(\left(\frac{x}{b}\right)^{0.5} + \left(\frac{a}{b}\right)^{0.5}\right)\right] \\ & + \left(\frac{x}{b}\right)^{0.5}\left(1 - \left(\frac{a}{b}\right)^{0.5}\right) + 0.5\left(1 - \frac{x}{b}\right) \\ & \times \ln\left[\left(1 + \left(\frac{x}{b}\right)^{0.5}\right) \Big/ \left(1 - \left(\frac{x}{b}\right)^{0.5}\right)\right]\Big\} \end{aligned} \tag{8.66}$$

The value when $a = 0$ is $a_{33} = \rho_w b^2 2/\pi$. When $a/b \to 1$, $a_{33} \to \infty$ because of the free surface,

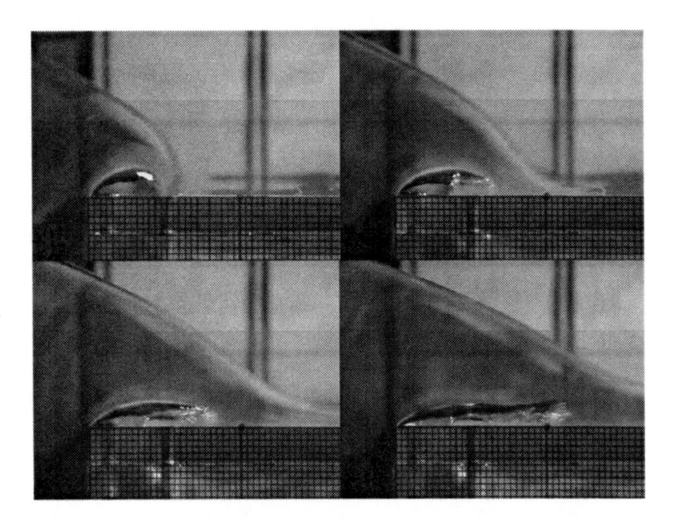

Figure 8.34. Two-dimensional water-on-deck experiments. Water impact with the deck and cavity formation during the initial stages of the water shipping (Greco 2001).

which is infinitely far away on the scale of $(b-a)$. The problem is then mathematically similar to solving the added mass problem with rigid free-surface condition. This gives infinite two-dimensional added mass in heave.

We should note that there is a conflict between assuming both U_n and p_1 to be constant between $x=a$ and b. So we must consider our analysis approximate from this point of view.

We substitute eq. (8.64) into eq. (8.63) and assume harmonic oscillations. This gives the natural frequency

$$\omega_n = \sqrt{\frac{8\gamma p_a}{\pi K \rho_w \Omega_0}}, \tag{8.67}$$

where K is defined by eq. (8.65). This equation also gives the time scaling; that is, we should present results in nondimensional form as a function of nondimensional time

$$t^* = t\sqrt{\frac{p_a}{\rho_w L^2}}, \tag{8.68}$$

where L is a length scale of the body. In other words, we must introduce the finite dimensions of the body.

Let us then study the scaling of the dynamic pressure p_1. We express then $U_n = U_{na}\cos\omega_n t$, where U_{na} is the initial normal velocity at $t=0$. This velocity is Froude scaled. By using eqs. (8.64) and (8.65), we can now write

$$\frac{p_1}{\sqrt{\rho_w g L p_a}} = -\sqrt{\frac{\pi}{8}K\gamma\left(\frac{b-a}{L}\right)^2\frac{L^2}{\Omega_0}}\frac{U_{na}}{\sqrt{gL}}\sin\omega_n t. \tag{8.69}$$

Let us then consider a model test based on Froude scaling. $U_{na}/\sqrt{gL}$, $(b-a)/L$, and Ω_0/L^2 would be the same in model and full scales. Eq. (8.69) says then that $p_1/\sqrt{\rho_w g L p_a}$ is the same in model and full scales. If we call L_m the model length and L_f the full-scale length, the pressure in full scale will be $(L_f/L_m)^{0.5}$ times the pressure in model scale. If Froude scaling of pressure had been used, the pressure in full scale would be L_f/L_m times the pressure in model scale. This means Froude scaling is clearly conservative when slamming pressures associated with air cushions are scaled. This was also documented numerically by Greco et al. (2003), who also showed that the linear behavior described in this section is appropriate in model tests. However, the oscillations of the air cushion had a strong nonlinear behavior in full scale.

Figure 8.34 shows another case in which an air cushion is created during impact. A plunging breaker hits the top of a deck structure. This can be an initial scenario for green water on deck (Barcellona et al. 2003). The results in Figure 8.34 are two-dimensional results by Greco (2001). The air cushion will, in this case, collapse into bubbles.

8.3.5 Impact of a fluid wedge and green water

Figure 8.35 shows theoretical results for slamming pressures on a rigid vertical wall due to an impacting fluid wedge with interior angle β and velocity V. The results are based on a similarity solution, which neglects gravity. This means that it does not need to be a vertical wall but can be any flat surface perpendicular to the impacting fluid wedge. When

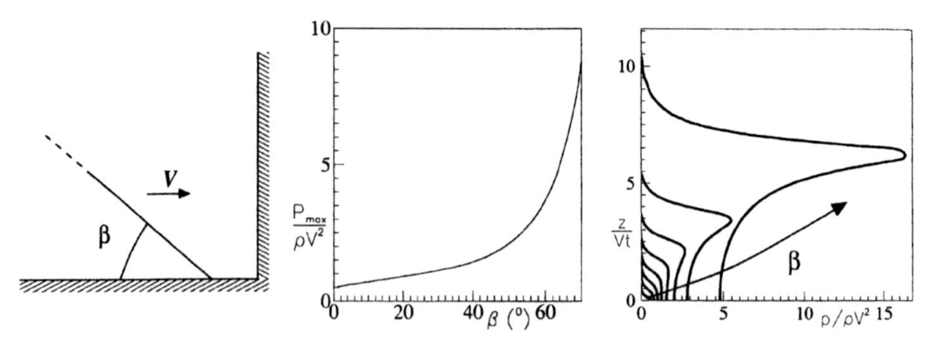

Figure 8.35. Left: sketch of the equivalent problem of a fluid (half) wedge impacting a flat wall at 90°. Center: maximum pressure on a wall due to the water impact. Right: pressure distribution along the vertical wall for $5^\circ \leq \beta \leq 75^\circ$ with increment $\Delta\beta = 10^\circ$. The results are numerically obtained by neglecting gravity and using the similarity solution by Zhang et al. (1996) (Greco 2001).

the interior angle β is close to 90°, we could obtain similar results by using a Wagner-type analysis.

The results in Figure 8.35 are of relevance, for instance, in studying the impact on a deck house of green water on deck (Greco 2001). A scenario causing green water is shown in Figure 8.36. The relative vertical motions between the ship and the waves cause a vertical wall of water around the bow. The behavior of the water later on is similar to the breaking of a dam. This causes water to flow with large velocities along the deck and to induce loading on the deck as well as on deck houses and equipment (Barcellona et al. 2003). The front of the water can locally be approximated as a fluid wedge with a small angle β. There are, of course, three-dimensional effects modifying this picture.

If the relative vertical velocity is not dominant in comparison with the relative longitudinal velocity,

Figure 8.36. Illustration of green water when the tanker *Siri* met Typhoon Judy southeast of Okinawa in 1963. The relative vertical motions between the ship and the waves caused a vertical wall of water around the bow. The behavior of the water later on was similar to the breaking of a dam. This caused water flowing with large velocity along the deck. Secondary slamming effects may occur when the water flows from the forecastle, shown on the picture, and hits the main deck (Photo: Per Meidel).

Figure 8.37. 2D and 3D experiments of green water on the deck of a stationary ship that is restrained from oscillating (Greco 2001, Barcellona et al. 2003).

the water may flow onto the deck in a manner similar to a plunging breaker (Greco 2001, see Figure 8.34). An extreme situation may be that a plunging breaker hits directly on a deck house in the forward part of the ship.

Figure 8.37 shows 2D and 3D experiments of green water on the deck of a stationary ship. The results in Figure 8.35 are directly applicable in the 2D case in the initial phase after the water has hit the vertical wall. This has been extensively studied numerically and experimentally by Greco (2001), who also described the strong interaction between the flow on the deck and exterior to the ship hull. The water in her case came initially as a plunging breaker hitting the front of the deck. The subsequent fluid motion on the deck resembled but was not equal to the flow due to the breaking of a dam.

Obviously, the results in Figure 8.35 cannot be valid for the entire time after the impact with the vertical wall, because the vertically moving fluid is influenced by gravity. The water near the wall will at some stage overturn, as illustrated in Figure 8.37. The overturning water will then impact on the underlying water, causing important pressure loading on the deck and the wall. Figure 8.37 shows the water after this impact of the overturning water. It illustrates also that the 3D flow situation is only qualitatively the same as for the 2D flow. The results in the figure are for a stationary and nonoscillating ship. However, large forward speed, wave-induced ship motions, and bow geometry typical for a high-speed vessel are expected to have a clear influence on the results.

Figure 8.38 shows experimental waterfront velocity along the deck centerline for three stationary ship models that are restrained from moving in head sea. If the initial height of the water above the deck, $H\text{-}f$, is set equal to 10 m, the waterfront velocity along the deck centerline once the flow is almost fully developed varies between 11 ms^{-1}, for $k_c a = 0.125$, and 17 ms^{-1}, for $k_c a = 0.225$. Here $k_c a$ means the incoming wave steepness. These values can be used to get a rough estimate of the maximum pressure on a vertical wall, associated with an initial water-superstructure impact. At the beginning of the impact, the waterfront velocity (impact velocity, V) and the angle β represent the impact parameters. Actually, if, as expected, β is sufficiently small (less than 40°),

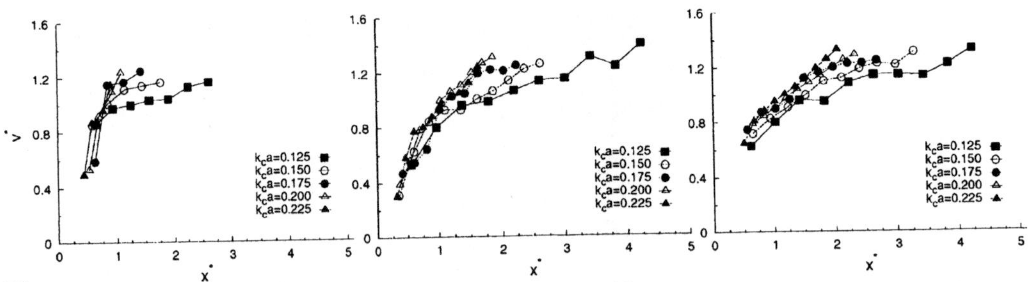

Figure 8.38. Experimental waterfront velocity $v^* = v/(g(H-f))^{0.5}$ along the deck centerline x^*-*coordinate* $= x/(H-f)$. $x = 0$ is the bow. The influence of incoming-wave steepness k_ca and bow shape are examined in head sea. Left : ESSO Osaka; center: circular bow; right: elliptical bow. The ship models are restrained from moving. H-f = initial height of the water above the deck at the bow (Barcellona et al. 2003).

the impact is dominated by the impact velocity only, and the pressure will depend on it as a square power. For $0^\circ < \beta < 40^\circ$, the maximum pressure varies between $1/2\rho V^2$ and $\approx 1.4\rho V^2$, respectively. For $k_ca = 0.225$, this means pressure of 145 kPa and 405 kPa, respectively. Both values are of concern for the superstructure, their time duration and spatial concentration being not very small relative to relevant local structural natural periods and structural dimensions. The angle β will be reduced as the water flows along the deck.

The superstructure of a high-speed vessel tends to be streamlined. This means it is more relevant to study impact against an inclined wall. Numerical results by Greco (2001) are presented in Figures 8.39 and 8.40. The initial conditions corresponding to the breaking of a dam with height h and an impacting fluid wedge with apex angle $\beta = 11^\circ$ and impacting velocity $V = 1.983\,(gh)^{0.5}$ were examined. The shallow-water solution to the dam-breaking problem differs from this (see Stoker 1958) and would give $V = 2\,(gh)^{0.5}$ and an apex angle equal to zero. This means the shallow-water solution does not solve the dam-breaking problem exactly.

Greco (2001) varied the slope of the wall α (see sketch in Figure 8.39) between 0° and 40°. The right-hand plot in Figure 8.39 shows the normal force acting on the wall for increasing values of α. In particular, as α increases, the force component decreases at a smaller rate, resulting in a weaker load for a given time. As an example, when $\alpha = 40^\circ$, at the end of the simulation, $F^{\alpha}_{n,\max}$ is about 50% of the value $F^{0}_{n,\max}$ obtained for the vertical wall ($\alpha = 0^\circ$). In general, the ratio $F^{\alpha}_{n,\max}/F^{0}_{n,\max}$ decreases almost linearly with α. The pressure values along the wall (Figure 8.40), and in particular the maximum pressure occurring at the position of the first impact, decreases as α increases. The difference among the pressure profiles reduces as time increases.

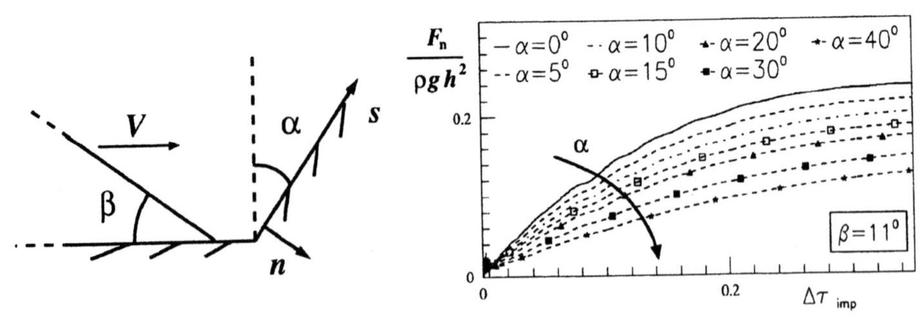

Figure 8.39. Water impacting against an inclined wall. Left: sketch of the problem. Right: time evolution of the normal force acting on the structure for increasing α. $\beta = 11^\circ$, $V = 1.983\,(gh)^{0.5}$, h = initial dam height. $\Delta\tau_{imp} = (t - t_{imp})(g/h)^{0.5}$. The results are obtained by numerically solving the exact dam-breaking problem (Greco 2001).

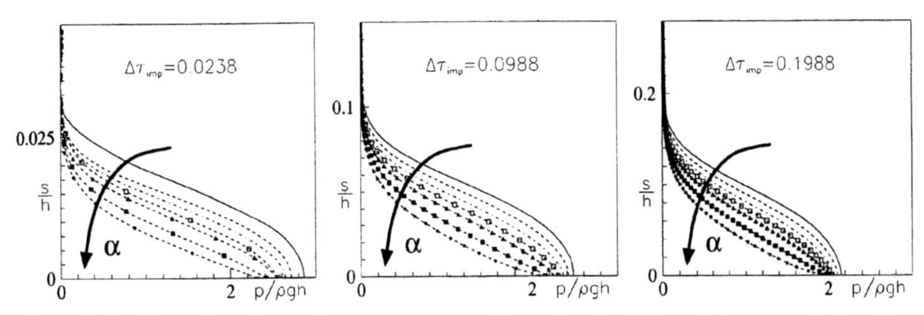

Figure 8.40. Water flow impacting a structure with angle $\beta = 11^\circ$ and impact velocity $V = 1.983\,(gh)^{0.5}$. The wall has an arbitrary slope α. Pressure (p) distributions, related to the values of α considered in Figure 8.39, are shown at three time instants after the impact. $\Delta\tau_{imp} = (t - t_{imp})(g/h)^{0.5}$, h = initial dam height, s = coordinate along the inclined wall defined in Figure 8.39 (Greco 2001).

8.4 Global wetdeck slamming effects

Slamming also causes global effects on the ship. For monohull vessels, these effects are associated with bow flare slamming, whereas catamarans and SES are dominated by wetdeck slamming effects. An SES that has suffered severe speed loss in heavy seas and in practice is off-cushion represents a risky scenario. Transient heave, pitch, and global vertical elastic vibrations are excited because of the wetdeck slamming. The dominant elastic vibrations in head sea are in terms of two-node longitudinal vertical bending. The phenomenon is called whipping and also induces global shear forces, bending moments, and stresses. Global longitudinal vertical bending is of concern for vessels of lengths larger than 50 m, but we also observe the effect on smaller vessels. Figure 8.41 shows a full-scale measurement of vertical accelerations at the bow of the 30 m–long Ulstein test catamaran in head sea conditions with significant wave height $H_{1/3} = 1.5$ m. The forward speed was 18 knots and the vessel was allowed to operate up to $H_{1/3} = 3.5$ m. Local slamming-induced bending stress in a longitudinal stiffener (located in the wetdeck at the bow area) corresponding to approximately half the yield stress was recorded at the same time. The high-frequency oscillations in Figure 8.41 have a period corresponding to the global two-node bending. Because the largest vertical accelerations in Figure 8.41 are about $2g$, a shipmaster may have avoided the situation during passenger transportation by reducing the speed and/or changing the course.

The natural period of the global two-node bending is of the order of 1 s when whipping matters. Because local hydroelastic slamming has typically a time scale of the order of 10^{-2}s, we can consider the structure locally rigid in the global structural analysis. There are also other modes to be considered in a practical evaluation. Our focus is on head sea and longitudinal vertical bending about a transverse axis.

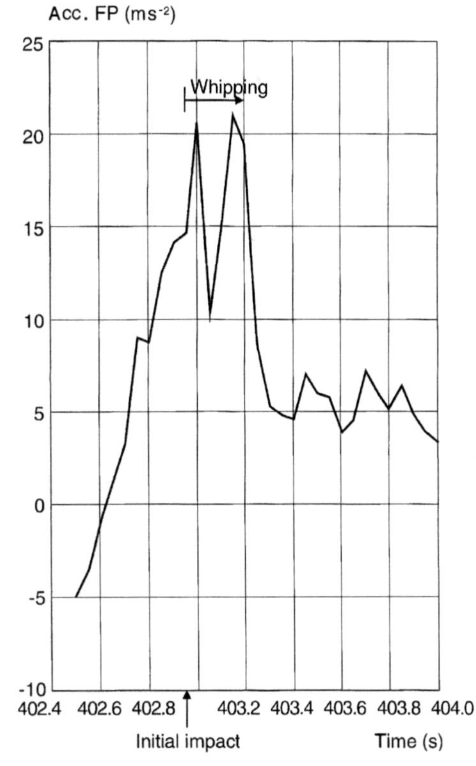

Figure 8.41. Measured vertical acceleration at the forward perpendicular (FP) of the Ulstein test catamaran; test no. 204. Significant wave height $H_{1/3} = 1.5$ m. Modal wave period $T_0 = 6$ s. Head sea. $U = 18$ knots (Faltinsen 1999).

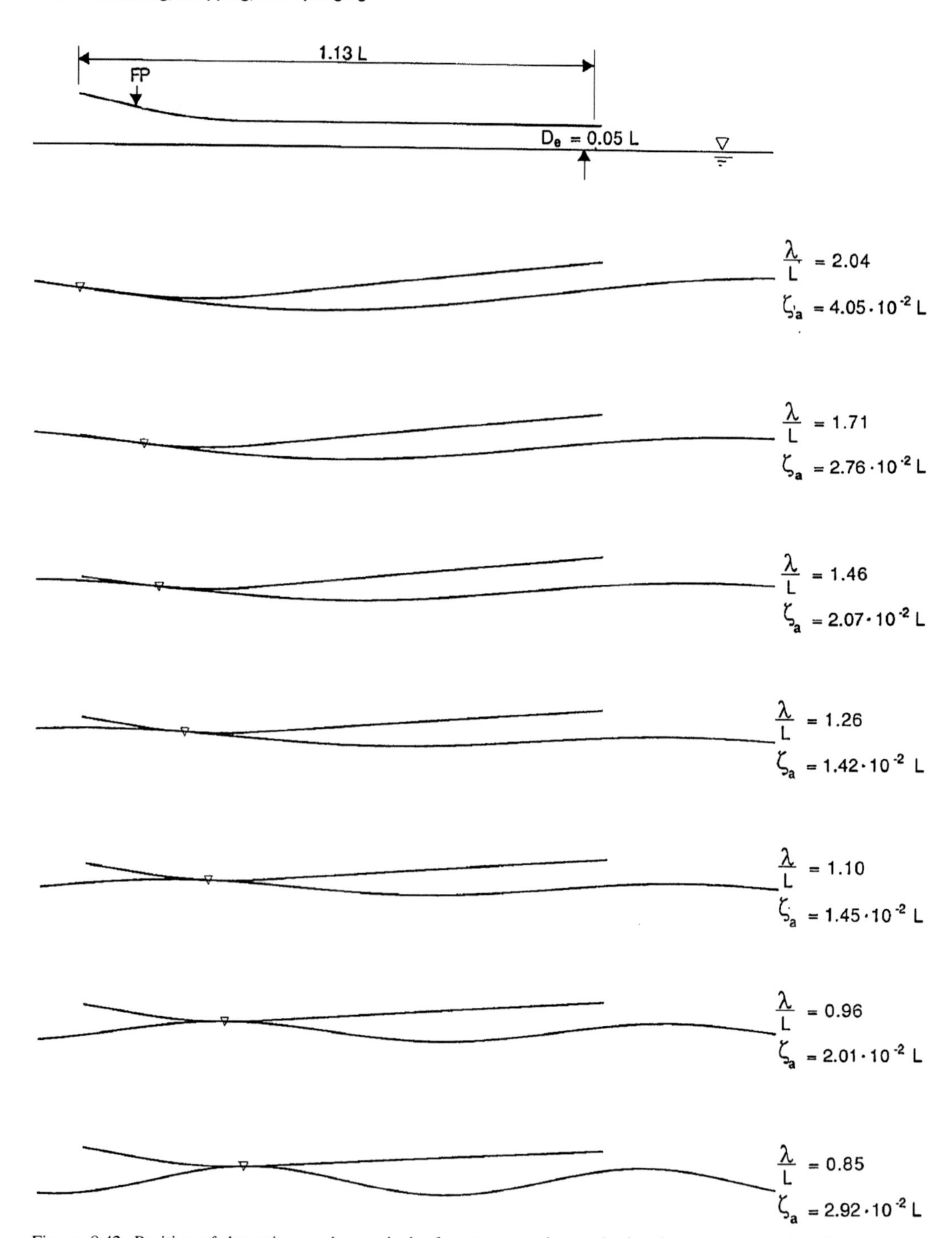

Figure 8.42. Position of slamming on the wetdeck of a catamaran in regular head sea waves as a function of wavelength λ. The figure shows a longitudinal cross section at the centerplane of the catamaran. The bow ramp is seen in the fore part. $Fn = 0.5$, $\zeta_a = \zeta_{slam}$ = lowest incident wave amplitude when slamming occurs, $L = L_{PP}$ = length between perpendiculars (Zhao and Faltinsen 1992).

It matters how the water hits the wetdeck. Figure 8.42 from Zhao and Faltinsen (1992) shows how the impact position depends on the wave period in regular head sea waves for a given catamaran and Froude number. The water always hits the forward part of the wet deck. This follows from the fact that the relative vertical motions of the catamaran are always largest in the forward part of the ship at forward speed in head sea. Interaction between the demihulls are often disregarded at high speed. However, this cannot be assumed at low speed. Generally, we cannot for all speeds and wave headings say that the water will always impact on the forward part of the wetdeck. Figure 8.42 shows that the longer the wavelengths are, the closer to the bow the initial impact occurs. The figure also presents the minimum wave amplitude ζ_a for slamming to occur for a given incident wavelength λ. This minimum wave amplitude is smallest for $\lambda/L = 1.26$ for the cases presented in Figure 8.42. The smaller the minimum wave amplitude, the larger the amplitude of the relative vertical motion divided by ζ_a. When the water does not initially hit at the end of the forward deck, the water surface has to be initially tangential to the wetdeck surface at the impact position. Let us consider a wetdeck with a plane transverse horizontal cross section, long-crested incident head waves, and a forward speed that is not small. The water surface at the initial impact position can then be approximated by the incident waves, and the flow due to slamming can be assumed two-dimensional in the longitudinal cross-sectional plane of the vessel. It can be shown by using Wagner's (1932) theory, that the initial slamming force F_3 is equal to

$$F_3 = 2\rho\pi V_R^2 RB. \tag{8.70}$$

Here V_R is the relative velocity normal to the deck surface at initial impact, R is the radius of curvature of the incident waves at the impact position, and B is the breadth of the wetdeck at the impact position. For linear incident regular waves, R can be expressed as $1/(k^2\zeta_a)$ at the wave crest. Here k and ζ_a are, respectively, wave number and wave amplitude. If V_R is proportional to ζ_a, it implies that F_3 is proportional to ζ_a.

Let us prove eq. (8.70). We then use eq. (8.43) and multiply it with B to get the vertical force, exchange V with $-V_R$, and assume constant V_R. We use eq. (8.53) for $c(t)$. Actually this was derived for a rigid body with parabolic shape hitting an initially horizontal free surface. However, following an analysis in which the impacting free surface has a parabolic shape and the body is flat gives the same expression for $c(t)$. This implies that

$$\frac{dc}{dt} = \sqrt{\frac{-V_R R}{t}}.$$

Here we should note that V_R is negative during the impact. The expression for dc/dt is initially infinite, but the product $c(dc/dt) = -2V_R R$ is initially finite. The final expression gives eq. (8.70).

If the water instead hits initially at the forward end of the deck, there will be a small angle α between the free surface and the deck surface. This implies initially zero slamming force. The increase in the slamming force on wedges is sensitive to α in a similar way as water entry forces on wedges are sensitive to the deadrise angle β for small β. If the wetdeck has a wedge-shaped transverse cross section, the slamming loads are for the same reason smaller than those for a plane horizontal wetdeck. It is also beneficial to have a bow ramp. This reduces the probability of slamming. A trim angle has a similar effect. The trim can be significantly increased (i.e., the bow rises) at Froude numbers larger than 0.35 (Molland et al. 1996). This is physically caused by the increased importance of the velocity square term in the steady Bernoulli's equation for the pressure relative to the hydrostatic pressure. Another way of saying this is that the increased trim is associated with the wavemaking of the ship in calm water. Nonlinear unsteady wave-body interaction will also cause a mean trim angle. However, Lugni et al. (2004) showed a small unsteady effect on the trim in their experimental studies. The same physical effects also cause a sinkage.

8.4.1 Water entry and exit loads

The global slamming analysis requires consideration of both the water entry and water exit phases. Because a Wagner method cannot be used during the water exit phase, we will base the slamming load analysis on a von Karman method. The following assumptions are made:

- Incident regular head sea waves act on a catamaran at forward speed
- The wetdeck has a plane horizontal transverse cross section

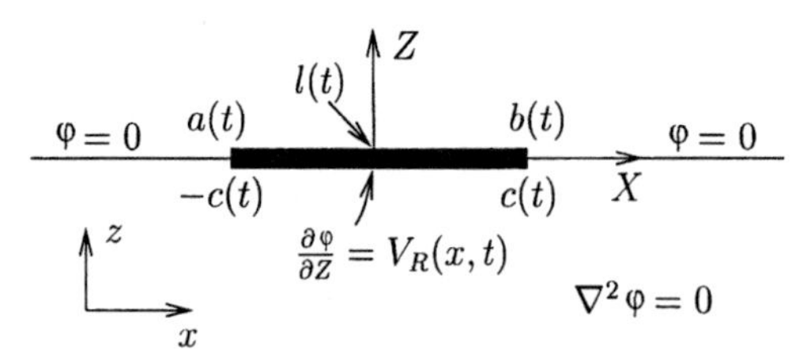

Figure 8.43. Two-dimensional boundary-value problem for velocity potential φ due to wetdeck slamming. $a(t)$, $b(t)$ and $l(t)$ are ship fixed x-coordinates. X-Z is the local 2D coordinate system on the wetted part of the deck (Ge 2002).

Because the initial impact occurs in the forward part of the wetdeck in the following analysis, we can assume that the free surface can be described by the incident waves. The wetted area with a von Karman method can then be found by examining the relative vertical displacement

$$\eta_R = \eta_B(x, t) - \zeta_a \sin(\omega_e t - kx) + h(x). \quad (8.71)$$

Here $h(x)$ is the time-independent wetdeck height above calm water and $\eta_B(x, t)$ is the vertical ship motion, which includes global elastic vibrations in addition to rigid body heave and pitch motions. The incident wave part of eq. (8.71) is consistent with eq. (7.21). If η_R is less than zero, slamming occurs.

Our assumptions imply that the flow caused by the impact can be assumed two-dimensional in the longitudinal cross-sectional planes. This would not be true if the wetdeck did not have a flat horizontal transverse cross section. Figure 8.43 illustrates the boundary-value problem that we have to solve for each time instant to find the velocity potential φ due to slamming. We note from Figure 8.43 that $2c(t)$ is the wetted length of the deck, X and Z are local coordinates with $X = x - l(t)$. Further, $b(t) - l(t) = l(t) - a(t) = c(t)$. The coordinates $x = a(t)$ and $x = b(t)$ follow by solving eq. (8.71) with $\eta_R = 0$. The free-surface condition $\varphi = 0$ is the same as the one we used earlier for slamming studies. The relative impact velocity V_R can be expressed as

$$V_R = \frac{\partial \eta_B}{\partial t} - U\left(\tau - \frac{\partial \eta_B}{\partial x}\right) - \omega_0 \zeta_a \cos(\omega_e t - kx). \quad (8.72)$$

Here $\eta_B = \eta_3 - x\eta_5$ and $\partial\eta_B/\partial x = -\eta_5$ in the case of no global elastic vibrations. The second term $U(\tau - \partial\eta_B/\partial x)$ in eq. (8.72) is the velocity component of U normal to the wetdeck. τ is the local time-averaged inclination of the wetdeck relative to the mean free surface. Positive τ and η_5 correspond to bow up. The angle τ expresses the local geometry, for instance, due to the bow ramp. It also includes the trim due to hydrostatic and steady forward speed–dependent hydrodynamic forces on the vessel in calm water. There is also a contribution due to time-averaged nonlinear hydrodynamic loads by unsteady wave-body interaction. The latter effect is normally neglected.

Because the wetted length is small relative to the incident wavelength, eq. (8.72) can be approximated on the wetted area as

$$V_R = V_1 + V_2 X. \quad (8.73)$$

This follows by keeping the constant and linearly varying terms of a Taylor expansion of V_R about $X = 0$. Because the flow associated with $V_2 X$ in eq. (8.73) is antisymmetric about $X = 0$, V_2 does not contribute to the vertical force. This means we can multiply eq. (8.43) by the breadth B of the wetdeck to get the vertical water entry and water exit forces on the wetdeck. We must just remember that in eq. (8.43), V is positive downward, that is, $V = -V_1$, and that during the water exit phase, we include only the added mass force.

Froude-Kriloff and hydrostatic forces on the wetdeck will also contribute, but they are generally smaller than the slamming and added mass forces previously described. It is a good approximation to consider incident linear waves, which we already have done when evaluating relative vertical motions and velocities. A clear understanding of the pressure distribution in incident waves is needed. This will be discussed in an Earth-fixed coordinate system. We must consider both $-\rho\partial\varphi/\partial t$ and $-\rho g z$ in the pressure calculation. When we derive the linear theory, we assume implicitly that φ is constant from the mean free-surface level to the instantaneous free-surface level. It is the hydrostatic component that makes the pressure atmospheric at the free surface. This is consistent with the dynamic free-surface condition $g\zeta + \partial\varphi/\partial t = 0$ on $z = 0$.

We have illustrated in Figure 3.5 how the two terms $-\rho\partial\varphi/\partial t$ and $-\rho g z$ give the total pressure variation with depth. We should, of course, note that $-\rho g z$ also includes the conventional hydrostatic pressure for $z \le 0$, but we need $-\rho g z$ to describe the dynamic pressure distribution in the free-surface zone. What this illustrates is that

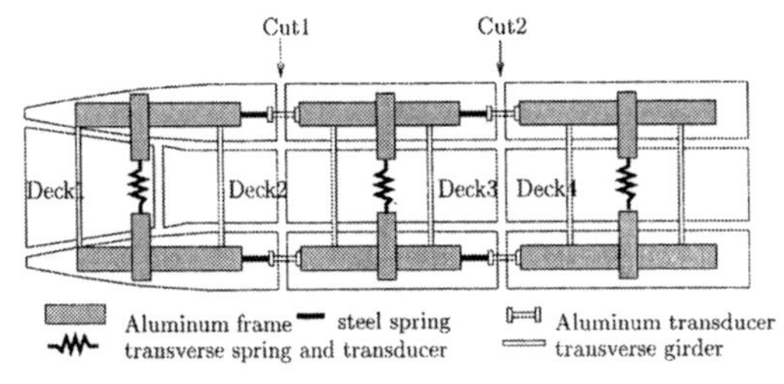

Figure 8.44. Outline of the experimental hull arrangements (top view) (Ge 2002).

pressure in the vicinity of the instantaneous free-surface level ζ behaves as $-\rho g(z-\zeta)+p_a$. We can use this information to evaluate the hydrodynamic pressure at the wetdeck, that is, at $z = \eta_B + h(x)$. Here η_B is the vertical ship motion and $h(x)$ is the time-independent wetdeck height above calm water. Assuming that the incident free-surface $\zeta = \zeta_a \sin(\omega_e t - kx)$ is higher than $\eta_B + h(x)$, we get the following "buoyancy" force:

$$F_{3,buoy} = \rho g B \int_{a(t)}^{b(t)} [\zeta_a \sin(\omega_e t - kx) - \eta_B - h(x)]\, dx. \tag{8.74}$$

8.4.2 Three-body model

Ge (2002) (see also Ge et al. 2005) studied numerically and experimentally wetdeck slamming-induced global loads on a catamaran in head sea deep-water regular waves. The vessel model is shown in Figure 8.44. The overall length is 4.1 m. Each side hull consists of three rigid sections. The hull sections are connected by steel springs and aluminum transducers longitudinally and transversely. These elastic connections then model the global elastic behavior of a catamaran. However, this can only be approximate. The wetdeck consists of four rigid flat sections. Deck 1, where wetdeck slamming mainly occurs, has a ramp angle of 3.72° with minimum and maximum heights of 0.34 m and 0.39 m, respectively, from the baseline. The draft at zero speed is 0.225 m and 0.220 m at FP and AP, respectively.

The catamaran in Figure 8.44 was theoretically modeled as three rigid bodies with longitudinal connections of elastic beams (Figure 8.45). The transverse flexibility between the two hulls was not accounted for. A reason for neglecting the transverse connecting springs and beams is that long-crested head sea waves are considered. The split moment (transverse vertical bending moment in Figure 7.48) will be small, but not zero because of different transverse centers for the hydrodynamic and structural inertia loads. The longitudinal vertical bending modes will then be dominant relative to transverse bending and torsional modes. Each rigid ship segment has two degrees of freedom, namely heave and pitch. So in total, there are six degrees of freedom in this system, and each degree of motion is referred to its local COG. The connecting beams are denoted as AB and CD, respectively.

The slamming-induced flow was described by a von Karman method as described in the previous section. Froude-Kriloff and hydrostatic forces were included when wetdeck loads were calculated. The hydrodynamic loads on the side hulls were described by a modification of the linear frequency-domain strip theory by Salvesen et al. (1970). Hydrodynamic hull interaction was neglected. The general equation system for the motion of the hull segments can be expressed as

$$\mathbf{M}_{gen}\ddot{\mathbf{r}} + \mathbf{B}_{gen}\dot{\mathbf{r}} + \mathbf{K}_{gen}\mathbf{r} = \mathbf{F}_{gen}(\mathbf{r}, \dot{\mathbf{r}}, \ddot{\mathbf{r}}, t). \tag{8.75}$$

Here $\mathbf{r}$ is the displacement matrix of this six-degrees-of-freedom system, containing the heave and pitch for each segment. $\mathbf{M}_{gen}$ and $\mathbf{B}_{gen}$ are the mass and damping matrices. Here mass refers to both the segment mass and the added mass. $\mathbf{K}_{gen}$ is the restoring (stiffness) matrix, including the hydrostatic restoring terms from the ship segments as well as the coupling terms from the spring beams. $\mathbf{F}_{gen}$ constitutes the forces due to wetdeck slamming and linear wave excitation loads on the side hulls.

We now describe how the elastic connections between the three hulls were accounted for and start with the static beam equation with zero loading, that is,

$$EI\frac{d^4 w}{dx^4} = 0, \tag{8.76}$$

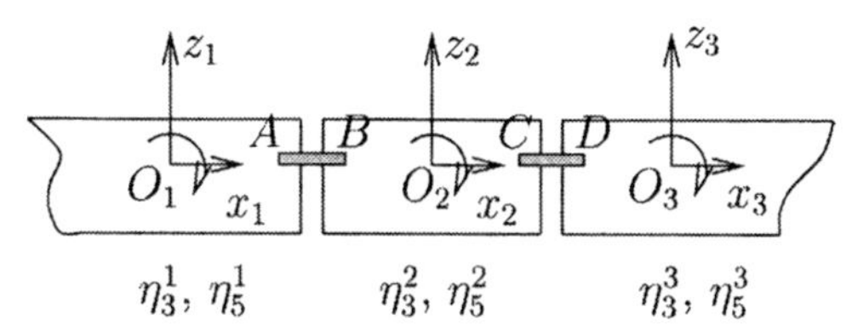

Figure 8.45. Degrees of freedom of segmented model (side view) (Ge 2002).

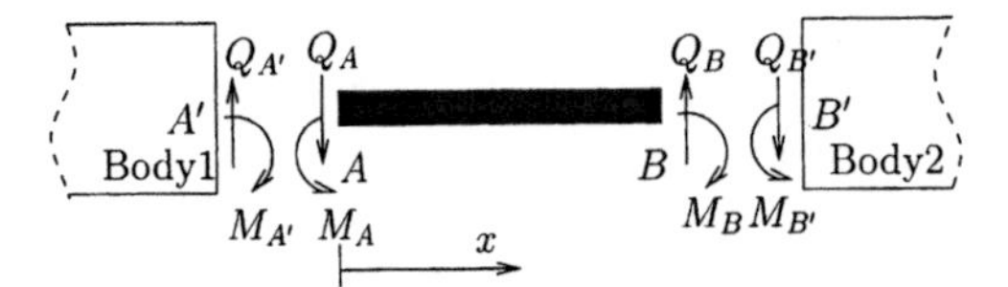

Figure 8.46. Elastic beam connection between two adjacent rigid-body segments (side view) (Ge 2002).

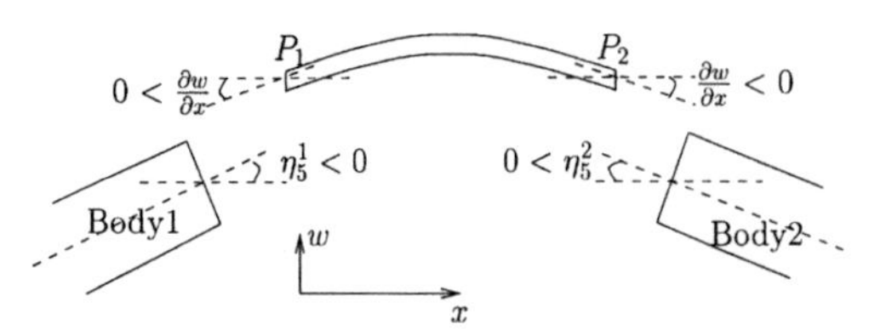

Figure 8.47. Rotational sign illustration for beam and adjacent bodies (Ge 2002).

where EI is the bending stiffness of a beam connection and w is the elastic deflection of this beam. The connecting beam AB between body 1 and body 2 is used to illustrate the procedure (Figure 8.46). $x = 0$ and L correspond to, respectively, point A and point B. Integration of eq. (8.76) gives

$$w(x) = \frac{1}{EI}\left(\frac{1}{6}ax^3 + \frac{1}{2}bx^2 + cx + d\right). \quad (8.77)$$

The boundary conditions of the beam require that the vertical and rotational displacements at the ends of A and B match those at the adjacent ends of body 1 and body 2. Hence,

$$\begin{aligned} w|_A &= \eta_3^1 - \overline{O_1A}\eta_5^1 \\ w|_B &= \eta_3^2 + \overline{O_2B}\eta_5^2 \\ \left.\frac{\partial w}{\partial x}\right|_A &= -\eta_5^1 \\ \left.\frac{\partial w}{\partial x}\right|_B &= -\eta_5^2 \end{aligned} \quad (8.78)$$

where η_3^i and η_5^i are heave at COG and pitch of body-number i. Further, $\overline{O_iX}$ denotes the distance from the beam end X to the local COG of the adjacent body-number i. One should notice the different sign definition between $\partial w/\partial x$ and η_5. This means $\partial w/\partial x = -\eta_5$, as illustrated in Figure 8.47. The coefficients in eq. (8.77) can then be expressed in terms of the six degrees of freedom of the three-body system as

$$\begin{pmatrix} a \\ b \\ c \\ d \end{pmatrix} = \frac{EI}{L^3}\begin{bmatrix} 12 & -6(L+2\overline{O_1A}) & -12 & -6(L+2\overline{O_2B}) \\ -6L & 2(2L+3\overline{O_1A})L & 6L & 2(L+3\overline{O_2B})L \\ 0 & -L^3 & 0 & 0 \\ L^3 & -\overline{O_1A}L^3 & 0 & 0 \end{bmatrix}\begin{pmatrix} \eta_3^1 \\ \eta_5^1 \\ \eta_3^2 \\ \eta_5^2 \end{pmatrix}. \quad (8.79)$$

The longitudinal distribution of vertical shear force $Q(x)$ and bending moment $M(x)$ at the right-hand side of the beam element, Figure 8.46, can be expressed as

$$\begin{aligned} Q(x) &= -EI\tfrac{\partial^3 w}{\partial x^3} = -a \\ M(x) &= -EI\tfrac{\partial^2 w}{\partial x^2} = -ax - b \end{aligned} \quad (8.80)$$

The sign definition is given in Figure 8.46. The constants a and b can now be expressed in terms of η_j^i by means of eq. (8.79). By evaluating eq. (8.80) at $x = 0$ and L, we find the shear forces Q_A and Q_B and the bending moments M_A and M_B at A and B. A similar formulation can be obtained for the second connecting beam between body 2 and body 3 with ends named C and D, respectively. The loads acting on the three rigid bodies due to the connecting beams can then be expressed as

$$\begin{pmatrix} F_3^1 \\ F_5^1 \\ F_3^2 \\ F_5^2 \\ F_3^3 \\ F_5^3 \end{pmatrix} = \begin{pmatrix} Q_A \\ M_A - Q_A\overline{O_1A} \\ Q_C - Q_B \\ M_C - M_B - Q_B\overline{O_2B} - Q_C\overline{O_2C} \\ -Q_D \\ -M_D - Q_D\overline{O_3D} \end{pmatrix} = \mathbf{kr}. \quad (8.81)$$

Here F_i^j with $i = 3{,}5$ represent heave force and pitch moment, respectively, and $j = 1{,}2{,}3$ mean body 1, body 2, and body 3. The vector $\mathbf{r}$ is $[\eta_3^1\ \eta_5^1\ \eta_3^2\ \eta_5^2\ \eta_3^3\ \eta_5^3]^T$, where the superscript T indicates matrix transposition. The matrix $\mathbf{k}$ is symmetric and is part of the restoring matrix in eq. (8.75).

There are six degrees of freedom and thus six pairs of eigenmodes and frequencies. The undamped natural frequencies and eigenmodes are obtained by setting $F_{gen} = 0$ and $B_{gen} = 0$ in eq. (8.75) and assuming harmonic time dependence $\exp(i\omega t)$. The two lowest modes with lowest natural frequencies are the coupled heave and pitch modes for the whole catamaran. This means modes that are very close to rigid-body modes for the whole catamaran. The important stiffnesses of these modes are the result of the hydrostatic restoring coefficients. The third and fourth modes

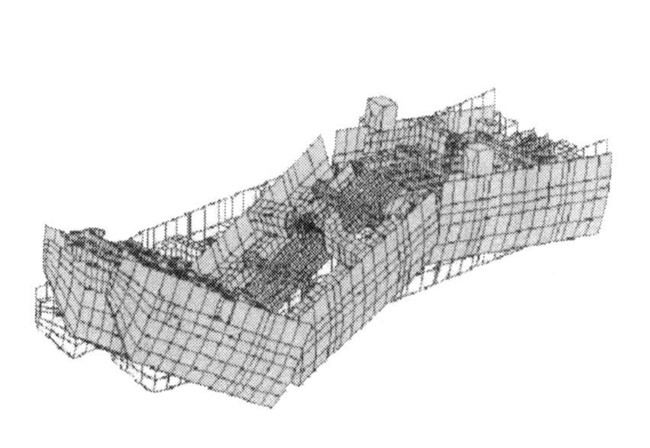
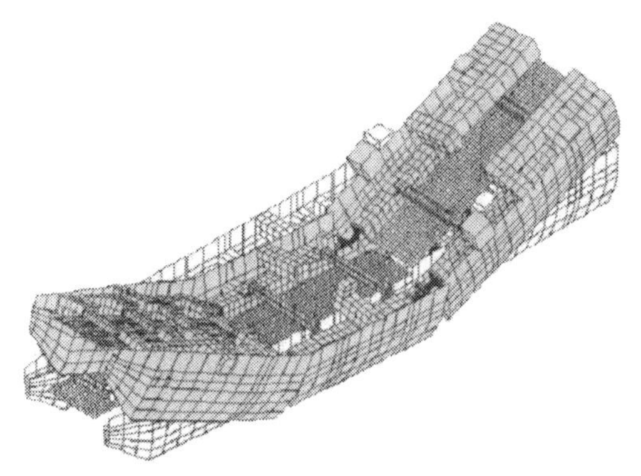

Three-noded vertical deflections Two-noded vertical deflections

Figure 8.48. Calculated shapes of eigenmodes for the three-body model shown in Figure 8.44 (Økland 2002).

are the two-node and three-node bending modes in the longitudinal vertical plane and are illustrated in Figure 8.48. The illustration is based on a finite-element model, which is really not necessary for finding the required modes for a segmented model such as this. However, a finite-element model is needed to find the modes for a real ship. The fifth and sixth modes (shear vibration modes) have very high natural frequencies relative to the other modes and are in reality highly structurally damped. They are not physically representative for the shear deformation of a true ship. This structural damping is not included in eq. (8.75). The experiments did not show any evidence of the fifth and sixth modes. The effects of them were in the theoretical model eliminated by using the modal decomposition method.

Figure 8.49 presents steady-state experimental and numerical vertical shear force (VSF) and vertical bending moment (VBM) at cut 1 (of Figure 8.44) in regular head sea waves for the most severe slamming case obtained for the model. The dominant contributions are the result of the two-node bending mode, but there are also noticeable rigid-body effects. The Froude number is $Fn = 0.29$, the wave period is $T = 1.8$ s, and the incident wave amplitude is 0.041 m. It is referred to as case 1114 in Ge's analysis. The results in Figure 8.49 include experimental error estimates (Table 8.4). The wave amplitude error is caused by

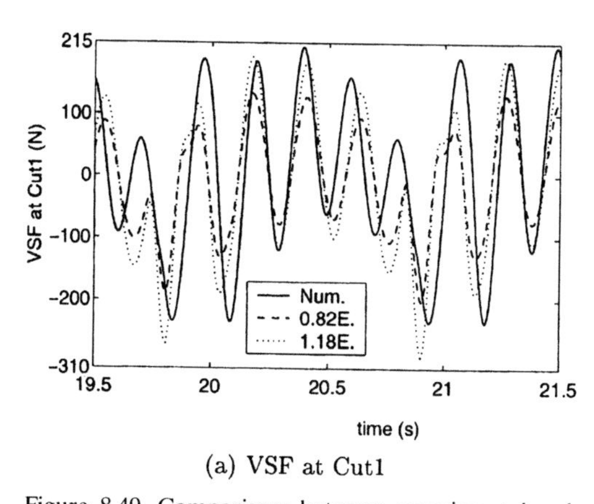

(a) VSF at Cut1

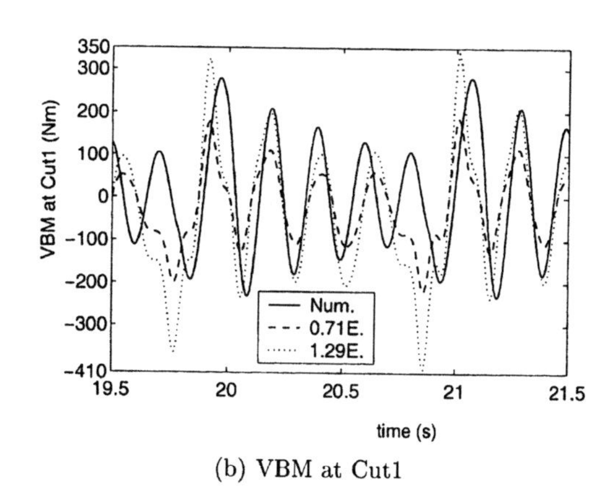

(b) VBM at Cut1

Figure 8.49. Comparisons between experimental and numerical values of vertical shear force (VSF) and vertical bending moment (VBM) at cut 1 (see Figure 8.44) in regular head sea waves for case 1114; The range of experimental data is based on the relative error in Table 8.4; E means that the experimental data are obtained by adding the values of the starboard and port sides. Values that are fractions of E are presented to give experimental error bands. Only three modes (heave, pitch, and two-node bending) were included (Ge 2002).

Table 8.4. *Relative error in global loads due to individual error sources as well as combined experimental relative error due to these error sources; case 1114 (Ge 2002)*

Error source	VSF cut 1 (N)	VBM cut 1 (Nm)	VSF cut 2 (N)	VBM cut 2 (Nm)
Speed	0.033	0.032	0.029	0.030
Wave amplitude	0.055	0.102	0.087	0.057
Seiching (sloshing)	≈0.0	≈0.0	≈0.0	≈0.0
Wave measurement	0.055	0.102	0.087	0.057
Roll, yaw, and sway	0.091	0.076	0.123	0.055
Sinkage	0.019	0.037	0.029	0.020
Trim	0.128	0.228	0.189	0.137
COMBINED	0.179	0.285	0.260	0.172

the change in the incident wave amplitude in the model basin along the sailing track of the model. Errors due to lateral motions were caused by the autopilot system and an unintended asymmetry in the mass distribution about the centerplane. The error due to trim is large and a consequence of the fact that the trim was not properly measured at forward speed.

Figure 8.50 shows fair agreement between the experimental and numerical slamming forces on deck 1 for case 1114. Because mass inertia force on deck 1 due to the vessel motions is included, there is also a non-zero force after the deck wetting has finished. However, the magnitude is not large relative to maximum slamming force. The wetdeck force is mainly upward during the entry phase, whereas it is mainly negative during the exit phase. The maximum value and the absolute value of the minimum wetdeck force have comparable magnitudes. The reason for the large negative force during the water exit is the added mass force. We can understand this added mass force from Figure 8.51, which shows the calculated wetted length and relative impact acceleration as a function of time. The water entry phase ends when the increase in the wetted length has stopped. Because the relative impact acceleration a_R is positive and the added mass force is $-\rho 0.5\pi c^2 B a_R$, we see that the force is negative. Because the wetted length, $2c$, is

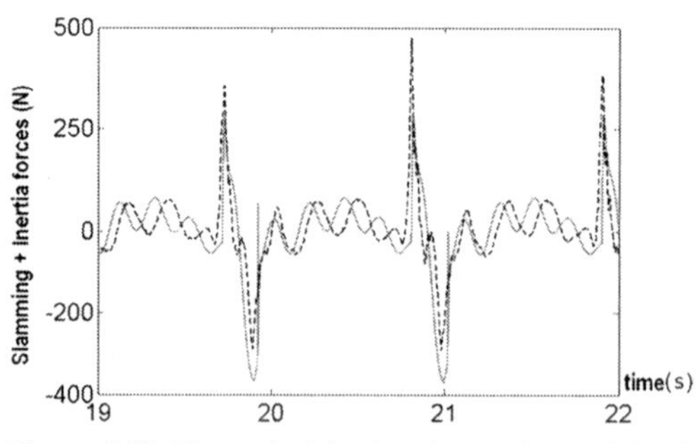

Figure 8.50. Theoretical (___) and experimental (-----) predictions of slamming force on deck 1 (see Figure 8.44) for case 1114; structural inertia force of deck 1 is included. Theoretical results are 0.07 s shifted left compared with the results in Figure 8.49. Only three modes (heave, pitch, and two-node bending) were included (Ge 2002).

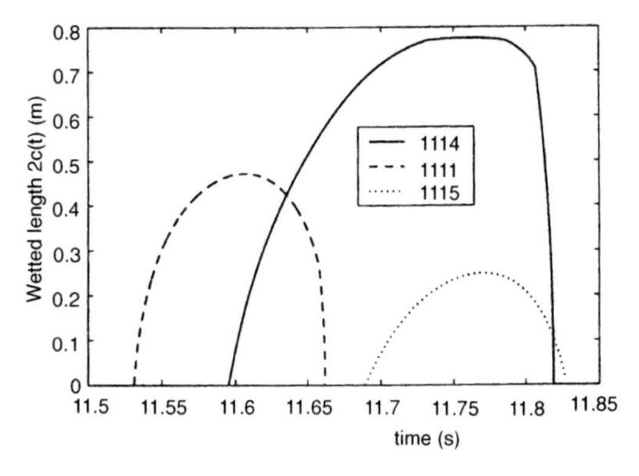

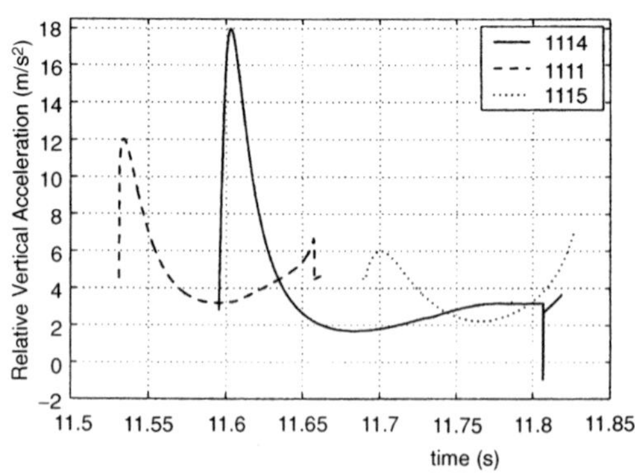

Figure 8.51. Calculated wetted length and relative impact acceleration at the midpoint of the wetted length for cases 1111, 1114, and 1115 (Ge 2002).

maximum when the water entry phase ends and c keeps a high value during a large part of the subsequent water exit phase, the added mass force is then largest.

Both the water entry and water exit phases are important for the global response. Ge (2002) documented by using alternative slamming models, such as the Wagner method and a Kutta condition method, that the global loads are not sensitive to the wetdeck load model. The Wagner model can be used only during the water entry phase and was therefore combined with a von Karman method during the water exit phase. The Kutta condition model assumes a smooth detachment of the flow at the aft intersection line between the free surface and the wetdeck. However, there is spray occurring at the forward intersection line. The Wagner and von Karman methods will lead to spray at both intersection lines. Because we consider an outer flow domain solution, the spray is not explicitly dealt with.

The theoretical duration T_d of the sum of the water entry and exit phase was 0.22 s for case 1114. The ratios between T_d and the natural periods T_N for the two coupled global heave and pitch modes were 0.22 and 0.25, whereas T_d/T_N was 1.1 for two-node bending. The maximum response is proportional to the force impulse if $T_d/T_N < \approx 0.25$ for a single excitation event (Clough and Penzien 1993). Because the water entry and exit phases give strongly canceling contributions to the force impulse, this gives one reason why the response due to two-node bending dominates over the response due to global heave and pitch modes.

The studies showed that the elastic vibrations gave significant contributions to the relative motions, impact velocities, and accelerations. The vibrations are more important for velocities than for motions and more important for accelerations than for velocities. This can be understood by first approximating the contribution to relative motion from the elastic vibrations as $A \sin(\omega_n t + \varepsilon)$. Here ω_n is the natural frequency for two-node vibration. The contribution to relative impact velocity becomes then $\omega_n A \cos(\omega_n t + \varepsilon)$. Because ω_n is clearly higher than the frequency of encounter with which the rigid-body response oscillates, we understand why the contributions from elastic vibrations are more important for velocities than for motions. A similar argument may be followed when comparing accelerations and velocities. The consequence of this is that we cannot consider the wetdeck slamming loads on a rigid ship and then later do an elastic response analysis. This has consequences both for the numerical analysis and the experiments.

In order to explain the differences between experimental and theoretical results in Figure 8.49, one must, for instance, investigate the importance of hull interaction. Eq. (7.58) shows that waves generated by one of the hulls will be incident to the other hull at a distance of $0.66L_{PP}$ from AP. However, using hull interaction in a strip theory model means that hull interaction must be considered over the whole length of the catamaran. This can give worse results than neglecting hull interaction. If we use eq. (7.64) for the piston mode resonance frequency for the midships section, as Ronæss (2002) did, we will find that the test case 1114 is sufficiently far away from piston mode resonance between the two hulls.

Because we have several frequencies, namely frequency of encounter and the natural frequencies of the elastic hull, Figure 8.49 illustrates that a time-domain model should be used instead of a frequency-domain model for the side hull hydrodynamics. Both the frequency of encounter and the natural frequency of two-node bending matter. The theoretical and experimental two-node bending modes show different oscillation periods. A better prediction of the oscillation period of the two-node bending mode is obtained by calculating the added mass of the side hulls at infinite frequency instead of frequency of encounter. 3D effects might also matter.

Nonlinear side hull loads should also have been considered. This would, for instance, alter the relative vertical motions and velocities in the impact region.

8.5 Global hydroelastic effects on monohulls

We will use the beam equation to describe the global hydroelastic effects on monohulls. There exist both the Timoshenko and Euler beam models. The Timoshenko model accounts for the shear deformation and rotational inertia, but is more complicated than the Euler beam model and does not predict much difference when it comes to bending moments. Alternatively, one can use finite element beam formulation. In the stiffness matrix of each beam element, the effect of shear

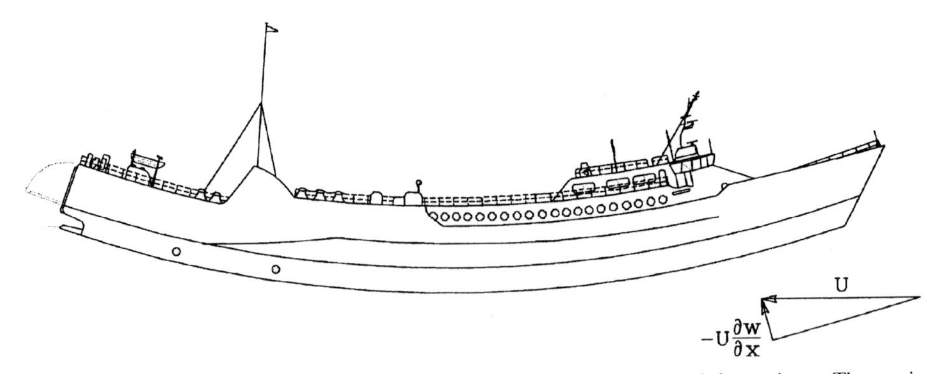

Figure 8.52. Exaggerated drawing of a ship vibrating with vertical two-node deformation w. The x-axis is in the direction of the inflow velocity U. U = ship speed.

deformation may be included without difficulties. The effect of shear deformation should be included, especially when higher modes are important.

We will use the Euler beam model, which can be expressed as

$$m(x)\frac{\partial^2 w}{\partial t^2} + \frac{\partial^2}{\partial x^2}\left[EI(x)\frac{\partial^2 w}{\partial x^2}\right] = f_3(x, t). \tag{8.82}$$

This equation assumes that the deformation is small or rather that $\partial w/\partial x \ll 1$, which should be sufficient for the bending of ship hull girders. Here x is the longitudinal coordinate of the ship, w is vertical deflection, $m(x)$ is the body mass per unit length, and $EI(x)$ is the bending stiffness. Further, f_3 is time-dependent vertical hydrodynamic force per unit length. This implies that we exclude static forces due to weight, mg, and buoyancy forces. This can be done by a separate analysis by also including steady hydrodynamic forces, which are important for high-speed vessels. Eq. (8.82) requires end conditions and initial conditions. The end conditions are zero shear force and bending moment at the forward and aft ends of the ship. By using the fact that the shear force and the bending moment are proportional to, respectively, $\partial^3 w/\partial x^3$ and $\partial^2 w/\partial x^2$ (see eq. (8.80)), we get that

$$\frac{\partial^3 w}{\partial x^3} = 0 \quad \text{and} \quad \frac{\partial^2 w}{\partial x^2} = 0 \text{ at the ends of the ship.} \tag{8.83}$$

We will first neglect excitation and express the contributions to f_3 due to linear hull vibrations. This means that we consider added mass, damping, and restoring loads in a similar, but generalized, way to that described for a rigid body. Because we are interested in oscillations with clearly higher frequencies than typical frequencies of encounter due to incident waves, it is appropriate to use the free-surface condition $\varphi = 0$ on the mean free surface $z = 0$. Here φ is the velocity potential due to the ship's vibrations. This implies that no waves are generated due to the vibrating ship. Because the ship can be assumed slender, φ satisfies the 2D Laplace equation in the transverse cross-sectional plane of the ship. Then we need body boundary conditions. As before, we use a coordinate system translating with the forward speed U of the ship (Figure 8.52). The ship speed appears in this coordinate system as a steady flow with velocity U in the x-direction. Let us consider a cross section of the vibrating ship. The vibrations cause a local angle $\partial w/\partial x$ of the ship relative to the x-axis. This angle implies that the steady flow with velocity U along the x-axis has a velocity component $-U\partial w/\partial x$ in the cross-sectional plane of the vibrating ship (Figure 8.52). In order to satisfy no flow through the hull surface, the ship must counteract this component of the incident flow. (We should note that $\partial w/\partial x$ is negative, as presented in Figure 8.52). In addition, we must account for the vibrating velocity $\partial w/\partial t$ in formulating that there is no flow through the hull surface. This gives the following linear body boundary condition

$$\frac{\partial \varphi}{\partial n} = n_3\left(\frac{\partial w}{\partial t} + U\frac{\partial w}{\partial x}\right) \quad \text{on } C(x). \tag{8.84}$$

Here $C(x)$ is the mean submerged cross-sectional curve of the hull surface. Further, $\mathbf{n} = (n_1, n_2, n_3)$ is, as usual, the normal vector to the hull surface with positive direction into the fluid. We have implicitly assumed a slender ship in formulating eq. (8.84). This means $n_1 \ll n_2$ and n_3, and $\partial/\partial n \approx n_2\partial/\partial y + n_3\partial/\partial z$.

A normalized velocity potential φ_3 is now introduced by

$$\varphi = \varphi_3\left(\frac{\partial w}{\partial t} + U\frac{\partial w}{\partial x}\right). \quad (8.85)$$

This means φ_3 is the velocity potential due to forced heave with unit velocity. The next step is to calculate the linear hydrodynamic pressure p on the hull. The effect of hydrostatic pressure is left out for the time being. We can write (see eq. (3.6))

$$p = -\rho\frac{\partial \varphi}{\partial t} - \rho U\frac{\partial \varphi}{\partial x}. \quad (8.86)$$

The interaction with the local steady flow has been neglected in formulating eqs. (8.84) and (8.86). The 2D vertical force f_3^{HD} on the hull due to the dynamic pressure, p given by eq. (8.86), induced by the hull vibrations can now be expressed as

$$f_3^{HD} = \rho \int\limits_{C(x)} n_3\left(\frac{\partial}{\partial t} + U\frac{\partial}{\partial x}\right) \times \left[\varphi_3\left(\frac{\partial w}{\partial t} + U\frac{\partial w}{\partial x}\right)\right] ds.$$

Interchanging integration and differentiation and using eq. (7.39) gives

$$f_3^{HD} = -\left(\frac{\partial}{\partial t} + U\frac{\partial}{\partial x}\right)\left[a_{33}\left(\frac{\partial w}{\partial t} + U\frac{\partial w}{\partial x}\right)\right], \quad (8.87)$$

where a_{33} is the 2D infinite-frequency added mass in heave. This equation has been derived by Lighthill (1960) in analyzing the swimming motion of a slender fish and by Newman (1977) in ship maneuvering studies. However, because ship maneuvering analysis uses a ship-fixed coordinate system, the velocity component $U\partial w/\partial x$ does not appear (see section 10.3).

Introducing the change of buoyancy due to the beam deflection, we get the following version of eq. (8.82):

$$(m + a_{33})\frac{\partial^2 w}{\partial t^2} + 2a_{33}U\frac{\partial^2 w}{\partial x \partial t} + U\frac{da_{33}}{dx}\frac{\partial w}{\partial t} + U^2\frac{\partial}{\partial x}\left(a_{33}\frac{\partial w}{\partial x}\right) + \rho g b w + \frac{\partial^2}{\partial^2 x}\left(EI\frac{\partial^2 w}{\partial x^2}\right) = f_3^{exc}. \quad (8.88)$$

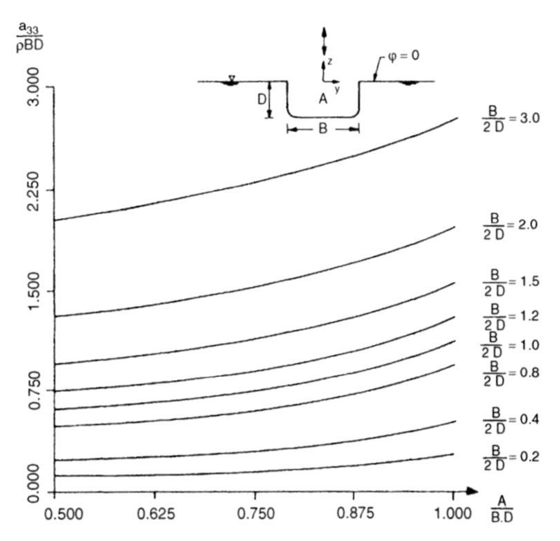

Figure 8.53. Two-dimensional added mass in heave a_{33} for Lewis form sections when the frequency of oscillation $\omega \to \infty$. Infinite water depth. A = cross-sectional area, B = beam, D = draft, φ = velocity potential (Faltinsen 1990).

Here b means sectional beam and f_3^{exc} is the hydrodynamic excitation load per unit length. The Lewis form technique is a simple way to estimate a_{33}. The expression for infinite frequency is

$$a_{33} = \rho\, 0.5\pi((a + aa_1)^2 + 3(aa_3)^2), \quad (8.89)$$

where

$$aa_1 = 0.5\,(0.5b - d)$$
$$aa_3 = -0.25(0.5b + d) + 0.25\sqrt{(0.5b + d)^2 - 8\,(2A/\pi - 0.5bd)}$$
$$a = 0.5(0.5b + d) - aa_3$$

Here A is the submerged cross-sectional area and d is the sectional draft. In order for a Lewis form to exist, it is necessary for $C_B = A/(bd)$ to be less than $\pi(0.5b/d + 2d/b + 10)/32$ and for C_B to be larger than $3\pi(2 - 0.5b/d)/32$ and $3\pi(2 - 2d/b)/32$ for, respectively, $0.5b/d \leq 1$ and $0.5b/d > 1$ (von Kerczek and Tuck 1969). Added mass results based on eq. (8.89) are presented in Figure 8.53. For cross sections with sharp corners – for instance, rectangular sections – the results are only approximate. Exact solutions for rectangular sections have been presented by Riabouchinski (1920). Barringer (1998) presented a curve fit to those results over the range $0 < \frac{2d}{b} < 2$. His formula is

$$\frac{a_{33}}{\rho\pi (b/2)^2} = 0.505589 + 0.26405 \left(\frac{2d}{b}\right)^{1/2} - 0.0251687 \frac{2d}{b} + 0.00104839 \left(\frac{2d}{b}\right)^{3/2} - 0.000014487 \left(\frac{2d}{b}\right)^2. \quad (8.90)$$

8.5.1 Special case: Rigid body

The beam equation is also applicable to high-frequency rigid-body oscillations due to heave and pitch. We can use the equation to derive global added mass and damping coefficients A_{ij} and B_{ij} for heave and pitch.

a) Forced heave $w = \eta_3$

By excluding the hydrostatic term, the hydrodynamic terms in eq. (8.88) are equal to

$$a_{33}\frac{d^2\eta_3}{dt^2} + U\frac{da_{33}}{dx}\frac{d\eta_3}{dt}. \quad (8.91)$$

We integrate this over the ship length L, assume the flow separates from the transom stern, and get by following the definition of added mass and damping

$$A_{33} = \int_L a_{33}\, dx \quad (8.92)$$

$$B_{33} = Ua_{33}(x_T). \quad (8.93)$$

Here x_T is the x-coordinate of the transom stern. The expression for B_{33} is the same as the one we have already derived in eq. (7.73).

We now take the pitch moment of eq. (8.91), that is, we multiply by $-x$ and integrate over the ship length. We then get

$$A_{53} = -\int_L x a_{33}\, dx \quad (8.94)$$

$$B_{53} = UA_{33} - Ux_T a_{33}(x_T). \quad (8.95)$$

B_{53} is a consequence of integration by parts. The term UA_{33} in B_{53} is associated with the Munk moment $U^2 A_{33}(\frac{d\eta_3/dt}{U})$ (see p. 197 in Faltinsen 1990).

b) Forced pitch $w = -x\eta_5$

By excluding the hydrostatic term, the hydrodynamic terms in eq. (8.88) are equal to

$$-xa_{33}\frac{d^2\eta_5}{dt^2} - 2a_{33}U\frac{d\eta_5}{dt} - U\frac{da_{33}}{dx}x\frac{d\eta_5}{dt} - U^2\frac{da_{33}}{dx}\eta_5. \quad (8.96)$$

By integrating this over the length of the ship, we get A_{35} and B_{35}. We should note that the last term is proportional to η_5, that is, it is a restoring term. We can always switch between including it as a restoring term or an added mass term. This is simply done by noting that $d^2\eta_5/dt^2 = -\omega^2\eta_5$. Because we have decided only to have hydrostatic effects in the restoring terms, we therefore include the last term in eq. (8.96) in the corresponding added mass term.

We now get

$$A_{35} = -\int_L x a_{33} dx + \frac{U^2}{\omega^2} a_{33}(x_T) \quad (8.97)$$

$$B_{35} = -UA_{33} - Ux_T a_{33}(x_T). \quad (8.98)$$

We note a certain similarity between B_{35} and B_{53}. If we disregard the end terms, then $B_{53} = -B_{35}$. We now take the pitch moment of eq. (8.96); that is, we multiply it by $-x$ and integrate over the ship length. This gives

$$A_{55} = \int_L x^2 a_{33}\, dx + \frac{U^2}{\omega^2}\left[A_{33} - x_T a_{33}(x_T)\right] \quad (8.99)$$

$$B_{55} = Ux_T^2 a_{33}(x_T). \quad (8.100)$$

These results for A_{ij} and B_{ij}, $i = 3, 5$, $j = 3, 5$ are the same as those obtained by the strip theory by Salvesen et al. (1970) when the frequency of oscillation is high and wave radiation damping is zero. However, when comparing, we should note that Salvesen et al. used a different coordinate system, with x pointing forward. This implies differences in the signs of some of the terms.

The hydrostatic term $\rho g b w$ in eq. (8.88) gives the restoring terms C_{jk} by following the same procedure as the one used in deriving the added mass and damping terms above. C_{33}, C_{35}, and C_{53} are the same as given by eq. (7.41). C_{55} will not include the term $\rho g \nabla (z_B - z_G)$ in eq. (7.41). However, this term is relatively small and is commonly neglected in strip theory calculations.

Further, by integrating the body mass term in eq. (8.88) as we did for the added mass and damping terms, we get the elements M_{33}, M_{35}, M_{53}, and M_{55} in the body mass matrix given by eq. (7.36).

The end conditions given by eq. (8.83) are automatically satisfied for rigid-body motions. So we now have shown that rigid-body motions are solutions of the presented beam equation.

8.5.2 Uniform beam

It is possible to derive analytical expressions for natural frequencies and eigenmodes if a uniform beam is assumed. This means m, a_{33}, I, and b are assumed independent of x. Because an eigenvalue problem is considered, there is no excitation; that is, f_3^{exc} is zero. We will consider undamped modes; that is, all terms associated with $\partial w/\partial t$ in eq. (8.88) are disregarded. Harmonic oscillations written as $\exp(i\omega t)$ are assumed. This leads to

$$EI\frac{\partial^4 w}{\partial x^4} + U^2 a_{33}\frac{\partial^2 w}{\partial x^2} - (\omega^2(m + a_{33}) - \rho g b)w = 0. \quad (8.101)$$

We choose $x = 0$ to be midships.

We have already seen that the beam equation can describe rigid-body heave and pitch motions and so exclude that in our discussion here. This means we focus on elastic vibrations. The fourth-order linear differential equation given by eq. (8.101) has solutions of the form $\sin(px)$, $\cos(px)$, $\sinh(px)$, and $\cosh(px)$. We focus on two-noded vibrations, that is, a solution that is symmetric about $x = 0$. This means we write $w = A\cos px + B\cosh px$. In order to satisfy the end conditions given by eq. (8.83), it is convenient to write

$$w = \frac{\cos px}{2\cos(0.5pL)} + \frac{\cosh px}{2\cosh(0.5pL)}, \quad (8.102)$$

where the condition $\partial^2 w/\partial x^2 = 0$ on $x = \pm 0.5L$ is directly satisfied. We also want to satisfy $\partial^3 w/\partial x^3 = 0$ on $x = \pm 0.5L$. This gives

$$\tan(0.5pL) = -\tanh(0.5pL). \quad (8.103)$$

This equation has an infinite number of solutions corresponding to the different modes. We are interested in the lowest modes, that is, the lowest value of solution of p, which we call p_1. The solution is approximately

$$p_1 = \frac{\pi}{2L} \cdot 3.01. \quad (8.104)$$

The corresponding mode shape expressed by eq. (8.102) is shown in Figure 8.54. The dynamic response is often on the order of centimeters for a real ship. The two-node mode will be the governing dynamic mode in whipping and springing. However, to include quasi-static response, more modes should be included, because the bending moment from this mode shape does not represent the vertical bending moment distribution along the ship for any load distribution. In order to find the corresponding eigenfrequency ω_n, we first substitute eq. (8.102) into eq. (8.101). We multiply eq. (8.101) with eq. (8.102) and integrate from $x = -0.5L$ to $x = 0.5L$. We can use the fact that

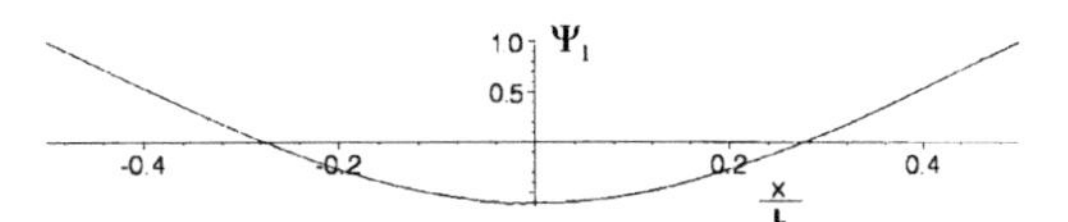

Figure 8.54. Eigenmode for two-node vertical vibration of a uniform beam.

$$\int_{-L/2}^{L/2} \Psi_1^2(x)\,dx \approx L/4,$$

where Ψ_1 is the same as eq. (8.102) with p given by eq. (8.104). Further, we need

$$\int_{-L/2}^{L/2} \frac{d^4\Psi_1}{dx^4}\Psi_1\,dx \approx p_1^4 L/4$$

$$\int_{-L/2}^{L/2} \frac{d^2\Psi_1}{dx^2}\Psi_1\,dx \approx -p_1^2 0.14L.$$

This means eq. (8.82) becomes

$$EIp_1^4\frac{L}{4} - U^2 a_{33}p_1^2 0.14L - \left(\omega_n^2(m + a_{33}) - \rho g b\right)\frac{L}{4} = 0$$

or that

$$\omega_n = \sqrt{\frac{EIp_1^4 - U^2 a_{33}p_1^2 0.56 + \rho g b}{m + a_{33}}}. \quad (8.105)$$

In order to assess the relative importance of the different terms in the nominator of eq. (8.105) and how ω_n varies with ship length, we need first to select representative values of EI. We consider a beam with height equal to the molded depth D (draft + freeboard) midships. The maximum stresses can be expressed as

$$\sigma = \frac{M}{I}\frac{D}{2}, \quad (8.106)$$

where M is vertical bending moment midships. This is specified in class society rules and is independent of material. The rule values are empirical values and account for still water and wave bending moment. By introducing the allowable stress σ_ℓ we get from eq. (8.106) that

$$EI = M\left(\frac{E}{\sigma_\ell}\right)0.5D. \quad (8.107)$$

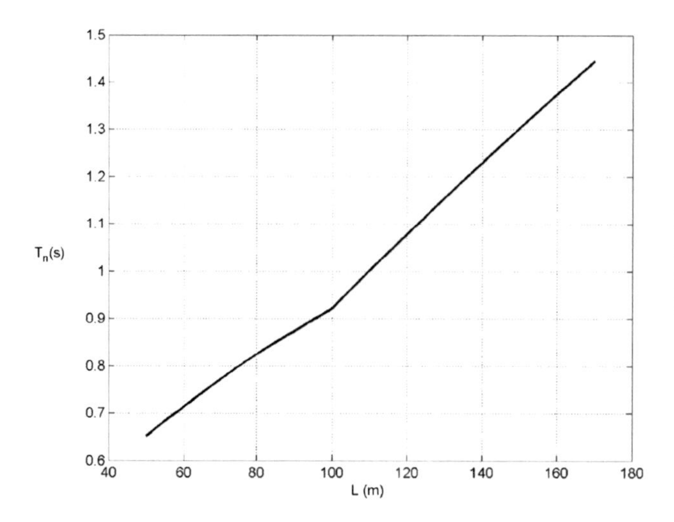

Figure 8.55. Natural period T_n for two-node vertical vibration as a function of ship length for a high-speed monohull vessel, based on uniform beam approximation.

We set the Young modulus, E, equal to 210 GPa and 70 GPa for, respectively, steel and aluminum. The yield stress σ_y is 235 MPa and 355 MPa for, respectively, normal-strength steel and high-strength steel. The allowable stress is, according to the DNV Rules for High-Speed and Light Craft, 175 MPa for normal-strength steel and $1.39 \cdot 175$ MPa=243 MPa for high-strength steel. σ_y is typically 220 MPa for aluminum used in the hull girder. The corresponding allowable global stress σ_ℓ is $0.89 \cdot 175$ MPa=156 MPa. This gives a variation of E/σ_ℓ between $0.4 \cdot 10^3$ and $1.2 \cdot 10^3$. We will use $E/\sigma_\ell = 10^3$ in the following example. Because ω_n is approximately proportional to $\sqrt{E/\sigma_\ell}$, E/σ_ℓ values of $0.4 \cdot 10^3$ and $1.2 \cdot 10^3$ will give, respectively, 37% lower and 10% higher ω_n than those we will estimate in the following text.

We express M in kilo-newton-meter (kNm) as

$$M = 0.3C_W L^2 BC_B, \tag{8.108}$$

where $C_W = 0.08L$ for $L < 100$ m and $C_W = 6 + 0.02L$ for $L > 100$ m. This is consistent with the hogging bending moment in DNV Rules for Classification of High-Speed, Light Craft and Naval Surface Vessels, July 2002, according to Class notation **R0** and **R1**. This implies that there is no reduction of C_W for restricted service. We note that C_W appears to have a dimension, but it should be interpreted as without dimension in eq. (8.108) for the bending moment to have correct dimension. This requires that L in the formula for C_W be given in meters.

We use the data in Table 7.1 as a basis for selecting dimensions. Mean values for beam and draft are used. That means the beam-to-draft ratio is 6.04 and the length-to-beam ratio is 5.68. The molded depth D is selected as 0.115 times the ship length. This corresponds to the mean value of the sum of the draft and the bow height given in Table 7.1. C_B is selected as 0.5. This is somewhat high according to Table 7.1. The reason is simply that eq. (8.89) is used to calculate a_{33}. This requires C_B to be above a minimum value. The EIp_1^4-term in the denominator of eq. (8.105) will be completely dominant, also if shear deformation had been included.

Because the EI-term is dominant, we can approximate eq. (8.105) as

$$\omega_n = \frac{22.4}{L^2}\left(\frac{EI}{m + a_{33}}\right)^{1/2}. \tag{8.109}$$

Figure 8.55 presents the natural period T_n for two-node vertical bending as a function of ship length between 50 m and 170 m. Because C_W has a kink at $L = 100$ m, T_n has a kink at $L = 100$ m. T_n varies nearly linearly for L below 100 m and for L above 100 m. T_n for $L = 100$ m is about 0.9 s. For shorter ships, the stiffness is commonly higher than required by the rules, hence the period would be less.

8.6 Global bow flare effects

Our focus is on head sea. Bow flare slamming affects heave and pitch motions and global elastic vibrations of the hull. The problem has to be solved in the time domain, and there will be coupling between rigid-body motions and elastic vibrations. We showed previously in the text that the beam equation, eq. (8.88), could describe rigid-body motions and elastic vibrations. However, this is a frequency-domain solution based on an infinite-frequency approximation. This means, for instance, that wave radiation damping, which is crucial for rigid-body resonant heave and pitch motions, is not accounted for. Let us say we accounted for a finite frequency and that wave radiation damping was included. Then this will cause inaccurate predictions of the elastic vibrations. For instance, we pointed this out in connection with natural frequency of two-node bending mode in the wetdeck slamming results by Ge (2002). This means that the linear part of the problem must be formulated in the time domain, as Cummins (1962) and Ogilvie (1964) did, as was described in section 7.3. This leads to another set of equations involving convolution integrals, which require information about either added mass or damping for all frequencies. We will not pursue this in this context and rather make the following major approximations:

a) Relative vertical motions and impact velocities are determined by assuming a rigid body.
b) The global elastic effects are in terms of two-node vertical bending vibration.
c) The elastic motions are decoupled from the rigid-body motions.
d) Rigid-body motions are not influenced by the slamming loads.

Our focus is on two-node bending vibration and we use eq. (8.88) to describe this. However, we can, as previously discussed, disregard the terms $U^2\partial\left(a_{33}\partial w/\partial x\right)/\partial x$ and $\rho g b w$ in the derivation of the natural frequency and mode shape. We will represent the solution in terms of undamped normal modes. This means we have to find the solution of

$$(m+a_{33})\frac{\partial^2 w}{\partial t^2}+\frac{\partial^2}{\partial x^2}\left(EI\frac{\partial^2 w}{\partial x^2}\right)=0 \quad (8.110)$$

with the boundary conditions given by eq. (8.83) and when w has the harmonic time dependence $\exp\left(i\omega t\right)$. We discussed this for a uniform beam, for which analytical solutions exist. There is an infinite number of modes, but we concentrate on the lowest mode Ψ_1 and corresponding wet natural frequency ω_1. The modes and natural frequencies have to be found by a numerical method for a realistic ship. We assume this has been done. Then we proceed with eq. (8.88) and represent the solution as $w=a_1\left(t\right)\Psi_1\left(x\right)$, where $a_1\left(t\right)$ is unknown at this stage. Because the damping will in reality be small and will not affect the maximum response due to a single transient loading, we neglect the damping terms, that is, the terms associated with $\partial w/\partial t$. Another matter is if the harmonic excitation frequency is in the vicinity of ω_1. The damping is then crucial. We now have the following version of eq. (8.88):

$$(m+a_{33})\Psi_1(x)\frac{d^2a_1}{dt^2}+a_1\frac{d^2}{dx^2}\left(EI\frac{d^2\Psi_1}{dx^2}\right)=f_3^{BF}(t). \quad (8.111)$$

The superscript BF means bow flare. We multiply now eq. (8.111) by $\Psi_1\left(x\right)$ and integrate over the length of the ship, that is,

$$\frac{d^2a_1}{dt^2}\left[\int_L(m+a_{33})\,\Psi_1^2(x)\,dx\right]+a_1\left[\int_L\Psi_1(x)\frac{d^2}{dx^2}\left(EI(x)\frac{d^2\Psi_1}{dx^2}\right)dx\right]=\int_L f_3^{BF}(t)\Psi_1(x)\,dx. \quad (8.112)$$

This can be rewritten by noting once more that eigenvalues are found by assuming harmonic oscillations and no excitation. This gives, by using eq. (8.110), that

$$-\omega_1^2\left[\int_L(m+a_{33})\Psi_1^2(x)\,dx\right]+\left[\int_L\Psi_1(x)\frac{d^2}{dx^2}\left(EI(x)\frac{d^2\Psi_1}{dx^2}\right)dx\right]=0. \quad (8.113)$$

Eq. (8.111) can therefore be written as

$$\frac{d^2a_1}{dt^2}+\omega_1^2a_1=F_V(t), \quad (8.114)$$

where

$$F_V(t)=\frac{\int_L f_3^{BF}(t)\Psi_1(x)\,dx}{\int_L(m+a_{33})\Psi_1^2(x)\,dx}. \quad (8.115)$$

Eq. (8.114) represents the response for a simple mass-spring system with a transient loading. We know from many textbooks (see, e.g., Clough and

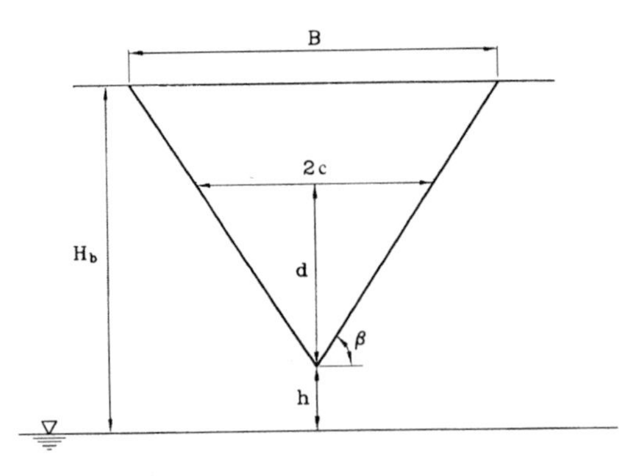

Figure 8.56. Wedge-shaped bow flare section. H_b = bow height.

Penzien 1993) that the ratio between the time duration of loading T_d and the natural period $T_1 = 2\pi/\omega_1$ is an important parameter. For instance, if $T_d/T_1 < \approx 0.25$, then $a_1(t)$ is proportional to the force impulse

$$I = \int_0^{T_d} F_V(t)\,dt, \qquad (8.116)$$

where $t = 0$ corresponds to the initial impact.

The bow flare slamming force can be derived by first generalizing eq. (8.87) by also accounting for the incident waves. We can write

$$f_3^{HD} = -\left(\frac{\partial}{\partial t} + U\frac{\partial}{\partial x}\right)(a_{33}V_R), \qquad (8.117)$$

where the relative impact velocity V_R is given by eq. (8.72) in the case of regular waves. The superscript HD means that f_3^{HD} is a hydrodynamic force per unit length. We earlier discussed slamming loads by neglecting the effect of $U\partial/\partial x$ in eq. (8.117). We then pointed out that the slamming term $V_R\, da_{33}/dt$ is set equal to zero during the water exit phase. In addition, we must account for hydrostatic and Froude-Kriloff forces.

We follow a von Karman approach and write $a_{33} = 0.5\rho\pi c^2 K_1$, where $2c$ is the instantaneous beam at the intersection between the incident wave and the cross section. Here $K_1 = 1$ if the local deadrise angle is small. Otherwise, K_1 has to be numerically determined by assuming an infinite-frequency free-surface condition. Values of a_{33} (or K_1) for wedges are presented in eq. (9.63) and Figure 9.27. We can express the vertical force per unit length f_3 due to the water as

$$\begin{aligned} f_3 = &-\rho\pi V_R\, c\frac{dc}{dt}K_1K_2 - \rho\frac{\pi}{2}c^2K_1\frac{dV_R}{dt} \\ &- U\frac{\partial}{\partial x}\left[\rho\frac{\pi}{2}c^2K_1V_R\right] \qquad (8.118) \\ &+ \rho g A(x,t) \end{aligned}$$

Here K_2 is 1 during the water entry phase and 0 during the water exit phase. Further, η_B is the vertical body motion, ζ is the incident wave elevation, and $A(x, t)$ is the submerged cross-sectional area. The last term in eq. (8.118) represents the hydrostatic and Froude-Kriloff forces. V_R and η_B in eq. (8.118) include both vertical elastic vibrations and rigid-body motions in the general case with coupling between all modes. We must then realize that eq. (8.118) also includes linear terms due to the body motions. These terms must not be accounted for twice.

In our simplified case, we have decoupled the rigid-body motions and elastic vibration and assumed that V_R and η_B are expressed by rigid motions. f_3 is then the same as f_3^{BF} in eq. (8.111).

Let us exemplify the different terms in eq. (8.118) by considering the wedge-formed cross section in Figure 8.56. We call H_b the bow height, and B is the beam at the deck. We consider incident head sea waves and express $\zeta - \eta_B$ as $A_R \sin\omega_e t$, where ω_e is the frequency of encounter and A_R is the relative motion amplitude. If we neglect the contribution from $U\eta_5$ in V_R, we can express V_R as $-\omega_e A_R \cos\omega_e t$. The instantaneous draft of the wedge is

$$d = A_R \sin\omega_e t - h, \qquad (8.119)$$

where h is the vertical distance between the apex of the wedge and mean free surface when there are no waves and body motions. Because d must be positive, $\sin\omega_e t > h/A_R$. Further, c in eq. (8.118) equals $d\cot\beta$, where β is the deadrise angle. This means $A(x)$ in eq. (8.118) is $d^2\cot\beta$.

We will have a ship length L=100 m in mind and use Figure 8.7 with $Fn = 0.53$ as a basis. This means that the ship speed is $U = 16.6\ \mathrm{ms}^{-1}$. We select the nondimensional mean wave period

$T_2\sqrt{g/L}$ as 2.4. This is close to where the RMS of relative vertical motion η_R has a maximum. This means that the zero-upcrossing period $T_2 = 7.7$ s. If a Pierson-Moskowitz (PM) wave spectrum is used, the modal (peak) period of the spectrum will be $T_0 = 1.408 T_2$, that is, $T_0 = 10.8$ s. In order to find the frequency of encounter, we use that $\omega_e = \omega_0 + U\omega_0^2/g$ for head sea and set $\omega_0 = 2\pi/T_0$. This gives $T_e = 2\pi/\omega_e = 5.4$ s. From Figure 8.7, we see that $\sigma_R = 0.95 H_{1/3}$ at the mean wave period and Froude number that we consider. The value of $H_{1/3}$ depends on the operational limitations and sea area. The most probable largest value follows from eq. (7.113) and is approximately $4\sigma_R$.

We illustrate the relative importance of the different slamming terms in eq. (8.118) by neglecting the $U\partial/\partial x$-term. There are many parameters to be selected. Table 7.1 gives some guidance. For instance, a 100 m–long monohull may have a beam between 15 and 18 m. The bow height H_b may be between 6 and 9 m. Operational limits could be $H_{1/3}$ between 4.5 and 6 m. Using $\sigma_R = 0.95 H_{1/3}$ leads to the most probable largest relative motion equal to about 23 m based on $H_{1/3} = 6$ m. This would certainly lead to green water on deck. However, there are several reasons why this estimate is too conservative. The calculations leading to Figure 8.7 are based on linear theory. A motion control system as described in section 7.1.3 is not accounted for. Nonlinearities tend to decrease heave and pitch. Further, the results of Figure 8.7 are for a catamaran. The large beam-to-draft ratio of a high-speed monohull is an advantage from a heave and pitch point of view relative to a catamaran, which has side hulls with a much smaller beam-to-draft ratio compared with a monohull. Further, too-large relative motions would lead to voluntary speed reduction due to green water. We will instead choose $A_R = H_b$. Then we prevent green water coming onto the deck. Green water or $A_R > H_b$ does also mean that we should reconsider the load expression. We assume $H_b = 9$ m, $A_R = 9$ m, $T_e = 5$ s, and a maximum sectional breadth $B = 8$ m (see Figure 8.56). Results for $\beta = 20°$ and $40°$ with $K_1 = 1$ in eq. (8.118) are shown in Figure 8.57. The hydrostatic and Froude-Kriloff forces are always positive and become increasingly significant with increasing β. The added mass force is always negative and is the reason the total force is negative in the water exit phase and also at the end of the water entry phase. The time instant when the water entry phase ends can be seen from the slamming force, which is zero in the water exit phase. The duration T_d of the force depends on β and is from 1.7 to 1.9 s in Figure 8.57. The representative natural period for two-node bending vibration is $T_1 = 1$ s, which makes $T_d/T_1 = 1.7 - 1.9$. This means that it is not an impulse type of transient load, which would require $T_d/T_1 < 0.25$. Because a representative natural period for heave and pitch would be 5 s, we note that it is closer to being an impulse type of response when the effects of bow flare loads on heave and pitch are considered. In the same way as Ge (2002) showed for wetdeck slamming, both the water entry and exit phase will matter for the response.

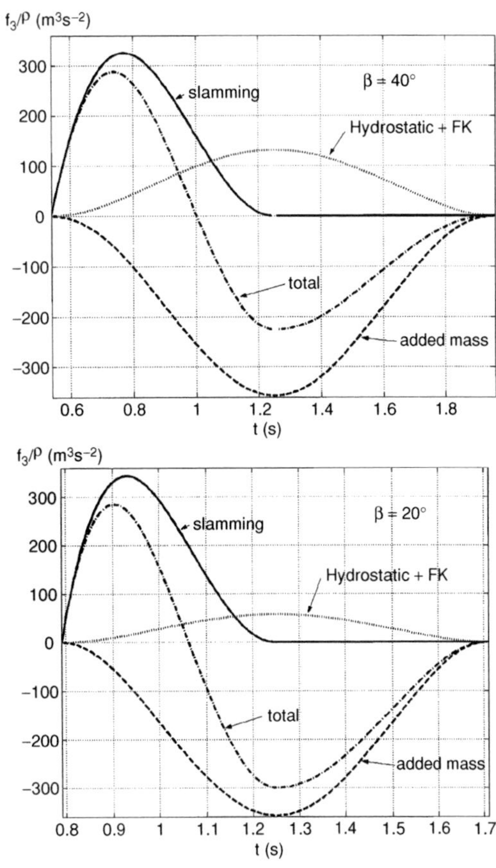

Figure 8.57. 2D bow flare slamming force on a wedge-shaped cross section. FK = Froude-Kriloff force, β = deadrise angle. Bow height is $H_b = 9$ m. Relative motion amplitude is $A_R = 9$ m. Sectional maximal breadth is $B = 8$ m (see Figure 8.56).

8.7 Springing

Springing is steady-state resonant elastic vibrations due to continuous wave loading. If slamming occurs frequently, springing may be difficult to distinguish from whipping. The reason is that the small damping causes slow decay of the whipping-induced response. Both springing and whipping in head sea are mainly related to the two-node bending of the hull. We now concentrate on springing; both linear and nonlinear wave loads have to be considered. Our focus is on head sea, but other headings may also matter. When the linear wave effects are important, the frequency of encounter ω_e is in the vicinity of the natural frequency ω_1 for two-node bending.

In addition to high-speed vessels, springing is a well-known effect for conventional ships. It is described for Great Lakes bulk carriers by Matthews (1967) and Cleary et al. (1971) and for large ocean-going ships of full form by Goodman (1971). Storhaug et al. (2003) documented that the springing may contribute to approximately 50% of the accumulated fatigue damage based on full-scale measurements of a 300 m–long bulk carrier.

Because springing is a resonant phenomenon, the damping is crucial. The hydrodynamic damping is the result of the terms $2a_{33}U\partial^2 w/\partial x\partial t$ and $U(da_{33}/dx)\partial w/\partial t$ in eq. (8.88). These terms have a small effect for conventional ships, and other damping mechanisms, such as structural, cargo, and viscous damping, must be considered as well. However, there are uncertainties related to the magnitude of the damping that should be used, and it may be necessary to determine it from full-scale measurements (Storhaug et al. 2003). The total damping may vary between 0.5% and 2% of the critical damping for ships at moderate speed. Because the total damping is small, strong amplification of the response occurs in resonant conditions.

In order to see what wave conditions cause significant springing, we note $\omega_e = \omega_0 + U\omega_0^2/g$ for head seas and consider a case in which the two-node natural frequency $\omega_1 = 2\pi$ rad/s and $U = 20$ ms^{-1}. Linear springing effects are important when $\omega_1 = \omega_e$. This gives a wave period $T_0 = 2\pi/\omega_0 = 4.1$ s, which means a wavelength of 26.2 m or about 4 wavelengths along a 100 m–long ship.

In reality, we have to consider a wave spectrum. It may then be more illustrative to present a "frequency-of-encounter" wave spectrum $S_e(\omega_e)$. We will show how this can be derived for head sea. It follows by energy consideration that

$$S_e(\omega_e)d\omega_e = S(\omega_0)\,d\omega_0. \tag{8.120}$$

This ensures that the areas under the wave spectrum and the frequency-of-encounter wave spectrum are the same. Eq. (8.120) assumes implicitly that there is a one-to-one correspondence between the wave frequency ω_0 and the frequency of encounter ω_e. This is, for instance, not true for all frequencies for following sea. Given $\omega_e = \omega_0 + U\omega_0^2/g$ for head seas, it follows that

$$S_e(\omega_e) = \frac{S(\omega_0)}{1 + 2U\omega_0/g} \tag{8.121}$$

and

$$\omega_0 = \frac{g}{2U}\left(-1 + \sqrt{1 + 4U\omega_e/g}\right). \tag{8.122}$$

We now use the modified Pierson-Moskowitz (PM) spectrum given by eq. (3.55) to illustrate $S_e(\omega_e)$. Instead of using the mean wave period T_1 as a parameter, we use the modal period T_0, which corresponds to the peak period of the spectrum. By using eqs. (3.59) and (3.60), we write $T_1 = 0.77\ T_0$ for a Pierson-Moskowitz spectrum. We now consider $U = 20$ ms^{-1} and $T_0 = 4.1$ s. Figure 8.58 shows $S_e(\omega_e)/H_{1/3}^2$ as a function of ω_e. As expected from our discussion of regular waves, the maximum spectral value occurs at $\omega_e = 2\pi$ rad/s. In order to see whether there is sufficient wave energy to excite springing when $\omega_e = \omega_1$, it is better to present the wave spectrum as a frequency-of-encounter wave spectrum like that illustrated for head sea. Obviously we should vary T_0 and $H_{1/3}$ in accordance with scatter diagrams showing the frequency of occurrence of T_0 and $H_{1/3}$ for given operational areas of the vessel.

Normally, the resonance frequency occurs in the high-frequency tail of the sea spectra. It is therefore of concern how the tail is represented. As seen from the PM spectrum, the decay is represented by ω^{-5}, whereas Torsethaugen's (1996) two-peaked wind- and swell-generated spectra may have ω^{-4}. The wave energy in the high-frequency tail is an important uncertainty for prediction of springing response.

A simple way to illustrate the presence of nonlinear wave excitation is to consider the quadratic

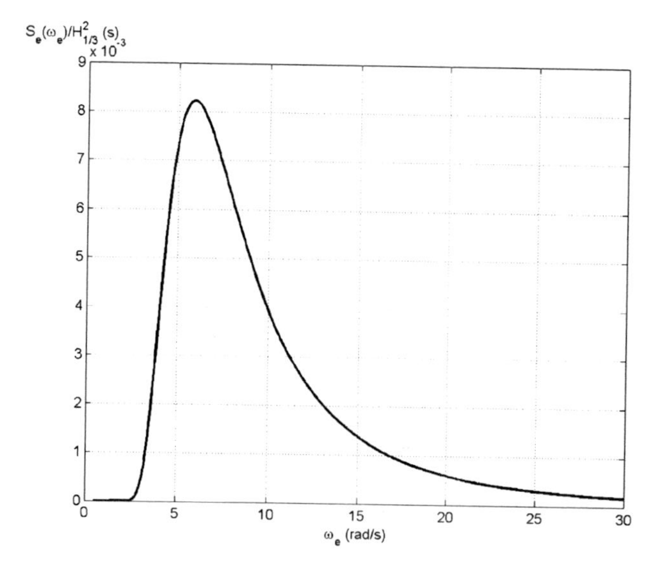

Figure 8.58. Frequency-of-encounter wave spectrum $S_e(\omega_e)$ for head sea. $U = 20\ \text{ms}^{-1}$. Modal wave period $T_0 = 4.1$ s.

velocity term in Bernoulli's equation for the fluid pressure. We can write this term as

$$-\frac{\rho}{2}(u^2 + v^2 + w^2) = -\frac{\rho}{2}\,|\nabla\varphi|^2, \quad (8.123)$$

where $\mathbf{u} = (u, v, w)$ is the fluid velocity vector. We emphasize that eq. (8.123) provides only one of the nonlinear effects. Other contributions may be equally important. They arise because we have to

a) Satisfy body boundary conditions on the exact wetted ship surface. Body boundary conditions are satisfied on the mean wetted surface in a linear theory.
b) Satisfy nonlinear free-surface conditions.
c) Integrate the pressure on the instantaneous position of the wetted surface of the ship. In a linear theory, we integrate the pressure component $-\rho\,(\partial\varphi/\partial t + U\partial\varphi/\partial x)$ over the mean wetted surface. (It is here assumed there is no interaction between local steady flow and unsteady flow.) In addition, we consider a linearized effect of the hydrostatic pressure $-\rho g z$.

We will return to eq. (8.123) and consider an idealized sea state consisting of two wave components ω_j and ω_k. Associated with ω_j and ω_k, there are two frequencies of encounter ω_{ej} and ω_{ek}. A linear approximation for the x-component of the velocity can be written as

$$u = A_j\cos(\omega_{ej}t + \varepsilon_j) + A_k\cos(\omega_{ek}t + \varepsilon_k), \quad (8.124)$$

so

$$\begin{aligned}-\frac{\rho}{2}u^2 = -\frac{\rho}{2}\Big[&\frac{A_j^2}{2} + \frac{A_k^2}{2} + \frac{A_j^2}{2}\cos(2\omega_{ej}t + 2\varepsilon_j)\\ &+ \frac{A_k^2}{2}\cos(2\omega_{ek}t + 2\varepsilon_k) \\ &+ A_jA_k\cos((\omega_{ej} - \omega_{ek})t + \varepsilon_j - \varepsilon_k)\\ &+ A_jA_k\cos((\omega_{ej} + \omega_{ek})t + \varepsilon_j + \varepsilon_k)\Big].\end{aligned} \quad (8.125)$$

This means that we have found the presence of pressure terms oscillating with $2\omega_{ej}$, $2\omega_{ek}$, $\omega_{ej} + \omega_{ek}$, and $\omega_{ej} - \omega_{ek}$. For a more realistic representation of the seaway, and considering the wave as the sum of N components of waves with different frequencies, we will find pressure terms with sum-frequencies $\omega_{ej} + \omega_{ek}$ $(k, j = 1, N)$. These nonlinear interaction terms produce sum-frequency excitation loads that may cause resonant springing. The energy in these waves is also significantly higher than in those that contribute to excitation at an encounter frequency corresponding to the springing frequency.

The depth decay of these pressure terms plays an important role. For instance, if we consider an analysis like the one in eq. (8.125), the depth decay of the pressure will be like $\exp((k_j + k_k)z)$, where k_j are wave numbers associated with ω_{ej}. However, there are terms associated with a second-order approximation of the velocity potential that decay much more slowly with depth. This can be illustrated with an example from pages 168 to 169 in Faltinsen (1990). Consider two linear regular waves of same frequency but with opposite propagation directions. The second-order pressure field will then oscillate with the sum-frequency and not decay with depth. Sea states with waves traveling in opposite directions with the same dominant frequencies are not typical in the open sea, but you may have swell from one direction and wind-generated sea from another.

Experiments in regular waves were carried out by Troesch (1984) on a model jointed amidships to measure both wave excitation and springing response. The dominant effect was linear, but he found a measurable springing excitation at $2\omega_e$ and sometimes $3\omega_e$. Let us illustrate what wave conditions cause $2\omega_e = \omega_1$ with $\omega_1 = 2\pi$ rad/s, $U = 20$ ms^{-1}, and head seas. This gives a wave period $T_0 = 6.2$ s, that is, a larger wavelength than that causing linear springing. The springing excitation with $2\omega_e$-oscillations will be approximately quadratic in wave amplitude. This discussion shows that linear springing excitation is important for small sea states and that nonlinear excitation due to sum-frequency effect is important for higher, but still small, sea states. Because small sea states frequently occur, springing is of concern from a fatigue point of view, whereas whipping is more common in larger sea states and is of more concern for ultimate strength. We must always have in mind that springing and whipping responses must be combined with rigid-body ship motion responses due to continuous wave loading.

There is not much literature on springing for high-speed ships. However, Jensen (1996) and Hansen et al. (1994, 1995) have studied the importance of springing for high-speed monohull ships. The quadratic strip theory described by Jensen and Pedersen (1978, 1981) and Jensen and Dogliani (1996) was used. Second-order nonlinear hydrodynamic effects are accounted for in an approximate way. An advantage from a computational efficiency and accuracy point of view is that a frequency-domain solution is used. Jensen (1996) found that nonlinear wave load effects on springing are important for a proper estimate of the fatigue damage, whereas Hansen et al. (1994) stated that springing is especially important for ships made of GRP (glass-reinforced plastic) or aluminum. In Jensen and Wang (1998), springing was found to be unimportant in more extreme sea conditions, whereas whipping was important for fatigue damage. Hermundstad (1995) and Hermundstad et al. (1995, 1997) showed comparisons between numerical predictions by a 2.5D high-speed theory and experimental results for a catamaran. Significant springing response was found in the transfer functions of the vertical bending moment due to (artificial) low springing frequency. Evidence of higher-order harmonics was also found.

8.7.1 Linear springing

We exemplify linearly excited two-node springing by using a uniform beam model. This means m, a_{33}, and EI are assumed constant along the ship. Further, f_3^{exc} in eq. (8.88) is approximated by only considering the linear Froude-Kriloff loads. This means

$$f_3^{exc} = -i\rho g \zeta_a e^{-kD} b e^{i(\omega_e t - kx)}, \qquad (8.126)$$

where it is understood that it is the real part that has physical meaning. Eq. (8.126) is consistent with the pressure given in Table 3.1. Because we operate with a linear system, it is convenient to use complex formulation, as in eq. (8.126). Using eq. (8.126) is a large simplification, but it enables us to derive analytical solutions for further discussions. We express the solution of eq. (8.88) as

$$w = a_1(t)\Psi_1(x).$$

Here Ψ_1 is the same as w expressed in eq. (8.102), with the constant $p = p_1$ given by eq. (8.104). In the discussion of the wet natural frequency ω_1 given by eq. (8.105), we noted that the restoring effect of the terms $U^2\partial\left(a_{33}\partial w/\partial x\right)/\partial x$ and $\rho g b w$ in eq. (8.88) can be neglected. This will therefore be done here. Then eq. (8.88) becomes

$$(m + a_{33})\Psi_1 \frac{d^2 a_1}{dt^2} + 2a_{33}U \frac{d\Psi_1(x)}{dx}\frac{da_1}{dt} + U\frac{da_{33}}{dx}\Psi_1(x)\frac{da_1}{dt} + EI\frac{d^4\Psi_1}{dx^4}a_1 = \bar{f}_3^{exc} e^{i\omega_e t}, \qquad (8.127)$$

where $\bar{f}_3^{exc} e^{i\omega_e t}$ is the same as eq. (8.126). Eq. (8.127) is solved by multiplying the equation by $\Psi_1(x)$ and integrating over the ship length. This is similar to what we did when we found ω_1. Because $\Psi_1(x)$ and $d\Psi_1/dx$ are, respectively, symmetric and antisymmetric about $x = 0$, $\int_{-L/2}^{L/2} (\Psi_1 d\Psi_1/dx)\, dx = 0$. This means the second

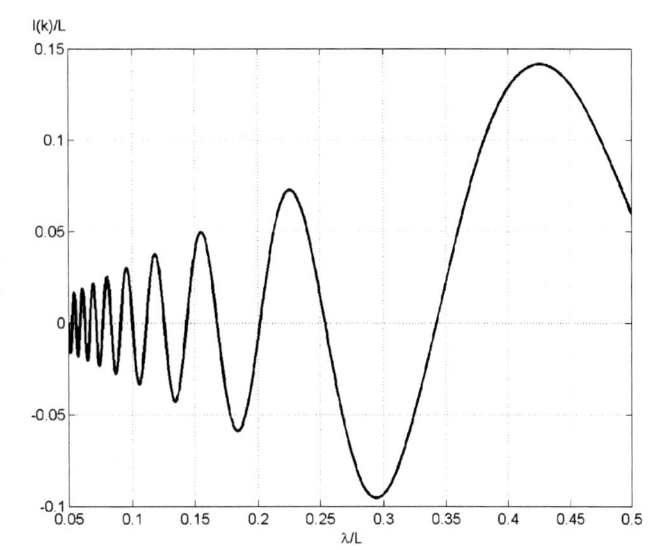

Figure 8.59. $\frac{I(k)}{L}$ given by eq. (8.135) as a function of λ/L. λ = incident wavelength, L = ship length.

term on the left-hand side of eq. (8.127) does not contribute. The third term involves the integral

$$\int_{-L/2}^{L/2} \frac{da_{33}}{dx}\Psi_1^2\,dx = a_{33}(x_T). \tag{8.128}$$

Here $x = x_T$ is the transom stern. Eq. (8.128) follows by partial integration and use of $\int_{-L/2}^{L/2}(\Psi_1\,d\Psi_1/dx)\,dx = 0$. The generalized excitation force involves the integral

$$I(k) = \int_{-L/2}^{L/2} e^{-ikx}\Psi_1\,dx = \int_{-L/2}^{L/2} \cos(kx)\Psi_1 dx. \tag{8.129}$$

Eq. (8.127) can now be written as

$$\frac{d^2a_1}{dt^2} + b_d\frac{da_1}{dt} + \omega_1^2 a_1 = F(k)\zeta_a e^{i\omega_e t}, \tag{8.130}$$

where

$$F(k) = \frac{-i\rho g e^{-kD}b}{0.25L(m+a_{33})}\int_{-L/2}^{L/2} \cos(kx)\Psi_1(x)\,dx \tag{8.131}$$

$$b_d = \frac{Ua_{33}}{0.25L(m+a_{33})}. \tag{8.132}$$

This means

$$a_1 = \frac{F(k)\zeta_a e^{i\omega_e t}}{-\omega_e^2 + b_d i\omega_e + \omega_1^2}. \tag{8.133}$$

The damping b_d can be assessed by comparing it with critical damping; that is, we consider the fraction ξ between damping and critical damping. Using eq. (7.9) gives

$$\xi = \frac{U}{0.5L(m/a_{33}+1)\,\omega_1}. \tag{8.134}$$

Let us choose $U = 20$ ms^{-1}, $L = 100$ m, $a_{33} = m$, and $\omega_1 = 2\pi$ rad/s. This gives a damping ratio of 0.03 (structural damping comes in addition). This means small damping and strong amplification of the response. $F(k)$ oscillates with the wavelength $\lambda = 2\pi/k$. To illustrate this, we rewrite $I(k)$ given by eq. (8.129) as

$$\frac{I(k)}{L} = \int_{-0.5}^{0.5} \cos(kLu) \times \left(\frac{\cos(p_1 Lu)}{2\cos(0.5p_1 L)} + \frac{\cosh(p_1 Lu)}{2\cosh(0.5p_1 L)}\right) du, \tag{8.135}$$

where $p_1 L = 0.5\pi \cdot 3.01$ according to eq. (8.104). $I(k)/L$ is plotted as a function of λ/L in Figure 8.59. We note the oscillatory values of $I(k)/L$ and that $I(k)/L$ becomes zero for certain values of λ/L. If these values for a given U correspond to when $\omega_e = \omega_1$, we get no excitation at the natural frequency. Because our theoretical model is strongly simplified, we must look only at the tendency, that is, that the excitation force tends to become small for certain wavelengths.

Even though Figure 8.59 shows a cancellation at a specific λ/L, a change of heading or speed may change the overall picture.

When regular waves of small wavelength relative to the ship length are propagating along the ship hull in head sea, the total wave elevation will decay along the ship. This 3D effect was accounted for in the linear springing studies by Skjørdal and Faltinsen (1980). For short waves compared with the ship length, this might be important. It may be understood from introducing in the integrand of eq. (8.131) a varying force along the ship. Hence any significant changes in forces at the ends and middle of the ship will then contribute, possibly significantly.

8.8 Scaling of global hydroelastic effects

When model testing with an elastic model is made in waves, we must check both Froude scaling and that the elastic properties are properly scaled. Froude scaling means that the frequency of the waves must satisfy that $\omega_0\sqrt{L/g}$ is the same in model scale and full scale. Further, the Froude number $U/\sqrt{Lg}$ is the same in model scale and full scale. Let us use the subscripts m and f to indicate model and full scale. This means Froude scaling gives

$$\omega_m = \omega_f(\Lambda_L)^{1/2}, \qquad (8.136)$$

where $\Lambda_L = L_f/L_m$ is the geometrical scale factor. In order to find out how to scale elastic properties associated with bending stiffness EI, we can use eq. (8.109) as a basis. We note that the mass m and a_{33} are proportional to ρL^2. This means

$$\omega\sqrt{\frac{\rho L^6}{EI}} \qquad (8.137)$$

is a nondimensional frequency associated with elastic vibrations due to bending stiffness. Formula (8.111) was based on a uniform beam, but eq. (8.137) is a general way to nondimensionalize the bending stiffness effect. Eq. (8.137) must be the same in model and full scale. This means

$$\omega_m = \omega_f\left[\Lambda_L^6 (EI)_m/(EI)_f\right]^{1/2}. \qquad (8.138)$$

In order to satisfy both eqs. (8.136) and (8.138), we must require that

$$(EI)_m = (EI)_f/\Lambda_L^5. \qquad (8.139)$$

We have not considered the shear rigidity kAG. However, this will lead to (Maeda 1991)

$$(kAG)_m = (kAG)_f/\Lambda_L^3. \qquad (8.140)$$

Because structural damping matters for springing, the scaling of structural damping must also be considered; that is, the ratio between damping b and critical damping must be the same in model and full scales. This gives $b_m = b_f\Lambda_L^{-2.5}$.

Neither the damping nor the scaling of the frequency is always properly assessed in model testing with flexible models. One reason for this may be that the quality of the waves for small wavelengths may be rather poor in a model basin. Hence one chooses to lower the eigenfrequency to have more control over the excitation. It is, however, then more difficult to come to valid conclusions for full-scale cases based on experiments only.

8.9 Exercises

8.9.1 Probability of wetdeck slamming

Consider an irregular sea state and head sea long-crested waves. According to Ochi (1964), the probability of bottom slamming can be written as

$$P(\text{slamming}) = \exp\left(-\left(\frac{V_{cr}^2}{2\sigma_v^2} + \frac{d^2}{2\sigma_r^2}\right)\right). \qquad (8.141)$$

Here d is the ship draft at the impact position and

$$V_{cr} = 0.093(gL)^{1/2}, \qquad (8.142)$$

where L is the ship length. V_{cr} is a threshold velocity for slamming to occur. Further, σ_v^2 and σ_r^2 are variances of, respectively, relative impact velocity (see eq. (8.72)) and relative vertical motion at the impact position. If $P(\text{slamming}) \geq 0.03$, a typical shipmaster will reduce the speed.

We will apply this to wetdeck slamming, where d then means the vertical distance between the wetdeck and the steady free-surface elevation.

Consider a 70 m–long catamaran by scaling the data for the catamaran in section 7.1.2. Assume the waves hit at the front end of the wetdeck. Use the data in Figure 8.7 for σ_v and σ_r, and consider a Froude number of 0.71.

a) How high must the significant wave height ($H_{1/3}^{lim}$) be for $P(\text{slamming}) \geq 0.03$ when $T_2 = 8$ s.

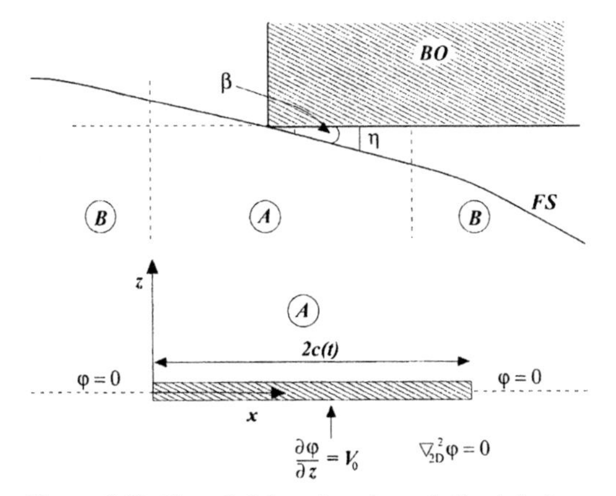

Figure 8.60. Top: Subdomains A and B at bottom impact. Bottom: Boundary-value problem for φ due to bottom-slamming in subdomain A. $x = 0$ front edge of bottom, $2c(t)$ = wetted length.

b) Consider different mean wave periods and $Fn = 0.71$. Discuss, based on Figure 8.7, whether the ship can satisfy $P(\text{slamming}) < 0.03$ for $H_{1/3}^{lim}$ by reducing the speed.

c) Express the slamming pressure as $p = 0.5\rho C_p V_R^2$, where V_R is the relative impact velocity and C_p is a constant slamming pressure coefficient. Generalize eq. (8.141) and express the probability that the slamming pressure is higher than a given value p.

8.9.2 Wave impact at the front of a wetdeck

A wave hits the front edge of a rigid wetdeck (Figure 8.60). The following assumptions are made :

- The free surface can be approximated as a straight line. There is an angle β between the free surface and wetdeck.
- The relative impact velocity V_0 is constant in time and space (positive V_0 is upward).

The flow due to the impact can be solved as in Wagner (1932) (see section 8.3.1) for two-dimensional impact of a rigid body on an initially calm free surface. The boundary-value problem for the velocity potential φ due to the impact is shown in Figure 8.60. The details of the jet flow at $x = 2c$ are not considered. An important difference from Wagner's analysis is that the body–free-surface intersection point corresponding to the front edge of the bottom ($x = 0$) does not change with time. The solution of the velocity potential φ on the wetted surface is

$$\varphi = -V_0\sqrt{x\,(2c - x)},\; 0 < x < 2c. \quad (8.143)$$

The wetted length $2c\,(t)$ can be found by the kinematic free-surface condition. It results in an integral equation that is solved in a way similar to the one Wagner used.

a) Show that

$$c = \frac{2V_0 t}{3\tan\beta}, \quad (8.144)$$

where $t = 0$ corresponds to initial impact.

b) Show that the free-surface elevation due to slamming is

$$\begin{aligned}\eta_s &= V_0 \int_0^t \frac{|x - c\,(t')|}{\sqrt{x\,(x - 2c)}}\,dt' - V_0 t \\ &= \frac{1}{2}\tan\beta\left[(c - 2x)\sqrt{\frac{x - 2c}{x}} + 2x\right] - V_0 t.\end{aligned} \quad (8.145)$$

We note that η_s is infinite at $x = 0$, which means that the finite vertical distance of the wetdeck at the front edge has to be considered (Faltinsen et al. 2004).

c) Express the impact pressure and vertical force on the wetdeck.

8.9.3 Water entry of rigid wedge

Consider water entry of a rigid wedge with deadrise angle β. The water entry velocity is

$$V = V_0 + V_1 t. \quad (8.146)$$

Consider the average pressure p_{av} from $y = y_i$ to y_{i+1} in Figure 8.18.

a) Show by using Wagner's theory that

$$p_{av} - p_a = \rho V^2 \frac{c\pi}{2\tan\beta} I_1 + \rho V_1 I_2, \quad (8.147)$$

where

$$I_1 = \frac{1}{y_{i+1} - y_i}\left(\sin^{-1}\left(\frac{y_{i+1}}{c}\right) - \sin^{-1}\left(\frac{y_i}{c}\right)\right) \quad (8.148)$$

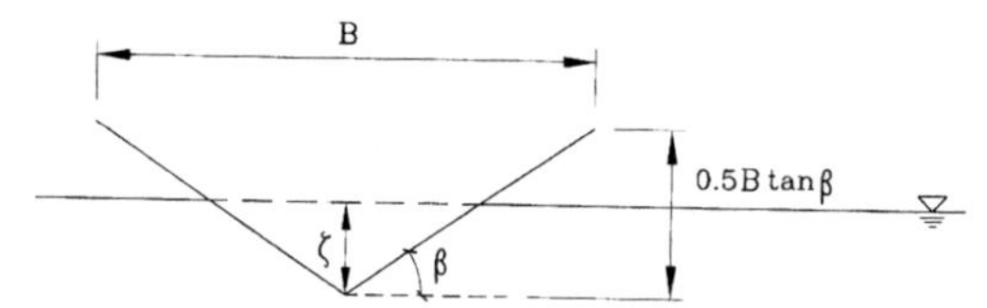

Figure 8.61. Water entry of a wedge.

$$I_2 = \frac{c^2}{(y_{i+1}-y_i)}\left\{\frac{y_{i+1}}{2c}\left(1-\frac{y_{i+1}^2}{c^2}\right)^{1/2} - \frac{y_i}{2c}\left(1-\frac{y_i^2}{c^2}\right)^{1/2} + 0.5\sin^{-1}\left(\frac{y_{i+1}}{c}\right) - 0.5\sin^{-1}\left(\frac{y_i}{c}\right)\right\} \quad (8.149)$$

when

$$c = \frac{\pi}{2\tan\beta}(V_o t + 0.5V_1 t^2) > y_{i+1} \quad (8.150)$$

and $V \geq 0, c \leq b$.

b) Express p_{av} for other time instants than those described in a) when $V \geq 0$.

c) Use $V = 1.8 - 6.8\,t\,(\text{ms}^{-1})$. Set $\tan\beta = 0.25$, and b in Figure 8.18 equal to 1.3 m. Make a computer program to solve this problem.

- How long does it take for the spray root to be at b, that is, $c = b$?
- Set $y_i = 0.317$ m and $y_{i+1} = 0.634$ m. When is the average pressure p_{av} largest?
- Judge the relative importance of hydroelasticity by using the structural data in section 8.3.3 in combination with Figure 8.20. Because Figure 8.20 is for constant water entry velocity V, you may use V at the instant when $c = y_i$.

8.9.4 Drop test of a wedge

a) Consider drop tests of a symmetric wedge with deadrise angle β and instantaneous submergence $\zeta(t)$ relative to undisturbed free surface. The maximum beam is B (Figure 8.61). The body mass per unit length is m. Use von Karman's theory with a flat plate approximation for the added mass and show that the differential equation for ζ when $\zeta < 0.5$ B $\tan\beta$ and the wetted length increases is

$$\frac{d^2\zeta}{dt^2} + \alpha\frac{d}{dt}\left(\zeta^2\frac{d\zeta}{dt}\right) + \gamma\zeta^2 = g, \quad (8.151)$$

where

$$\alpha = 0.5\rho\pi\cot^2\beta/m \quad (8.152)$$

$$\gamma = \rho g\cot\beta/m. \quad (8.153)$$

Further, $d\zeta/dt = V = (2gh)^{0.5}$ when $\zeta = 0$. Here h is the initial height of the wedge apex above the horizontal free surface. It is common to set the slamming part of the hydrodynamic force equal to zero when the wetted length decreases. How would you then modify eq. (8.151)?

b) What do the expressions look like if Wagner's theory is used? Discuss your choice of buoyancy force.

c) We want to do a drop test with constant water entry velocity. However, this can only be obtained approximately in an experiment. Let us consider an example in which $B = 1$ m and $\beta = 20°$ in Figure 8.61. The initial water entry velocity is 1 ms^{-1}. Use Wagner's theory and consider when the spray root has a vertical distance $0.5B\tan\beta$ from the apex of the wedge. We want the water entry velocity to be 0.999 ms^{-1} at that time. How large a body mass per unit length must be used? You may want to make a computer program to solve this problem.

8.9.5 Generalized Wagner method

Zhao et al. (1996) presented a generalized Wagner method in which the exact body boundary condition is satisfied. However, the dynamic free-surface condition is the same as in Wagner's outer flow domain solution. All terms in Bernoulli's equation are included except the hydrostatic pressure term. Further, predicted pressures less than atmospheric pressure are set equal to atmospheric pressure.

A further simplification will be made in this exercise. We will use the flat plate approximation by Wagner to find the velocity potential. This means that we do not satisfy the exact body boundary condition. Now study water entry of a wedge with constant entry velocity by using the idea of Zhao et al. (1996) about keeping all tems in Bernoulli's equation except the hydrostatic pressure term.

Derive an expression for the vertical water entry force as a function of the deadrise angle and compare the results with Figure 8.23.

8.9.6 3D flow effects during slamming

Consider the water impact of an ellipsoid of revolution (spheroid) with a horizontal axis. Assume the water entry velocity is constant and use a von Karman method with a flat plate approximation to calculate water impact loads. Assess the importance of 3D flow effects on the slamming loads by using both eq. (8.45) and a strip theory formulation to calculate heave-added mass.

8.9.7 Whipping studies by a three-body model

The three-body model used in the theoretical and experimental studies of wetdeck slamming-induced global loads on a catamaran in head sea regular waves is described in section 8.4.2. The hydrodynamic loads on the side hulls were calculated by strip theory. Hydrodynamic interaction between the side hulls was neglected.

a) Consider now the high-frequency model used in section 8.5.1 and apply this to each of the three rigid bodies. Derive added mass and damping coefficients in heave and pitch for each of the three bodies, as was done for a single body in section 8.5.1.

b) Explain why it is necessary to account for end terms at both the front and aft ends of the second and third bodies.

8.9.8 Frequency-of-encounter wave spectrum in following sea

Because there is not a one-to-one correspondence between wave frequency ω_0 and frequency of encounter ω_e for all values of ω_e in following seas, eq. (8.120) is not generally valid for following sea. Suppose that ω_e should always be a positive quantity and use that $\omega_e = \left|\omega_0 - U\omega_0^2/g\right|$ for following sea. Show for $\omega_e < 0.25g/U$ that

$$S_e(\omega_e) = \frac{S(\omega_1) + S(\omega_2)}{\sqrt{1 - \frac{4U\omega_e}{g}}} + \frac{S(\omega_3)}{\sqrt{1 + \frac{4U\omega_e}{g}}}, \tag{8.154}$$

where

$$\omega_{1,2} = \frac{g}{2U}\left[1 \pm \sqrt{1 - \frac{4U\omega_e}{g}}\right] \tag{8.155}$$

and ω_3 is the solution of ω_0 following from

$$\omega_e = \frac{\omega_0^2}{g}U - \omega_0. \tag{8.156}$$

Hint: Write

$$S_e(\omega_e) = S(\omega_1)\left|\frac{d\omega_0}{d\omega_e}\right|_{\omega=\omega_1} + S(\omega_2)\left|\frac{d\omega_0}{d\omega_e}\right|_{\omega=\omega_2} + S(\omega_3)\left|\frac{d\omega_0}{d\omega_e}\right|_{\omega=\omega_3} \tag{8.157}$$

Discuss the phase velocities of the wave components ω_i, $i = 1, 2, 3$ relative to the ship velocity U. Show for $\omega_e > 0.25g/U$ that

$$S_e(\omega_e) = \frac{S(\omega_3)}{\sqrt{1 + \frac{4U\omega_e}{g}}}. \tag{8.158}$$

What is now the phase velocity relative to the ship velocity?

We note from eq. (8.154) that $S_e(\omega_e)$ is singular at $\omega_e = 0.25g/U$. Is $S_e(\omega_e)$ integrable?

8.9.9 Springing

Faltinsen (1972) analyzed head sea regular waves on a restrained ship when the wavelength was small relative to the ship length. He showed that the total velocity potential at the ship could be expressed as

$$\varphi = \varphi_0 e^{i\pi/4}\Psi(y, z; x)(x + 0.5L)^{-1/2}. \tag{8.159}$$

Here φ_0 is the incident wave potential and Ψ is a function depending on the cross-sectional shape, ship speed, and wave number. Further, $x = -0.5L$ corresponds to the ship bow. Because the expression is singular at the bow, it is, strictly speaking, not valid there, but we are going to use the expression over the whole ship length.

Consider the generalized wave excitation force $F(k)$ for linear springing of a ship with constant cross section. This was expressed as eq. (8.131) when only the Froude-Kriloff loads were considered, which led to wavelengths in which $F(k)$ was zero as shown in Figure 8.59.

Show how this behavior is modified when the x-dependence given by eq. (8.159) is used to calculate the generalized force $F(k)$. It would be too much work to calculate Ψ, so just set Ψ to be an unspecified constant in the calculations.

9 Planing Vessels

9.1 Introduction

Planing vessels are used as patrol boats, sportfishing vessels, service craft, ambulance craft, recreational craft, and for sport competitions. The Italian vessel *Destriero*, designed by Donald L. Blount and Associates, won in 1992 the Blue Riband Award for the fastest transatlantic passage without refueling. The average speed was 53.1 knots. The vessel is a 67 m–long planing monohull equipped with gas turbines and three waterjets with a combined horsepower of 60,000. Use of hydrodynamic test facilities was an important part of the design. However, the amount of research on planing vessels is relatively small, considering the large amount of different recreational craft that exist. Our focus is on monohull vessels, but catamaran types also exist. Most of the planing vessels have lengths smaller than 30 m. Recreational craft are typically smaller than that.

A vessel is planing when the length Froude number $Fn > 1.2$ (Savitsky 1992). However, $Fn = 1.0$ is also used as a lower limit for planing; that is, we cannot set a clear line of demarcation between planing and nonplaning conditions just by referring to the Froude number. During planing, the weight of the vessel is mainly supported by hydrodynamic pressure loads, with buoyancy having less importance. The hydrodynamic pressure both lifts the vessel and affects the trim angle.

Figure 9.1 shows a typical high-speed planing hull. In order to avoid negative pressures relative to atmospheric pressure on the hull at high speed, it is essential to have flow separation at the transom and along the sides. Pressures less than atmospheric pressure may result in dynamic instabilities of the vessel (Müller-Graf 1997). Flow separation along the sides is usually accomplished by using a hard chine. Further, the longitudinal shape (buttock lines) must not be convex aft of the bow sections (Savitsky 1992). A buttock line is the contour of a longitudinal section parallel to the centerplane. A typical deadrise angle is 10° to 15° for a hard-chine monohull. Up to a 25° deadrise angle is used for high-speed offshore hulls.

A double-chine hull is shown in Figure 9.2. The flow will separate from the lower chine at high speed, whereas the upper chine provides a large local beam at low speed. This is beneficial from a transverse hydrostatic stability point of view at zero speed. The position of the upper chine must be selected to avoid the separating flow from the lower chine at high speed from reattaching to the hull. Grigoropoulos and Loukakis (2002) presented a systematic series of double-chine hull forms with wide transom, warped planing surface and fine bow, developed in the Laboratory for Ship and Marine Hydrodynamics of the National Technical University of Athens (NTUA).

Stepped planing monohulls and catamarans have been successful in offshore racing. Stepped hulls were originally used on flying boats for stabilization during takeoff. Figure 9.3 illustrates a monohull with one step. The vessel is equipped with a free-surface–piercing propeller and ventilating rudder. Trim tabs are placed at the transom to optimize the trim angle. A stepped planing hull has hard chines. The step is placed aft of where the main hydrodynamic lift occurs. This means some distance aft of where the flow separation from the chines starts. In the main text, we will be more clear about what we mean by "some distance." The flow will separate from the step at high speed, and the afterbody will be partially ventilated, reducing the wetted surface and hence the resistance without significantly affecting the hydrodynamic lift. The ventilated length is shortest along the keel. It increases with speed and is dependent on the height of the keel in the flow attachment area above the keel before the step. This is explained further in section 9.2.3. A rough estimate is that the ventilated length may be between $0.5B$ and B, where B is the beam of the vessel. Sometimes more than one step is used. Trim control can also be obtained by placing a hydrofoil at the stern (Clement and Koelbel 1992). This hydrofoil typically carries 10% of the weight. Clement and Koelbel (1992) discuss how to design stepped planing hulls for practical load-carrying purposes. When the step is wet at lower speeds, vortex separation occurs at the step. This increases the resistance relative to no step. The viscous drag

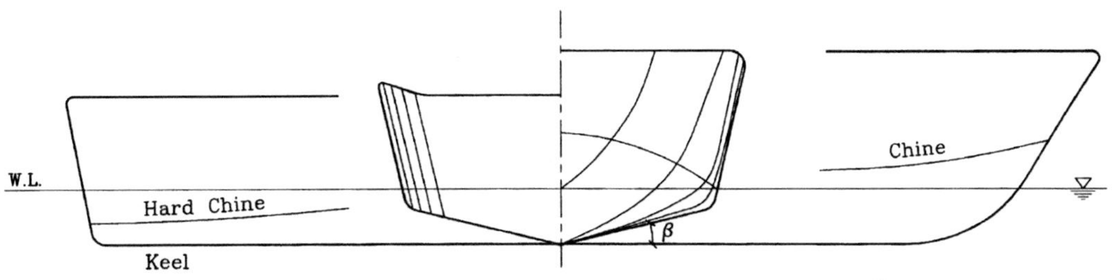

Figure 9.1. Typical high-speed planing hull. Series 62 (Savitsky 1992).

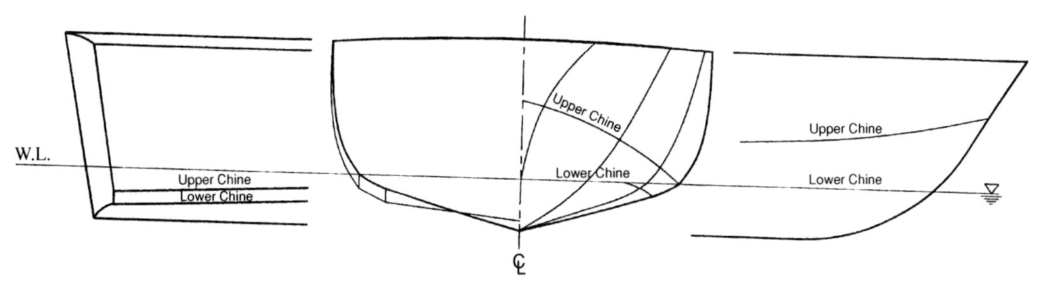

Figure 9.2. Double-chine hull (Savitsky 1992).

associated with a wet step is, roughly speaking, proportional to the step height. Our following discussion does not assume a stepped planing hull. The only exception is in section 9.2.3 dealing with the steady behavior of a stepped planing hull on a straight course.

Transom tabs (flaps) that can be automatically controlled (see section 7.1.3), are also used to optimize the trim angle. This is beneficial for resistance and dynamic stability in heave and pitch (porpoising). Automatically controlled flaps may also reduce the vertical ship motions.

Factors in selecting hull forms from a hydrodynamic point of view are

- Heel stability at zero speed in calm water
- Resistance and propulsion in calm water
- Maneuvering in calm water
- Broaching in following sea
- Steady and dynamic stability at high speed in calm water
- Wave-induced vertical accelerations and roll
- Deck wetness
- Slamming

An important consideration at high speed is the possibility of cavitation and ventilation on, for instance, the rudder and propulsion unit. This may cause unexpected behavior. We cannot deal with all these issues from a theoretical point of view and must rely on experiments.

Steady and dynamic stability are of major concern for planing vessels. The loss of steady heel restoring moment with forward speed is discussed

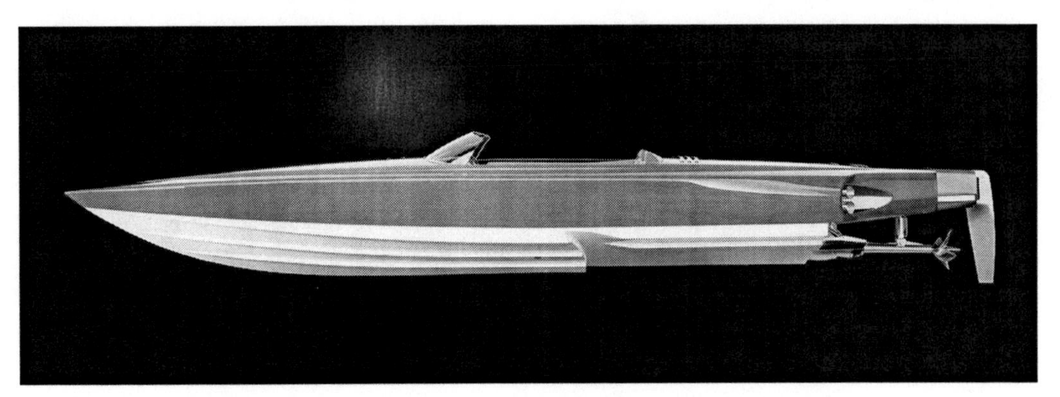

Figure 9.3. The Alpha-Z stepped planing hull designed by Michael Peters.

in section 7.7. Figure 7.40 shows three examples of dynamic instabilities at planing speed. These are "chine walking," "cork-screw" pitch-yaw-roll oscillations, and "porpoising." Cork-screwing is considered in section 10.9.3. Porpoising is a periodic, bounded, vertical plane motion that a planing hull may exhibit at certain speeds. Dynamic instabilities occur in calm water in the absence of excitation. Linear stability analysis shows when this unstable behavior is possible. This behavior is a function of speed, and there is a lower speed limit for porpoising to occur. Common practice suggests that forward movement of the longitudinal center of gravity (lcg) reduces porpoising instabilities, but this may not always be the case. However, one can just lower the speed to avoid the problem. In reality, one often sees porpoising of small pleasure boats, not necessarily with disastrous results. Perhaps some people enjoy the bumpy ride! However, it may lead to structural damage when the motions are so severe that the hull is thrown out of water and subsequently impacts on the water. Porpoising may also result in diving (tripping over the bow).

There are other dynamic stability problems to be considered. Examples are directional stability in calm water and broaching in calm water and following waves. Katayama (2002) has reported the occurrence of bow-diving and transverse porpoising during experiments in calm water. Bow-diving was detected when the model was at high speed and rapidly accelerated. "Aerodynamic pitch up" occurs only for lightweight and very high–speed planing boats. Serious accidents have occurred during racing. An aerodynamic lift with its center of pressure in the bow region may cause the craft to be airborne. The vessel may either flip over or slam back onto the water surface. It is a more serious problem for catamarans than for monohulls because of the large wetdeck area of catamarans.

Milburn (1990) reports an unexpected and unexplained anomaly in planing boat performance that occurred during high-speed trials of the U.S. Coast Guard's 47-foot Motor Life Boat (MLB) prototype. A sudden roll motion, almost like a submarine "snap roll," occurred at speeds greater than 20 knots when the rudder was deflected quickly to 20° or more. It is possible that cavitation and ventilation may have influenced this behavior.

The following description starts with steady flow, that is, how to predict the rise, trim, and resistance of a planing vessel on a straight course in calm water. Steady heel stability and some of the many dynamic instability phenomena occurring at high speed are then considered. A detailed analysis of porpoising and wave-induced vertical motions in head sea is given. The final section deals with maneuvering which also is discussed in Chapter 10.

9.2 Steady behavior of a planing vessel on a straight course

The steady behavior of a planing vessel on a straight course in calm water is a function of how the trim moment, vertical force, and longitudinal force on the hull depend on trim angle, draft (rise), and speed. The pressure can be divided into hydrodynamic and hydrostatic pressures. The hydrodynamic pressure can, to a large extent, be described by potential flow and by neglecting gravity. Flow separation from the chines and the transom stern strongly influences the pressure load distribution and is essential for the water flow to lift and trim the vessel. The lift force is approximately proportional to the trim angle. If the vessel has hard chines, the separation lines along the hull are well defined along the chines. Calculations can then be made by neglecting the effect of the viscous boundary layer on the pressure distribution. This may not be true for a planing vessel with round bilges. The separation lines may then be dependent on laminar or turbulent flow conditions in the boundary layer.

Water resistance components are discussed in Chapter 2. There is a viscous and a residual resistance component scaling with Reynolds number and Froude number, respectively. In addition comes the appendage resistance. The residual resistance includes the wave (pattern) resistance associated with the far-field waves. The residual resistance accounts also for plunging waves and spray generated along the hull. This must be considered in combination with the fact that the planing hull is a lifting surface and a lift-induced resistance is present. However, we can clearly identify only the wave pattern resistance component of the residual resistance.

It is common to talk specifically about a spray resistance for planing vessels. However, it is difficult to identify. It is caused by a pressure and viscous friction component. The spray is affected by surface tension and hence the Weber number.

However, this does not mean that the spray resistance is strongly affected by Weber number. The reason is that the behavior of the spray has a small influence on the flow along the hull causing the spray. The pressure component of the resistance is implicitly accounted for in the residual resistance. In the case of hard chines, the viscous component is associated with an increased wetted hull surface due to the jet flow originating at the spray roots forward of flow separation from the chines. Spray rails are used to minimize this effect.

The wave pattern resistance cannot be neglected at lower planing speeds. The results in Figure 4.14 for a Wigley hull show, for instance, that the wave resistance matters for $Fn = 2$.

The counterpart to resistance is propulsion. This is discussed in Chapter 2. Figure 2.1 shows examples of propulsion systems used for planing vessels. The figure also illustrates appendages that cause resistance. Examples are rudders, struts, and inclined propeller shafts.

The air resistance and the added resistance in wind and waves must also be accounted for. The added resistance in waves can partly be explained as in section 7.5; that is, it is the result of the vessel's ability to generate unsteady waves. However, there is also an important effect due to interaction between the steady and unsteady flows. This effect increases with forward speed. A theoretical model including this effect was presented by Faltinsen et al. (1991a). Good agreement was shown with experimental results up to $Fn = 1.14$ for a round-bilge hull. However, no experimental results were available for higher planing speeds.

The hydrodynamic behavior of a vessel at non-planing speed is important. The results presented for semi-displacement vessels in Chapters 2 and 4 are then relevant. For instance, the trim angle starts to be influenced by the hydrodynamic flow along the vessel when the Froude number is larger than about 0.35. Further, the generation of transverse waves may cause a maximum in the wave resistance for Froude numbers around 0.5. An example of a resistance hump (maximum) is illustrated in Figure 6.15 for a hydrofoil vessel that typically uses a planing hull. The vessel must have sufficient horsepower to overcome a possible hump and to reach planing condition. Because the resistance is influenced by the trim, either trim tabs or interceptors can be used to counteract the trim and minimize the resistance. An optimum trim angle

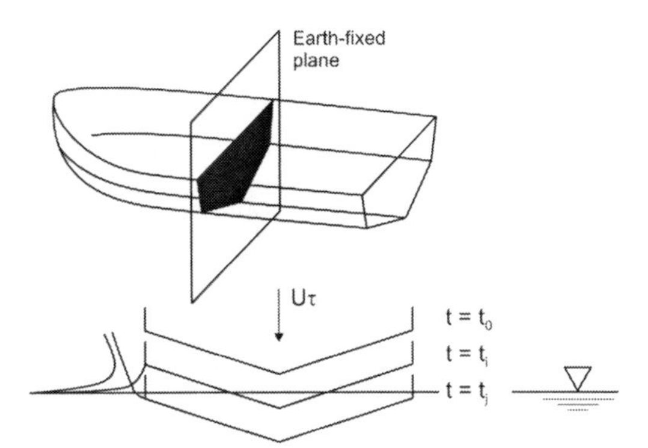

Figure 9.4. Illustration of how a 2D water entry analysis can be used in steady flow analysis of a planing vessel. When the planing craft passes through an Earth-fixed plane, the problem is similar to 2D water entry of a body with changing form. U = ship speed, τ = trim angle (rad) (Zhao et al. 1997).

at planing speed from a resistance point of view is sometimes said to be about 4°. However, this must depend on the hull form. Ikeda et al. (1993) experimentally studied the effect of small and large trim tabs of a hard-chine hull in the Froude number range from 0.7 to 1.2. The trim tab angle varied from 0° to 20° for the small trim tab. The lowest resistance for $Fn < 0.8$ was obtained with a trim tab angle of 20°, whereas no trim tab or no trim tab angle gave the lowest resistance for $Fn > 1.0$. Similar results were obtained with the larger trim tab. Trim tabs are particularly effective in reducing the hump trim angle of vessels with low length-to-beam ratio (Savitsky 1992).

9.2.1 2.5D (2D + t) theory

Figure 9.4 illustrates how a 2D water-entry analysis can be used in a 2.5D (2D + t) analysis of steady flow relative to a ship-fixed coordinate system. An Earth-fixed cross-plane that is initially ahead of the ship is considered. We will follow the fluid particles in this cross-plane. These fluid particles are assumed initially not to know that the ship is coming. This is an approximation. At a certain time $t = t_0$, a cross section of the ship that is not in the water is penetrating the Earth-fixed cross-plane (see Figure 9.4). This ship cross section is above the water and will not influence the fluid particles in the Earth-fixed cross-plane. However, as time evolves, other ship cross sections as illustrated at $t = t_i$ and t_j in Figure 9.4 are penetrating the Earth-fixed cross-plane and are submerged

Figure 9.5. Drop test of a wedge with a 10° deadrise angle and a breadth B = 0.28 m. The wedge is free falling. The pictures show snapshots of the water entry at time instants $t = 0$, 0.01, 0.0219, 0.344, and 0.0625 s. $t = 0$ is when the wedge first hits the free surface. Downward velocity V of the wedge is presented in Figure 9.6. (Photo by Olav Rognebakke.)

in the water. When $t = t_j$, the flow has separated from the chines. The vertical velocity of the ship's cross section is for small τ equal to $U\tau$, where τ is the local trim angle in radians. The analysis of the flow in the studied Earth-fixed cross-plane is therefore the same as the flow due to water entry of a 2D body with changing form and downward velocity $V = U\tau$.

Let us therefore first concentrate on water entry of 2D bodies, particularly wedges with chines (knuckles). Figure 9.5 shows how the free-surface elevation looks in experimental drop tests of a wedge at different time instants. The corresponding water entry velocity V as a function of time is presented in Figure 9.6. Our analysis of steady performance of a planing vessel requires that V is constant with time. However, this was not achieved in the experiments. Figure 9.5 shows that the water initially separates from the chines (knuckles) tangentially to the wedge surface. The water rises almost vertically close to the wedge, with resulting plunging breakers.

Figure 9.7 illustrates calculated free-surface elevation during the water entry of a wedge with deadrise angle 30° and knuckles (hard chines). The calculations were done by a boundary element method (BEM). Potential flow of an incompressible fluid was assumed, and the exact free-surface conditions without gravity were satisfied. However, the method numerically cuts off parts of

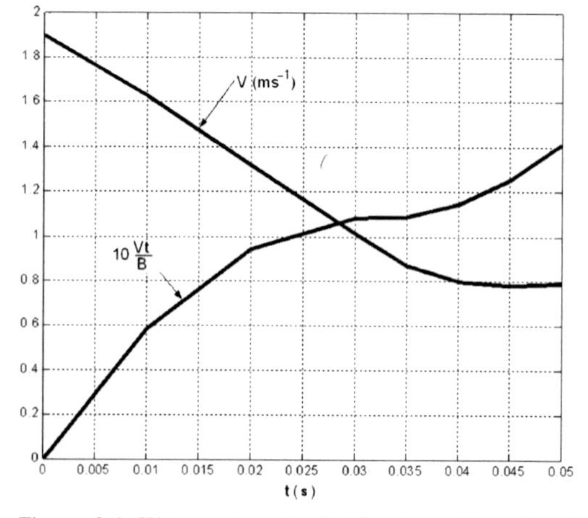

Figure 9.6. Water entry velocity V and nondimensional velocity Vt/B as a function of time for the experiments presented in Figure 9.5. B = beam.

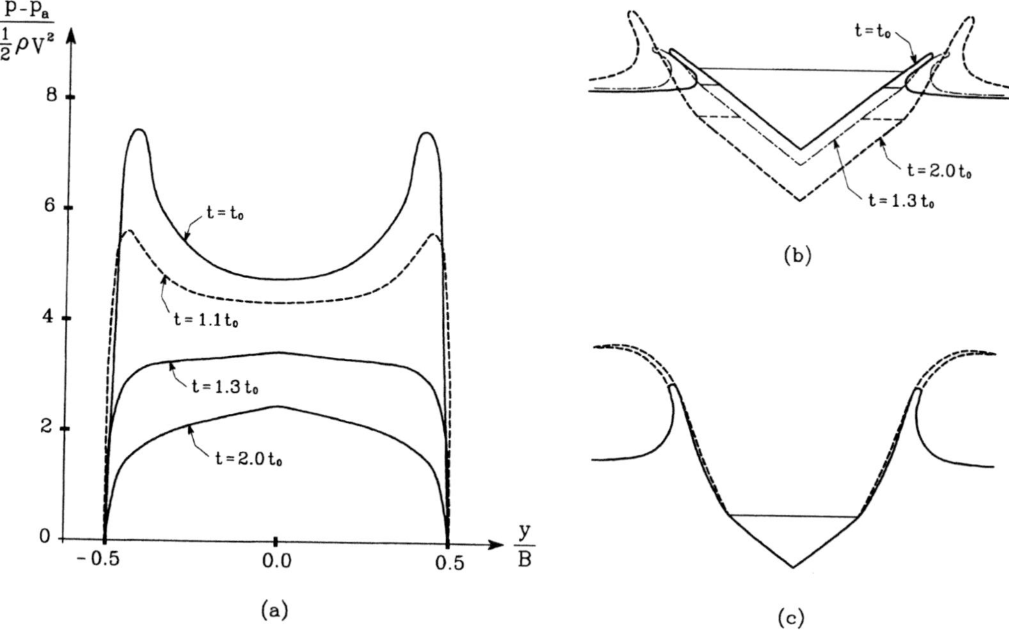

Figure 9.7. The pressure (p) distribution and free-surface elevation during the water entry of a wedge with deadrise angle 30° and chines (knuckles), calculated by a fully nonlinear potential flow solution without gravity. V is constant drop velocity, p_a is atmospheric pressure, ρ is mass density of the water, and B is breadth of the wedge. y is horizontal coordinate on the body surface. t_0 is the time instance when the spray roots of the jets reach the separation points (chines). (a) Pressure distribution at selected time instants after flow separation from the chines. (b) Free-surface elevation at selected time instants after flow separation from the chines. (c) Comparison of free-surface elevation between theory and experiments. $t = 2.9_0$; ————, theory; - - - - - - - -, experiments by Greenhow and Lin 1983 (Zhao et al. 1996).

the spray, as we see in the figure. This does not have a significant effect on the flow outside the neglected spray domain. This is confirmed in Figure 9.7 by comparing with the experimental results by Greenhow and Lin (1983). The flow at this relatively large deadrise angle shows also almost vertical jets, that is, similar to those in Figure 9.5. Figure 9.7 also shows the resulting pressure distribution on the wedge. What is of interest in the corresponding analysis of steady flow around the ship is the resulting vertical force. The case of a deadrise angle of 20° is presented in Figure 8.24. A major contribution to the vertical force occurs ahead of the flow separation from the chines. Because the separation line has to be known in the 2.5D analysis by Zhao et al. (1996), rounded bilges with flow separation are difficult to handle. We will come back later to how this vertical force following from the water entry analysis can be used in calculating rise, trim, and resistance of a planing vessel.

Let us now illustrate how a free-surface elevation, as in Figures 9.5 and 9.7, appears in a ship-fixed coordinate system. The coordinate transformation $x = Ut$ is then introduced. Here x is a body-fixed longitudinal coordinate of the ship. The positive x-direction is from the bow toward the stern. $t = 0$ corresponds to the initial time of water entry of a ship cross section into the water in the previously mentioned Earth-fixed cross-plane. The x-coordinate of this ship cross section is then zero according to $x = Ut$.

This is illustrated in Figure 9.8 by showing the free-surface elevation for four sections: A, B, C, and D. How this looks depends on

$$\frac{Vt}{B} = \frac{(U\tau)(x/U)}{B} = \frac{\tau x}{B}. \tag{9.1}$$

There is a hole in the water at the last section, D. This will in reality disappear at some distance behind the ship, and a rooster tail as in Figure 4.18

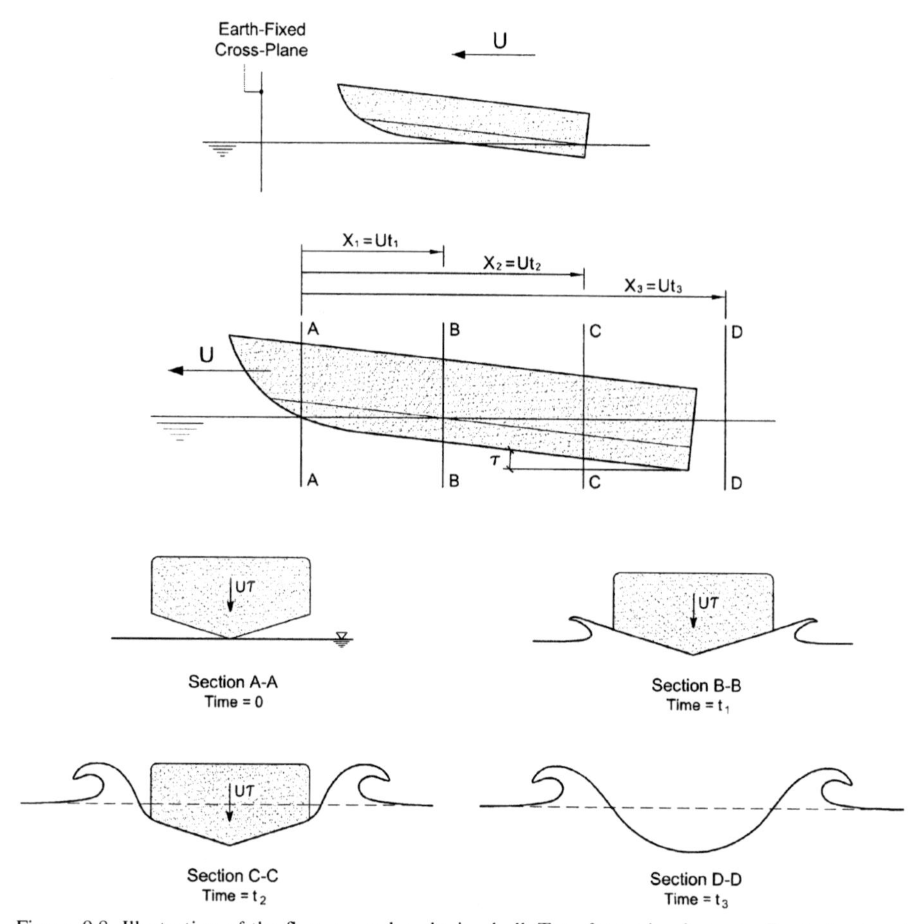

Figure 9.8. Illustration of the flow around a planing hull. Transformation between Earth-fixed and ship-fixed coordinate systems (see also Figures 9.5 and 9.7).

will appear. This is affected by gravity. Further, gravity will pull down the spray appearing at cross section C. The presence of gravity will cause far-field waves and therefore wave pattern resistance.

A simplified version of the 2.5D theory will first be used to explain qualitatively why a transom stern and hard chines are important in creating hydrodynamic lift and trim moment. The basis is eq. (8.44), which will be expressed in a coordinate system following the forward velocity of the vessel. Eq. (8.44) then gives the following vertical force per unit length on a cross section:

$$f_3 = U\frac{d}{dx}(a_{33}U\tau) \tag{9.2}$$

It is here used that $x = Ut$. Further, a_{33} is the two-dimensional infinite-frequency added mass in heave of the cross section. If the keel line is expressed as $z(x)$ where z is positive upward, then the local trim angle $\tau = -dz/dx$. If the keel is a straight line, as it is for the prismatic hull forms studied by Savitsky (1964) (see section 9.2.2), the local trim angle is the same as the global trim angle of the vessel.

Eq. (9.2) gives a vertical force on a cross section only as long as $a_{33}\tau$ varies with x at the cross section. When flow separation from the chines has occurred, neither a_{33} nor τ change for a prismatic planing hull; that is, f_3 is zero. By comparing this with Figure 8.24 and exchanging the abscissa from Vt/B to $\tau x/B$, as in eq. (9.1), we see that this zero force is only qualitatively true at some distance downstream of where the chine wetting starts.

Integrating eq. (9.2) over the whole length of the vessel gives the hydrodynamic lift force

$$L = U^2 \tau a_{33}(x_T), \qquad (9.3)$$

where x_T means the x-coordinate of the transom stern. If the vessel had a pointed aft end, then $a_{33}(x_T)$ is zero; that is, there is no hydrodynamic lift on the vessel. This demonstrates the importance of the transom stern to create lift. Eq. (9.3) shows, for finite $\tau a_{33}(x_T)$, that the lift force increases with the square of the speed. Because $a_{33}(x_T)$ is proportional to B^2, the lift will also have this proportionality factor according to the simplified method. Further, $a_{33}(x_T)$ will decrease with increasing deadrise angle.

Eq. (9.3) illustrates that the vessel must have a trim angle to experience a hydrodynamic lift. This is caused by the trim moment on the hull. In order to generate a trim moment causing the bow to rise, it is essential that the center of pressure of the lift is forward of the center of gravity during the acceleration to a constant velocity U. This is achieved when the flow separation from the chines starts close to the bow and $a_{33}\tau$ does not change downstream from where chine wetting starts. If τ does not vary, that is, the keel and buttock lines are straight and parallel, constant $a_{33}\tau$ means constant wetted cross section. This is achieved by both the geometry of the cross section and the fact that the hard chines force the flow to separate.

Because the 2D vertical force expressed by eq. (9.2) is a consequence of hydrodynamic pressure loads, the equation predicts that negative hydrodynamic pressures on a cross section occur when $d(a_{33}\tau)/dx$ is negative. The total pressure is the sum of hydrostatic, hydrodynamic, and atmospheric pressures. Negative pressures relative to atmospheric pressure may result in dynamic instabilities of the vessel (Müller-Graf 1997). Let us assume that the speed is sufficiently high for the hydrostatic pressure to be negligible. For instance, if the cross sections are wedge-formed with constant deadrise angles, eq. (9.2) shows that negative pressures relative to atmospheric pressure occur when $b^2\tau$ decreases with increasing x, that is, toward the stern. Here b is the local beam. If the keel line is convex, τ decreases with increasing x. Our slender-body theory results are therefore consistent with Savitsky's (1992) warning against using convex keel and buttock lines aft of the bow sections. If the planing surface is warped, the deadrise angle varies with x and influences a_{33}. If we use the results for a_{33} of wedges, then a_{33}/b^2 will increase with decreasing β (see Figure 9.27). This qualitative analysis shows that negative hydrodynamic pressures can be avoided by a proper design of the keel and buttock lines, local beam, and deadrise angles. However, the following simplifications in the analysis should be remembered.

- Eq. (9.2) is approximate.
- The considered pressure has been averaged over a cross section by weighing the pressure with the z-component of the normal surface vector.
- The effect of propulsors have not been considered. A propeller may cause negative hydrodynamic pressures on the hull.

If a trim tab is considered hydrodynamically as an integrated part of the hull, eq. (9.2) can be used to approximate the trim moment caused by a trim tab. However, 3D flow effects matter in the close vicinity of the stern.

The following text returns to the more exact 2.5D theory by Zhao et al. (1997) and demonstrates that it matters that there is non-zero vertical force aft of the cross section where flow separation from the chines starts.

9.2.2 Savitsky's formula

Zhao et al. (1997) compared their 2.5D method with the empirical formula by Savitsky (1964) for the lift, drag, and center of pressure coefficients for a prismatic and chine wetted planing hull. The formula is based on extensive experimental data. The lift coefficient can be written as

$$C_{L\beta} = C_{L0} - 0.0065\beta C_{L0}^{0.60}. \qquad (9.4)$$

Here

$$C_{L\beta} = \frac{F_{L\beta}}{0.5\rho U^2 B^2}$$

and

$$C_{L0} = \frac{F_{L0}}{0.5\rho U^2 B^2} = \tau_{\text{deg}}^{1.1}\left(0.012\lambda_W^{0.5} + 0.0055\lambda_W^{2.5}/Fn_B^2\right). \qquad (9.5)$$

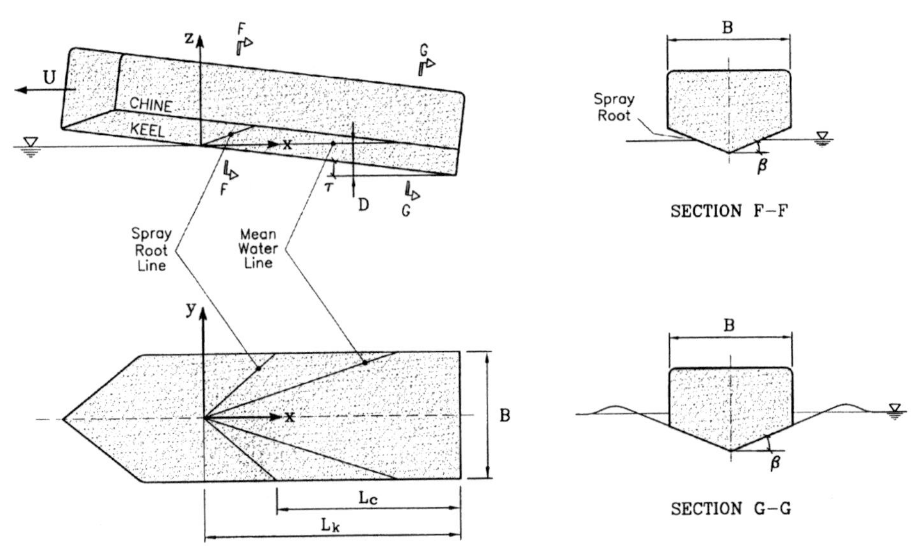

Figure 9.9. Coordinate system (x, y, z) and symbols used in a prismatic planing boat analysis (Savitsky 1964).

Further,

C_{L0} = lift coefficents for zero deadrise angle ($\beta = 0^\circ$)
$C_{L\beta}$ = lift coefficient
F_{L0} = lift force for zero deadrise angle ($\beta = 0^\circ$)
$F_{L\beta}$ = lift force
λ_W = mean wetted length-to-beam ratio
$\tau_{\deg}$ = trim angle of planing area in degrees
τ = trim angle of planing area in radians
β = angle of deadrise of planing surface in degrees
B = beam of planing surface.
Fn_B = $U/(gB)^{0.5}$

We note that the beam is used as a length parameter in the Froude number. A reason is that the beam is fixed whereas a longitudinal length, such as the wetted keel length, is not known before the equations of equilibrium for vertical force and trim moment for a given speed are solved.

Eq. (9.4) is valid for $2^\circ \leq \tau_{\deg} \leq 15^\circ$ and $\lambda_W \leq 4$. Figure 9.9 defines the hull geometry and the angles β and τ. The mean wetted length-to-beam ratio λ_W is equal to $0.5(L_K + L_C)/B$ (see Figure 9.9). L_K and L_C are, respectively, the keel and chine wetted lengths. Savitsky's formula assumes a prismatic hull form; for example, the deadrise angle is constant along the craft.

Ikeda et al. (1993) experimentally studied a series of hard-chine hulls (Figure 9.10). The length-to-beam ratios varied between 3 and 6. The deadrise angles of the aft part of the series B vessels were kept constant while the deadrise angles of the series A ships became zero at the transom stern. If the deadrise angle in Savitsky's formula were chosen at the forward section midway between the first keel wetted section and the first chine wetted section, Ikeda et al. show that Savitsky's formula can be applied to their nonprismatic hulls for length Froude numbers larger than about 0.9. This is valuable information, but we should be careful in generalizing this finding. Resistance, rise, and trim from systematic series of model experiments have also been presented by Clement and Blount (1963), Keuning and Gerritsma (1982), and Keuning et al. (1993).

We note from Savitsky's formula that the lift goes to zero when the trim angle goes to zero. The trim angle plays a role similar to that of the angle of attack in hydrofoil theory. Further, part of the lift decreases linearly with increasing deadrise angle β.

The resistance component R_P due to the pressure force in this case is simply

$$R_P = F_{L_\beta}\tau. \tag{9.6}$$

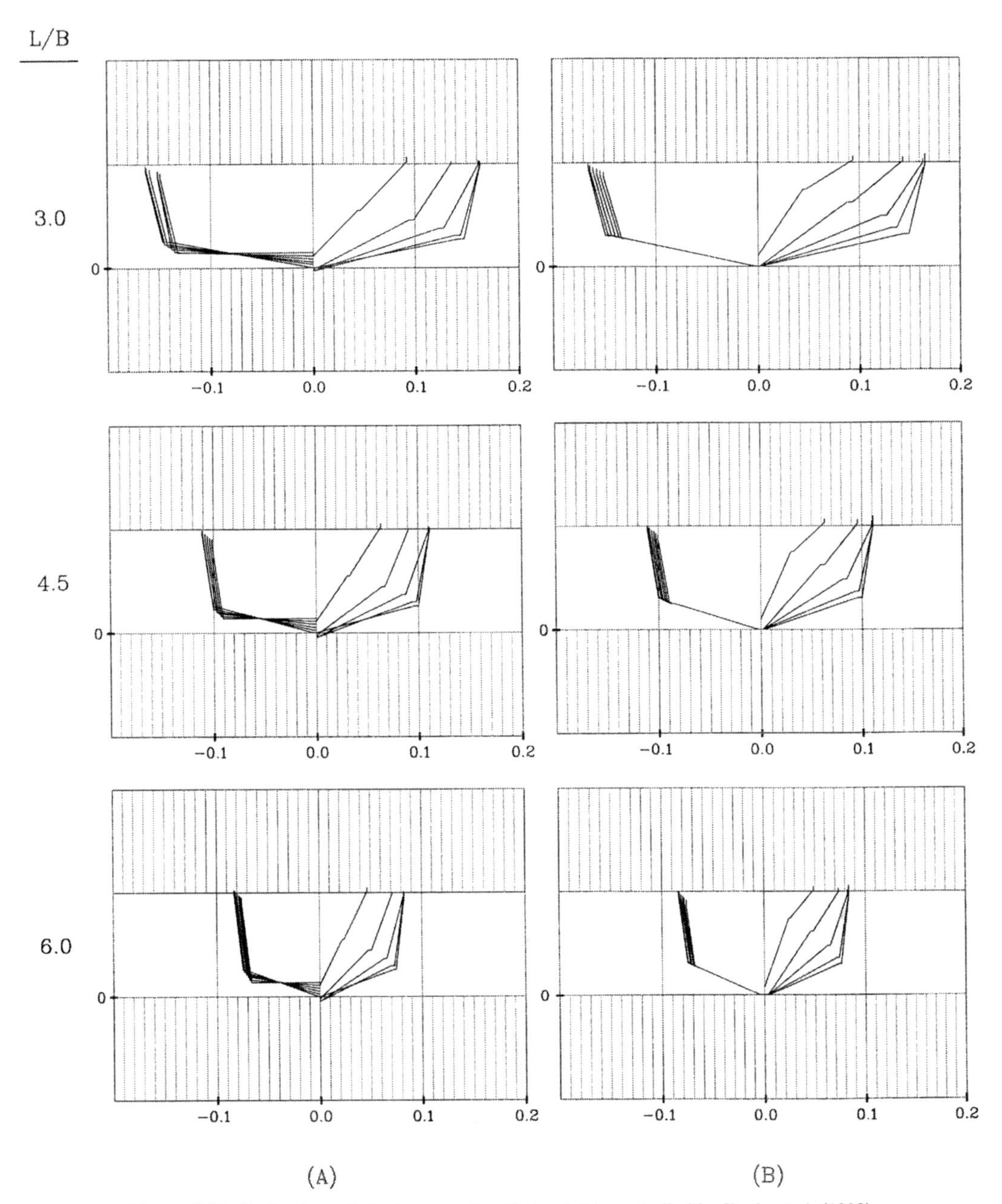

Figure 9.10. Body plans of planing vessels with hard chines studied by Ikeda et al. (1993).

Here τ is in radians. The longitudinal position of center of pressure is expressed by

$$\frac{l_p}{\lambda_W B} = 0.75 - \frac{1}{5.21 Fn_B^2/\lambda_W^2 + 2.39}, \quad (9.7)$$

where l_p is the distance measured along the keel from the transom stern to the center of pressure of the hydrodynamic force. The parts of the force and moment expression obtained by letting $Fn_B \to \infty$ in eqs. (9.5) and (9.7) are the result of the hydrodynamic lift. Hydrostatic loads and the effect of free-surface wave generation are implicitly included in the formula.

Figure 9.11 shows comparisons between the empirical formula and the numerical results for the lift coefficient and the center of pressure. Infinite Froude number Fn_B is assumed in eqs. (9.5) and (9.7). Because there is a simple relationship

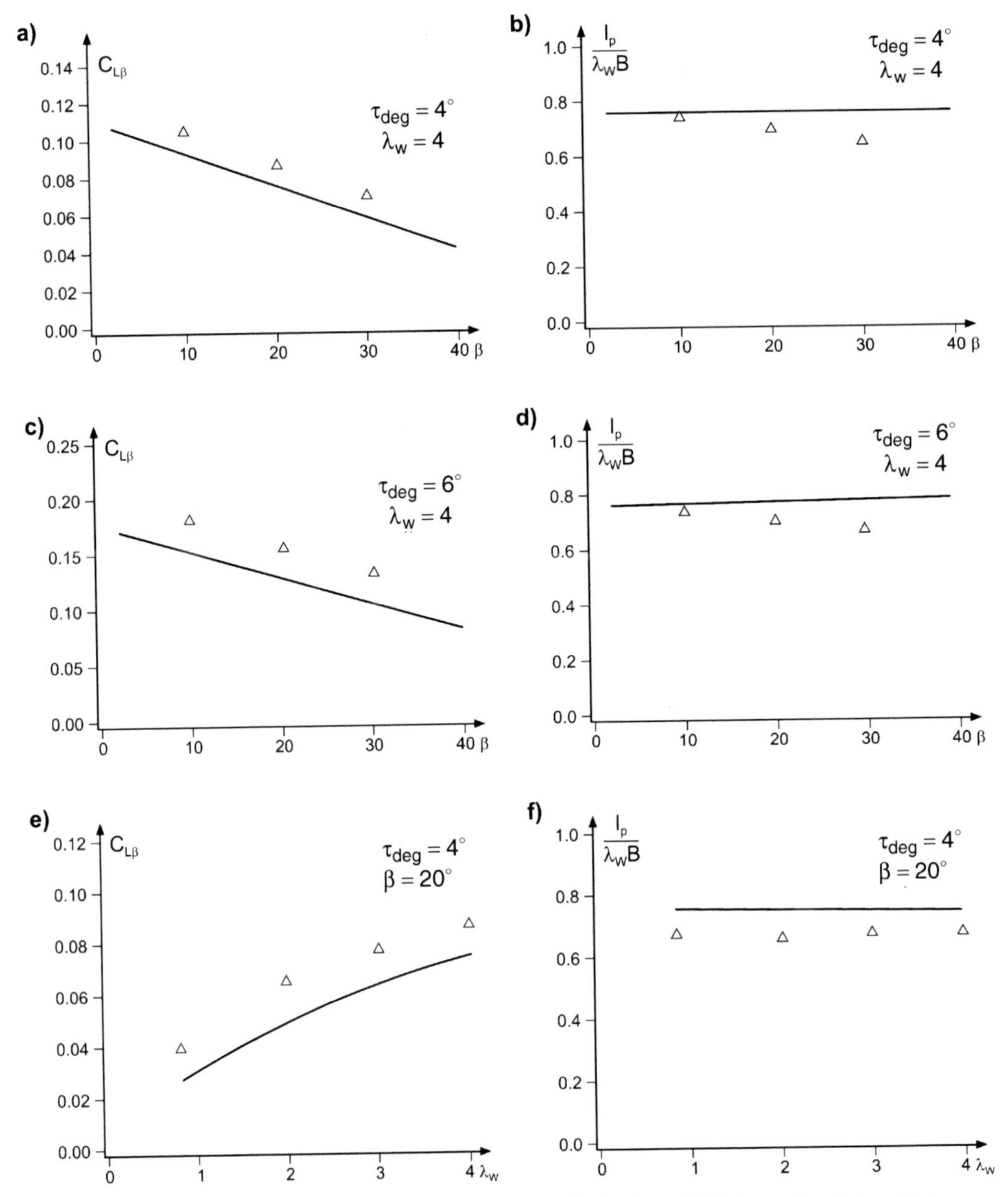

Figure 9.11. Comparison between the empirical formula by Savitsky (1964) (———) and the numerical results (Δ) for the lift coefficient $C_{L\beta}$ and the center of pressure at infinite Froude number. β is the angle of deadrise of planing surface, l_p is the distance along the keel from the transom stern to the center of pressure, τ_{deg} is the trim angle of the planing area in degrees, and λ_W is the mean wetted length-to-beam ratio (Zhao et al. 1997).

(see eq. (9.6)) between R_P and $F_{L\beta}$, results are not presented for R_P. There is a reasonable agreement between the theory and the empirical formula. The results are presented either as a function of deadrise angle β or as a function of λ_W. The detailed theoretical force distribution along the ship when $\beta = 20^{\circ}$ can be obtained approximately from Figure 8.24 by setting $V = U\tau$ and $x = Ut$. This means that the horizontal axis in Figure 8.24 becomes $\tau x/B$. The maximum force value is at $x = L_K - L_C$ (see the definition in Figure 9.9), that is, when the spray roots are at the chines and flow separation has started. The theoretical force distribution up to $x = L_K - L_C$ is the same for all λ_W. The difference in theoretical values for different values of λ_W is simply a matter of how large the maximum of $\tau x/B$ (or Vt/B in Figure 8.24) is. Even if the local forces aft of the chine position where flow separation starts are smaller than those ahead of the separation point, they are not negligible. The λ_W-dependence of the results demonstrates that. This means that the effect of flow separation on the hydrodynamic loads is important. The theory overpredicts $C_{L\beta}$ and underpredicts l_p relative to the empirical formula. The difference between theory and the empirical formula does not vary much with λ_W. A possible reason for the difference is that three-dimensional flow effects are of some importance from $x = 0$ to $x = L_K - L_C$. This is likely because $(L_K - L_C)/B$ varies from 0.63 to 2.66 in the numerical results presented in Figure 9.11. The values of $(L_K - L_C)/B$ imply a rapid change of the flow in the longitudinal direction from $x = 0$ to $x = L_K - L_C$. The general trend is that $(L_K - L_C)/B$ decreases with decreasing deadrise angle β for fixed value of trim angle τ. If β is fixed, then $(L_K - L_C)/B$ decreases with increasing value of τ. If the three-dimensional flow effects are mainly in the bow region, it can explain why the differences between theory and the empirical formula does not vary much with λ_W. Lai's (1994) three-dimensional numerical results are also an indication of the presence of three-dimensional effects.

Alternative flow description in the bow region

For sections in which flow separation from the chines does not occur, we may use the similarity solution or the Wagner's solution for water entry of wedges with constant velocity to estimate the vertical force. We start with Wagner's solution, that is, eq. (8.52), to estimate where flow separation occurs. Eq. (8.52) expresses the wetted half beam of the wedge as a function of time. This can be expressed as a function of the longitudinal vessel coordinate x by noting that $Vt = x\tau$ (see the previous discussion and Figure 9.9). This implies that low separation from the chines will start at $x = x_S = L_K - L_C$, where x_S satisfies

$$\frac{B}{2} = \frac{\pi}{2\tan\beta} x_S \tau. \tag{9.8}$$

Because $\lambda_W = 0.5 L_K/B$ with $L_C = 0$, flow separation from the chine does not occur when

$$\lambda_W < 0.5 \frac{\tan\beta}{\pi\tau}. \tag{9.9}$$

For instance, if $\beta = 20^\circ$ and $\tau_{\text{deg}} = 4^\circ$, eq. (9.9) gives $\lambda_W < 0.83$. We could alternatively have used the similarity solution to find x_S. We introduce then $z_{\max}$ defined in Figure 8.22. This means that the vertical distance between the point of maximum pressure and the keel is $Vt + z_{\max}$. The horizontal coordinate of this point in the outer domain solution, in which we do not see the details of the spray roots, is then c. It follows by geometry that $c = (z_{\max} + Vt)/\tan\beta$. The flow separation at the chine will then start at x_S, which satisfies

$$\frac{B}{2} = \left(1 + \frac{z_{\max}}{Vt}\right)\frac{Vt}{\tan\beta} = \left(1 + \frac{z_{\max}}{Vt}\right)\frac{x_S\tau}{\tan\beta}. \tag{9.10}$$

We can determine $z_{\max}/Vt$ by means of Table 8.3. Eq. (9.10) implies that $\pi/2$ on the right-hand side of eq. (9.8) is replaced by $1 + z_{\max}/Vt$. We see that this factor is 1.5087 for $\beta = 20^\circ$ and this will cause a larger x_S-value than Wagner's solution.

The vertical force distribution along the hull can then be approximated as

$$\frac{F_3^{2D}}{\rho V^3 t} = K, \tag{9.11}$$

where K depends only on β (see Table 8.3). F_3^{2D} means the same as F_3 in Table 8.3. We now introduce x and U in eq. (9.11). This means

$$F_3^{2D} = \rho K U^2 \tau^3 x. \tag{9.12}$$

The total vertical force follows by integration and is

$$F_3 = \rho K U^2 \tau^3 0.5 x^2. \tag{9.13}$$

For $x = L_K$ and no flow separation, that is, $L_C = 0$, this means that

$$C_{L\beta} = \frac{F_3}{0.5\rho U^2 B^2} = K\tau^3 4\lambda_W^2. \tag{9.14}$$

If, for instance, $\tau_{\text{deg}} = 4^\circ$, $\beta = 20^\circ$, and $\lambda_W = 0.83$, this gives $C_{L\beta} = 0.04$, which agrees with the numerical results in Figure 9.11e). We may note that eqs. (9.4) and (9.14) have a very different parametric dependence for small λ_W. However, it may be that Savitsky's formula was not intended for small λ_W. Actually, because none of the cross sections is then wetted over the breadth B, B is an

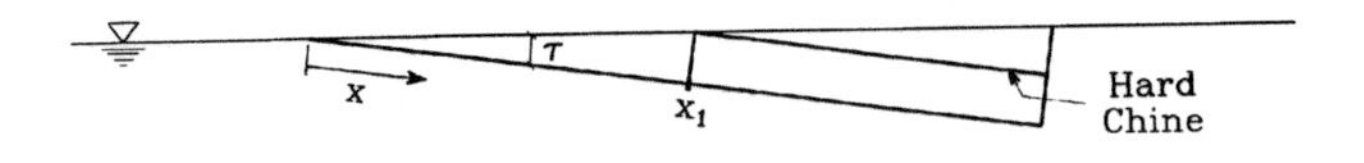

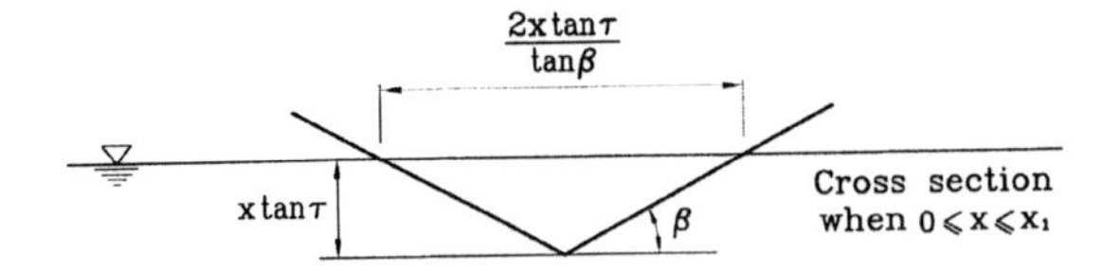

Figure 9.12. Definition of variables used in the calculation of the hydrostatic vertical force on a prismatic planing vessel.

unphysical length dimension to use for nondimensionalizing the force.

We can also use eq. (9.12) to find the center of pressure of the force. The corresponding x-coordinate is

$$x_C = \frac{1}{F_3}\int_0^{L_K} x F_3^{2D}\, dx = \frac{2}{3}L_K. \qquad (9.15)$$

We write this similar to eq. (9.7) and note that $\lambda_W = 0.5L_K/B$ and $l_p = L_K - x_C$. This gives

$$\frac{l_p}{\lambda_W B} = \frac{2}{3}, \qquad (9.16)$$

which agrees with the corresponding numerical value in Figure 9.11f).

Gravity effects

Gravity has to be accounted for at a finite Froude number for the planing hull. There are, in principle, two effects: hydrostatic pressure and generation of gravity waves. However, the latter effect is considered small in the following discussion. The hydrostatic pressure contribution is evaluated by considering the hull volume ∇ below the intersection between the mean free surface and the hull in its planing condition.

We use Figure 9.12 to illustrate the calculations. An x-axis along the keel is introduced where $x = 0$ and $x = x_1$ correspond to where the keel and hard chines, respectively, intersect the mean free surface. The cross-sectional area $A(x)$ below the mean free surface between $x = 0$ and $x = x_1$ can be expressed as

$$A(x) = \frac{x^2 \tan^2 \tau}{\tan \beta}.$$

The hull volume from $x = 0$ to x_1 below the mean free surface is then

$$\mathrm{Vol}_1 = \int_0^{x_1} A(x)\, dx = \frac{1}{3}x_1^3 \frac{\tan^2 \tau}{\tan \beta}.$$

The hull volume below the hard chines from $x = x_1$ to the transom is simply

$$\mathrm{Vol}_2 = (L_K - x_1)0.25B^2 \tan \beta.$$

Then we have to add the hull volume between the mean free surface and a plane between the hard chines from $x = x_1$ to the transom. The final answer is

$$\nabla = x_1^3 \tan^2 \tau/(3\tan\beta) + (L_K - x_1)\,0.25B^2 \tan\beta + 0.5(L_K - x_1)^2 \tan \tau \cdot B,$$

where $x_1 = 0.5B\tan\beta/\tan\tau$. Writing the vertical force as $F_{HS} = \rho g \nabla$ gives

$$C_{LHS} = \frac{F_{HS}}{0.5\rho U^2 B^2} = \frac{2}{Fn_B^2}\cdot\frac{\nabla}{B^3}. \qquad (9.17)$$

This assumes the wetted hull surface is below the mean waterplane, but because the dry hull surface above the chines is vertical, it does not contribute to vertical forces. Further, a correction for a dry transom stern has a negligible effect on the vertical force. We want to stress that what we are doing is approximate and that the generation of free-surface waves should have been analyzed simultaneously with the lifting effect. The effect of hydrostatic pressure would then have been included. However, we continue with our simplifications.

Another effect is a suction pressure at the transom stern. This is caused by the flow separation from the transom stern and the fact that the pressure has to be atmospheric at the transom stern. The consequence is a small loading in the vicinity of the transom stern. This will be accounted for by using a smaller L_K in the expression for ∇. Reducing L_K somewhat arbitrarily to $0.5B$ correlates well with Savitsky's formula. This is illustrated in Figure 9.13, in which $C_{L\beta}$ and C_{LHS} are presented as functions of $1/Fn_B^2$ for $\beta = 10^\circ$, $\tau_{\deg} = 4^\circ$, and $\lambda_W = 3$. The value of $C_{L\beta}$ for $1/Fn_B^2 = 0$

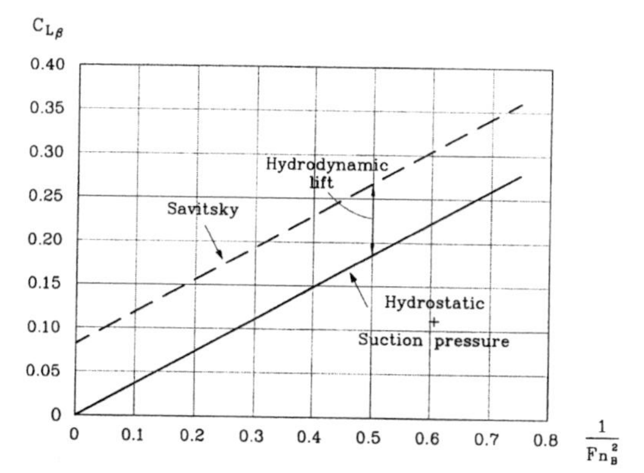

Figure 9.13. Comparison between Savitsky's lift coefficient $C_{L\beta}$ and the lift coefficient C_{LHS} due to suction pressure at the transom stern and the hydrostatic pressure. Prismatic planing hull. $\beta = 10^\circ$, $\tau_{deg} = 4^\circ$, and $\lambda_W = 3$ (Faltinsen 2001).

is the hydrodynamic lift. Because $C_{L\beta}$ and C_{LHS} are nearly parallel to increasing $1/Fn_B^2$, it suggests that the steady lift force on a planing hull can be divided into hydrodynamic lift, buoyancy force, and a suction pressure loading at the transom stern. It means that gravity wave generation has a minor relative importance on the lift. However, because we somewhat arbitrarily reduced L_K with $0.5B$ to reach our conclusion, we cannot be sure about that. In order to be more precise in this matter, we would need a numerical tool that includes the effect of wave generation and at the same time is able to predict the details of the flow at the transom.

9.2.3 Stepped planing hull

The strategy behind the design of a stepped planing hull is to reduce the viscous resistance by decreasing the wetted hull surface area while maintaining a high hydrodynamic lift force. This can be achieved if the flow separates from a step (see Figure 9.3) and ventilates the aft part of the hull in an area where the hydrodynamic pressures are small for the same planing hull without a step. Because the vertical hydrodynamic force per unit length has a maximum at which flow separation from the chines starts (see the previous discussion of results by the 2.5D method in section 9.2.1), it means the step must be placed some distance aft of this location.

The flow separation from the step raises two important questions:

- What is the condition for the flow to separate at the step and cause ventilation aft of the step?
- What is the length of the ventilated area of the hull?

These questions can be answered by first neglecting the hull aft of the step. The answer to the first question can then be found from experimental investigations by Doctors (2003), who studied the condition for the transom of a monohull to be dry. The most important parameter is the draft Froude number $Fn_D = U/(gD)^{1/2}$, where D is the draft at the transom measured relative to the calm water level. This means D accounts for the rise and trim of the vessel at the considered speed. An estimate of the condition for separation with ventilation at the step is given as

$$\frac{U}{(gD_S)^{1/2}} > 2.5 \tag{9.18}$$

based on Doctors (2003). Here D_S is the draft at the step accounting for the trim and rise of the vessel.

In order to answer the second question about the length of the ventilated hull area aft of the step, we can use the empirical formula by Savitsky (1988) for the centerline free-surface profile aft of the transom stern of a prismatic planing hull. If the separated flow from the step reattaches to the aft hull, it will do so first at the centerline. In the local coordinate system (X, Z) defined in Figure 9.14, the free-surface profile can be expressed as

$$\frac{Z}{B} = C_1\left(\frac{X}{B}\right)^2 - C_2\left(\frac{X}{B}\right)^{2.44} + C_3\left(\frac{X}{B}\right), \tag{9.19}$$

where

$$C_1 = 0.02064\left[\frac{\tau_{deg}^{0.7}}{Fn_B^{0.6}}\right]^2 \tag{9.20}$$

$$C_2 = 0.00448\left[\frac{\tau_{deg}^{0.7}}{Fn_B^{0.6}}\right]^{2.44} \tag{9.21}$$

$$C_3 = 0.0108\,\lambda_W\,\tau_{deg}^{0.34}. \tag{9.22}$$

Here B is the beam of the planing surface, τ_{deg} is the trim angle in degrees, Fn_B is the Froude number with the beam as a length parameter, and λ_W is the mean wetted length-to-beam

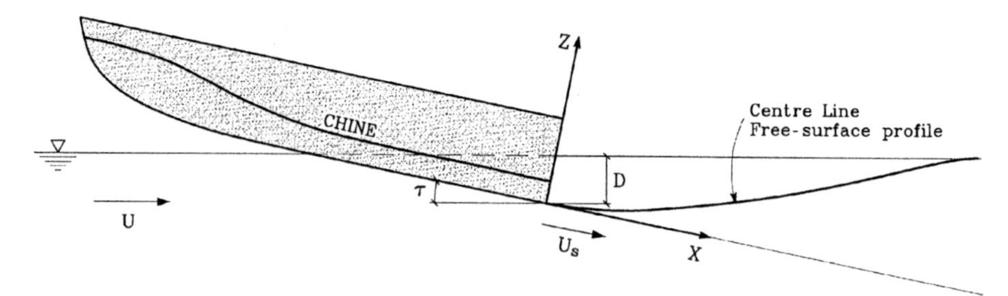

Figure 9.14. Definition of the parameters and coordinate system used in the analysis of the centerline free-surface profile aft of the transom stern of a planing vessel. U = forward speed, U_S = tangential flow velocity at the transom stern, D = draft at the transom.

ratio. According to Savitsky (1988), eq. (9.19) is valid for $6^\circ \leq \tau_{deg} \leq 14^\circ$, $3 \leq Fn_B \leq 6$, $10^\circ \leq \beta \leq 30^\circ$, $0.34 \tan\beta/\tan\tau \leq \lambda_W \leq 3$, and $0 \leq (X/B) \leq 6$. Here β means the deadrise angle.

Examples of predictions based on eq. (9.19) are shown in Figure 9.15, which illustrates that the length of the hollow aft of the transom increases with Fn_B for given values of λ_W and trim. This is also consistent with the results in Figure 4.23 for length Froude numbers between 0.5 and 0.8.

For a given step height, that is, a given value of Z/B, we can use eq. (9.19) to find if and where the flow reattaches to the aft hull. Let us say Z/B is 0.02. Figure 9.15 shows the the corresponding ventilated length X along the keel for different beam Froude numbers Fn_B. The trim angle τ and the mean wetted length-to-beam ratio λ_W are assumed constant in Figure 9.15. These values will in reality be a function of Fn_B. The results also give guidance in the use of several steps. For instance, if the longitudinal distance between the first step and the transom is much longer than the predicted ventilated length, one or more steps may be introduced. A procedure such as this must also consider the fact that the reattached flow will cause additional pressure loads on the vessel that influence the trim and rise. This is an area requiring further research. A possibility is to consider the wetted hull surface abaft the reattachment as a high-aspect–ratio planing surface. However, the inflow from the separated flow from the step should be accounted for. One suggestion is to introduce a relative trim angle between the vessel's trim and the incident free-surface profile at the reattachment.

Local analytical solution near the transom

We will show how we can analytically derive the free-surface profile at the centerline aft of the

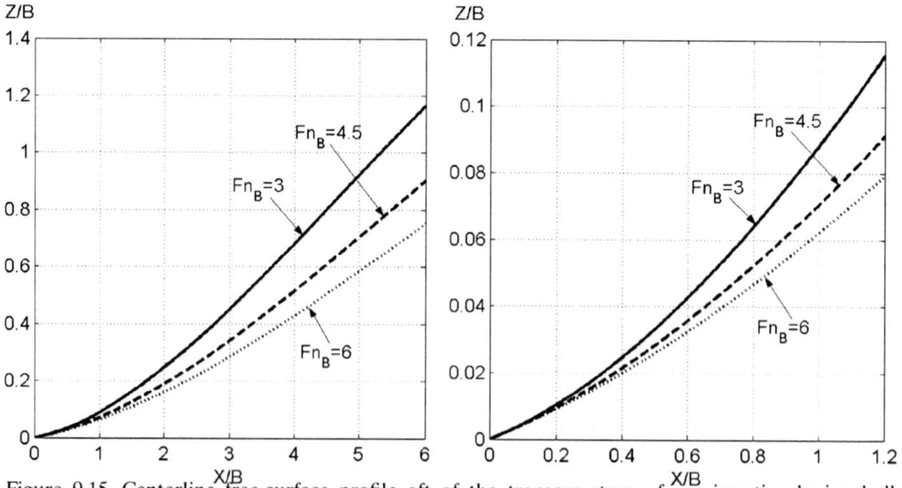

Figure 9.15. Centerline free-surface profile aft of the transom stern of a prismatic planing hull (Savitsky 1988). The right figure gives a detailed view of a part of the left figure. Coordinate system (X, Z) is defined in Figure 9.14. $\tau_{deg} = 6^\circ$, $\lambda_W = 2$.

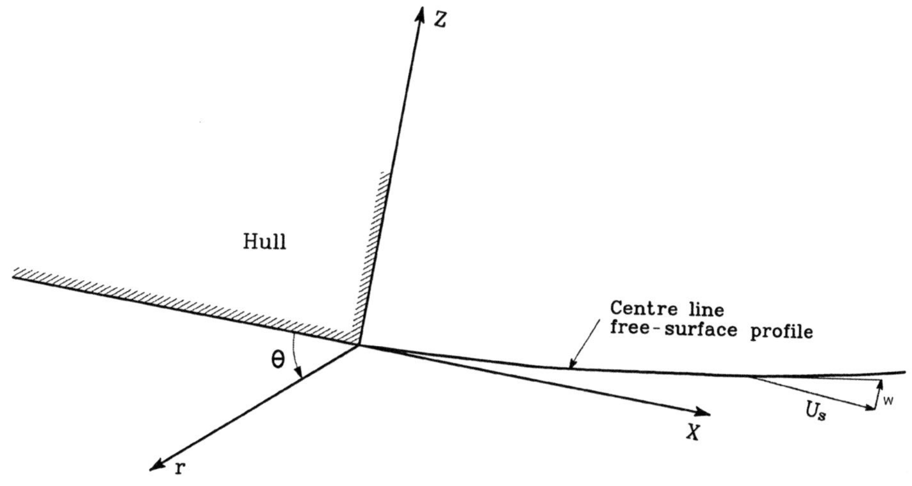

Figure 9.16. Detailed view of the flow separation area at the centerline of the transom stern (see Figure 9.14). (r, θ) = polar coordinates. The Z-component of the water velocity at the free surface is w. U_s is a first approximation of the X-component of the water velocity.

transom. A local solution in the vicinity close to the separation point at the transom is studied (see Figure 9.16). The flow is assumed two-dimensional in the X-Z-plane. Polar coordinates (r, θ) as defined in Figure 9.16 are introduced. The local solution of the velocity potential Φ is expressed as

$$\Phi = U_S X + Ar^n \cos n\theta. \tag{9.23}$$

Here r is assumed to be small and U_S, A, and n are presently unknown constants. U_S represents the tangential velocity at the transom (see Figure 9.14). We can check that the body boundary condition is satisfied by evaluating the velocity component $v_\theta = r^{-1}\partial\Phi/\partial\theta$ at $\theta = 0$. This gives $v_\theta = 0$ for $\theta = 0$, that is, no flow through the body boundary. Then we must satisfy the dynamic free-surface condition by using Bernoulli's equation for the pressure p and require that the pressure is atmospheric on the free surface, that is, $p = p_a$. Using eq. (3.5) with $\partial\Phi/\partial t = 0$ gives

$$\frac{\rho}{2}[(U_S + u)^2 + w^2] - \rho g(D - Z) = \frac{\rho}{2}U^2. \tag{9.24}$$

Here D is the draft at the transom (see Figure 9.14) and u and w are the X- and Z-components of the flow velocity due to the velocity potential component $Ar^n \cos n\theta$ in eq. (9.23). Here u and w are small relative to U_S. We will satisfy this dynamic free-surface condition approximately at $\theta = \pi$ (or $Z = 0$ for $X > 0$). Because u/U_S and w/U_S are small, the lowest-order terms in eq. (9.24) are

$$\frac{\rho}{2}U_S^2 = \rho g D + \frac{\rho}{2}U^2,$$

that is,

$$U_S = \sqrt{2gD + U^2}. \tag{9.25}$$

The next order term in eq. (9.24) becomes $\rho U_S u = 0$ on $Z = 0$. We can write on $Z = 0$

$$u\Big|_{\theta=\pi} = \frac{\partial}{\partial r}Ar^n \cos n\theta\Big|_{\theta=\pi}.$$

This means $\cos n\pi = 0$ or $n\pi = 0.5\pi, 1.5\pi, \ldots$. Because r is small, we must choose the lowest possible n. However, we must disregard $n = 1/2$, which gives infinite velocity at $r = 0$. That cannot be permitted because the flow must leave smoothly from the transom. Hence $n = 3/2$. At $Z = 0$, that is, $\theta = \pi$, we then have

$$w = \frac{1}{r}Ar^{3/2}\frac{d}{d\theta}\cos\frac{3}{2}\theta\Big|_{\theta=\pi} = \frac{3}{2}Ar^{1/2}. \tag{9.26}$$

A first approximation to the free-surface profile can be obtained by noting that the free surface is a streamline. This means (see Figure 9.16) that the free-surface slope dZ/dX satisfies approximately

$$\frac{dZ}{dX} = \frac{w}{U_S}. \tag{9.27}$$

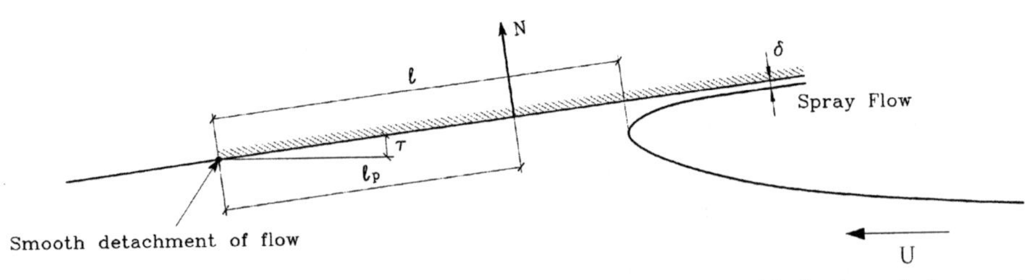

Figure 9.17. 2D flat planing surface with angle of attack τ. N is the normal hydrodynamic force, and ℓ_p is the distance measured along the plate from the trailing edge to the center of pressure.

Substituting w given by eq. (9.26) into eq. (9.27) and solving the differential equation gives

$$Z = \frac{A}{U_S} X^{3/2}. \tag{9.28}$$

The constant A can be determined by matching with the flow outside of the transom stern. This will not be pursued. Eq. (9.28) appears at first glance to be very different from eq. (9.19). However, it turns out that eqs. (9.28) and (9.19) have a very similar behavior even for quite large values of X/B, despite the fact that X was assumed to be small in our local flow analysis. We can then determine A by a least squares fit of eqs. (9.19) and (9.28), for, let us say, X/B between 0 and 3. This means we introduce

$$I = \int_0^{3B} dX \left(C_1 \left(\frac{X}{B}\right)^2 - C_2 \left(\frac{X}{B}\right)^{2.44} + C_3 \left(\frac{X}{B}\right) - \frac{A}{U_S} X^{3/2} \right)^2. \tag{9.29}$$

The least squares fit requires that I is a minimum; that is, A is determined by requiring $\partial I/\partial A = 0$. This gives

$$A = \frac{4U_S}{81B^4} \int_0^{3B} dX \left(C_1 \left(\frac{X}{B}\right)^2 - C_2 \left(\frac{X}{B}\right)^{2.44} + C_3 \left(\frac{X}{B}\right) \right) X^{3/2}. \tag{9.30}$$

A more qualitative and subjective estimate of A was obtaining by directly comparing eq. (9.28) with eq. (9.19). This was done by plotting eq. (9.28) with different values of A obtained by eq. (9.30). This gave

$$Z = 0.05B \frac{\tau_{\text{deg}}}{Fn_B} \left(\frac{X}{B}\right)^{3/2}. \tag{9.31}$$

This expression does not fit equally well for all trim angles, Froude numbers, and λ_W. However, it illustrates in a more simple way how the free-surface profile depends on τ and Fn_B.

9.2.4 High-aspect–ratio planing surfaces

Because the ratio between the beam and wetted keel length is typically small, planing vessels can in most situations be considered as low-aspect–ratio planing surfaces. If we look upon a trim tab as an appendage, this can be considered a high-aspect–ratio planing surface. Further, in extreme cases, the planing vessel can be supported by the water flow only in a small area at the stern. We also then have a high-aspect–ratio planing surface. Another scenario could be the reattached flow due to separation from a step on a planing vessel.

When the aspect ratio is high, we can approximate the flow close to the planing surface as two-dimensional in a vertical plane parallel to the forward motion of the planing surface. The flow is illustrated in Figure 9.17. There is a forward jet flow resulting in spray at the front wetted part of the planing surface. The flow leaves tangentially at the trailing edge and the presence of the planing surface causes a local uprise of the water. Gravity waves will be generated for a finite Froude number. When the Froude number is very high and the effect of gravity can be neglected, this problem has been studied for small trim angles τ by Wagner (1932). If we neglect the details of the jet flow and assume a small angle of attack τ, the body boundary condition can be transferred to a horizontal line of length l, as defined in Figure 9.17. This is similar to what we did in the outer domain solution of Wagner's slamming model in section 8.3.1. We can impose the same free-surface condition $\varphi = 0$, where φ is the velocity potential due to the planing surface. The flow is assumed to leave

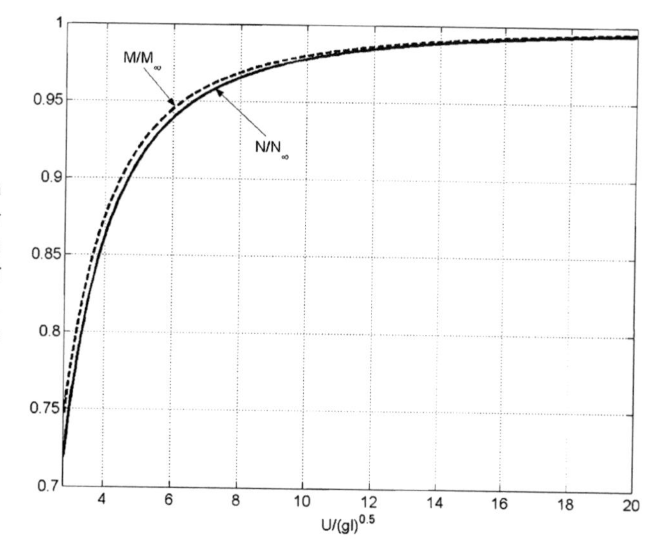

Figure 9.18. Effect of Froude number on theoretical hydrodynamic steady linear normal force N and pitch moment M about the center of a 2D planing flat plate of length ℓ (see Figure 9.17). N_∞ and M_∞ are values of N and M at infinite Froude number.

tangentially and smoothly from the trailing edge of the flat plate, that is, like a Kutta condition in foil theory is imposed on the trailing edge. Then we create an image body above the free surface, and we have the flow around a 2D foil in infinite fluid. Because in our problem we integrate pressure forces on only one side of the plate, we get half the lift of a flat foil in infinite fluid. This gives a lift coefficient of $\pi\tau$ for a flat planing surface. The lift force will act three quarters of the chord length from the trailing edge.

We argue in Chapter 6 that there is no drag due to steady 2D potential flow past a foil in infinite fluid. We explain that this is true even for a flat plate. This is associated with a finite suction force acting at the leading edge. However, because there is no flow around the leading edge in the planing problem, there is a drag force equal to the lift force times the angle of attack. When the Froude number is infinite, the drag is physically caused by the jet (spray) flow in front of the planing surface.

If the free-surface condition $\varphi = 0$ is assumed, we can also handle a finite-aspect–ratio planing surface by making the analogy to foil theory and using the methods described in Chapter 6. This analogy is not dependent on a flat foil; for example, we can use camber to create lift on the planing surface. Sedov (1965) presented a comprehensive presentation on two-dimensional steady planing. The linear wave generation problem for a flat plate in deep water was analyzed in detail. High Froude–number asymptotic formulas for the linear normal hydrodynamic force N (see Figure 9.17) and hydrodynamic pitch moment M about the center of the plate were presented. When the Froude number $Fn = U/\sqrt{\ell g}$ is larger than 2.8, the expressions

$$\frac{N}{0.5\rho U^2 \ell\tau} = \pi\left[1 - \left(\pi + \frac{4}{\pi}\right)\frac{0.5}{Fn^2}\right] \qquad (9.32)$$

$$\frac{8M}{\rho\pi U^2\ell^2\tau} = 1 - \frac{8+3\pi^2}{3\pi}\frac{0.5}{Fn^2} \qquad (9.33)$$

agree well with Sedov's exact solutions. Figure 9.18 shows how N/N_∞ and M/M_∞ vary with Fn. The subscript ∞ refers to infinite Froude number, that is, the case analyzed by Wagner (1932). When Fn is larger than 4.25, N and M differ less than 10% from the results at infinite Froude number.

The results for N can be used to calculate the sum of wave and spray resistance. This resistance is simply $N\tau$, that is, the normal force component opposite the forward motion direction of the plate.

Sedov (1965) also considered 2D nonlinear steady planing of a flat plate at infinite Froude number (see also Green 1936). The normal hydrodynamic force N and longitudinal distance ℓ_p of the center of pressure from the trailing edge can be expressed as

$$\frac{N}{0.5\rho U^2\ell} = \frac{2\pi}{\cot(0.5\tau)+\pi+\tan(0.5\tau)\ln\left[\cot^2(0.5\tau)-1\right]} \qquad (9.34)$$

$$\frac{\ell_p}{\ell} = \frac{1+0.5\cos\tau+2(1-\cos\tau)\ln 2+0.5\pi\sin\tau}{(1-\cos\tau)\ln\left[2\cos\tau/(1-\cos\tau)\right]+1+\cos\tau+\pi\sin\tau}. \qquad (9.35)$$

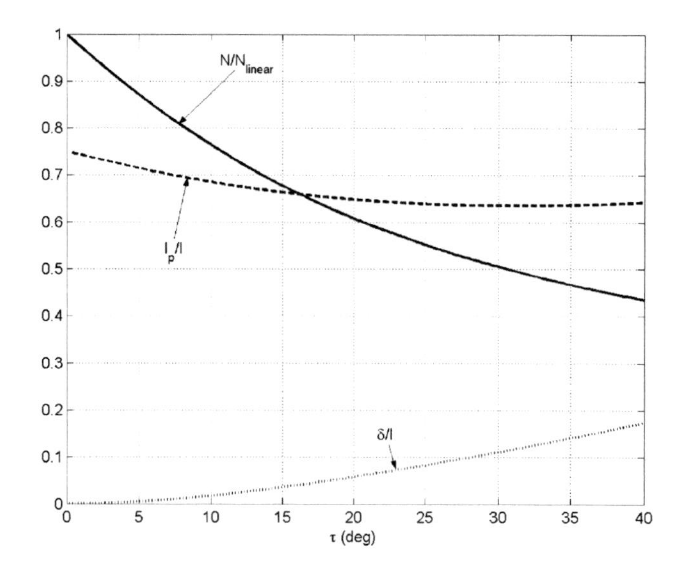

Figure 9.19. Theoretical hydrodynamic normal force N, distance ℓ_p of center of pressure from trailing edge, and jet thickness δ as a function of the trim angle τ for steady flow past a 2D planing flat plate of length ℓ at infinite Froude number (see Figure 9.17). N_{linear} is the value of N based on linear theory.

The linear values of N and ℓ_p are, respectively, $N_{linear} = 0.5\rho U^2 \ell \pi \tau$ and $\ell_p/\ell = 0.75$. Kochin et al. (1964) presented the following formula for the jet thickness δ:

$$\frac{\ell}{\delta} = \frac{1}{\pi}\left[\frac{1+\cos\tau}{1-\cos\tau} + \pi\frac{\sin\tau}{1-\cos\tau} + \ln\frac{2\cos\tau}{1-\cos\tau}\right]. \tag{9.36}$$

The results for N, ℓ_p, and δ are presented in Figure 9.19 as a function of τ. Nonlinearities start to matter for quite small trim angles (angles of attack). This is contrary to what happens for a 2D foil in infinite fluid. The lift is then linearly dependent on the angle of attack for much larger angles (Figure 2.17). Figure 9.19 shows that the jet thickness is very small relative to ℓ for small τ. However, this is not true for τ larger than 10°.

Because a 2D infinite Froude number theory predicts infinite free-surface elevation at infinity, there is no reference height for the planing surface. This can be circumvented by assuming a high-aspect–ratio planing surface and letting the 2D flow analysis be valid in the near field. The far field requires a 3D flow analysis (Shen and Ogilvie 1972).

9.3 Prediction of running attitude and resistance in calm water

By running attitude, we mean the trim and rise (steady heave or negative sinkage) of the vessel. We will exemplify how this can be predicted by Savitsky's formula for prismatic hull forms. This includes prediction of resistance and needed horsepower.

9.3.1 Example: Forces act through COG

A case similar to the one presented by Savitsky (1964) is considered. The following data are given:

Mass, M		27,000 kg
lcg	29 ft	8.84 m
Beam, B	14 ft	4.27 m
Deadrise angle, β	10°	
Ship speed	40 knots	20.58 ms^{-1}
Fn_B	3.18	

The frictional force R_V and propeller thrust T act through COG. This is illustrated in Figure 9.20, where N means the potential flow force discussed in section 9.2. The wetted length and running trim follow by satisfying vertical force and pitch moment equilibrium. The power requirement can then be investigated. Savitsky (1964) did this by extensive use of graphs. However, a computer program can easily be made to determine the effective horsepower. In the following, we will go through the necessary steps.

Step 1. Average wetted length-to-beam ratio λ_W

Because the forces act through COG, ℓ_p (see eq. (9.7)) has to be equal to lcg. Eq. (9.7) determines then the average wetted length-to-beam

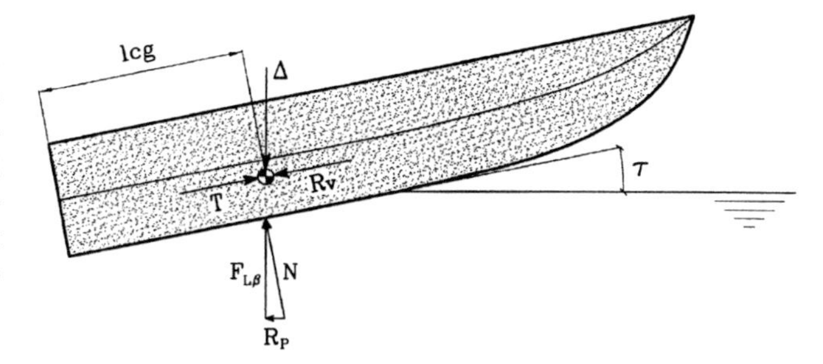

Figure 9.20. Prismatic planing hull in which all forces pass through COG. N is the force due to hydrodynamic pressures acting on the wetted hull. This has a vertical component $F_{L\beta}$ and a longitudinal component R_P. Δ = vessel weight, T = thrust from propulsion unit, R_V = viscous frictional force on the hull (Savitsky 1964).

ratio λ_W. The unknown λ_W is therefore the solution of the nonlinear equation

$$\frac{lcg}{\lambda_W B} - 0.75 + \frac{1}{\left[5.21 Fn_B^2/\lambda_W^2 + 2.39\right]} = 0.$$

This requires a numerical solution. The answer is $\lambda_W = 3.43$.

Step 2. Trim angle τ

We now use the fact that the lift force calculated by eq. (9.4) balances the weight of the vessel, that is,

$$C_{L\beta} = Mg/(0.5\rho U^2 B^2). \tag{9.37}$$

Using the mass density of salt water at 15°C, that is, $\rho = 1026\,\text{kg/m}^3$ gives $C_{L\beta} = 0.067$. Eqs. (9.4) and (9.5) determine then the trim angle τ by solving the equation

$$0.067 - (C_{L0} - 0.0065\beta C_{L0}^{0.6}) = 0.$$

The solution is $\tau_{\text{deg}} = 2.21°$.

Step 3. Wetted length

The length $x_s = L_K - L_C$ along the keel from the intersection between the keel and the mean free surface until chine wetting starts is first defined. This can be obtained by eq. (9.8), in which τ is given in radians. The chine wetted length L_C and keel wetted length L_K follow from

$$\lambda_W = 0.5(L_K + L_C)/B = 0.5(x_s + 2L_C)/B. \tag{9.38}$$

This means $L_C = \lambda_W B - 0.5x_s = 11.5\,\text{m}$ and $L_K = 2\lambda_W B - L_C = 17.7\,\text{m}$. The consequence is that the draft of the keel at transom is

$$D = L_K \sin\tau = 0.68\,\text{m}. \tag{9.39}$$

Step 4. Effective horsepower (EHP)

The resistance of the vessel will now be calculated. The frictional resistance on the hull is obtained by eqs. (2.3), (2.4), and (2.86). Using that the kinematic viscosity coefficient $\nu = 1.19 \cdot 10^{-6}\,\text{m}^2/\text{s}$ for saltwater at 15°C gives a Reynolds number $Rn = UL_K/\nu = 3.1 \cdot 10^8$. The average hull roughness (AHR) will be set equal to $150 \cdot 10^{-6}$ m. This gives $C_F = 1.78 \cdot 10^{-3}$ and $\Delta C_F = 3.7 \cdot 10^{-4}$.

The wetted area is divided into two parts. The wetted area S_1 from the bow ($x = 0$) up to where chine wetting starts ($x = x_s$) is chosen as the hull area below the spray root. Actually, this wetted area part may be larger. It follows by introducing $d(x)$ as the vertical distance from the spray root to the keel at x and the discussion before eq. (9.10), so that

$$S_1 = 2\int_0^{x_S} \frac{d(x)}{\sin\beta} dx = \frac{2}{\sin\beta}\int_0^{x_S}\left(1 + \frac{z_{\max}}{Vt}\right) x\tau\, dx$$
$$= \frac{\tau}{\sin\beta}\left(1 + \frac{z_{\max}}{Vt}\right) x_s^2.$$

Using eq. (9.10) to express x_S gives

$$S_1 = \frac{\tan^2\beta}{\sin\beta}\left(\frac{B^2}{4(1 + z_{\max}/Vt)\tau}\right). \tag{9.40}$$

The wetted area from $x = x_s$ to the transom is simply

$$S_2 = \frac{B}{\cos\beta} L_C. \tag{9.41}$$

The total wetted area is by setting $1 + z_{\max}/Vt = 0.5\pi$ equal to $S = S_1 + S_2 = 63.3\,\text{m}^2$. It follows that the longitudinal frictional force is $R_V = 29603\,N$. We have to add the lift-induced resistance given by eq. (9.6). However, other resistance components will be neglected in this example. The total longitudinal drag force is then $R_T = 39825\,N$. This gives a drag-lift ratio of 0.15. The needed power is $R_T \cdot U = 820$ kW. By dividing this by 0.7457 or multiplying by 1.36, we obtain the effective horsepower $EHP = 1115\,Hp$.

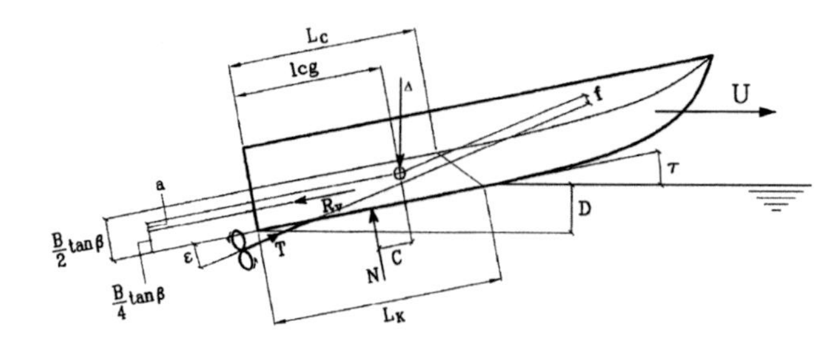

Figure 9.21. Prismatic planing hull in which forces do not pass through COG (Savitsky 1964).

9.3.2 General case

Figure 9.21 illustrates a general case in which the frictional force R_V and propeller thrust T do not act through COG. The viscous component of drag R_V is assumed to act parallel to keel line, half-height between the keel and chine lines. Why the viscous drag force acts like this is difficult to justify. The distance between R_V and COG measured normal to R_V is called a. The thrust line has an inclination angle ε relative to the keel. The angle ε on small vessels with outboards and stern drives is adjustable and may often be negative, lifting the bow up. The negative ε may cause a large thrust spray seen behind the vessel. The distance between thrust line and COG measured normal to shaft line is f. N is the result of pressure forces due to potential flow. The distance between N and COG measured normal to N is c. We can set up the following equations for force and moment equilibrium.

Vertical equilibrium of forces:

$$Mg = N\cos\tau + T\sin(\tau+\varepsilon) - R_V\sin\tau. \quad (9.42)$$

Horizontal equilibrium of forces:

$$T\cos(\tau+\varepsilon) = R_V\cos\tau + N\sin\tau. \quad (9.43)$$

Pitch moment equilibrium:

$$Nc + R_V a - Tf = 0. \quad (9.44)$$

These equations can be rearranged as follows. Balance of forces along the keel line is first considered. N will then not have a component in this direction and we can write

$$T\cos\varepsilon = Mg\sin\tau + R_V. \quad (9.45)$$

Assuming $\cos\varepsilon \approx 1$ in eq. (9.45) and substituting eq. (9.45) into eq. (9.42) gives

$$\begin{aligned} Mg &= N\cos\tau + Mg\sin\tau\sin(\tau+\varepsilon) \\ &\quad + R_V\sin(\tau+\varepsilon) - R_V\sin\tau \\ &\approx N\cos\tau + Mg\sin\tau\sin(\tau+\varepsilon). \end{aligned}$$

This means

$$N = \frac{Mg(1-\sin\tau\sin(\tau+\varepsilon))}{\cos\tau}. \quad (9.46)$$

We now substitute eqs. (9.45) and (9.46) into eq. (9.44). This results in

$$Mg\left\{\frac{(1-\sin\tau\sin(\tau+\varepsilon))c}{\cos\tau} - f\sin\tau\right\} + R_V(a-f) = 0. \quad (9.47)$$

A computational procedure to find trim angle, wetted length, and so forth can now be set up. This is different and more involved than the procedure presented in section 9.3.1.

The first step is to assume a trim angle τ. Because the weight, ship speed, and beam are given, $C_{L\beta}$ is known. This is similar to step 2 in section 9.3.1. The difference now is that we have assumed τ and used eq. (9.4) to determine the average wetted length-to-beam ratio. We note that Fn_B then is given. We now proceed with calculations similar to those in steps 3 and 4 in section 9.3.1. This determines R_V. We determine c by noting that lcg is given and ℓ_p is found by eq. (9.7). Because f, a, and ϵ are known, the left-hand side of eq. (9.47) can now be evaluated.

By repeating this procedure for different assumed values of the trim angle, we will find which trim angle will satisfy eq. (9.47). By using a computer program, we can easily do this for many assumed τ-values and accurately determine trim moment equilibrium. We have then solved the problem, that is, found the equilibrium position for the vessel. Effective horsepower can be evaluated in a way similar to the one used in section 9.3.1.

Savitsky (1964) presented an example with values of M, lcg, B, β, and U similar to those in section 9.3.1. In addition, $a = 1.39ft(0.42m)$, $f = 0.50ft$ (0.15 m), and $\varepsilon = 4°$. This caused small differences relative to $a = f = \varepsilon = 0$, that is, the case in section 9.3.1.

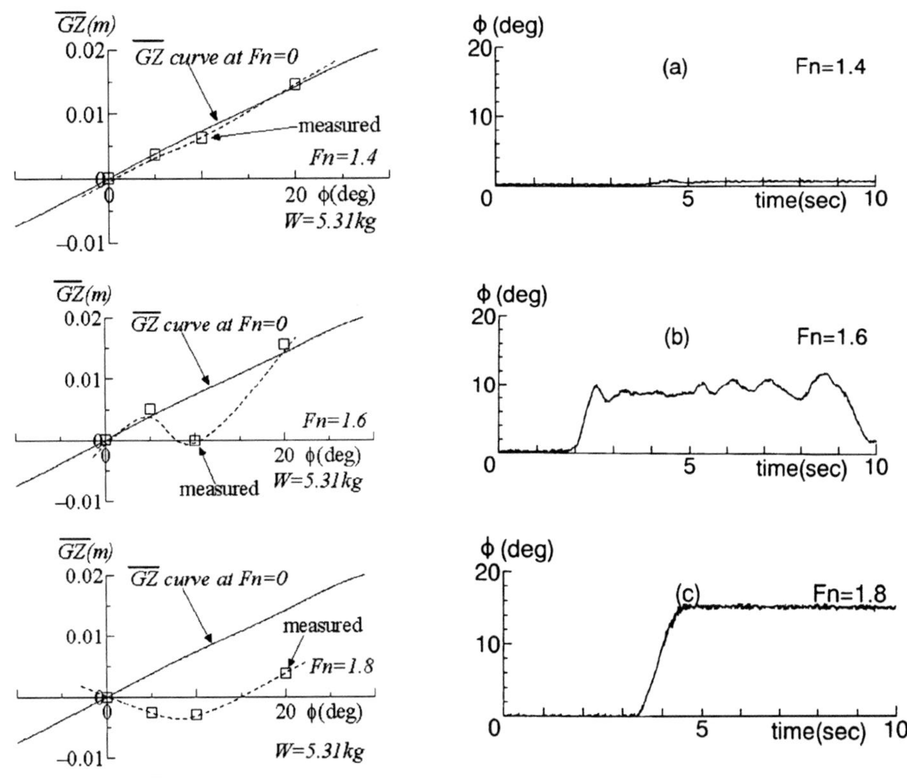

Figure 9.22. $\overline{GZ}$-curve and experimental time histories of roll motion of ship TB45 at a trim angle of 2°, center of gravity of 0.074 m, and ship weight of 5.31 kgf for several advance speeds: (a) $Fn = 1.4$, (b) $Fn = 1.6$, (c) $Fn = 1.8$ (Ikeda and Katayama 2000a).

9.4 Steady and dynamic stability

Steady heeling stability of a vessel depends on the steady heel (roll) restoring moment $W \cdot \overline{GZ}$ about the center of gravity of the vessel as a function of the steady heel angle ϕ. Here W is the weight of the vessel and $\overline{GZ}$ is the moment arm. $\overline{GZ}$ at zero speed is a result of the hydrostatic pressure. However, as the ship speed increases, the influence of the hydrodynamic pressure on $\overline{GZ}$ increases. This has been discussed in section 7.7 for round-bilge monohulls.

A 2.5D theory can be applied to a heeled hard-chine planing vessel and used to study steady transverse stability. This can be done similarly to the way steady vertical forces and trim moments were analyzed in section 9.2.1. This means that results from water entry of a heeled 2D section with constant entry velocity are used. Xu et al. (1998) used a vortex distribution method to analyze water entry of a heeled 2D section with flow separation from hard chines. The deadrise angle is assumed small and boundary conditions are transferred to a horizontal line. The possibility of transverse flow separation from the keel is incorporated. A method as this does not include the hydrostatic pressure, rudders, propulsion, and possible effect of cavitation and ventilation.

Ikeda and Katayama (2000 a) presented measured $\overline{GZ}$ of a planing vessel on a straight course in calm water as a function of the heel angle ϕ. The TB45 model (see Figure 9.10) with a length of 1 m was used. Figure 9.22 shows their results for $Fn = 0, 1.4, 1.6,$ and 1.8 at a trim angle $\tau_{\text{deg}} = 2°$. The $\overline{GZ}$-curve at zero speed is close to a straight line in the presented ϕ-range between 0° and 30°. However, $\overline{GZ}$ may at high speed be negative in this ϕ-range, depending on the length Froude number, Fn, and the distance $\overline{KG}$ between the keel and the center of gravity. Figure 9.22 shows, for $\overline{KG} = 0.074$ m and $Fn = 1.8$, that $\overline{GZ}$ is negative for a large range of ϕ.

Values of ϕ corresponding to $\overline{GZ} = 0$ represent equilibrium positions. This gives $\phi = 0^\circ$ and 15° as the two equilibrium positions for $\overline{KG} = 0.074$ m and $Fn = 1.8$. An equilibrium position is statically stable if small deviations in ϕ from the equilibrium position cause a heel moment that restores the vessel to equilibrium. Analyzing the $\overline{GZ}$-curve in the vicinity of $\phi = 0^\circ$ and 15° gives that $\phi = 0^\circ$ and 15° correspond to, respectively, unstable and stable equilibrium positions. This is also evident from the roll (heel) time history presented in Figure 9.22 for $Fn = 1.8$. The vessel is initially in an upright position ($\phi = 0^\circ$). It then quickly heels to an angle $\phi = 15^\circ$, and the vessel continues to move forward with this statically stable heel angle.

The behavior at $Fn = 1.6$ and $\overline{KG} = 0.074$ m is clearly different from $Fn = 1.8$ (see Figure 9.22). The $\overline{GZ}$-curve then only becomes negative in a small vicinity of the heel angle 10°. There are two static equilibrium positions for non-zero heel. The vessel will in this condition heel over to a mean heel angle of about 10° at the beginning of the run. However, an oscillatory heel starts to develop ("chine-walking"). Because the oscillation amplitude increases with time, the system is dynamically unstable. If the craft has a $\overline{GZ}$-curve, as shown for $Fn = 1.6$, and there are waves and wind present or the vessel makes a maneuver, a scenario may be that the vessel rapidly heels over to the opposite side. This may cause a dangerous situation.

The $\overline{GZ}$-curve depends strongly on the trim angle at high speed. If we generalize this to unsteady motions, it means that the roll-restoring moment is a function of pitch. This can lead to Mathieu instabilities (see section 7.7.1). Ikeda and Katayama (2000a) demonstrated that this was possible with a forced pitching period equal to half the natural roll period.

Lewandowski (1997) presents semi-empirical formulas for how the metracentric height $\overline{GM}$ of prismatic (hard-chine) vessels varies with beam, deadrise angle, trim angle, transom draft, and speed at planing conditions. Here $\overline{GM} = d\overline{GZ}/d\phi$ at zero heel angle ϕ. A minimum $\overline{GM}$ occurring between volumetric Froude number $F_V = U/\sqrt{g\nabla^{1/3}}$ of 2 and 3 is characteristic for the studied Series 62 hulls. Dynamic instability of coupled sway-roll-yaw motions of prismatic planing vessels is studied by Lewandowski (1997) by means of linear stability analysis and semi-empirical formulas for the hydrodynamic coefficients. Systematic calculations are reported, and maximum $\overline{KG}$ for stable behavior is presented. A simple method to check the transverse dynamic stability of a proposed design is also presented by Lewandowski (1997). Dynamic stability analysis of coupled sway-roll-yaw motions is discussed further in section 10.9.3.

Blount and Codega (1992) presented design guidelines on how to avoid dynamic transverse instabilities of hard-chine craft operating at speeds greater than 25 knots. The criteria were based on full-scale trials and expressed in terms of the hull loading factor $A_p/\nabla^{2/3}$ and the dimensionless lcg-parameter $(CA_p - lcg)/L_p$. Here A_p is the project area of the planing bottom bounded by the chine and transom, CA_p is the centroid of A_p, lcg is the longitudinal position of center of gravity measured from the transom, and L_p is the projected chine length. Boats with observed dynamic transverse instability had $A_p/\nabla^{2/3} \le 5.8$ and $(CA_p - lcg)/L_p \le 0.03$.

Figure 7.41 also illustrates how the static stability in roll is influenced by the forward speed. For instance, the heel (roll) angle of craft A is 13.4° at Froude number 1.6 and 3° at zero speed. Werenskiold (1993) showed that 3° of the 13.4° could be explained by the hydrodynamic pressure on the hull without the effect of the propeller and rudder. The effect of using the rudder for yaw compensation is 3.4°. The remaining 4° is caused by the dynamic pressure generated by the propeller. It is possible that cavitation and ventilation are contributing factors, for instance, by influencing the pressure distribution in the propeller tunnels.

Werenskiold (1993) says the following about the four vessels presented in Figure 7.41, which have a relatively large influence of forward speed on static roll stability:

Craft A: "being unstable at speeds above 25 to 30 knots and having broaching problems."

Craft B: "having no problems with broaches, however, has to slow down in following seas in order to maintain steering control."

Craft C: "broaching can be provoked, in forward trim conditions, in particular."

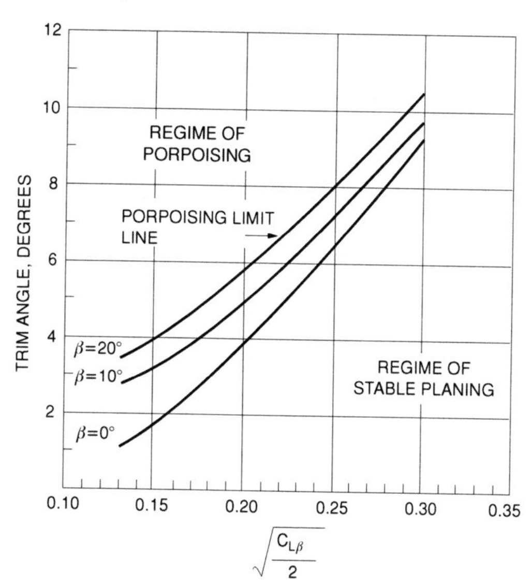

Figure 9.23. Porpoising limits for prismatic planing hulls (Savitsky 1964). $C_{L\beta} = Mg/\left(0.5\rho U^2 B^2\right)$, where M is the vessel mass; U = vessel speed; B = beam.

Craft D: "before modification of rudder design and trim condition the craft had severe roll stability and broaching problems. Ventilation of aft body was experienced in hard turns and in waves, with measured change of negative bottom pressure to zero, giving a violent upward kick."

This suggests that a vessel with unsatisfactory steady heel stability on a straight course is an indication of dynamic stability problems. If this is true, it will facilitate how guidelines can be formulated for the stability of high-speed vessels. One could, for instance, require model tests such as those in the example in Figure 7.41 and start with a heel angle of 3° at zero speed. Then one could require that the vessel have less than, for instance, 8° heel at maximum operating speed. This would ideally require a depressurized towing tank to scale cavitation properly. However, there is not sufficient documentation showing that the outlined procedure is a reflection of all possible stability problems. In the following text, we describe in more detail porpoising, which is one of the many dynamic instability problems that may occur.

9.4.1 Porpoising

Porpoising is unstable coupled heave and pitch motions. Design guidelines are available for predicting and avoiding porpoising (Blount and Codega 1992). Figure 9.23 can be used to evaluate the risk of porpoising. It is based on comprehensive experimental studies by Day and Haag (1952) and was presented by Savitsky (1964). Figure 9.23 presents limit curves for stability for different deadrise angles β. If the combination of the trim angle and the lift coefficient corresponds to a point above the limit curve for a given β, porpoising will occur. The horizontal coordinate $(C_{L\beta}/2)^{1/2}$ in Figure 9.23 is 0.18 for the example in section 9.3.1. Because the trim angle $\tau_{\text{deg}} = 2.21°$ and $\beta = 10°$, no porpoising occurs. Because $C_{L\beta}$ is proportional to the weight and is a constant, Figure 9.23 shows that it is beneficial to lower the trim angle if porpoising occurs. This can, for instance, be achieved by moving the longitudinal center of gravity forward or by using trim tabs. The effect of trim tabs on porpoising has been experimentally studied by Celano (1998). Later, we will see that there is a lower speed limit for porpoising to exist, so we can also lower the speed to avoid the problem.

Inception of porpoising can be found by a linear stability analysis. Small perturbations from the steady equilibrium position are then dynamically examined. There are no excitations, for instance, due to wave loads. If a small initial perturbation is given to the system and the motions grow with time, the system is unstable. A nonlinear stability analysis is needed to get a measure of how large the unstable motions may be. Nonlinearities occur, for instance, because of the hydrodynamic loads on the craft. Further, the vessel's speed and thrust may vary as a consequence of porpoising. Our analysis assumes linearity.

Let us first define a coordinate system (x, y, z) that does not oscillate with the ship (see Figure 9.24). It is steady relative to the steady forward motion of the ship. When the ship is in the steady equilibrium position, the origin coincides with COG. z is vertical and positive upward. x is horizontal and positive in the aft direction of the ship. The time-dependent motions are denoted η_k, where η_3 means the vertical motions (heave) of COG and η_5 is the pitch rotation in radians. Both the trim angle τ and pitch angle are positive when the bow goes up. The vertical distance between COG and the keel is called vcg, and lcg is the longitudinal center of gravity measured from the transom stern. The vertical position z_{wl} of COG above the mean water surface when the ship is not

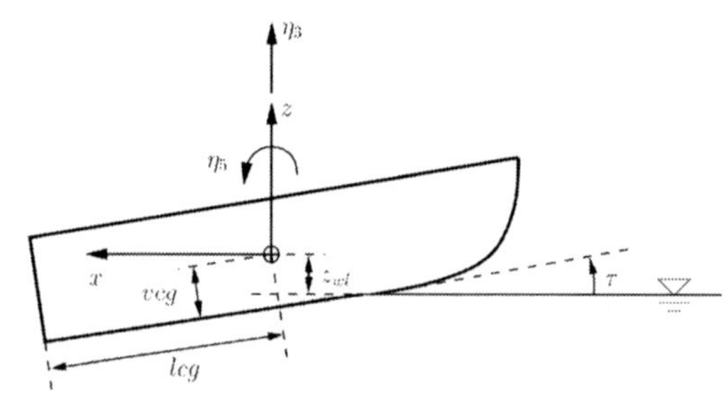

Figure 9.24. Coordinate system (x, y, z) moving with forward speed of a ship and fixed relative to mean oscillatory position of the ship. Definitions of heave (η_3), pitch (η_5), and positions lcg and vcg of the center of gravity (COG) relative to the ship. z_{wl} is the height of COG above mean water surface with no heave.

oscillating will be needed later in the analysis. It follows from Figure 9.24 that

$$z_{wl} = vcg \cdot \cos\tau - (L_K - lcg)\sin\tau. \quad (9.48)$$

Here L_K is the wetted keel length.

The linear coupled equations of motion in heave and pitch have the same structure as eq. (7.35) for semi-displacement vessels. By setting the excitation force in heave and pitch equal to zero it follows that

$$\begin{aligned}(M + A_{33})\frac{d^2\eta_3}{dt^2} + B_{33}\frac{d\eta_3}{dt} + C_{33}\eta_3 + A_{35}\frac{d^2\eta_5}{dt^2}\\ + B_{35}\frac{d\eta_5}{dt} + C_{35}\eta_5 = 0\\ A_{53}\frac{d^2\eta_3}{dt^2} + B_{53}\frac{d\eta_3}{dt} + C_{53}\eta_3 + (I_{55} + A_{55})\\ \times\frac{d^2\eta_5}{dt^2} + B_{55}\frac{d\eta_5}{dt} + C_{55}\eta_5 = 0,\end{aligned} \quad (9.49)$$

where M is the vessel mass and I_{55} is the vessel moment of inertia in pitch with respect to the coordinate system defined in Figure 9.24. The added mass (A_{jk}), damping (B_{jk}), and restoring coefficient (C_{jk}) are separately discussed below. An important difference from the analysis of a semi-displacement vessel is that the added mass and damping coefficients for the planing vessel are assumed to be frequency independent by using the high-frequency free-surface condition. The damping is then the result of hull-lift effects. Further, the restoring coefficients in the case of a semi-displacement vessel were a consequence of the change of buoyancy forces on the hull. An important contribution for a planing vessel is the result of changes in steady hydrodynamic lift force and trim moment due to heave and pitch motions. This means that the restoring coefficients increase strongly with forward speed. The added mass coefficients will have a smaller dependence on forward speed. A consequence is that natural periods in heave and pitch will decrease with forward speed. This was documented experimentally by Katayama et al. (2000).

Our theoretical presentation of added mass and damping will be highly simplified. Because the final porpoising analysis shows that porpoising is sensitive to the hydrodynamic coefficients, a more accurate method should be developed. One possibility is to include dynamic effects in the 2.5D analysis described for steady flow in section 9.2.1. A more engineering-type approach would be to adopt the frequency-domain strip theory by Salvesen et al. (1970). However, a strip theory should be questioned for Froude numbers larger than 0.4 to 0.5.

Restoring force and moment

We consider now a constant displacement in heave and angular pitch rotation and evaluate the corresponding steady vertical force and pitch moment. Eqs. (9.4), (9.5), and (9.7) will be used even if τ and the mean wetted length-to-beam ratio λ_W may be larger than the prescribed domain of validity. This means τ_{deg} is replaced by $\tau_{\text{deg}} + \eta_5 180/\pi$ in eq. (9.5). A requirement is that the instantaneous trim angle must be positive for eq. (9.5) to be mathematically valid. Further, it is necessary to know how λ_W changes with η_3 and η_5. Troesch's (1992) procedure (Figure 9.25) will then be partly followed. We write first AB in Figure 9.25 as

$$AB = vcg - \frac{z_{w\ell} + \eta_3}{\cos(\tau + \eta_5)}.$$

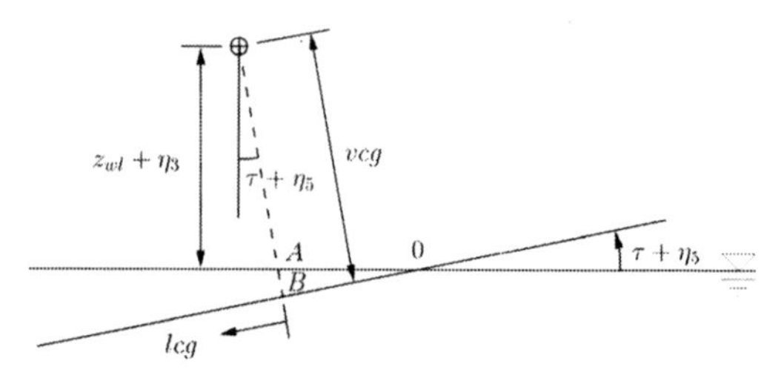

Figure 9.25. Instantaneous position of the center of gravity (COG) and keel.

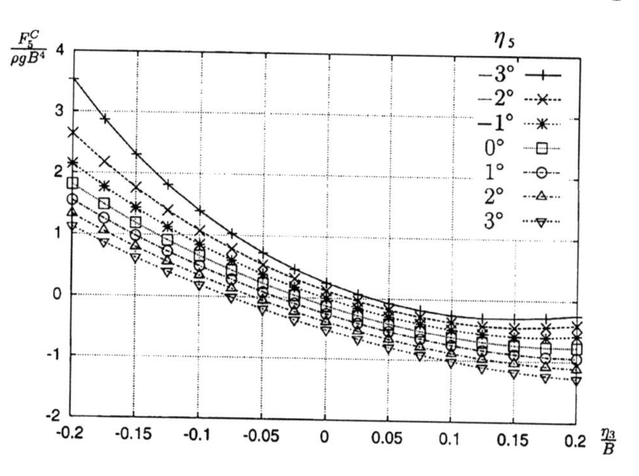

Figure 9.26. Steady vertical force F_3^C and steady pitch moment F_5^C about COG as a function of given vertical displacement η_3 and pitch angle η_5. $\lambda_W = 3$ for $\eta_3 = 0$ and $\eta_5 = 0$. $Fn_B = 4.35$, $\tau_{\text{deg}} = 6°$, $\beta = 20°$, $lcg/B = 2.13$, $vcg/B = 0.25$, $M/\rho B^3 = 1.28$.

This means BO in Figure 9.25 is

$$BO = \frac{AB}{\tan(\tau + \eta_5)} = \frac{vcg}{\tan(\tau + \eta_5)} - \frac{(z_{w\ell} + \eta_3)}{\sin(\tau + \eta_5)}.$$

This gives

$$L_K = lcg + \frac{vcg}{\tan(\tau + \eta_5)} - \frac{(z_{wl} + \eta_3)}{\sin(\tau + \eta_5)}. \quad (9.50)$$

Negative L_K means simply that the boat is out of the water, that is, $L_K = 0$. By using eq. (9.10) with τ replaced by $\tau + \eta_5$, we have

$$L_C = L_K - x_s = L_K - \frac{0.5 B \tan\beta}{(1 + z_{\max}/Vt)(\tau + \eta_5)}. \quad (9.51)$$

If eq. (9.51) gives negative L_C, L_C is zero. The instantaneous value of $\lambda_W = 0.5\,(L_K + L_C)/B$ follows now from eqs. (9.50) and (9.51). The use of the formula will be illustrated with an example in which $\lambda_W = 3$ with no oscillatory motions, $\tau_{\text{deg}} = 6°$, $vcg/B = 0.25$, $lcg/B = 2.13$, $M/\left(\rho B^3\right) = 1.28$, and $Fn_B = 4.35$. The resulting vertical force F_3^c and pitch moment F_5^c about COG are presented in Figure 9.26 as a function of heave for different values of pitch. We should note that the vertical force is in balance with the weight when η_3 and η_5 are zero. However, there is a non-zero pitch moment when $\eta_3 = \eta_5 = 0$. This means viscous and propeller forces must contribute to pitch moment equilibrium. However, we will neglect time-varying viscous and propeller loads in the analysis.

The linearized restoring coefficients in heave and pitch are obtained by

$$C_{jk} = -\left.\frac{\partial F_j^c}{\partial \eta_k}\right|_0, \quad j = 3, 5 \text{ and } k = 3, 5, \quad (9.52)$$

where 0 means the static equilibrium position, that is, corresponding to $\eta_3 = \eta_5 = 0$. C_{jk} can be obtained either analytically or by numerical differentiation. The analytical derivation is shown below.

Using $\lambda_W = 0.5(L_K + L_C)/B$ and eqs. (9.50) and (9.51) gives

$$\left.\frac{\partial \lambda_W}{\partial \eta_5}\right|_0 = \frac{-vcg/B}{\sin^2\tau} + \frac{z_{wl}/B}{\sin^2\tau}\cos\tau + \frac{0.25\tan\beta}{(1 + z_{\max}/Vt)\tau^2} \tag{9.53}$$

$$\left.\frac{\partial \lambda_W}{\partial \eta_3}\right|_0 = -\frac{1}{\sin\tau}\frac{1}{B}. \tag{9.54}$$

Here τ must be evaluated in radians. An important part of the static forces is C_{L0} (see eq. (9.5)). Differentiating C_{L0} results in

$$\left.\frac{\partial C_{L0}}{\partial \eta_5}\right|_0 = 1.1\left(\frac{180}{\pi}\right)^{1.1}\tau^{0.1}\left[0.012\lambda_0^{0.5} + 0.0055\lambda_0^{2.5}/Fn_B^2\right] + \tau_{\text{deg}}^{1.1}\left[0.006\lambda_0^{-0.5} + 0.01375\lambda_0^{1.5}/Fn_B^2\right]\left.\frac{\partial \lambda_W}{\partial \eta_5}\right|_0 \tag{9.55}$$

$$\left.\frac{\partial C_{L0}}{\partial \eta_3}\right|_0 = \tau_{\text{deg}}^{1.1}\left[0.006\lambda_0^{-0.5} + 0.01375\lambda_0^{1.5}/Fn_B^2\right]\left.\frac{\partial \lambda_W}{\partial \eta_3}\right|_0. \tag{9.56}$$

Here λ_0 is the value of λ_W at the static equilibrium position. Using the expression for vertical static force given by eqs. (9.4) and (9.5) leads to

$$\frac{C_{33}}{0.5\rho U^2 B} = -B\left.\frac{\partial C_{L\beta}}{\partial \eta_3}\right|_0 = -B\left.\frac{\partial C_{L0}}{\partial \eta_3}\right|_0\left[1 - 0.0039\beta C_{L0}^{-0.4}\right] \tag{9.57}$$

$$\frac{C_{35}}{0.5\rho U^2 B^2} = -\left.\frac{\partial C_{L\beta}}{\partial \eta_5}\right|_0 = -\left.\frac{\partial C_{L0}}{\partial \eta_5}\right|_0\left[1 - 0.0039\beta C_{L0}^{-0.4}\right]. \tag{9.58}$$

The pitch moment about COG can be expressed as

$$\frac{F_5^c}{0.5\rho U^2 B^3} = \left(\frac{\ell_p}{B} - \frac{\ell cg}{B}\right)C_{L\beta}, \tag{9.59}$$

where ℓ_p is given by eq. (9.7). We can write

$$\frac{1}{B}\left.\frac{\partial \ell_p}{\partial \lambda_W}\right|_0 = \left[0.75 - \frac{15.63 Fn_B^2/\lambda_0^2 + 2.39}{(5.21 Fn_B^2/\lambda_0^2 + 2.39)^2}\right]. \tag{9.60}$$

It follows that

$$\frac{C_{53}}{0.5\rho U^2 B^2} = -\left[\frac{1}{B}\frac{\partial \ell_p}{\partial \lambda_W}B\frac{\partial \lambda_W}{\partial \eta_3}C_{L\beta} + \left(\frac{\ell_p}{B} - \frac{\ell cg}{B}\right)B\frac{\partial C_{L\beta}}{\partial \eta_3}\right]_0 \tag{9.61}$$

$$\frac{C_{55}}{0.5\rho U^2 B^3} = -\left[\frac{1}{B}\frac{\partial \ell_p}{\partial \lambda_W}\frac{\partial \lambda_W}{\partial \eta_5}C_{L\beta} + \left(\frac{\ell_p}{B} - \frac{\ell cg}{B}\right)\frac{\partial C_{L\beta}}{\partial \eta_5}\right]_0. \tag{9.62}$$

The restoring coefficients for the case presented in Figure 9.26 are then

$$\frac{C_{33}}{\rho g B^2} = 3.978 \qquad \frac{C_{35}}{\rho g B^3} = -11.734$$
$$\frac{C_{53}}{\rho g B^3} = 6.327 \qquad \frac{C_{55}}{\rho g B^4} = 7.227.$$

These restoring coefficients show strong coupling between heave and pitch. This coupling effect is important for the occurrence of porpoising. Ikeda and Katayama (2000b) showed that porpoising did not occur if the coupled restoring coefficients were set equal to zero in their analysis of a personal watercraft.

Added mass in heave and pitch

The added mass calculations will be based on a high-frequency free-surface condition and strip theory. The two-dimensional added mass coefficient in heave a_{33} for a wedge is then an essential part for a prismatic planing hull. An analytical solution of a_{33} has been presented by many researchers. One version is (Faltinsen 2000)

$$a_{33} \equiv \rho d^2 K = \frac{\rho d^2}{\tan\beta}\left[\frac{\pi}{\sin\beta}\frac{\Gamma(1.5 - \beta/\pi)}{\Gamma^2(1 - \beta/\pi)\Gamma(0.5 + \beta/\pi)} - 1\right]. \tag{9.63}$$

Here d is the draft, which is equal to $0.5b\tan\beta$. Here b is the beam of the wedge. Further, Γ is the gamma function and K is by definition $a_{33}/\rho d^2$. Eq. (9.63) is graphically presented in Figure 9.27.

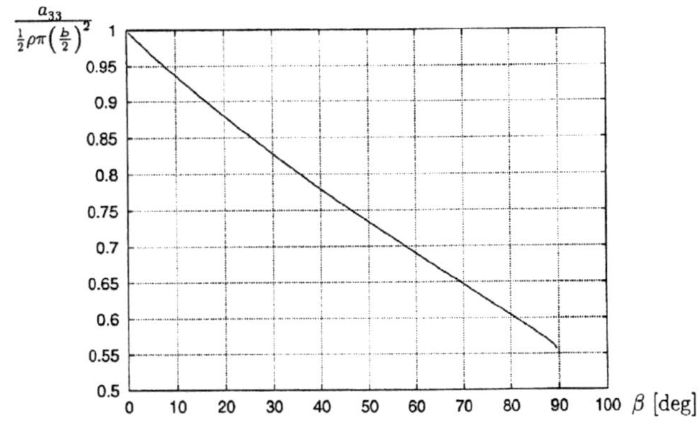

Figure 9.27. Two-dimensional infinite-frequency added mass in heave a_{33} for a wedge with deadrise angle β. $\rho =$ mass density of fluid, $b =$ beam.

An x'-axis is introduced along the keel with $x' = 0$ corresponding to the intersection between the keel and the free surface. Positive x' is aftward. The hull is divided into two parts. The first part is from $x' = 0$ until the chine wetting starts at $x' = x_s$. The second part is where the chine is wetted. The superscripts 1 and 2 will be used to indicate the added mass contributions from the two hull parts. From now on, the apostrophe in x' is neglected; that is, we write it as x. This must not be confused with the x-axis shown in Figure 9.24.

A. Hull part until chine wetting starts at x_s

We can argue as in the discussion before eq. (9.10) and write the draft d from the spray root as

$$d = \left(1 + \frac{z_{max}}{Vt}\right) x\tau, \tag{9.64}$$

where z_{max} is defined in Figure 8.22. We should recall the definition of added mass (see section 7.2.1) and start out with the contribution to heave-added mass, that is,

$$A_{33}^{(1)} = \rho K \left(1 + \frac{z_{max}}{Vt}\right)^2 \tau^2 \int_0^{x_s} x^2\, dx.$$

K is defined by eq. (9.63), and x_S can be expressed by eq. (9.10) as

$$x_s = \frac{B}{2} \frac{\tan\beta}{(1 + z_{max}/Vt)\tau}. \tag{9.65}$$

$(1 + z_{max}/Vt)$ can be obtained from Table 8.3 for any deadrise angle β. When $\beta \to 0$, the value is asymptotically equal to $\pi/2$, that is, the same as Wagner's theory. It follows now by integration that

$$\frac{A_{33}^{(1)}}{\rho B^3} = \frac{K}{24} \frac{\tan^3\beta}{(1 + z_{max}/Vt)\tau}. \tag{9.66}$$

The contribution to coupled added mass in heave and pitch is

$$A_{35}^{(1)} = A_{53}^{(1)} = -\rho K \int_0^{x_s} d^2 (x - x_G)\, dx,$$

where

$$x_G = L_K - lcg. \tag{9.67}$$

L_K is the keel wetted length, and lcg is the longitudinal position of the center of gravity from the transom stern measured along the keel. This gives

$$\frac{A_{35}^{(1)}}{\rho B^4} = \frac{A_{53}^{(1)}}{\rho B^4} = \frac{A_{33}^{(1)}}{\rho B^3} \frac{x_G}{B} - \frac{K}{64} \frac{\tan^4\beta}{(1 + z_{max}/Vt)^2 \tau^2}. \tag{9.68}$$

The contribution to added mass in pitch is

$$A_{55}^{(1)} = \rho K \int_0^{x_s} d^2 (x - x_G)^2\, dx.$$

This means

$$\frac{A_{55}^{(1)}}{\rho B^5} = \frac{K}{160} \frac{\tan^5\beta}{(1 + z_{max}/Vt)^3 \tau^3} - \frac{K}{32} \frac{x_G}{B} \frac{\tan^4\beta}{(1 + z_{max}/Vt)^2 \tau^2} + \left(\frac{x_G}{B}\right)^2 \frac{A_{33}^{(1)}}{\rho B^3}. \tag{9.69}$$

B. Hull part with chine wetting

The integration limits in the added mass expressions are now from x_s to L_K. We find that

$$\frac{A_{33}^{(2)}}{\rho B^3} = C_1 \frac{\pi}{8} \frac{L_C}{B}, \tag{9.70}$$

where

$$C_1 = \frac{2\tan^2\beta}{\pi} K \tag{9.71}$$

and L_C is the chine wetted length. Further,

$$\frac{A_{35}^{(2)}}{\rho B^4} = \frac{A_{53}^{(2)}}{\rho B^4} = -C_1 \frac{\pi}{16}\left[\left(\frac{L_K}{B}\right)^2 - \left(\frac{x_s}{B}\right)^2\right] + \frac{x_G}{B}\frac{A_{33}^{(2)}}{\rho B^3} \quad (9.72)$$

and

$$\frac{A_{55}^{(2)}}{\rho B^5} = \frac{C_1 \pi}{24}\left(\left(\frac{L_K}{B}\right)^3 - \left(\frac{x_s}{B}\right)^3\right) - \frac{C_1 \pi}{8}\left(\frac{x_G}{B}\right)\left(\left(\frac{L_K}{B}\right)^2 - \left(\frac{x_s}{B}\right)^2\right) + \left(\frac{x_G}{B}\right)^2 \frac{A_{33}^{(2)}}{\rho B^3}. \quad (9.73)$$

Damping in heave and pitch

First, forced heave velocity is considered. We could apply the analysis in section 7.2.7 that led to hull-lift damping in heave. However, this was based on a linearized high-frequency free-surface condition. The nonlinear dynamic and kinematic free-surface conditions matter for a planing vessel, as was discussed in section 9.2. A quasi-steady approach will instead be followed. *Quasi-steady* means that the heave velocity causes a change in the angle of attack and changes the steady lift force. The steady analysis will be based on Savitsky's (1964) empirical formula. However, parts of the formula account for hydrostatic effects. We are only interested in the lifting part of the force and moment. This follows by letting $Fn_B \to \infty$ in the expressions for C_{L0} and l_p given by eqs. (9.5) and (9.7). The hydrodynamic lifting force coefficient is therefore

$$C_{L\beta} = C_{L0} - 0.0065\beta_{\text{deg}} C_{L0}^{0.60},$$

where

$$C_{L0} = \tau_{\text{deg}}^{1.1} 0.012 \lambda_W^{0.5} = \left(\frac{180}{\pi}\right)^{1.1} \tau^{1.1} 0.012 \lambda_W^{0.5}.$$

The index deg indicates if the angle should be given in degrees. Otherwise, it is given in radians. The argument is now like the one in section 7.2.8. Because of the heave velocity, there is a change in the angle of attack (trim)

$$\alpha = -\frac{d\eta_3}{dt}/U.$$

This causes a vertical force

$$F_3 = -\frac{\rho}{2}U^2 B^2 \frac{\partial C_{L\beta}}{\partial \tau}\frac{\dot{\eta}_3}{U}, \quad (9.74)$$

where

$$\frac{\partial C_{L\beta}}{\partial \tau} = \frac{\partial C_{L0}}{\partial \tau}\left[1 - 0.0039\beta_{\text{deg}} C_{L0}^{-0.4}\right] \quad (9.75)$$

and

$$\frac{\partial C_{L0}}{\partial \tau} = \left(\frac{180}{\pi}\right)^{1.1} 0.0132\tau^{0.1}\lambda_W^{0.5}. \quad (9.76)$$

When this force is moved to the left-hand side of the equations of motions, we can identify a damping coefficient B_{33}. This can be expressed as

$$\frac{B_{33}}{\rho B^3 (g/B)^{1/2}} = 0.5 Fn_B \frac{\partial C_{L\beta}}{\partial \tau}. \quad (9.77)$$

The forced heave velocity also causes a pitch moment F_5 about COG, which can be found by using eq. (9.7) with $Fn_B \to \infty$. (Note that eq. (9.7) gives the moment arm ℓ_p about the transom). The result is

$$F_5 = F_3(0.75\lambda_W B - lcg), \quad (9.78)$$

where F_3 is the vertical force for $Fn_B \to \infty$. When this moment is moved to the left-hand side of the equations of motion, we can identify a damping coefficient B_{53}. The result is

$$\frac{B_{53}}{B_{33} B} = 0.75\lambda_W - lcg/B. \quad (9.79)$$

The damping coefficients and B_{35} and B_{55} are found by studying forced pitch velocity $d\eta_5/dt$ about COG and calculating corresponding vertical force and pitch moment about COG that is 180° out of phase with $d\eta_5/dt$. However, there is no simple expression like Savitsky's to rely on. The analysis will instead be simplified as in section 8.5.1. This means that $B_{55} = Ux_T^2 a_{33}(x_T)$ and $B_{35} = -UA_{33} - Ux_T a_{33}(x_T)$ (see eqs. (8.100) and (8.98)).

Porpoising stability analysis

The stability analysis implies that nontrivial solutions of eq. (9.49) are studied. Possible solutions can be expressed as

$$\eta_j = \eta_{ja} e^{st}, \quad j = 3 \text{ and } 5. \quad (9.80)$$

Here η_j is a complex function. This is convenient in the solution of linear differential equations. It is the real part of eq. (9.80) that has physical meaning. s in eq. (9.80) can be written as

$$s = \alpha + i\omega, \qquad (9.81)$$

where α and ω are real and i is the complex unit. It means that

$$\eta_j = \eta_{ja} e^{\alpha t} e^{i\omega t}, \quad j = 3 \text{ and } 5, \qquad (9.82)$$

where η_{ja} is a complex constant. In the stability analysis, we are interested in whether the solutions decay or increase with time. Because $\exp(i\omega t)$ oscillates with time, stability or instability is expressed by the term $\exp(\alpha t)$. If the real part α of s is negative, then the solution decays with time and the solution is stable. If α is positive, then the solution is unstable. There exists more than one solution of s, so it must be ensured that the real part of all possible s-values is negative for the system to be stable. To find s, we introduce first eq. (9.80) into eq. (9.49). This gives

$$\begin{aligned}
&[s^2(M + A_{33}) + sB_{33} + C_{33}]\eta_{3a} \\
&\quad + [s^2 A_{35} + sB_{35} + C_{35}]\eta_{5a} = 0 \\
&[s^2 A_{53} + sB_{53} + C_{53}]\eta_{3a} + [s^2(I_{55} + A_{55}) \\
&\quad + sB_{55} + C_{55}]\eta_{5a} = 0.
\end{aligned} \qquad (9.83)$$

Because the right-hand side of this equation system is zero and we are interested in nontrivial solutions, the only possibility is that the coefficient determinant is equal to zero. This means

$$\begin{aligned}
&[s^2(M + A_{33}) + sB_{33} + C_{33}][s^2(I_{55} + A_{55}) \\
&\quad + sB_{55} + C_{55}] \\
&- [s^2 A_{53} + sB_{53} + C_{53}][s^2 A_{35} + sB_{35} + C_{35}] = 0.
\end{aligned} \qquad (9.84)$$

Eq. (9.84) can be rewritten as

$$A's^4 + B's^3 + C's^2 + D's + E' = 0, \qquad (9.85)$$

where

$$\begin{aligned}
A' &= (M + A_{33})(I_{55} + A_{55}) - A_{53}A_{35} \\
B' &= (M + A_{33})B_{55} + B_{33}(I_{55} + A_{55}) \\
&\quad - A_{53}B_{35} - A_{35}B_{53} \\
C' &= (M + A_{33})C_{55} + B_{33}B_{55} \\
&\quad + C_{33}(I_{55} + A_{55}) - A_{53}C_{35} \\
&\quad - B_{53}B_{35} - C_{53}A_{35} \\
D' &= B_{33}C_{55} + C_{33}B_{55} - B_{53}C_{35} - C_{53}B_{35} \\
E' &= C_{33}C_{55} - C_{53}C_{35}
\end{aligned} \qquad (9.86)$$

Analytical solutions of eq. (9.85) can be found and the real parts of the solutions studied. There exist four solutions of s. Because the coefficients A', B', C', D', and E' are real, two of the solutions are complex conjugates.

An alternative way to study the stability is to use Routh-Hurwitz stability criterion (Dorf and Bishop 1998). The requirement for a stable system is

$$\begin{aligned}
&\frac{B'}{A'} > 0, \quad \frac{D'}{A'} > 0, \quad \frac{E'}{A'} > 0, \\
&\frac{B'C'D' - A'D'^2 - B'^2E'}{A'^3} > 0.
\end{aligned} \qquad (9.87)$$

A third way of studying the eigenvalue s requires that the second-order differential equations (9.49) are rewritten into a system of first-order differential equations. The heave velocity $u_3 = d\eta_3/dt$ and the angular pitch velocity $u_5 = d\eta_5/dt$ are then introduced as new variables. This gives the following equation system:

$$\begin{Bmatrix} \dfrac{d\eta_3}{dt} \\ \dfrac{d\eta_5}{dt} \\ \dfrac{du_3}{dt} \\ \dfrac{du_5}{dt} \end{Bmatrix} = \{K\} \begin{Bmatrix} \eta_3 \\ \eta_5 \\ u_3 \\ u_5 \end{Bmatrix}. \qquad (9.88)$$

It is left as an exercise (see section 9.7.5) to derive the K-matrix. We now substitute in eq. (9.88), $\eta_3 = \eta_{3a}e^{st}$, $\eta_5 = \eta_{5a}e^{st}$, $u_3 = u_{3a}e^{st}$, and $u_5 = u_{5a}e^{st}$ and get

$$(K - sI)\begin{Bmatrix} \eta_{3a} \\ \eta_{5a} \\ u_{3a} \\ u_{5a} \end{Bmatrix} = 0. \qquad (9.89)$$

Here I is the identity matrix with non-zero elements equal to 1 only on the diagonal. Eq. (9.89) shows that the problem of determining s is the same as determining the eigenvalues s of the matrix K. The advantage of following this procedure is that standard computer subroutines may be used. If the number of degrees of freedom is increased – for instance, if coupled surge, heave, and pitch are considered – this procedure is more convenient than following the approach that led to the characteristic equation (9.85).

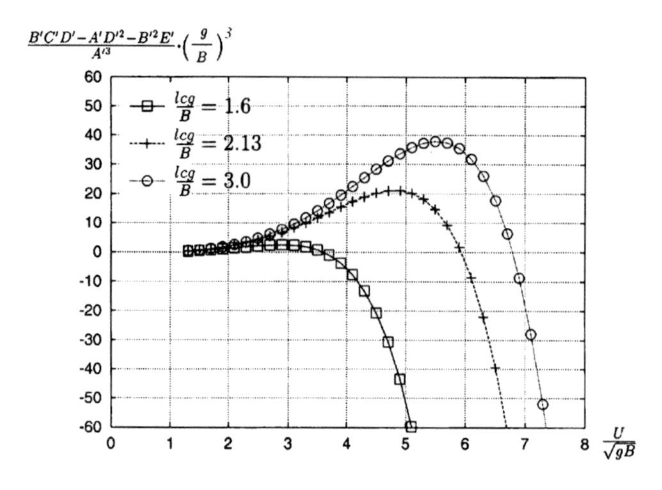

Figure 9.28. Porpoising stability of prismatic planing hull studied by the stability parameter $(B'C'D' - A'D'^2 - B'^2E')/A'^3$ (see eq. (9.87)) as a function of beam Froude number and influence of lcg. Negative stability parameter means instability. Theoretical estimates.

Example: Porpoising stability

The theoretical framework will be illustrated by examples. The base case is a prismatic hull form with average wetted length-to-beam ratio $\lambda_W = 4$, deadrise angle $\beta_{\text{deg}} = 20^{\text{o}}$, trim angle $\tau_{\text{deg}} = 4^{\text{o}}$, $lcg/B = 2.13$, $vcg/B = 0.25$, $M/(\rho B^3) = 1.28$, pitch radius of gyration $= 1.3\ B$. Hydrodynamic coefficients are theoretically calculated as previously described. The following parameter study is meant only to illustrate trends, and one must be careful in making quantitative conclusions.

The stability analysis is based on eq. (9.87). B'/A', D'/A', and E'/A' are always positive in the studied case. So it is the last condition in eq. (9.87) that decides if the planing vessel is stable or not in heave and pitch. Therefore, the following figures present graphs of the stability parameter $(B'C'D' - A'D'^2 - B'^2E')/A'^3$.

Figure 9.28 shows the stability parameter as a function of the beam Froude number for three different lcg/B ratios. It has not been accounted for that the parameters depend on each other. For instance, changing lcg influences τ and λ_W (see section 9.2). It confirms the common experience that porpoising can be avoided by reducing the speed and/or moving the COG forward in the ship.

In order to demonstrate how sensitive the stability parameter is to the hydrodynamic coefficients, extrapolated added mass and damping coefficients from Troesch's (1992) experiments will be used. Our base case, in which $\beta = 20^{\text{o}}$, $\lambda_W = 4$, $lcg/B = 2.13$, and $\tau_{\text{deg}} = 4^{\text{o}}$, is studied. Table 9.1 shows the predicted values of added mass and damping for our base case with $Fn_B = 5.0$. The big differences are for A_{35} and A_{53}, but A_{35} is small.

Table 9.1. *Added mass* (A_{jk}) *and damping* (B_{jk}) *in heave and pitch for a prismatic planing hull with* $\beta = 20^\circ$, $\lambda_w = 4$, $lcg/B = 2.13$, $\tau_{\text{deg}} = 4^\circ$, and $Fn_B = 5.0$.

	Simplified theory	Empirical formula (Troesch, unpublished)
$A_{33}/(\rho B^3)$	1.28	1.28
$A_{35}/(\rho B^4)$	−0.31	−0.04
$A_{53}/(\rho B^4)$	−0.31	−1.63
$A_{55}/(\rho B^5)$	1.66	1.48
$B_{33}/[\rho B^3(g/B)^{0.5}]$	3.53	4.15
$B_{35}/[\rho B^4(g/B)^{0.5}]$	−10.09	−8.90
$B_{53}/[\rho B^4(g/B)^{0.5}]$	3.07	4.21
$B_{55}/[\rho B^5(g/B)^{0.5}]$	7.84	10.70

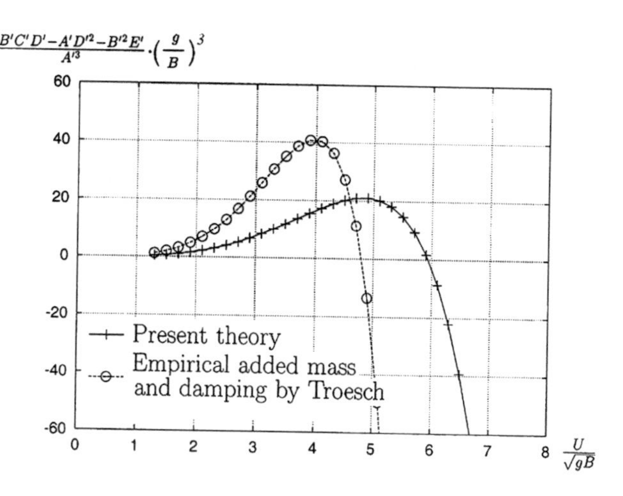

Figure 9.29. Porpoising stability of a prismatic planing hull studied by the stability parameter $(B'C'D' - A'D'^2 - B'^2E')/A'^3$ (see eq. (9.87)) as a function of beam Froude number. Influence of changing added mass and damping coefficients according to the empirical formula by Troesch (unpublished). Negative stability parameter means instability.

Figure 9.29 presents the stability results by using our theory for added mass and damping as well as an unpublished empirical formula for added mass and damping by Troesch. The same restoring coefficients are used in both cases. The figure shows that using only theory gives porpoising when $U/(gB)^{0.5} > 5.9$. When Troesch's empirical formulas are used, porpoising occurs when $U/(gB)^{0.5} > 4.8$. Actually, Troesch and Falzarano (1993) used a B_{55}-value that was less than 5% higher than that of the empirical formula by Troesch. The other hydrodynamic coefficients were the same. That gave the $U/(gB)^{1/2} = 5.0$ to be a marginally stable case. This illustrates the sensitivity to the hydrodynamic coefficients. One reason is that the different components of the stability parameter counteract each other and have absolute values that are an order of ten times larger than the stability parameter. The first component, $B'C'D'/A'^3$, has a sign opposite to the two other components, $-A'D'^2/A'^3$ and $-B'^2E'/A'^3$. A consequence is that relatively small differences in each component will have a large influence on the stability parameter.

9.5 Wave-induced motions and loads

Seakeeping operability criteria for fast small craft are listed in Tables 1.1 and 1.2. These criteria are related to accelerations, roll, slamming, and deck wetness. Our focus is on vertical vessel motions, which also include vertical accelerations as a part of the analysis.

Fridsma (1969, 1971) presented systematic experimental studies of motions and accelerations of planing monohulls in head waves. The dependence on parameters such as forward speed, deadrise angle, longitudinal position of the center of gravity, and trim angles was investigated. The results were analyzed and presented as design "charts" for designers.

A planing craft in waves may show strong nonlinear behavior. Reasons are the strong variations in the wetted area and a nonvertical hull surface at the intersection between the body surface and the water surface, which result in an increase of the trim and the rise of the vessel due to the wave-induced motions of the vessel. This was also experimentally confirmed by Katayama et al. (2000). The length Froude number was varied between 2 and 5 in their study. The nonlinear effects have, generally speaking, a larger influence on accelerations than on motions.

Both regular and irregular jumping of a planing vessel may occur in a seaway. Katayama et al. (2000) investigated this systematically by means of model tests in incident regular waves. As the Froude number increases, the limiting wave height for jumping to occur decreases.

A scenario as in Figure 9.30 may lead to jumping in which the speed is high enough for the vessel to jump into the air and subsequently fall down and impact on the water. The resulting slamming loads may cause important structural effects and leads to large vertical accelerations of the vessel, which may affect equipment onboard.

Figure 9.31 shows computer simulations by Lin et al. (1995) of vertical accelerations at the center of gravity of a 57-foot (17.4-m) high-speed patrol boat in head sea and sea state 5 (see Table 3.5). The ship speed is 40 knots; that is, the length Froude number is 1.58. The vessel position

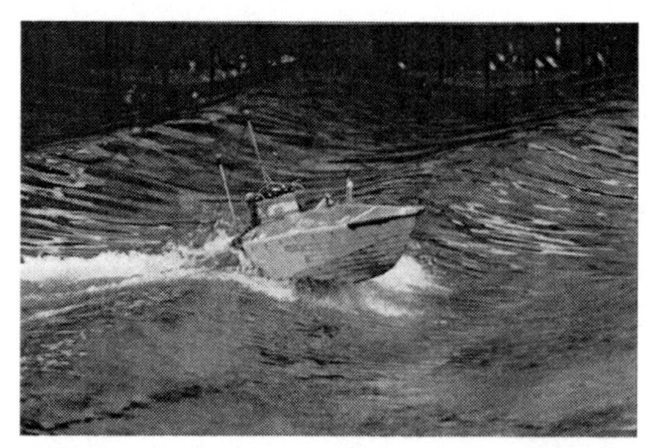

Figure 9.30. Model testing of planing vessel in waves at MARINTEK.

relative to waves is also illustrated for selected time instants.

The boat is out of the water at time 13.13 s. The ship therefore has a downward acceleration equal to the acceleration of gravity, g.

The ship reenters the water at time 13.33 s. A large wave then approaches, and the acceleration rises from $-g$ to $+1.3g$ in 0.2 s. This is followed by a very large upward acceleration close to $6g$.

At the time step $t = 14.86$ s, the vessel is once more in the air and has a downward acceleration equal to g.

9.5.1 Wave excitation loads in heave and pitch in head sea

The following analysis assumes no jumping. Including jumping would imply that global slamming loads as discussed in Chapter 8 must be incorporated. We start with studying the wave excitation loads in heave and pitch in head sea. An important aspect is the strong interaction between the steady and unsteady flows. This is of a similar nature to that discussed in section 9.4.1 for restoring forces and moments due to heave and pitch.

Generalized Froude-Kriloff loads

When studying wave-induced motions, we have to introduce wave excitation loads on the right-hand side of eq. (9.49). The left-hand side is unchanged. We will limit ourselves to head sea regular waves and assume that the wave steepness is small so that linear wave theory can describe the incident waves. Further, the incident wavelength λ is assumed long relative to the ship length. This simplifies the analysis. The incident instantaneous wave elevation is written as

$$\zeta = \zeta_a \sin(\omega_e t - kx). \tag{9.90}$$

Here $\omega_e = \omega_0 + kU$ is the frequency of encounter, ω_0 is the frequency of the wave in an Earth-fixed coordinate system, and $k = 2\pi/\lambda = \omega_0^2/g$ is the wave number. Because k is assumed small (or λ long), we can write along the ship

$$\begin{aligned} \zeta &= \zeta_a \sin\omega_e t \cos kx - \zeta_a \cos\omega_e t \sin kx. \\ &\approx \zeta_a \sin\omega_e t - xk\zeta_a \cos\omega_e t \end{aligned} \tag{9.91}$$

We have expanded $\cos kx$ and $\sin kx$ in a series in kx and neglected terms of order k^2. The first term, $\zeta_a \sin\omega_e t$, represents a spatial uniform vertical motion along the ship, that is, similar to the heave motion. The difference is that the water moves and not the ship. So if the water goes up, it corresponds to negative heave. The last term, $-xk\zeta_a \cos\omega_e t$, can be compared with the vertical motion due to pitch; that is, $k\zeta_a \cos\omega_e t$ is like a pitch angle.

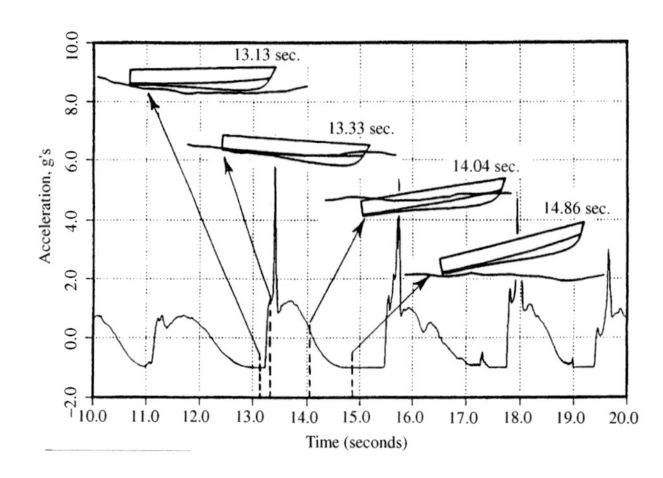

Figure 9.31. Computer simulations of vertical accelerations in sea state 5 for a 57-foot patrol boat measured in terms of acceleration of gravity, g. The length Froude number is 1.58 (Lin et al. 1995).

Actually, $-k\zeta_a \cos\omega_e t$ is the incident wave slope at $x = 0$. This similarity with heave and pitch will be used when the wave excitation loads are formulated. However, let us first recall Figure 3.5. This figure expresses the fact that the total pressure in the vicinity of the free surface has hydrostatic depth dependence relative to the instantaneous free surface with atmospheric pressure at the free surface. Normally when we talk about the hydrostatic pressure $-\rho g z$, $z = 0$ corresponds to the mean free surface. How far down in the fluid this hydrostatic depth dependence has relevance depends on the wavelength. This can be seen by noting that the pressure below the mean free surface is composed of one part that is exponentially decaying like $\rho g \zeta_a \exp(kz)$ for deep-water waves and another part that is $-\rho g z$ (the normal hydrostatic pressure). Here $z = 0$ means mean free surface; see Figure 3.5. The larger the wavelength $\lambda = 2\pi/k$ is, the slower $\exp(kz)$ decays with depth and the more appropriate it is to approximate $\exp(kz)$ with 1. This implies that the pressure in incident long waves at the ship hull has a hydrostatic depth dependence relative to the instantaneous wave elevation.

We have now a sufficient basis to formulate what we call the generalized Froude-Kriloff wave excitation loads on the ship. Froude-Kriloff loads originally meant that the presence of the ship does not influence the pressure distribution. Generalized Froude-Kriloff loads will now mean that we consider quasi-steady hydrodynamic loads on the ship; that is, we consider all pressure terms in the steady Bernoulli equation for different instantaneous positions of the incident waves. We apply, then, Savitsky's formulas for the steady vertical force and pitch moment by making the similarity between the vertical incident wave motion along the ship and the steady heave and pitch motions. The word *steady* for motions is stressed because the velocities and accelerations due to the incident waves are not accounted for. We will come back to that later, when the diffraction potential is introduced. This implies that the generalized Froude-Kriloff wave excitation loads can be formulated like the restoring force and moment (see section 9.4.1). We can write the vertical generalized Froude-Kriloff force as

$$F_3^{FK} = C_{33}\zeta_a \sin\omega_e t + C_{35}k\zeta_a \cos\omega_e t. \quad (9.92)$$

The generalized Froude-Kriloff pitch moment becomes

$$F_5^{FK} = C_{53}\zeta_a \sin\omega_e t + C_{55}k\zeta_a \cos\omega_e t. \quad (9.93)$$

Similar terms are used by Troesch and Falzarano (1993) in their study of wave-induced motions of planing boats in long incident deep-water waves. We should once more note that C_{jk} accounts not only for the hydrostatic pressure term, but all pressure terms given by the steady Bernoulli equation.

We should recall that we have assumed long wavelengths relative to the ship length. Let us try to quantify what this means. This is done by a strip theory approach by assuming that the phasing of the vertical loads F_3^{2D} per unit length along the ship can be expressed as eq. (9.90). This means that F_3^{2D}

$$F_3^{2D} = A(\sin\omega_e t \cos kx - \cos\omega_e t \sin kx), \quad (9.94)$$

where A is a constant depending on the cross-sectional form. Let us assume constant cross-sectional form along the ship and that $x = 0$ corresponds to the midship. L is the length of the ship. By integrating the following vertical force F_3 and pitch moment F_5 on the ship, this gives

$$F_3 = \frac{2A}{k}\sin\left(\frac{kL}{2}\right)\sin\omega_e t \quad (9.95)$$

$$F_5 = A\left[-\frac{L}{k}\cos\left(\frac{kL}{2}\right) + \frac{2}{k^2}\sin\left(\frac{kL}{2}\right)\right]\cos\omega_e t. \quad (9.96)$$

If the load phasing is approximated as in eq. (9.91), we get the following approximation:

$$F_3^{LW} = AL\sin\omega_e t \quad (9.97)$$

$$F_5^{LW} = A\frac{kL^3}{12}\cos\omega_e t. \quad (9.98)$$

The superscript LW means long-wavelength approximation. We have plotted both F_3/F_3^{WL} and F_5/F_5^{WL} as a function of λ/L in Figure 9.32. For instance, if we want to make less than a 10% error in using the phase approximation given by eq. (9.91), then λ/L must be larger than 4. For this wavelength to be sufficiently small to evaluate resonant heave and pitch motions depends on the natural frequencies ω_n and the Froude number. This can be examined by eq. (7.46) by replacing ω_{n3} with ω_n. We should recall that there are two natural frequencies for coupled heave and pitch motions. Their values follow from the

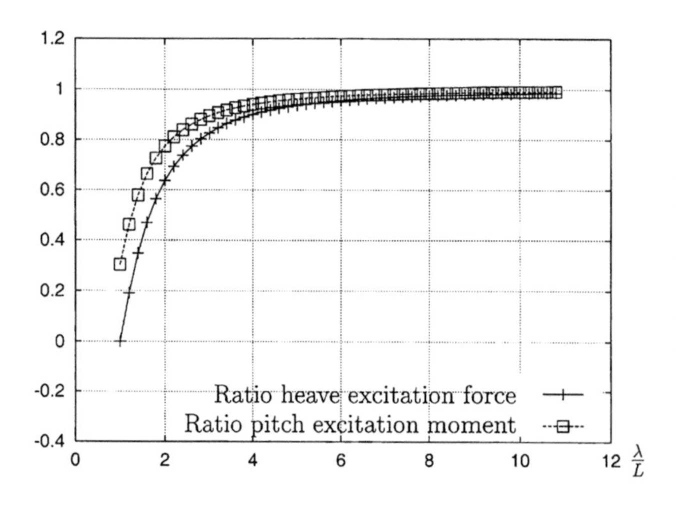

Figure 9.32. Ratio between "exact" heave force and pitch moment and long-wavelength approximation of heave force and pitch moment as a function of wavelength (λ)-to–ship length (L) ratio. Head sea, constant cross section, and strip theory are assumed.

analysis in section 9.4 (see eqs. (9.80) and (9.81)) and depend on ship length as well as beam, trim angle, deadrise angle, lcg, and Froude number. Troesch and Falzarano (1993) presented predicted highest natural frequency for different conditions of a prismatic planing hull. The values varied from $\omega_n(L/g)^{0.5} = 2.96$ to 6.24.

We have chosen three cases of nondimensional frequencies and present in Figure 9.33 the relationship between λ/L and length Froude number $Fn = U/(Lg)^{0.5}$ that gives resonance. We note that the higher $\omega_n(L/g)^{0.5}$ is, the lower λ/L is for a given Fn. In the first case, $\omega_n(L/g)^{0.5} = 3.53$, which corresponds to $Fn = 2.51$ in Troesch and Falzarano's calculations. By comparing Figure 9.33 with Figure 9.32, we note that the phase approximation given by eq. (9.91) can be used to describe the wave excitation loads at the natural frequency. The second case is $\omega_n(L/g)^{0.5} = 4.48$, which corresponds with $Fn = 1.72$. This gives, according to Figure 9.33, $\lambda/L = 3.45$, which by Figure 9.32 means that the phase approximation causes a relative error of 13% in the heave approximation force and 8% in the pitch excitation load. The third case is $\omega_n(L/g)^{0.5} = 6.24$, which corresponds with $Fn = 2.5$. According to Figure 9.33, this gives $\lambda/L = 3.24$, which by Figure 9.32 means that the phase approximation causes a relative error of 14% in the heave excitation force and 9% in the pitch excitation moment.

Diffraction loads

Using what we called the generalized Froude-Kriloff approximation implies that the effects of the incident wave velocities and accelerations on the ship have not been considered. The first step in accounting for this is to find a velocity potential φ_7 that, together with the incident wave

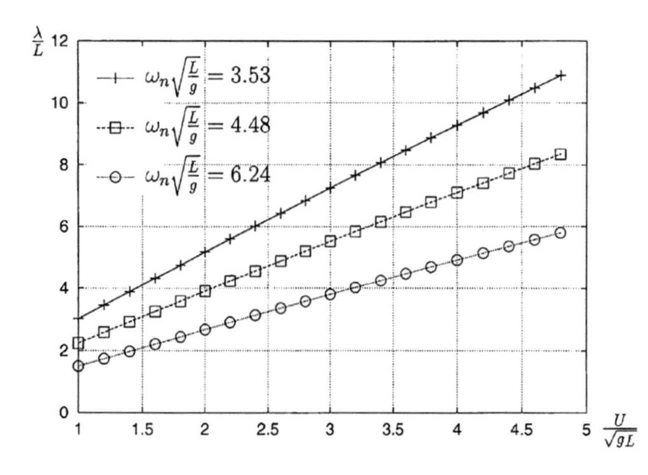

Figure 9.33. Relationship between wavelength-to–ship length ratio, λ/L, and length Froude number, $U/(Lg)^{0.5}$, for different nondimensionalized natural frequencies, $\omega_n(L/g)^{0.5}$, of heave and pitch in head sea.

velocity potential, causes zero flow through the body boundary. The ship is restrained from oscillating. This will be done by strip theory using a long-wavelength approximation and letting φ_7 satisfy the high-frequency free-surface condition $\varphi_7 = 0$, similar to what we did for the added mass and damping problem. Strip theory implies that φ_7 satisfies a 2D Laplace equation in a cross-sectional plane of the ship. We can formally write the body boundary condition as

$$\frac{\partial \varphi_7}{\partial n} = -w n_3, \qquad (9.99)$$

where $\partial/\partial n$ means the normal derivative in a cross-sectional plane of the ship. The positive normal direction is into the fluid. Further, n_3 is the z-component of the normal vector and w is the vertical fluid velocity of the incident waves. By a long-wavelength approximation, this can be written as

$$\begin{aligned} w &= \omega_0 \zeta_a e^{kz} \cos(\omega_e t - kx) \\ &= \omega_0 \zeta_a e^{kz} \left[\cos \omega_e t \cos kx + \sin \omega_e t \sin kx\right] \\ &\approx \omega_0 \zeta_a \cos \omega_e t + \omega_0 \zeta_a kx \sin \omega_e t \\ &= V_3 - x V_5 \end{aligned} \qquad (9.100)$$

where

$$V_3 = \omega_0 \zeta_a \cos \omega_e t \qquad (9.101)$$

$$V_5 = -\omega_0 k \zeta_a \sin \omega_e t. \qquad (9.102)$$

Here V_3 corresponds to the vertical velocity of the incident free surface at $x = 0$ and V_5 is the angular velocity of the incident wave slope at $x = 0$. We can make analogies between V_3 and heave velocity and between V_5 and pitch velocity. The boundary-value problem resembles, therefore, the forced heave and pitch problem, but there is no angle of attack term $U\eta_5$. The solution at a ship cross section can then formally be written as

$$\varphi_7 = -(V_3 - x V_5)\varphi_3, \qquad (9.103)$$

where φ_3 is the strip theory velocity potential due to forced heave with unit heave velocity. It is the same φ_3 as the one used in section 7.2.7. The pressure is now similar to that in eq. (7.67), that is,

$$\begin{aligned} p &= -\rho \frac{\partial \varphi_7}{\partial t} - \rho U \frac{\partial \varphi_7}{\partial x} \\ &= \rho \varphi_3 \frac{\partial}{\partial t}(V_3 - x V_5) + \rho U \frac{\partial}{\partial x}\left[(V_3 - x V_5)\varphi_3\right]. \end{aligned} \qquad (9.104)$$

We leave it as an exercise to show that the resulting vertical force F_3^D and pitch moment F_5^D can be expressed as

$$\begin{aligned} F_3^D = {} & A_{33}\frac{\partial V_3}{\partial t} + A_{35}\frac{\partial V_5}{\partial t} + U a_{33}(x_T) V_3 \\ & - U x_T a_{33}(x_T) V_5 \end{aligned} \qquad (9.105)$$

$$\begin{aligned} F_5^D = {} & A_{53}\frac{\partial V_3}{\partial t} + A_{55}\frac{\partial V_5}{\partial t} + [-U x_T a_{33}(x_T) \\ & + U A_{33}]\, V_3 + \left[U x_T^2 a_{33}(x_T) + U A_{35}\right] V_5. \end{aligned} \qquad (9.106)$$

Let us compare the phasing of these force and moment terms with corresponding generalized Froude-Kriloff terms in eqs. (9.92) and (9.93). Because $\partial V_3/\partial t$ is 180° out of phase with the incident wave elevation at $x = 0$ and $\partial V_5/\partial t$ is 180° out of phase with $k\zeta_a \cos \omega_e t$, it means that each $A_{jk}\partial V_k/\partial t$-term in eqs. (9.105) and (9.106) is 180° out of phase with corresponding C_{jk}-terms in eqs. (9.92) and (9.93) and causes a reduction in the wave excitation.

The terms associated with the V_k-terms in F_3^D and F_5^D can formally be denoted B_{jk}^D. Each $B_{jk}^D V_k$-term in eqs. (9.105) and (9.106) is 90° out of phase with corresponding $A_{jk}\partial V_k/\partial t$ terms in eqs. (9.105) and (9.106). It means they cause an increase in the wave excitation and change the phasing.

Summary

We have shown that the linear vertical wave excitation force F_3 and wave excitation pitch moment F_5 about COG ($x = 0$) in regular head sea waves described with free-surface elevation

$$\zeta = \zeta_a \sin(\omega_e t - kx) \qquad (9.107)$$

can, by a long-wavelength approximation, be expressed as

$$F_3 = F_{3s}\zeta_a \sin \omega_e t + F_{3c}\zeta_a \cos \omega_e t \qquad (9.108)$$

$$F_5 = F_{5s}\zeta_a \sin \omega_e t + F_{5c}\zeta_a \cos \omega_e t. \qquad (9.109)$$

Here F_{jc} and F_{js} are expressed by the added mass (A_{jk}), damping (B_{jk}), and restoring coefficients (C_{jk}) as

$$F_{3s} = C_{33} - A_{33}\omega_0\omega_e - B_{35}^D \omega_0 k \qquad (9.110)$$

$$F_{3c} = C_{35}k - A_{35}\omega_0\omega_e k + B_{33}^D \omega_0 \qquad (9.111)$$

$$F_{5s} = C_{53} - A_{53}\omega_0\omega_e - B_{55}^D \omega_0 k \qquad (9.112)$$

$$F_{5c} = C_{55}k - A_{55}\omega_0\omega_e k + B_{53}^D \omega_0, \qquad (9.113)$$

where $B_{33}^D = B_{33}$ and $B_{53}^D = B_{53}$ according to linear theory. Further,

$$B_{35}^D = B_{35} + UA_{33} \quad (9.114)$$

$$B_{55}^D = B_{55} + UA_{35}. \quad (9.115)$$

Further,

$$\omega_e = \omega_0 + kU, \quad k = \omega_0^2/g. \quad (9.116)$$

9.5.2 Frequency-domain solution of heave and pitch in head sea

We will first assume linear theory and a steady-state solution (frequency-domain solution) and consider a prismatic planing hull in regular head sea waves. Because the cross-sectional shape is wedge-formed and the local draft is small in the bow region, small wave-induced motions can make large changes in the instantaneous water-plane area and submerged volume. This implies that linear theory may have limited value in practice. Assuming a steady-state solution implies that all transient effects have died out. This means that the real parts α of the eigenvalues s (see eqs. (9.81) and (9.82)) are negative. Another way of saying this is that a linear steady-state solution does not exist when porpoising instability occurs.

The left-hand sides of the equations of coupled heave and pitch motions are the same as those in eq. (9.49). The right-hand sides of the first and second equation are, respectively, F_3 and F_5 given by eqs. (9.108) and (9.109). When solving the equations, it is convenient to use complex notation, which means we write

$$F_j = \zeta_a(F_{jc} - iF_{js})e^{i\omega_e t}, \quad j = 3, 5, \quad (9.117)$$

where F_{jc} and F_{js} are given by eqs. (9.110) through (9.113). When operating with a complex quantity as eq. (9.117), it is always understood that it is the real part that has physical meaning. We see that the real part of eq. (9.117) is

$$\begin{aligned} &\text{Re}\left[\zeta_a(F_{jc} - iF_{js})e^{i\omega_e t}\right] \\ &= \text{Re}\left[\zeta_a(F_{jc} - iF_{js})(\cos\omega_e t + i\sin\omega_e t)\right] \\ &= F_{jc}\zeta_a\cos\omega_e t + F_{js}\zeta_a\sin\omega_e t \end{aligned}$$

We write the motions as

$$\eta_j = (\eta_{Rj} + i\eta_{Ij})e^{i\omega_e t}, \quad j = 3, 5. \quad (9.118)$$

Inserting this into the equations of motions and dividing by the common factor $\exp(i\omega_e t)$ on the left- and right-hand sides of the equations, give the following two complex equations:

$$\begin{aligned} &\left[-\omega_e^2(M + A_{33}) + i\omega_e B_{33} + C_{33}\right]\left[\eta_{R3} + i\eta_{I3}\right] \\ &\quad + \left[-\omega_e^2 A_{35} + i\omega_e B_{35} + C_{35}\right] \\ &\quad \times \left[\eta_{R5} + i\eta_{I5}\right] = \zeta_a(F_{3c} - iF_{3s}) \\ &\left[-\omega_e^2 A_{53} + i\omega_e B_{53} + C_{53}\right]\left[\eta_{R3} + i\eta_{I3}\right] \\ &\quad + \left[-\omega_e^2(I_{55} + A_{55}) + i\omega_e B_{55} + C_{55}\right] \\ &\quad \times \left[\eta_{R5} + i\eta_{I5}\right] = \zeta_a(F_{5c} - iF_{5s}) \end{aligned} \quad (9.119)$$

We can solve these linear algebraic equations for the unknowns, directly with complex unknowns or by dividing each equation into a real and imaginary part. The latter eq. (9.119) gives four equations with four unkowns, η_{R3}, η_{I3}, η_{R5}, and η_{I5}, which can be solved by standard computer subroutines. The transfer functions in heave and pitch are given as, respectively,

$$\frac{|\eta_j|}{\zeta_a} = \frac{(\eta_{Rj}^2 + \eta_{Ij}^2)^{1/2}}{\zeta_a}, \quad j = 3, 5. \quad (9.120)$$

The phases ε_j for heave and pitch relative to the wave elevation at $x = 0$, that is, COG, can be obtained as described in section 7.2. Results based on a frequency-domain solution are presented later in the chapter.

9.5.3 Time-domain solution of heave and pitch in head sea

We remarked in the beginning of this chapter that nonlinearities may matter because of large wave-induced motions. Also, if porpoising instabilities occur, nonlinearities will cause bounded solutions. A linear stability analysis will give an unbounded unstable solution as time goes on. The latter is a result of the exponential growth in the $\exp(\alpha t)$-term of eq. (9.82) when the real part α of the eigenvalue s is positive. Accounting for nonlinearities implies that we must solve the equations of motions in the time domain. There also exist nonlinear frequency-domain solutions, for example, the common procedures in analyzing second-order wave-induced motions of offshore structures (Faltinsen 1990). However, a frequency-domain solution assumes a steady-state solution and cannot handle unstable solutions. Further, nonlinear frequency-domain solutions are convenient when we can use a perturbation analysis with the wave amplitude as a small parameter to find higher-order hydrodynamic loads. This will, in practice, require that the hull surface is vertical at the free surface. The latter is not true for planing hulls. So we proceed with a nonlinear time-domain solution. Dealing with all nonlinear hydrodynamic forces and moments is beyond the present state of the art. So we must simplify and try to include what are the most important nonlinearities. According to Troesch and Falzarano (1993), the most important nonlinearities are the result of the restoring forces and moments and not added mass and damping. However, because we explained the generalized Froude-Kriloff excitation forces and moments as restoring forces and moments (see eqs. (9.92) and (9.93)), we will also include that load part in the nonlinear analysis. The approximate nonlinear equations for the coupled heave and pitch motions in the time domain are expressed as

$$(M + A_{33})\frac{d^2\eta_3}{dt^2} + A_{35}\frac{d^2\eta_5}{dt^2} = F_3 \qquad (9.121)$$

$$A_{53}\frac{d^2\eta_3}{dt^2} + (I_{55} + A_{55})\frac{d^2\eta_5}{dt^2} = F_5, \qquad (9.122)$$

where

$$\begin{aligned} F_3 = {} & F_3^c - F_{03}^c + (F_{3s} - C_{33})\zeta_a \sin\omega_e t \\ & + (F_{3c} - C_{35}k)\zeta_a \cos\omega_e t \\ & - B_{33}\frac{d\eta_3}{dt} - B_{35}\frac{d\eta_5}{dt} \end{aligned} \qquad (9.123)$$

$$\begin{aligned} F_5 = {} & F_5^c - F_{05}^c + (F_{5s} - C_{53})\zeta_a \sin\omega_e t \\ & + (F_{5c} - C_{55}k)\zeta_a \cos\omega_e t. \\ & - B_{53}\frac{d\eta_3}{dt} - B_{55}\frac{d\eta_5}{dt} \end{aligned} \qquad (9.124)$$

Here F_{03}^c and F_{05}^c mean the values of F_3^c and F_5^c for zero η_3, η_5, and ζ_a. F_3^c is the vertical steady force as calculated by eq. (9.4) and accounting for the change in trim due to pitch and wave slope and submergence due to heave, pitch, and wave elevation. The latter is found by calculating keel wetted length L_K and chine wetted length L_C as in eqs. (9.50) and (9.51). We must, of course, require positive L_K. Negative predicted L_K means the ship is out of the water. In reality, this occurs at high speed. If this should be included in our analysis, it requires that slamming loads also be considered when the vessel subsequently impacts on the water surface.

We now replace η_3 by $\eta_3 - \zeta_a \sin\omega_e t$ and η_5 by $\eta_5 - k\zeta_a \cos\omega_e t$ (see eq. (9.91) and follow the discussion in which the analogy between the incident wave elevation at $x = 0$ and heave and between the incident wave slope at $x = 0$ and pitch was made). Savitsky's formula sets the requirement that the trim angle be positive. This means in the quasi-steady analysis, that

$$\tau + \eta_5 - k\zeta_a \cos\omega_e t \geq 0$$

where τ is the steady trim angle in radians. This represents a practical limitation for steepnesses or wave heights for the present nonlinear theory. This limitation is dependent on the wavelength. For instance, when the wavelength is larger than the wavelength with large and resonant motions, η_5 tends to be in phase with the incident wave slope at $x = 0$. This means that it is easier to meet the requirement for long wavelengths for a given wave amplitude than for smaller wavelengths relative to the resonant condition. In the latter case, one should also note that $k\zeta_a$ increases with decreasing wavelength, λ. F_5^c is the steady pitch moment about $x = 0$ found by F_3^c and the moment arm given by eq. (9.7). (Note that ℓ_p is the moment arm about the transom, so we have to subtract lcg from ℓ_p to get the moment arm about $x = 0$). Heave, pitch, and wave elevation are accounted for in a similar way as for F_3^c. Linear analysis then gives

$$\begin{aligned} F_3^c - F_{03}^c &= -C_{33}\eta_3 - C_{35}\eta_5 + C_{33}\zeta_a \sin\omega_e t \\ &\quad + C_{35}k\zeta_a \cos\omega_e t \\ F_5^c - F_{05}^c &= -C_{53}\eta_3 - C_{55}\eta_5 + C_{53}\zeta_a \sin\omega_e t \\ &\quad + C_{55}k\zeta_a \cos\omega_e t. \end{aligned}$$

Further, F_{js} and F_{jc} in eqs. (9.123) and (9.124) define the wave excitation loads as given by eqs. (9.110) through (9.113). We have subtracted the C_{jk} (linear generalized Froude-Kriloff) part of F_{js} and F_{jc} because that is accounted for in a nonlinear way in F_j^c. Why we have moved the linear damping terms to the right-hand sides of eqs. (9.121) and (9.122) becomes evident when we solve the equations.

When numerically time integrating eqs. (9.121) and (9.122), it is convenient to rewrite the equations into a set of first-order differential equations. We then define, as in section 9.4, $d\eta_3/dt = u_3$ and $d\eta_5/dt = u_5$ and introduce u_3 and u_5 in eqs. (9.121) and (9.122). We then reformulate these two

Table 9.2. *Calculated real (α) and imaginary (ω) terms of eigenvalues s for coupled linear heave and pitch of prismatic planing hull with $\beta = 20^\circ$, $\lambda_W = 4$, $lcg/B = 2.13$, $\tau_{deg} = 4^\circ$, and $U/\sqrt{gB} = 3.0$*

Eigenvalue no.	$\alpha\sqrt{\frac{B}{g}}$	$\omega\sqrt{\frac{B}{g}}$
1	−0.12	+1.91
2	−0.12	−1.91
3	−0.86	+0.67
4	−0.86	−0.67

equations into explicit expressions for du_3/dt and du_5/dt. This gives, then, the following four first-order differential equations:

$$\begin{aligned} \frac{d\eta_3}{dt} &= u_3 \\ \frac{d\eta_5}{dt} &= u_5 \\ \frac{du_3}{dt} &= [F_3(I_{55} + A_{55}) - F_5 A_{35}]/D \\ \frac{du_5}{dt} &= [(M + A_{33})F_5 - A_{53} F_3]/D, \end{aligned} \tag{9.125}$$

where

$$D = (M + A_{33})(I_{55} + A_{55}) - A_{53} A_{35}.$$

Eq. (9.125) is now in a convenient form for standard procedures for numerical time integration, but initial conditions have to be given. In our analysis, we have used a Runge-Kutta method of fourth order.

9.5.4 Example: Heave and pitch in regular head sea

A prismatic planing hull in regular head sea waves is considered, and the linear transfer functions of heave and pitch are calculated. The average wetted length-to-beam ratio $\lambda_W = 4.0$, the trim angle $\tau_{deg} = 4^\circ$, the deadrise angle $\beta_{deg} = 20^\circ$, the beam Froude number $U/(gB)^{0.5} = 3.0$, $lcg/B = 2.13$, $vcg/B = 0.25$, the ship mass $M = 1.28\rho B^3$, and the pitch radius of gyration r_{55} with respect to COG is $1.3B$. The theoretical procedure to find transfer functions is described in section 9.5.2. Three different methods are used to predict added mass, damping, and wave excitation loads.

Transfer functions (or steady-state solutions) have no physical meaning if the system is dynamically unstable, that is, porpoising occurs. However, Figure 9.29 shows that the coupled heave and pitch motions will be stable when $U/(gB)^{0.5} = 3.0$. This is true both for our theoretical method and for Troesch's empirical method. The stability is determined by the eigenvalues s (see eq. (9.81)). The four eigenvalues s calculated by our theory gives nondimensionalized real (α) and imaginary (ω) parts, as presented in Table 9.2. We can combine the eigenvalues into two sets, in which the eigenvalues are complex conjugates for each set. When studying the system response, it is sufficient to consider positive imaginary parts. Because all real parts are negative, Table 9.2 confirms that porpoising instability does not occur. If the ratio between the real and imaginary parts is small, strong amplification of the transfer function occurs when the system is excited with a frequency ω_e equal to ω. The ratio between the absolute values of α and ω can then be approximated as the ratio between damping and critical damping of the eigenmode.

The transfer functions in heave and pitch are presented in Figures 9.34 and 9.35. The peak in the transfer functions predicted by our theory

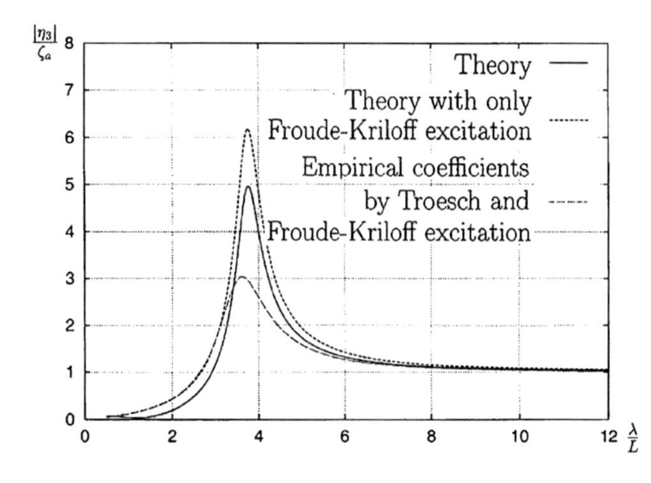

Figure 9.34. Transfer function for wave excited heave (η_3) in regular head sea waves with incident wave amplitude ζ_a. $\lambda =$ wavelength, $L =$ average wetted ship length, $B =$ beam, prismatic planing hull with $L/B = 4.0$, $\tau_{deg} = 4^\circ$, $\beta = 20^\circ$, $lcg/B = 2.13$, and $U/\sqrt{gB} = 3.0$.

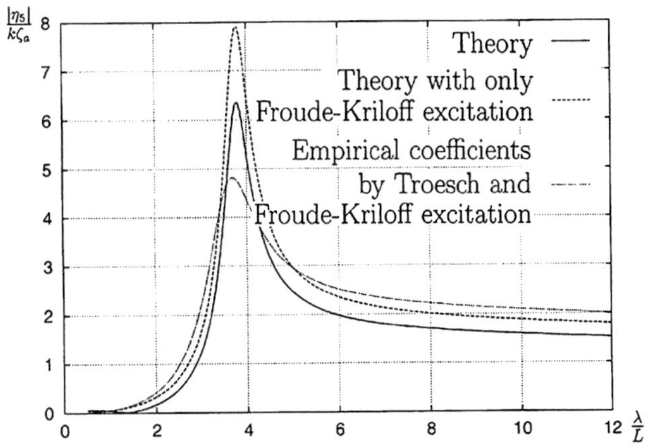

Figure 9.35. Transfer function for wave excited pitch (η_5) in regular head sea waves with incident wave amplitude ζ_a. λ = wavelength, k = wave number, L = average wetted ship length, B = beam. Prismatic planing hull with $L/B = 4.0$, $\tau_{deg} = 4^\circ$, $\beta = 20^\circ$, $lcg/B = 2.13$, and $U/\sqrt{gB} = 3.0$.

corresponds to $\omega_e(B/g)^{0.5} = 1.91$, that is, the absolute value of the imaginary part of the first two eigenvalues presented in Table 9.2. Because $|\alpha|/\omega$ is only 0.06 in this case, it supports the strong amplification of the response at this natural frequency. The two other eigenvalues in Table 9.2 have nondimensionalized absolute values of the imaginary part $\omega(B/g)^{0.5}$ equal to 0.67. An incident wavelength $\lambda = 1.4L$ will give a frequency of encounter corresponding to this ω. The corresponding response is not presented in Figures 9.34 and 9.35. However, because these eigenmodes are highly damped, that is, α/ω is large, we will not observe the response as a peak in the transfer function.

As expected, Figure 9.34 shows that heave amplitude approaches incident wave amplitude for large λ/L-values, but the pitch amplitude presented in Figure 9.35 is not close to the incident wave slope for the largest presented λ/L value.

Figures 9.34 and 9.35 show a clear influence of using the complete expressions of the wave excitation loads relative to a generalized Froude-Kriloff approximation. However, we should recall that the wave excitation loads are based on a long-wavelength approximation and that we earlier used Figure 9.32 to assess the error in doing so. For instance, when $\lambda/L = 4$, this error is less than 10%. Using Troesch's empirical added mass and damping coefficients shows a clear influence on the results around the peaks of the transfer functions. Anyway, the results around the peaks of the transfer functions will have limited practical applicability. For realistic incident wave amplitudes, the wave-induced motions will be so large that nonlinearities matter. We can, for instance, easily see that if we examine the restoring forces and moments as a function of heave and pitch, that is, similar to Figure 9.26. Troesch and Falzarano (1993) included the nonlinearities due to the restoring forces and moments but kept linear added mass, damping, and wave excitation loads. Using linearized added mass and damping terms was partly experimentally justified. Because the exact nonlinear restoring forces and moments were used, a time-domain formulation had to be used. Troesch and Falzarano (1993) demonstrated in their test cases that increasing wave amplitude caused a reduction in $|\eta_3|/\zeta_a$ and $|\eta_5|/\zeta_a$ when strong amplification of the response occured for small incident wave amplitudes.

We will demonstrate this by using the nonlinear time-domain method described in section 9.5.3. This differs from the nonlinear method by Troesch and Falzarano (1993) in that nonlinear generalized Froude-Kriloff forces are also included. Further, we will use the hydrodynamic coefficients and wave excitation formulations consistent with our theory. We use the same planing hull conditions as those used in Figures 9.34 and 9.35. The time-domain solutions will contain transient effects initially. These are disregarded and the steady-state solution is studied. We concentrate on the mean values and the part oscillating with frequency ω_e. In addition, there will be higher harmonics. This means we study

$$\eta_3 \approx \eta_3^0 + \eta_3^1 \sin(\omega_e t + \varepsilon_3) \qquad (9.126)$$

and

$$\eta_5 \approx \eta_5^0 + \eta_5^1 \sin(\omega_e t + \varepsilon_5). \qquad (9.127)$$

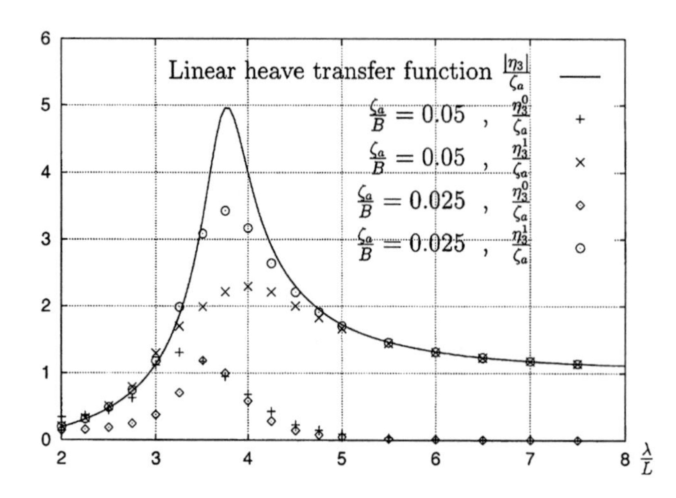

Figure 9.36. Influence of nonlinearities on wave excited heave of a prismatic planing hull in regular head sea waves with wavelength λ and wave amplitude ζ_a. Steady-state conditions. η_3^0 is the mean heave and η_3^1 is the amplitude of the heave component oscillating with the frequency of encounter. L = average wetted ship length, B = beam, $L/B = 4.0$, $\tau_{deg} = 4°$, $\beta = 20°$, $lcg/B = 2.13$, and $U/\sqrt{gB} = 3.0$.

Here η_j^0 and η_j^1, $j = 3, 5$ are, respectively, the constant offsets and amplitudes of harmonically varying motions with frequency ω_e. The results for η_j^0 and η_j^1 are presented in Figures 9.36 and 9.37 for two incident wave amplitudes: $\zeta_a/B = 0.025$ and 0.05.

The transfer functions predicted by the linear theory are also presented. These were evaluated from both the frequency- and time-domain solutions. We note a clear influence of the nonlinearities around resonance. It is worth noting the large steady offset values caused by the dynamic effect, so the ship is, on average, lifted up at COG and has, on average, an increase in the trim angle. The amplitudes of both the heave and pitch motions oscillating with ω_e decrease with increasing wave amplitudes. Figures 9.36 and 9.37 clearly illustrate the importance of nonlinearities in describing the wave-induced motions of planing hulls. The effect of nonlinearities would have been even stronger if vertical accelerations had been considered.

The objective of the previous analysis is to demonstrate important physical effects in an analysis of wave-induced heave and pitch motions. Because the method has not been validated, that is, compared with model tests, we cannot claim that all physical effects are properly accounted for. Further, we must not forget that simplifications were made in evaluating the added mass and damping and that a long-wavelength approximation was assumed in calculating the wave excitation loads.

Irregular seas must be considered in order for an analytical method to be of value in the design. Further, the method must be able to predict that a planing vessel can jump out of the water and cause important impact loads. This has increased significance with increasing forward speed and wave height.

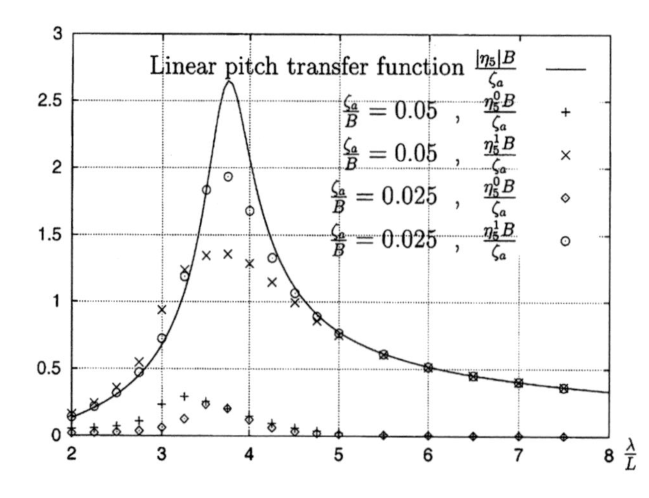

Figure 9.37. Influence of nonlinearities on wave excited pitch of a prismatic planing hull in regular head sea waves with wavelength λ and wave amplitude ζ_a. Steady-state conditions. η_5^0 is the mean pitch and η_5^1 is the amplitude of the pitch component oscillating with the frequency of encounter. L = average wetted ship length, B = beam, $L/B = 4.0$, $\tau_{deg} = 4°$, $\beta = 20°$, $lcg/B = 2.13$, and $U/\sqrt{gB} = 3.0$.

Table 9.3. *Main parameters of a personal watercraft*

Length (m)	L	0.630
Breadth (m)	B	0.223
Depth (m)	D	0.100
Draft (m)	d	0.055
Ship weight (kgf)	W	5.796
$\overline{KG}$ (m)		0.107
LCG from transom (m)		0.255
Deadrise angle (degrees)		22

Reprinted from *Contemporary Ideas on Ship Stability* (ISBN 0080436528), 2000, Ikeda et al., pp. 449–495, "Stability of a planning craft in turning motion," with permission from Elsevier.

9.6 Maneuvering

Ikeda et al. (2000a,b) have presented extensive experimental results of steady hydrodynamic forces and moments acting on a one-fourth–scale model of a personal watercraft that is obliquely towed at high speed in calm water. The hull has hard chines and a duct for waterjet, but no impeller. These results can be used to analyze the steady-state turning of the vessel. However, this also requires information on the longitudinal and transverse thrust provided by the waterjet propulsion of the full-scale craft.

The main parameters of the model are presented in Table 9.3. Longitudinal, transverse, and vertical hydrodynamic forces as well as hydrodynamic moments in heel, trim, and yaw were measured as a function of the rise H, the heel angle ϕ, and the yaw angle ψ. The length Froude number was varied between 2 and 4.4. The coordinate system used in the experiments is presented in Figure 9.38. The x- and y-axes are parallel to the mean free surface and rotate with the yaw angle of the vessel.

Figure 9.39 illustrates the transverse forces that act on the vessel during steady-state turning. The centrifugal force on the vessel must balance the transverse hydrodynamic force on the hull and the transverse thrust due to the waterjet propulsion system. Further, the figure defines

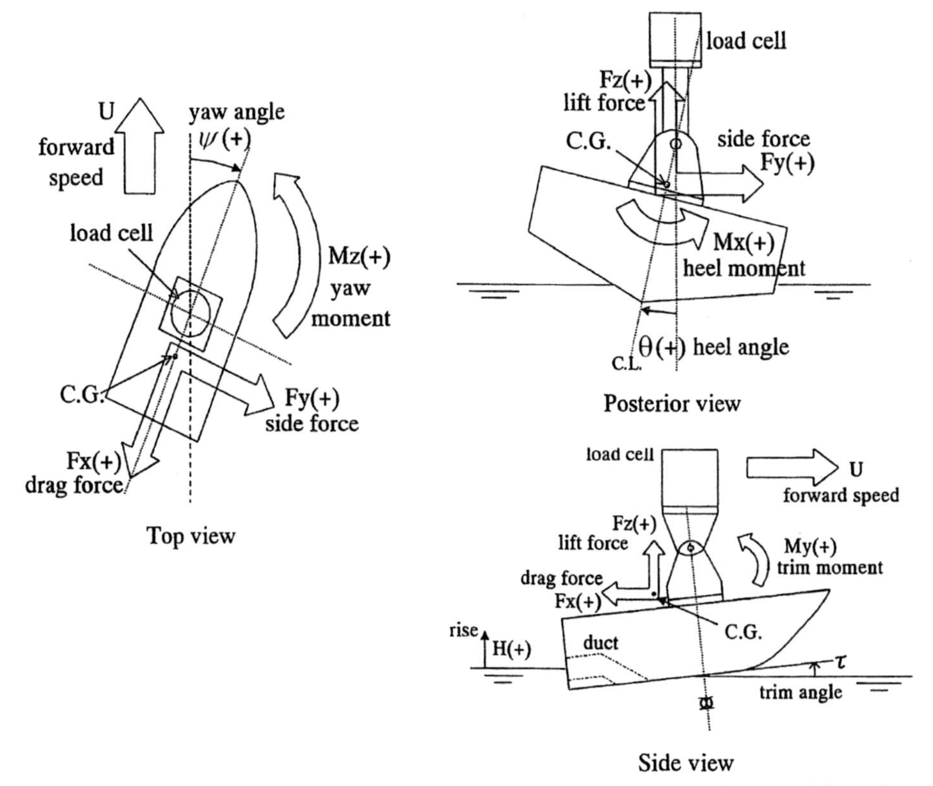

Figure 9.38. Schematic views of experimental setup and coordinate system. (Reprinted from *Contemporary Ideas on Ship Stability* (ISBN 0080436528), 2000, Ikeda et al., pp. 449–495, "Stability of a planning craft in turning motion," with permission from Elsevier.)

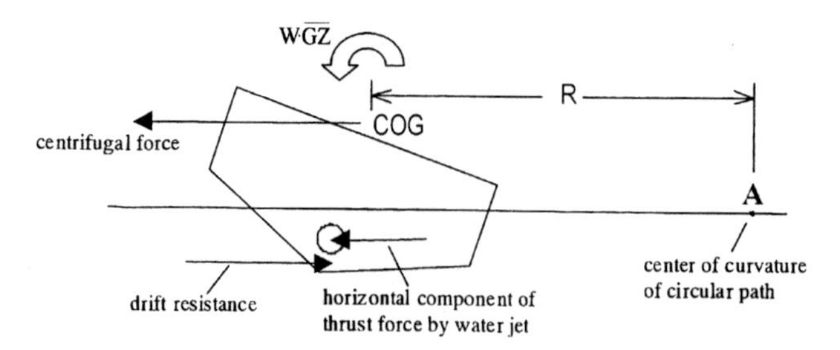

Figure 9.39. Forces acting on a planing hull during steady turning motion. (Reprinted from *Contemporary Ideas on Ship Stability* (ISBN 0080436528), 2000, Ikeda et al., pp. 449–495, "Stability of a planning craft in turning motion," with permission from Elsevier.)

an uprighting moment $W \cdot \overline{GZ}$ about the center of gravity, including the effect of both the hydrodynamic and hydrostatic pressures acting on the hull. The uprighting/heel moment due to the waterjet thrust must also be considered. Here W means the weight of the vessel. Values of $\overline{GZ}$ for a rise of 20 mm are presented in Figure 9.40 as a function of the heel angle ϕ for different trim angles at a side slip (yaw) angle of $\psi = 20^\circ$. The length Froude number is 2.0. The large negative values of $\overline{GZ}$ are a consequence of the hydrodynamic pressure. The results indicate that the vessel will have a large inboard heel angle during steady-state turning with a side slip angle $\psi = 20^\circ$ and a rise of 20 mm in model scale. This is true for all the trim angles τ indicated in the figure. For instance, if we consider zero trim angle and neglect the effect of the transverse waterjet thrust, which we do not know, then Figure 9.40 gives a heel angle of 20°. We see that this equilibrium heel angle is statically stable by first considering a small increment to the equilibrium angle. $W \cdot \overline{GZ}$ is then positive, which means the vessel experiences a static restoring heel moment forcing the vessel back to the

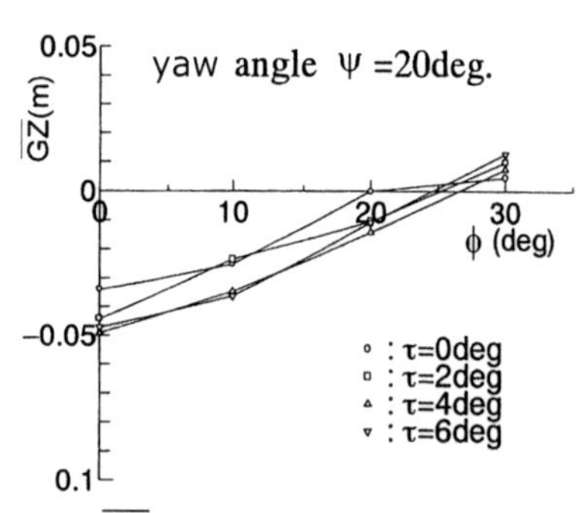

Figure 9.40. $\overline{GZ}$-curve when the vessel rise is H = 20 mm. The length Froude number is 2.0. (Reprinted from *Contemporary Ideas on Ship Stability* (ISBN 0080436528), 2000, Ikeda et al., pp. 449–495, "Stability of a planning craft in turning motion," with permission from Elsevier.)

equilibrium angle. Similarly, we may consider a small decrement to the equilibrium heel angle and the static restoring heel moment. The highest heel angle for $\tau_{\mathrm{deg}} = 0^\circ, 2^\circ, 4^\circ,$ and 6° is obtained for $\tau_{\mathrm{deg}} = 4^\circ$ and is $\phi = 27^\circ$. How these results depend on the forward speed is not presented. A vessel will not always heel inboard during turning, as we have indicated in the previous cases.

The experimental results show that steady hydrodynamic forces and moments on a planing hull can be significantly influenced by rise, trim, and yaw. However, in order to predict, for instance, what trim angle causes minimum resistance during a steady turn, we need to solve the six equations given by the balance in forces and moments for given longitudinal and transverse thrusts from the waterjet propulsion system. This has not been pursued. The experimental results also show that a strong restoring moment in yaw is acting on the hull. This suggests good directional stability.

The possible effect of cavitation and ventilation at high speed was not investigated. This would require a depressurized towing tank. It is well known for some high-speed vessels that cavitation and ventilation can cause undesirable behavior.

The previously found statically stable inboard heel angle does not need to be dynamically stable. Dynamic instabilities during maneuvering of a vessel on a straight course are discussed in Chapter 10. Further, the analysis assumes a constant ship speed. The results in Figure 5.5 show that an SES in a turning maneuver may suffer from a large speed loss. The physical reasons may be different, for instance, because of air coming into the waterjet system and affecting the thrust. In any case, it should be expected that the smaller the turning radius, the larger the speed loss.

Ikeda et al. (2000a,b) used their experimental data based on PMM tests and oblique towing tests to study maneuvering of a planing vessel. They demonstrated that maneuvering can cause violent

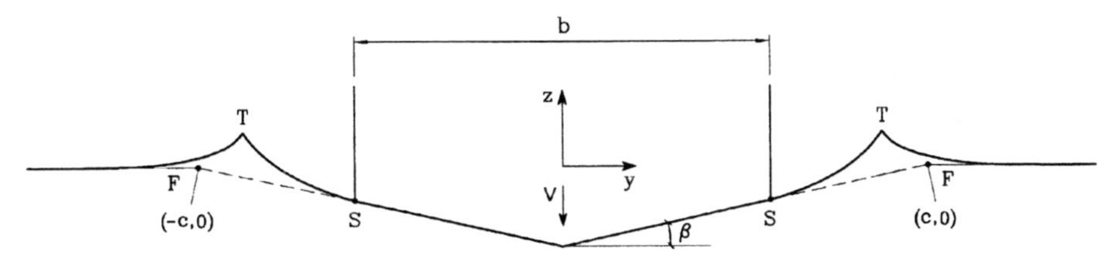

Figure 9.41. Water entry of 2D cross section with deadrise angle β and constant water entry velocity V. Flow separation from the chines.

roll, heave, and pitch motions when the natural frequencies of heave and pitch motions are twice the roll natural frequency and the maneuvering motions have the same frequency as the roll natural frequency.

The maneuvering characteristics at nonplaning speeds must also be considered. For instance, if the Froude number is low so that a hollow in the water aft of the transom does not occur, the maneuvering characteristics may be quite different from those at high speed. Maneuvering at low speed with a wetted transom will cause vortex shedding at the transom. This implies a transverse force and yaw moment component that oscillates with the vortex shedding frequency. The result can be an oscillatory vessel behavior during turning and on a straight course.

9.7 Exercises

9.7.1 2.5D theory for planing hulls

A 2.5D theory can, to a large extent, explain the lift and the trim moment that a planing vessel experiences on a straight course in calm water. It is then essential that flow separation from the chines be incorporated. In this exercise, we examine whether a simplified theory can explain the nature of the vertical force after flow separation from the bilges has occurred.

The basis of a 2.5D theory is water entry of a 2D section with constant entry velocity V. We consider the scenario in Figure 9.41 after the flow separation has occurred (see also Figure 9.5), showing a picture of the free surface. The pressure on the free surface has to be atmospheric. If we consider the free surface from S to T in Figure 9.41, we could have considered this surface as a rigid surface with the special property that the pressure is atmospheric there. Let us use this description to construct a simplified model. We consider, then, water entry of a wedge. This imaginary wedge has an instantenous beam $2c(t)$, as shown in the figure. However, the physical wedge part of the impacting body has a breadth $b < 2c(t)$. We then pretend that the part of this imaginary wedge from S to F is an approximation of a part of the free surface. For $|y| > c$, we use the free-surface condition $\varphi = 0$ typically used in impact problems (see the Wagner slamming model in section 8.3.1). We then use Wagner's model to represent the solution of the velocity potential φ on the impacting wedge with breadth $2c(t)$, that is,

$$\varphi = -V(c^2 - y^2)^{1/2}, \quad |y| < c(t) \quad (9.128)$$

(see eq. 8.41). We will now represent the water entry force F_3 on the physical wedge as in eq. (8.44), that is,

$$F_3 = V\frac{da_{33}}{dt}, \quad (9.129)$$

where

$$a_{33} = \rho \int_{-0.5b}^{0.5b} (c^2 - y^2)^{1/2} dy. \quad (9.130)$$

a) Choose $\beta = 20^\circ$ and calculate this water entry force from the time when flow separation from the bilges occurs. Use Wagner's method to find $c(t)$. Compare F_3 with the results in Figure 8.24. Initially after flow separation, there is a resemblance in the results. However, the disagreement increases as times goes on. There are two reasons for that. The free-surface part SF becomes more and more different from the true free-surface shape (see Figure 9.5). Further, the pressure is not atmospheric at SF.

b) Discuss qualitatively why F_3 in reality must be smaller than predicted by this simplified approach.

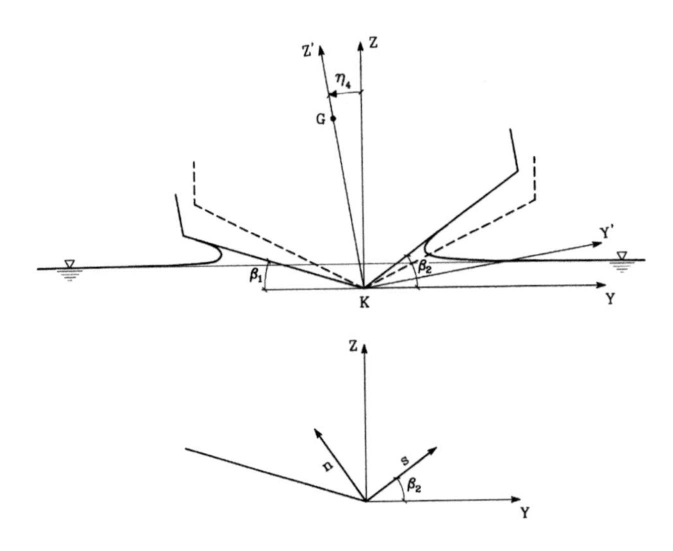

Figure 9.42. Coordinate systems used in the analysis of the steady hydrodynamic heel moment acting on a prismatic planing vessel.

9.7.2 Minimalization of resistance by trim tabs

Consider as a basis a prismatic planing hull with mass $M = 27000\,\text{kg}$, $lcg = 8.84\,\text{m}$, beam 4.27 m, and a deadrise angle $\beta = 10°$. The propeller thrust and the frictional force on the vessel attack through the center of gravity is illustrated in Figure 9.20.

Assume that the vessel is equipped with a horizontal trim tab attached to the stern and with a breadth equal to the beam. The deadrise angle of the planing hull at the stern is modified and set equal to zero. We assume this change of the hull does not influence the following analysis. The trim tab is hinged to the bottom of the stern (see Figure 7.4 for an illustration of a possible arrangement). In the analysis, one should consider the trim tab as an appendage that does not influence the flow at the hull of the vessel.

a) Combine Savitsky's formula for loads on the bare hull with the force acting on the trim tab to find which trim angle causes the lowest resistance on the hull for a vessel speed of 40 knots. What trim moment must the trim tab generate to obtain this minimum resistance?

(Hint: Follow the procedure in section 9.3.1, except that step 1 has to be modified and be the last step.)

b) Assume that the vertical force on the trim tab can be obtained by strip theory and linear lifting theory for a flat plate. You must, of course, account for the fact that there is air on the top of the trim tab. What combination of trim tab angle and trim tab length is needed to generate the trim moment found as part of question a)?

c) Decide on a length of the trim tab that is consistent with strip theory; that is, the length should be much smaller than the breadth. Estimate the drag force on the trim tab due to both viscous and potential flow forces. Use the same procedure as the one in section 9.3.1 to calculate viscous drag forces. Discuss what length should be used in defining the Reynolds number and note that there is no unique answer when one is following this approximate procedure.

d) Repeat the calculations under question a) for other vessel speeds. The lowest ship speed should correspond to approximately length Froude number 0.9. Discuss the relative importance of viscous resistance.

9.7.3 Steady heel restoring moment

We will discuss the contribution of hydrodynamic pressure to $\overline{GZ}$ for a prismatic planing vessel with deadrise angle β. The vessel has a steady heel angle η_4. Figure 9.42 shows a cross section of the vessel and defines the body-fixed coordinate axis Y' and Z'. When $\eta_4 = 0$, the Y'- and Z'-axes coincide with the Y- and Z-axes, respectively. The angles between the wedge surfaces and the Y-axis are $\beta_1 = \beta - \eta_4$ and $\beta_2 = \beta + \eta_4$ on the two sides of the centerplane.

A 2.5D theory can be applied similarly to the way steady vertical forces and trim moments were analyzed in section 9.2.1. This means that results from water entry of a heeled 2D section with constant entry velocity are used. Our approach will be simplified. The wetted surface and pressure

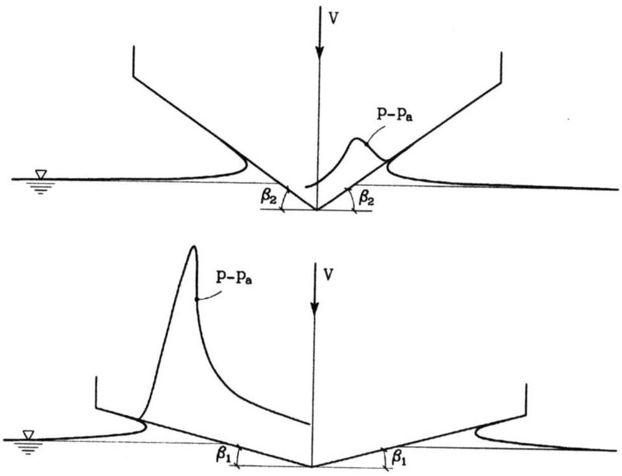

Figure 9.43. Water entry of symmetric wedges with constant entry velocity V and deadrise angles β_1 and β_2. Resulting pressure distribution p is shown for a given time instant.

distribution will be obtained by analyzing symmetric water entry of wedges with deadrise angles β_1 and β_2. This is illustrated for constant entry velocity V in Figure 9.43. Because β_1 is smaller than β_2, the wetted surface and pressure loads are for a given time instant t, largest for the wedge with deadrise angle β_1. The symmetric wedge results for β_1 and β_2 will be applied to the left- and right-hand sides of the heeled planing surface, respectively. This asymmetry in the pressure distribution about the centerplane creates a hydrodynamic heel moment about the center of gravity (COG). The sign of the heel moment depends on the distance $\overline{KG}$ between the keel and COG.

Let us elaborate more by combining the 2.5D theory with Wagner's theory until chine wetting occurs. Use eqs. (8.42) and (8.52), where β is either β_1 or β_2. Positive Y is first considered. This means that $\beta = \beta_2$. We introduce the coordinate system (s, n) with the s-axis along the wedge surface (see Figure 9.42).

The heel moment about COG due to the pressure distribution for positive Y will be expressed in terms of the heel moment about the keel. Explain that we can write

$$F_{4G} = F_{4K} + \overline{KG} \cdot F_{Y'}, \qquad (9.131)$$

where F_{4G} and F_{4K} are heel moments about the COG and keel, respectively. Further, $F_{Y'}$ is the force component along the body-fixed Y'-axis.

We will first express the time-dependent heel moment during water entry of the heeled section. Show that the heel moment about the keel due to the pressure distribution for positive Y can be expressed as

$$f_{4K}^{\beta_2} = \frac{\rho V c^2 \frac{dc}{dt}}{\cos^2 \beta_2}. \qquad (9.132)$$

(Hint: Introduce s as an integration variable.)

We consider, then, the force component $f_{Y'}^{\beta_2}$ along the Y'-axis. Show that

$$f_{Y'}^{\beta_2} = -f_n^{\beta_2} \sin \beta, \qquad (9.133)$$

where $f_n^{\beta_2}$ is the hydrodynamic force due to the pressure distribution for positive Y-values acting in the normal direction n (see Figure 9.42).

Show that

$$f_n^{\beta_2} = \frac{\rho V c \frac{dc}{dt}}{\cos \beta_2} \frac{\pi}{2}. \qquad (9.134)$$

We now introduce the 2.5D theory. This means $Vt = \tau x$ and $V = U\tau$ (see eq. 9.1). Integrate eq. (9.132) from $x = 0$ to $x = x_s = B \tan \beta_2 / (\pi \tau)$ (see eq. (9.8)), that is, where chine wetting starts. Show that the resulting contribution to the heel moment about the keel can be expressed as

$$F_{4K}^{\beta_2} = \frac{\rho U^2 B^3 \tau}{24 \cos^2 \beta_2}. \qquad (9.135)$$

Is it correct to stop the integration when the chine wetting starts?

Do the same for the normal force, and start with eq. (9.134). Show that the integration from $x = 0$ to $x = x_s$ gives

$$F_n^{\beta_2} = \frac{\rho U^2 B^2 \pi \tau}{16 \cos \beta_2}. \qquad (9.136)$$

Follow a similar procedure for the pressure distribution for negative Y. Show that the total heel moment about COG due to the hydrodynamic pressure can then be expressed as

$$F_{4G} = \rho U^2 B^3 \tau \left[\frac{1}{24} \left(\frac{1}{\cos^2 \beta_2} - \frac{1}{\cos^2 \beta_1} \right) - \frac{\overline{KG}}{B} \left(\frac{\pi}{16} \right) \left(\frac{\sin \beta}{\cos \beta_2} - \frac{\sin \beta}{\cos \beta_1} \right) \right]. \quad (9.137)$$

We must also consider the effect of the hydrostatic pressure on the heel moment. Discuss how you would do that by modifying the results for zero speed.

9.7.4 Porpoising

Consider a prismatic planing hull with deadrise angle $\beta = 20°$, trim angle $\tau_{\text{deg}} = 4°$, average wetted length-to-beam ratio $\lambda_W = 4$, $lcg/B = 2.13$, $vcg/B = 0.25$, $M/(\rho B^3) = 1.28$, and pitch radius of gyration $= 1.3B$. The added mass and damping coefficients will be approximated as

$$\begin{aligned}
&\frac{A_{33}}{\rho B^3} = 1.3; \quad \frac{B_{33}}{\rho B^3 (g/B)^{1/2}} = 0.7 Fn_B + 0.5 \\
&\frac{A_{53}}{\rho B^4} = -0.3 Fn_B - 0.1; \quad \frac{B_{53}}{\rho B^4 (g/B)^{1/2}} = 1.1 Fn_B - 0.5 \\
&\frac{A_{55}}{\rho B^5} = 1.3; \quad \frac{B_{55}}{\rho B^5 (g/B)^{1/2}} = 2.1 Fn_B - 0.3 \\
&\frac{A_{35}}{\rho B^4} = 0.1 Fn_B - 0.5; \quad \frac{B_{35}}{\rho B^4 (g/B)^{1/2}} = -1.8 Fn_B + 0.6
\end{aligned} \quad (9.138)$$

Use the procedure based on Savitsky's formula and described in the main text to calculate the restoring coefficients C_{33}, C_{35}, C_{53}, and C_{55}.

a) By using Routh-Hurwitz stability criterion, study for which beam Froude numbers Fn_B porpoising occurs.

b) Repeat the stability analysis by systematically varying the hydrodynamic coefficients as follows:

Case 1: C_{35} and C_{53} are zero.

Case 2: use either twice the heave or pitch damping B_{33} and B_{55} given by eq. (9.138).

c) Assume that lcg differs from $2.13B$, which is the basis of the coefficients given in eq. (9.138). This changes the steady position of the vessel. However, neglect this fact in the following derivation. Express added mass and damping coefficients in heave and pitch about this different center of gravity in terms of the expressions in eq. (9.138).

(Hint: Remember that added mass and damping coefficients are a consequence of either vertical hydrodynamic forces or pitch moments due to either forced heave or pitch motions.)

9.7.5 Equation system of porpoising

In the analysis of porpoising, we reformulated the equations of motions into a system of first-order differential equations (see eq. 9.88). Show that the K-matrix has the following elements:

$$\begin{aligned}
&K_{11} = 0 \quad K_{12} = 0, \quad K_{13} = 1, \quad K_{14} = 0 \\
&K_{21} = 0 \quad K_{22} = 0, \quad K_{23} = 0, \quad K_{24} = 1 \\
&K_{31} = -\tfrac{1}{D} C_{33}(I_{55} + A_{55}) + \tfrac{1}{D} C_{53} A_{35} \\
&K_{32} = -\tfrac{1}{D} C_{35}(I_{55} + A_{55}) + \tfrac{1}{D} C_{55} A_{35} \\
&K_{33} = -\tfrac{1}{D} B_{33}(I_{55} + A_{55}) + \tfrac{1}{D} B_{53} A_{35} \\
&K_{34} = -\tfrac{1}{D} B_{35}(I_{55} + A_{55}) + \tfrac{1}{D} B_{55} A_{35} \\
&K_{41} = -\tfrac{1}{D} C_{53}(M + A_{33}) + \tfrac{1}{D} C_{33} A_{53} \\
&K_{42} = -\tfrac{1}{D} C_{55}(M + A_{33}) + \tfrac{1}{D} C_{35} A_{53} \\
&K_{43} = -\tfrac{1}{D} B_{53}(M + A_{33}) + \tfrac{1}{D} B_{33} A_{53} \\
&K_{44} = -\tfrac{1}{D} B_{55}(M + A_{33}) + \tfrac{1}{D} B_{35} A_{53}
\end{aligned} \quad (9.139)$$

where

$$D = (M + A_{33})(I_{55} + A_{55}) - A_{53} A_{35}.$$

9.7.6 Wave-induced vertical accelerations in head sea

Use the planing vessel described in exercise 9.7.4 and a linear frequency-domain solution for head sea as described in section 9.5. The added mass and damping coefficients are described by eq. (9.138), and the restoring coefficients should be based on the same procedure as the one in section 9.4.1.

a) Decide on a high Froude number when the vessel is not porpoising.

b) Calculate the transfer function of the heave accelerations of the center of gravity (COG) of the vessel.

c) Assume long-crested head sea waves in a short-term sea state that can be described by the Pierson-Moskowitz wave spectrum (see eq. (3.55)). Calculate nondimensional standard deviations $\sigma_{3a} B/(H_{1/3} g)$ as a function of nondimensional wave period $T_1 \sqrt{g/B}$. Here σ_{3a} is the standard deviation of vertical accelerations at COG. This can be calculated as described in section 7.4.1.

d) Select a beam B of the vessel. Set the operability-limiting criterion for standard deviation (*RMS*) of vertical accelerations equal to 0.275 g (see Table 1.1). Find the limiting value of the significant wave height $H_{1/3}$ as a function of mean wave period T_1 for the vessel to operate according to this criterion.

e) The previous analysis was based on linear theory. We saw in section 9.5.4 that nonlinear effects were important in resonant conditions. Discuss for which values of T_1 nonlinear effects have a large influence.

f) The helmsman will reduce the speed if the vertical accelerations become too large (voluntary speed reduction). Use the operability-limiting criterion for σ_{a3} as previously described as a basis for the helmsman to make this decision. Select a T_1 and $H_{1/3}$ where this criterion is not satisfied.

How much must the ship speed be reduced in order to satisfy the operability-limiting criterion?

10 Maneuvering

10.1 Introduction

High standards for maneuverability are required for high-speed vessels to operate, particularly in congested areas, where emergency maneuvers may be necessary to avoid collisions. An important aspect is training of personnel who operate the vessels. A maneuvering simulator is then a useful tool. This requires mathematical models that reflect the very different physical features of the various categories of high-speed vessels. The maneuvering characteristics of a vessel are documented in terms of turning circle maneuver, zigzag (Z) maneuver and crash astern test. The IMO (International Maritime Organization) maneuvering criteria from 2002 for ships longer than 100 m are described in Table 10.1.

Figures 10.1 and 10.2 define a turning circle maneuver and a zigzag maneuver, respectively. The course changing ability of the ship is expressed by the turning circle maneuver. The ability to bring the ship to a straight course is determined by the zigzag maneuver. The crash astern test is illustrated in Figure 10.20. Later, in section 10.5, we give examples of a turning circle maneuver, zigzag maneuver, and crash astern test.

The hydrodynamics clearly differ between high-speed maneuvering on one hand and low-speed maneuvering and dynamic positioning on the other hand. Let us assume calm water conditions and divide the hydrodynamic loads on the hull into potential and viscous flow effects, as we did for ship resistance (see Chapter 2). However, an interaction exists in reality between these two effects. The potential flow causes added mass and damping effects, as we have shown for wave-induced motions of semi-displacement vessels in Chapter 7. Maneuvering and steering control action typically occur with a much lower frequency than important linear wave encounter frequencies. One exception may be for small planing vessels. However, by disregarding this scenario, we may assume the added mass and damping coefficients are not frequency dependent, as they are in the seakeeping problem. This means no wave radiation damping. The potential flow damping is therefore zero at zero speed. However, there is potential flow damping at forward speed. This is influenced by steady wave generation when the Froude number of a monohull or multihull vessel is larger than approximately 0.2. Part of the damping at forward speed may be categorized as hull-lift damping.

The viscous flow effect must be handled differently if we go from low- to high-speed maneuvering. At low forward speed, we can use the cross-flow principle. This means the effect of forward speed on the cross-flow at sections along the ship is neglected. Increasing forward speed will have the effect that the cross-flow separates more easily at a cross section in the stern than it does in the bow.

The effectiveness of the propulsion and steering devices also depends on the forward speed. Thrusters are effective only for speeds lower than approximately 2 to 3 ms^{-1}. The waterjet, azimuth thrusters, and podded propulsion may be used in the whole speed range, whereas rudders require a certain threshold speed, let us say, larger than 2 ms^{-1}, to be efficient. However, the slip stream from the propeller can be used to decrease this threshold speed.

Notations and coordinate systems differ in seakeeping and maneuvering. To some extent, we will follow conventions adopted in seakeeping and used in the previous chapters. It is convenient in ship maneuvering to use a body-fixed coordinate system, whereas in the case of seakeeping problems, a nonoscillatory coordinate system is used following the ship in its forward motion on a straight course. This is referred to as the reference-parallel frame in the literature on maneuvering (Fossen 2002). In order to minimize confusion due to different coordinate systems and notations, in section 10.2 we show the link between the maneuvering and seakeeping conventions.

Ship maneuvering is traditionally handled in calm water conditions. When the Froude number is moderate, that is, less than approximately 0.2, it is normally sufficient to study coupled surge, sway, and yaw of the vessel. However, turning (change of heading operation) of a vessel may lead to important roll (heel). This is certainly an effect of high

Table 10.1. *IMO maneuvering criteria from 2002 for ships longer than 100 m at service speed*

From the turning circle maneuver:
- Turning ability at 35° rudder angle or maximum permissible rudder angle at test speed
  - Advance $< 4.5 L_{PP}$
  - Tactical diameter $< 5 L_{PP}$
- Course initiating ability at 10° rudder angle
  - Traveled distance $< 2.5 L_{PP}$ at 10° change of heading

From the zigzag maneuver:
- Course checking ability
  - 10°/10° Z-maneuver
    First overshoot $\leq 10^\circ$ if $L_{PP}/U < 10\,\mathrm{s}$
    First overshoot $\leq 20^\circ$ if $L_{PP}/U \geq 30\,\mathrm{s}$
    First overshoot $< (5 + 0.5 L_{PP}/U)^\circ$ for $10\,\mathrm{s} \leq L_{PP}/U < 30\,\mathrm{s}$
    Second overshoot $\leq 25^\circ$ if $L_{PP}/U < 10\,\mathrm{s}$
    Second overshoot $\leq 40^\circ$ if $L_{PP}/U \geq 30\,\mathrm{s}$
    Second overshoot $\leq (17.5 + 0.75 L_{PP}/U)^\circ$ if $10\,\mathrm{s} \leq L_{PP}/U \leq 30\,\mathrm{s}$
  - 20°/20° maneuver
    First overshoot $< 25^\circ$
- Stopping ability
  - Track reach $< 15 L_{PP}$
    May be modified for large displacement ships

L_{PP} is the length between perpendiculars. Turning circle and zigzag (Z) maneuvers are defined in Figures 10.1 and 10.2.

forward speed. The steady restoring heel moment $-Mg\,\overline{GZ}$ is one of several effects to account for. Here M means the vessel mass, and the moment arm $\overline{GZ}$ is influenced by both the hydrodynamic and hydrostatic pressure. Further, ship maneuvering leads to increased resistance and speed loss, as is illustrated in Figure 5.5 for an SES.

Practical situations exist in which there is a need to combine maneuvering and seakeeping. Broaching of a ship in following sea is an example. This is initiated by directional instability and may lead, in extreme cases, to the ship turning into a beam sea condition relative to the waves. This may then cause capsizing of the vessel. Figure 10.3 shows an example of the full-scale measurements of broaching of a 30 m–long catamaran in quartering sea. We notice the large yaw angle and the attempt to use a large rudder angle to steer

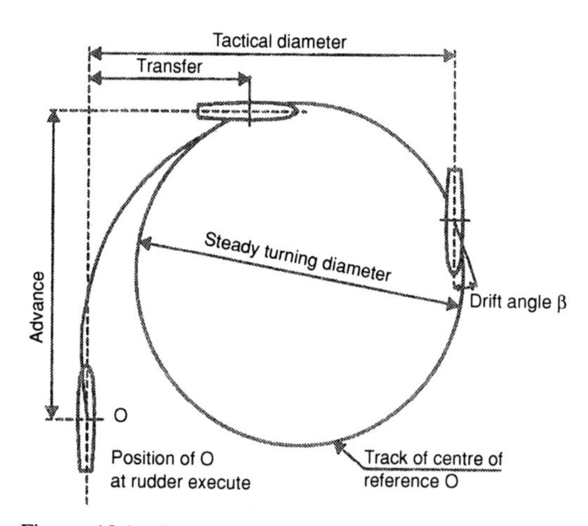

Figure 10.1. Description of the turning circle maneuver (Hooft and Nienhuis 1994).

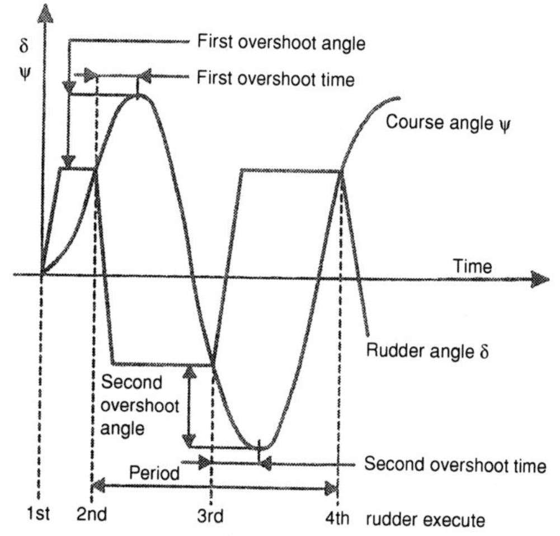

Figure 10.2. Description of the zigzag (Z) maneuver (Hooft and Nienhuis 1994).

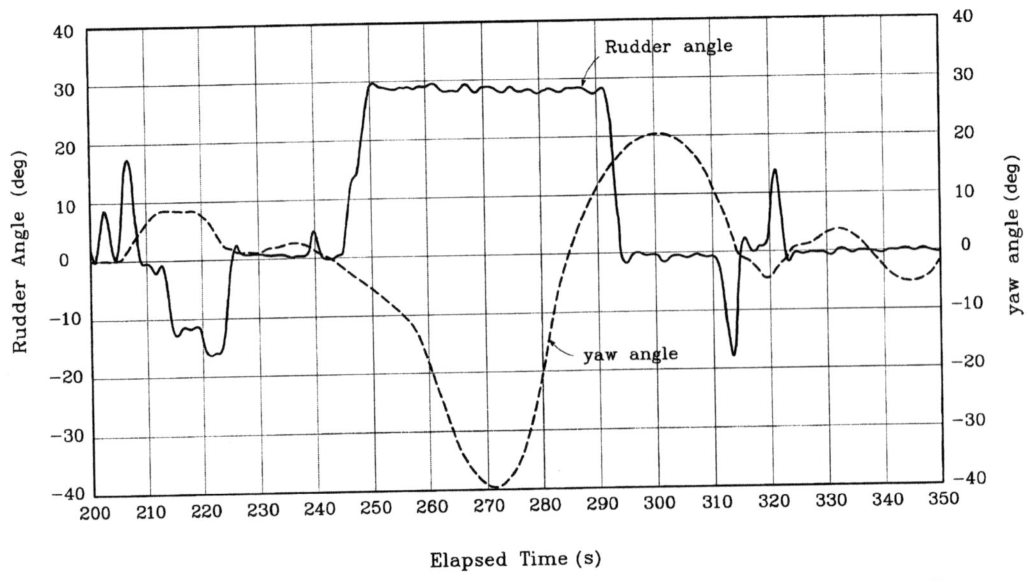

Figure 10.3. Illustration of broaching of a 30 m–long catamaran in quartering sea with 120° heading. The ship speed is 18 knots. Wave condition: $H_{1/3} = 1.5$ m, $T_0 = 10$ s. Note the large yaw angles and the attempt to steer the vessel by using a large rudder angle. The results are from full-scale tests by MARINTEK.

the vessel. Analysis of broaching is discussed by Vassalos et al. (2000) and in section 10.10.3.

When the dominant frequency of encounter between the waves and the ship is low, as it is for broaching, we must analyze seakeeping and maneuvering simultaneously. The wave-induced flow will change the flow along the hull, as well as the inflow to the propulsion and steering devices. This affects the ability to control the vessel. The greater difficulties involved can be partially reduced in the analysis because the wave-induced flow problem can be highly simplified for low frequencies.

The frequency domain associated with maneuvering is generally much lower than dominating wave encounter frequencies. As a first approximation, we can then consider combined seakeeping and maneuvering as a two-scale problem. This means the seakeeping is analyzed for a given wave heading assuming the ship is on a straight course. The nonlinear interaction between the incident waves and the flow due to the ship, for instance, caused by wave-induced ship motions, will result in mean and slowly varying horizontal forces and yaw moments on the ship (Salvesen 1974, Faltinsen et al. 1980). Both linear and nonlinear wave loads must be included in the ship maneuvering analysis. The mean wave force in surge is what we earlier called added resistance in waves. Slowly varying wave loads have been extensively studied in the context of stationkeeping (Faltinsen 1990). Because the wave-induced ship motions are strongly speed dependent, the forward speed has an important effect on the mean and slowly varying wave loads. In addition to wave loads, we must consider mean and slowly varying wind loads. Ship maneuvering performance in waves is, for instance, analyzed by Kijima and Furukawa (2000).

Studies of ship maneuvering require an integrated analysis of maneuvering, steering devices, propulsion, and resistance. Cavitation and ventilation on the hull, rudders, and propulsion units may lead to directional instability. This was a contributing factor to the large yaw angles of the catamaran shown in Figure 10.3.

The limited size of indoor basins makes it difficult to do maneuvering tests of high-speed vessels at maximum operating speed. An alternative is to use sheltered outdoor lakes and basins. Another possibility is to use a planar motion mechanism (PMM, see Crane et al. 1989) in a towing tank to determine hydrodynamic coefficients in a mathematical model for the ship motions. Use of purely theoretical methods to predict ship maneuvering is still under development. A difficulty is to account properly for turbulent flow; for example see section 10.6.

There are also important aspects, such as interactions between ships and the effect of laterally restricted water. These issues will not be considered in detail. Tuck and Newman (1974) theoretically studied the interaction between two ships of different lengths and speeds when they are on a parallel course in deep water. Norrbin (1971), Fujino (1976), Brix (1993), and Crane et al. (1989) discussed the maneuvering of a ship in restricted water.

We start by studying maneuvering in deep water at moderate speed and in water of infinite lateral extent by using linear theory. This means that it is sufficient to analyze coupled sway and yaw. The next steps are to consider the effect of finite water depth and high Froude number by linear theory in yaw and sway. However, nonlinearities due to viscous flow will matter. This is handled by using empirical drag formulas. Two main categories are considered. The forward speed is much lower than the transverse speed in the first case, and the cross-flow principle is used. It is demonstrated that the viscous forces depend on many flow parameters. In the second case, the forward speed dominates over the transverse speed. Coupled surge, sway and yaw, and control means are also studied. We then discuss models for analyzing maneuvering in six degrees of freedom, which is needed at high-speed conditions and is exemplified for hydrofoil vessels. Coupled sway-roll-yaw motions of a monohull are studied in detail for moderate speed by introducing a slender body theory. High-speed cases are also discussed, including unstable oscillatory sway-roll-yaw motions ("corkscrewing").

10.2 Traditional coordinate systems and notations in ship maneuvering

The traditional body-fixed coordinate system (x, y, z) used in analyzing ship maneuvering is shown in Figure 10.4. The origin is in the center of gravity with the z-axis downward and the x-axis forward. The hydrodynamic force components along the x- and y-axes are called X and Y. N is the hydrodynamic turning (yaw) moment about the z-axis. u and v are the longitudinal and lateral components of the ship velocity, respectively. r is the yaw angular velocity and ψ is the yaw angle.

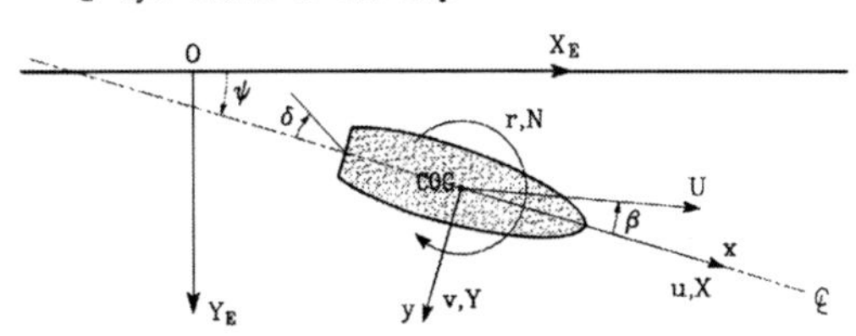

Figure 10.4. Traditional body-fixed coordinate system (x, y, z) used in ship maneuvering. The z-axis is downward. X and Y are hydrodynamic force components along the x- and y-axes. N is the hydrodynamic yaw (turning) moment about the z-axis. u and v are the longitudinal and lateral components of ship velocity. r is yaw angular velocity (Fujino 1976).

Further, δ is the rudder angle. Positive directions are indicated by arrows.

The linearized equations of motions in sway and yaw of a ship advancing with mean forward speed U in water of infinite horizontal extent are expressed as

$$M\left(\frac{dv}{dt} + Ur\right) = Y_{\dot{v}}\frac{dv}{dt} + Y_v v + Y_{\dot{r}}\frac{dr}{dt} + Y_r r + X_{\dot{u}}Ur + Y_\delta \delta \quad (10.1)$$

$$I_{66}\frac{dr}{dt} = N_{\dot{v}}\frac{dv}{dt} + N_v v + N_{\dot{r}}\frac{dr}{dt} + N_r r + N_\delta \delta. \quad (10.2)$$

Here M is the ship mass, I_{66} is the mass moment of inertia in yaw with respect to COG, and U means a constant forward speed. We note the ship mass term MUr appearing in eq. (10.1). This is a consequence of using a body-fixed coordinate system, which will be explained in section 10.3. The time derivatives are indicated by a dot – for example, $\dot{v}$. The subscripts on Y mean that we should take partial derivative of Y with respect to the variable in the subscript – for example, $\dot{Y}_{\dot{r}}\frac{dr}{dt} = \frac{\partial Y}{\partial \dot{r}}\frac{dr}{dt}$. The subscripts on N and X have a similar meaning. However, this has no practical consequence for our presentation. We can just look upon $Y_{\dot{v}}$ and other derivatives as constants. The term $X_{\dot{u}}Ur$ in eq. (10.1) is also a consequence of using a body-fixed coordinate system. $X_{\dot{u}}$ is in eq. (10.159) derived as minus the added mass in surge. Søding (1982) approximated this as

$$X_{\dot{u}} = -A_{11} \approx -2.7\rho\nabla^{5/3}/L^2, \quad (10.3)$$

where ∇ is the displaced volume of water. The added mass in surge A_{11} is small relative to the body mass. For instance, if the average beam-to-draft and length-to-beam ratios in Table 7.1 are combined with a block coefficient of 0.35, eq. (10.3) gives that the added mass in surge is only 4% of the ship mass $M = \rho\nabla$.

It is common to operate with nondimensional coefficients and variables. We will follow Society of Naval Architects and Marine Engineers (SNAME) standards (Crane et al. 1989) and first divide eq. (10.1) by $0.5\rho L^2 U^2$, where L is the ship length. The equation can then be re-expressed as

$$M'(\dot{v}' + r') = Y'_{\dot{v}}\dot{v}' + Y'_{v}v' + Y'_{\dot{r}}\dot{r}' + Y'_{r}r' + X'_{\dot{u}}r' + Y'_{\delta}\delta,$$

where the symbol $'$ (prime) has been introduced to indicate non-dimensional values. These can be expressed as

$$v' = \frac{v}{U};\quad \dot{v}' = \frac{dv}{dt}\frac{L}{U^2};\quad r' = \frac{rL}{U};\quad \dot{r}' = \frac{dr}{dt}\frac{L^2}{U^2}$$
$$M' = \frac{M}{0.5\rho L^3};\quad Y'_v = \frac{Y_v}{0.5\rho L^2 U};\quad Y'_r = \frac{Y_r}{0.5\rho L^3 U}$$
$$Y'_{\dot{v}} = \frac{Y_{\dot{v}}}{0.5\rho L^3};\quad X'_{\dot{u}} = \frac{X_{\dot{u}}}{0.5\rho L^3}:\quad Y'_{\dot{r}} = \frac{Y_{\dot{r}}}{0.5\rho L^4};$$
$$Y'_{\delta} = \frac{Y_{\delta}}{0.5\rho L^2 U^2} \tag{10.4}$$

Dividing eq. (10.2) with $0.5\rho L^3 U^2$ gives

$$I'_{66}\dot{r}' = N'_{\dot{v}}\dot{v}' + N'_{v}v' + N'_{\dot{r}}\dot{r}' + N'_{r}r' + N'_{\delta}\delta,$$

where

$$I'_{66} = \frac{I_{66}}{0.5\rho L^5};\quad N'_v = \frac{N_v}{0.5\rho L^3 U};\quad N'_r = \frac{N_r}{0.5\rho L^4 U}$$
$$N'_{\dot{v}} = \frac{N_{\dot{v}}}{0.5\rho L^4};\quad N'_{\dot{r}} = \frac{N_{\dot{r}}}{0.5\rho L^5};\quad N'_{\delta} = \frac{N_{\delta}}{0.5\rho L^3 U^2} \tag{10.5}$$

There are other ways to introduce nondimensional coefficients and variables. Fossen (2002) presented a different system, which has become common to use in automatic control.

In order to describe the position of the vessel in the Earth-fixed coordinate system (X_E, Y_E), we need two additional kinematic equations describing the coordinates $X_E(t)$ and $Y_E(t)$ of the center of gravity (COG) (see notations in Figure 10.4). One of them is

$$\frac{dY_E(t)}{dt} = v\cos\psi + u\sin\psi = U\sin(\psi - \beta), \tag{10.6}$$

where U is the velocity of COG. The drift angle β is defined by

$$v = -U\sin\beta. \tag{10.7}$$

The other kinematic equation is

$$\frac{dX_E(t)}{dt} = u\cos\psi - v\sin\psi. \tag{10.8}$$

When v/U and ψ are small, we can approximate $dY_E/dt \approx U(\psi - \beta)$, $v \approx -U\beta$, and $dX_E/dt \approx U$. Further, the heading angle ψ is related to the yaw angular velocity r by

$$\frac{d\psi}{dt} = r. \tag{10.9}$$

We can then integrate eqs. (10.1), (10.2), (10.6), (10.8) and (10.9) in time with given initial conditions to find the time-dependent position of the vessel. This obviously requires knowledge of the steering force and moment, which may be influenced by the use of an autopilot system.

We are going to use another body-fixed coordinate system with x- and z-axes pointing in opposite directions to the coordinate system described above. The equations for sway and yaw velocities and accelerations will be expressed in this coordinate system as

$$M\left(\frac{d^2\eta_2}{dt^2} - U\frac{d\eta_6}{dt}\right) = -A_{22}\frac{d^2\eta_2}{dt^2} - B_{22}\frac{d\eta_2}{dt} - A_{26}\frac{d^2\eta_6}{dt^2} - B_{26}\frac{d\eta_6}{dt} + A_{11}U\frac{d\eta_6}{dt} + F_2^S \tag{10.10}$$

$$I_{66}\frac{d^2\eta_6}{dt^2} = -A_{62}\frac{d^2\eta_2}{dt^2} - B_{62}\frac{d\eta_2}{dt} - A_{66}\frac{d^2\eta_6}{dt^2} - B_{66}\frac{d\eta_6}{dt} + F_6^S. \tag{10.11}$$

Here $\dot{\eta}_2$, $\ddot{\eta}_2$, $\dot{\eta}_6$, and $\ddot{\eta}_6$ are consistent with the symbols in Figure 7.11, except that we refer to transverse velocity $\dot{\eta}_2$ of the center of gravity of the ship in a body-fixed coordinate system. Further, F_2^S and F_6^S are sway force and yaw moment due to a steering device such as a rudder, interceptor, or waterjet system. The coordinate system with definitions of positive $\dot{\eta}_2$, $\dot{\eta}_6$, force F_2, and moment F_6 is shown in Figure 10.5. A_{jk} and B_{jk} are what we defined as added mass and damping coefficients in eq. (7.39). When determining the relationship between A_{jk} and B_{jk} and the derivative terms

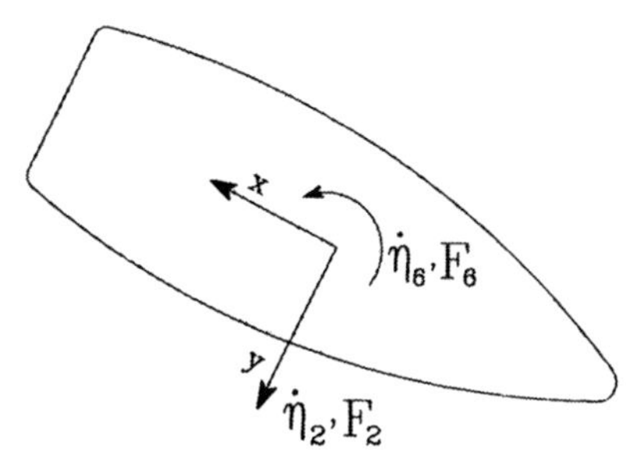

Figure 10.5. Body-fixed coordinate system (x, y, z) based on seakeeping terminology. The z-axis is positive upward.

$Y_{\dot{v}}, Y_v, Y_{\dot{r}}, Y_r, N_{\dot{v}}, N_v, N_{\dot{r}}$, and N_r in eqs. (10.1) and (10.2), we must be careful with the different coordinate systems and with what is positive sway force, yaw moment, sway, and yaw. We start with transverse hydrodynamic hull force due to sway velocity. In the two different systems, this can be written as

$$F_2 = -A_{22}\ddot{\eta}_2 - B_{22}\dot{\eta}_2; \quad Y = Y_{\dot{v}}\dot{v} + Y_v v. \quad (10.12)$$

From Figures 10.4 and 10.5, we see that the positive direction for $\dot{\eta}_2$ is the same as for v. The same is true for F_2 and Y. This gives

$$Y_{\dot{v}} = -A_{22}; \quad Y_v = -B_{22}. \quad (10.13)$$

Then we consider hydrodynamic yaw moment on the hull due to sway velocity. In the two different systems, this can be written as

$$F_6 = -A_{62}\ddot{\eta}_2 - B_{62}\dot{\eta}_2; \quad N = N_{\dot{v}}\dot{v} + N_v v. \quad (10.14)$$

Because F_6 has positive direction opposite that of N whereas the sway velocity has the same direction, it follows that

$$N_{\dot{v}} = A_{62}; \quad N_v = B_{62}. \quad (10.15)$$

Then we consider the transverse force and yaw moment due to yaw velocity $\dot{\eta}_6$. Because $\dot{\eta}_6$ has the opposite positive direction to r, and recalling what are positive force and moment, it follows that

$$Y_{\dot{r}} = A_{26}; \quad Y_r = B_{26} \quad (10.16)$$

$$N_{\dot{r}} = -A_{66}; \quad N_r = -B_{66}. \quad (10.17)$$

10.3 Linear ship maneuvering in deep water at moderate Froude number

The slender body theory, leading to eq. (8.87) and describing infinite-frequency added mass and damping in heave and pitch, can be generalized to ship maneuvering at moderate speed. By *infinite frequency* is meant a very high frequency so that added mass is frequency independent. Linear sway (η_2) and yaw (η_6) motions are assumed. An essential difference from analysis of wave-induced motions is that now ship velocities are relative to a body-fixed coordinate system. This means that the body boundary condition is

$$\frac{\partial \varphi}{\partial n} = n_2(\dot{\eta}_2 + x\dot{\eta}_6) \quad \text{on } C(x). \quad (10.18)$$

Here $C(x)$ is the mean submerged cross-sectional surface per unit length and φ is the velocity potential caused by the ship. In contrast to the corresponding seakeeping case, see eq. (8.84), no forward speed term is present in the body boundary condition in the maneuvering case. The free-surface condition is approximated as a rigid wall condition; that is, $\frac{\partial \varphi}{\partial z} = 0$ on $z = 0$. This implies the assumption of a moderate Froude number. It follows that eq. (8.87) is replaced by

$$f_2^{HD} = -\left(\frac{\partial}{\partial t} + U\frac{\partial}{\partial x}\right)[a_{22}(\dot{\eta}_2 + x\dot{\eta}_6)], \quad (10.19)$$

where f_2^{HD} is the 2D horizontal force on the hull and a_{22} is the low-frequency 2D added mass in sway. By *low frequency* is meant that the rigid free-surface condition is satisfied.

For a broad class of cross sections, the 2D added mass in sway for a ship cross section of a monohull in deep water can be approximated as

$$a_{22} = \rho\, 0.5\pi\,[(a + aa_1)^2 + 3(aa_3)^2]. \quad (10.20)$$

We note this is the same as the expression for infinite-frequency added mass in heave a_{33} given by eq. (8.89). However, coefficients a, a_1, and a_3 differ in the two cases. When using eq. (10.20), we must set b equal to twice the draft of the ship and d equal to half the beam of the ship in the formula presented in eq. (8.89). We show this by means of Figure 10.6 and by assuming a 2D body oscillating in heave at infinite frequency. The free-surface condition is then $\varphi = 0$ on the mean free surface, $z = 0$. Here, φ is the velocity potential. The coordinate system in Figure 10.6 is consistent with the seakeeping coordinate system defined in

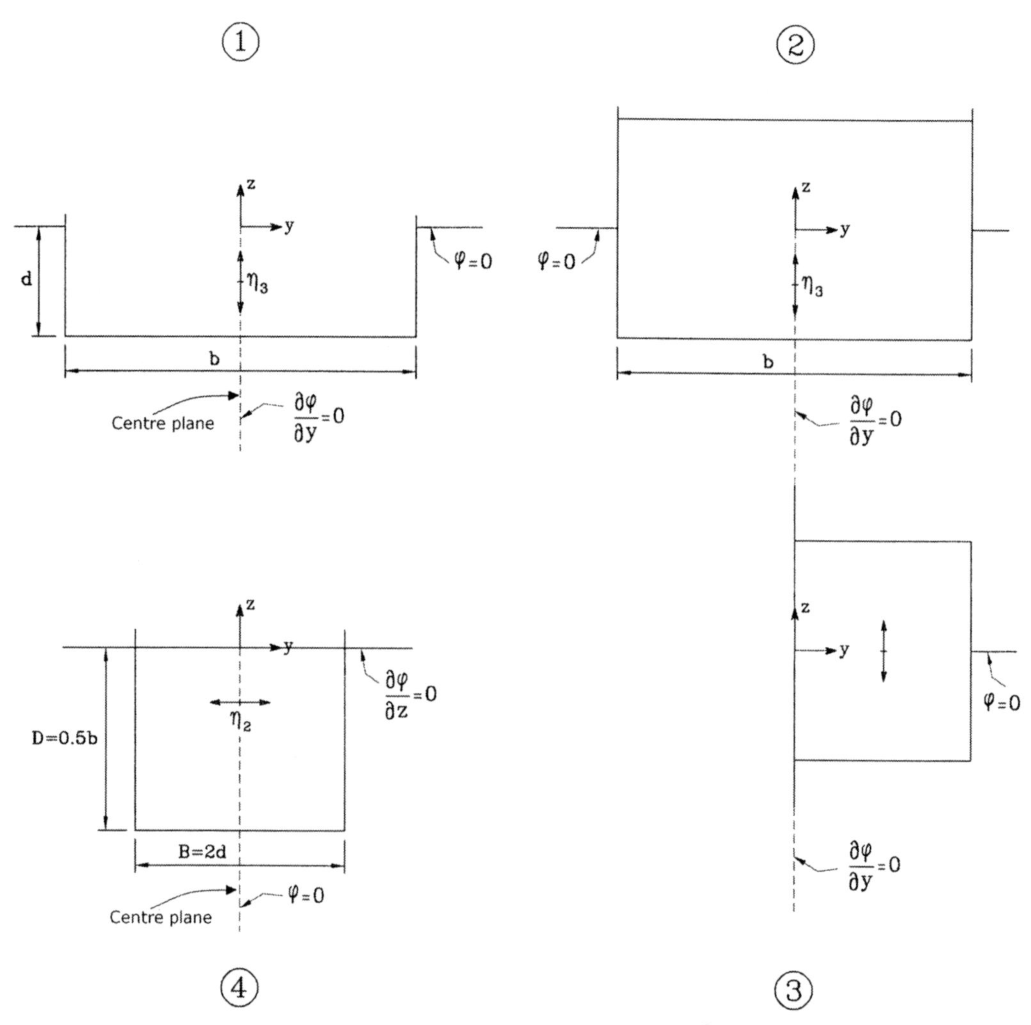

Figure 10.6. Illustration of how an infinite-frequency heave problem ① can be transformed into a low-frequency sway problem ④ by first considering a double-body problem ② and then cutting this double-body problem into two ③. The connection between ③ and ④ is just the result of a 90° rotation and of the introduction of a new coordinate system (y, z) in picture ④.

Figure 7.11. Because the body is symmetric about the centerplane and we consider forced heave motion, the flow is symmetric about the centerplane. We then generate a double body, in which there is an image body above the mean free surface. The flow due to this double body will be antisymmetric about the mean free surface. This is consistent with the free-surface condition $\varphi = 0$. This means that the double body creates the same flow as the original body. Then we consider a new body by dividing the double body into two single bodies by making a cut along the centerplane. Because there is no flow through the centerplane, it behaves similarly to a rigid wall. Then we rotate the new single body by 90° in the clockwise direction and rename forced heave motion as forced sway motion. The rigid free-surface condition is implicitly taken care of. The new body, then, has a beam that is twice the draft of the original body and a draft that is half the beam of the original body. In this way, we have obtained the problem that we want to study, that is, a low-frequency sway problem.

We now use eq. (10.19) to derive the global added mass and damping coefficients A_{jk} and B_{jk} in sway and yaw ($j, k = 2,6$). We assume that the flow separates from the transom stern. A consequence is that the velocity potential is continuous at the transom stern. The first step is to integrate the equation over the length of the ship. This gives the following hydrodynamic sway force

$$F_2^{HD} = -\left[\int_L a_{22}\,dx\ddot{\eta}_2 + \int_L a_{22}x\,dx\ddot{\eta}_6 + Ua_{22}(x_T)\dot{\eta}_2 + Ux_T a_{22}(x_T)\dot{\eta}_6\right]. \quad (10.21)$$

Here x_T is the x-coordinate of the transom stern. Then we take the yaw moment F_6^{HD} based on eq. (10.19). This means the equation is multiplied by x and then integrated over the ship length. This involves the integral

$$\int_L x\frac{\partial}{\partial x}[a_{22}(\dot{\eta}_2 + x\dot{\eta}_6)dx] = x_T a_{22}(x_T)(\dot{\eta}_2 + x_T\dot{\eta}_6) - \int_L a_{22}(\dot{\eta}_2 + x\dot{\eta}_6)dx.$$

The result is

$$F_6^{HD} = -\left\{\int_L xa_{22}\,dx\ddot{\eta}_2 + \int_L a_{22}x^2\,dx\ddot{\eta}_6 + U\left\{\left[x_T a_{22}(x_T) - \int_L a_{22}\,dx\right]\dot{\eta}_2 + \left[x_T^2 a_{22}(x_T) - \int_L xa_{22}\,dx\right]\dot{\eta}_6\right\}\right\}. \quad (10.22)$$

Let us study steady flow with $\dot{\eta}_6 = 0$ as a special case. Because the flow is steady, $\ddot{\eta}_2$ and $\ddot{\eta}_6$ are zero. The transverse force and yaw moment are then equal to

$$F_2^{HD} = -U\,a_{22}(x_T)\dot{\eta}_2 \quad (10.23)$$

$$F_6^{HD} = -U\left[x_T a_{22}(x_T) - \int_L a_{22}\,dx\right]\dot{\eta}_2 \quad (10.24)$$

according to eqs. (10.21) and (10.22), respectively. Looking at eq. (10.23), we may believe that the force acts at the transom stern. However, eq. (10.19) shows that the transverse load distribution f_2^{HD} is non-zero when $\partial a_{22}/\partial x$ is non-zero. Let us consider an extreme case, such as the one in Figure 10.7, to illustrate this. A ship with a long nearly parallel middle and aft part extending until the transom stern is considered. The ship has circular cross sections with axes in the mean free surface. The cross-sectional area is given as

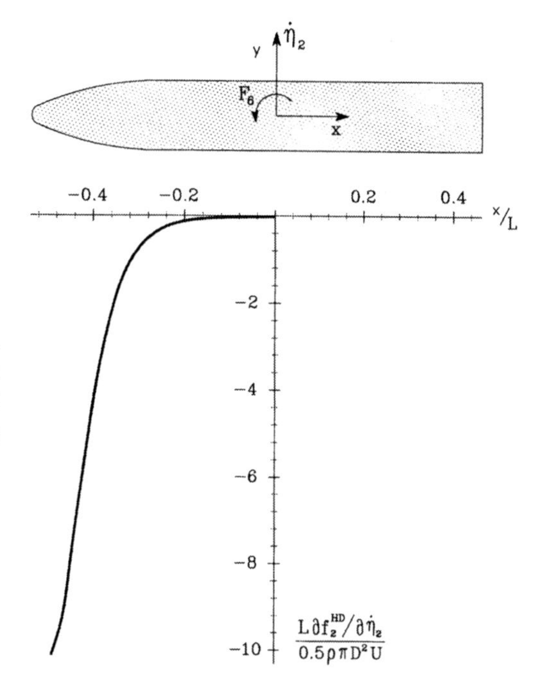

Figure 10.7. Prediction of steady transverse force distribution $\partial f_2^{HD}/\partial\dot{\eta}_2$ according to slender body theory for a ship maneuvering with moderate forward speed U and small transverse ship speed $\dot{\eta}_2$ relative to U. The ship has a long parallel middle and aft body extended until the transom stern with circular cross sections and cylinder axis in mean free surface. D = draft, L = ship length.

$$S(x) = 0.5\pi D^2 \tanh[k(0.5 + x/L)], \quad (10.25)$$

where D is the draft and L is the ship length. The parameter k in eq. (10.25) is chosen as 10. The two-dimensional added mass is $a_{22} = \rho S(x)$. By using eq. (10.19), we get

$$\frac{(\partial f_2^{HD}/\partial\dot{\eta}_2)L}{0.5\rho\pi D^2 U} = -L\frac{\partial}{\partial x}\tanh[10(0.5 + x/L)]. \quad (10.26)$$

This force distribution is presented in Figure 10.7. It shows that the resulting transverse force per unit sway velocity $\partial f_2^{HD}/\partial\dot{\eta}_2$ is negative and acts in the bow part. The center of pressure of F_2^{HD} can be found from

$$\ell_{\dot{\eta}2} = \frac{\int_{-L/2}^{L/2} dx\, x\,\partial f_2^{HD}/\partial\dot{\eta}_2}{\int_{-L/2}^{L/2} dx\,\partial f_2^{HD}/\partial\dot{\eta}_2}. \quad (10.27)$$

This gives $\ell_{\dot\eta_2} = -0.43L$, which means the force acts $0.07L$ from FP. The steady transverse force and yaw moment due to a positive sway velocity are negative and positive, respectively, in this particular case. However, the sign of the steady yaw moment for positive sway velocity depends on the relative magnitude of the two terms $x_T a_{22}(x_T)$ and $\int_L a_{22}\,dx$ in eq. (10.24). Eq. (10.23) shows that the steady transverse force can never be positive for positive sway velocity.

Let us say that the flow did not separate at the transom stern or the ship had a pointed end. We would then have a force contribution from the stern canceling the force contribution from the bow. This would result in a zero force in the potential flow case (D'Alembert's paradox). However, the corresponding potential flow yaw moment is non-zero. This is equal to $U\int_L a_{22}\,dx\,\dot\eta_2$ according to eq. (10.24) and is a slender-body approximation of what is called the Munk moment. The exact Munk moment, see eq. (10.159), with the traditional coordinate system and notation used in ship maneuvering (see Figure 10.4) is $(A_{11} - A_{22})vu$. The difference in sign is a consequence of different coordinate systems and definitions of positive velocities. The Munk moment causes a destabilizing (broaching) moment on a ship with steady oblique translatory motion. The Munk moment will be significant for high forward speed. However, lifting and viscous effects will, to some extent, counteract this.

Let us perform a similar analysis by studying constant forced yaw velocity $\dot\eta_6$. By using eq. (10.19), we get

$$\frac{\partial f_2^{HD}/\partial\dot\eta_6}{0.5\rho\pi D^2 U} = -\frac{\partial}{\partial x}\left\{x\tanh\left[10\left(0.5 + x/L\right)\right]\right\}. \tag{10.28}$$

This force distribution is presented in Figure 10.8. It shows that $\partial f_2^{HD}/\partial\dot\eta_6$ is not concentrated in the bow part as $\partial f_2^{HD}/\partial\dot\eta_2$ was. The center of pressure of resulting transverse force F_2^{HD} can be found from

$$\ell_{\dot\eta_6} = \frac{\int_{-L/2}^{L/2} dx\, x\, \partial f_2^{HD}/\partial\dot\eta_6}{\int_{-L/2}^{L/2} dx\, \partial f_2^{HD}/\partial\dot\eta_6}. \tag{10.29}$$

This gives $\ell_{\dot\eta_6} = 0.44L$, which means the transverse force acts $0.06L$ from AP. If $\dot\eta_6$ is positive, the resulting transverse force and yaw moment are negative.

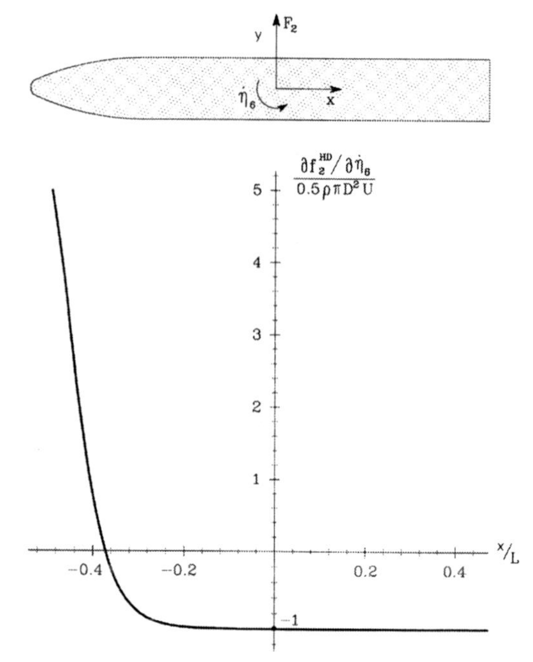

Figure 10.8. Prediction of steady transverse force distribution $\partial f_2^{HD}/\partial\dot\eta_6$ according to slender body theory for a ship maneuvering with moderate forward speed U and yaw velocity $\dot\eta_6$. The ship has a long parallel middle and aft body extended until the transom stern with circular cross sections and cylinder axis in mean free surface. D = draft, L = ship length.

10.3.1 Low-aspect–ratio lifting surface theory

Eqs. (10.23) and (10.24) can be applied to give the lift force L and moment M on a low-aspect–ratio lifting surface in infinite fluid. This can be done by introducing an image body about the mean free surface and considering the double body as the lifting surface in infinite fluid. If we consider a rectangular plane planform in infinite fluid with span s and chord length ℓ, we get

$$L = \rho U^2 0.25\pi s^2 \alpha. \tag{10.30}$$

Here α is the angle of attack or $-\dot\eta_2/U$ using the notation in this chapter. Eq. (10.30) gives, then, a lift coefficient

$$C_L = 0.5\pi\Lambda\alpha. \tag{10.31}$$

The aspect ratio Λ is defined as s^2/A (see eq. (2.87)), where the planform area A is $s\ell$ in this case. The trailing vortex sheet enters this analysis indirectly through the condition that the velocity potential is continuous at the trailing edge. This condition means the downwash due to the trailing vortex sheet is negligible. Eq. (10.24) gives a moment about $x = 0$ for the rectangular lifting surface that can be expressed as

$$M = -\rho U^2 0.125\pi s^2 \ell\alpha. \tag{10.32}$$

10.3.2 Equations of sway and yaw velocities and accelerations

Let us go back to eqs. (10.21) and (10.22). Introducing the definition of added mass and damping coefficients given by eq. (7.39), we have

$$A_{22} = \int_L a_{22}\,dx; \quad B_{22} = Ua_{22}(x_T) \tag{10.33}$$

$$\begin{aligned} A_{62} &= \int_L xa_{22}\,dx; \\ B_{62} &= -UA_{22} + Ux_T a_{22}(x_T) \end{aligned} \tag{10.34}$$

$$A_{26} = \int_L xa_{22}\,dx; \quad B_{26} = Ux_T a_{22}(x_T) \tag{10.35}$$

$$\begin{aligned} A_{66} &= \int_L x^2 a_{22}\,dx; \\ B_{66} &= -UA_{62} + Ux_T^2 a_{22}(x_T). \end{aligned} \tag{10.36}$$

The expressions of coupling coefficients between sway and yaw show that $A_{26} = A_{62}$, whereas $B_{26} \neq B_{62}$. If we want to express the hydrodynamic forces and moments in terms of the derivative term Y_v and similar terms, as in eqs. (10.1) and (10.2), we can use eq. (10.16) in combination with eqs. (10.33) to (10.36). However, we must recall that the x-axis in Figure 10.4 is pointing in the opposite direction with respect to the one we have used above.

Let us compare eqs. (10.33) to (10.36) with the results in eq. (10.159). The latter case does not consider lifting effects, which is the same as saying that the end terms in eqs. (10.33) to (10.36) are not present. We must also linearize eq. (10.159). This means that u is replaced with U in the expressions. There is a term $-A_{11}u\dot{\psi}$ in the expression for the transverse force Y given by eq. (10.159). However, noting that A_{11} is negligible in a slender body theory gives consistent results.

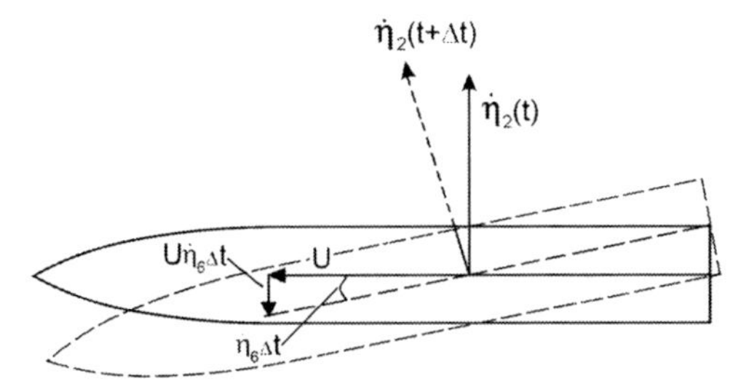

Figure 10.9. Ship velocity components in the body-fixed coordinate system (see Figure 10.5) decomposed in an inertial system in order to derive ship accelerations in an inertial system. The time difference Δt is assumed to be small; however, this is exaggerated in the figure.

Because we are using a body-fixed coordinate system and not an inertial system, it will be evident below that the ship mass terms will be different. We use Figure 10.9, in which the sway velocity at the COG of the ship at time t and $t + \Delta t$ is shown. If we refer to the ship coordinate system at time t, the sway velocity at time $t + \Delta t$ will be approximately

$$\dot{\eta}_2(t + \Delta t) - U\dot{\eta}_6\Delta t$$

for small Δt. This means the sway acceleration in the inertial system is $\ddot{\eta}_2 - U\dot{\eta}_6$ or, in other words, that the ship inertia term is

$$M(\ddot{\eta}_2 - U\dot{\eta}_6).$$

This gives the linearized equations of sway and yaw accelerations of a maneuvering ship as follows:

$$\begin{aligned} &(M + A_{22})\frac{d^2\eta_2}{dt^2} + Ua_{22}(x_T)\frac{d\eta_2}{dt} + A_{26}\frac{d^2\eta_6}{dt^2} \\ &\quad + U[-M + x_T a_{22}(x_T)]\frac{d\eta_6}{dt} = F_2^S \\ &A_{62}\frac{d^2\eta_2}{dt^2} + U[x_T a_{22}(x_T) - A_{22}]\frac{d\eta_2}{dt} \\ &\quad + (I_{66} + A_{66})\frac{d^2\eta_6}{dt^2} \\ &\quad + U[-A_{62} + x_T^2 a_{22}(x_T)]\frac{d\eta_6}{dt} = F_6^S. \end{aligned} \tag{10.37}$$

Here F_2^S and F_6^S are control forces in sway and yaw moments due to, for instance, a rudder or a water-jet propulsion system. We should note that there are also hydrodynamic forces on the rudder due to sway and yaw. These have not been explicitly included in the formulation of the equations.

10.3.3 Directional stability

By setting F_2^S and F_6^S equal to zero in eq. (10.37) and assuming a solution of the form $\exp(st)$, we can study the directional stability of the ship. This means we write $\dot{\eta}_2 = \dot{\eta}_{2a}\exp(st)$ and $\dot{\eta}_6 = \dot{\eta}_{6a}\exp(st)$, where the dot means time derivative. This gives

$$\begin{aligned}
&[s(M + A_{22}) + Ua_{22}(x_T)]\dot{\eta}_{2a}\\
&\quad+\{sA_{26} + U[-M + x_T a_{22}(x_T)]\}\dot{\eta}_{6a} = 0\\
&\{sA_{62} + U[x_T a_{22}(x_T) - A_{22}]\}\dot{\eta}_{2a}\\
&\quad+\{s(I_{66} + A_{66}) + U[-A_{62} + x_T^2 a_{22}(x_T)]\}\dot{\eta}_{6a} = 0.
\end{aligned} \qquad (10.38)$$

We may write eq. (10.38) in matrix form as

$$\begin{bmatrix} s(M + A_{22}) + Ua_{22}(x_T) & sA_{26} + U[-M + x_T a_{22}(x_T)] \\ sA_{62} + U\left[x_T a_{22}(x_T) - A_{22}\right] & s(I_{66} + A_{66}) + U\left[-A_{62} + x_T^2 a_{22}(x_T)\right] \end{bmatrix} \begin{bmatrix} \dot{\eta}_{2a} \\ \dot{\eta}_{6a} \end{bmatrix} = \begin{bmatrix} 0 \\ 0 \end{bmatrix}.$$

Because the right-hand side of this equation system is zero and we are interested in the nontrivial solutions, these can be obtained by enforcing the coefficient determinant to be zero. This means

$$As^2 + Bs + C = 0, \qquad (10.39)$$

where

$$\begin{aligned}
C &= U^2[-MA_{22} + x_T a_{22}(x_T)(M + A_{22})\\
&\quad - a_{22}(x_T)A_{26}]\\
B &= U[a_{22}(x_T)(I_{66} + A_{66} + x_T^2(M + A_{22}))\\
&\quad - 2A_{26}x_T a_{22}(x_T)]\\
A &= (M + A_{22})(I_{66} + A_{66}) - A_{26}^2
\end{aligned} \qquad (10.40)$$

Here we have used the symmetry property $A_{26} = A_{62}$. The solutions of eq. (10.39) are two, $s_j(j = 1, 2)$, and generally complex. This means $s_j = \mathrm{Re}(s_j) + i\,\mathrm{Im}(s_j)$. The requirement for a stable system is that both $\mathrm{Re}(s_1)$ and $\mathrm{Re}(s_2)$ must be less than zero.

Formally, the solutions s_j can be written as

$$s_j = \frac{-B \pm (B^2 - 4AC)^{1/2}}{2A}. \qquad (10.41)$$

We will assume in the following that A and B are positive. We see this is true if A_{26} is negligible. In fact, $A_{26} = 0$ for a ship with fore and aft symmetry and with the longitudinal position of center of gravity (COG) at midships.

We note that

$$\mathrm{Re}(s_2) = \mathrm{Re}\left\{\frac{1}{2A}\left[-B - (B^2 - 4AC)^{1/2}\right]\right\} < 0$$

while

$$\mathrm{Re}(s_1) = \mathrm{Re}\left\{\frac{1}{2A}\left[-B + (B^2 - 4AC)^{1/2}\right]\right\}$$

$$\text{is}\begin{cases} <0 & \text{for } C > 0 \\ >0 & \text{for } C < 0 \end{cases}.$$

This means that directional stability of the ship requires $C > 0$ or

$$|x_T|\,a_{22}(x_T) > \frac{MA_{22}}{M + A_{22}} \qquad (10.42)$$

by using eq. (10.40) with $A_{26} = 0$. The same condition was derived by Newman (1977). Eq. (10.42) shows the advantage of moving COG forward, resulting in large $|x_T|$-values. Further, $a_{22}(x_T)$ is, roughly speaking, proportional to the square of the draft at the transom stern. Having a skeg that causes an increase in the local draft at the aft part of the ship will obviously improve the directional stability. The same is true with aft trim that lowers the stern. Contrarily, if the ship has a pointed aft end, the ship is directionally unstable. We should note that this stability condition is speed independent, but it has been derived by assuming moderate speed, let us say, a Froude number less than about 0.2. From this discussion, condition eq. (10.42) has to be used for qualitative guidelines. In order to quantify the directional stability of the ship, we must examine the eigenvalues.

In order to express the stability condition in a way that is more general and independent of slender body theory, we should start by setting δ equal to zero in eqs. (10.1) and (10.2) and assuming that v and r have the time dependence $\exp(st)$. This leads to A, B, and C in eq. (10.40) being replaced by

$$\begin{aligned}
A &= (M - Y_{\dot{v}})(I_{66} - N_{\dot{r}}) - N_{\dot{v}}Y_{\dot{r}}\\
B &= -(M - Y_{\dot{v}})N_r - Y_v(I_{66} - N_{\dot{r}})\\
&\quad - N_{\dot{v}}[Y_r - (M - X_{\dot{u}})U] - N_v Y_{\dot{r}}\\
C &= Y_v N_r - N_v[Y_r - (M - X_{\dot{u}})U]
\end{aligned} \qquad (10.43)$$

Because A and B in practice are positive, the stability condition is $C > 0$ or that

$$\frac{N_r}{Y_r - (M - X_{\dot{u}})U} - \frac{N_v}{Y_v} > 0. \qquad (10.44)$$

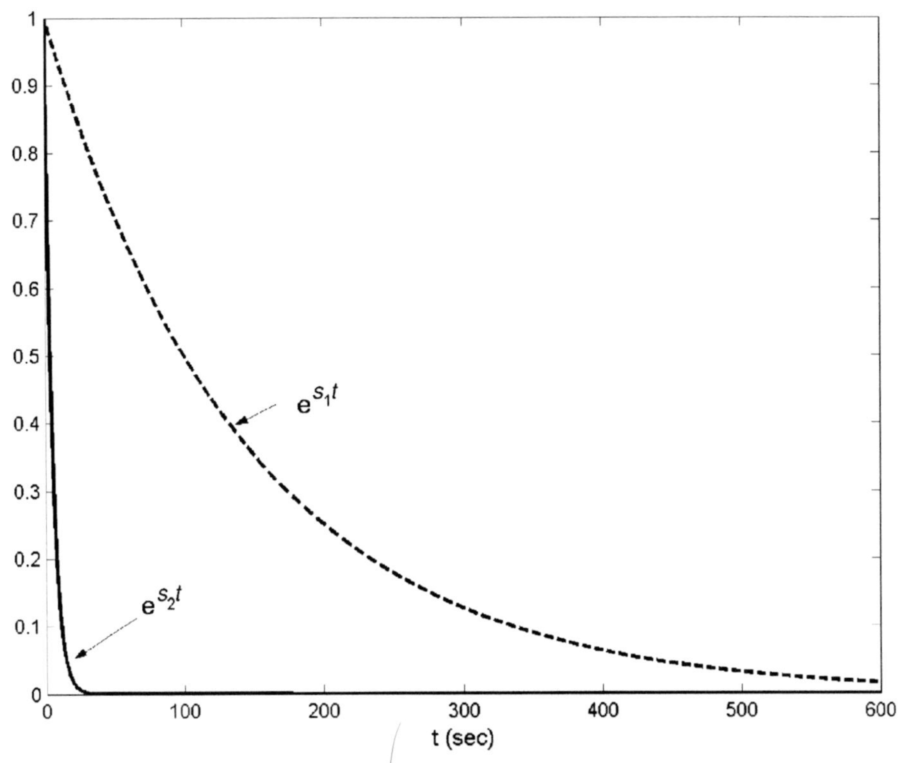

Figure 10.10. Time decay of eigensolutions for sway and yaw of the ship described by eq. (10.25) in deep water. Ship length $L = 100$ m. Ship speed $U = 10\ \mathrm{ms}^{-1}$.

It is not an advantage with a too-stable ship. It is then difficult to change the course. One should notice that unstable ships can be stabilized by feedback control using the autopilot.

10.3.4 Example: Directional stability of a monohull

We consider the ship described by eq. (10.25) with $k = 10$ and center of gravity at $x = 0$. The vessel mass M is simply $\rho\nabla$, where ∇ is the displaced volume of water. The vessel moment of inertia in yaw I_{66} is expressed as Mr_{66}^2 with $r_{66} = 0.25L$. Here L is the length of the waterline. The x-coordinate of the transom stern is $L/2$. Calculations of the eigenvalues s_1 and s_2 give

$$\frac{s_1 L}{U} = -0.06; \quad \frac{s_2 L}{U} = -2.31. \qquad (10.45)$$

This means the ship is directionally stable.

We choose as an example $L = 100$ m and $U = 10\ \mathrm{ms}^{-1}$. The time-dependent parts $\exp(s_j t)$ of the eigensolutions, $j = 1, 2$ are plotted in Figure 10.10. From the graph, it takes more than ten minutes for the eigensolution associated with s_1 to die out.

10.3.5 Steady-state turning

We will apply the linear maneuvering equations, that is, eqs. (10.1) and (10.2), to steady-state turning. This means the center of gravity (COG) of the vessel follows a circular path with radius R. The velocity of the COG is tangential to the circular path and has a constant magnitude U (Figure 10.11). The yaw angular velocity r and the velocity component v along the body-fixed y-axis also remain constant in time. Eqs. (10.1) and (10.2) can then be expressed as

$$-Y_v v + [(M - X_{\dot{u}})U - Y_r]r = Y_\delta \delta \qquad (10.46)$$

$$-N_v v - N_r r = N_\delta \delta. \qquad (10.47)$$

Positive rudder angle δ is defined in Figure 10.4. This means the rudder angle needed for the turning maneuver shown in Figure 10.11 is negative. Further, we consider the rudder as an appendage

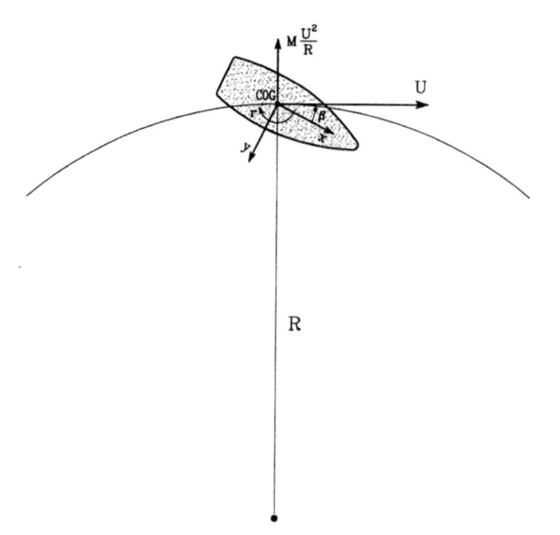

Figure 10.11. Steady-state turning of a ship with constant speed U along a circular track with radius R. The centrifugal force MU^2/R is indicated.

and approximate N_δ as $-|x_T|\,Y_\delta$, where x_T is the x-coordinate of the transom stern of the vessel. By using that $Rr = U$ and neglecting the surge added mass $-X_{\dot{u}}$, we can express eq. (10.46) as

$$M\frac{U^2}{R} = Y_v v + Y_r r + Y_\delta \delta. \quad (10.48)$$

This expresses that the centrifugal force MU^2/R on the vessel is balanced by the hydrodynamic transverse hull force $Y_v v + Y_r r$ and the rudder force $Y_\delta \delta$.

Solving eqs. (10.46) and (10.47) gives

$$v = \frac{-N_r + |x_T|\left[(M - X_{\dot{u}})\,U - Y_r\right]}{Y_v N_r + N_v\left[(M - X_{\dot{u}})\,U - Y_r\right]} Y_\delta \delta \quad (10.49)$$

$$r = \frac{Y_v\,|x_T| + N_v}{Y_v N_r + N_v\left[(M - X_{\dot{u}})\,U - Y_r\right]} Y_\delta \delta. \quad (10.50)$$

Here the denominator $Y_v N_r + N_v[(M - X_{\dot{u}})U - Y_r]$ is the same as C given by eq. (10.43). Because $C > 0$ is a requirement for a directionally stable ship, we must also require this in the analysis here. It follows from eqs. (10.13), (10.15), (10.16), and (10.17) and slender body theory expressed by eqs. (10.33) through (10.36) that

$$Y_v = -Ua_{22}(x_T) \quad (10.51)$$

$$N_v = -UA_{22} + U|x_T|a_{22}(x_T) \quad (10.52)$$

$$Y_r = U|x_T|a_{22}(x_T) \quad (10.53)$$

$$N_r = U\int_L xa_{22}\,dx - Ux_T^2 a_{22}(x_T). \quad (10.54)$$

Here $a_{22}(x)$ means the 2D added mass in sway and A_{22} is the 3D added mass in sway for the vessel. Further, x_T is the x-coordinate of the transom. The x-integration in eq. (10.54) is along the x-axis defined in Figure 10.5. However, this integral is small and will be neglected in the following expressions. Further, $X_{\dot{u}}$ will be neglected. Substituting the slender body theory expressions into eqs. (10.49) and (10.50) gives

$$v = \frac{U\,|x_T|\,MY_\delta\delta}{C} \quad (10.55)$$

$$r = \frac{-U A_{22} Y_\delta \delta}{C}, \quad (10.56)$$

where

$$\begin{aligned} C &= U^2\left\{x_T^2 a_{22}^2(x_T) - \left[A_{22} - |x_T|a_{22}(x_T)\right]\right. \\ &\quad \left.\times\left[M - |x_T|a_{22}(x_T)\right]\right\} \quad (10.57) \\ &= U^2\left[-MA_{22} + |x_T|a_{22}(x_T)(M + A_{22})\right]. \end{aligned}$$

Because Y_δ is proportional to U^2, eqs. (10.55) and (10.56) show that v and r are proportional to U. Because Y_δ is positive, C must be positive, and δ is negative, eq. (10.55) shows that v is negative and r is positive. The latter is consistent with the fact that the bow is pointed into the turn, as illustrated in Figure 10.11.

10.3.6 Multihull vessels

The procedure can be easily generalized to a catamaran by accounting for the effect of hull interaction when evaluating the 2D "low-frequency" added mass in sway. Figure 10.12 shows 2D added mass results for semi-submerged circular demihulls with axes in the mean free surface in equilibrium position. The hydrodynamic interaction between the demihulls reduces the added mass.

Kaplan et al. (1981) applied a slender body theory like this to an SES by also accounting for viscous drag force and moment. This was done by using the cross-flow principle (see section 10.6.1) with constant drag coefficient along the ship. Kaplan et al. (1981) ignored the hydrodynamic hull interaction and the fact that the free surface is lower inside the cushion than outside. However, they reported satisfactory agreement with experiments. Kaplan et al. (1981) also studied the effect of a ventral fin or rudder, as in Figure 7.5. This was considered as an appendage in the mathematical modeling.

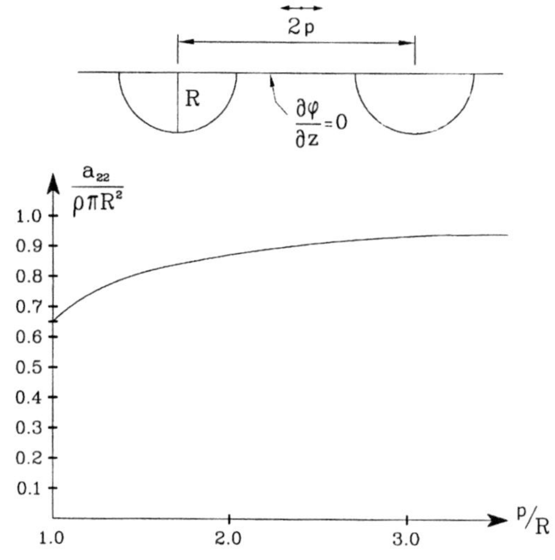

Figure 10.12. 2D "low-frequency" sway-added mass a_{22} for a catamaran with semi-submerged circular demihulls in infinite fluid depth, based on Greenhow and Li (1987). The value of $a_{22}/(\rho\pi R^2)$ when $p/R = 1$ is $\pi^2/6 - 1$.

10.3.7 Automatic control

If a vessel is directionally unstable, it can be stabilized by automatic control. Here, we will describe a control system based on the use of a PID (proportional integral derivative) regulator in combination with rudders. The control system acts by reducing the error,

$$e = \psi - \psi_d f(t), \tag{10.58}$$

where ψ is the actual yaw angle (see Figure 10.4) and ψ_d is the yaw angle desired by the helmsman. In eq. (10.58), $f(t)$ is a filter function, so that the desired yaw angle is not introduced abruptly. We now express the rudder angle as

$$\delta = -K_p e - K_i \int_0^t e\,dt - K_d \dot{e}, \tag{10.59}$$

where K_p, K_i, and K_d are positive constants set by the system. Here K_p and K_d will act as restoring and damping coefficients, respectively, in the mathematical model. K_i eliminates static offset (bias). The subscripts $_p$, $_i$, and $_d$ are associated with the name of the regulator, that is, PID or that p refers to *proportional*, i to *integral*, and d to *derivative*. More details of a control system design such as this can be found in Fossen (2002).

10.4 Linear ship maneuvering at moderate Froude number in finite water depth

If we base the calculations on a slender body, we can use the same expressions for A_{jk} and B_{jk} as those in section 10.3. The important difference is that the two-dimensional added mass in sway is influenced by the water depth. Our discussion will be based on a constant water depth and a rigid sea floor.

Figure 10.13 presents the ratio between "low-frequency" sway added mass A_{22} at water depth h and A_{22} at $h \to \infty$ as a function of water depth–to-draft ratio h/D. The figure includes results for both low-speed ships and two-dimensional sections. We have used the same notation for the 2D and 3D added mass. $A_{22}(h)/A_{22}(\infty)$ is clearly a function of the cross-sectional shape.

The results in Figure 10.13 show a clear increase in $A_{22}(h)/A_{22}(\infty)$ when h/D decreases. This strong change in hydrodynamic coefficients with depth influences the directional stability of a vessel. For instance, a vessel may start by being directionally stable in infinite depth, then become unstable in a certain h/D range and then stable again at shallow depth (Fujino 1976).

The trim and sinkage of a vessel affect the ratio between the water depth and the local draft along the ship. The directional stability will therefore be influenced. We discussed the importance of sinkage and trim in shallow water for a ship on a straight course in section 4.5.6.

When h/D becomes close to 1, three-dimensional flow effects become pronounced (Newman 1969). This causes a strong flow around the ship ends.

10.5 Linear ship maneuvering in deep water at high Froude number

Previous analysis is also relevant for high-speed vessels, for instance, when moving at moderate speed in their approach to harbors. However, Chapman (1976) shows that the hydrodynamic coefficients for maneuvering at high speed are quite different from those for maneuvering at moderate speed and should be properly evaluated. The differences are connected with the divergent wave system generated by the ship during maneuvering. The high-speed analysis can be carried out numerically by using a 2.5D (2D+t) theory in a

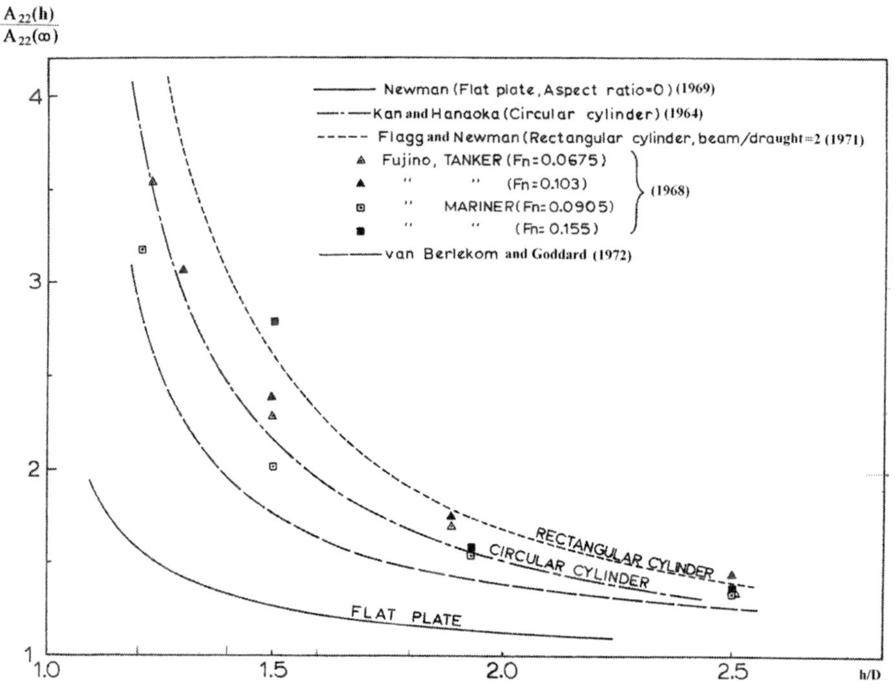

Figure 10.13. Comparison of the shallow-water effects on the sway-added mass A_{22} obtained by two-dimensional calculations and by experiments. We normally use A_{22} as the 3D sway-added mass for the vessel. However, because both 2D and 3D results are presented, A_{22} is used as a common notation. h = water depth, D = draft (Fujino 1976).

way similar to the one we described for wave resistance in section 4.3.4 and for wave-induced motions in section 7.2.12.

van den Brug et al. (1971) presented extensive experimental results for hydrodynamic sway force and yaw moment on vertical surface-piercing flat plates at high Froude numbers. Chapman (1976) demonstrated good agreement between these experiments and his 2.5D numerical results (Figure 10.14). The results show a clear Froude number dependency. When the Froude number is small, there is fair agreement with the slender body theory presented in section 10.3. The results illustrate that the maneuvering model must account for the Froude number dependency at high Froude numbers.

Ishiguro et al. (1993) studied the maneuvering properties of the high-speed vessel "Super Slender Twin Hull" (SSTH). The waterjet version is shown in Figure 10.15. PMM (planar motion mechanism) tests were used to obtain hydrodynamic coefficients in sway and yaw. Nondimensional hydrodynamic derivative terms Y_v, Y_r, N_v, and N_r are presented as a function of Froude number in Figure 10.16. The derivative terms are nondimensionalized by dividing by corresponding values at Froude number 0.184. The results show a very clear Froude number dependence. How this influences the directional stability as a function of Froude number can be illustrated by means of Figure 10.17, in which

$$C' = \frac{N_r'}{Y_r' - M' + X_{\dot{u}}'} - \frac{N_v'}{Y_v'} \tag{10.60}$$

is plotted as function of Froude number for the vessel with and without skegs. The notation m_x' is used instead of $-X_{\dot{u}}'$ in Figure 10.17. The nondimensional coefficients on the right-hand side of eq. (10.60) are defined in eqs. (10.4) and (10.5). Figure 10.17 shows that the vessel without skegs becomes directionally unstable for $Fn > 0.25$ in the considered Froude number range up to $Fn = 0.735$.

A skeg was mounted at the stern of each demihull. The ratio between total projected area of both skegs divided by the product of ship length and draft was 0.02. Each skeg had a rectangular shape with an aspect ratio of 1.5. The skegs caused the

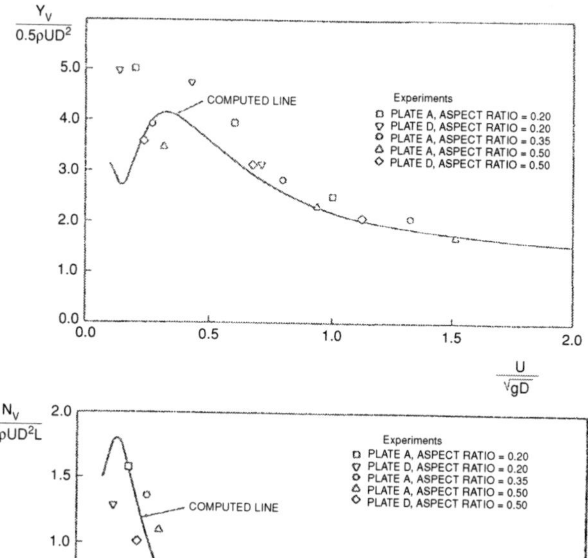

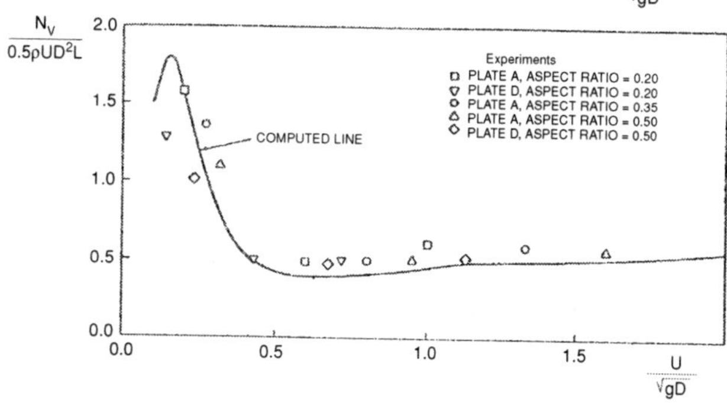

Figure 10.14. Theoretical and experimental hydrodynamic derivative terms Y_v and N_v as a function of the draft Froude number $U/\sqrt{gD}$ for flat plates with different aspect ratios D/L. D draft, L = plate length. The yaw moment N is with respect to an axis through the mid-chord (midships). The experiments are by van den Brug et al. (1971) (Chapman 1976).

vessel to be directionally stable up to $Fn = 0.37$. The benefit of using skegs at moderate speed from a directionality stability point of view is also evident from eq. (10.42). The skegs increase $a_{22}(x_T)$.

A vessel at high speed will heel during turning. This means that the coupled effect of sway, yaw, and roll (heel) on the directional stability should be investigated. This is discussed further in section 10.9.3.

Ishiguro et al. (1993) also presented full-scale results with a 30 m–long version of the vessel. The results were compared with simulated results. Model test results were used for the hydrodynamic coefficients in the simulation model. Figure 10.18 shows turning test results with a ship speed of 12 knots. One meter per second corresponds to 1.944 knots. Even though the full-scale trial results may be influenced by current, wind, and waves, there is satisfactory agreement between the

Waterjet Version

Model data: L_{pp} : 2.000 m
B(Twin Hull) : 0.333 m
draught : 0.051 m

Figure 10.15. Waterjet version of the high-speed vessel "Super Slender Twin Hull" (SSTH), shown together with length dimensions used in PMM model tests. L_{PP} is the length between perpendiculars (Ishiguro et al. 1993).

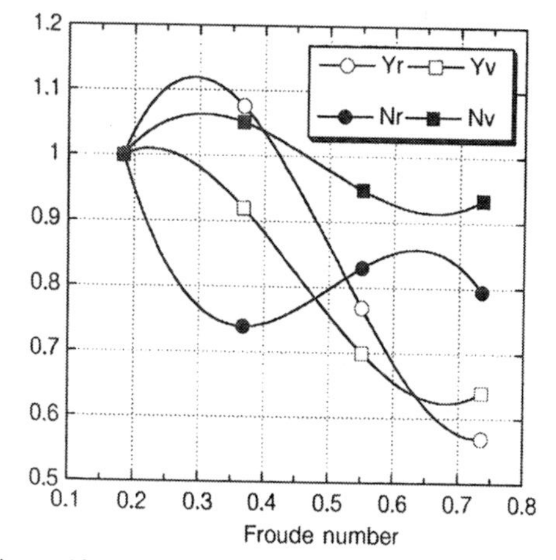

Figure 10.16. Length Froude number dependence of nondimensional hydrodynamic derivative terms Y_r, Y_v, N_r, N_v obtained by PMM tests of the twin hull vessel SSTH shown in Figure 10.15. The hydrodynamic derivative terms are normalized with respect to corresponding values at Froude number 0.184 (Ishiguro et al. 1993).

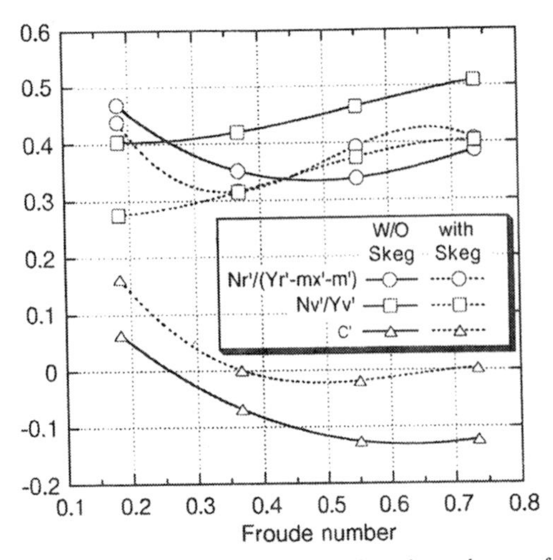

Figure 10.17. Length Froude number dependence of combinations of nondimensional hydrodynamic derivative terms $Y_v', Y_r', N_v', N_r', m_x'$ of the twin-hull vessel SSTH shown in Figure 10.15. C' is defined in eq. (10.60) and expresses the directional stability of the vessel. If $C' > 0$, the vessel is stable, and if $C' < 0$, the vessel is unstable (Ishiguro et al. 1993).

simulated and measured results. The turning diameter is about four ship lengths. Figures 10.19 and 10.20 show similarly good agreement between simulations and measurements for a zigzag maneuver test and a crash astern test. The ship speed is about 9 knots in the zigzag maneuver test. The maximum rudder angle is 20°, with a resulting maximum yaw angle that is about 30% larger. The initial speed is 24 knots in the crash astern test, and the vessel stops moving ahead after about three ship lengths. The documented performance can be related to the IMO maneuvering criteria for ships longer than 100 m discussed in connection with Table 10.1 and Figures 10.1 and 10.2.

10.6 Nonlinear viscous effects for maneuvering in deep water at moderate speed

The previous sections consider hydrodynamic loads that can be described by linear potential flow theory. This section focuses on nonlinear viscous loads due to sway and yaw velocities.

10.6.1 Cross-flow principle

First we will use the cross-flow principle to evaluate the transverse viscous force and yaw moment on the ship. This requires the transverse component of the ship velocity to be larger than the forward ship speed. Thus, in dynamic positioning and low speed maneuvering, this assumption is good. The cross-flow principle assumes that (i) the flow separates because of the cross-flow past the ship, (ii) the longitudinal velocity components do not influence the transverse forces on a cross section, and (iii) the transverse force on a cross section is mainly the result of separated flow effects on the pressure distribution around the ship. This means we can write the transverse viscous force F_2^v on the ship as

$$F_2^v = -\frac{1}{2}\rho \int_L [C_D(x)|\dot{\eta}_2 + x\dot{\eta}_6| \times (\dot{\eta}_2 + x\dot{\eta}_6) D(x)]\, dx, \tag{10.61}$$

where the integration is over the ship length L. Here, $C_D(x)$ is the drag coefficient for the cross-flow past an infinitely long cylinder, with the cross-sectional area of the ship at the longitudinal coordinate x. $D(x)$ is the sectional draft. Further, $|\dot{\eta}_2 + x\dot{\eta}_6|$ is equal to $(\dot{\eta}_2 + x\dot{\eta}_6)$ and $-(\dot{\eta}_2 + x\dot{\eta}_6)$ when $\dot{\eta}_2 + x\dot{\eta}_6$ is positive and negative, respectively. This behavior of $|\dot{\eta}_2 + x\dot{\eta}_6|$ implies that $|\dot{\eta}_2 + x\dot{\eta}_6|\,(\dot{\eta}_2 + x\dot{\eta}_6)$ cannot be explicitely expressed in terms of $\dot{\eta}_2^2$, $\dot{\eta}_2\dot{\eta}_6$ and $\dot{\eta}_6^2$ in a simple way.

The viscous yaw moment due to the cross-flow is

$$F_6^v = -\frac{1}{2}\rho \int_L [C_D(x)|\dot{\eta}_2 + x\dot{\eta}_6| \times (\dot{\eta}_2 + x\dot{\eta}_6) D(x)x]\, dx. \tag{10.62}$$

C_D-values for ship sections

In order to improve the predictions by eqs. (10.61) and (10.62), we need to know more about C_D-values. It is difficult to do this by theoretical means only. In the following text, we discuss what the important parameters are that influence C_D when the ship has zero forward speed. Important factors are, for instance, free-surface effects, beam-to-draft ratio, bilge radius, Reynolds number, hull roughness, Keulegan-Carpenter (KC) number and three-dimensional flow effects. The KC number describes the effect of oscillatory ship motions. Let us consider a 2D section with transverse velocity $V_a \sin((2\pi/T)t + \varepsilon)$, where V_a is

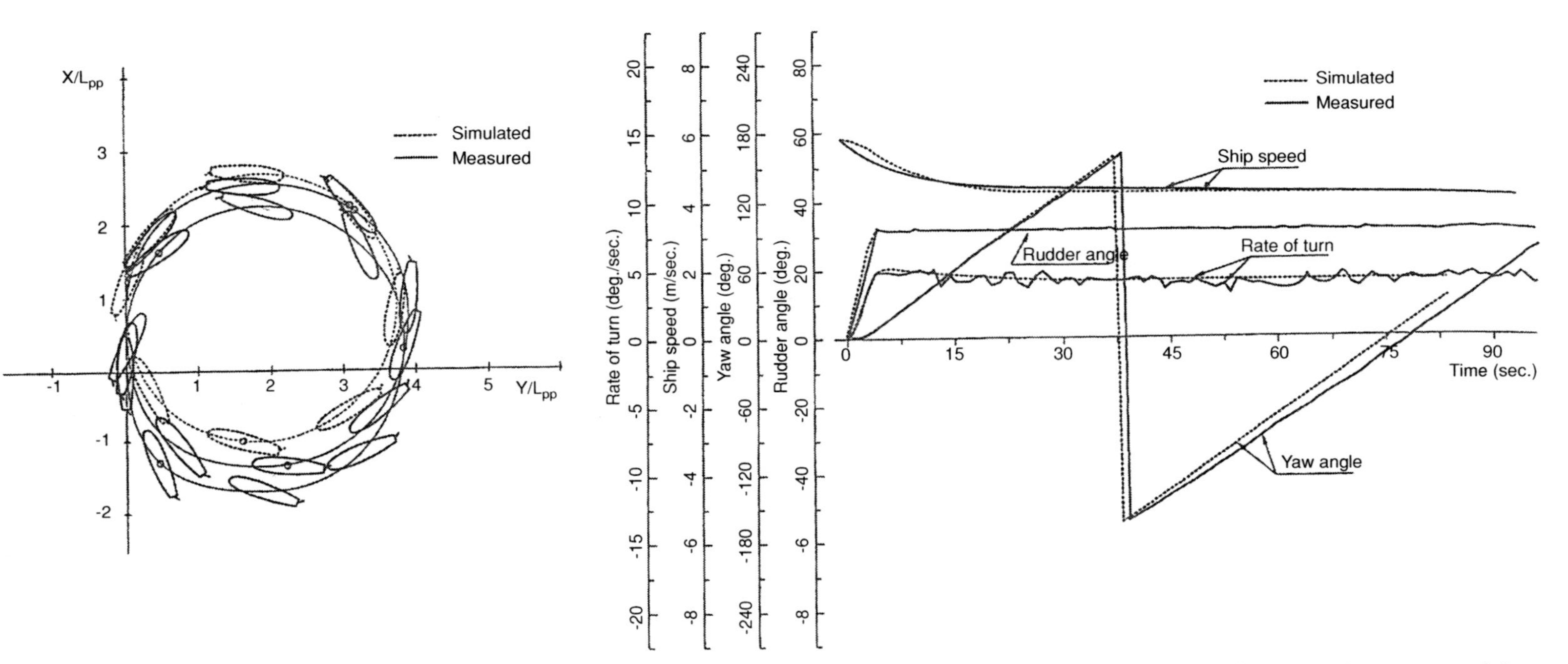

Figure 10.18. Full-scale turning tests with a 30 m–long version of the twin-hull vessel SSTH. Simulated and measured trajectories and time history records are compared. L_{PP} is the length between perpendiculars (Ishiguro et al. 1993).

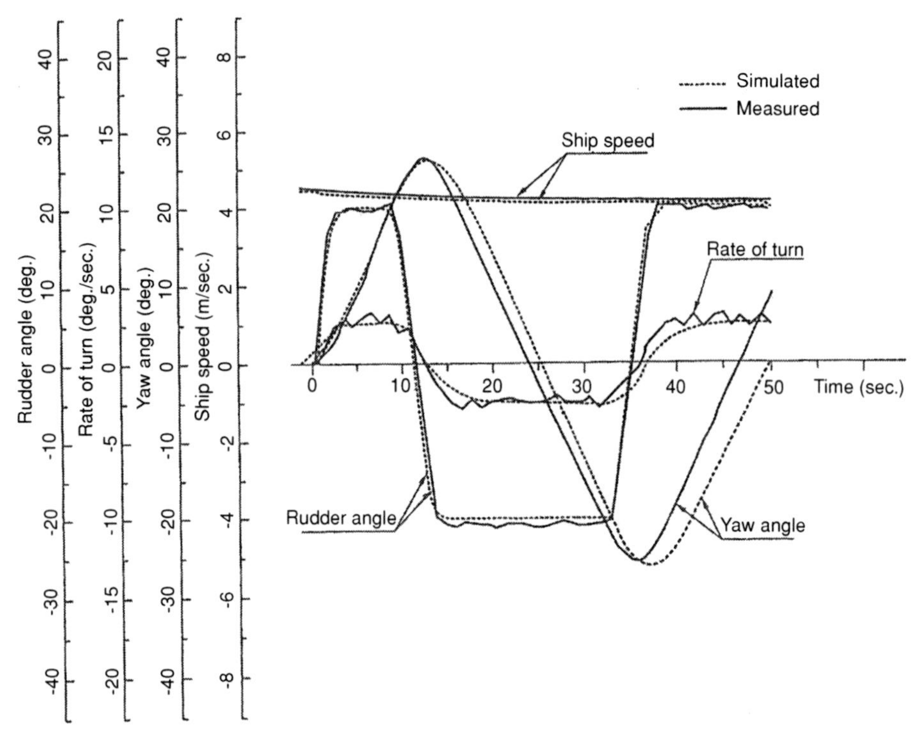

Figure 10.19. Full-scale zigzag maneuver test with a 30 m–long version of the twin-hull vessel SSTH. Simulated and measured time history records are compared (Ishiguro et al. 1993).

the velocity amplitude, T is the oscillation period, and ε is the phase. Then KC can be defined as $KC = V_a T / L_c$, where L_c is a characteristic length of the cross section, such as the beam of the section. How KC affects C_D is discussed by Faltinsen (1990) and Sarpkaya and Isaacson (1981). In the following part of this section, we also consider that the horizontal velocity of the cross section is time independent. This is the same as studying a fixed cross section in a current and, therefore, this

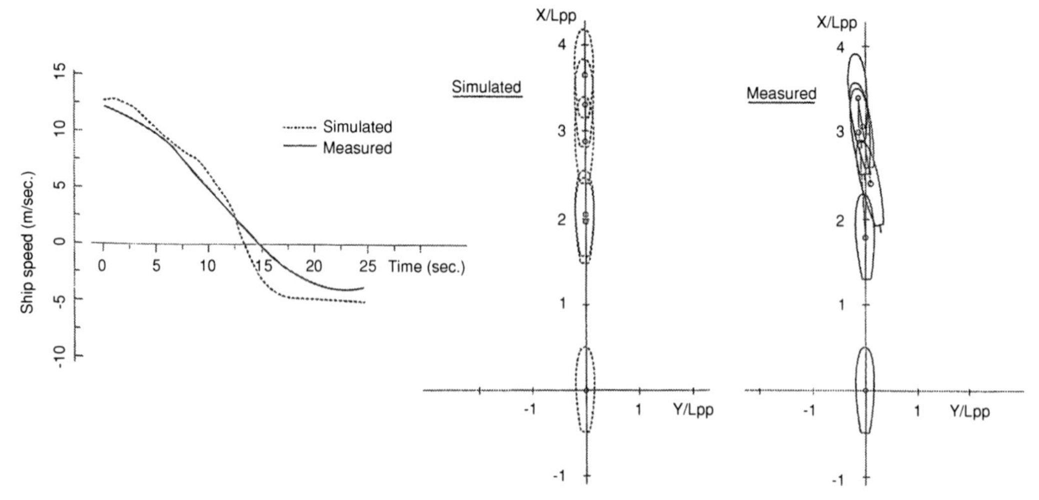

Figure 10.20. Full-scale crash astern test with a 30 m–long version of the twin-hull vessel SSTH. Simulated and measured time history records are compared (Ishiguro et al. 1993).

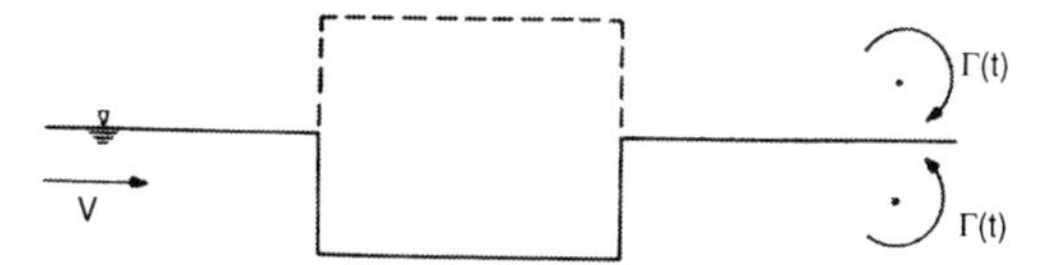

Figure 10.21. Simple vortex system with an image flow above the free surface so that the rigid free-surface condition $\partial\varphi/\partial z = 0$ on $z = 0$ is satisfied. $\Gamma(t)$ = circulation, V = inflow (ambient) velocity.

problem is considered. Two-dimensional flow is first assumed. Three-dimensional effects are considered at the end.

1. Free-surface effects

The free surface at moderate Froude number tends to act as an infinitely long splitter plate. There is, of course, a difference because of the boundary layer on the splitter plate. However, there is no cross-flow either at the splitter plate or at the free surface. Hoerner (1965) refers to C_D-values for bodies with splitter plates of finite length in steady incident flow. The splitter plate causes a clear reduction of the drag coefficient.

A simple explanation of why the free-surface presence affects the drag coefficient can partly be given by means of Figure 10.21. The shed vorticity is represented by one single vortex of strength Γ, which is a function of time. To account for the free-surface effect, one has to introduce an image vortex. This ensures zero normal velocity on the free surface (see Figure 2.13 for a more complete picture of vortices). If the splitter plate (free surface) had not been there, instabilities would cause a Karman vortex street to develop behind the double body. The image vortex illustrated in Figure 10.21 has a stronger effect on the motion of the real vortex than the vortices in a Karman vortex street behind the double body have on each other. Because there is a connection between the velocities of the shed vortices and the force on the body, we can understand why the free surface influences the drag coefficient.

In the case of oscillating ambient flow at low amplitudes, the eddies will stay symmetric for the double body without a splitter plate. This means the free surface has the same effect in this case. However, we should note that the drag coefficient for ambient oscillatory flow with small amplitude is larger than that for steady incident (ambient) flow.

2. Beam-to-draft ratio effects

Experimental results by Tanaka et al. (1982) show only a small effect of the height-to-length ratio on the drag coefficient for two-dimensional cross sections of rectangular forms. One exception is for small height-to-length ratios. If one translates the results to ship cross sections, it implies that the beam-to-draft ratio B/D has a small influence on the drag coefficient when $B/D > 0.8$.

3. Bilge radius effects

Experimental results by Tanaka et al (1982) show a strong effect of the bilge radius r on the drag coefficient. The bilge radius influence on C_D appears as $C_D = C_1 e^{-kr/D} + C_2$, where C_1 and C_2 are constants of similar magnitude and D is the draft. As an example, k may be 6. Therefore, the increase of bilge radius will cause a substantial decrease of the drag coefficient. This effect is less relevant for high-speed hulls.

4. Effect of laminar or turbulent boundary-layer flow

The classical results for steady-state flow past a circular cylinder in infinite fluid and with a constant incident flow show the existence of a critical Reynolds number. The boundary layer flow is laminar below the critical Reynolds number. In the supercritical and transcritical ranges, the boundary-layer flow is turbulent. As a consequence, the location of the separation points is substantially different in the subcritical and transcritical Reynolds number ranges. A further consequence is a difference in drag coefficient. For a smooth cylinder, the critical Reynolds number is $2{\cdot}10^5$. In this case, Reynolds number is defined as VD/ν, where V is the ambient cross-flow velocity, D is the diameter, and ν is the kinematic viscosity coefficient. By increasing the roughness of the cylinder surface, the critical Reynolds number will decrease. One often finds a situation in which model tests have to be performed in the subcritical range, whereas the full-scale situation is in the transcritical range. However, when the separation occurs from sharp corners one would expect less-severe scale effects.

Aarsnes (1984, see also Aarsnes et al. 1985) has shown that the drag coefficient may be substantially different depending on laminar or turbulent separation. This is also evident from Delany and Sorensen's (1953) results. Aarsnes's results are for

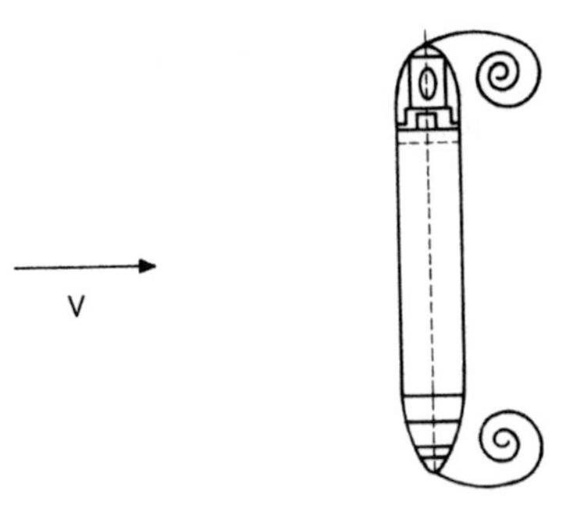

Figure 10.22. Sketch of the vertical vortex system at ship ends present in steady incident cross-flow past the ship. V is inflow (ambient) velocity.

ship cross sections. There are, in general, significant differences between the drag coefficient in subcritical and transcritical flows. The reason is that the flow separates more easily in the subcritical regime. In subcritical flow conditions, the boundary layer is laminar and the flow separates at the "leading" bilge. In contrast, in turbulent boundary layers, which occur in the transcritical regime, the flow can sustain a larger adverse pressure gradient without separating. This is the reason there is no separation at the leading bilge for transcritical flow. If separation occurs at both corners, the drag coefficient is, roughly speaking, twice the value we have when separation occurs at one corner only.

5. Three-dimensional effects

Aarsnes (1984) pointed out that three-dimensional effects at the ship ends will reduce the drag force compared with a pure strip theory approach. One way of taking this into account would be to use a reduced effective incident flow at ship ends, as predicted in a qualitative way by Aarsnes. Physically, the reduced inflow velocity is caused by the eddies at the ship ends (Figure 10.22). The effective reduced inflow can be translated into a reduction factor of the two-dimensional drag coefficient. The reason is that the forces on a cross section are proportional to the square of the local inflow velocity, that is, v_L^2, and that the drag coefficient is normalized by the square of the global inflow (ambient) velocity, that is, V^2. The reduction factor is simply v_L^2/V^2. This reduction factor is exemplified in Figure 10.23 and should be multiplied by the two-dimensional results to obtain the correct local two-dimensional solution. The local two-dimensional results can be added together by a strip theory approach to find the three-dimensional results.

10.6.2 2D+t theory

When the forward speed of the ship is high relative to the transverse velocity component along the vessel, it is relevant to consider a 2D + t approach, where t indicates the time variable. In order to explain the procedure, we start out with a 2D time-dependent cross-flow past a fixed circular cylinder of radius R in infinite fluid. High Reynolds number and laminar boundary layer (subcritical) flow are assumed. The fluid is at rest at time $t = 0^-$. At time $t = 0$, an ambient flow velocity V is assumed that will thereafter remain constant in time. Under these conditions, it takes a nondimensional time $Vt/R = 0.351$ (Schlichting 1979) before flow separation starts. Separation occurs first at the downstream stagnation point on the cylinder. The separation points then move rapidly to the vicinity of separation points for steady-state conditions. The importance of *pressure drag forces* relative to *viscous shear forces* increases with time after the flow separation has started. Figure 10.24 shows the experimental flow path of particles in the wake of an impulsively started cylinder for $Vt/R = 2,3,4,$ and 5. The Reynolds number is $VD/\nu = 5000$. We note that the wake is symmetric; that is, no asymmetric vortex shedding has developed. As time increases in Figure 10.24, we see the development of two large symmetric eddies in the wake.

Sarpkaya (1966) presented experimental results for drag coefficients C_D for a circular cylinder

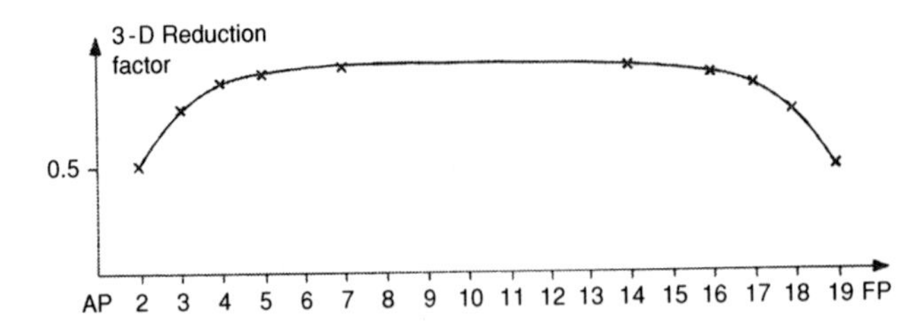

Figure 10.23. Examples of the three-dimensional reduction factor of local drag coefficient due to the vertical vortex system at ship ends described in Figure 10.22. FP and AP mean forward and aft perpendicular, respectively (Adapted from Aarsnes et al. 1985).

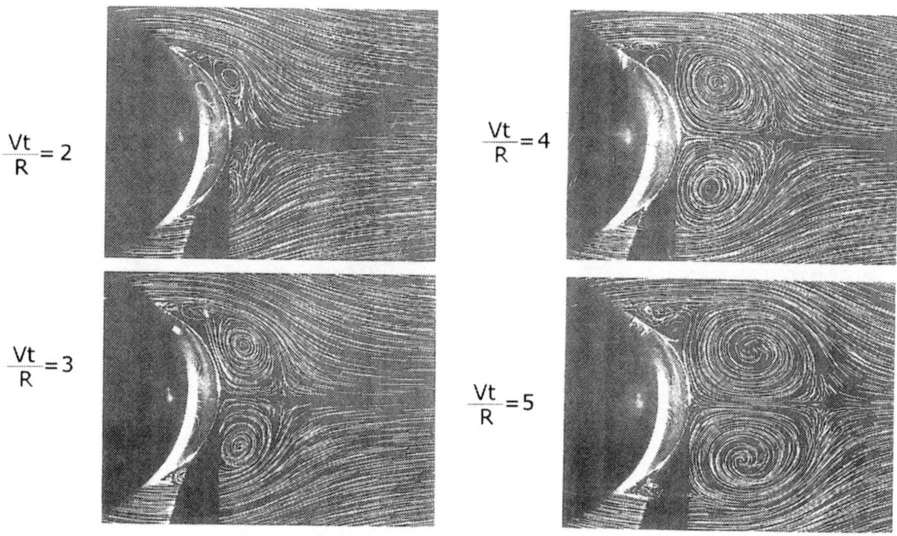

Figure 10.24. Early stage of development of the wake behind an impulsively started cylinder. $Rn = VD/\nu = 5000$ (Bouard and Coutanceau 1980).

in infinite fluid during nearly impulsively started (transient) laminar flow (Figure 10.25). Because there is a finite acceleration time during the experiments, hydrodynamic forces in phase with acceleration are present initially. Morison's equation (Morison et al. 1950) with a mass coefficient $C_M = 2$ was used to subtract the acceleration-dependent term from the experimental force record.

The Morison equation expresses the transverse force per unit length on a nonmoving circular cylinder as

$$f = \rho \frac{\pi D^2}{4} C_M a_1 + \frac{\rho}{2} C_D D |V| V, \qquad (10.63)$$

where D is the cylinder diameter. V and a_1 are the undisturbed (ambient) cross-flow fluid velocity and acceleration at the cylinder axis, respectively. The force direction coincides with the direction of the ambient fluid velocity.

The results in Figure 10.25 are presented as a function of

$$\frac{s}{R} = \int_{t_0}^{t} V \, dt / R. \qquad (10.64)$$

Here s corresponds to the longitudinal motion of a particle in the incident flow during the time interval $t - t_0$, where t_0 corresponds to the initial time of flow separation. This is a function of dV/dt. The C_D-value has a maximum value around $s/R = 8$. The wake flow is symmetric, as the one shown in Figure 10.24 up to that time instant. Antisymmetric vortex shedding will start at a later stage. This is because symmetric wake flow is unstable. Small perturbations of the flow can cause the wake flow to be asymmetric even before $s/R = 8$. The time it takes for C_D to reach its maximum value and the time it takes to reach steady value are sensitive to the ambient flow velocity evolution (Sarpkaya and Shoaff 1979).

The C_D-value during steady-state conditions contains mean and time-dependent parts. The time-dependent part is small relative to the mean part and oscillates with twice the vortex shedding (Strouhal) frequency (Faltinsen 1990).

In order to facilitate the use of these experimental data in the subsequent 2D+t analyses, we assume an impulsively started laminar flow and make the following approximation of the experimental data:

$$C_D = p_1 t'^5 + p_2 t'^4 + p_3 t'^3 + p_4 t'^2 + p_5 t' + p_6, \qquad (10.65)$$

where $t' = (Vt/R - 0.351) \geq 0$. The nondimensional p_i-coefficients are given as $p_1 = 2.4805 \cdot 10^{-7}$, $p_2 = -3.647 \cdot 10^{-5}$, $p_3 = 1.9058 \cdot 10^{-3}$, $p_4 = -4.4173 \cdot 10^{-2}$, $p_5 = 4.3146 \cdot 10^{-1}$, and $p_6 = 7.3386 \cdot 10^{-2}$. When $0 \leq Vt/R \leq 0.351$, we set C_D equal to zero. Strictly speaking, the above representation of C_D is not correct. The experimental data are not for an impulsively started flow, and

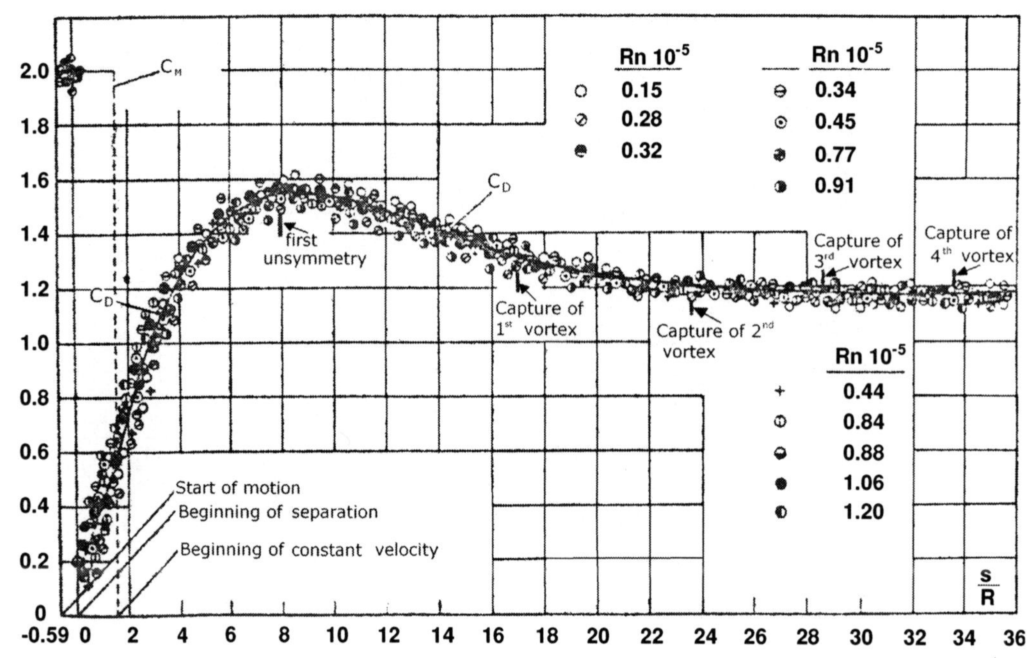

Figure 10.25. Transient drag coefficient C_D for a circular cylinder in nearly impulsively started laminar flow. The particle motion s of the ambient flow is defined by eq. (10.64). Experiments by Sarpkaya (1966).

C_D will differ from zero until flow separation starts. We should also recall that maximum C_D may in reality occur for a smaller s/R-value than that shown by Sarpkaya (1966) (see Figure 10.25). These effects can be illustrated by numerical simulations reported by Koumoutsakas and Leonard (1995). Initially, their predicted C_D-value is large and friction and pressure drag have equal importance. When flow separation is established, pressure drag is dominant. The results by Koumoutsakas and Leonard (1995) show a maximum C_D-value close to $Vt/R = 4$, that is, earlier than that shown in Figure 10.25.

Let us now consider a circular cylinder of length L and forward speed U in infinite fluid. We have made a streamlined bow to avoid flow separation at the bow when this body has a straight course (Figure 10.26). We then give to the body a constant sway velocity at time $t = 0$. If we see the flow from the body reference frame, there is then an incident cross-flow velocity V. If we see the motion of the shed vorticity from an Earth-fixed coordinate system, the vorticity will mainly move in the transverse direction of the body. The consequence is that different cross sections of the body experience different vorticity fields in the cross-sectional plane and therefore different C_D-values. Figure 10.27 illustrates how the vortex pattern looks according to a 2D+t theory.

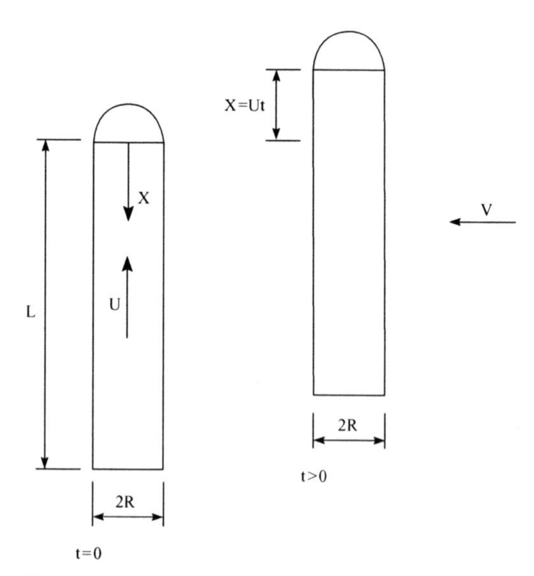

Figure 10.26. A slender body in infinite fluid consisting of a bow part and a circular cylindrical part with radius R and length L. The forward speed is U. At time $t \geq 0$, there is a constant incident cross-flow velocity V.

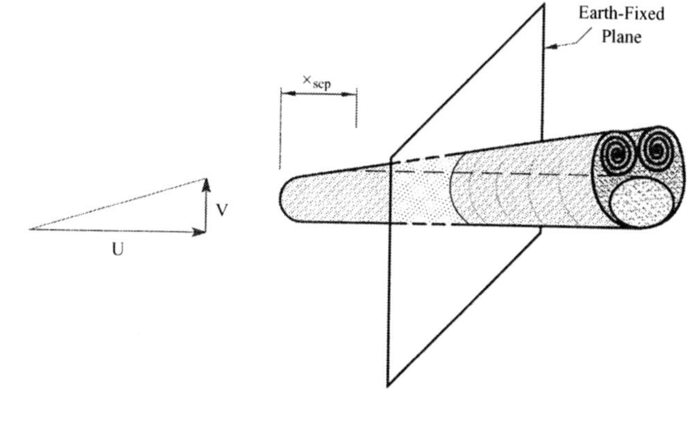

Figure 10.27. 2D+t analysis of a cylinder with constant forward velocity U and incident cross-flow velocity V. The cross-flow starts to separate at a longitudinal distance x_{sep} from the front end.

Let us quantify the transverse force and moment on the body by disregarding the streamlined bow part and applying eq. (10.65) for the circular cylinder part. We define a body-fixed longitudinal x-axis with $x = 0$ in the front of the circular cylinder and with positive x pointing toward the stern (see Figure 10.26). Then we consider time t and the cross section at $x = Ut$. The vorticity field at that cross section must be the same as that developed during the cross-flow of the cylinder with incident velocity V at the same time t. Because the body is moving forward, the vorticity field in this cross-plane of the ship cannot further develop. This means we can obtain the C_D-value on this cross section by using eq. (10.65) with

$$s' = \frac{Vt}{R} = \frac{V}{R}\frac{x}{U} \tag{10.66}$$

and by using the assumption that $C_D = 0$ when $0 \le s' < 0.351$. We then get a total transverse drag force on the body that is equal to

$$\begin{aligned} F_2 &= \rho V^2 R \int_0^L C_D\left(\frac{V}{R}\frac{x}{U}\right) dx \\ &= \rho V U R^2 \int_0^{s_1} C_D(s')ds', \end{aligned} \tag{10.67}$$

where

$$s_1 = \frac{VL}{UR}. \tag{10.68}$$

If we had used the cross-flow principle, then the mean drag force would have been

$$F_2^{CF} = \rho 1.2 V^2 RL \tag{10.69}$$

based on using a steady-state mean C_D-value of 1.2. This gives

$$\frac{F_2}{F_2^{CF}} = \frac{\int_0^{s_1} C_D(s')ds'}{1.2 s_1}. \tag{10.70}$$

This ratio is presented in Figure 10.28. When $VL/(UR) < 5$, the transverse drag force is less than about 60% of the drag force based on the cross-flow principle. For instance, if $L/R = 20$, this means $V/U < 0.25$. We should realize that this approach does not account for damping forces due to potential flow contributions. The latter effect is included in the linear theory presented in section 10.3 and is the result of changing cross-sectional area. The results indicate that a cross-flow approach as expressed by eq. (10.61) and with drag coefficients for steady ambient flow will overestimate the transverse damping for increasing forward speed. We can also calculate the yaw moment about the midpoint of the cylinder. This will be zero according to the cross-flow principle. Using the 2D+t approach, we have

$$\begin{aligned} F_6 &= \rho V^2 R \int_0^L C_D\left(\frac{V}{R}\frac{x}{U}\right) x\, dx - \frac{L}{2}F_2 \\ &= \rho U^2 R^3 \int_0^{s_1} C_D(s')s'\, ds' - \frac{L}{2}F_2. \end{aligned} \tag{10.71}$$

This gives

$$\frac{F_6}{F_2 L} = \frac{\int_0^{s_1} C_D(s')s'ds'}{s_1 \int_0^{s_1} C_D(s')ds'} - 0.5. \tag{10.72}$$

The results are presented in Figure 10.28. We note that the moment approaches zero when $VL/(UR) \to \infty$. This is consistent with the cross-flow principle. Further, the center of pressure of

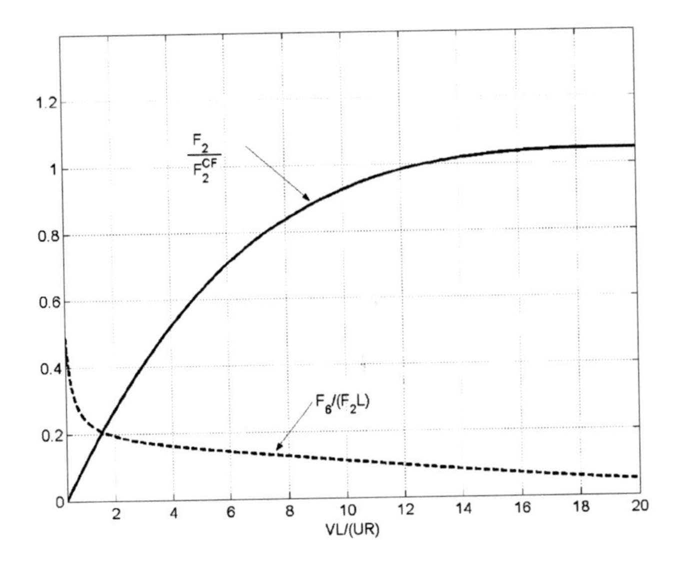

Figure 10.28. Sway force F_2 and yaw moment F_6 about COG of a circular cylinder with forward speed U and incident cross-flow velocity V in infinite fluid and with subcritical boundary-layer flow. Only the viscous force after flow separation is accounted for. R = cylinder radius, L = cylinder length, F_2 and F_6 are estimated by a 2D+t approach, F_2^{CF} = sway force based on cross-flow principle. Note that the horizontal axis with $VL/(UR)$ does not start from zero.

the drag force moves toward the transom stern when $VL/(UR) \to 0$.

The longitudinal distribution of C_D as a function of x/R is presented in Figure 10.29 for $V/U = 0.05$ and 0.2. This is obtained by using eq. (10.66) and eq. (10.65) for positive C_D-values and setting C_D equal to zero for other x-values. In reality, there will be a transverse viscous force for all x/R-values. When the flow separation has not occurred, the transverse viscous force will have a component that is linearly dependent on V for laminar flow.

For the results presented in Figure 10.29, the wake behind the cylinder is always symmetric, that is, like in Figure 10.24. If we interpret the cylinder as a double body consisting of a semicircular cylinder and its image above the free surface, it means that our results will implicitly account for free-surface effects for moderate Froude number, that is, when a rigid free surface is appropriate.

We want to emphasize that the results presented are for laminar boundary-layer flow. Turbulent boundary-layer flow will change the separation points, and this has a large influence on C_D-values.

We could use the 2D+t analysis qualitatively to estimate when the cross-flow principle is appropriate. Let us somewhat arbitrarily require C_D to have the 2D steady-state value for $x/R > 2$. Further, we assume that we are close to the steady-state C_D-value when $Vt/R > 10$, that is, $Vx/UR > 10$. This gives that V/U must be larger than 5.

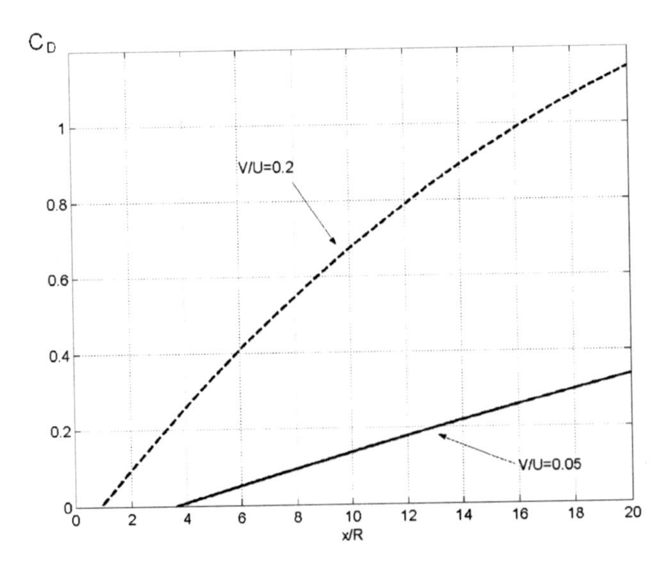

Figure 10.29. Longitudinal distribution of drag coefficient C_D along the body shown in Figure 10.26 for different ratios V/U between transverse and longitudinal body velocities. Only the viscous force after flow separation is accounted for. R = cylinder radius, $x = 0$ is at the front end of the circular cylinder. The results assume cross-flow with subcritical boundary layer.

The transverse ship velocity in reality will be oscillatory. A characteristic time scale T of the oscillations may be from half a minute to three minutes. If the ship has zero forward speed, The Keulegan-Carpenter (KC) number reflects the important effect of shed vorticity becoming incident to the ship during an oscillation cycle. If the ship has a forward speed, the shed vorticity at the forward part of the ship can be incident to an aft cross section of the vessel. Let us make a simplified analysis and assume $V = V_a \cos \omega t$, where $\omega = 2\pi/T$. The vorticity will as a first approximation be convected with V in the Earth-fixed coordinate system. We consider the shed vorticity at the forward part of the vessel corresponding to $t = 0$. This shed vorticity will be incident to the ship after the time $t = T/4$. The vessel has then moved forward a length $UT/4$. If this length is larger than the ship length L, we can be sure that this shed vorticity will not be incident to a cross section of the vessel. We therefore introduce

$$KC_{UL} = UT/L \tag{10.73}$$

as a parameter. It has a similar physical meaning as the KC number for cross-flow past a stationary object. We have denoted it as KC_{UL} to indicate that the velocity U and the length L are the characteristic velocity and length. The ratio between $KC = V_a T/(2R)$ and KC_{UL} is proportional to the angle of attack V_a/U which is an important parameter.

We could have done a 2D+t analysis similar to the one we did for sway by studying the combined sway and yaw motions of the vessel. However, for this to be of any practical use, we have to consider all the flow parameters mentioned previously. Further, we have to account for the change of the cross section shape along the ship. As a result, the 2D+t method should solve the time-dependent 2D Navier-Stokes equations for cross sections varying with time.

Nonaka (1993) combined a 2D+t method with a vortex method to represent the cross-flow wake. A difficulty is to correctly predict the separation points with a vortex method. Nonaka (1993) chose instead to specify where separation occurs. Oblique towing of realistic ships at moderate Froude number was considered. Good agreement between numerical and experimental results for transverse force and yaw moment up to a drift angle β (see Figure 10.4) of 20° was documented for both finite and infinite water depths. Obviously one can also numerically solve the maneuvering problem by the 3D Navier-Stokes equation (see e.g., Fujino 1996).

10.6.3 Empirical nonlinear maneuvering models

One way to introduce nonlinear terms into a maneuvering model as described by eqs. (10.1) and (10.2) is to use a Taylor expansion in terms of the variables of interest. The notation used in eqs. (10.1) and (10.2) is based on a Taylor expansion of the hydrodynamics sway force Y and yaw moment N in $v, r,$ and δ. However, only linear terms are kept in eqs. (10.1) and (10.2). If we now consider the results in Figure 10.28, we should interprete F_2 as Y and F_6 as N. A formal Taylor expansion in v gives terms like

$$Y = Y_0 + Y_v v + Y_{vv} v^2 + Y_{vvv} v^3 + \cdots \tag{10.74}$$

This requires that $Y_{vv} = \frac{1}{2}\partial^2 Y/\partial v^2 |_{v=0}$ and $Y_{vvv} = \frac{1}{6}\partial^3 Y/\partial v^3 |_{v=0}$. The transverse viscous force is antisymmetric in v. It means that Y_0 and Y_{vv} are zero. The lowest-order nonlinear term is the result of Y_{vvv}. The results in Figure 10.28 should be supplemented by a linear viscous term in V as well as a linear potential flow term, as predicted by slender body theory. This can be generalized to account for the viscous influence of r on Y as well as the viscous influence of v and r on the yaw moment N.

An alternative way to express the influence of viscous flow is to write

$$Y = Y_v v + Y_r r + Y_{v|v|}\, v\,|v| + Y_{v|r|}\, v\,|r| + Y_{|v|r}\, |v|\, r + Y_{r|r|}\, r\,|r| \tag{10.75}$$

$$N = N_v v + N_r r + N_{v|v|}\, v\,|v| + N_{v|r|}\, v\,|r| + N_{|v|r}\, |v|\, r + N_{r|r|} r|r|. \tag{10.76}$$

This implies that N and Y are antisymmetric with respect to v and r. Eqs. (10.75) and (10.76) are more consistent than eq. (10.74) regarding how drag formulas are expressed. However, it is not clear if eqs. (10.75) and (10.76) are the physically correct way to represent N and Y. Even if we assumed the valid cross-flow formulation given by eqs. (10.61) and (10.62), it would not be straightforward to find the link with eqs. (10.75) and (10.76), except in the case in which either v or r is equal to zero.

The coefficients in eqs. (10.75) and (10.76) have to be experimentally determined, for example by PMM (planar motion mechanism) tests. A difficulty is to account properly for nondimensional flow parameters. Examples are the influence of Reynolds number, the Keulegan-Carpenter numbers, and the Froude number. Further, cavitation and ventilation may also matter for high-speed vessels.

Before starting to estimate all nonlinear terms, it is important to have an idea of their influence on ship motion, that is, what accuracy is needed in estimating the different coefficients. If $v' = v/U$ and $r' = rL/U$ are small, as they would be for a ship at high speed, the effect of viscous flow on the linear terms Y_v, Y_r, N_v, and N_r is the most important.

10.7 Coupled surge, sway, and yaw motions of a monohull

We have in the previous sections assumed constant forward speed and discussed coupled sway and yaw motions. The effect of varying longitudinal vessel velocity is now investigated. A monohull at moderate speed and equipped with a propeller and rudder is considered.

Søding (1984) formulated the longitudinal component of Newton's second law in the body-fixed coordinate system as

$$M(\dot{u} - v\dot{\psi}) = X_{\dot{u}}\dot{u} - R_T(u) + (1-t)\,T(u,n) + X_{vv}v^2 + X_{v\psi}v\dot{\psi} + X_{\psi\psi}\dot{\psi}^2 + X_{\delta\delta}\delta^2. \quad (10.77)$$

The notation is consistent with Figure 10.4, and derivative terms such as $X_{\dot{u}}$, X_{vv}, ... are used to express longitudinal hydrodynamic forces on the hull and the rudder. u and v are the longitudinal and transverse velocity of the center of gravity of the vessel in the body-fixed coordinate system. Further,

- $\dot{\psi} = r$ = yaw angular velocity
- R_T = ship resistance
- t = thrust deduction coefficient
- T = propeller thrust based on open-water propeller characteristics
- n = number of propeller revolutions per second
- δ = rudder angle

The mass term $-Mv\dot{\psi}$ on the left-hand side of eq. (10.77) is a consequence of formulating Newton's second law in the body-fixed coordinate system. This can be shown similarly to how the term $-MUd\eta_6/dt$ in eq. (10.37) was derived by means of Figure 10.9. A more general derivation of the mass terms in a body-fixed coordinate system is presented in section 10.9, in which we consider motions in six degrees of freedom.

If we assume an incompressible fluid with irrotational motion, no circulation, and a rigid free-surface condition, the coefficients X_{vv}, $X_{v\psi}$, and $X_{\psi\psi}$ in eq. (10.77) can be theoretically determined by infinite fluid results by Kochin et al. (1964). We can show the equivalence to the infinite fluid problem by introducing an image ship with respect to the free surface (see Figure 2.9). Because the ship velocity is in a horizontal plane and the rigid free-surface condition is assumed, the flow around the double body consisting of the ship and the image ship will correctly describe the flow around the ship. Using general expressions presented by Kochin et al. (1964) (see section 10.10.2) and accounting for the fact that the x-z–plane is a symmetry plane for the hull gives, according to eq. (10.159),

$$X_{vv} = 0, \quad X_{v\psi} = A_{22}, \quad X_{\psi\psi} = A_{26}. \quad (10.78)$$

Here A_{22} and A_{26} are the low-frequency added mass in sway and coupled added mass between sway and yaw, respectively. A_{22} and A_{26} can, for instance, be calculated by a 3D boundary element method (BEM) or by a strip (slender body) theory, as shown in section 10.3. If the ship has fore-and-aft symmetry about the y-z–plane, A_{26} or $X_{\psi\psi}$ will be zero according to potential flow. Because A_{22} is the order of magnitude of the ship mass, we see from eq. (10.77) that the term $X_{v\psi}v\dot{\psi}$ on the right-hand side is of equal importance as the mass term $-Mv\dot{\psi}$ on the left-hand side. Norrbin (1971) reported that experimental values of $X_{v\psi}$ may be as low as 20% to 50% of the theoretical value based on eq. (10.78). To what extent this is true for a high-speed slender hull at moderate speed needs experimental documentation.

A main cause of speed loss in a turning motion is the terms $-Mv\dot{\psi}$ and $X_{v\psi}v\dot{\psi}$. In order to see that they cause a speed loss, we keep both terms on the right-hand side of eq. (10.77). We can write $u = R\dot{\psi}$, where R is the radius of curvature of the ship's path. Further, $v \approx -u\beta$ according to eq. (10.7) for small drift angles β. This gives the following two terms on the right-hand side of the modified eq. (10.77):

$$(M + X_{v\dot{\psi}})v\dot{\psi} = -(M + X_{v\dot{\psi}})\frac{u^2}{R}\beta. \qquad (10.79)$$

Here $u^2\beta/R$ is the x-component of the centrifugal acceleration. Because the ship proceeds with the bow pointing inward in a steady turn, β is positive (see Figure 10.11). This means that $(M + X_{v\dot{\psi}})v\dot{\psi}$ causes a resistance, that is, a speed loss. The term $X_{\delta\delta}\delta^2$ in eq. (10.77) is also important to consider in assessing the speed loss due to maneuvering.

We need to determine v, $\dot{\psi}$, and δ in order to solve eq. (10.77) for u. Two additional equations follow by considering the transverse component of Newton's second law and the equation for yaw angular momentum. These equations are discussed in detail for linear sway and yaw in section 10.3. Nonlinear terms in viscous transverse force and yaw moment due to sway and yaw velocities are handled in section 10.6. If the control system for maneuvering is based on a PID regulator, the rudder angle is expressed by eq. (10.59).

In the following section, we further discuss the coupled surge-sway-yaw equations.

10.7.1 Influence of course control on propulsion power

Søding (1984) studied the influence of course control on the propulsion power $P = 2\pi nQ$ for a monohull at moderate speed. Here Q is the propeller torque, which is a function of the instantaneous values of u and n. In order to estimate the added power, we need to calculate the time averages of the time fluctuations in u and n. Eq. (10.77) is used as a part of the analysis. The calm water resistance was estimated by empirical formula by Auf'm Keller (1973). X_{vv}, $X_{v\dot{\psi}}$, and $X_{\dot{\psi}\dot{\psi}}$ were experimentally obtained. The rudder control algorithm was here expressed as a PD (proportional derivative) controller according to

$$\delta = a\psi + b\dot{\psi}, \qquad (10.80)$$

and the influence of the control factors a and b was studied (see Fossen (2002) about more details concerning automatic control). The deviations from a straight course were assumed to be small so that the linearized coupled sway and yaw equations could be applied. A time-harmonic excitation force in sway and yaw was assumed. The linearized coupled sway and yaw equations (see eqs. (10.1) and (10.2)) can, by using eq. (10.80), be expressed in matrix form as

$$\mathbf{A}\ddot{\mathbf{x}} + \mathbf{B}\dot{\mathbf{x}} + \mathbf{C}\mathbf{x} = \hat{\mathbf{s}}e^{i\omega t}. \qquad (10.81)$$

Here

- $\mathbf{x}$ = motion vector = $[y, \psi]^T$. Further, y and ψ follow from the equations $\dot{y} = v$ and $\dot{\psi} = r$, respectively.
- $\hat{\mathbf{s}} = [\hat{Y}_S, \hat{N}_S]^T$, where $\hat{Y}_S$ and $\hat{N}_S$ are complex amplitudes of environmental external lateral force and yaw moment.

The superscript T indicates matrix transposition. Further,

$$\mathbf{A} = \begin{bmatrix} M - Y_{\dot{v}} & -Y_{\dot{r}} \\ -N_{\dot{v}} & I_{66} - N_{\dot{r}} \end{bmatrix} \qquad (10.82)$$

$$\mathbf{B} = \begin{bmatrix} -Y_v & (M - X_{\dot{u}})u - Y_r - bY_\delta \\ -N_v & -N_r - bN_\delta \end{bmatrix} \qquad (10.83)$$

$$\mathbf{C} = \begin{bmatrix} 0 & -aY_\delta \\ 0 & -aN_\delta \end{bmatrix}. \qquad (10.84)$$

In eq. (10.83), we use u instead of U. This is consistent with linear theory.

Søding (1984) determined the hydrodynamic derivative terms in eq. (10.81) by the method described by Søding (1982). This means that the hull coefficients were derived by slender body theory with correction factors. The rudder force is determined by the lifting line theory that accounts for the propeller slip stream and hull-rudder interaction.

Assuming a steady-state response, $\mathbf{x} = \hat{\mathbf{x}}e^{i\omega t}$ gives that the solution of eq. (10.81) can be expressed as

$$\hat{\mathbf{x}} = \left[-\mathbf{A}\omega^2 + i\omega\mathbf{B} + \mathbf{C}\right]^{-1}\hat{\mathbf{s}}. \qquad (10.85)$$

Here the superscript -1 indicates matrix inversion.

The presence of the restoring matrix $\mathbf{C}$ due to the control factor a in combination with the mass matrix $\mathbf{A}$ means that the control algorithm introduces natural periods for the sway and yaw motions. The control factor b has a damping effect and can be used to minimize the resonance.

When there is no excitation $\hat{\mathbf{s}}$, the ship is assumed to have a straight course along the

X_E-axis with the heading ψ equal to zero (see Figure 10.4). The longitudinal speed is $u = u_0$, and the number of propeller revolutions per second is $n = n_0$. When $\hat{\mathbf{s}}$ is different from zero, the resistance R_T and propeller thrust T is expressed by the Taylor expansion

$$R_T(u) = R_T(u_0) + \left.\frac{dR_T}{du}\right|_{u=u_0} (u - u_0) + \cdots \quad (10.86)$$

$$T(u, n) = T(u_0, n_0) + \left.\frac{\partial T}{\partial u}\right|_{\substack{u=u_0\\n=n_0}} (u - u_0) + \left.\frac{\partial T}{\partial n}\right|_{\substack{u=u_0\\n=n_0}} (n - n_0) + \cdots \quad (10.87)$$

The derivative terms dR_T/du, $\partial T/\partial u$, and $\partial T/\partial n$ can be obtained by numerically differentiating the known curves of $R_T(u)$ and $T(u, n)$.

In order to find the influence of course control on the propulsion power, we need to find the time average eq. (10.77). Using eqs. (10.86) and (10.87) and introducing $\Delta u = u - u_0$ and $\Delta n = n - n_0$ give

$$\begin{aligned} M\left(\bar{\dot{u}} - \overline{v\dot{\psi}}\right) &= X_{\dot{u}}\bar{\dot{u}} - R_T(u_0) - \frac{dR_T}{du}\overline{\Delta u} + (1-t) \\ &\quad \times \left[T(u_0, n_0) + \frac{\partial T}{\partial u}\overline{\Delta u} + \frac{\partial T}{\partial n}\overline{\Delta n}\right] \\ &\quad + X_{vv}\overline{v^2} + X_{v\dot{\psi}}\overline{v\dot{\psi}} + X_{\dot{\psi}\dot{\psi}}\overline{\dot{\psi}^2} + X_{\delta\delta}\overline{\delta^2}, \end{aligned} \quad (10.88)$$

where the bar over the different terms indicates time average. For instance,

$$\bar{\dot{u}} = \frac{1}{T_e}\int_0^{T_e} \dot{u}\, dt = \frac{1}{T_e}\left[u(T_e) - u(0)\right], \quad (10.89)$$

where T_e is the excitation period. Because a steady-state solution is considered, $u(T_e) = u(0)$; that is, $\bar{\dot{u}} = 0$. The terms $\overline{v^2}$, $\overline{v\dot{\psi}}$, and $\overline{\dot{\psi}^2}$ can be found by using $\hat{\mathbf{x}}$ given by eq. (10.85) with $u = u_0$. The term $\overline{\delta^2}$ follows by using eq. (10.80). It should be noted that it is only the real part of $\hat{\mathbf{x}}e^{i\omega t}$ that has physical meaning. This real part must be taken before time averaging the quadratic expressions in v and $\dot{\psi}$. Let us show how $\overline{v\dot{\psi}}$ can be expressed by writing y as $y_a \cos(\omega t + \varepsilon_y)$ and ψ as $\psi_a \cos(\omega t + \varepsilon_\psi)$. This gives

$$\begin{aligned} \overline{v\dot{\psi}} &= \omega^2 y_a \psi_a \overline{\sin(\omega t + \varepsilon_y)\sin(\omega t + \varepsilon_\psi)} \\ &= \omega^2 y_a \psi_a \frac{1}{2}\overline{[\cos(\varepsilon_y - \varepsilon_\psi) - \cos(2\omega t + \varepsilon_y + \varepsilon_\psi)]} \\ &= \omega^2 y_a \psi_a \frac{1}{2}\cos(\varepsilon_y - \varepsilon_\psi). \end{aligned}$$

The same can be done with the other terms. Eq. (10.88) can be simplified by using the fact that the resistance balances the propeller thrust at $u = u_0$ and $n = n_0$; that is, $-R_T(u_0) + (1-t)\,T(u_0, n_0) = 0$.

There are two unknowns, $\overline{\Delta u}$ and $\overline{\Delta n}$, in eq. (10.88). An additional equation is therefore needed. The term $\overline{\Delta u}$ expressing the time average of the longitudinal speed oscillations can be found by considering the vessel speed dX_E/dt along the Earth-fixed X_E-axis in Figure 10.4. This is expressed by eq. (10.8). The time average of $dX_E(t)/dt$ is u_0. Further, noting that ψ is assumed to be small and Taylor expanding $\cos\psi$ and $\sin\psi$ in eq. (10.8) gives

$$\begin{aligned} u_0 &= \overline{(u_0 + \Delta u)\left(1 - \frac{1}{2}\psi^2\right)} - \overline{v\psi} \\ &\approx u_0 + \overline{\Delta u} - \frac{1}{2}u_0\overline{\psi^2} - \overline{v\psi}. \end{aligned}$$

This means

$$\overline{\Delta u} = \frac{1}{2}u_0\overline{\psi^2} + \overline{v\psi}. \quad (10.90)$$

Eq. (10.88) with using balance between the resistance and propeller thrust at $u = u_0$ and $n = n_0$ determines $\overline{\Delta n}$, that is,

$$\begin{aligned} \overline{\Delta n} = \Big[&-M\overline{v\dot{\psi}} + \frac{dR_T}{du}\overline{\Delta u} - (1-t) \\ &\times \frac{\partial T}{\partial u}\overline{\Delta u} - X_{vv}\overline{v^2} - X_{v\dot{\psi}}\overline{v\dot{\psi}} - X_{\dot{\psi}\dot{\psi}}\overline{\dot{\psi}^2} \\ &- X_{\delta\delta}\overline{\delta^2}\Big] \Big/ (1-t)\frac{\partial T}{\partial n}. \end{aligned} \quad (10.91)$$

When $\overline{\Delta u}$ and $\overline{\Delta n}$ are known, the added propulsion power is determined from the propeller torque Q as a function of u and n. Q is known from the open-water propeller characteristics. The propulsion power P is $2\pi nQ$. The mean added propulsion power is

$$\overline{\Delta P} = \frac{\partial P}{\partial u}\overline{\Delta u} + \frac{\partial P}{\partial n}\overline{\Delta n}, \quad (10.92)$$

where $\partial P/\partial u$ and $\partial P/\partial n$ are numerically evaluated at $n = n_0$ and $u = u_0$.

Søding (1984) applied the described procedure in a case study with the Mariner ship. Representative wind gust loads for Beaufort 8 to 9 were used to express the complex excitation vector $\hat{\mathbf{s}}$ in eq. (10.81). The excitation period T_e was varied between 27 s and 566 s, and the influence of the control factors a and b in eq. (10.80) was investigated. The added propulsion power had a typical resonant behavior that depended on a, b, and the

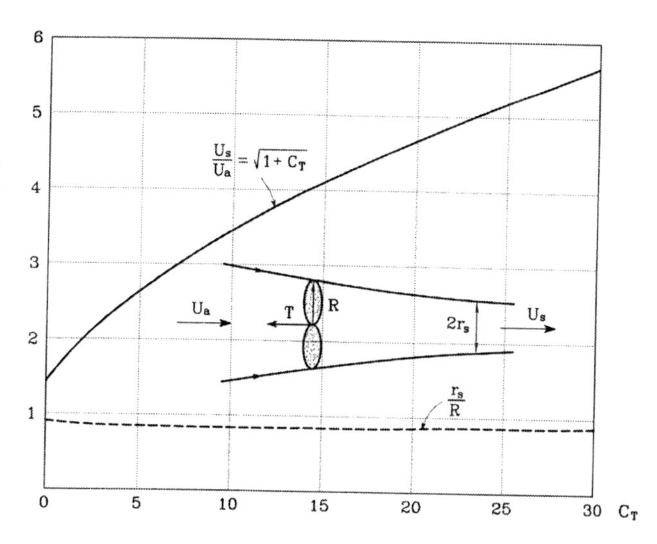

Figure 10.30. The axial slip stream velocity U_S behind a propeller in infinite fluid divided by the inflow velocity U_a to the propeller as a function of the thrust coefficient $C_T = T/(0.5\rho U_a^2 \pi R^2)$, where T is the propeller thrust and R is the propeller radius. Also shown is the slip stream radius r_s. The calculations are done by actuator disc theory, assuming axially symmetric flow. Note that the horizontal axis with C_T does not start with zero.

excitation period T_e. This is a consequence of our comment following eq. (10.85) and the fact that $\overline{\Delta u}$ and $\overline{\Delta n}$ are functions of ψ and v. The added propulsion power could be as much as the order of 20% of the power in calm water conditions without wind. Søding (1984) also showed how to generalize the procedure to include the effect of a wind gust spectrum. However, we will not discuss this here.

10.8 Control means

Figures 2.1 and 2.2 show examples of propeller-rudder arrangements for high-speed vessels. The rudders in Figure 2.2 are twisted and adapted to the propeller slip stream. Cavitation and ventilation are important concerns for high-speed rudders. This was discussed in section 6.10 and related to a rudder-foil system on a foil catamaran. It can also be demonstrated by Figures 6.58 and 10.14 that free-surface wave generation, that is, Froude number, has an important effect on a free surface–piercing rudder.

If waterjet propulsion is used to steer the vessel (see Figure 2.56), the consequences are reduced thrust and reduced maximum operating speed. An alternative is to use interceptors as described in section 7.1.3. This is more efficient at Froude numbers larger than 0.3 to 0.4. Another alternative is to use a high-speed rudder as illustrated in Figure 7.5. The effect of the hull must then be considered in evaluating the steering force and moment. If the hull boundary layer is neglected and the hull surface is approximated as horizontal, the hull can be accounted for by introducing an image rudder about the horizontal hull surface, that is, similar to the one illustrated for a ship at low Froude number in Figure 2.9. The lift coefficient can be estimated by considering a fictitious foil in infinite fluid consisting of the rudder and the image rudder. Because this increases the aspect ratio relative to the rudder in infinite fluid, the lift coefficient C_L of the rudder is increased because of the presence of the hull. Procedures on how to calculate C_L are described in Chapter 6; see for example, eq. (6.130). We must also account for the change in the inflow velocity due to the waves generated by the vessel. The ship wave calculations are then done by neglecting the presence of the rudder.

Let us consider as another case a rudder operating behind a propeller. The rudder is in the centerplane of the vessel. We use the coordinate system with velocities, forces, moments, and rudder angle δ defined in Figure 10.4. The centroid of the rudder has a longitudinal coordinate x_R.

We can use eqs. (2.142), (2.143), and (2.144) to show how the inflow velocity U_S to a rudder behind a propeller increases over the height $2r_s$. This is illustrated in Figure 10.30, which shows U_S/U_a and r_S/R as a function of the thrust-loading coefficient $C_T = T/(0.5\rho\pi R^2 U_a^2)$. Here U_a, R, and T are the inflow velocity to the propeller, the propeller radius, and the propeller thrust, respectively. We can express U_a as $U(1 - \bar{w})$, where $\bar{w}$ is the mean wake fraction and U is the vessel velocity. Because a rudder provides steering force and moment through lift and the lift force and moment increase with the square of the incident flow velocity in noncavitating and nonventilating conditions,

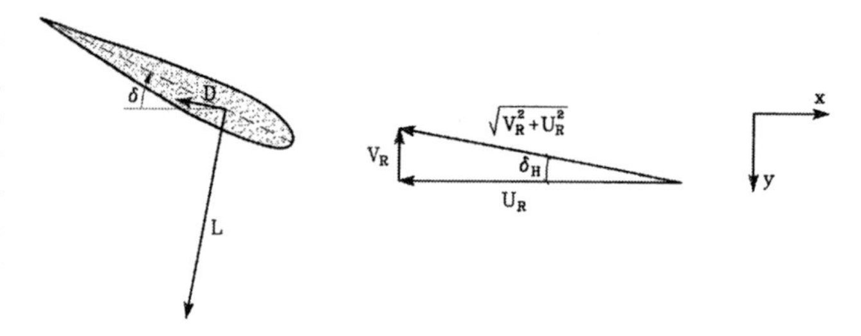

Figure 10.31. An all-movable rudder with a rudder angle δ as defined in Figure 10.4. There is an incident flow velocity $\sqrt{u_R^2 + v_R^2}$ to the rudder with an angle of attack δ_H. L and D are the rudder lift and drag force, respectively (see also Figure 2.16). The body-fixed coordinate system (x, y) is consistent with Figure 10.4.

we can understand that the propeller slip stream is beneficial for the effectiveness of the rudder. However, the calculations in Figure 10.30 neglect the swirling flow in the propeller slip stream and do not consider the effect of the free surface and the vessel. Further, the analysis requires the rudder to be, let us say, at least the order of a propeller radius behind the propeller (Søding 1982).

We illustrate how to evaluate forces and moments on the vessel due to an all-movable rudder behind a propeller. As seen from the rudder, there is an incident flow velocity $\sqrt{u_R^2 + v_R^2}$ with components u_R and v_R along the body-fixed negative x- and y-axes, respectively (Figure 10.31). Here u_R will vary along the rudder axis and is U_s if we consider a rudder cross section within the propeller slip stream. Otherwise u_R is U_a. We introduce an averaged u_R^2 instead of using a varying u_R^2. One way of doing this is by using the rudder planform area parts inside and outside the propeller slip stream as weighting factors, that is,

$$u_R = \sqrt{\left[A_{RS} U_S^2 + (A_R - A_{RS})\, U_a^2\right] \big/ A_R}, \quad (10.93)$$

where

A_R = rudder planform area

A_{RS} = rudder planform area within propeller slip stream

The transverse (v_R) inflow velocity component along the negative y-axis as observed from the rudder can be expressed as

$$v_R = \gamma_v v + \gamma_r r x_R, \quad (10.94)$$

where x_R is the x-coordinate of the rudder centroid and γ_v and γ_r are flow rectification factors due to the hull and the propeller (Ankudinov et al. 1993). Neglecting this effect means that $\gamma_v = \gamma_r = 1$.

The incident flow causes an angle of attack δ_H relative to the rudder (see Figure 2.16 for a definition of angle of attack). We can write (see Figure 10.31)

$$\delta_H = \arctan(v_R/u_R).$$

The total or effective angle of attack δ_e of an all-movable rudder must also include the rudder angle, that is,

$$\delta_e = \delta - \delta_H. \quad (10.95)$$

The resulting force on the rudder can be decomposed into a lift (L) and a drag (D) component (see Figure 10.31). The lift is perpendicular to the inflow velocity direction and can be expressed as $0.5\rho A_R C_{LR}(u_R^2 + v_R^2)$, where C_{LR} is the lift coefficient of the rudder. The drag force, which is in line with the inflow direction, can be expressed as $0.5\rho A_R C_{DR}(u_R^2 + v_R^2)$, where C_{DR} is the drag coefficient.

The hydrodynamic longitudinal (X_R) and transverse (Y_R) force components and yaw moment (N_R) due to the rudder in the body-fixed coordinate system can then be expressed as

$$\begin{aligned} X_R &= \frac{\rho}{2} A_R \left(u_R^2 + v_R^2\right)\left(-C_{LR}\sin\delta_H - C_{DR}\cos\delta_H\right) \\ Y_R &= \frac{\rho}{2} A_R \left(u_R^2 + v_R^2\right)\left(C_{LR}\cos\delta_H + C_{DR}\sin\delta_H\right) \\ N_R &= Y_R x_R. \end{aligned} \quad (10.96)$$

Whicker and Fehlner (1958) did an extensive experimental study of the lift and drag C_{LR} and C_{DR} for rudders as a function of the angle of attack. A spatially uniform inflow velocity along the rudder span was considered. Investigations were made of the following: three aspect ratios, 1, 2, and 3; five section shapes; two tip shapes, faired and square; and three sweep angles (−8, 0, 11). Aspect ratios and sweep angles are defined by Figure 6.3. Tests were made in a low-speed wind tunnel, and the models were mounted on a ground board. If the rudder is free surface–piercing, then the ground board will simulate the effect of the free surface for small and moderate Froude numbers.

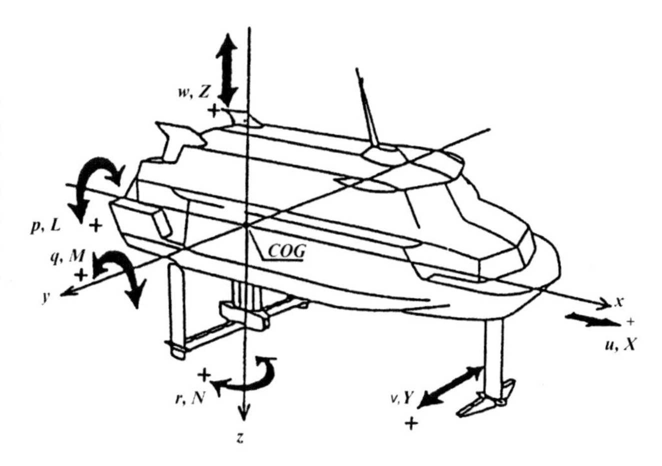

Figure 10.32. Body-fixed coordinate system (x, y, z). The center of gravity (COG) has velocity $\mathbf{V} = (u, v, w)$, and the angular velocity of the vessel is $\mathbf{\Omega} = (p, q, r)$. The external forces and moments with respect to COG acting on the vessel are, respectively, (X, Y, Z) and (L, M, N). All components are in the (x, y, z)-system. The Earth-fixed coordinate system (X_E, Y_E, Z_E) is shown in Figure 10.4 (Saito et al. 1991).

Rudder lift and drag at small angles of attack

If we assume that the effective angle of attack δ_e and the incident flow angle of attack δ_H are small (see Figure 10.31), we can linearize the transverse rudder force given by eq. (10.96). Noting that $\cos\delta_H \approx 1$, $\sin\delta_H \approx \delta_H$ and that $v_R \ll u_R$ gives $Y_R \approx 0.5\rho A_R u_R^2 C_{LR}$, where C_{LR} is a function of the effective angle of attack δ_e. We will consider a symmetric rudder profile; that is, the camber is zero (see definition of *camber* in Figure 6.3). This means C_{LR} is zero when δ_e is zero. A Taylor expansion of C_{LR} gives $C_{LR} \approx dC_{LR}/d\delta_e|_{\delta_e=0}\,\delta_e$. This implies that the transverse rudder force Y_R can be approximated as

$$Y_R = \frac{\rho}{2} A_R u_R^2 C\ell_\delta(\delta - \delta_H), C\ell_\delta = \frac{dC_{LR}}{d\delta_e}|_{\delta_e=0}. \tag{10.97}$$

This means that the coefficient Y_δ in eq. (10.1) can be expressed as $0.5\rho A_R u_R^2 C\ell_\delta$. N_δ in eq. (10.2) is simply $x_R Y_\delta$ according to eq. (10.96). For instance, if we use eq. (2.97) to express C_L, then $C\ell_\delta$ is $2\pi/(1 + 2/\Lambda)$, where Λ is the aspect ratio (see eq. (2.87)). However, eq. (2.97) is based on infinite fluid, and we must modify it to account for hull–free surface interaction.

When it comes to the drag on the rudder, we will concentrate on the effect of the rudder angle and once more consider small values of δ_e and δ_H. We approximate X_R given by eq. (10.96) in a way similar to the one we used for Y_R. This gives

$$X_R = -0.5\rho A_R C_{DR} u_R^2. \tag{10.98}$$

Part of C_{DR} can be accounted for by considering the rudder as a lifting surface. If we approximate the rudder as an elliptical foil in infinite fluid with uniform incident flow, lifting line theory gives that C_{DR} can be expressed by eq. (2.98). However, the actual rudder shape, the presence of the propeller slip stream, the hull, and the free surface must be considered. Viscous flow will also contribute to C_{DR}. It causes, for instance, a viscous resistance in a similar way as the viscous resistance on the ship hull. If we concentrate on the lift-induced drag, C_{DR} is proportional to δ_e^2 for small δ_e (see e.g., eq. (2.98)). We can write $C_{DR} = C_{DIR}\delta_e^2$, and eq. (10.98) gives that the coefficient $X_{\delta\delta}$ in eq. (10.77) can be expressed as

$$X_{\delta\delta} = -0.5\rho A_R C_{DIR} u_R^2. \tag{10.99}$$

Additional specific details relevant to rudders are considered by Søding (1982), Crane et al. (1989), and Brix (1993).

10.9 Maneuvering models in six degrees of freedom

10.9.1 Euler's equation of motion

It is not sufficient to consider only surge, sway, and yaw in a maneuvering analysis of a high-speed vessel at high speed. If the vessel makes a turn, it will heel (bank), as illustrated for a hydrofoil vessel in Figure 6.20 and for a planing vessel in Figure 9.39. Further, a hydrofoil vessel with a fully submerged foil system needs to be controlled in heave, pitch, and roll. Nonlinear effects may also matter. We will therefore formulate the Euler equations of motions for a rigid body and use a hydrofoil vessel in foilborne condition, as illustrated in Figure 10.32, as an example. A body-fixed coordinate system (x, y, z) is introduced with origin in the center of gravity (COG) of the vessel. The x-z–plane is assumed to be a symmetry plane of the vessel. Further, the velocity vector $\mathbf{V}$ of the COG has components (u, v, w) and the angular velocity vector $\mathbf{\Omega}$ of the vessel has components (p, q, r) in the body-fixed coordinate system. The translatory motions of the vessel are referred to an Earth-fixed coordinate system (X_E, Y_E, Z_E), as illustrated in Figure 10.4. The Earth-fixed and body-fixed coordinate systems coincide at initial time. A yaw angle Ψ, pitch angle Θ, and roll angle Φ of the vessel must also be introduced. These angles are called Euler angles. Because the angles are finite, it matters in which order they are executed. The usual order is yaw, pitch, and roll, as illustrated in Figure 10.33. We introduce a coordinate system (x_1, y_1, z_1) with origin in COG and imagine that the vessel is first oriented so that the x_1-, y_1-, and z_1-axes are parallel to, respectively, the X_E-, Y_E- and Z_E-axes.

The *first step* is to rotate the vessel in yaw about the z_1-axis. That brings the x_1- and y_1-axes to the x_2- and y_2-axes. The z_2-axis is the same as the z_1-axis.

The *second step* is to rotate the vessel in pitch about the y_2-axis. That brings the x_2- and z_2-axes to the x_3- and z_3-axes. The y_3-axis is the same as the y_2-axis.

The *final step* is to rotate the vessel in roll about the x_3-axis. That brings the coordinate axis to the x-, y-, and z-axes.

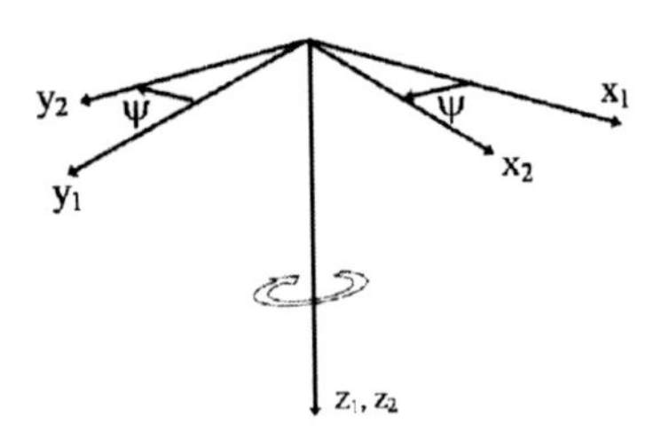

Step 1 : Rotation in yaw ψ about z1 – axis

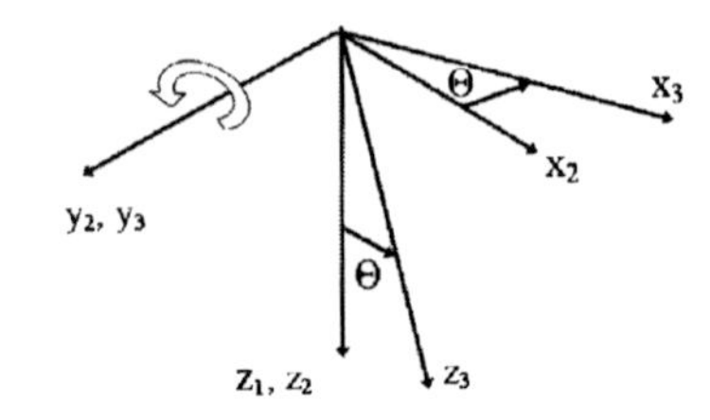

Step 2 : Rotation in pitch Θ about y2 – axis

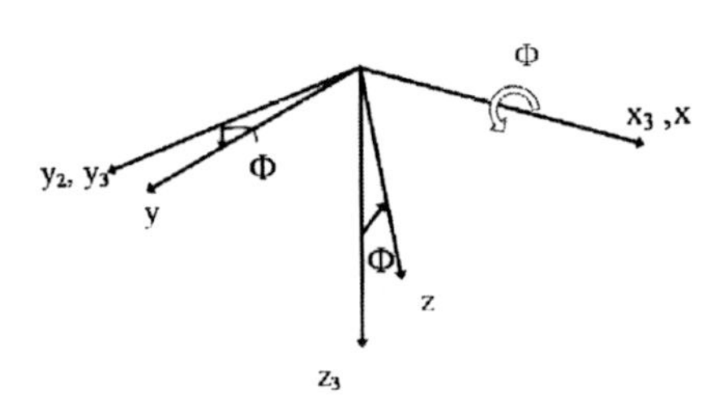

Step 3 : Rotation in roll Φ about x3 – axis

Figure 10.33. The order of the rotation of the Euler angles Φ, Θ, and Ψ. We start with a coordinate system (x_1, y_1, z_1) that has origin in the center of gravity of the vessel and axes parallel to the Earth-fixed coordinate system (X_E, Y_E, Z_E) at initial time. When the rotations are finished, we end up with the body-fixed coordinate system (x, y, z).

Having defined the necessary variables describing the kinematics of the vessel, we will now apply Newton's second law. Because Newton's law refers to acceleration in an inertial system, we must be careful in differentiating the body-fixed velocity components of the vessel with respect to time. We illustrated that for linear sway and yaw velocities in connection with Figure 10.9. We can write

$$\frac{d\mathbf{V}}{dt} = \frac{du}{dt}\mathbf{i} + \frac{dv}{dt}\mathbf{j} + \frac{dw}{dt}\mathbf{k} + u\frac{d\mathbf{i}}{dt} + v\frac{d\mathbf{j}}{dt} + w\frac{d\mathbf{k}}{dt}. \tag{10.100}$$

Here $\mathbf{i}$, $\mathbf{j}$, and $\mathbf{k}$ are unit vectors along the body-fixed coordinate axes x, y, and z, respectively. This means that the unit vectors change with time relative to an inertial system. Let us start with showing how $d\mathbf{k}/dt$ can be calculated. By definition,

$$\frac{d\mathbf{k}}{dt} = \lim_{\Delta t \to 0} \frac{\mathbf{k}(t + \Delta t) - \mathbf{k}(t)}{\Delta t}. \tag{10.101}$$

Figure 10.34 shows a first approximation of $\mathbf{k}$ at time $t + \Delta t$. $\mathbf{k}$ has then moved because of the angular velocity vector $\mathbf{\Omega}$ of the vessel to a new position in the Earth-fixed coordinate. As a first approximation, the head of the $\mathbf{k}$-vector follows a circular path about the angular velocity vector $\mathbf{\Omega}$ (see Figure 10.34). The length AB of the circular path from t to $t + \Delta t$ can be approximated as

$$AB = \sin\alpha \cdot |\mathbf{\Omega}|\, \Delta t. \tag{10.102}$$

Here $|\mathbf{\Omega}|$ means the magnitude of the vector $\mathbf{\Omega}$ and α is the angle between the $\mathbf{\Omega}$-vector and the $\mathbf{k}$-vector at time t (see Figure 10.34).

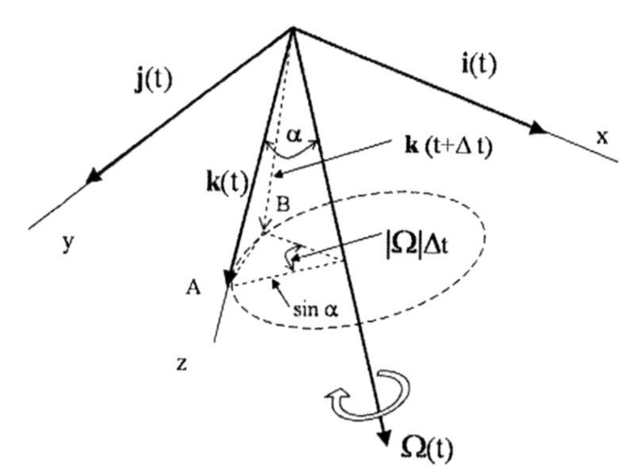

Figure 10.34. Rotating body-fixed frame of reference (x, y, z) with unit vectors $\mathbf{i}, \mathbf{j}, \mathbf{k}$. $\boldsymbol{\Omega}$ is the angular velocity vector of the vessel. The unit vector $\mathbf{k}$ is shown at time t and $t + \Delta t$, where Δt is small.

It follows from the definition of a vector cross-product that

$$|\boldsymbol{\Omega}(t) \times \mathbf{k}(t)| = |\boldsymbol{\Omega}(t)| \cdot |\mathbf{k}(t)| \sin \alpha. \qquad (10.103)$$

Here $|\mathbf{k}|$ is equal to one. Further, the vector $\boldsymbol{\Omega}(t) \times \mathbf{k}(t)$ has, as a first approximation, the direction of the vector $\mathbf{k}(t + \Delta t) - \mathbf{k}(t)$ with the head in point B and the origin in point A. Using eq. (10.102) and (10.103) gives, then,

$$\mathbf{k}(t + \Delta t) - \mathbf{k}(t) \approx \boldsymbol{\Omega}(t) \times \mathbf{k}(t)\Delta t. \qquad (10.104)$$

Using eq. (10.101) results in

$$\frac{d\mathbf{k}}{dt} = \boldsymbol{\Omega} \times \mathbf{k}. \qquad (10.105)$$

By a similar analysis, we will find that

$$\frac{d\mathbf{i}}{dt} = \boldsymbol{\Omega} \times \mathbf{i}, \quad \frac{d\mathbf{j}}{dt} = \boldsymbol{\Omega} \times \mathbf{j}. \qquad (10.106)$$

Eqs. (10.105) and (10.106) then give that eq. (10.100) can be rewritten as

$$\frac{d\mathbf{V}}{dt} = \frac{du}{dt}\mathbf{i} + \frac{dv}{dt}\mathbf{j} + \frac{dw}{dt}\mathbf{k} + \boldsymbol{\Omega} \times \mathbf{V}. \qquad (10.107)$$

By decomposing eq. (10.107), it follows from Newton's second law that

$$\begin{aligned} M(\dot{u} + qw - rv) &= X - Mg \sin \Theta \\ M(\dot{v} + ru - pw) &= Y + Mg \cos \Theta \sin \Phi \\ M(\dot{w} + pv - qu) &= Z + Mg \cos \Theta \cos \Phi \end{aligned} \qquad (10.108)$$

Here M is the mass of the vessel and (X, Y, Z) are the hydrodynamic and aerodynamic external forces acting on the vessel. The other terms on the right-hand side of eq. (10.108) follow by decomposing the weight Mg of the vessel along the body-fixed coordinate axis. We can show that by means of Figure 10.33. The gravitational acceleration acts along the z_1-axis. Going to the second drawing in Figure 10.33, we see that the acceleration of gravity will have components $-g \sin \Theta$ and $g \cos \Theta$ along the x_3- and z_3-axes, respectively. There is no component along the y_3-axis. We then use the last drawing in Figure 10.33. Because the x-axis coincides with the x_3-axis, the x-component of g is $-g \sin \Theta$. This is consistent with the first equation in eq. (10.108). Decomposing the z_3-component $g \cos \Theta$ along the y- and z-axes then gives the terms in the second and third equations of eq. (10.108).

We also need to consider external moments about the x-, y-, and z-axes. These are related to the time derivative of the moment of momentum, that is,

$$\frac{d}{dt}\int \mathbf{r} \times (\mathbf{V} + \boldsymbol{\Omega} \times \mathbf{r})\, dM = \mathbf{H}, \qquad (10.109)$$

where $\mathbf{r} = x\mathbf{i} + y\mathbf{j} + z\mathbf{k}$ and dM is the mass of an infinitesimally small structural element located at (x, y, z). The integration in eq. (10.109) is over the whole structure. Further, $\mathbf{H}$ is the external moment vector with components along the x-, y-, and z-axes. When differentiating with respect to time in eq. (10.109), we must once more note that the unit vectors along the body-fixed coordinate axis vary with time; see eqs. (10.105) and (10.106). The details of the derivation are given in Etkin (1959). By assuming symmetry about the xz-plane, we have that

$$\begin{aligned} I_{44}\dot{p} - (I_{55} - I_{66})\,qr - I_{64}(\dot{r} + pq) &= L \\ I_{55}\dot{q} - (I_{66} - I_{44})\,rp - I_{64}(r^2 - p^2) &= M \\ I_{66}\dot{r} - (I_{44} - I_{55})\,pq - I_{64}(\dot{p} - qr) &= N \end{aligned} \qquad (10.110)$$

Here I_{jj} is the moment of inertia of the jth mode and I_{jk} is the product of inertia with respect to the coordinate system (x, y, z). The expressions are the same as those presented in eq. (7.37). Further, L, M, and N are the external moments about the x-, y-, and z-axes, respectively. We must obviously not confuse M with the vessel mass in this context. If we limit ourselves to a vessel with constant forward speed and small sway and yaw velocities, then eqs. (10.108) and (10.110) are consistent with eqs. (10.1) and (10.2).

We have not said too much about the external forces and moments. They are functions of the unknowns (u, v, w) and (p, q, r) in eqs. (10.108) and (10.110) as well as the unknown orientation of the vessel. It is not an easy task to find consistent nonlinear expressions for external forces and moments. If the hydrodynamic problem is solved in the body-fixed coordinate system and is based on potential flow theory, it would, for instance, require that Bernoulli's equation, which is expressed in an inertial system, is transferred to a body-fixed coordinate system. Additional terms will then appear (Kochin et al. 1964).

In the body-fixed coordinate system, the hydrodynamic pressure can be expressed as

$$\frac{p}{\rho} + \frac{\partial \Phi}{\partial t} + \frac{1}{2}(\nabla\Phi)^2 - \nabla\Phi\cdot(\mathbf{V} + \boldsymbol{\Omega} \times(x\mathbf{i} + y\mathbf{j} + z\mathbf{k})) - gz_1 = C. \tag{10.111}$$

Here Φ is the velocity potential and $\partial\Phi/\partial t$ is calculated in the moving coordinate system, that is, for a point rigidly connected with the body-fixed coordinate system $Oxyz$. The constant C can be determined by evaluating eq. (10.111) on the free surface far away from the body, where there is no flow disturbance and the pressure is atmospheric (see section 3.2.1). We note that eq. (10.111) differs from eq. (3.5), which is the Bernoulli equation for the pressure in an inertial coordinate system. Final expressions of nonlinear, nonlifting and nonviscous hydrodynamic forces and moments on a maneuvering body in infinite fluid are given in section 10.10.2.

In order to solve the Euler equations, we need to derive equations that describe the orientation of the vessel. Velocity components U_i, V_i, W_i of the COG along the x_i-, y_i-, and z_i-axes in Figure 10.33 are then introduced. Here the subscript i goes from 1 to 3. We consider the following differential equations describing the vessel position in the Earth-fixed coordinate system (X_E, Y_E, Z_E):

$$\frac{dX_E}{dt} = U_1, \quad \frac{dY_E}{dt} = V_1, \quad \frac{dZ_E}{dt} = W_1. \tag{10.112}$$

By using the first drawing in Figure 10.33, we see that

$$\begin{aligned} U_1 &= U_2 \cos\Psi - V_2 \sin\Psi \\ V_1 &= U_2 \sin\Psi + V_2 \cos\Psi \\ W_1 &= W_2 \end{aligned}$$

This can be shown by noting that the rotation occurs in a plane, that is, similar to Figure 7.10 and resulting eqs. (7.16) and (7.17). The previous expression can also be expressed as

$$\begin{bmatrix} U_1 \\ V_1 \\ W_1 \end{bmatrix} = \mathbf{A} \begin{bmatrix} U_2 \\ V_2 \\ W_2 \end{bmatrix},$$

where the matrix $\mathbf{A}$ is

$$\mathbf{A} = \begin{bmatrix} \cos\Psi & -\sin\Psi & 0 \\ \sin\Psi & \cos\Psi & 0 \\ 0 & 0 & 1 \end{bmatrix}.$$

The second drawing gives

$$\begin{bmatrix} U_2 \\ V_2 \\ W_2 \end{bmatrix} = \mathbf{B} \begin{bmatrix} U_3 \\ V_3 \\ W_3 \end{bmatrix},$$

where the matrix $\mathbf{B}$ is

$$\mathbf{B} = \begin{bmatrix} \cos\Theta & 0 & \sin\Theta \\ 0 & 1 & 0 \\ -\sin\Theta & 0 & \cos\Theta \end{bmatrix}.$$

The third drawing gives the relationships

$$\begin{bmatrix} U_3 \\ V_3 \\ W_3 \end{bmatrix} = \mathbf{C} \begin{bmatrix} u \\ v \\ w \end{bmatrix},$$

where the matrix $\mathbf{C}$ is

$$\mathbf{C} = \begin{bmatrix} 1 & 0 & 0 \\ 0 & \cos\Phi & -\sin\Phi \\ 0 & \sin\Phi & \cos\Phi \end{bmatrix}.$$

Using these relationships gives that eq. (10.112) can be expressed as

$$\begin{bmatrix} \frac{dX_E}{dt} \\ \frac{dY_E}{dt} \\ \frac{dZ_E}{dt} \end{bmatrix} = \mathbf{ABC} \begin{bmatrix} u \\ v \\ w \end{bmatrix}$$

or that

$$\begin{aligned} \frac{dX_E}{dt} &= u\cos\Theta\cos\Psi \\ &\quad + v(\sin\Phi\sin\Theta\cos\Psi - \cos\Phi\sin\Psi) \\ &\quad + w(\cos\Phi\sin\Theta\cos\Psi + \sin\Phi\sin\Psi) \\ \frac{dY_E}{dt} &= u\cos\Theta\sin\Psi + v(\sin\Phi\sin\Theta\sin\Psi \\ &\quad + \cos\Phi\cos\Psi) \\ &\quad + w(\cos\Phi\sin\Theta\sin\Psi - \sin\Phi\cos\Psi) \\ \frac{dZ_E}{dt} &= -u\sin\Theta + v\sin\Phi\cos\Theta + w\cos\Phi\cos\Theta. \end{aligned} \tag{10.113}$$

Finally, we need differential equations for Θ, Φ, and Ψ. These can be expressed as (see Etkin 1959)

$$\frac{d\Theta}{dt} = q\cos\Phi - r\sin\Phi$$
$$\frac{d\Phi}{dt} = p + q\sin\Phi\tan\Theta + r\cos\Phi\tan\Theta \qquad (10.114)$$
$$\frac{d\Psi}{dt} = (q\sin\Phi + r\cos\Phi)\sec\Theta, \qquad \cos\Theta \neq 0$$

We note that the angular velocity transformation has a singularity for $\Theta = \pm 90^\circ$. Other formulations may remove this (Fossen 2002). However, the Euler formulation holds for practical purposes. We then have presented 12 nonlinear differential equations given by eqs. (10.108), (10.110), (10.113), and (10.114) with the 12 unknowns: u, v, w, p, q, r, X_E, Y_E, Z_E, Θ, Φ, and Ψ. In order to solve these equations numerically we need to specify initial conditions and express the external forces and moments acting on the vessel. Practical procedures that partly rely on empirical knowledge are presented by van Walree (1999) and Saito et al. (1991) for a hydrofoil vessel. This means that we must consider hydrodynamic loads on the foils, struts, appendages, and propulsion units in combination with aerodynamic loads on the vessel. The effect of rudders and foil flaps in combination with an automatic control system must be incorporated. As long as the vessel is maneuvering in calm water, a quasi-steady approach for hydrodynamic and aerodynamic forces can be followed. However, Saito et al. (1991) emphasize the importance of nonlinear saturation effects due to ventilation and cavitation on the foils and struts.

As another example, consider an SES in a turn. There is the danger of air leakage from the cushion as a consequence of the heel (bank) angle. This can be handled in a way similar to that described in Chapter 5.

10.9.2 Linearized equation system in six degrees of freedom

We will use the Euler equation of motion as a basis and consider small time-dependent deviations from a steady upright equilibrium position. The vessel is assumed to be on a straight course with forward velocity U in the steady condition. The unsteady part of u and the unsteady velocity components v and w are assumed small relative to U. We use the notations ϕ, θ, and ψ for unsteady roll, pitch, and yaw, respectively. The order in which these small angles are executed in the linear case does not matter. Using a Taylor expansion in the equation for $d\Theta/dt$ in eq. (10.114) means that

$$\frac{d\Theta}{dt} = q\cos\Phi - r\sin\Phi$$
$$= q\left(1 - \frac{1}{2}\Phi^2 + \cdots\right) - r\left(\Phi + \cdots\right).$$

Keeping only linear terms in q, r, Θ, and Φ gives $d\theta/dt = q$, where we have used the small letter θ for Θ to indicate a linear quantity. Following a similar procedure for the equations for $d\Phi/dt$ and $d\Psi/dt$ in eq. (10.114) implies $d\phi/dt = p$, $d\psi/dt = r$. Further, qw, rv, pw, and pv can be neglected in eq. (10.108). Those terms are products between small quantities. Further, $ru \approx rU$ and $qu \approx qU$. We can also use the approximations $\sin\Theta \approx \theta$, $\sin\Phi \approx \phi$, $\cos\Theta \approx 1$, and $\cos\Phi \approx 1$. There appears, then, the term Mg, representing the vessel weight on the right-hand side of the third equation in eq. (10.108). This is a steady term that will be balanced by a steady term in the z-component of the external force. Keeping linear unsteady terms in eqs. (10.108) and (10.110) gives, therefore,

$$M\dot{u} = X_1 - Mg\theta \qquad (10.115)$$

$$M(\dot{v} + U\dot{\psi}) = Y_1 + Mg\phi \qquad (10.116)$$

$$M(\dot{w} - U\dot{\theta}) = Z_1 \qquad (10.117)$$

$$I_{44}\ddot{\phi} - I_{46}\ddot{\psi} = L_1 \qquad (10.118)$$

$$I_{55}\ddot{\theta} = M_1 \qquad (10.119)$$

$$I_{66}\ddot{\psi} - I_{64}\ddot{\phi} = N_1. \qquad (10.120)$$

The subscript 1 used for X_1, Y_1, Z_1, L_1, M_1, and N_1 means the linear unsteady part of X, Y, Z, L, M, and N, respectively. By Taylor expansion, as in eqs. (10.1) and (10.2), we can now express the forces and moments acting on the vessel as linear expressions in the motion velocity and acceleration variables, as well as in terms of the unsteady control angles, such as the rudder angle δ in eqs. (10.1) and (10.2). However, for a hydrofoil vessel with fully submerged foils, it is not sufficient with a rudder angle controlling sway and yaw only.

We need to control the vessel in heave, roll, pitch, and yaw. This means more than one control angle. Schmitke and Jones (1972) and Hamamoto et al. (1993) have presented more details on this linear equation system for a surface-piercing hydrofoil craft.

As a consequence of linearity and symmetry properties of the body, the coupled equations for surge, heave, and pitch given by eqs. (10.115), (10.117), and (10.119) will be uncoupled from the coupled equations for sway, roll, and yaw given by eqs. (10.116), (10.118), and (10.120) for a vessel with a mean upright position. The reason is that the hydrodynamic pressure associated with surge-heave-pitch and sway-roll-yaw are, respectively, symmetric and antisymmetric with respect to the centerplane. Saito et al. (1991) have described how the linear equation system is combined with a control system for a hydrofoil vessel with a fully submerged foil system. Equations similar to those derived above are used in analyzing the dynamic behavior of airplanes. A thorough presentation of this is given by Etkin (1959).

In the next section, we study in more detail coupled sway-roll-yaw motions of a monohull. The procedure can be easily generalized to a catamaran.

10.9.3 Coupled sway-roll-yaw of a monohull

We will generalize the linear slender body theory presented in section 10.3 to include roll. A body-fixed coordinate system as shown in Figure 10.5 is used. Further, the z-axis is positive upward, with $z = 0$ in the mean free surface. The center of gravity of the vessel has coordinates $(0, 0, z_G)$.

The body-boundary condition given by eq. (10.18) for coupled sway-yaw motions has to be modified to account for roll. The ship velocity due to sway, roll, and yaw has the velocity components $\dot{\eta}_2 - z\dot{\eta}_4 + x\dot{\eta}_6$ and $y\dot{\eta}_4$ along the y- and z-axes, respectively. The body-boundary condition becomes then

$$\frac{\partial \varphi}{\partial n} = n_2 (\dot{\eta}_2 + x\dot{\eta}_6) + (-zn_2 + yn_3)\,\dot{\eta}_4 \quad \text{on } C(x). \tag{10.121}$$

The free-surface condition is $\partial\varphi/\partial z = 0$ on $z = 0$. This means no wave generation and is relevant for Froude numbers up to approximately 0.2. Further, φ satisfies a 2D Laplace equation in the yz-plane.

We focus on roll and denote the velocity potential due to roll as $\varphi_4\dot{\eta}_4$. Because the flow problem is linearly dependent on $\dot{\eta}_2$, $\dot{\eta}_4$, and $\dot{\eta}_6$, we can study separately the $\dot{\eta}_4$-dependent term on the right-hand side of eq. (10.121). It follows then that φ_4 satisfies

$$\frac{\partial \varphi_4}{\partial n} = n_4 \quad \text{on } C(x), \tag{10.122}$$

where

$$n_4 = -zn_2 + yn_3. \tag{10.123}$$

Two-dimensional added mass in roll a_{44} and coupled added mass between sway and roll a_{24} are needed as a part of the analysis. We will show how a_{44} and a_{24} are related to φ_4. The procedure described in section 7.2.1 is then followed, and the hydrodynamic pressure due to roll velocity without the presence of forward speed is considered, that is,

$$p = -\rho\frac{\partial}{\partial t}(\varphi_4\dot{\eta}_4) = -\rho\varphi_4\ddot{\eta}_4. \tag{10.124}$$

This gives a 2D horizontal force,

$$-\int_C pn_2\,dS = \rho\left(\int_C \varphi_4 n_2\right)\ddot{\eta}_4. \tag{10.125}$$

The coupled added mass between sway and roll is by definition (see eq. (7.39)) equal to

$$a_{24} = -\rho\int_C \varphi_4 n_2\,dS. \tag{10.126}$$

Similarly, we consider the roll moment $-\int_C pn_4\,dS$. This gives

$$a_{44} = -\rho\int_C \varphi_4 n_4\,dS. \tag{10.127}$$

We will also need the coupled added mass between roll and sway, that is, a_{42}. This follows by analyzing the pressure $-\rho\partial(\varphi_2\dot{\eta}_2)/\partial t$ due to forced sway velocity and the resulting roll moment, that is,

$$a_{42} = -\rho\int_C \varphi_2 n_4\,dS. \tag{10.128}$$

We will show by using Green's second identity that $a_{42} = a_{24}$. We start out with eq. (6.23) and set $\psi = \varphi_2$ and $\varphi = \varphi_4$. Because φ_2 and φ_4 both satisfy the 2D Laplace equation, it follows that

$$\int_{C+S_F+S_\infty} \varphi_2\frac{\partial\varphi_4}{\partial n}\,dS = \int_{C+S_F+S_\infty} \varphi_4\frac{\partial\varphi_2}{\partial n}\,dS. \tag{10.129}$$

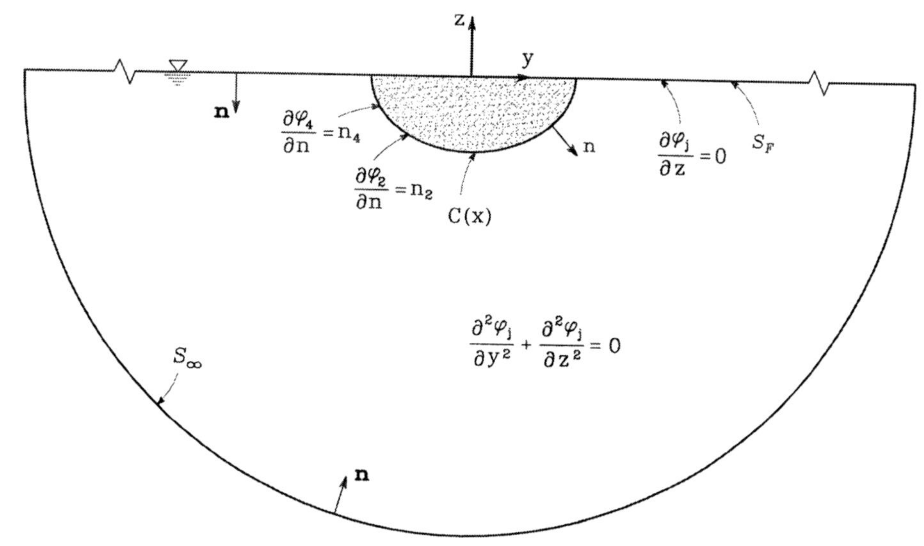

Figure 10.35. Surfaces C, S_F, and S_∞ used in applying Green's second identity to show symmetry properties of 2D cross-coupling terms in sway- and roll-added mass. φ_j, $j = 2$ and 4 are velocity potentials due to unit sway and roll velocity, respectively.

Here, $C + S_F + S_\infty$ is a closed surface, S_∞ is a control surface at infinity, and S_F is the mean free surface (Figure 10.35). Because both φ_4 and φ_2 satisfy a rigid free-surface condition, that is, $\partial\varphi_j/\partial z = 0$, $j = 2{,}4$ on $z = 0$, the integration along S_F does not contribute to eq. (10.129). Further, φ_2 and φ_4 go to zero at infinity and it can be shown that the integrals along S_∞ are zero. If we use that $\partial\varphi_4/\partial n = n_4$. and $\partial\varphi_2/\partial n = n_2$ on C, it follows that

$$\int_C \varphi_2 n_4 \, dS = \int_C \varphi_4 n_2 \, dS. \qquad (10.130)$$

Using eqs. (10.126) and (10.128) gives that $a_{42} = a_{24}$.

The added mass coefficient can be determined by, for instance, a boundary element method (BEM). Approximate expressions can be obtained by the Lewis form technique. Grim (1955) showed that

$$a_{44} = \frac{\rho\pi}{8}\left(\frac{b}{2}\right)^4 \times \left\{ \frac{128}{\pi^2} \frac{\left[a_1^2(1+a_3)^2 + \frac{8}{9}a_1 a_3(1+a_3) + \frac{16}{9}a_3^2\right]}{(1+a_1+a_3)^4} \right\} \qquad (10.131)$$

The Lewis form coefficients a_1 and a_3 are defined in connection with eq. (8.89) in terms of the cross-sectional area A, beam b, and draft d. Requirements for a Lewis form to exist are also given in connection with eq. (8.89).

Figure 10.36 presents nondimensional values of a_{24} and a_{44} as a function of b/d when $A/(bd) = 0.6$. We note that a_{44} becomes small and has a minimum for b/d close to 2. a_{24} changes sign close to this minimum value of a_{44}.

We consider now the 2D hydrodynamic roll moment f_{44}^{HD} and sway force f_{24}^{HD} on a cross section of the vessel due to roll velocity when the vessel has a constant forward speed. This can be derived in a similar way as eqs. (8.87) and (10.19), that is,

$$f_{44}^{HD} = -\left(\frac{\partial}{\partial t} + U\frac{\partial}{\partial x}\right)[a_{44}\dot{\eta}_4] \qquad (10.133)$$

$$f_{24}^{HD} = -\left(\frac{\partial}{\partial t} + U\frac{\partial}{\partial x}\right)[a_{24}\dot{\eta}_4]. \qquad (10.134)$$

Further, there exist 2D hydrodynamic roll moments f_{42}^{HD} and f_{46}^{HD} on a cross section as the result of sway and yaw, respectively. They can be

$$a_{24} = a_{42} = \rho\frac{8}{3d}\left(\frac{b}{2}\right)^4 \frac{\left[a_1(1-a_1)(1+a_3) + a_1 a_3(1+a_3)0.6 + a_3(1-a_1)0.8 - 1.714a_3^2\right](1-a_1+a_3)}{(1+a_1+a_3)^4}. \qquad (10.132)$$

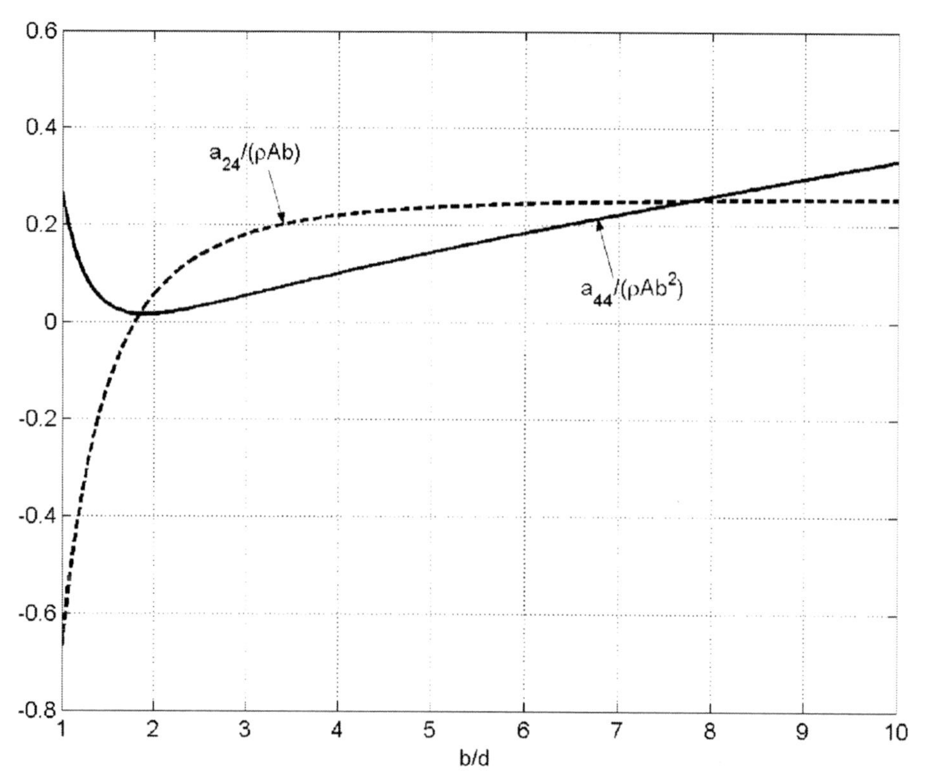

Figure 10.36. Two-dimensional added mass in roll a_{44} and coupled added mass a_{24} between sway and roll for Lewis form cross sections with beam b, draft d, and cross-sectional area A. The calculations are with rigid free-surface conditions and when $A/(bd) = 0.6$. The coefficients are defined with respect to a coordinate system y-z with origin in the mean free surface and z positive upward. The z-axis is in the centerline (see Figure 10.6).

expressed as

$$f_{42}^{HD} = -\left(\frac{\partial}{\partial t} + U\frac{\partial}{\partial x}\right)[a_{42}\dot{\eta}_2] \qquad (10.135)$$

$$f_{46}^{HD} = -\left(\frac{\partial}{\partial t} + U\frac{\partial}{\partial x}\right)[a_{42}x\dot{\eta}_6]. \qquad (10.136)$$

Integrating eqs. (10.133), (10.135), and (10.136) along the length L of the ship gives the hydrodynamic roll moment

$$F_4^{HD} = -\left[\int_L a_{42}\,dx\ddot{\eta}_2 + \int_L a_{44}\,dx\ddot{\eta}_4 + \int_L a_{42}x\,dx\ddot{\eta}_6 + Ua_{42}(x_T)\dot{\eta}_2 + Ua_{44}(x_T)\dot{\eta}_4 + Ux_T a_{42}(x_T)\dot{\eta}_6\right]. \qquad (10.137)$$

Here x_T is the x-coordinate of the transom stern. Further, eq. (10.134) gives the following hydrodynamic sway force due to roll:

$$F_{24}^{HD} = -\left[\int_L a_{24}\,dx\ddot{\eta}_4 + Ua_{24}(x_T)\dot{\eta}_4\right]. \qquad (10.138)$$

Then we take the yaw moment F_{64}^{HD} based on eq. (10.134). This means the equation is multiplied by x and then integrated over the ship length. We get in a way similar to how eq. (10.22) was derived that

$$F_{64}^{HD} = -\left\{\int_L xa_{24}\,dx\ddot{\eta}_4 + U\left[x_T a_{24}(x_T) - \int_L a_{24}\,dx\right]\dot{\eta}_4\right\}. \qquad (10.139)$$

Introducing the definition of added mass and damping coefficients given by eq. (7.39), we have by using eqs. (10.137) through (10.139)

$$A_{24} = \int_L a_{24}\,dx; \quad B_{24} = Ua_{24}(x_T) \tag{10.140}$$

$$A_{42} = \int_L a_{42}\,dx; \quad B_{42} = Ua_{42}(x_T) \tag{10.141}$$

$$A_{44} = \int_L a_{44}\,dx; \quad B_{44} = Ua_{44}(x_T) \tag{10.142}$$

$$A_{46} = \int_L a_{42}x\,dx; \quad B_{46} = Ux_T a_{42}(x_T) \tag{10.143}$$

$$A_{64} = \int_L a_{24}x\,dx; \quad B_{64} = -UA_{24} + Ux_T a_{24}(x_T). \tag{10.144}$$

The added mass and damping coefficients in the linear sway-roll-yaw maneuvering equations of a slender ship can then, according to slender body theory, be expressed by eqs. (10.33) through (10.36) and eqs. (10.140) through (10.144). However, this approach does not account for viscous effects, which have to be experimentally or empirically determined.

When formulating the linear maneuvering equations, we must also consider the effect of hydrostatic and steady hydrodynamic pressure and the fact that the origin of our coordinate system is not in the center of gravity (COG).

Because the vertical force on the hull due to the hydrostatic and steady hydrodynamic pressures balances the weight Mg of the vessel, we can disregard the $Mg\phi$ term in eq. (10.116). However, there are additional linear restoring terms due to roll. They are a consequence of the fact that the hydrostatic and steady hydrodynamic pressure distribution on the hull is a function of the heel angle. This is further discussed in the following text.

Because $\dot{\eta}_2 - z_G\dot{\eta}_4$ is the sway velocity of COG, the sway equation (10.116) can with the present notation be expressed as

$$\begin{aligned}(M + A_{22})\frac{d^2\eta_2}{dt^2} &+ B_{22}\frac{d\eta_2}{dt}\\ &+ (-Mz_G + A_{24})\frac{d^2\eta_4}{dt^2} + B_{24}\frac{d\eta_4}{dt}\\ &+ C_{24}\eta_4 + A_{26}\frac{d^2\eta_6}{dt^2}\\ &+ (-MU + B_{26})\frac{d\eta_6}{dt} = F_2^S.\end{aligned} \tag{10.145}$$

The restoring coefficient C_{24} is a result of the steady hydrodynamic pressure and will be small at moderate speed when our slender body theory is valid. F_2^S is the control force in sway.

Eqs. (10.118) and (10.120) can be modified as

$$\begin{aligned}&(-Mz_G + A_{42})\frac{d^2\eta_2}{dt^2} + B_{42}\frac{d\eta_2}{dt}\\ &+ (I_{44} + A_{44})\frac{d^2\eta_4}{dt^2} + B_{44}\frac{d\eta_4}{dt} + C_{44}\eta_4\\ &+ (-I_{46} + A_{46})\frac{d^2\eta_6}{dt^2}\\ &+ (Mz_G U + B_{46})\frac{d\eta_6}{dt} = F_4^S\end{aligned} \tag{10.146}$$

$$\begin{aligned}&A_{62}\frac{d^2\eta_2}{dt^2} + B_{62}\frac{d\eta_2}{dt} + (-I_{64} + A_{64})\frac{d^2\eta_4}{dt^2}\\ &+ B_{64}\frac{d\eta_4}{dt} + C_{64}\eta_4 + (I_{66} + A_{66})\frac{d^2\eta_6}{dt^2}\\ &+ B_{66}\frac{d\eta_6}{dt} = F_6^S,\end{aligned} \tag{10.147}$$

where the moment of inertia I_{jj} and product of inertia I_{jk} are with respect to the coordinate system (x, y, z) and defined by eq. (7.37). F_4^S and F_6^S are the control roll and yaw moments, respectively. We should note that there are also hydrodynamic forces and moments on a rudder due to sway, roll, and yaw (see section 10.8). These should be added to the left-hand sides of eqs. (10.145) through (10.147). The restoring coefficient C_{44} can be expressed as

$$C_{44} = \rho g \nabla \overline{GM} + C_{44}^D, \tag{10.148}$$

where ∇ is the displaced volume and the transverse metacentric height $\overline{GM}$ is defined by eq. (7.41). The restoring coefficients C_{44}^D and C_{64} are a result of the steady hydrodynamic pressure and are small at moderate speed. How C_{44} varies with U for the different vessels is illustrated indirectly in Figure 7.41. The static heel angle can be approximated as F_4/C_{44}, where F_4 is a given constant heel moment. Because the heel angle increases with increasing Froude number, C_{44}^D is negative and decreases with increasing Froude number for all the cases presented in Figure 7.41. However, this trend is not universal for all vessels. Lewandowski (1997) presented experimental results for hard-chine planing hulls showing a different trend as a function of Froude number at planing speeds. For instance, a minimum heel restoring moment occurring between volumetric Froude numbers

$F_v = U/\sqrt{g\nabla^{1/3}}$ of 2 and 3 was characteristic for the studied Series 62 hulls.

The restoring coefficients C_{24} and C_{64} are important when the steady heel of a semi-displacement vessel generates large asymmetric bow waves. Müller-Graf (1997) has described a scenario such as this. The bow waves become substantially higher at the immersed side than at the emerged side of the hull. The center of pressure of the steady side force will be close to the bow. For instance, if the vessel heels to the port, the steady yaw moment will turn the bow to starboard; that is, the largest hydrodynamic pressures occur on the hull side where the bow waves are largest. Eqs. (10.145) through (10.147) are consistent with Lewandowski (1997) and Haarhoff and Sharma (2000). However, they used standard maneuvering nomenclature and did not express the added mass and damping coefficients by slender body theory.

By setting F_2^S, F_4^S, and F_6^S equal to zero in the linear maneuvering equations, assuming a solution of the form $\exp(st)$, we can study the coupled sway-roll-yaw stability of the ship. This means we write $d\eta_2/dt = \dot{\eta}_{2a}\exp(st)$, $\eta_4 = \eta_{4a}\exp(st)$, and $d\eta_6/dt = \dot{\eta}_{6a}\exp(st)$. This leads to a linear equation system for $\dot{\eta}_{2a}$, η_{4a}, and $\dot{\eta}_{6a}$ with the unknown $\dot{\eta}_{2a}$, η_{4a}, and $\dot{\eta}_{6a}$ on the left-hand side. Because the right-hand side of this equation system is zero and we are interested in the nontrivial solutions, these can be obtained by enforcing the coefficient determinant to be zero. This is similar to what we did in analyzing eq. (10.37). The condition for nontrivial solutions is then

$$\det\begin{vmatrix} (M+A_{22})s+B_{22} & (-Mz_G+A_{24})s^2+B_{24}s+C_{24} & A_{26}s+(-MU+B_{26}) \\ (-Mz_G+A_{42})s+B_{42} & (I_{44}+A_{44})s^2+B_{44}s+C_{44} & (-I_{46}+A_{46})s+(Mz_GU+B_{46}) \\ A_{62}s+B_{62} & (-I_{64}+A_{64})s^2+B_{64}s+C_{64} & (I_{66}+A_{66})s+B_{66} \end{vmatrix} = 0. \quad (10.149)$$

This gives the fourth-degree polynomial equation

$$A's^4 + B's^3 + C's^2 + D's + E' = 0. \quad (10.150)$$

This is similar to eq. (9.85), used in the porpoising stability analysis. However, the coefficients A', B', C', D', and E' are obviously different. The stability can be studied by means of the Routh-Hurwitz stability criterion given by eq. (9.87). Haarhoff and Sharma (2000) studied the coupled sway-roll-yaw stability and stated that three of the four stability criteria given by eq. (9.87) are fulfilled for realistic conventional hull forms and that it is only necessary to study if $E' > 0$ to ensure stability. Even though we cannot be sure about this for any high-speed monohull at any speed, we will limit the discussion to a study of E'.

It follows from eqs. (10.149) and (10.150) that the stability criterion is

$$\begin{aligned} E' = {}& \rho g\nabla\overline{GM}\left[B_{22}B_{66} - B_{62}\left(-MU + B_{26}\right)\right] \\ & + \left\{B_{22}\left[C_{44}^D B_{66} - C_{64}\left(Mz_GU + B_{46}\right)\right]\right. \\ & - C_{24}\left[B_{42}B_{66} - B_{62}\left(Mz_GU + B_{46}\right)\right] \\ & \left. + \left(-MU + B_{26}\right)\left[B_{42}C_{64} - B_{62}C_{44}^D\right]\right\} > 0. \end{aligned} \quad (10.151)$$

When the ship speed is small or moderate, let us say the Froude number Fn is less than 0.2, we can use the slender body theory to estimate the damping coefficients. Because the restoring coefficients C_{24}, C_{44}^D, and C_{64} are negligible for small Fn, eq. (10.151) gives

$$\rho g\nabla\overline{GM}[B_{22}B_{66} - B_{62}(-MU + B_{26})] > 0 \quad (10.152)$$

as a criterion for directional stability of a slow ship. Because $\overline{GM}$ is obviously positive because of static heel stability requirements, eq. (10.152) is consistent with the criterion for dynamic sway-yaw stability discussed in section 10.3.3. When the ship speed increases, the restoring coefficients C_{24}, C_{44}^D, and C_{64} due to heel get increased importance in determining the dynamic stability. This is particularly true if $\overline{GM}$ is low. We see that by studying how the terms in eq. (10.151) depend on the forward speed U. If the previously described slender body theory is used, the damping coefficients are proportional to U. This is based on neglecting the generation of free-surface waves, that is, the effect of Froude number. The restoring coefficients C_{24}, C_{44}^D, and C_{64} depend on the steady hydrodynamic pressure, which is proportional to U^2 when the generation of free-surface waves is negligible. Even though we are interested in Froude numbers in which wave generation matters, we will as a first approximation assume that B_{jk} is proportional to U and C_{24}, C_{44}^D, and C_{64} are proportional to U^2. This means that the first term in E' involving $\overline{GM}$ is proportional to U^2 and the rest of the expression for E' is proportional to U^4. It implies that the first term gets reduced importance with increased speed, particularly if $\overline{GM}$ is small.

Eq. (10.151) indicates that a possible scenario is a ship that is stable at low Froude number and high $\overline{GM}$, but becomes unstable with increasing speed and/or decreasing metacentric height (Haarhoff and Sharma 2000). This is consistent with what has been reported in the literature, that is, Eda (1980). Haarhoff and Sharma (2000) also described another less likely but not impossible scenario in which the vessel is unstable at low Froude number and high $\overline{GM}$ but can become stable with increasing speed and/or decreasing transverse metacentric height.

We need experimental results of B_{jk} and C_{jk} as a function of Froude number in order to quantify the dynamic coupled sway-roll stability of a vessel at high speed by means of eq. (10.151). Because B_{jk} and C_{jk} are functions of trim and sinkage, we must also know how the trim and sinkage vary with the Froude number.

Lewandowski (1997) has presented semi-empirical added mass, damping, and restoring coefficients for prismatic planing vessels at planing speed. The coefficients are expressed as functions of the beam and deadrise angle, and the speed, trim angle, and transom draft, which can be determined by Savitsky's (1964) method (see section 9.2.2). Formulas for appendages are also provided. Lewandowski (1997) recommends that all four Routh-Hurwitz stability criteria based on eq. (10.150) be examined. Further, studies of dynamic loss of transverse stability should not be restricted to the highest operating speed. Systematic calculations are reported, and maximum distance between the keel and the center of gravity $\overline{KG}$ ($\overline{KG}_{max}$) for stable behavior is presented. Comparisons are made with experimental results of a free-running radio-controlled model representing a 22.5-m hard-chined patrol boat. The volumetric Froude number varied between 1.3 and 4, and experimental $\overline{KG}_{max}$ was about 60% of the beam B. The linear stability analysis shows generally good agreement with observed capsizes. Because a linear stability analysis assumes small deviations from the equilibrium position, it cannot predict the details of the highly nonlinear capsize process. The linear stability analysis can only suggest when a dangerous situation may occur. A linear instability does not always lead to capsizing. Lewandowski's systematic stability calculations for Series 62 models at planing speed show that the ratio $\overline{KG}_{max}/B$ increases with either decreasing length-to-beam ratio or decreasing deadrise angle. A simple method to check the transverse dynamic stability of a proposed design is also presented by Lewandowski (1997). More experimental data for hydrodynamic coefficients are needed at planing speed in order to bridge the gap between zero and planing speeds. If the vessel is turning at high speed, the resulting mean heel angle implies that coupled motions in six degrees of freedom must be considered in the dynamic stability analysis.

Unstable oscillatory sway-roll-yaw motions are known to operators of high-speed vessels as "cork-screwing" and make it difficult to steer the vessel. A more dangerous situation is "calm water broaching" (Müller-Graf 1997). This is a nonoscillatory sway-roll-yaw instability occurring at higher speeds than the oscillatory "cork-screw" instabilities. The loss of steady restoring moment in heel with speed causes a sudden list of the vessel to one side followed by a violent yaw to the other side. The consequence may be capsizing. The stopped vessel may be overrun by the stern wave system and may create a dangerous situation for small craft (Müller-Graf 1997). Calm water broaching is the main reason round-bilge hulls should not operate beyond a Froude number of 1.2 (Lavis 1980).

10.10 Exercises

10.10.1 Course stability of a ship in a canal

Let us consider a ship moving in a canal with constant cross section, and let the coordinate axis OX_E in Figure 10.4 be along the centerline of the canal. The presence of the canal introduces a side force and yaw moment on the ship, which by linearization can be expressed as the terms $Y_{Y_E}Y_E(t)$ and $N_{Y_E}Y_E(t)$ appearing on the right-hand side of eqs. (10.1) and (10.2), respectively.

Use the Routh-Hurwitz criterion (see eq. (9.87)) to express the conditions for course stability of the ship in terms of the coefficients in the equations of motions of the vessel in the canal.

10.10.2 Nonlinear, nonlifting, and nonviscous hydrodynamic forces and moments on a maneuvering body

Kochin et al. (1964) presented expressions for the nonlinear, nonlifting, and nonviscous hydrodynamic force and moment on a maneuvering body in infinite fluid. A body-fixed Cartesian coordinate system with origin in O is introduced. The velocity vector of an arbitrary point M of the body is expressed as

$$\mathbf{u} = \mathbf{V} + \boldsymbol{\Omega} \times \mathbf{r}, \qquad (10.153)$$

where $\mathbf{V}$ is the velocity vector of the point O, $\mathbf{r}$ is the radius vector from M to O, and $\boldsymbol{\Omega}$ is the vector of the angular velocity of rotation of the body. We express

$$\mathbf{V} = (V_1, V_2, V_3), \quad \boldsymbol{\Omega} = (V_4, V_5, V_6), \qquad (10.154)$$

where V_1, V_2, and V_3 mean the components of $\mathbf{V}$ along the body-fixed coordinate axis. V_4, V_5, and V_6 have similar meaning for $\boldsymbol{\Omega}$. We introduce

$$B_j = \sum_{k=1}^{6} A_{jk} V_k, \quad j = 1, \ldots 6. \qquad (10.155)$$

Here A_{jk} means added mass coefficients. These can be calculated by, for instance, a 3D boundary element method (BEM). We define the vectors

$$\mathbf{B} = (B_1, B_2, B_3), \quad \mathbf{I} = (B_4, B_5, B_6). \qquad (10.156)$$

The hydrodynamic force vector $\mathbf{F}$ acting on the body is

$$\mathbf{F} = -\frac{\partial \mathbf{B}}{\partial t} - \boldsymbol{\Omega} \times \mathbf{B} \qquad (10.157)$$

and the hydrodynamic moment vector $\mathbf{M}$ with respect to the body-fixed coordinate system is

$$\mathbf{M} = -\frac{\partial \mathbf{I}}{\partial t} - \boldsymbol{\Omega} \times \mathbf{I} - \mathbf{V} \times \mathbf{B}. \qquad (10.158)$$

a) Consider a ship that maneuvers in the horizontal plane. The water has infinite depth and horizontal extent. The Froude number is assumed moderate so that the rigid free-surface condition applies. Use eqs. (10.157) and (10.158) to express consistently with Figure 10.5 the yaw moment and the longitudinal and transverse forces on the ship in terms of the yaw angular velocity and the longitudinal and lateral components of ship velocity. The expressions should account for the fact that the hull is symmetric about the body-fixed xz-plane defined in Figure 10.4. You must explain why this causes some of the coupled added mass coefficients A_{jk} to be zero. (Hint: Start with how added mass was explained in section 7.2.1 and discuss symmetry and antisymmetry of the flow with respect to the xz-plane due to forced surge, sway, and yaw velocity.)

The answer is

$$\begin{aligned} X &= -A_{11}\dot{u} + (A_{22}v + A_{26}\dot{\psi})\dot{\psi} \\ Y &= -A_{22}\dot{v} - A_{26}\ddot{\psi} - A_{11}u\dot{\psi} \qquad (10.159) \\ N &= -A_{62}\dot{v} - A_{66}\ddot{\psi} + (A_{11} - A_{22})vu - A_{26}u\dot{\psi}. \end{aligned}$$

Here X, Y, N, u, v, and $r = \dot{\psi}$ are consistent with Figure 10.4 and A_{jk} are the low-frequency added mass coefficients for the ship.

Modify eq. (10.159) by introducing lifting terms consistent with linear slender body theory.

b) Why can we not apply the infinite fluid results by Kochin et al. (1964) to the roll of a ship moving at moderate Froude number? Consider now the free-surface condition that the velocity potential due to body motion is zero on the mean free surface. Use the infinite fluid results to derive hydrodynamic transverse force, roll, and yaw moment due to roll.

10.10.3 Maneuvering in waves and broaching

We will consider the hydrodynamic loads during maneuvering of a monohull at moderate speed in linear regular deep-sea waves with a wavelength λ that is long relative to the cross dimensions of the vessel.

a) Show by making a coordinate transformation of the results in Table 3.1 that the incident wave potential can be represented as

$$\varphi_0 = \frac{g\zeta_a}{\omega_0} e^{kz} \cos(\omega_e t - kx\cos\beta - ky\sin\beta + \varepsilon) \qquad (10.160)$$

in the body-fixed coordinate system defined in Figure 10.5 with $z = 0$ in the mean free surface and positive z upward. Here

$$\omega_e = \omega_0 + ku\cos\beta, k = \frac{\omega_0^2}{g} = \frac{2\pi}{\lambda}, \qquad (10.161)$$

where u is the forward speed component of the vessel along the negative x-axis. Further, β is the wave propagation direction measured relative to the x-axis, so that $\beta = 0^\circ$ is head sea and $\beta = 180^\circ$ is following sea.

b) The wave-induced hydrodynamic loads can be divided into Froude-Kriloff and diffraction loads (see section 7.2). We consider first the Froude-Kriloff loads, that is, the loads following by integrating the pressure loads of the incident waves. Consider the horizontal Froude-Kriloff force f_2^{FK} per unit length on a cross section of the vessel. Show that this can be approximated as

$$f_2^{FK} = \rho A(x) a_2 \qquad (10.162)$$

for long wavelengths relative to the cross-sectional dimension. Here $A(x)$ is the mean submerged cross-sectional area and

$$a_2 = \omega_0^2 \sin\beta e^{kz_m} \zeta_a \cos(\omega_e t - kx\cos\beta + \varepsilon), \qquad (10.163)$$

where z_m is the z-coordinate of the centroid of $A(x)$.

(Hint: Use the generalized Gauss theorem given by eq. (2.205).)

c) Linear sway and yaw motions and small ω_e are assumed. Explain that the horizontal diffraction force follows from modifying eq. (10.19) as

$$f_2 = -\left(\frac{\partial}{\partial t} + U\frac{\partial}{\partial x}\right)\left[a_{22}\left(\dot\eta_2 + x\dot\eta_6 - v_2\right)\right], \qquad (10.164)$$

where

$$v_2 = \omega_0 \zeta_a e^{kz_m} \sin\beta \sin(\omega_e t - kx\cos\beta + \varepsilon). \qquad (10.165)$$

Which part is the diffraction force per unit length? You must argue that the diffraction potential satisfies a rigid free-surface condition.

d) Express the wave-induced transverse force and yaw moment on the ship.

e) We will now study broaching in following waves at zero frequency of encounter. Broaching is caused by directional instability. We express β as $\pi - \eta_6$, where η_6 is the yaw angle. Hydrodynamic forces and moments that are linear in η_6 will be derived. Show that v_2 given by eq. (10.165) can then be approximated as

$$v_2 \approx \omega_0 \zeta_a e^{kz_m} \eta_6 \sin(\varepsilon + kx).$$

Express ε as $\pi/2 + ka$ and explain that $x = -a$ corresponds to a wave crest.

In order to study the directional stability in waves, you must first formulate the linear equations of motions in yaw and sway. When doing this, you should assume a PD controller (see eq. (10.80)) and include rudder forces and moment.

Derive the characteristic equation that determines the eigenvalues.

Does the directional stability depend on the position of the vessel in waves?

10.10.4 Linear coupled sway-yaw-roll motions of a monohull at moderate speed

Coupled sway-roll-yaw was analyzed in section 10.9.3 by using seakeeping nomenclature and a coordinate system that differs from the traditional maneuvering coordinate system.

a) Consider the body-fixed coordinate system (x, y, z) and the nomenclature for velocity, angular velocity, and external forces and moments defined in Figure 10.32. Start out with the slender body results for added mass and damping given by eqs. (10.33) through (10.36) and eqs. (10.140) through (10.144), and express the hydrodynamic derivative terms $Y_{\dot p}$, Y_p, $L_{\dot v}$, L_v, $L_{\dot p}$, L_p, $L_{\dot r}$, L_r, $N_{\dot p}$, and N_p.

(Hints: You must transfer the added mass (A_{jk}) and damping (B_{jk}) coefficients to a coordinate system with origin in the center of gravity. This means you must consider forced velocities, forces, and moments with respect to this coordinate system. You must then realize the differences in positive signs of velocities, forces, and moments between Figure 10.32 and the coordinate system used in section 10.9.3.

b) Consider a ship in steady state turning with heel. Generalize the analysis in section 10.3.5 to include the effect of heel and derive an expression for the yaw rate. Discuss why E' given by eq. (10.151) is an important part of the analysis.

10.10.5 High-speed motion in water of an accidentally dropped pipe

Pipes may be accidentally dropped from offshore platforms used for oil and gas production. Pipe impact may cause damage to risers and subsea equipment. The drop height above the mean water level may vary from 20 to 60 m. The orientation

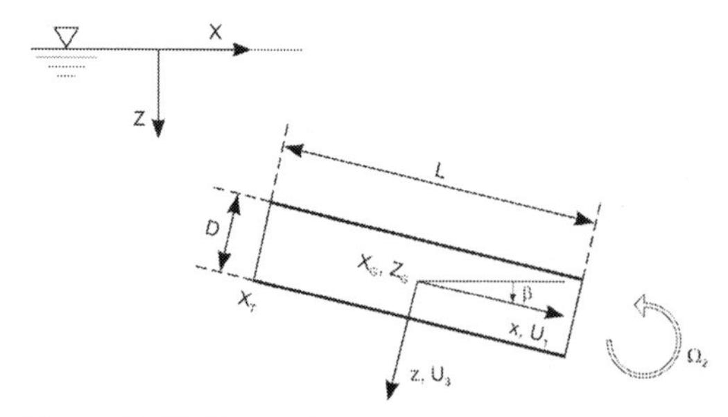

Figure 10.37. Nomenclature and coordinate systems of motions of a slender pipe in the Earth-fixed X-Z–plane. The body-fixed coordinate system (x, z) has origin in the center of gravity of the pipe. U_1 and U_3 are translatory velocity components of the pipe along the x- and z-axes, respectively. Ω_2 is the rate of turn of the pipe.

of the pipe when it hits the water surface is an unknown parameter. The water entry of the pipe will subsequently change the orientation and velocity of the pipe. We will focus on the next phase, when the pipe is completely submerged and has no influence on the free surface. The situation is illustrated in Figure 10.37. We assume 2D flow in the global vertical X-Z–plane. The X- and Z-coordinates of the center of gravity of the pipe are denoted X_G and Z_G. We introduce a body-fixed coordinate system (x, y, z) as illustrated in Figure 10.37. The origin is in the center of gravity of the pipe. The pipe has translatory velocity components U_1 and U_3 along the x- and z-axes, respectively. The angle of the pipe axis relative to the X-axis is denoted β, and the angular velocity of the pipe about the y-axis is called Ω_2.

We can then set up the following relationships:

$$\frac{dX_G}{dt} = U_1 \cos\beta + U_3 \sin\beta \quad (10.166)$$

$$\frac{dZ_G}{dt} = U_3 \cos\beta - U_1 \sin\beta \quad (10.167)$$

$$\frac{d\beta}{dt} = \Omega_2. \quad (10.168)$$

We will first assume the pipe has end caps. The following three equations follow from Newton's second law:

$$(M + A_{11})\frac{dU_1}{dt} = F_{Dx} - (M - \rho\nabla)\, g \sin\beta + F_{x?} \quad (10.169)$$

$$(M + A_{33})\frac{dU_3}{dt} = -|U_1|\, U_3 a_{33T} + U_1 (x_T a_{33T} + M)\,\Omega_2 + F_{Dz} + (M - \rho\nabla)\, g\cos\beta + F_{z?} \quad (10.170)$$

$$(I_{55} + A_{55})\frac{d\Omega_2}{dt} = U_1 (A_{33} + x_T a_{33T}) \times U_3 - x_T a_{33T}\Omega_2 |U_1| + M_{Dy} + M_{z?}. \quad (10.171)$$

Here the longitudinal viscous force can be expressed as

$$F_{Dx} = -0.5\rho C_F \pi DLU_1 |U_1| - \frac{\rho}{8}\pi C_{Dx} D^2 U_1 |U_1|, \quad (10.172)$$

where

$$C_F = 0.0015 + \left(0.30 + 0.015\left(\frac{2L}{D}\right)^{0.4}\right) Rn^{-1/3}. \quad (10.173)$$

The C_F-value assumes turbulent axisymmetric flow along a smooth surface (White 1972). Rn means the Reynolds number. Further, C_{Dx} is a base drag coefficient that may be set equal to 0.65 (Hoerner 1965). F_{Dz} and M_{Dy} in eqs. (10.170) and (10.171) also represent viscous loads. The mass of the pipe is called M, and we will assume uniform mass distribution. In eqs. (10.169) through (10.171), we have not explicitly expressed all the mass and added mass terms. The latter terms are simply denoted $F_{x?}$, $F_{z?}$, $M_{z?}$.

a) Explain the different terms in eqs. (10.169) through (10.171). Expressions for $F_{x?}$, $F_{z?}$, and $M_{z?}$ should be presented. Use the cross-flow principle to formulate the viscous terms F_{Dz} and M_{Dy} in eq. (10.170) and eq. (10.171), respectively. One should make a special effort to explain why $|U_1|$ appears in the equations.

b) Assume now that the pipe has no end caps and there is a flow through the pipe. Consider the special case of steady incident flow with a constant small angle of attack relative to the cylinder axis and assume that the inner diameter of the pipe is equal to the outer diameter.

Show by slender body theory that the lift force with interior flow is twice the lift force on the same pipe with end caps.

APPENDIX

Units of Measurement and Physical Constants

The fundamental units of measurement in mechanics are mass, length, and time. The basic units in the SI system are kilograms (kg), meters (m), and seconds (s) for mass, length, and time, respectively. Special names are given to derived units in the SI system. The unit force is one newton (N), which is equal to one kilogram-meter per second squared ($kgms^{-2}$). One pascal (Pa) is the unit of pressure and stress. This is the same as one newton per meter-squared (Nm^{-2}). The unit of work or energy is one joule (J), that is, one newton-meter (Nm). The unit power is one watt (W), equal to one joule per second (Js^{-1}). Prefixes denote decade factors, such as kilo- (k) for 10^3, mega- (M) for 10^6, giga- (G) for 10^9,tera- (T) for 10^{12}, centi- (c) for 10^{-2}, milli- (m) for 10^{-3}, and micro- (μ) for 10^{-6}.

Table A.1 presents the relationship between the SI system and some other commonly used measurement systems. Table A.2 lists values of density and viscosity of water and air, whereas Table A.3 shows how the vapor pressure varies with the temperature. The standard acceleration of gravity (g) is equal to 9.80665 ms^{-2}. The standard atmospheric pressure at sea level is 1.01325×10^5 Nm^{-2}. The surface tension of the interface between air and water varies between 0.076 Nm^{-1} and 0.071 Nm^{-1} for the temperature range 0° to 30°C. Representative values for speed of sound in water and air are 1500 ms^{-1} and 340 ms^{-1}, respectively.

Table A.1. *Conversion factors for different units of measurement*

Quantity	SI unit	Other unit	Inverse factor
Length	1 m	3.281 feet (ft)	0.3048 m
	1 km	0.540 nautical miles	1852 m
Area	1 m^2	10.764 ft^2	0.0929 m^2
Volume	1 m^3	35.315 ft^3	0.0283 m^3
	1 m^3	264.2 gallons (US)	0.00379 m^3
	1 m^3	220.0 gallons (UK)	0.00455 m^3
Velocity	1 ms^{-1}	3.281 fts^{-1}	0.305 ms^{-1}
	1 ms^{-1}	1.944 knots	0.514 ms^{-1}
Mass	1 kg	2.205 pounds	0.454 kg
	1000 kg	0.984 tons (long)	1016 kg
	1000 kg	1 tonne (metric)	1000 kg
Force	1 N	0.225 pound-force	4.448 N
	1 N	0.1020 kg-force (kgf)	9.807 N
	1 MN	102.0 tonne-force	9807 N
	1 MN	100.4 ton-force	9964 N
Pressure	1 Nm^{-2}	0.000145 psi (pounds per square inch)	6895 Nm^{-2}
	1 Nm^{-2}	10^{-5} bar	100 kNm^{-2}
Energy	1 J	0.738 foot-pounds	1.356 J
	1 W	0.00134 horsepower	745.7 W

Table A.2. *Mass density (ρ) and kinematic viscosity (ν) of water and air*

	Freshwater		Saltwater (salinity 3.5%)		Dry air	
Temperature	$\rho(kg\ m^{-3})$	$\nu \cdot 10^6(m^2s^{-1})$	$\rho(kg\ m^{-3})$	$\nu \cdot 10^6(m^2s^{-1})$	$\rho(kg\ m^{-3})$	$\nu \cdot 10^6(m^2s^{-1})$
0°C	999.8	1.79	1028.0	1.83	1.29	13.2
5°C	1000.0	1.52	1027.6	1.56	1.27	13.6
10°C	999.7	1.31	1026.9	1.35	1.25	14.1
15°C	999.1	1.14	1025.9	1.19	1.23	14.5
20°C	998.2	1.00	1024.7	1.05	1.21	15.0

Table A.3. *Vapor pressure of water for various temperatures (Breslin and Andersen 1994)*

Temperature		Vapor pressure,
°C	°F	$p_V(Nm^{-2})$
0	32	610.8
5	41	871.8
10	50	1227.1
15	59	1704.0
20	68	2336.9
25	77	3166.6
30	86	4241.4
35	95	5622.2
40	104	7374.6

References

Aarsnes, J. V., 1984, Current forces on ships. Dr.ing. thesis, Report UR-84-39, Dept. of Marine Hydrodynamics, Nor. Inst. Technol., Trondheim.

Aarsnes, J. V., Faltinsen, O. M., Pettersen, B., 1985, Application of a vortex tracking method to current forces on ships. In *Proc. Conf. Separated Flow around Marine Structures*, pp. 309–46, Trondheim: Nor. Inst. Technol.

Abbott, J. H., Doenhoff, A. E. von, 1959, *Theory of Wing Sections*, New York: Dover Publications, Inc.

Abramowitz, M., Stegun, I., 1964, *Handbook of Mathematical Functions with Formulas, Graphs and Mathematical Tables*, New York: Dover Publications Inc.

Abramson, N., 1974, Structural dynamics of advanced marine vehicles, In *Int. Symp. Dynamics of Marine Vehicles and Structures in Waves*, ed. R. E. D. Bishop, W. G. Price., pp. 344–57, London: Mechanical Engineering Publications Ltd.

Adegeest, L. J. M., 1995, Nonlinear hull girder loads, Ph.D. thesis, Delft University of Technology, Faculty Mech. Eng. and Mar. Tech., Delft.

Allison, J., 1993, Marine waterjet propulsion, *Trans. SNAME*, **101**, 275–335.

Anderson, J. D., 2001, *Fundamentals of Aerodynamics*, third edition, New York: McGraw-Hill Book Company.

Andrewartha, M., Doctors, L., 2001, How many foils? A study of multiple hydrofoil configurations. In *Proc. FAST2001*, Vol. 3, pp. 79–86, London: The Royal Institution of Naval Architects.

Ankudinov, V., Kaplan, P., Jacobsen, B. K., 1993, Assessment and principal structure of the modular mathematical model for ship maneuverability prediction and real-time maneuvering simulations, In *Proc. MARSIM'93*, St. John's, Newfoundland.

Arai, M., Myanchi, T., 1998, Numerical study of the impact of water on cylindrical shells, considering fluid-structure interactions, In *Proc. PRADS'98*, ed. M. C. W. Oosterveld, S. G. Tan, pp. 59–68, London and New York: Elsevier Applied Science.

Armand. J. L., Cointe, R., 1986, Hydrodynamic impact analysis of a cylinder, In *Proc. Fifth Int. Offshore Mech. and Arctic Engng. Symp.*, Vol. 1, pp. 609–34, ASME.

Auf'M Keller, W. H., 1973, Extended diagrams for determining the resistance and required power for single-screw ships, *Intern. Shipb. Progr.*, **20**, 133–42.

Baarholm, R. J., 2001, Theoretical and experimental studies of wave impact underneath decks of offshore platforms, Dr. Ing thesis, Dept. of Marine Hydrodynamics, NTNU, Trondheim.

Baba, E., 1969, Study on separation of ship resistance components, Mitsubishi Technical Bulletin, No. 59.

Bailey, D. S., 1976, The NPl high speed round bilge displacement hull series, Maritime Technology Monograph No. 4, London, UK: RINA.

Baird, N., 1998, *The World Fast Ferry Market*, Melbourne, Australia: Baird Publications.

Bal, S., Kinnas, S. A., Lee, H., 2001, Numerical analysis of 2-D and 3-D cavitating hydrofoils under a free surface, *J. Ship Res.*, **45**, 1, 34–49.

Barcellona, M., Landrini, M., Greco, M., Faltinsen, O. M., 2003, An experimental investigation of bow water shipping, *J. Ship Res.*, **47**, 4, 327–46.

Barringer, I. E., 1998, The hydrodynamics of ship sections entering and exiting a fluid, Ph.D. thesis, Dept. of Mathematics, Brunel University.

Batchelor, G. K., 1967, *An Introduction to Fluid Dynamics*, Cambridge: Cambridge University Press.

Beek, T. van, 1992, Application limits for propellers at high speeds, In *Hydrodynamics: Computations, Model Tests and Reality*, ed. H. J. J. van den Boom, pp. 121–32, Amsterdam: Elsevier Science Publishers BV.

Berlekom, W. B. van, Goddard, T. A., 1972, Maneuvering of large tankers, *Trans. SNAME*, **80**, 264–98.

Berstad, A. J., Faltinsen, O. M., Larsen, C. M., 1997, Fatigue crack growth in side longitudinals, In *Proc. NAV&HSMV*, pp. 5.3–15, Naples: Dipartimento Ingeneria-Università di Napoli "Federico II."

Berstad, A. J., Larsen, C. M., 1997, Fatigue crack growth in the hull structure of high speed vessels, In *Proc. FAST'97*, ed. N. Baird, Vol. 1, pp. 255–62, South Yarra, Victoria, and London: Baird Publications.

Bertram, V., 1999, Numerical investigation of steady flow effects in three-dimensional seakeeping computations. In *Proc. 22nd Symposium on Naval Hydrodynamics*, Washington D.C.: Office of Naval Research–Dept. of the Navy.

Bertram, V., Iwashita, H., 1996, Comparative evaluation of various methods to predict seakeeping of fast ships, *Schiff & Hafen*, **48**, 6, 54–8.

Besch, P. K., Liu, Y-N, 1972, Bending flutter and torsional flutter of flexible hydrofoil struts, In *Proc. Ninth Symposium on Naval Hydrodynamics*, ed. R. Brard, A. Castera, Vol. 1, pp. 343–400, Arlington, Va.: Office of Naval Research–Department of the Navy.

Bethwaite, F., 1996, *High Performance Sailing*, Shrewsbury, England: Waterline.

Beukelman, W., 1991, Slamming on forced oscillating wedges at forward speed, Part I: Test results, Rep. no. 888, Delft University of Technology, Ship Hydromechanics Laboratory, Netherlands.

Billingham, J., King, A. C., 2000, *Wave Motion*, Cambridge: Cambridge University Press.

Birkhoff, G., Zarantonello, E. H., 1957, *Jets, Wakes and Cavities*, New York: Academic Press Inc.

Bishop, R. E. D., Price, W. G., 1979, *Hydroelasticity of Ships*, Cambridge: Cambridge University Press.

Bisplinghoff, R. L., Ashley, H., Halfman, R. L., 1996, *Aeroelasticity*, New York: Dover Publications.

Blevins, R. D., 1990, *Flow Induced Vibration*, Malabar, Florida: Krieger Publishing Company.

Blok, J. J., Beukelman, W., 1984, The high speed displacement ship systematic series hull forms, *Trans. SNAME*, **92**, 125–50.

Blount, D. L., 1997, Design of propeller tunnels for high-speed craft, In *Proc. FAST'97*, ed. N. Baird, Vol. 1, pp. 151–6, South Yarra, Victoria, and London: Baird Publications.

Blount, D. L., Codega, L. T., 1992, Dynamic stability of planing boats, *Marine Technology*, **29**, 1, 4–12.

Bouard, R., Coutanceau, M., 1980, The early stage of development of the wake behind an impulsively started cylinder for $40 < Re < 10^4$, *J. Fluid Mech.*, **101**, 3, 583–607.

Bowden, B., Davison, N., 1974, Resistance increments due to hull roughness associated with form factor extrapolation methods, NPL Ship Division Report TM 380.

Brennen, C. E., 1995, *Cavitation and Bubble Dynamics*, Oxford and New York: Oxford University Press.

Breslin, J. P., 1957, Application of ship-wave theory to the hydrofoil of finite span, *J. Ship Res.*, **1**, 1, 27–55.

Breslin, J. P., 1958, Discussion of the paper by K. L. Wadlin, In *Proc. Second Symp. on Naval Hydrodynamics*, pp. 434–40, Washington, D.C.: Office of Naval Research–Department of the Navy.

Breslin, J. P., 1994, Hydrofoil ships – fantasies, facts and fysiks, DCAMM Anniversary Volume, Danish Centre for Appl. Math. and Mech., Techn. Univ. of Denmark, Lyngby.

Breslin, J. P., Andersen, P., 1994, *Hydrodynamics of Ship Propellers*, Cambridge: Cambridge University Press.

Brix, J., 1993, *Maneuvering Technical Manual*, Hamburg: Seehafen Verlag GmbH.

Brizzolara, S., 2003, Hydrodynamic analysis of interceptors with CFD methods, In *Proc. FAST'2003*, ed. P. Cassella, Vol. III, Session E, pp. 49–56. Naples: Dipartimento Ingegneria Navale–Universitè di Napoli "Federico II."

Brug, J. B. van den, Beukelman, W., Prins, G. J., 1971, Hydrodynamic forces on a surface piercing flat plate, Report no. 325, Shipbuilding Laboratory, Delft University of Technology, Delft, The Netherlands.

Bryant, J. P., 1983, Waves and wave groups in deep water, In *Nonlinear Waves*, ed. L. Debnath, Ch. 6, pp. 100–15, Cambridge: Cambridge University Press.

Buckingham, E., 1915, Model experiments and the forms of empirical equations, *Trans. ASME*, **37**, 263–96.

Carlton, J. S., 1994, *Marine Propellers and Propulsion*, Oxford: Butterworth-Heineman.

Carstensen, C., 1983, Beitrag zur Berechnung von ebenen Einlaufströmungen, Dissertation, Technische Universität Berlin, D83, März.

Casanova, R. L., Latorre, R., 1992, The achievements of high performance in marine vehicles over the period 1970–1990, In *Proc. HPMV'92*, pp. O/A61–O/A66, Alexandria, Va.: American Society of Naval Engineers.

Cebeci, T., 2004, *Analysis of Turbulent Flows*, second revised and expanded edition, Oxford: Elsevier.

Celano, T., 1998, The prediction of porpoising inception for modern planing craft, *Trans. SNAME*, **106**, 269–92.

Chapman, R. B., 1972, Hydrodynamic drag of semisubmerged ships, *J. of Basic Eng., Trans. ASME*, **94 Series D**, 4, 879–84.

Chapman, R. B., 1976, Free surface effects for yawed surface-piercing plates, *J. Ship Res.*, **20**, 3, 125–36.

Chezhian, M., 2003, Three-dimensional analysis of slamming, Dr.ing thesis, Dept. of Marine Technology, NTNU, Trondheim, Norway.

Cleary, W. A., Robertson, J. B., Yagle, R. A., 1971, The results and significance of strength studies of Great Lakes bulk ore carrier, *Edward L. Ryerson*, In *SNAME Symp. on Hull Stresses in Bulk Carriers in the Great Lakes and Gulf of St. Lawrence Wave Environment*, Paper G, Jersey City, N.J.: The Society of Naval Architects and Marine Engineers.

Clement, E. P., Blount, D. L., 1963, Resistance tests of a systematic series of planing hull forms, *Trans. SNAME*, **71**, 201–77.

Clement, E. P., Koelbel, J. G., 1992, Optimized design for stepped planing monohulls and catamarans, In *Proc. HPMV'92*, pp. PC35–PC44, Alexandria, Va.: American Society of Naval Engineers.

Clough, R. W., Penzien, J., 1993, *Dynamics of Structures*, second edition, New York: McGraw-Hill, Inc.

Cohen, S., Blount, D., 1986, Research plan for the investigation of dynamic instability of small high-speed craft, *Trans. SNAME*, **94**, 197–214.

Cointe, R., 1991, Free surface flows close to a surface-piercing body, In *Mathematical Approaches in Hydrodynamics*, ed. T. Miloh, pp. 319–34, Philadelphia: Society for Industrial and Applied Mathematics.

Colagrossi, A., Lugni, C., Landrini, M., Graziani, G., 2001, Numerical and experimental transient tests for ship seakeeping, *Int. Journal Off. and Ocean Struct.*, **11**, 67–73.

Crane, C. L., Eda, H., Landsburg, A., 1989, Controllability, In *Principles of Naval Architecture,* Vol. III, Chapter IX, ed. E. V. Lewis, Jersey City, N.J.: The Society of Naval Architects and Marine Engineers.

Cummins, W. E., 1962, The impulse response function and ship motions, *Schiffstechnik*, **9**, 47, 101–9.

Cusanelli, D. S., Karafiath, G., 1997, Integrated wedge-flap for enhanced powering performance, In *Proc. FAST'97*, ed. N. Baird, Vol. 2, pp. 751–64, South Yarra, Victoria, and London: Baird Publications.

Day, J. P., Haag, R. J., 1952, *Planing Boat Porpoising*, New York: Webb Institute of Naval Architecture.

Day, S., Clelland, D., Nixon, E., 2003, Experimental and numerical investigation of "Arrow" Trimarans, In *Proc. FAST 2003*, ed. P. Casella, Vol. III, Session D2, pp. 23–36, Naples: Dipartimenta Ingegneria navale, Universtà di Napoli "Federico II."

Delany, N. K., Sorensen, N. E., 1953, Low-speed drag of cylinders of various shapes. Washington, D.C.: NACA Technical Note 3038.

Divitiis, N. de, Socio, L. M. de, 2002, Impact of floats on the water, *J. Fluid Mech.*, **471**, 365–79.

Dobrovol'skaya, Z. N., 1969, On some problems of fluid with a free surface, *J. Fluid Mech.*, **36**, 4, 805–29.

Doctors, L. J., 1978, Hydrodynamic power radiated by a heaving and pitching air-cushion vehicle, *J. Ship Res.*, **22**, 2, 67–79.

Doctors, L. J., 1992, The use of pressure distributions to model the hydrodynamics of air-cushion vehicles and surface effect ships, In *Proc. HPMV'92*, pp. SES56–SES72, Alexandria, Va.: American Society of Naval Engineers.

Doctors, L. J., 2003, Hydrodynamics of the flow behind a transom stern, In *Proc. Twenty-Ninth Israel Conference on Mechanical Engineering*, Paper 20–1, 11 pp., Haifa, Israel.

Doctors, L. J., Day, A. H., 1997, Resistance prediction for transom-stern vessels, In *Proc. FAST'97*, ed. N. Baird, Vol. 2, pp. 743–50, South Yarra, Victoria, and London: Baird Publications.

Doctors, L. J., Sharma, S. D., 1972, The wave resistance of an air-cushion vehicle in steady and accelerated motion, *J. Ship Res.*, **16**, 4, 248–60.

Dorf, R. C., Bishop, R. H., 1998, *Modern Control Systems*, Menlo Park, Calif.: Addison Wesley Longman, Inc.

Doyle, R., Whittaker, T. J. T., Elsasser, B., 2001, A study of fast ferry wash in shallow water, In *Proc. FAST2001*, Vol. 1, pp. 107–20, London: The Royal Institution of Naval Architects.

Eda, H., 1980, Rolling and steering performance of high speed ships – simulation studies of yaw-roll-rudder coupled instability, In *Proc. 13th Symp. on Naval Hydrodynamics*, pp. 115–31, Washington, D.C.: Office of Naval Research–Department of the Navy.

Etkin, B., 1959, *Dynamics of Flight: Stability and Control*, New York: John Wiley & Sons, Inc.

Falck, S., 1991, Seakeeping of foil catamarans. In *Proc. FAST'91*, ed. K. O. Holden, O. M. Faltinsen, T. Moan, Vol. 1, pp. 209–21, Trondheim: Tapir Publishers.

Faltinsen, O. M., 1972, Wave forces on a restrained ship in head-sea wave, In *Proc. Ninth Symp. on Naval Hydrodynamics*, ed. R. Brard and A. Castera, vol. 2, pp. 1763–844. Arlington, Va.: Office of Naval Research–Department of the Navy.

Faltinsen, O. M., 1983, Bow flow and added resistance of slender ships at high Froude number and low wavelengths, *J. Ship Res.*, **27**, 160–71.

Faltinsen, O. M., 1990, *Sea Loads on Ships and Offshore Structures*, Cambridge: Cambridge University Press.

Faltinsen, O. M., 1997, The effect of hydroelasticity on slamming, *Phil. Trans. R. Soc. Lond. A*, **355**, 575–91.

Faltinsen, O. M., 1999, Water entry of a wedge by hydroelastic orthotropic plate theory, *J. Ship Res.*, **43**, 3, 180–193.

Faltinsen, O. M., 2000, Water impact in ship and ocean engineering, In *Proc. Fourth Int. Conf. on Hydrodyn.*, ed. Y. Goda, M. Ikehata, K. Suzuki, pp. 17–36, Yokohama: ICHD2000 Local Organizing Committee.

Faltinsen, O. M., 2001, Steady and vertical dynamic behaviour of prismatic planning hulls, In *Proc. 22nd Intern Conf. HADMAR 2001*, pp. 89–104, Varna, Bulgaria: Bulgarian Ship Hydrodynamics Centre.

Faltinsen, O. M., Helmers, J. B., Minsaas, K. J., Zhao, R., 1991a, Speed loss and operability of catamarans and SES in a seaway, In *Proc. FAST'91*, ed. K. O. Holden, O. M. Faltinsen, T. Moan. Vol. 2, pp. 709–25, Trondheim: Tapir Publishers.

Faltinsen. O. M., Hoff, J. R., Kvålsvold, J., Zhao, R., 1992, Global wave loads on high-speed catamarans, In *Proc. PRADS'92*, ed. J. B. Caldwell, G. Ward, Vol. 1, pp. 1.360–1.375, London and New York: Elsevier Applied Science.

Faltinsen, O. M., Holden, K. O., Minsaas, K. J., 1991b, Speed loss and operational limits of high-speed marine vehicles, In *Proc. IMAS'91 – High Speed Marine Transportation*, pp. 13–21, London: The Institute of Marine Engineers.

Faltinsen, O. M., Kvålsvold, J., Aarsnes, J. V., 1997, Wave impact on a horizontal elastic plate, *J. Mar. Sci. Technol.*, **2**, 2, 87–100.

Faltinsen, O. M., Landrini, M., Greco, M., 2004, Slamming in marine applications, *J. Eng. Math.*, **48**, 187–217.

Faltinsen, O. M., Minsaas, K., Liapis, N., Skjørdal, S. O., 1980, Prediction of resistance and propulsion of a ship in a seaway, In *Proc. 13th Symp. on Naval Hydrodynamics*, ed. T. Inui, pp. 505–30. Tokyo: The Shipbuilding Research Association of Japan.

Faltinsen, O. M., Pettersen, B., 1983, Vortex shedding around two-dimensional bodies at high Reynolds number, In *Proc. 14th Symp. on Naval Hydrodynamics*, pp. 1171–213, Washington, D.C.: National Academy Press.

Faltinsen, O. M., Svensen, T. E., 1990, Incorporation of seakeeping theories on CAD, In *Proc. of Int. Symp. CFD and CAD in Ship Design*, ed. G. van Oortmersen, pp. 147–64, Amsterdam: Elsevier Science Publishers, B.V.

Faltinsen, O. M., Zhao, R., 1991a, Numerical predictions of ship motions at high forward speed, *Phil. Trans. R. Soc. Lond. A*, **334**, 241–52.

Faltinsen, O. M., Zhao, R., 1991b, Flow prediction around high-speed ships in waves, In *Mathematical Approaches in Hydrodynamics*, ed. T. Miloh, pp. 265–88, Philadelphia: Society for Industrial and Applied Mathematics.

Faltinsen, O. M., Zhao, R., 1998, Water entry of ship sections and axisymmetric bodies, In *AGARD Report 827 High Speed Body Motions in Water*, pp. 24-1–24-11, Neuilly-Sur-Seine, Cedex, France: AGARD/NATO.

Feifel, M. W., 1981, Advanced numerical methods for hydrofoil systems design and experimental verification, In *Proc. Third Int. Conf. on Num. Ship Hydrodynamics*, ed. J-C. Dern, H. J. Haussling, pp. 365–74, Paris: Bassin d'Essais des Carénes.

Fischer, H., Matjasic, K., 1999, The Hoverwing technology – bridge between WIG and ACV, In *RTO Meeting Proc. 15 – Fluid Dynamics Problems of Vehicles Operating Near or in the Air-Sea Interface*, pp. 30-1–30-7, Neuilly-Sur-Seine, Cedex, France: Res. and Techn. Org., NATO.

Flagg, C. N., Newman, J. N., 1971, Sway added-mass coefficients for rectangular profiles in shallow water, *J. Ship Res.*, **15**, December, 257–65.

Fontaine, E., Faltinsen, O. M., Cointe, R., 2000, New insight into generation of ship bow waves, *J. Fluid Mech.*, **421**, 15–38.

Førde, M., Ørbekk, E., Kubberud, N., 1991, Computational fluid dynamics applied to high speed craft with special attention to water intake for water jets, In *Proc. FAST'91*, ed. K. O. Holden, O. M. Faltinsen, T. Moan., Vol. 1, pp. 69–89, Trondheim: Tapir Publishers.

Fossen, T. I., 2002, *Marine Control Systems: Guidance, Navigation and Control of Ships, Rigs and Underwater Vehicles*, Trondheim: Marine Cybernetics AS.

Fridsma, G., 1969, A systematic study of the rough-water performance of planing boats, Report 1275, Davidson Laboratory, Stevens Institute of Technology, Hoboken, N.J.

Fridsma, G., 1971, A systematic study of the rough-water performance of planing boats; part 2, irregular waves, Davidson Laboratory, Stevens Institute of Technology, Hoboken, N.J.

Froude, W. W., 1877, Experiments upon the effect produced on the wave-making resistance of ships by length of parallel middle body, *Trans. Inst. of Naval Arch.*, London, UK.

Fujino, M., 1968, Experimental studies on ship maneuverability in restricted waves – part I, *Intern. Shipbuilding Progr.*, **15**, 168, 279–301.

Fujino, M., 1976, Maneuverability in restricted waters: state of the art, report no. 184, Dept. of Nav. Arch. and Mar. Eng., The University of Michigan, Ann Arbor, Mich.

Fujino, M., 1996, Keynote lecture: prediction of ship manoeuvrability: state of the art, *Marine Simulation and Ship Manoeuvrability*, ed. M. S. Chislett, pp. 371–87, Rotterdam: Balkema.

Garrett, R., 1987, *The Symmetry of Sailing: The Physics of Sailing for Yachtsmen*, London: Adlard Coles.

Gawn, R. W. L., 1953, Effect of pitch and blade width on propeller performance, *Trans. RINA*, **95**, 157–93.

Ge, C., 2002, Global hydroelastic response of catamarans due to wetdeck slamming, Dr.Ing thesis, Dept. of Marine Technology, NTNU, Trondheim.

Ge, C., Faltinsen, O. M., Moan, T., 2005, Global hydroelastic response of catamarans due to wetdeck slamming, *J. Ship Res.*, **49**, 1, 24–42.

Gerritsma, J., Beukelman, W., 1972, Analysis of the resistance increase in waves of a fast cargo ship, *Intern. Shipbuilding Progr.*, **19**, 217, 285–93.

Geurst, J. A., 1960, Linearized theory for fully cavitated hydrofoils, *Intern. Shipbuilding Progr.*, **7**, 65, 17–27.

Giesing, J. P., 1968, Nonlinear two-dimensional unsteady potential flow with lift, *J. Aircraft*, **5**, 2, 135–43.

Goodman, R. A., 1971, Wave excited main hull vibration in large bulk carriers and tankers, *Trans. RINA*, **113**, 167–84.

Gradshteyn, I., Ryzhik, I., 1965, *Tables of Integrals Series and Products*, fourth ed., London and New York: Academic Press.

Graff, W., Kracht, A., Weinblum, G. P., 1964, Some extensions of DW Taylor Standard Series, *Trans SNAME*, **72**, 374–403.

Greco, M., 2001, A two-dimensional study of green water loading. Dr.Ing thesis, Dept. of Marine Hydrodynamics, NTNU, Trondheim.

Greco, M., Landrini, M., Faltinsen, O. M., 2003, Local hydroelastic analysis on a VLFS with shallow draft, In *Proc. Hydroelasticity in Marine Technology*, ed. R. Eatock Taylor, pp. 201–14, Oxford: Dept. of Eng. Science, University of Oxford.

Green, A. E., 1936, Note on the gliding of a plate on the surface of a stream, *Proc. Cambridge Phil.Soc.*, **32**, 248–52.

Greenhow, M., 1986, High- and low-frequency asymptotic consequences of the Kramers-Kronig relations, *J. Eng. Math.*, **20**, 293–306.

Greenhow, M., Li, Y., 1987, Added masses for circular cylinders near or penetrating fluid boundaries – review, extension and application to water-entry, exit and slamming, *Ocean Engng.*, **14**, 4, 325–48.

Greenhow, M., Lin, W., 1983, *Non-linear Free Surface Effects: Experiments and Theory*, Report No. 83-19, Dept. Ocean Engn., Cambridge, Mass: Mass. Inst. Technol.

Grigoropoulos, G. J., Loukakis, T. A., 2002, Resistance and seakeeping characteristics of a systematic series in the pre-planing condition (part I), *Trans. SNAME*, **110**, 77–113.

Grim, O., 1955, Die hydrodynamischen Kräfte beim Rollversuch, *Schiffstechnik*, **3**, 14/15, 147–51.

Haarhoff, S., Sharma, S. D., 2000, A note on the influence of speed and metacentric height on the yaw-rate stability of displacement ships, Intern. Workshop on Ship Maneuvering at the Hamburg Ship Model Basin, Hamburg, Germany, October 10–11.

Hackmann, D., 1979, Written discussion to Jensen, J. J. and Pedersen, P. T. (1978).

Halstensen, S. O., Leivdal, P. A., 1990, The development of the SpeedZ Propulsion System, In *Seventh International High Speed Surface Craft Conference*, Kingston upon Thames: High Speed Surface Craft Ltd.

Hama, F. R., Long, J. D., Hegart, J. C., 1956, On transition from laminar to turbulent flow, University of Maryland, Technical Note BN-81, AFOSR-TN-56-381.

Hama, F. R., 1957, An efficient tripping device, *J. Aeronautical Sciences*, March.

Hamamoto, M., Inoue, K., Kato, R., 1993, Turning motion and directional stability of surface piercing hydrofoil craft, In *Proc. FAST'93*, ed. K. Sugai, H. Miyata, S. Kubo, H. Yamato, Vol. 1, pp. 807–18, Tokyo: The Society of Naval Architects of Japan.

Hansen, P. F., Jensen, J. J., Pedersen, T. P., 1994, Wave-induced springing and whipping of high-speed vessels, In *Proc. Hydroelasticity in Marine Technology*, ed. O. M. Faltinsen, C. M. Larsen, T. Moan, K. Holden, N. S. Spidsøe, pp. 191–204, Rotterdam and Brookfield: A. A. Balkema.

Hansen, P. F., Jensen, J. J., Pedersen, T. P., 1995, Long term springing and whipping stresses in high speed vessels, In *Proc. FAST'95*, ed. C. F. L. Kruppa, Vol. 1, pp. 473–85, Berlin and Hamburg: Schiffbautechnische Gesellschaft e.V.

Haugen, E. M., 1999, Hydroelastic analysis of slamming on stiffened plates with application to catamaran wetdeck, Dr.ing. thesis, Dept. of Marine Hydrodynamics, NTNU, Trondheim.

Haugen, E. M., Faltinsen, O. M., Aarsnes, J. V., 1997, Application of theoretical and experimental studies of wave impact to wetdeck slamming, In *Proc. FAST'97*, ed. N. Baird, Vol. 1, pp. 423–30, South Yarra, Victoria, and London: Baird Publications.

Havelock, T. H., 1908, The propagation of groups of waves in dispersive media with application to waves on water produced by a travelling disturbance, *Proc. Royal Soc., London, Series A*, **LXXXI**, 398–430.

Havelock, T. H., 1963, *Collected Papers*, ed. C. Wigley, Washington, D.C.: Office of Naval Research.

Hayman, B., Haug, B., Valsgård, S., 1991, Response of fast craft hull structures to slamming loads, In *Proc. FAST'91*, ed. K. O. Holden, O. M. Faltinsen, T. Moan, Vol. 1, pp. 381–398, Trondheim: Tapir Publishers.

Henry, C. J., Dugundji, J., Ashley, H., 1959, Aeroelastic stability of lifting surfaces in high-density fluids, *J. Ship Res.*, **2**, 4, 10–21.

Hermundstad, O. A., 1995, Theoretical and experimental hydroelastic analysis of high speed vessels, Dr.ing thesis, Dept. of Marine Structures, NTNU, Trondheim.

Hermundstad, O. A., Aarsnes, J. V., Moan, T., 1995, Hydroelastic analysis of a flexible catamaran and comparison with experiments, In *Proc. FAST'95*, ed. C. F. L. Kruppa, Vol. 1, pp. 487–500, Berlin and Hamburg: Schiffbautechnische Gesellschaft e.V.

Hermundstad, O. A., Aarsnes, J. V., Moan, T., 1997, Hydroelastic analysis of high speed catamaran in regular and irregular waves, In *Proc. FAST'97*, ed. N. Baird, Vol. 1, pp. 447–54, South Yarra, Victoria, and London: Baird Publications.

Hinze, J. O., 1987, *Turbulence*, second edition, New York: McGraw-Hill Book Company.

Hoerner, S. F., 1965, *Fluid Dynamic Drag*, Published by the author.

Holling, H. D., Hubble, E. N., 1974, Model resistance data of Series 65 Hullforms applicable to hydrofoils and planing craft, National Ship Research and Development Centee Report No. 4121, Bethesda, Md.

Hooft, J. P., Nienhuis, U., 1994, The prediction of the ship's maneuverability in the design stage, *SNAME Transactions*, **102**, 419–45.

Hough, G. R., Moran, J. P., 1969, Froude number effects on two-dimensional hydrofoils, *J. Ship Res.*, **13**, 1, 53–60.

Howison, S. D., Ockendon, J. R., Wilson, S. K., 1991, Incompressible water-entry problems at small deadrise angles, *J. Fluid Mech.*, **222**, 215–30.

Hughes, G., 1954, Friction and form resistance in turbulent flow, and a proposed formulation for use in model and ship correlation. *Transactions of the Institution of Naval Architects*, **96**, 314–76.

Huse, E., 1972, Pressure fluctuations on the hull induced by cavitating propellers, Norwegian Ship Model Experiment Tank Publications, No. 111, March, Trondheim.

Ikeda, Y. Katayama, T., 2000a, Stability of high speed craft, In *Contemporary Ideas on Ship Stability*, ed. D. Vassalos, M. Hamamoto, A. Papanikolaou, D. Molyneux, pp. 401–9, Oxford: Elsevier Science Ltd.

Ikeda, Y., Katayama, T., 2000b, Porpoising oscillations of very-high-speed marine craft, *Phil. Trans. R. Soc. Lond. A*, **358**, 1905–15.

Ikeda, Y., Katayama, T., Okumura, H., 2000a, Characteristics of hydrodynamics derivatives in maneuverability equations for super-high-speed planing hulls, In *Proc. Tenth Int. Offshore and Polar Engineering Conf.*, Vol. 4, pp. 434–44.

Ikeda, Y., Okumura, H., Katayama, T., 2000b, Stability of a planing craft in turning motion, In *Contemporary Ideas on Ship Stability*, ed. D. Vassalos, M. Hamamoto, A. Papanikolaou, D. Molyneux, pp. 449–95, Oxford: Elsevier Science Ltd.

Ikeda, Y., Yokomizo, K., Hamasaki, J., Umeda, N., Katayama, T., 1993, Simulation of running attitude and resistance of a high-speed craft using a database of hydrodynamic forces obtained by fully captive model experiments, In *Proc. FAST'93*, ed. K. Sugai, H. Miyata, S. Kubo, H. Yamata, Vol. 1, pp. 583–94, Tokyo: The Society of Naval Architects of Japan.

Inukai, Y., Horiuchi, K., Kinoshita, T., Kanou, H., Itakura, H., 2001, Development of a single-handed hydrofoil sailing catamaran, *J. Mar. Sci. Technol.* **6**, 1, 31–41.

Ishiguro, T., Uchida, K., Manabe, T., Michida, R., 1993, A study on the maneuverability of the Super Slender Twin Hull, In *Proc. FAST'93*, ed. K. Sugai, H. Miyata, S. Kubo, H. Yamata, Vol. 1, pp. 283–94, Tokyo: The Society of Naval Architects of Japan.

Iwashita, H., Nechita, M., Colagrossi, A., Landrini, M, Bertram, V., 2000, A critical assessment of potential flow models for ship seakeeping, In *Proc. Fourth Osaka Colloquium on Seakeeping Performance of Ships*, pp. 37–46. Osaka, Japan: Dept. of Naval Architecture and Ocean Engineering, Osaka University.

Jensen, J. J., 1996, Wave-induced hydroelastic response of fast monohull ships, CETENA Seminar on Hydroelasticity for Ship Structural Design, Genova: CETENA.

Jensen, J. J., Dogliani, M., 1996, Wave-induced ship hull vibrations in stochastic seaways, *Marine Structures*, **9**, 3/4, 353–87.

Jensen, J. J., Pedersen, P. T., 1978, Wave-induced bending moments in ships – a quadratic theory, *Trans. RINA*, **121**, 151–65.

Jensen, J. J., Pedersen, P. T., 1981, Bending moments and shear forces in ships sailing in irregular wave, *J. Ship Res.*, **24**, 4, 243–51.

Jensen, J. J., Wang, Z., 1998, Wave induced hydroelastic response of a fast monohull displacement ship, In *Proc. Second Int. Conf. on Hydroelasticity in Marine Technology*, ed. M. Kashiwagi, W. Koterayama, M. Okkusu, pp. 411–27, Fukuoka, Japan: RIAM, Kyushu University.

Johnston, R. J., 1985, Hydrofoils, *Naval Engineers Journal*, **97**, 2, 142–99.

Kaiho, T., 1977, A new method for solving surface-piercing strut problems, Ph.D. thesis, Dept. of Nav. Arch. and Mar. Eng., The University of Michigan, Ann Arbor, Mich.

Kan, M., Hanaoka, T., 1964, Analysis for the effect of shallow water upon turning (in Japanese), *J. Soc. Nav. Arch. Japan*, **115**, 49–55.

Kaplan, P., Bentson, J., Davis, S., 1981, Dynamics and hydrodynamics of surface-effect ships, *Trans. SNAME*, **89**, 211–47.

Kapsenberg, G. K., Brizzolara, S., 1999, Hydroelastic effects of bow flare slamming on a fast monohull, In *Proc. FAST'99*, pp. 699–708, Jersey City, N.J.: The Society of Naval Architects and Marine Engineers.

Karman, T. von, 1929, The impact on seaplane floats during landing, NACA, Tech. Note No. 321, Washington, D.C.

Karman, T. von, 1930, Mechanische Ähnlichkeit und Turbulenz, *Nachr. Ges. Wiss. Goett, Math-Phys. Kl.*, 58–76.

Kashiwagi, M., 1993, Heave and pitch motions of a catamaran advancing in waves, In *Proc. FAST'93*, ed. K. Sugai, H. Miyata, S. Kubo, H. Yamata, Vol. 1, pp. 643–55, Tokyo: The Society of Naval Architects of Japan.

Katayama, T., 2002, Experimental techniques to assess dynamic unstability of high-speed planing craft, non-zero heel, bow-diving, porpoising and transverse porpoising, In *Proc. Sixth Int. Ship Stability Workshop*, Jersey City, N.J.: The Society of Naval Architects and Marine Engineers.

Katayama, T., Hinami, T., Ikeda, Y., 2000, Longitudinal motion of a super high-speed planing craft in regular head waves, In *Proc. Fourth Osaka Colloquium on Seakeeping Performance of Ships*, pp. 214–20. Osaka, Japan: Dept. of Naval Architecture and Ocean Engineering, Osaka University.

Kato, H., 1996, Cavitation, In *Advances in Marine Hydrodynamics*, ed. M. Ohkusu, Ch. 5, pp. 233–77, Southampton: Computational Mechanics Publications.

Kerczek, C. von, Tuck, E. O., 1969, The representation of ship hulls by conformal mapping functions, *J. Ship Res.*, **13**, 4, 284–98.

Kerwin, J. E., 1991, Hydrofoils and propellers. Lecture notes, Dept. of Ocean Engineering, MIT, Cambridge, Massachusetts.

Kerwin, J. E., Lee, C-S., 1978, Prediction of steady and unsteady marine propeller performance by numerical lifting-surface theory, *Trans. SNAME*, **86**, 218–53.

Keuning, J. A., Gerritsma, J., 1982, Resistance tests of a series planing hull forms with 25 degrees deadrise angle, *Intern. Shipbuilding Progr.*, **29**, 337, 222–49.

Keuning, J. A., Gerritsma, J., Terwisga, P. F. van, 1993, Resistance tests of a series planing hull forms with 30 degrees deadrise angle and a calculation method based on this and similar systematic series, *Intern. Shipbuilding Progr.*, **40**, 424, 333–82.

Kijima, K., Furukawa, Y., 2000, Ship maneuvering performance in waves, in *Contemporary Ideas on Ship Stability*, ed. D. Vassalos, N. Hamamoto, A. Papanikolaous, D. Molyneux, pp. 435–48, Amsterdam: Elsevier Science Ltd.

Kinnas, S. A., 1996, Theory and numerical methods for the hydrodynamic analysis of marine propulsors, In *Advances in Marine Hydrodynamics*, ed. M. Okkusu, Ch. 6, pp. 279–323, Southampton: Computional Mechanics Publications.

Kinsman, B., 1965, *Wind Waves*, Englewood Cliffs, N.J.: Prentice-Hall Inc.

Klotter, K., 1978, *Technische Schwingungslehre. Erster Band: Einfache Schwinger. Teil A: Lineare Schwingungen*, Berlin, Heidelberg and New York: Springer-Verlag.

Kochin, N. E., Kibel, I. A., Roze, N. V., 1964, *Theoretical Hydromechanics*, New York: Interscience Publishers.

Koehler, B. R., Kettleborough, 1977, Hydrodynamics of a falling body upon a viscous incompressible fluid, *J. Ship Res.*, **20**, 190–8.

Kotik, J., Mangulis, V., 1962, On the Kramers-Kronig relations for ship motions, *Intern. Shipbuilding Progr.*, **9**, 97, 183–94.

Koumoutsakas, P., Leonard, A., 1995, High-resolution simulations of the flow around an impulsively started cylinder using vortex methods, *J. Fluid Mech.*, **296**, 1–38.

Koushan, K., 1997, Beitrag Zum Kanaleinfluss bei Tragflügelversuchen, Dr.ing thesis, Technische Universität Berlin.

Krasny, R., 1987, Computation of vortex sheet roll-up in the Trefftz plane, *J. Fluid Mech.*, **184**, 123–55.

Kruppa, C., 1990, Propulsion systems for high-speed marine vehicles, Second Conference on High-Speed Marine Craft, Oslo: Norwegian Society of Chartered Engineers.

Kruppa, C., 1991, On the design of surface piercing propellers, Seventh GE-US Symposium Hydroacoustics, Part II, Hamburg, Germany.

Kruppa, C. F. L., 1992, Testing surface piercing propellers, In *Hydrodynamics: Computation, Model Tests and Reality*, ed. H. J. J. van den Boom, pp. 107–14, Amsterdam: Elsevier Science Publishers B.V.

Kuchemann, D., 1978, *The Aerodynamic Design of Aircraft*, Oxford: Pergamon Press.

Kutta, W. M., 1910, Über eine mit den Grundlagen des Flugsproblems in Beziehung stehende zweidimensionale Strömung. *Sitzungsberichte der Königlischen Bayerschen Akademie der Wissenschaften*. (This paper reproduced Kutta's unpublished thesis of 1902).

Kvålsvold, J., 1994, Hydroelastic modelling of wetdeck slamming on multihull vessels, Dr.ing. thesis, Dept. of Marine Hydrodynamics, Nor. Inst. Technol., Trondheim.

Kvålsvold, J., Faltinsen, O. M, 1995, Hydroelastic modelling of wetdeck slamming on multihull vessels, *J. Ship Res.*, **39**, 225–29.

Kvålsvold, J., Faltinsen, O. M., Aarsnes, J. V., 1995, Effect of structural elasticity on slamming against wetdecks of multihull vessels, In *Proc. PRADS'95*, ed. H. Kim, J. W. Lee, **1**, 1684–99, Seoul: The Society of Naval Architects of Korea.

Lai, C., 1994, Three-dimensional planing hydrodynamics based on a vortex lattice method, Ph.D. thesis, Dept. of Nav. Arch. and Mar. Eng., The University of Michigan, Ann Arbor, Mich.

Landau, L. D., Lifshitz, E. M., 1959, *Fluid Mechanics*, Oxford: Pergamon Press.

Larsson, L., Baba, E., 1996, Ship resistance and flow computations, *Advances in Marine Hydrodynamics*, ed. M. Ohkusu, pp. 1–75, Southampton: Computational Mechanics Publication.

Larsson, L., Eliasson, R., 2000, *Principles of Yacht Design*, Camden, Maine: International Marine.

Latorre, R., Miller, A., Philips, R., 2003, Drag reduction on a high speed trimaran, In *Proc. FAST'03*, ed. P. Casella, Vol. 1, Session A1, pp. 87–92, Naples: Dipartimento Ingegneria Navale–Università di Napoli "Federico II."

Lavis, D. R., 1980, The development of stability standards for dynamically supported craft, a progress report, In *Proc. of the High Speed Surface Craft Exhibition and Conference*, pp. 384–94, Brighton, Sussex, UK: Kalerghi Publications.

Lee, C. S., 1977, A numerical method for the solution of the unsteady lifting problem of rectangular and elliptic hydrofoil, master's thesis, Dept. of Ocean Engineering, MIT, Cambridge, Mass.

Lee, W. T., Bales, S. L., 1984, Environmental data for design of marine vehicles, In *Ship Structure Symposium '84*, pp. 197–209, New York: The Society of Naval Architects and Marine Engineers.

Lefandeux, F., 1999, New advances in sailing hydrofoils, In *RTO Meeting Proc. 15. Fluid Dynamics Problems of Vehicles Operating Near or in the Air-Sea Interface*, pp. 15-1–15-14, Neuilly-Sur-Seine Cedex, France: Research and Technology Organization/ NATO.

Leonard, J. W., 1988, *Tension Structures, Behaviour and Analysis*, New York: McGraw-Hill Book Company.

Lewandowski, E. M., 1997, Transverse dynamic stability of planing craft, *Marine Technology*, **34**, 2, 109–18.

Lewis, R. I., 1996, *Turbomachinery Performance Analysis*, London: Arnold.

Lighthill, M. J., 1951, A new approach to thin airfoil theory, *The Aeronautical Quarterly*, **III**, 193–210.

Lighthill, M. J., 1960, Note on the swimming of slender ship, *J.Fluid Mech.*, **9**, 304–17.

Lin, W-M., Meinhold, M. J., Salvesen, N., 1995, SIMPLAN2, simulation of planing craft motions and load, Report SAIC-95/1000, SAIC, Annapolis, Md.

Lord Kelvin (Sir William Thompson), 1887, On ship waves, *Proc. Inst. Mech. Eng.*, London, UK.

Lugni, C., Colagrossi, A., Landrini, M., Faltinsen, O. M., 2004, Experimental and numerical study of semi-displacement monohull and catamaran in calm water and incident waves, In *Proc. 25th Symposium on Naval Hydrodynamics*, Washington D.C.: Dept. of the Navy–Office of Naval Research.

Lugt, H. J., 1981, Numerical modelling of vortex flows in ship hydrodynamics, a review, In *Proc. Third Int. Conf. on Numerical Ship Hydrodynamics*, ed. J-C. Dern, H. J. Hausling, pp. 297–316, Paris: Bassin d'Essais des Carènes.

Lunde, J. K., 1951, On the linearized theory of wave resistance for displacement ships in steady and accelerated motions, *Trans. SNAME*, **59**, 25–85.

Maeda, H., 1991, Modelling techniques for dynamics of ships, *Phil. Trans. R. Soc. Lond. A*, **334**, 307–17.

Malakhoff, A., Davis, S., 1981, Dynamics of SES bow seal fingers, *AIAA Sixth Marine Systems Conf.*, AIAA – 81-2087.

Manen, J. D. van, Oossanen, P. van, 1988, Resistance, propulsion and vibration, In *Principles of Naval Architecture*, ed. E. V. Lewis, Vol. II, Chapter VI, Jersey City, N.J.: The Society of Naval Architects and Marine Engineers.

Marchaj, C. A., 2000, *Aero-hydrodynamics of Sailing*, St. Michaels, Md: Tiller.

Maruo, H., 1963, Resistance in waves, *60th Anniversary Series SNA Japan*, **8**, 67–102.

Masilge, C., 1991, Konzeptien und Analyse eines interierten Strahlantriebes mit einem rotationssymmetrischen Grenzchichteinlauf, Dissertation, Technische Universität Berlin.

Maskell, E. C., 1972, On the Kutta-Joukowski condition in two-dimensional unsteady flow, Roy. Aircraft Establishment, Fanborough, Techn. Memo Aero 1451.

Matthews, S. T., 1967, Main hull girder loads on a Great Lakes bulk carrier, In *Proc. SNAME Spring Meeting*, pp. 11.1–11.32, Jersey City, N.J.: The Society of Naval Architects and Marine Engineers.

Meek-Hansen, B., 1990, Damage investigation on diesel engines in high speed vehicles, In *Proc. Fifth International Congress on Marine Technology Athens '90*, pp. 309–403, Athens: Hellenic Institute of Marine Technology.

Meek-Hansen, B., 1991, Engine running conditions during high speed marine craft operations, In *Proc. FAST'91*, ed. K. O. Holden, O. M. Faltinsen, T. Moan, Vol. 2, pp. 861–76, Trondheim: Tapir Publishers.

Mei, C. C., 1983, *The Applied Dynamics of Ocean Surface Waves*, New York: John Wiley & Sons. Revised printing (1989), Singapore: World Scientific.

Meyer, J. R., Wilkins, J. R. Jr., 1992, Hydrofoil development and applications, In *Proc. HPMV'92*, pp. HF1–HF24, Alexandria, Va.: American Society of Naval Engineers.

Michell, J. M., 1898, The wave resistance of a ship, *Phil. Mag., London, Series 5*, **45**, 106–23.

Milburn, D., 1990, Numerical model of 47'MLB high speed turns, USCG R&D Center Report.

Milne-Thomson, L. M., 1996, *Theoretical Hydrodynamics*, Mineola, N.Y.: Dover Publications, Inc.

Minsaas, K. J., 1993, Design and development of hydrofoil catamarans in Norway, In *Proc. FAST'93*, ed. K. Sugai, H. Miyata, S. Kubo, H. Yamato, Vol. 1, pp. 83–99, Tokyo: The Society of Naval Architects of Japan.

Minsaas, K. J., 1996, Flow studies with a pitot inlet in a cavitation tunnel, 20th ITTC Workshop on Waterjets, Supplement to the Report of the Waterjet Group, 21st ITTC, Trondheim, Norway.

Minsaas, K. J., Thon, H. J., Kauczynski, W., 1986, Influence of ocean environment on thrusters performance. In *Proc. Int. Symp. Propeller and Cavitation*, supplementary volume, pp. 142–42. Shanghai: The Editorial Office of Shipbuilding of China.

Molin, B., 1999, On the piston mode in moonpools, In *Proc. 14th Int. Workshop on Water Waves and Floating Bodies*, ed. R. F. Beck, W. W. Schultz, pp. 103–6, Ann Arbor, Mich.: Dept. of Nav. Arch. and Mar. Eng., The University of Michigan.

Molland, A. F., Wellicome, J. F., Couser, P. R., 1996, Resistance experiments on a systematic series of high speed displacement catamaran hull forms: Variation of length-displacement ratio and breadth-draught ratio, *Trans. RINA*, **138 pt A**, 55–72.

Mørch, J. B., 1992, Aspects of hydrofoil design with emphasis on hydrofoil interaction in calm water, Dr.ing. thesis, Div. of Marine Hydrodynamics, NTNU, Trondheim.

Mørch, H. J. B., Minsaas, K. J., 1991, Aspects of hydrofoil design with emphasis on hydrofoil interaction in calm water, In *Proc. FAST '91*, ed. K. O. Holden, O. M. Faltinsen, T. Moan, Vol. 1, pp. 143–61, Trondheim: Tapir Publishers.

Morison, J. R., O'Brien, M. P., Johnson, J. W., Schaaf, S. A., 1950, The force exerted by surface waves on piles, *Pet. Trans.*, **189**, 149–54.

Moulijn, J. 2000, Added resistance due to waves of surface effect ships, Ph. D. thesis, Technical University of Delft, The Netherlands.

Müller-Graf, B., 1991, The effect of an advanced spray rail system on resistance and development of spray of semi-displacement round bilge hulls, In *Proc. FAST'91*, ed. K. O. Holden, O. M. Faltinsen, T. Moan, Vol. 1, pp. 125–41, Trondheim: Tapir Publishers.

Müller-Graf, B., 1994, Spritzleisten und Staukeile-Massnahmen zur Verbesserung der hydrodynamischen Eigenschaften von Motorbooten (Spray rails and wedges – an effective tool to improve the hydrodynamic characteristics of motorboats). In *Proc. of the 15th Symp. on Yacht Design and Yacht Building*, 28–29 Oct. 1994, pp. 11–65, Hamburg, Germany: Hamburger Messe und Congress GmbH und Deutcher Boots und Schiffbauer Verband.

Müller-Graf, B., 1997, Dynamic stability of high speed small craft, WEGEMT Association Twenty-Fifth School Craft Technology, Athens, Greece: Dept. of Nav. Arch. and Mar. Eng., National Technical University of Athens.

Müller-Graf, B, 1999a, Widerstand und hydrodynamische Eigenschaften der schnellen Knickspant-Katamarane der VWS-Serie'89 (Resistance and hydrodynamic characteristics of the VWS Hard Chine Hull Catamaran Series '89). In *Proc. German 20. Symposium Yachtenwurf und Yachtbau*, 5.–6. November 1999, pp. 47–165, Hamburg, Germany: Hamburg Messe und Congress GmbH und Deutscher Boots- und Schiffbauer-Verband e.V.

Müller-Graf, B, 1999b, Leistingsbedarf und Propulsionseigenschaften der schnellen Knickspantkatamarane der VWS-Serie'89 (Power requirements and propulsive characteristics of the VWS Hard Chine Hull Catamaran Series '89). In *Proc. German 20. Symposium Yachtenwurf und Yachtbau*, 5.–6. November 1999, pp. 167–257, Hamburg, Germany: Hamburg Messe und Congress GmbH und Deutscher Boots- und Schiffbauer-Verband e.V.

Müller-Graf, B., Schmiechen, M., 1982, On the stability of semidisplacement craft, In *Proc. of Second Intern. Conf. on Stability of Ships and Ocean Vehicles*, pp. 67–76, Tokyo: The Society of Naval Architects of Japan.

Myrhaug, D., 2004, Lecture notes in oceanography: winds, waves, Trondheim: Dept. of Marine Technology, NTNU.

Nakatake, K. Ando, J., Kataoka, K., Yoshitake, A., 2003, A simple surface panel method "SQCM" in ship hydrodynamics, In *Proc. Int. Symp. on Naval Architecture and Ocean Engineering*, pp. 23/1–11, Shanghai: School of Naval Architecture and Ocean Engineering, Shanghai Jiao Tong University, China.

Nakos, D., 1990, Ship wave patterns and motions by a three-dimensional Rankine panel method, Ph.D. thesis, Dept. of Ocean Engineering, MIT, Cambridge.

Newman, J. N., 1962, The exciting forces on fixed bodies in waves, *J. Ship Res.*, **6**, 4, 10–7.

Newman, J. N., 1969, Lateral motion of a slender body between two parallel walls, *J. Fluid Mech.*, **39**, 1, 97–115.

Newman, J. N., 1977, *Marine Hydrodynamics*, Cambridge: The MIT Press.

Newman, J. N., 1978, The theory of ship motions, *Advances in Applied Mechanics*, **18**, 221–82.

Newman, J. N., 1987, Evaluation of the wave-resistance Green function: part 1 – the double integral, *J. Ship Res.*, **31**, 2, 79–90.

Newman, J. N., Sclavounos, P., 1980, The unified theory of ship motions, In *Proc. 13th Symp.on Naval Hydrodynamics*, ed. T. Inui, pp. 373–97, Tokyo: The Shipbuilding Research Association of Japan.

Newton, R. N., Rader, H. A., 1961, Performance data of propellers for high speed craft, *Trans. RINA*, **103**, 2, 93–129.

Nicholson, K., 1974, Some parametric model experiments to investigate broaching-to, In *Int. Symp. Dynamics of Marine Vehicles and Structures in Waves*, ed. R. E. D. Bishop, W. G. Price, pp. 160–6, London: Mechanical Engineering Publications Ltd.

Nikuradse, J., 1930, Untersuchungen über turbulente Strömungen in nicht kreisförmigen Rohren, *Ing. – Arch.*, **1**, 306–32.

Nikuradse, J., 1933, Strömungsgesetze in rauhen Rohren, *Forschungsheft*, **361**, Berlin: VDI-Verlag.

Nonaka, K., 1993, Estimation of hydrodynamic forces acting on a ship in maneuvering motion, In *Proc. MARSIM'93*, pp. 437–45, St. John's, Newfoundland.

Nordenstrøm, N., 1973, A method to predict long-term distributions of waves and wave-induced motions and loads on ships and other floating structures, DNV Publications No 81, Det Norske Veritas, Høvik, Norway.

Nordenstrøm, N., Faltinsen, O. M., Pedersen, B., 1971, Prediction of wave-induced motions and loads for catamarans, In *Proc. Offshore Technology Conference*, Paper No. OTC1418, Vol. 2, pp. 13–58, Richardson, Tex.: Offshore Technology Conference Inc.

NORDFORSK, 1987, The Nordic Cooperative Project, Seakeeping performance of ships, In *Assessment of a Ship Performance in a Seaway*, Trondheim, Norway: MARINTEK.

Norrbin, N. H., 1971, Theory and observation on the use of a mathematical model for ship maneuvering in deep and confined waters, SSPA Report No. 68, Gothenborg.

NS-ISO 2631-31. utgave November 1985 (Figure 1 side 6).

Nwogu, O., 1993, An alternative form of Boussinesq equations for nearshore wave propagation, *J. of Waterway, Port, Coastal and Ocean Engineering*, **119**, 6, 618–38.

Ochi, M. K., 1964, Prediction of occurrence and severity of ship slamming at sea, In *Proc. Fifth Symp. on Naval*

Hydrodynamics, pp. 545–96. Washington, D.C.: Office of Naval Research–Department of the Navy.

Ochi, M. K., 1982, Stochastic analysis and probability distribution in random seas, *Advances in Hydroscience*, **13**, 217–375.

Ogilvie, T. F., 1964, Recent progress towards the understanding and prediction of ship motions. In *Proc. Fifth Symp. on Naval Hydrodynamics*, pp. 3–128. Washington, D.C.: Office of Naval Research–Department of the Navy.

Ogilvie, T. F., 1969a, Lecture notes for the course Naval Hydrodynamics I, Dept. of Nav. Arch. and Mar. Eng., The University of Michigan, Ann Arbor, Mich.

Ogilvie, T. F., 1969b, Oscillating pressure fields on a free surface, Rep. no 030, Dept. of Nav. Arch. and Mar. Eng., The University of Michigan, Ann Arbor, Mich.

Ogilvie. T. F., 1972, The wave generated by a fine ship bow, In *Ninth Symp. Naval Hydrodynamics*, ed. R. Brard and A. Castaro, Vol. **2**, pp. 1483–525, Washington, D.C.: National Academy Press.

Ogilvie, T. F., 1978, End effects in slender-ship theory, In *Proc. Symp. on Applied Mathematics, dedicated to the late Prof. Dr. R. Timman*, ed. A. J. Hermans, M. W. C. Oosterveld, pp. 119–39, Delft: Delft University Press.

Ohkusu, M., 1969, On the heaving motion of two circular cylinders on the surface of a fluid, Reports of Research Institute for Applied Mechanics, Vol. XVII, No. 58, Kyushu University, Japan.

Ohkusu, M., 1996, Hydrodynamics of ships in waves, In *Advances in Marine Hydrodynamics*, ed. M. Ohkusu, Chapter 2, pp. 77–132, Southampton: Computational Mechanics Publications.

Ohkusu, M., Faltinsen, O. M., 1990, Prediction of radiation forces on a catamaran at high Froude number, In *Proc. 18th Symp. on Naval Hydrodynamics*, pp. 5–19, Washington, D.C.: National Academy Press.

Økland, O., 2002, Numerical and experimental investigation of whipping in twinhull vessels exposed to severe wet deck slamming, Dr.ing. thesis, Dept. of Marine Technology, NTNU, Trondheim.

Papanikolaou, A., 2002, Developments and potential of Advanced Marine Vehicles Concepts, *Bulletin of the KANSAI Society of Naval Architects*, **55**, 50–4.

Prandtl, L., 1933, Neuere Ergebnisse der Turbulenzforschung, *Z. Ver. Dtch. Ing.*, **77**, 5, 105–14, (Translated as NACA Tech. Mem. 720).

Prandtl, L., 1956, *Strömungslehre*, Braunschweig: Friedr. Vieweg & Sohn.

Riabouchinski, D., 1920, Sur la resistance des fluids, *Congres Intern. des Math*, Strasbourg, pp. 568–85, Toulouse; Henri Villat, Librairie de l'Université.

Rognebakke, O. F., Faltinsen, O. M., 2003, Coupling of sloshing and ship motions, *J. Ship Res.*, **47**, 3, 208–21.

Ronæss, M., 2002, Wave induced motions of two ships advancing on a parallel course, Dr. Ing. Thesis, Dept. of Marine Hydrodynamics, NTNU, Trondheim.

Rose, J. C., Kruppa, C., 1991, Surface piercing propellers – methodical series model test results, In *Proc. FAST'91*, ed. K. O. Holden, O. M. Faltinsen, T. Moan, Vol. 2, pp. 1129–47, Trondheim: Tapir Publishers.

Rose, J. C., Kruppa, C., Koushan, K., 1993, Surface piercing propellers – propeller/hull interaction, In *Proc. FAST'93*, ed. K. Sugai, M. Miyata, S. Kubo, H. Yamata, Vol. 1, pp. 867–81, Tokyo: The Society of Naval Architects of Japan.

Rouse, H., 1961, *Fluid Mechanics for Hydraulic Engineers*, New York: Dover Publications, Inc.

Saito, Y., Oka, M., Ikebuchi, K., Asao, M., 1991, Rough water capabilities of fully submerged hydrofoil craft "Jetfoil," In *Proc. FAST'91*, ed. K. O. Holden, O. M. Faltinsen, T. Moan, Vol. 2, pp. 1013–28, Trondheim: Tapir Publishers.

Salvesen, N., 1974, Second-order steady-state forces and moments on surface ships in oblique waves, In *Int. Symp. Dynamics of Marine Vehicles and Structures in Waves*, ed. R. E. D. Bishop, W. G. Price, pp. 212–26, London: Mechanical Engineering Publications.

Salvesen, N., Tuck, E. O., Faltinsen, O. M., 1970, Ship motions and sea loads, *Trans. SNAME*, **78**, 250–87.

Sarpkaya, T., 1966, Separated flow about lifting bodies and impulsive flow about cylinders, *AIAA Journal*, **44**, 414–20.

Sarpkaya, T., Shoaff, R. L., 1979, A discrete-vortex analysis of flow about stationary and transversely oscillating circular cylinders, Tech. Rep. NPS-69 SL 79011, Nav. Postgrad. Sch. Monterey, Calif.

Sarpkaya, T., Isaacson, M., 1981, *Mechanics of Wave Forces on Offshore Structures*, New York: Van Nostrand Reinhold Company.

Savitsky, D., 1964, Hydrodynamic design of planing hulls, *Marine Technology*, **1**, 1, 71–96.

Savitsky, D., 1988, Wake shapes behind planing hull forms, In *Proc. Int. High-Performance Vehicle Conf.*, pp. VII, 1–15, Shanghai: The Chinese Society of Naval Architecture and Marine Engineering.

Savitsky, D., 1992, Overview of planing hull developments, In *Proc. HPMV'92*, pp. PC1–PC14, Alexandria, Va.: American Society of Naval Engineers.

Schlichting, H., 1979, *Boundary-Layer Theory*, New York: McGraw-Hill Book Company.

Schmitke, R. T., Jones, E. A., 1972, Hydrodynamics and simulation in the Canadian hydrofoil program, In *Proc. Ninth Symp. on Naval Hydrodynamics*, ed. R. Brard and A. Castera, Vol. 1, pp. 293–342, Arlington, Va.: Office of Naval Research–Department of the Navy.

Schultz-Grunow, F, 1940, Neues Reibungswiderstandsgesetz für glatte Platten, *Luftfahrtforschung*, **17**, 239–46 (Translated as NACA Tech. Mem. 986).

Schwartz, L. W., 1974, Computer extension and analytic continuation of Stokes' expansion for gravity waves, *J. Fluid Mech.*, **62**, 553–78.

Sclavounos, P. D., 1987, An unsteady lifting line theory, *J. Eng. Math.*, **21**, 201–26.

Sclavounos, P., 1996, Computation of wave ship interactions, In *Advances in Marine Hydrodynamics*, ed. M. Ohkusu, Ch. 4, pp. 177–231, Southampton: Computational Mechanics Publications.

Sclavounos, P. D., Borgen, H., 2004, Seakeeping analysis of a high-speed monohull with a motion control bow hydrofoil, *J. Ship. Res.*, **28**, 2, 77–117.

Scolan, Y.-M., Korobkin, A. A., 2001, Three-dimensional theory of water impact, part 1, inverse Wagner problem, *J. Fluid Mech.*, **440**, 293–326.

Scolan, Y.-M., Korobkin, A. A., 2003, On energy arguments applied to slamming of elastic body, In *Proc. Hydroelasticity in Marine Technology*, ed. R. Eatock Taylor, pp. 175–83, Oxford: Dept. of Eng. Science, University of Oxford.

Sedov, I., 1940, On the theory of unsteady planing and the motion of a wing with vortex separation, NACA Technical Memorandum 942, 53 pp., Washington, D.C.

Sedov, I., 1965, *Two-Dimensional Problems in Hydrodynamics and Aerodynamics*, New York: Interscience Publishers.

Sfakiotakis, M., Lane, D. M., Davies, J. B. C., 1999, Review of fish swimming modes for aquatic locomotion, *IEEE Journal of Oceanic Engineering*, **24**, 2, 237–52.

Shen, Y. T., Ogilvie, T. F., 1972, Nonlinear hydrodynamic theory for finite-span planing surface, *J. Ship Res.*, **16**, 3–20.

Shen, Y. T., Eppler, R., 1979, Section design for hydrofoil wings with flaps, *J. Hydrodynamics*, **13**, 2, 39–45.

Shen, Y. T., 1985, Wing sections for hydrofoils, part 3: experimental verifications, *J. Ship Res.*, **29**, 1, 39–50.

Skjørdal, S., Faltinsen, O. M., 1980, A linear theory of springing, *J. Ship. Res.*, **24**, 2, 74–84.

Skomedal, N., 1985, Application of a vortex tracking method to three-dimensional flow past lifting surfaces and blunt bodies, Dr.ing thesis, Dept. of Marine Hydrodynamics, Nor. Inst. Technol., Trondheim.

Skorupka, S., Le Coz, D., Perdon, P., 1992, Performance assessment of the surface effect ship AGNES 200, DCN Bassin d'Essais des Carénes Translation, Paris, France.

Søding, H., 1982, Prediction of ship steering capabilities, *Schiffstechnik*, **29**, 3–29.

Søding, H., 1984, Influence of course control on propulsion power, *Schiff & Hafen/Kommandobrücke*, **3**, 63–8.

Søding, H., 1997, Drastic resistance reductions in catamarans by staggered hulls, In *Proc. FAST'97*, ed. N. Baird, Vol. 1, pp. 225–30, South Yarra, Victoria, and London: Baird Publications.

Sørensen, A., 1993, Modelling and control of SES dynamics in the vertical plane, Dr.ing. thesis, ITK-report 1993:7-W, Nor. Inst. Technol., Trondheim.

Sorensen, R. M., 1993, *Basic Wave Mechanics: For Coastal and Ocean Engineers*, New York: John Wiley & Sons Inc.

Steen, S., 1993, Cobblestone effect on SES, Dr.ing. thesis, Dept. of Marine Hydrodynamics, Nor. Inst. Technol., Trondheim.

Stoker, J. J., 1958, *Water Waves. The Mathematical Theory with Applications*, New York: John Wiley & Sons Inc.

Storhaug, G., 1996, SWATH project: seakeeping and wave load analysis of a SWATH, revision 2, DNV report 96-0174, Det Norske Veritas, Høvik, Norway.

Storhaug, G., Vidic-Perunovic, J., Rüdinger, F., Holtsmark, G., Helmers, J. R., Gu, X., 2003, Springing/whipping response of a large ocean going vessel – a comparison between numerical simulations and full scale measurements, In *Proc. Hydroelasticity in Marine Technology*, ed. R. Eatock Taylor, pp. 117–29, Oxford: Dept. of Eng. Science, University of Oxford.

Stratford, B. S., 1959, An experimental flow with zero pressure friction throughout its region of pressure rise, *J. Fluid Mech.*, **5**, 1, 17–35.

Svenneby, E. J., Minsaas, K. J., 1992, Foilcat 2900, Design and performance, In *Proc. Third Conf. on High-Speed Marine Craft*, paper no 6, Oslo: Norwegian Society of Chartered Engineers.

Takaishi, Y., Matsumoto, T., Ohmatsu, S., 1980, *Winds and Waves of the North Pacific Ocean 1964–1973. Statistical Diagrams and Tables*, Tokyo: Ship Research Institute.

Takaki, M., Iwashita, H., 1994, On the estimation methods of the seakeeping qualities for the high speed vessel in waves, applications of ship motion theory to design, *11th Marine Dynamics Symposium*, Tokyo: Soc. Naval Arch. of Japan.

Takemoto, H., 1984, Some considerations on water impact pressure, *J. Soc. Naval Arch. Japan*, **156**, 314–22.

Tanaka, N., Ikeda, Y., Nishino, K., 1982, Hydrodynamic viscous force acting on oscillating cylinders with various shapes. In *Proc. Sixth Symp. of Marine Technology*, The Society of Naval Architects of Japan. (Also Rep. Dep. Nav. Arch., University of Osaka Prefecture, no. 407, Jan. 1983).

Tatinclaux, J. C., 1975, On the wave resistance of surface effect ships, *Trans. SNAME*, **83**, 51–66.

Taylor, T. E., Kerwin, J. E., Scherer, J. O., 1998, Waterjet pump design and analysis using a coupled lifting-surface and RANS procedure, *Int. Conf. on Waterjet Propulsion, Latest Development*, London: The Royal Institution of Naval Architects.

Terwisga, T. van, 1991, The effect of waterjet-hull interaction on thrust and propulsive efficiency. In *Proc. FAST'91*, ed. K. O. Holden, O. M. Faltinsen, T. Moan, Vol. 2, pp. 1149–67, Trondheim: Tapir Publishers.

Terwisga, T. van, 1992, On the prediction of the powering characteristics of hull-waterjet systems, In *Hydrodynamics: Computations, Model Tests and Reality*, ed. H. J. J. van den Boom, pp. 115–20, Amsterdam: Elsevier Science Publishers B.V.

Theodorsen, T., 1935, General theory of aerodynamic instability and mechanism of flutter, NACA Report 496.

Todd, F. H., 1967, Resistance and propulsion. In *Principles of Naval Architecture*, ed. J. P. Comstock, pp. 228–462. New York: Society of Naval Architects and Marine Engineers.

Torsethaugen, K., 1996, Model for a doubly peaked wave spectrum, Rep. no. STF22 A96204, SINTEF Civil and Environmental Engineering, Trondheim, Norway.

Tregde, V., 2004, Aspects of ship design; optimization of aft hull with inverse geometry design, Ph.D thesis, Dept. of Marine Technology, NTNU, Trondheim.

Triantafyllou, M. S., Triantafyllou, G. S., 1995, An efficient swimming machine, *Scientific American*, March, 40–8.

Troesch, A. W., 1984, Effects of nonlinearities on hull springing, *Marine Technology*, **21**, 4, 356–63.

Troesch, A. W., 1992, On the hydrodynamics of vertically oscillating planing hulls, *J. Ship Res.*, **36**, 4, 317–31.

Troesch, A. W., Falzarano, J. M., 1993, Modern nonlinear dynamical analysis of vertical plane motion of planing hulls, *J. Ship Res.*, **37**, 3, 189–99.

Tuck, E. O., 1966, Shallow water flow past slender bodies, *J. Fluid Mech.*, **26**, 89–95.

Tuck, E. O., 1988, A strip theory for wave resistance, In *Proc. Third Int. Workshop on Water Waves and Floating Bodies*, ed. F. T. Korsmeyer, pp. 169–74, Cambridge, Mass.: Dept. of Ocean Engineering, MIT.

Tuck, E. O., Lazauskas, L., 1998, Optimum spacing of a family of multihulls, *Ship Technology Research*, **45**, 180–95.

Tuck, E. O., Lazauskas, L., 2001, Free-surface pressure distributions with minimum wave resistance, *ANZIAM Journal*, **43**, E75–E101.

Tuck, E. O., Newman, J. N., 1974, Hydrodynamic interactions between ships, In *Tenth Symp. on Naval Hydrodynamics*, ed. R. D. Cooper, S. W. Doroff, pp. 35–70, Arlington, Va.: Office of Naval Research–Department of the Navy.

Tucker, M. J., Challenor, P. G., Carter, D. J. T., 1984, Numerical simulation of a random sea, a common error and its effect upon wave group statistics, *Applied Ocean Research*, **6**, 2, 118–22.

Tucker, M. J., Pitt, E. G., 2001, *Waves in Ocean Engineering*, Elsevier Ocean Engineering Book Series, Vol. 5, ed. R. Bhattacharya, M. E. McCormick, Amsterdam: Elsevier.

Tulin, M. P., 1953, Steady two-dimensional cavity flows about slender bodies, David Taylor Model Basin, Rep. 834, Washington D.C.

Tulin, M., Landrini, M., 2000, Breaking waves in the ocean and around ships, In *Proc. 23rd Symp. on Naval Hydrodynamics*, Washington, D.C.: National Academy Press.

Ulstein, T., 1995, Nonlinear effects of a flexible stern seal bag by cobblestone oscillations of an SES, Dr.ing. thesis, Dept. of Marine Hydrodynamics, NTNU, Trondheim.

Ulstein, T., Faltinsen, O. M., 1996, Hydroelastic analysis of a flexible bag-structure, In *Proc. 20th Symp. on Naval Hydrodynamics*, pp. 702–21, Washington, D.C.: National Academy Press.

Ulstein, T., Faltinsen, O. M., 1996, Two-dimensional unsteady planing, *J. Ship Res.*, **40**, 3, 200–10.

Vanden-Broeck, J.-M., 1980, Nonlinear stern waves, *J. Fluid Mech.*, **96**, 3, 603–11.

Vassalos, D., Hamamoto, M., Papanikolaou, D, Molyneux, D., 2000, *Contemporary Ideas on Ship Stability*, Oxford: Elsevier Science Ltd.

Venning, E., Haberman, W. L., 1962, Supercavitating propeller performance, *Trans. SNAME* **70**, 354–417.

Vugts, J. H., 1968, Cylinder motions in beam waves, Nederlands Ship Research Centre, TNO, Delft.

Wadlin, K. L., 1958, Mechanics of ventilation inception, In *Proc. Second Symp. on Naval Hydrodynamics*, pp. 425–46, ed. P. Eisenberg, Washington, D.C.: Office of Naval Research–Department of the Navy.

Wagner, H., 1925, Über die Enstehung des Auftriebes von Tragflügeln, *Z. Angew. Mech.* **5**, 1, 17–35.

Wagner, H., 1932, Über Stoss- und Gleitvorgänge an der Oberfläche von Flüssigkeiten, *Zeitschr. f. Angew. Math und Mech*, **12**, 4, 193–235.

Wahab, R., Swaan, W. A., 1964, Course keeping and broaching in following waves, *J. Ship Res.*, **7**, 4, 1–15.

Walderhaug, H., 1972, *Ship Hydrodynamics, Basic Course* (in Norwegian), Trondheim: Tapir Publishers.

Walree, F. van, Yamaguchi, K., 1993, Hydrofoil research: model tests and computations, In *Proc. FAST'93*, ed. K. Sugai, H. Miyata, S. Kubo, H. Yamato, Vol. 1, pp. 791–806, Tokyo: The Society of Naval Architects of Japan.

Walree, F. van, 1999, Computational methods for hydrofoil craft in steady and unsteady flow, Ph.D. thesis, Technical University of Delft, Delft.

Walree, F. van, Luth, H. R., 2000, Scale effects on foils and fins in steady and unsteady flow, *RINA Conf. on Hydrodynamics of High Speed Craft*, November, article no. 15, p. 8, London, UK.

Wehausen, J. H., Laitone, E. H., 1962, Surface waves, in *Handbuch der Physik*, ed. S. Flügge, Ch. 9, Springer-Verlag.

Wehausen, J. H., 1973, The wave resistance of ships, *Advances in Applied Mechanics*, **13**, 93–245.

Weissinger, J., 1942, The lift distribution of swept back wings, Translated in NACA TM1120.

Werenskiold, P., 1993, Methods for regulatory and design assessment of planing craft dynamic stability, In *Proc. FAST'93*, ed. K. Sugai, H. Miyata, S. Kubo, H. Yamato, Vol. 1, pp. 883–94, Tokyo: The Society of Naval Architects of Japan.

Whicker, L. F., Fehlner, L. F., 1958, Free stream characteristics of a family of low aspect all movable control surfaces for application to ship design, DTNSRDC Report No. 933, Washington D.C.

White, F., 1972, An analysis of axisymmetric turbulent flow past a long cylinder, *Journal of Basic Engineering*, **94**, 200–6.

White, F. M., 1974, *Viscous Fluid Flow*, New York: McGraw-Hill Book Company.

Whittaker, T., Elsässer, B., 2002, Coping with the wash. The nature of wash waves produced by fast ferries, *Ingema*, **11**, 40–4.

Wigley, W. G. S., 1942, Calculated and measured wave resistance on a series of forms defined algebraically, The prismatic coefficient and angle of entrance being varied independently, *Trans RINA*, **84**, 52–74.

Xu, L., Troesch, A. W., Vorus, W. S., 1998, Asymmetric vessel and planing hydrodynamics, *J. Ship Res.*, **42**, 3, 187–98.

Yamakita, K., Itoh, H., 1998, Sea trial test results of the wear characteristics of SES bow seal fingers, In *Proc. Hydroelasticity in Marine Technology*, ed. M. Kashiwagi, W. Koteryama, M. Ohkusu, pp. 471–6, Fukuoka, Japan: RIAM, Kyushu University.

Yamamoto, Y., Ohtsubo, H., Kohno, Y., 1984, Water impact of wedge model, *Journal of the Soc. Nav. Arch. Japan*, **155**, 236–45.

Yang, Q., 2002, Wash and wave resistance of ships in finite water depth, Dr.ing. thesis, Dept. of Marine Hydrodynamics, NTNU, Trondheim.

Yang, Q., Faltinsen, O. M., Zhao, R., in press, Green function of steady motion in finite water depth, *J. Ship Res.*

Zhang, S., Yue, D. K. P., Tanizawa, K., 1996, Simulation of plunging wave impact on a vertical wall, *J. Fluid Mech.*, **327**, 221–54.

Zhao, R., Faltinsen, O. M, 1992, Slamming loads on high-speed vessel, In *Proc. 19th Symp. on Naval Hydrodynamics*, Washington, D.C.: National Academy Press.

Zhao, R., Faltinsen. O. M., 1993, Water entry of two-dimensional bodies, *J. Fluid Mech.*, **246**, 593–612.

Zhao, R., Faltinsen, O. M., Aarsnes, J. V., 1996, Water entry of arbitrary two-dimensional sections with and without flow separation, In *Proc. 21st Symp. on Naval Hydrodynamics*, pp. 408–23, Washington, D.C.: National Academy Press.

Zhao, R., Faltinsen, O. M., Haslum, H., 1997, A simplified non-linear analysis of a high-speed planing craft in calm water, In *Proc. FAST'97*, ed. N. Baird, Vol. 1, pp. 431–8, South Yarra, Victoria, and London: Baird Publications.

Index

Breinigsville, PA USA
07 October 2010
246812BV00003B/2/P

9 780521 178730